AF841403

724...
Botany
(R)

JOHNSON · FLOWERING PLANTS

ARTHUR M. JOHNSON

The Taxonomy of the Flowering Plants

with 478 figures

Reprint 1977

J. CRAMER
In der A.R. Gantner Verlag Kommanditgesellschaft
FL-9490 VADUZ

Printed in Germany

by Strauss & Cramer GmbH, D-6945 Hirschberg II
ISBN 3-7682-1169-X

TAXONOMY
OF THE
FLOWERING PLANTS

The Century Biological Series
Robert Hegner, Editor

Ralph H. Wetmore, Associate Editor for Botany

TAXONOMY
OF THE
FLOWERING PLANTS

By
ARTHUR MONRAD JOHNSON, Ph.D.
ASSISTANT PROFESSOR OF BOTANY
UNIVERSITY OF CALIFORNIA AT LOS ANGELES

Illustrations by the Author

THE CENTURY CO.
NEW YORK LONDON

COPYRIGHT, 1931, BY THE CENTURY CO.
ALL RIGHTS RESERVED, INCLUDING THE
RIGHT TO REPRODUCE THIS BOOK, OR
PORTIONS THEREOF, IN ANY FORM. 311

PRINTED IN U.S.A.

DEDICATED TO MY WIFE

WITH PROFOUND ADMIRATION

By nature I mean the natural out-of-doors—the snow and the rain, the sky, the plants, the animals, the garden and the orchard, the running brooks, and every landscape that is easy of access and undefiled.
The landscape is as important as any object that it contains, and the human sentiment is more important than either.
I would preach the surface of the earth because we walk on it.
Do you wish to study botany? There are weeds in your dooryard or trees on your lawn. You say that they are not interesting: that is not their fault.

L. H. Bailey, *The Outlook to Nature.*

PREFACE

This volume is the outgrowth of the author's experience in teaching classes in taxonomy, and has been in preparation for a good many years. Much time and thought have been devoted to the perfecting of it and as it now leaves the author's hands he will feel adequately rewarded if it will serve some useful purpose among the ranks of those who have the teaching of taxonomy at heart.

There is much in the volume that is not original with the author, as any botanist will see, but if a modest claim to originality may be made it lies in the method of approach to and in the presentation of the subject and in the use *in toto* of original drawings from the author's pen. As the volume is intended primarily for the student and the instructor, it does not pretend to offer any new scheme of classification of the flowering plants, but rather attempts to adapt for purposes of helpful instruction the best and most practical that has been proposed by the most eminent scholars in the field. The author has endeavored consistently to adhere to the thesis that simplification by the method of deduction from the general, or the known, or the obvious, offers a safer and saner method of approach to a subject like taxonomy than does an inductive method, although the opportunity is not slighted to take advantage of the inductive method when the occasion seems appropriate.

The system followed is in the main that of Engler and Prantl, but incorporates in addition certain very practical features of the systems of Warming-Potter, Wettstein, and Rendle. The sequence of the orders and families is that of Engler and Gilg's *Syllabus der Pflanzenfamilien*. Needless to say, the author has drawn freely upon these sources and the numerous others mentioned in the bibliography, as well as numerous monographs and other technical papers, and due recognition is hereby given of this fact. In the preparation of the contents a great amount of herbarium material has been examined, supplemented by wide studies of the vegetation in the field, covering the greater part of the United States and adjacent Canada and the Mexican border. Access to a wealth of exotic plants adorning private and public gardens of California and representing a considerable number of tropical and subtropical families has contributed a great deal of information applicable to the subject as well as contributing materially to the crystallizing and perfecting of the general plan.

The preparation of the drawings for the illustrations has extended over many years, during which time considerably over a thousand plates of drawings have been made from which to select illustrative types. With few exceptions these drawings have all been made from living specimens. That there are many gaps and inadequacies, as far as the illustrations are concerned, no one can appreciate more than the author, but those that are offered are intended to be sufficiently adequate to illustrate those features that are diagnostic and of most importance in the groups they represent. It has not seemed necessary that every family should be represented by a figure. Rather, advantage has been taken in several places of the peculiarly instructive evolutionary tendencies to be found there to point out these tendencies by a series of figures of the different types. This has been deemed more instructive and helpful to the student, not merely as taxonomic information, but as opening up to him those facts upon which phylogenetic theories and principles are based. And this method is deemed to be all the more effective in as much as it demonstrates vividly the fact that evolutionary tendencies do occur within limited groups, and that the student has before him the twigs of a phylogenetic branch.

As regards nomenclature no attempt has been made to conform strictly to any particular practice, and it is fully realized that there are many places at which criticism may be directed. However, the aim has been to use the names that a student will be most likely to meet with in standard American floras, and which will reduce confusion to a minimum. With this in mind resort has been had to Engler and Gilg's *Syllabus* (9th and 10th editions), Gray's *Manual* (7th edition), Britton and Brown's *Illustrated Flora,* Jepson's *Manual of the Flowering Plants of California,* Piper's *Flora of Washington,* Piper and Beattie's *Flora of Washington and Adjacent Idaho,* and Bailey's *Manual of Cultivated Plants.*

Needless to add, the preparation of a work of this sort has involved other factors than time, material, facilities, and application. These other factors are the many friends and associates in one place or another who have contributed of those illimitably valuable things that so much lighten one's effort. What to say to them properly in appreciation does not come trippingly on the tongue. But the author owes them one and all a large debt. For those things that words but feebly express yet are easily understood, the author owes a special debt of appreciation to his wife, to all his colleagues on the botany staff of the Department of Biology of the University of California at Los Angeles, to Professor B. L. Robinson, Professor M. L. Fernald, and Dr. I. M. Johnston of the Gray Herbarium, to Professor R. H. Wetmore of the Department of Botany, Harvard, and to Professor N. L. Huff of the Department of Botany of the University of Minnesota. Professor Wetmore's interest, demonstrated practically in more ways than one, is something that the author takes special pleasure in remembering.

To his many students, past and present, the author desires also to express his sincere appreciation for all that they have meant to him in the development of this volume.

It is hoped that this volume may find favor in courses in taxonomy designed especially for students in agriculture, forestry, and horticulture, where a simple and direct method of presenting the main principles of classification in the shortest time is of more utilitarian value than phylogenetic theories, offering at the same time a method of presentation that conforms to the generally accepted system of classification.

ARTHUR MONRAD JOHNSON.

UNIVERSITY OF CALIFORNIA AT
LOS ANGELES.
December, 1930.

INTRODUCTION

Many years ago the writer became intrigued with the idea that the teaching of taxonomy, old and dignified science, offered possibilities of improvement commensurate with its dignity. As experience grew with the years and with contacts in various places and with different methods, the idea impressed itself more firmly until it assumed the form expressed in this volume. That taxonomy *can* be taught with a minimum of the objectionable features that have attached to the subject and that are too well known to call for mention here, the author was long ago convinced, and he was also convinced that the only way to remedy the situation, so far as his own laboratory and his own students were concerned, was to break away from orthodox methods. This was promptly done, and with gratifying results.

But in breaking away from the orthodox the first thought that came to mind was what to substitute and how. Having observed the subject taught both pitifully and well it was easy to see how not to do it. One course only seemed plain and that was to strip the subject of its entanglements.

Another thought occurred and that was that phylogeny ought not to be made a fetish. Assuming that the average student really desires to know something about the subject and that he comes with that attitude, it seemed only fair that he should be given an opportunity to learn in the simplest and speediest way those things that are fundamental and which will help him to find his own way, and to think phylogenetically, rather than to hedge him about at the very outset with stuff that is likely to turn him against the subject and is an insult to his intelligence. Few students are sufficiently equipped at the outset to appreciate, much less understand, phylogeny, and when a student picks up this volume and that to read on the subject, or strives to follow a ponderous lecture, he is bewildered, discouraged. When even the best of authorities are not agreed, some caution is necessary. If there be any subject within the scope of biology in which hyperboles are so likely to take wing it is the subject of phylogeny. And why it should be necessary to involve the student in the intricate ramifications of a phylogenetic tree before he is able to climb, is something the author has never been quite able to comprehend. True, phylogeny is important, and the author does not neglect it; it has its place, and that place is given due recognition.

Still another thought insinuated itself and that was that the great majority of students who register for a course in taxonomy probably will never practise the subject professionally: that what they are seeking most is a knowledge of how to go about it to know the plants around them. To this majority phylogeny is perhaps interesting enough, if properly presented, but not very practical.

It is a well-known axiom that there is no short-cut to knowledge. Neither is there to a knowledge of the flowering plants. No useful or practical knowledge can ever be acquired without a preliminary thorough grounding in fundamentals. Nor can any really useful knowledge be inculcated by attempts to elementarize. To elementarize to extremes, as is often done, is to rob a subject of the essence of it, often at the risk of garbling the facts. And to strip a science of its technical language is to strip it of its clarity. The author has never in his experience found it necessary to do so. Strip law, strip the automobile, strip the airplane of its technical language, and what would the lawyer and the mechanician do? It is difficult to comprehend how a teacher who lacks a command of the technical language of his subject can be much of an inspiration to his students. Distaste for or dislike of the technical, of whatever nature it may be, is not inherent in the average student but has its roots in the mental attitude of the teacher. And such mental attitude is often reflected in other activities.

Naturally, in offering a work of this sort, some advertence should be made to the principles that underlie it. These are set forth in Part I, and an effort has been made throughout the pages of Part II to show the application of these principles, particularly in the taxonomy of certain groups. For this purpose those groups have been selected which offer special opportunities for demonstrating the application of the principles, and for the study of which adequate material can usually be obtained. Special emphasis is placed on those points, methods, and devices by which, only, satisfactory results can be attained. Among the various devices, major emphasis is placed upon the median longitudinal section of the flower, for reasons stated in the text. Measurements are disregarded as a waste of precious time and as of little meaning to the student, except in technical investigations.

A word may be in place in regard to laboratory material for class use. The ideal condition of course obtains when fresh specimens are available at all times. But if a course of study is to follow some logical sequence, at least as regards some of the more important aspects of the subject, a certain amount of material must necessarily be of the dried or preserved kind, the dried being moistened as required. The author is fully aware of the objections to using material of this sort, but as the objections are puerile, and as such material can be so prepared as to reduce the objectionable features to a bagatelle, they may be disregarded. If the proper spirit be there, the student will not take cognizance of such trifles.

Associated with the study of taxonomy is the matter of the field excursion, a subject on which much can be said. A field-trip should be something more than a pleasure jaunt, although the pleasurable features are desirable and essential. The success of a field-trip depends much upon the instructor's enthusiasm—of the dignified sort—and his knowledge of the flora. But a field-trip may also be made a waste of time and effort that were better employed in the laboratory. What the author has endeavored to do is to put into practical application in the field the principles of classification inculcated in the laboratory, that is, an effort is made to classify the plants found in flower according to the categories presented in the chart (Part II), so that the student may in this way learn to place with some assurance any strange plant that he may come across. The student's time is well spent in studying this classification, as he will thus acquire information and resourcefulness that will enable him to help himself. He will then learn how to know the wild flowers.

A subject that can hardly be dissociated from taxonomy is that of laboratory drawing. This may be made and often is made a discouraging procedure, mere routine that leads nowhere. The main characteristic of the orthodox system is its uselessness. It is not only destructive of the student's patience, but it defeats the very purpose for which it was intended. The statement seems paradoxical, but this system is anything but scientific. Precious hours are wasted by the student in drawing or trying to draw something which is beyond his present ability and which, when he is done with it, is meaningless for the purpose in view—if there ever were any purpose in view. As a specific example, take the student's drawing of a buttercup. If this can be distinguished from his drawing of a blackberry blossom or an apple blossom, it is exceptional. What useful knowledge has the student acquired in making such drawings? What taxonomic principles has he learned? Better far that he had spent the time studying the flower so as to be ready to construct a floral diagram and the median longitudinal section of it. Worse still will be his effort to represent the "habit" sketch—the most useless of all inflictions upon the student. Is it not possible that the average student possesses a mind sufficiently capable of retaining the image of the "habit" without being afflicted with the tedium of delineating it? And is the habit always so important that it should require delineation? If something seems required about it, why not pick out those points that are diagnostic and let the rest go? Or may not a diagrammatic drawing in line suffice to tell the story? After all, is it not the taxonomic *story* that we would tell, in the simplest pictorial rhetoric? In demanding such work of students, are we not overlooking the very thing we pride ourselves on exemplifying—the scientific method?

Opposed to this orthodox idea of laboratory drawing in taxonomy courses is the idea that only such features need be drawn as really illustrate the important, the *diagnostic,* characters. And after the preliminary studies of the parts of the flower only these diagnostic features need be labeled—that is, pointed out. In delineating these characters, their relation to this or that organ of the flower should be clearly shown, and the simpler the drawing the better for clearness. However, a drawing may be superlatively simple yet fail to show the very thing that is important, as, for example, the insertion of the stamens or the gynœcium in the flowers of a willow—for the very reason that the subtending bract may have been so drawn as to conceal completely the very feature we need to know.

As will be pointed out in the text, certain structural features of flowers are at all times of fundamental importance in connection with classification, Briefly, these are, in the order in which they may be observed in examining a flower: (1) the symmetry of the flower; (2) the number of parts; (3) the relation of the ovary to the receptacle—that is, superior or inferior, and the associated conditions of hypogyny, perigyny, and epigyny; (4) the structure of the ovary; (5) the structure of the fruit. The first can be shown by means of a diagrammatic or semidiagrammatic figure; the second by the floral diagram and the floral formula; the third by the median (longitudinal) section; the fourth by the transverse section (diagrammatic if desired) of the ovary, in association with the median section; the fifth by median and transverse sections (diagrammatic if desired). The median section shows the relations of the parts one to the other, and for this reason should be made the most important figure, whenever possible, in a plate of drawings, and should be made sufficiently large to bring out the structural plan without confusing the details. It may be made a diagrammatic drawing entirely or semidiagrammatic, as desired, the important consideration being always the positional relations of the floral organs. A sufficient number of such drawings are illustrated in the figures in the text to convey an idea of their character. Such drawings tell the story, simply and effectively, without encumbering appendages that have no bearing on the case. Aside from its utility in these respects, this method of drawing will at the same time train the student in a very important feature of all scientific work, namely to see what is and what is not essential to the immediate problem.

Like all other scientific work, botanical drawing depends for its success upon the development of a technique, and successful technique depends upon the proper utilization of adequate instruments. Skill will come with practice, and no special deftness is required. For ordinary laboratory purposes a medium hard pencil—well pointed—a pencil divider, a small metric ruler, and a good bevel-edged eraser—always kept beveled on one end—are all that is required. Colored pencils are sometimes useful in making certain types of diagrammatic drawings, and India ink may be used if desired, but requires a certain amount of skill in using. The "crow-quill" pen has long been in vogue in biological laboratories, but it is an instrument which even in the hand of a skillful draftsman has definite limitations. Yet in biological laboratories it is made to do all sorts of work for which it was never intended. In the hands of an amateur it is impossible: it tends to tightness and woolliness in drawing and inhibits the development of that sure, bold, and vigorous stroke that is the earmark of good draftsmanship. Outside of cytological laboratories stippling has little to recommend it. Aside from its tediousness, it is flat and lifeless and unless used with thoughtful consideration of the principles of light and shade results in nothing but hopeless crudities. Covering a surface with stippling is not shading: it is merely changing its color. No device of drawing is more useful to the student than the simple line, and none possesses such unlimited possibilities.

CONTENTS

PART I

FLOWER ANALYSIS: FUNDAMENTAL PRINCIPLES

CHAPTER I

THE SCIENCE OF TAXONOMY

Taxonomy is that branch of biology that deals with the orderly arrangement of plants and animals. It is one of the oldest of the sciences and one of the most important. The name *taxonomy* is derived from two Greek words τάξις, meaning "order" and νόμος, "law." While the term *classification* likewise conveys the idea of orderly arrangement of organisms in groups, taxonomy is concerned more particularly with the principles of classification. Taxonomy arrives at its principles by the method of analysis, comparison, and contrast, and by the method of synthesis attempts to reduce to order what otherwise appears chaotic. Since we are here dealing with plants, and since botanical taxonomy attempts to reduce the living plants of the world to an orderly system of classification, this science is also known as *systematic botany*.

It is only by observation of the structure of plants that we shall arrive at an understanding of the principles on which taxonomy rests. But our observations should not be confined to the laboratory nor to the herbarium only; they should be supplemented by extensive studies in the field, for only in this way shall we learn to appreciate the multitudinous variations which plants exhibit, and only in this way shall we learn to appreciate the problems that confront the taxonomist on every hand. Plants are not static things: they are living organisms, the products of an evolution that has extended over a long geological past, and they are still being subjected to the vicissitudes of a changing environment. Little wonder that no two individuals of the same species are exactly alike.

Aristotle long ago (384-322 B.C.) and his pupil Theophrastus (372-287 B.C.) classified plants as *trees, shrubs,* and *herbs,* and this simple classification is the one we all first become aware of in our youth. It is plain that this classification is based on "life-form," and that the structure of the flower is ignored. But a similar point of view we also find in *Genesis,* where we may read (Chap. I:11), "Let the earth bring forth grass, the *herb* yielding seed, and the fruit *tree* yielding fruit after his kind, whose seed is in itself, upon the earth; and it was so." (The italics are ours.) Here we note the recognition of the particular *group* of plants we know as grasses. But we might go further back into antiquity, to the dawn of articulate speech, and find there, if we could, the beginnings of the naming of things, some sort of classification in other words, for, as Greene * has so well put it, "Botany, as certainly as geography, had its initiative in primal man's distinguishings and separatings between objects appertaining to the world of plants. The fact that in the rudest and simplest dialects of primitive peoples there exist group names for botanical entities establishes this. It is improbable that there ever was a primitive language, other than that of some arctic tribe, in which there did not occur words equivalent to tree, bush, grass, or to trunk, branch, leaf, fruit, root; and every one of these is the *name, not of an individual object, but of a group of like objects.* Each is a *general—a generic*—name, and each testifies most clearly to *observation, comparison, reflection, generalization,* and also either *the invention of a new word,* or else *the more extended application of an old one,* which, in as far as science is concerned, amounts to the same thing." (The italics are ours.) We may pause, in passing, to reflect that this *generic* concept which Greene here so clearly emphasizes is one that will confront the student at every step in his pursuit of taxonomic problems.

For centuries following Aristotle and Theophrastus systematic botany and the study of plants in nature languished, until in the beginning of the sixteenth century and on through the seventeenth we come to the *herbalists*—the gatherers of medicinal herbs—who produced large tomes or *herbals,* in which were described and pictured the herbs they gathered. While some of these herbals showed a return to original observation, others contained much material copied from ancient works, even back to Aristotle. Among the more notable of these sixteenth and seventeenth century herbalists were Tragus, Cordus, Fuchsius, Brunfels, Cæsalpino, Bauhin, Dodonæus. "The books of Tragus and Cordus," says Greene (*loc. cit.*), "abound in new and original descriptions," while Brunfels and Fuchsius (Fuchs) "sought to render identification of remedies more easy and certain by supplying engravings of the plants." Cæsalpino, an Italian physician, studied fruits and seeds to the extent of using them in a scheme of classification, but he did not distinguish between monocotyledonous and dicotyledonous seeds. Dodonæus and Bauhin were probably the first to use leaf characters as a basis of classification, but their neglect of genera led to an artificial and heterogeneous group-

* Greene, E. L., *Landmarks of Botanical History* (Smithsonian Institution, 1909).

ing of the plants they used. At least their work showed an attempt to arrive at some system by means of comparisons. However, both these writers are recognized more particularly as being the first to use a binomial system of nomenclature.

In the beginning of the eighteenth century the Englishman John Ray wrote a history of plants, *Historia Plantarum* (1703), in which a simple classification of plants is presented, the major categories of which, however, are the Aristotelian herbs and trees; but here for the first time we find an approach to a natural system in his distinguishing between *"Monocotyledones"* and *"Dicotyledones."* With Tourneforte, a contemporary of John Ray, we find a systematic attempt to recognize that the structure of the flower is of more fundamental importance in classification than are vegetative characters. This led Tourneforte to recognize genera, to which he gave descriptions. He was also the last of the old school to adhere to the Aristotelian classification of plants as herbs, shrubs, and trees.

Modern systematic botany, particularly as it applies to the use of the binomial system of nomenclature and the systematic description of species, dates from the system of Linnæus (Carl von Linné, 1707-1778), at first a practising physician in Stockholm and later professor of natural history at the University of Upsala. Linnæus divided the plant kingdom into twenty-four classes, based upon the number of stamens, carpels, or gynœcia, their union, and their presence or absence in the flower. Thus his different classes became, according to the number of stamens, *Monandria, Diandria, Triandria,* etc. The flowerless plants—the algæ, fungi, mosses, and ferns—he placed last, in the Class *Cryptogamia*—plants with concealed reproductive organs. Obviously, from the modern viewpoint, Linnæus's system is an artificial one—as he himself recognized. In 1753 Linnæus published his *Species Plantarum* in which he described all species of plants known up to that time and in which he adopted the binomial system of nomenclature. This work has become a classic in botanical literature and has, by general agreement among botanists, been adopted as the basic reference in the scientific naming of plants. This volume was followed in 1754 by the *Genera Plantarum,* in which Linnæus describes the genera then known to him. Considering the nature and inadequacy of much of the material Linnæus had at his disposal to work with, and the slowness of communication, not to mention the chaotic condition of botanical knowledge at the time, one must marvel at the amount of work he accomplished; and not the least of what he did was concerned with the describing and naming of a large mass of American plants brought or sent to him by collectors; one cannot open a volume on our American flora without the eye's falling upon the initial letter of his name appearing after the name of some genus or species, as *Rosa carolina* L., *Lilium superbum* L.

Since the time of Linnæus many systems of classification have been proposed, all having as their main object the grouping of plants in a natural system, that is, according to their affinities. The rapid advances in the branch sciences of morphology, genetics, and phytogeography have contributed a wealth of new evidences regarding the probable relationships of plants, and the result is reflected in all the systems of classification that have been proposed.

In 1789, Antoine Laurent de Jussieu proposed a system in which he recognized three major groups of plants which he called the *Acotyledones, Monocotyledones,* and *Dicotyledones.* In the first he included the algæ, the fungi, the mosses, and ferns, and the "Naiades." The *Dicotyledones* he grouped as *Apetalæ, Monopetalæ, Polypetalæ,* and *Diclines Irregulares*—the first three obviously grouped on characters of the corolla, the last on the sexuality of the flowers (unisexual), and general absence of a corolla. The *Monocotyledones* and the *Dicotyledones* were further subdivided into classes based on *hypogyny, perigyny,* and *epigyny,* thus giving prominence to the positional relations of the floral organs—a noteworthy step in advance toward a natural system of classification.

With the system of A. P. De Candolle (1819) we find anatomical features for the first time used as a basis of classification. De Candolle divided the plant kingdom into two major groups based on the presence or absence of *vascular bundles.* The first group he named *Vasculares*—plants with vascular bundles; the second he named *Cellulares*—vascular bundles absent. The *Cellulares* he subdivided into two categories according to the presence or absence of leaves. His *Vasculares* are significant in that he here distinguished those plants in which the vascular bundles are arranged "in a ring," the *Exogenæ* as he called them, and which we of to-day know as plants possessing exogenous growth in thickness. Opposed to these were his *Endogenæ* or plants with "scattered" vascular bundles, that is, characterized by endogenous growth in thickness, namely the Monocotyledons. But he went further, taking into account the features of the perianth and placing in one group (*Monochlamydeæ*) all forms with a "simple" or undifferentiated perianth, and in another (*Diplochlamydeæ*) all forms in which the perianth is differentiated into calyx and corolla. Polypetaly and sympetaly, as well as hypogyny, perigyny, and epigyny, formed the bases of his more minor categories.

Stephen Endlicher (1836-1840) formed two major groupings, the *Thallophyta* or plants not differentiated into stem and root, and the *Cormophyta* or plants with stem and root. In the former he placed the algæ, the fungi, and the lichens; in the latter all other plants. But in 1827, the Englishman Robert Brown had published his discovery of the naked ovules of the cycads and the conifers and had established the Gymnosperms as an independent group—a discovery which is reflected in Endlicher's classification, where we find these plants as a subordinate group in his subdivision *Acramphibrya*—plants with "stems growing both at the point and circumference"—of the *Cormophyta.* But he neglected to consider the Angiosperms as one comprehensive group, rather subdividing them as *Apetala*—with a perianth single or absent, *Gamopetala*—with a gamopetalous (sympetalous) perianth, and *Dialypetala*—with polypetalous corolla. The Monocotyledons he named *Amphibrya,* and gave them coördinate rank with the *Acramphibrya* and the *Acrobrya* (mosses, liverworts, ferns).

Coming to Brongniart (1843) we find two major divisions, the *Cryptogamæ* or plants without flowers, and the *Phanerogamæ* or plants with flowers. These groups and names are still recognized at the present day. But Brongniart made a notable advance in making the Monocotyledons and the Dicotyledons coördinate subdivisions of the *Phanerogamæ.* However, he did not see the full significance of the characters of the embryo of the Gymnosperms, which led him to give this group coördinate rank with the Angiosperms under the *Dicotyledones.*

In 1845, in England, John Lindley proposed a system of classification in which he distinguished "Asexual or Flowerless Plants" and "Sexual or Flowering Plants." The former we recognize as the *Cryptogamæ,* the latter as the *Phanerogamæ,* of Brongniart. Lindley established seven classes, two of which, *Thallogens*—plants with "stem and leaves undistinguishable," and *Acrogens*—plants with "stem and leaves distinguishable," he included in the asexual or flowerless group. Of the five remaining classes that comprised his sexual or flowering plants he distinguished (1) the *Rhizogens* as plants having a "fructification springing from a thallus," and (2) the remaining four as having a "fructification springing from a stem." These four he separated chiefly on anatomical characters of the stem: the *Endogens* having the "wood of stem youngest in the center . . . and always confused"; the *Dictyogens* having the "wood of stem, when perennial, arranged in a circle with central pith"; the *Gymnogens* and the *Exogens* having the "wood of stem youngest at the circumference, always concentric, and cotyledons two or more." The *Gymnogens* were distinguished as plants with naked seeds (*Gymnospermæ!*) and the *Exogens* as plants with enclosed seed-vessels. Thus Lindley used vegetative, anatomical, fruiting, and seed characters as the bases of his classification.

With the rapid strides in morphological botany at the time came the system of Alexander Braun, in 1864, in Germany. Braun proposed three major categories, the *Bryophyta,* comprising the algæ, fungi, lichens, charas, and mosses; the *Cormophyta,* comprising the ferns, horsetails, and clubmosses; and the *Anthophyta,* comprising the *Gymnospermæ* and *Angiospermæ*—with coördinate ranking. The *Anthophyta* he subdivided into *Monocotyledones* and *Dicotyledones* and the *Dicotyledones* into *Apetalæ, Sympetalæ,* and *Eleutheropetalæ.* The main feature of Braun's system as far as the higher plants are concerned is the coördinate ranking he gave to the Gymnosperms and Angiosperms under the *Anthophyta*—the *Spermatophyta* of the modern grouping—and also his distinguishing of the cycads (*Frondosæ*) and the conifers (*Acerosæ*) of the Gymnosperms.

Between the years 1862 and 1893 came the system of Bentham and Hooker, which appeared in their monumental work, *Genera Plantarum,* published in England. This scholarly work held the stage of systematic botany until the appearance of Engler and Prantl's *Die Natürlichen Pflanzenfamilien* in the 90's. The completeness of the system precludes a detailed discussion, but in its essential features the higher plants were grouped in three major categories—the *Dicotyledones, Gymnospermæ,* and *Monocotyledones.* These were subdivided into "series," "cohorts," and "orders," the cohorts corresponding to what are now known as orders, and the orders to what we now know as families. The *Dicotyledones* were subdivided into three groups, of which two—the *Polypetalæ* and the *Gamopetalæ*—are based on the configuration of the corolla, and the third, the *Monochlamydeæ,* on the nature of the perianth. The *Gymnospermæ* comprised three orders —*Gnetaceæ, Coniferæ, Cycadaceæ.* The *Monocotyledones* comprised seven series based on the characters of the perianth, gynœcium, and seeds. In all 200 orders (families) were recognized. The system of Bentham and Hooker thus epitomizes the progress of the studies of floral morphology up to that time, and their *Genera Plantarum* is still a standard source of reference in all taxonomic work of the present day. And side by side with the *Species Plantarum* of Linnæus, the *Genera Plantarum* of Bentham and Hooker has become a classic of systematic botany.

With Eichler (1883) came the system of classification that is generally in vogue at the present day, particularly with reference to the major categories. Eichler's major groups are the *Cryptogamæ* and the *Phanerogamæ*. The *Cryptogamæ* he subdivided into the *Thallophyta*—comprising the *Algæ* and the *Fungi;* the *Bryophyta*—comprising the *Hepaticæ* (liverworts) and the *Musci* (mosses); and the *Pteridophyta*—comprising the *Equisetinæ* (horsetails), the *Lycopodinæ* (clubmosses), and the *Felicinæ* (ferns). His *Phanerogamæ* comprised the *Gymnospermæ* and the *Angiospermæ*. The *Angiospermæ* he grouped as *Monocotyleæ* and *Dicotyleæ,* the latter comprising two subclasses, the *Choripetalæ* and the *Sympetalæ*. The next lower subdivisions he called *Reihen* ("series").

Then in 1892 came the epochal system of Adolph Engler, the distinguished German botanist who for many years was professor of botany at the University of Berlin and director of the Berlin Botanical Garden. This system has superseded that of Bentham and Hooker and is now in general use among botanists the world over. It is the system adhered to in the present volume, particularly as regards the concepts of orders and families and their sequence. The system is the basis of *Die Natürlichen Pflanzenfamilien* of Engler and Prantl, a work of many volumes and the result of the coöperation of many specialists. A complete outline of Engler's system will be found in the *Syllabus der Pflanzenfamilien* by Engler and Gilg, brought up to date and amplified in the ninth and tenth editions published in 1924 in one volume. For students of taxonomy this is an exceedingly useful reference volume, possessing the great advantage of having the minor categories keyed out so that one may grasp at once the diagnostic characters and the phylogenetic positions. And under each such category are listed the more important genera and species, often with a brief characterization accompanied by notes on distribution. Engler divides the plant kingdom into thirteen major divisions (*Abteilungen*), the thirteenth division comprising the *Gymnospermæ* and the *Angiospermæ,* that is, the seed plants—which he calls the *Embryophyta Siphonogama* or embryo-producing plants in which fertilization is effected by means of a pollen-tube. The Dicotyledons are divided into two subclasses, the *Archichlamydeæ* or plants with a primitive floral envelop and the *Metachlamydeæ* or plants with a modified floral envelop. In all Engler established eleven orders of Monocotyledons and forty-one of Dicotyledons.

In contrast to Engler's system stands that of Richard Wettstein, professor of botany in the University of Vienna. Wettstein's system appears in his comprehensive volume, *Handbuch der Systematischen Botanik,* the last edition of which appeared in 1924. In this system Wettstein divides the plant kingdom into seven major categories or phyla (*Stämme*) each of which is subdivided into classes. For the seed plants (Gymnosperms and Angiosperms) he adopted Braun's name *Anthophyta*—as possessing priority over the name *Spermatophyta,* which connotes *sperm* or male reproductive cell. At the same time the name *Anthophyta* (meaning *flower* + *plant*) embodies Wettstein's conception of the Gymnosperms as being essentially flowering plants. A further departure from Engler's system is seen in the position assigned to the Monocotyledons, namely, following the Dicotyledons, thereby conceiving them as derived from the Dicotyledons —through the *"Polycarpicæ,"* the *Ranales* of Engler. The Dicotyledons Wettstein subdivides, as did Warming,* as *Choripetalæ* and *Sympetalæ*. The *Choripetalæ* or flowering plants with a corolla of distinct petals he very conveniently subdivides into *Monochlamydeæ* (flowers with undifferentiated perianth) and *Dialypetalæ* (flowers with perianth differentiated into calyx and corolla). In the main features otherwise, as for example the position of the *Verticillatæ, Fagales,* etc., the system parallels that of Engler. But in the interpretation of orders some notable departures will be observed. Wettstein recognizes thirty-nine orders of Dicotyledons and nine of Monocotyledons. In the discussion of his principles he takes occasion to call attention to the probable significance of serum diagnoses in phylogenetic classification. On the whole Wettstein's system is suggestive and probably comes closer to a correct interpretation of the phylogenetic connections of the flowering plants than does Engler's.

In America, Asa Gray (1810-1888), who has been called the father of American botany and who was professor of botany and director of the botanical garden at Harvard, became the chief exponent of Bentham and Hooker's system. Gray's *Lessons* and *Manual* are classics of American botany. The seventh edition of Gray's *Manual* was rearranged and modified in accordance with the systems of Eichler and of Engler and Prantl by Professors Benjamin Lincoln Robinson and Merritt Lyndon Fernald of Harvard.

Another classic of American Botany, the *Illustrated Flora of the Northern States and Canada and the British Possessions,* by Nathaniel Lord Britton, Professor of Botany in Columbia University and until recently director of the New York Botanical Garden, and the Honorable Addison Brown, President of the New York Botanical

* Warming-Potter, *Systematic Botany,* English edition.

Garden, appeared in 1896, in three volumes. The outstanding feature of these volumes is the illustration which accompanies the description of each species. The system followed is in the main that of Engler and Prantl. A smaller work, in one volume—the *Manual of the Flora of the Northern States and Canada,* by Professor Britton, appeared in 1907.

In 1894 appeared a notable system of classification of the flowering plants by Professor Charles E. Bessey, then head of the department of botany in the University of Nebraska. The system was modeled somewhat on that of Bentham and Hooker. A revised and amplified form of the system appeared in 1915. Bessey divided the flowering plants into two classes, the *Alternifoliæ* (*Monocotyledoneæ*) and the *Oppositifoliæ* (*Dicotyledoneæ*), on the basis of the position of the primary (embryonic) leaves. Of the *Alternifoliæ* he made two subclasses, the *Strobiloideæ,* which comprised most of the plants below the irises, and the *Cotyloideæ,* which comprised the irises and the other monocotyledonous groups above them, besides the *"Vallisneriaceæ"* (tape-grasses). He made a similar subdivision of the *Oppositifoliæ*—into *Strobiloideæ* ("cone flowers") and *Cotyloideæ* ("cup flowers"). These were further subdivided into superorders and orders. The orders and their families he construed as forming three phylogenetic series radiating from the dicotyledonous order *Ranales,* thus at the same time conceiving the Monocotyledons as derived from the Ranalian complex, and as constituting one of the three evolutionary series. The two series (*Strobiloideæ* and *Cotyloideæ*) of the Dicotyledons are shown to have evolved from a hypogynous, apocarpous, apopetalous, polycarpellate type of flower to a syncarpous, dicarpellate, sympetalous, epigynous type. We have thus a phylogenetic tree of three main branches radiating from the polycarpellate, apocarpous, hypogynous, Ranalian base. While Bessey's contribution embodies some admirable features from the phylogenetic standpoint, the system has not received wide adoption. The names *Alternifoliæ, Oppositifoliæ, Strobiloideæ,* and *Cotyloideæ,* like Engler's *Embryophyta Siphonogama,* while expressing a morphological condition, are rather cumbersome, and unless one possesses a professional knowledge of the field it is rather difficult to grasp the distinction at once between a "cone flower" and a "cup flower." From the teaching and learning standpoints the system has the disadvantage of being rather involved and rendering it difficult to place quickly in its proper category any plant that may be in hand, unless one is thoroughly grounded in the numerous types of flowers which a satisfactory manipulation of the system requires. Furthermore, the "apocarpous" as opposed to the "syncarpous," the "hypogynous" as opposed to the "epigynous," the "actinomorphic" as opposed to the "zygomorphic," and the "polypetalous" as opposed to the "sympetalous" categories are so involved in the phylogenetic tree ordinarily presented in explanation of the system as to be difficult to disentangle. To the student of phylogeny, however, the system is of special interest.

Under the authorship of Alfred Barton Rendle, Keeper of the Department of Botany, British Museum, there has recently appeared (1925, 1930) an excellent work on *The Classification of the Flowering Plants* (in two volumes), which in the main embodies the system of Engler and Prantl but which offers the added advantage of having the choripetalous Angiosperms subdivided into the two very convenient categories, *Monochlamydeæ* and *Dialypetalæ,* while the *Sympetalæ* are subdivided into the equally convenient groups *Pentacyclicæ* (flowers with five whorls) and *Tetracyclicæ* (flowers with four whorls). The *Tetracyclicæ* are again conveniently subdivided into the two easily recognizable groups *Superæ* and *Inferæ,* according as the ovary is superior or inferior, respectively. Rendle here follows Warming (*loc. cit.*) and Wettstein (*loc. cit.*) with respect to the major categories. The characters of these categories are very easily grasped and enable the student to place a plant in its respective category with comparative ease. As a workable system, in which technical problems of phylogeny may be set aside for the specialist, and in which the major taxonomic groups are made to stand out clearly, Rendle's system is superior. It is the system adhered to in the present volume, as regards major groupings.

The latest system to attract attention is that of J. Hutchinson, of the Royal Botanic Gardens, Kew. Hutchinson's system, as published in his *The Families of Flowering Plants* (1926), covers, however, only the Dicotyledons. In his system Hutchinson retains the *Archichlamydeæ* and the *Metachlamydeæ* (*Sympetalæ*). In the former he recognizes two phylogenetic lines, one radiating from the *Magnoliales* and comprising forms in which the arborescent habit is predominant, the other, in which the herbaceous habit is predominant, radiating from the *Ranales.* In thus separating the (primitive) arborescent forms from the herbaceous, Hutchinson departs from Engler and Gilg and from Wettstein, who retain them in the *Ranales* and *Polycarpicæ,* respectively. From these "polypetalous" groups of the *Archichlamydeæ* evolutionary lines are traced through the *Metachlamydeæ,* terminating in several climax groups, among which are the *Apocynales, Rubiales, Asterales,* and *Lamiales.* Hutchinson also regards the Monocotyledons as derived from the Ranalian complex. Besides, several new order names are proposed. In his own words, his system is "a phylogenetic system based on the assumption that plants *with sepals and petals,* associated with other floral characters regarded as also primitive, are more *ancient phylogenetically* than plants without sepals or petals." *"Free parts* are regarded as *primitive,* and *connate* or *adnate* as *more recent,"* and *"the spiral* arrangement of parts is more primitive than the *cyclic* (whorled), and *numerous free* stamens are earlier than the *few* or connate. Also the hermaphrodite precedes the unisexual flower," and *"resemblances* are emphasized rather than *differences."* The last clause is significant as emphasizing a viewpoint that needs to be given thoughtful consideration in taxonomy.

CHAPTER II

NOMENCLATURE, GENUS AND SPECIES, PHYLOGENY, HOMOLOGY AND ANALOGY

By *nomenclature* is meant the technical or scientific names and naming of plants (and animals) and of the groups to which they belong. Obviously nomenclature must be reduced to a system or confusion would result. The *binomial system* of nomenclature now in general use in the biological sciences dates from Linnæus. According to this system the technical name of a plant consists of two parts, the *generic* name, or the name of the genus to which the plant belongs, and the *specific* name, which sets it apart from other species of the same genus. Thus the technical name of the red oak is written *Quercus rubra*. In more formal writing the name, an abbreviation of the name, or the initial letter of the name of the author who first published it is added, as *Quercus rubra* L., meaning the *Quercus rubra* of Linnæus. If a variety of the species has been described, the technical name is written as a trinomial, as *Carex intumescens Fernaldii* or *Carex intumescens* var. *Fernaldii,* or more formally *Carex intumescens* Rudge, var. *Fernaldii* Bailey. If a species is transferred to another genus, the name or abbreviation of the name of the author of the original name of the species is placed in parentheses followed by the name or abbreviation of the name of the author who made the transfer (Example: *Agrostis hyemalis* (Walt.) B.S.P., from *Cornucopia hyemalis* Walt.).

The generic name may be derived from any source, but when derived from any other language than Latin it must be written in the Latin form, in the nominative singular. The specific name may be (1) an adjective, (2) a noun in the Latin genitive, or (3) a noun in apposition (Hitchcock)—*Quercus rubra, Quercus Michauxii, Myrica Gale,* respectively. In the best practice technical names are made as euphonious and as convenient to manipulate as possible.

Nomenclature cannot be separated from classification, for every group or category must have a name, and the name is intended to convey some particular feature of the group, which sets it off from other groups, thus at once classifying it. In order that system may prevail in nomenclatorial practice, an international code of nomenclature has been established for the guidance of taxonomists. There is, besides, an American code in use among some American systematists. According to the international code specific names that are nouns in apposition or proper nouns are written with a capital letter, as *Zephyranthes Atamasco, Carex Fraseri.*

A full classification of *Quercus rubra* with the corresponding nomenclature, according to Engler and Gilg's *Syllabus der Pflanzenfamilien,* is written as follows:

Division *Embryophyta Siphonogama,*
Subdivision *Angiospermæ,*
Class *Dicotyledoneæ,*
Subclass *Archichlamydeæ,*
Order *Fagales,*
Family *Fagaceæ,*
Genus *Quercus,*
Species *rubra.*

Genus and species. Such names as *rose, oak, willow, cherry,* designate different *kinds* of plants. These are *generic* names, in English. In technical nomenclature these names become *Rosa, Quercus, Salix, Prunus.* But when we wish to distinguish one rose from another, one oak from another, and so on, we say *white* rose, *red* oak, *black* willow, etc.; in other words we designate a *particular* kind of rose, oak, willow. These names, then, are *specific* names, as used in English. Translated into technical language the specific names of these plants become *alba, rubra, nigra,* and the technical names of the plants become *Rosa alba, Quercus rubra, Salix nigra.* But the technical name is not always a direct translation of the common name: for the "pin cherry," for instance, the technical name is *Prunus pennsylvanica.* The advantage of the technical name over a common name is seen in the use of the appellation "pigweed," for example. This name is applied to at least two different species belonging not only to different genera but to different families. Hence in using this name some characterization of the plant is necessary before its identity can be determined. But the use of the technical name, *Amaranthus retroflexus* or *Chenopodium album,* at once establishes the identity of the species, whichever it may be.

The species is the unit of classification. The term may be variously defined. But what we have to deal with here are taxonomic species, and in this sense a species may be defined as a group of individuals that are fundamentally and uniformly alike, particularly with respect to the more deeply seated characters such as appear in the organs of the flower. Variations in vegetative characters or in shape and size of petals, for example, may ordinarily be regarded as of minor importance. Such characters may serve to differentiate subspecies or varieties or forms. Often it is very difficult to delimit a species precisely, yet if every minor variation were to be made the basis of specific distinction, there would be no end to the number of species that might be made. A species, therefore, is a concept, not something absolute and inelastic. Just where to draw the line rests upon the good judgment of the taxonomist. But for taxonomic purposes a line must be drawn somewhere, and preferably broadly. And in drawing the line differences should not be allowed to obscure resemblances.

Genera, like species, are concepts, that is they are aggregations, as nearly as can be determined, of closely related species. The same general principles of delimitation hold for genera as for species. Many genera—*Rosa, Salix, Quercus, Prunus, Ribes, Geranium*—seem rather well defined as far as our experience goes. However, with wider acquaintance the lines may seem to break down. Thus the genus *Ribes,* as broadly and ordinarily construed, comprises currants and gooseberries. The question is: Are the gooseberries sufficiently distinct from the currants to warrant separating them as a genus apart? That has been done by some authorities. In some systematic works, therefore, the gooseberries will be found listed under the genus *Grossularia.* Similarly *Prunus* comprises the cherries, plums and, in a still broader sense, peaches, almonds, and apricots. If *Prunus* shall be limited to the cherries alone, new generic names must be found for the other groups. Indeed the line between cherries and plums is not so easily drawn. The tan-bark oak, as the common name implies and as the fruit at once suggests, is an oak to the uninitiated, yet it goes by the technical name *Lithocarpus densiflora.* Similar instances could be multiplied without end.

When a genus consists of but one species, it is said to be *monotypic.* But a family may also be monotypic. These conditions are fulfilled by the lopseed, *Phryma Leptostachya,* a species of Eastern parts of the country (p. 557). Some genera, on the other hand, at present contain a large number of species, in some cases running into the hundreds (*Senecio, Euphorbia, Carex*), and it is a matter of opinion whether or not such genera should be broken up as a matter of convenience. But convenience is better served, no doubt, by larger units in the case of genera.

Phylogeny. The ultimate aim of taxonomy or of systematic botany is to arrange plants in a "natural" series, that is, in a series based on relationships. Such a series will be an evolutionary series, and will show the progressive steps from a primitive or generalized type to a specialized one. If the present living organisms have arisen by spontaneous creation, no such evolutionary series can be shown, for obviously there can be no relationship, although there might be similarities in form, etc. Graphically represented, an evolutionary series would take the form of a tree, but a tree which, owing to the limitations of the sheet of paper on which it might be drawn, would have all its branches lying in the same plane.

That branch of biology which deals with the relationships of organisms is called *phylogeny* (*phylo,* "race" + *geny,* "to give birth to"). Stated in other words phylogeny is race-history, or the developmental history of the race. The term is synonymous with *genealogy.* Many of the data of phylogeny are deduced from a study of the developmental history of the individual or *ontogeny* (*onto,* "a thing which exists" + *geny*). And, as Von Baer explained it, the developmental history of the individual recapitulates the developmental history of the race, or "ontogeny repeats phylogeny."

The main sources from which the evidences of the racial affinities of plants have been drawn are *paleobotany* and *comparative morphology.** While the fossil record has yielded much valuable information, there yet remain gaps to be closed. Unfortunately the tissues of many plants are of such nature as to have precluded preservation as fossils. The morphological record includes a great mass of evidence derived from intensive studies of the form and structure of the reproductive organs and reproductive elements, and of anatomy and embryology, and of the form and structure of the flower, fruit, and seed of the flowering plants.

Biologists are generally agreed that the existing living organisms have sprung from some simple, probably unicellular, ancestral stock. But whether they have sprung from one solitary cell that originated in one particular spot or from different cells that originated in widely separated places, it is not possible to determine. That organisms developed from simple to complex forms appears certain. Probably the first organisms originated in the sea and by stages came to inhabit the land, where the more unstable environment brought about differentiated forms that became the ancestral stocks of the present diverse forms. If it could be shown that the flowering plants, for instance, have descended in direct line from some primitive simple ancestral stock, the evolutionary series, or phylogenetic tree, would be very simple—merely a linear series, or

* Recently another source, that of *serum diagnosis,* is yielding some significant evidences, but the method has not as yet been widely applied.

trunk, without branches. Such an origin would be *monophyletic.* On the other hand, if, as is the more likely, the various groups have sprung from different ancestral stocks, their origin would be *polyphyletic.* But any phylogenetic tree that we might construct will be more or less of an abstraction, a concept, based on our interpretation of the evidences of the possible affinities. The points from which the branches of the phylogenetic tree depart from the main trunk are often very obscure owing to the incompleteness of our knowledge. Perhaps we might better conceive the evolutionary series as a shrub of numerous sprouts or stems springing from the ground but all connected beneath the surface in some way we do not see.

Homology and analogy. The study of comparative morphology involves an understanding of what is meant by homologous and analogous parts. Many organs of plants are similar in appearance but different as to origin. Such organs are *analogous.* Thus the leaf of a moss may resemble very strikingly the leaf of the water-weed, *Elodea* (*Philotria*) *canadensis,* for instance; indeed the resemblance is such that to the popular mind this water-weed sometimes goes by the name "ditch-*moss.*" Many other small-leaved flowering plants are called "mosses" for the same reason. However, while the leaves of the moss and the water-weed are similar in appearance and also in function, nevertheless they are of widely different morphological origin—being a gametophytic organ in the moss and a sporophytic organ in the water-weed. Similarly the showily colored leaves and bracts of the inflorescence of the poinsettia and other euphorbias, and the purple bracts of the bougainvillæa, are commonly mistaken for petals, although they are entirely extraneous to the flower; they may be analogous to but not homologous with petals. On the other hand, the leaves of the water-weed, and of a grass, a lily, or an oak, although dissimilar in appearance, are *homologous* organs—since they are of similar origin, belonging to the sporophytic generation. The tubers of the potato plant and the tubers of the Jerusalem artichoke are homologous organs, both being stems, but the potato tuber and the tuber of the sweet potato are analogous organs—the one a stem, the other a root. From these examples it will be seen that analogous organs are more or less similar in appearance, while homologous organs may or may not be similar in appearance.

CHAPTER III

THE FLOWER

The flower is the characteristic organ of the Angiosperms. We know from experience that there are many kinds of flowers, differing widely in color, size, shape, etc., and we have learned to associate each kind with a particular kind of plant; in fact, when we think of any particular plant it is its flower that comes first to mind. Often, too, we think of the flower as being the only colored and showy part of the plant, although other parts of the plant, such as bracts, leaves, and fruits, may be equally showy, or even more so. And, generally, when we think of the flowers of a sunflower, aster, dandelion, or thistle, it is the flower-cluster that comes to mind, rather than the individual flowers of the cluster; that is because the individual flowers in such clusters are very small, as is often the case in many other flowering plants—so small in fact, in hundreds of cases, that their structure is more or less of a mystery. Some flower-clusters, composed of thousands of very small flowers, often grow to enormous size, as in the palms, while such gigantic clusters as those of the yuccas, aloes, and agaves are composed of quite large flowers. In the duckweeds of ditches and ponds, the smallest of all flowering plants, the flowers are minute and very simple, while in contrast to these are the gigantic flowers of the rafflesias of Sumatra, which in one species measure a meter—over three feet—across, and are the largest flowers known.

It is the structure of the flower that forms the basis of classification of the flowering plants, and it is on differences in its structure that all major and minor groupings are based. However, it should be remembered that fruit, seed, and vegetative organs play an important and significant rôle.

In spite of the fact that flowers appear to differ so widely, in outward appearance at least, nevertheless they are all constructed on the same general plan. Differences have been brought about by modifications of the general plan, and it is in these numerous modifications that the systematic botanist seeks for evidences of affinities or relationships and attempts to tie together all those forms that appear to possess similar modifications or structure, arranging them in an orderly series, according to the degree of modification. The ultimate aim, in fact, of any scheme of classification is to arrange all plants in an orderly series according to affinities, proceeding from the more primitive to the more specialized, so as to give us a sort of

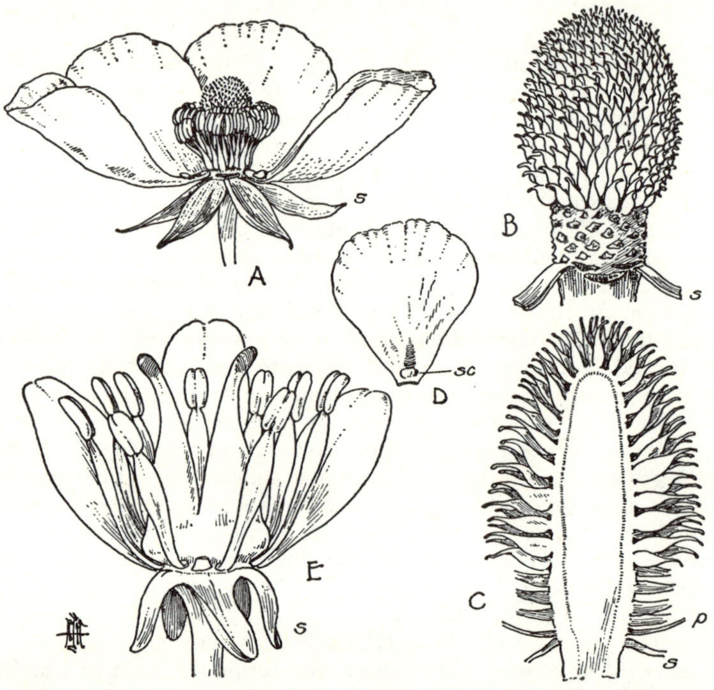

FIG. 1.—*A,* The flower of a buttercup (*Ranunculus asiaticus*), with one petal removed; *B,* the gynoecium of the same; *C,* the gynoecium of the same in median section; *D,* a petal of the same, with a scale (*sc*) at the base; *p,* petals; *s,* sepals; *E,* the flower of a saxifrage (*Saxifraga*), with two petals removed; *p,* petals; *s,* sepals.

picture of the whole course of evolution—a phylogenetic tree of the plant kingdom. But, as has already been pointed out, a phylogenetic tree has its limitations, especially since it is obviously impossible in the two-dimensional plane of a sheet of paper to demonstrate the probable course of events in the evolution of all modern plants. Plants as we see them to-day are at the most merely the ultimate branches or twigs of any phylogenetic tree that we could construct; their ancestry is buried in the sands of the past. Much evidence has already been brought to light, however, that has enabled students of phylogeny to reconstruct, in part at least, the phylogenetic tree, although the probable connections of the branches are as yet subject to differences of opinion. With our

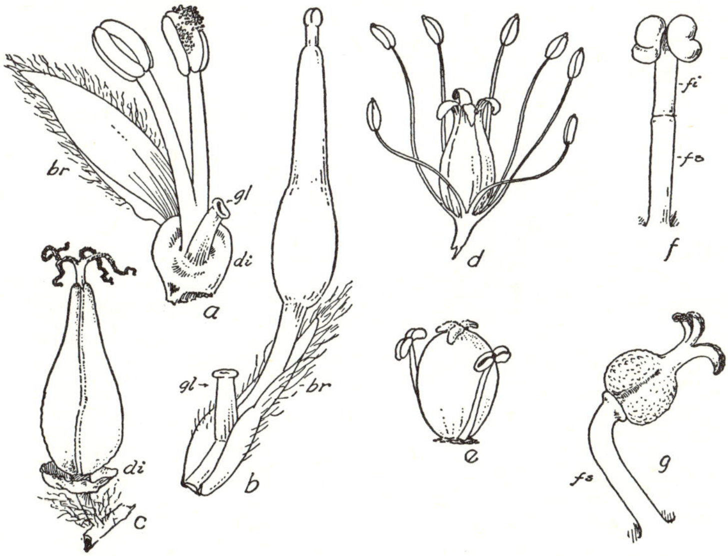

FIG. 2.—Flowers without a perianth (naked): *a,* staminate, *b,* carpellate, flowers of a willow; *br,* bract; *gl,* gland; *di,* disk; *c,* carpellate flower of a poplar; *di,* disk; *d,* flower of lizard's-tail (*Saururus*); *e,* flower of a pepper (*Piper*); *f,* staminate and *g,* carpellate flowers of *Euphorbia; fi,* filament; *fs,* flower-stalk (pedicel).

present knowledge no system of classification can hope to be absolute or final, but before we can comprehend any system, as far as the flowering plants are concerned, we must acquaint ourselves thoroughly with floral structures over as wide a field as possible and with the structure, life-habits, and distribution of plants in general. The more we can know about a plant, the more fascinating does the study of it become.

Morphologically a flower is a modified shoot. The parts of a flower are the spore-leaves or *sporophylls,* which are differentiated into *stamens* and *carpels;* and the floral envelop or *perianth,* which is commonly differentiated into *sepals* and *petals.* These organs are arranged spirally or in whorls on the axis of the shoot, or *floral axis,* called the *receptacle* (Fig. 1). The flower-stalk, when present, is called the *pedicel;* if the pedicel is absent, the flower is *sessile.*

The stamens and the carpels are known as the *essential organs* of the flower because they have to do with reproduction. Collectively the sepals are termed the *calyx* and the petals the *corolla.* Similarly the stamens constitute the *andrœcium* and the carpels the *gynœcium.*

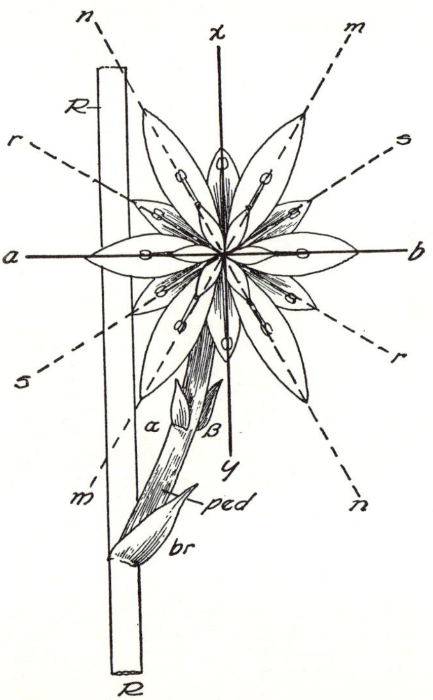

FIG. 3.—Schematic representation of radial symmetry in a flower: *xy,* median plane; *ab,* transverse plane; *mm, nn, rr, ss,* oblique planes; *R,* axis (*rachis*) of inflorescence; *br,* floral bract; *ped,* pedicel; *α* and *β,* bracteoles.

When all the organs are present, the flower is *complete;* but if either the calyx or the corolla is wanting, it is *incomplete. Naked* flowers are those that have neither calyx nor corolla. A flower is *bisexual, perfect,* or *hermaphroditic,* when both stamens and carpels are present; but when either one or the other set of these organs is wanting, the flower is *unisexual* or *imperfect.* Unisexual flowers are either *staminate* or *carpellate* (*pistillate*), according to which set of these organs is present. In either case rudiments of the other set are sometimes present (Figs. 1, 2).

With respect to their organization, flowers are *primitive* or *generalized, modified* or *specialized* (Fig. 1). Intergrading forms are numerous, since evolution has not proceeded in the same direction nor accomplished the same result in all cases. That is to say, a flower may be specialized in one respect, yet retain primitive characters.

In primitive flowers the organs are usually spirally arranged and numerous or of indefinite number, but with increasing specialization the organs tend to become whorled or *cyclic* and to be reduced to a definite and constant number. Complete suppression of a member or of an entire whorl or set of organs is of common occurrence.

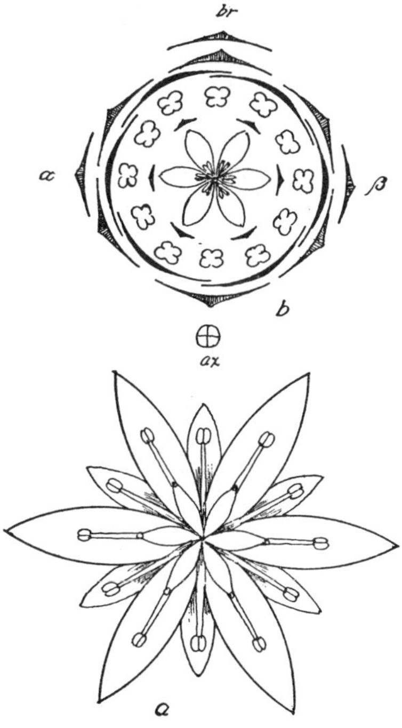

FIG. 4.—The radially symmetrical flower (*a*) of a stonecrop and its floral diagram (*b*); between the stamens and carpels in *b* is a whorl of small scales; *ax,* floral axis; *br,* floral bract; α and β, bracteoles.

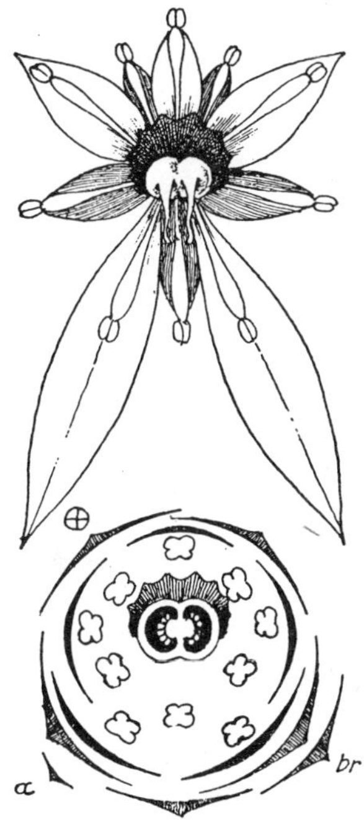

FIG. 5.—A bilaterally symmetrical flower of a saxifrage (*Saxifraga sarmentosa*) and its floral diagram; shaded portion the gland; *br,* floral bract; α, bracteole.

With respect to *symmetry,* flowers are generally *radially* or *bilaterally symmetrical,* these types being most prominently expressed by the corolla. Radially symmetrical flowers are usually termed *regular* or *actinomorphic,* bilaterally symmetrical forms *irregular* or *zygomorphic* (Figs. 3, 4, 5). In correlation with these evolutionary tendencies, the floral axis or receptacle tends to become more and more shortened and, besides, may develop in such a way as to modify materially the general structural plan of the flower. The general tendency in floral evolution appears to be towards the attainment of stability of floral design; but many forms have not as yet attained stability, and in such cases we may expect to find often considerable variation, especially as to number of parts, shape of petals, etc. Forms of this sort often present some puzzling and difficult taxonomic problems. It should be understood, too, in this connection, that simplicity of organization does not necessarily imply primitiveness: the flower of a valerian, for instance (Fig. 396), appears simple enough, having a sympetalous (simple!) corolla, one style, and sometimes only one stamen, but what is calyx, and what is ovary? As contrasted with the flower of a magnolia, a buttercup, or a columbine, it appears, indeed, to be very simple, but is it as primitive? Many questions of this sort will confront a student of taxonomy, and experience will show that what often appears very simple on the surface may prove to be just the opposite. Searching observation and careful weighing of evidence are nowhere more important than in the taxonomy of the flowering plants.

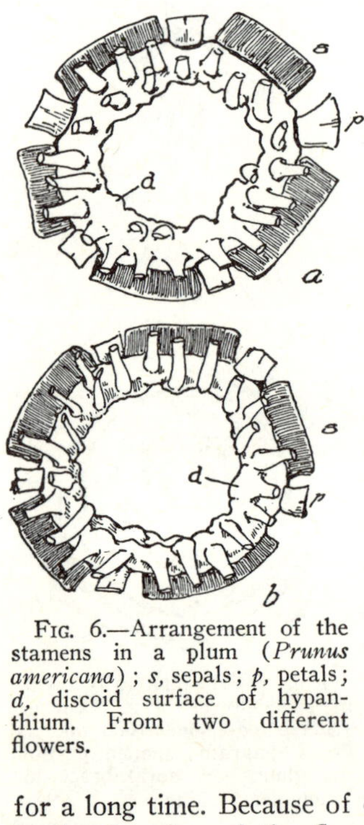

FIG. 6.—Arrangement of the stamens in a plum (*Prunus americana*); *s*, sepals; *p*, petals; *d*, discoid surface of hypanthium. From two different flowers.

Expansion of the flower. The period of expansion or opening of the flower is known as *anthesis*. Some flowers burst open very suddenly and may last only a short time, while others may open more slowly and may last for a long time. Because of differences in the time of flowering and in the duration of the flowering period, we have a procession of flowers of different kinds from early spring to late autumn. The sudden bursting into bloom of a cherry orchard, for instance, and the almost equally sudden vanishing of the "flowers" is a familiar phenomenon. In warm countries, however, some plants may bloom all the year round. Dandelions (and other composites) appear to bloom for a long time, but this is because the "dandelion" is a cluster of numerous very small flowers arranged in a close spiral, blossoming commencing with the outermost flower of the spiral and proceeding inwards until the last or innermost flower of the cluster, or head as it is called, has opened. A similar but more striking example of this sort is the gigantic sunflower of the gardens. In these forms the fruits of the outermost flowers may be ripe before the innermost flowers commence to bloom. Among some of our Northern trees and

FIG. 7.—*a*, empirical diagram of the flower of a lily; *b*, theoretical diagram of the flower of an orchid (*Cypripedium*).

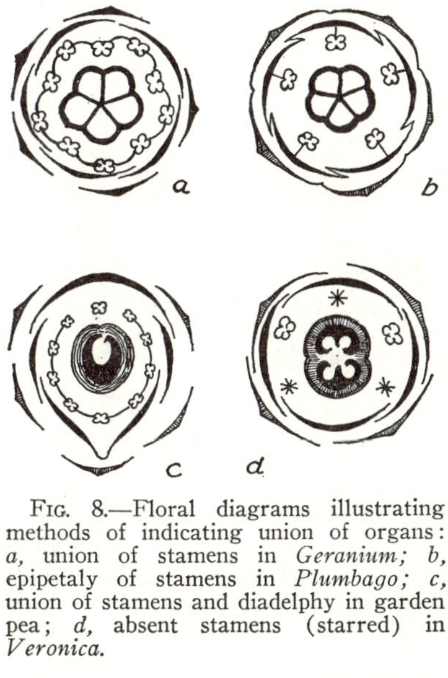

FIG. 8.—Floral diagrams illustrating methods of indicating union of organs: *a*, union of stamens in *Geranium; b*, epipetaly of stamens in *Plumbago; c*, union of stamens and diadelphy in garden pea; *d*, absent stamens (starred) in *Veronica.*

shrubs, the flowers appear before the leaves; in others they appear with the leaves, and in still others they appear after the leaves. All three conditions may be observed among the willows. Plums produce their flowers before the leaves, while their cousins, the cherries, produce them with or after the leaves. The showy flower of *Lewisia rediviva* of the Northwest appears months after the leaves have disappeared.

That branch of botany which deals with the relations between climate and the flowering period of plants is called *phenology.*

Fragrance of flowers. Fragrance is usually associated with flowers, but many flowers are without fragrance, and in many cases fragrance or distinctive odor emanates from the vegetative parts, as in the skunk-cabbage, sweet-flag, catnip, and the eucalypts. Characteristic odors may also emanate from fruits, especially of the fleshy sort. Fragrance or odor may be used as a key character if especially distinctive; at least it serves as a recognition-mark of the particular plants.

The color of flowers. The color of flowers is sometimes used as a basis of classification and within limits is convenient for the purpose in view; but a system of classification based on color alone can only be more or less artificial. The color-pattern of a flower is nevertheless a helpful guide in identification in many cases. "Albino" flowers may sometimes appear in species with flowers normally of another color (*Epilobium angustifolium*).

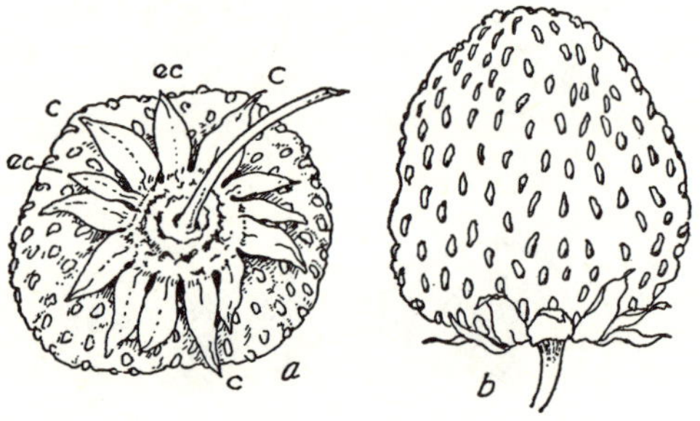

FIG. 9.—Fruits of the strawberry, showing epicalyx, *ec*, and calyx, *c*.

Insertion of the floral organs. The floral organs occupy definite portions or zones of the floral axis or receptacle—the order of sequence downwards being gynœcium, andrœcium, corolla, calyx. This is the sequence in all cases and is especially evident in hypogynous flowers: perigyny and epigyny do not alter the situation, whatever the modifications of the flower may be (Figs. 1, 68).

Arrangement (phyllotaxy) of the floral organs. In flowers of the primitive type all the organs are spirally arranged or *acyclic*. In advanced types the arrangement is whorled or *cyclic*. Intermediate forms in which some of the organs are acyclic and others cyclic are common, especially among the lower orders (*Ranales*). Persistence of the acyclic condition is confined especially to the gynœcium and the andræcium.

Number of members in a whorl. In this connection many variations will be found. A whorl consisting of but one member is *monomerous,* and when it has two members it is *dimerous;* higher numbers are expressed by the terms *tri-, tetra-, penta-,* to *polymerous* (many or numerous), as the case may be. Whorls having the same number of members are *isomerous,* but if they vary in the number of their members they are *heteromerous.*

Oligomery and pleiomery. When a whorl has fewer than the typical number of members it is *oligomerous,* and when it has more than the typical number it is *pleiomerous.* For example, in the harebell (*Campanula*) the whorls are typically pentamerous, but in some species of Campanula the gynœcium is trimerous and hence oligomerous with respect to the other whorls. And in the apple, which typically has a pentamerous gynœcium, a pleiomerous condition of six or seven carpels will sometimes be found.

Number of whorls in flowers. Suppression may involve not only individual members of a whorl but often entire whorls. Flowers having five whorls of floral organs—one of carpels, two of stamens, one of petals, and one of sepals—are *pentacyclic.* They are typical of the subdivision *Pentacyclicæ* (the five-whorled group) of the *Metachlamydeæ* and of certain groups of the *Archichlamydeæ* and of the Monocotyledons. A *tetracyclic* or four-whorled condition is characteristic of the *Tetracyclicæ* of the *Metachlamydeæ* and of certain groups of the *Archichlamydeæ* and of the Monocotyledons. *Mono-, di-, tri-,* and *polycyclic* flowers will also be met with.

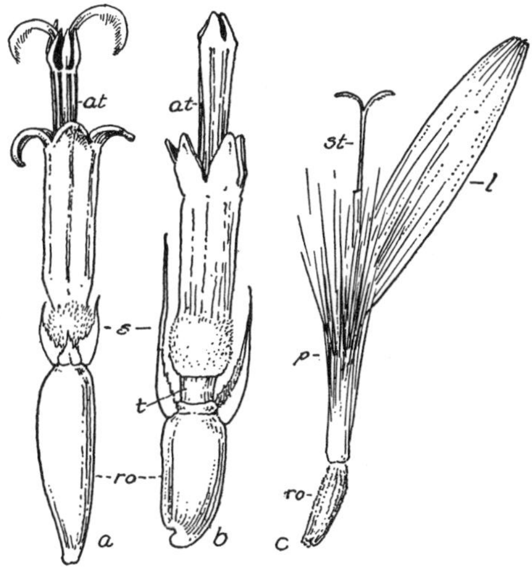

FIG. 10.—Flowers of composites: *a, b,* disk-flowers of *Helianthus orgyalis* (*a*) and *H. strumosus* (*b*); ligulate flower (*c*) of *Tragopogon porrifolius; l,* ligule; *st,* style; *at,* anther-tube; *s,* scale-like or chaffy and *p,* bristly pappus; *t,* stipe supporting corolla and stamens; *ro,* fused receptacle and ovary.

Oligotaxy and pleiotaxy. Reduction in the number of whorls of the floral organs is called *oligotaxy,* while an increase in the number is called *pleiotaxy.* Pleiotaxy may be exemplified by the common "doubling" of some flowers. Oligotaxy may be observed in the *Primulaceæ,* where one whorl of stamens has been suppressed in several cases. (See also *Scitamineæ, Microspermæ, Iridaceæ.*)

Cleistogamy. Flowers that open and permit cross-pollination by wind or insects or other agency are called *chasmogamous.* But besides these some plants produce, at the same time or later, another set of flowers that never open: these are called *cleistogamous* flowers. Cleistogamous flowers are usually inconspicuous, are sometimes produced underground, and differ in certain respects from the chasmogamous flowers of the same species. The presence of cleistogamous flowers is often of taxonomic value, and at the same time serves as a useful recognition mark of the species. Cleistogamous flowers occur among the violets, in some catchflies (*Silene*), in the grasses *Danthonia spicata* and *D. sericea,* and in many other species of different families. In a number of other cases flowers that are normally chasmogamous will, under certain conditions, remain closed; these have been called *pseudocleistogamous* flowers. They have been reported for the dandelion, portulaca, and many others.

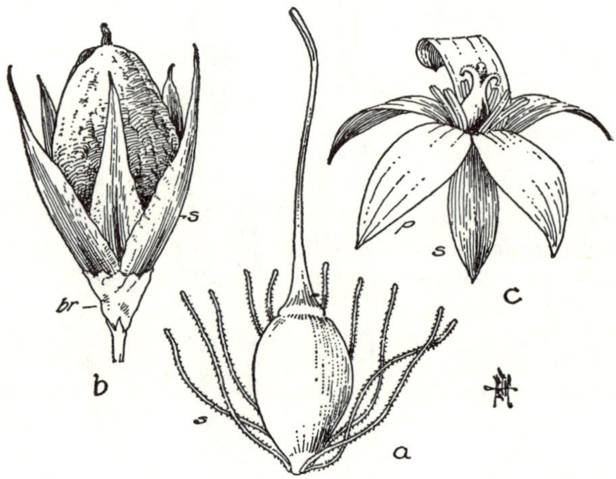

FIG. 11.—*a,* Gynoecium and bristles (perianth) in a sedge (*Rhynchospora alba*); *b,* the fruit and *s* perianth in a rush (*Juncus inflexus*); *s,* outer segments (sepals) of perianth, *br,* bract, and *c,* the flower of a trillium; *p.* petals; *s,* sepals.

The floral formula. For representing the numerical relations of the floral organs the *floral formula* is used. Thus the numerical relations in *Phlox* may be written *S*5 *P*5 *A*5 *G*5, the letters representing the different sets of organs and the numerals the number of members in each set. To show the number of whorls in each set the formula may be written thus: $S2 + 2\ P4\ A2 + 4\ G2$. An absent whorl is represented by O, as $S5\ P5\ A0 + 5\ G5$; in this case the inner whorl of stamens is suppressed. Coalescence of the organs of any set may be indicated by bracketing the proper numerals, as $S5\ P(5)\ A5\ G3$, the corolla in this case being sympetalous. Epipetalous stamens may be represented thus: $S5\ (P5\ A5)\ G3$ or $S5\ P5\ A5\ G3$. Superior and inferior ovary may be indicated by drawing a dash below or above the symbol for gynœcium, as the case may

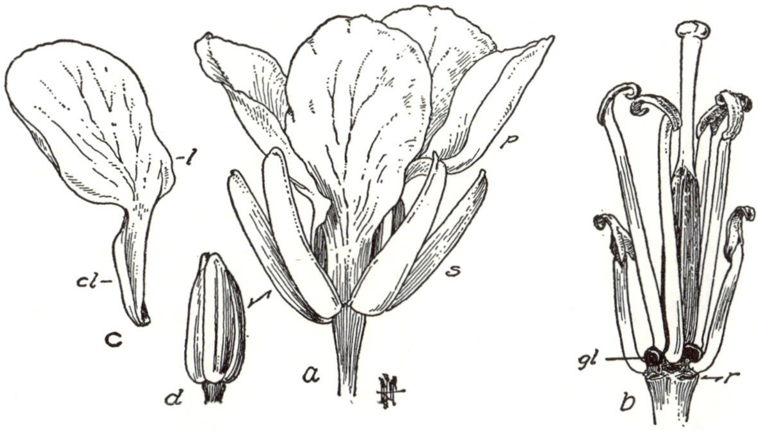

FIG. 12.—*a,* The flower of a mustard; *b,* the same with petals and sepals removed; *c,* a petal; *cl,* claw, *l,* limb; *s,* sepals; *p,* petals; *gl,* glands; *r,* receptacle.

be, as $S5\ (P5)\ A5\ G5$, $S5\ (P5)\ A5\ G2$. Syncarpy may be indicated by bracketing: $(G2)$, and the numerous carpels of an apocarpous gynœcium by the symbol for infinity, ∞, as $G\infty$.

As the floral formula is merely an arbitrary shorthand method of representing the numerical relations of the floral organs, it follows that other letters than the above and other types of formulas may be devised, the chief consideration being simplicity of expression. As a time-saver and a concise mode of statement of the numerical relations, it is a very convenient tool.

The floral diagram. What the floral formula expresses symbolically the floral diagram expresses graphically, but it has the further advantage of picturing in a simple way the arrangement or phyllotaxy of the floral organs. The floral organs are delineated as they appear *in situ* in a transverse section of the flower, certain conventional details being used to distinguish, for instance, petals from sepals, and to show other special features. For accurate work it is always advisable to cut transverse sections of a young or unopened flower-bud, as these will show the prefloration of the calyx and corolla exactly. If the stamens are numerous, their arrangement can be made out by cutting off the filaments close to their base, leaving the stubs to indicate the position (Fig. 6).

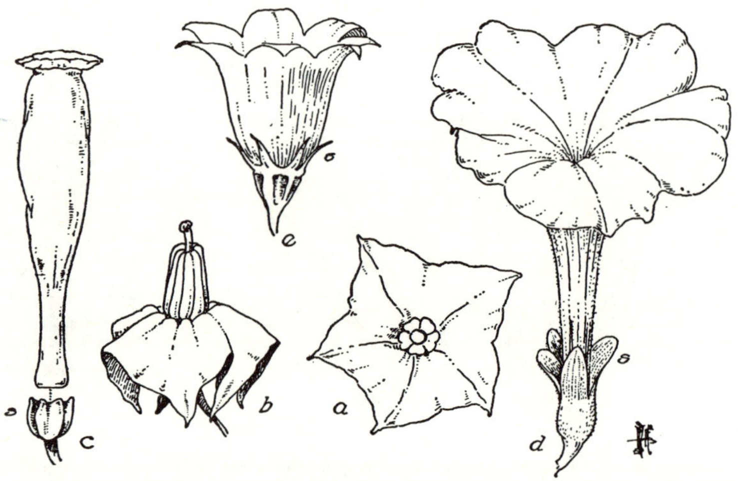

FIG. 13.—Sympetalous (metachlamydeous) flowers: *a,* flower of the potato plant, viewed from above; *b,* the same in side view; *c,* the flower of *Fabiana;* *d,* the flower of *Petunia;* *e,* the flower of the harebell (*Campanula rotundifolia*); *s,* calyx segments (sepals).

Two kinds of floral diagrams are usually made use of, namely, the *empirical* and the *theoretical.* The empirical diagram takes into account only the organs that are actually present in the flower; the theoretical diagram, on the other hand, attempts to replace in their proper relations organs that appear to have been suppressed, that is, it attempts to reconstruct the ancestral floral type (Fig. 7). In this way the evolutionary history and the probable relationships of the species may be expressed.

From the floral diagram we see, also, the type of symmetry of the flower, and by various conventional methods we may indicate sympetaly, epipetaly of the stamens, etc., etc. (Fig. 8).

The median section. The median (longitudinal) section of the flower is the most useful tool at our disposal in the study of floral structures. It, more than any other feature or device, at once shows us the phylogenetic connections of the flower. The relative ease with which it can be made, and the features that it reveals, enable us to place with reasonable facility many a strange flower that may come to hand. Of course the numerical relations of the floral organs must also be known, but such decisive features as hypogyny, perigyny, and epigyny, or superior or inferior ovary, can only be clearly demonstrated by the median section. For the use and significance of the median section see the subject of application of the system on a succeeding page, and figures in the text.

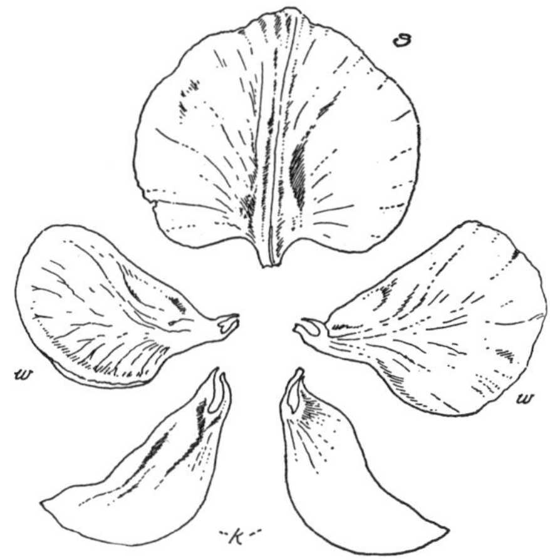

FIG. 14.—Parts of the corolla of the sweet pea: *s*, standard; *w*, wing petals; *k*, keel petals.

For convenience of reference, a summary of the probable steps in the evolution of flowers and the characters of a primitive flower are added below. These items constitute the so-called "canons" of taxonomy or systematic botany and should be familiar to every student. They are the criteria upon which modern phylogenetic systems of classification are based.

THE DIRECTION OF EVOLUTION OF FLOWERS

1. From many to few parts.
2. From spiral (acyclic) to whorled (cyclic) arrangement of the parts.
3. From free or distinct to united parts.
4. From hypogyny to perigyny and epigyny, or from superior to inferior ovary.
5. From complete to perfect forms.
6. From bisexual to unisexual forms—monœcious and diœcious.

7. From an undifferentiated perianth to one differentiated into calyx and corolla.
8. From many to few seeds.
9. From insect (or animal) pollination to wind pollination.
10. And with respect to the plant: from woody to herbaceous.

Characters of a Primitive Flower

1. Relatively numerous floral organs.
2. Floral organs spirally arranged.
3. Perianth undifferentiated or a gradual transition from calyx to corolla and even into stamens.
4. More or less gradual transition from bracts to perianth leaves.
5. Floral organs all free or distinct, not united.
6. Radial or actinomorphic symmetry.
7. Stamens leaf-like and connective exceeding anthers.
8. Absence of style.
9. Numerous ovules and seeds.
10. Small embryo in copious endosperm.
11. Two cotyledons or more.
12. Relatively long floral axis (receptacle).

And as to vegetative characters of the plant:

13. Entire leaves.
14. Tree-like growth-form.
15. Aquatic habit.

CHAPTER IV

CALYX AND COROLLA

Calyx. In all cases where the perianth or floral envelop is clearly differentiated the lowermost set of organs is the calyx. Usually the calyx is green, but in many flowers it is colored (petaloid), *e.g.,* in *Delphinium, Aconitum, Anemone, Fuchsia, Iris.* The sepals or segments of the calyx are commonly distinct, but many cases will be encountered in which they are more or less united (gamosepalous). In regular (actinomorphic) flowers the sepals are generally similar, while in irregular (zygomorphic) flowers they are often more or less modified and dissimilar (*Leguminosæ,* orchids). In some flowers of the primitive type (water-lilies) there is a gradual transition from sepals to petals (compare with the magnolias). The calyx is an ephemeral organ in the *Papaveraceæ;* consequently the bud stage of the flower may be required for the proper study of the calyx characters. While the calyx generally drops off sooner or later after anthesis, there are numerous cases in which it persists and may appear as a more or less prominent part of the fruit—especially in fruits of the spurious (receptacular) type such as the apple, pomegranate, cranberry, blueberries, *Philadelphus.* A persistent calyx is also characteristic of the strawberry, blackberries, tomato, *Datura,* and many others. In the mulberries the persistent calyx forms the succulent part of the separate fruitlets. By way of contrast such forms as the dandelion and the bush-honeysuckle (*Diervilla Lonicera*) may be noted (Figs. 9, 392).

The sepals or segments of the calyx are commonly of the same number as the petals, but instances will be met with in which the number is less, as in *Monolepis* and *Atriplex* of the *Chenopodiaceæ,* where only one sepal is present. Two sepals occur in *Portulacaceæ, Papaveraceæ, Fumariaceæ.* In some cases an *epicalyx* of supernumerary "sepals" occurs (strawberry) (Fig. 9). It is also important to distinguish between calyx and involucre or involucel in such genera as *Hepatica, Althæa,* and *Monardella,* for example. In *Hepatica* the corolla is wanting but instead the calyx is petaloid, and closely approximated to the calyx below is the calyx-like involucre. In *Althæa* there is an involucel of several bracts. In *Monardella,* on the other hand, the flowers are aggregated in a head, and the calyx of each flower is tubular (gamosepalous), but the entire head is subtended by a number of involucral bracts. A more advanced type of head (capitulum) subtended by numerous green involucral bracts is seen in the composites. These involucral bracts of the composites are often mistaken for the calyx, but it should be remembered that the head or capitulum of the composites is an inflorescence—*i.e.,* a cluster of numerous flowers, each with its own calyx (pappus), and that the involucral bracts belong to the inflorescence and not to the individual flowers (Figs. 407, K; 408, 412).

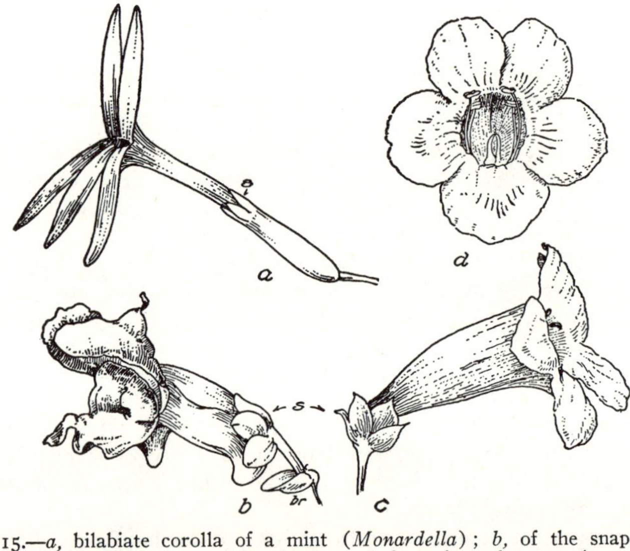

FIG. 15.—*a,* bilabiate corolla of a mint (*Monardella*); *b,* of the snapdragon (*Antirrhinum*); *c,* of *Pentstemon; d,* face view of *c; s,* calyx.

A petaloid calyx is present in *Caltha* (Marsh-Marigold), *Anemone, Anemonella, Clematis,* and other genera. (Fig. 187).

In composites and several other groups the calyx takes the form of bristles or scale-like or teeth-like structures; these are either deciduous or persistent. The following genera exemplify some of the different types: *Solidago, Aster, Helianthus, Madia, Taraxacum, Hieracium, Dipsacus, Diervilla, Campanula* (Figs. 10, 392, 408, 414, 415).

In the Monocotyledons a clearly differentiated green calyx is present, *e.g.*, in *Sagittaria, Tradescantia, Trillium, Cypripedium.* Often, however, while more or less differentiated from the corolla, the calyx of Monocots is petaloid, as in *Lilium, Iris, Strelitzia.* In *Scirpus* and some other sedges the perianth is modified to bristles (Figs. 11, 424, 456, 462, 468, 472, 476).

It is evident that in all cases where the calyx possesses distinctive characters, as in the forms mentioned above, it at once becomes a useful guide to the identity of a species or genus, and at the same time these characters can be used to advantage in classification or keying.

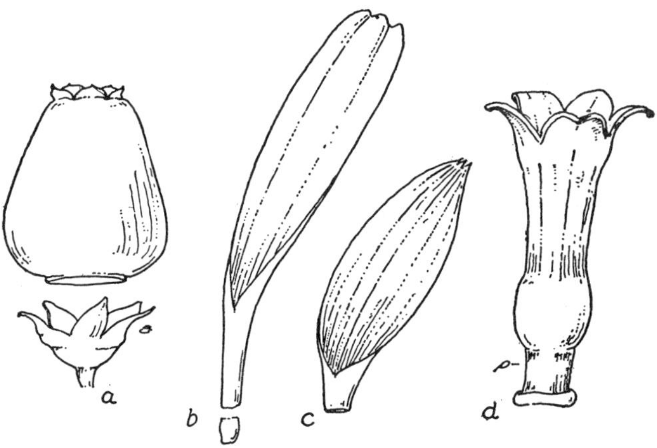

FIG. 16.—Various types of sympetalous corollas: *a,* in the winter-green (*Gaultheria procumbens*); *b,* ligulate corolla in a composite (*Encelia californica; c,* ligulate corolla in a sunflower (*Helianthus*); *c,* corolla of a disk-flower of a composite (*Helianthus*); *p,* stipe; *s,* calyx.

Corolla. In general the corolla is the most conspicuous organ of the flower; however, it is often small and inconspicuous or even absent. The showy colorings and the many remarkable modifications which it displays are thought to be directly associated with insect pollination. The commonest colorings are yellow, white, blue, and red, these occurring as solid colors or variously mottled.

When the corolla is wanting the flowers are described as *apetalous* (see *Anemone, Clematis, Thalictrum*). When the petals are distinct, the corolla is *apopetalous, polypetalous,* or *choripetalous*—these terms being synonymous. This type of corolla is characteristic of the *Di-*

alypetalæ of the *Archichlamydeæ* (p. 233; Figs. 1, 12). A more specialized type is the *sympetalous* (*gamopetalous*) corolla characteristic of the *Metachlamydeæ* (*Sympetalæ, Gamopetalæ*) (Fig. 13). Thus it will be noted that, according to the scheme of classification being followed in these studies, the type of corolla is of major importance in the classification of the flowering plants. While the sympetalous corolla is commonly spoken of as consisting of "united

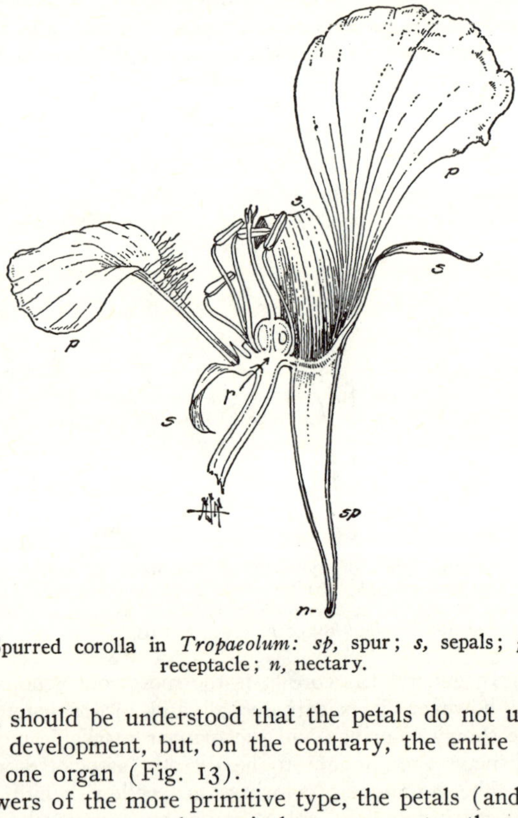

FIG. 17.—Spurred corolla in *Tropaeolum: sp,* spur; *s,* sepals; *p,* petals; *r,* receptacle; *n,* nectary.

petals," it should be understood that the petals do not unite in the course of development, but, on the contrary, the entire corolla develops as one organ (Fig. 13).

In flowers of the more primitive type, the petals (and the sepals as well) have a more or less spiral arrangement; otherwise, in the advanced types, the arrangement is cyclic (whorled). Relatively numerous or an indefinite number of petals is also characteristic of many primitive flowers (Fig. 185).

Petals are often differentiated into a characteristic *limb* and a basal *claw,* these parts corresponding in a way to the blade and stalk of a leaf (Fig. 12). Sympetalous corollas are often differentiated into a more or less prominent *corolla-tube* which is crowned above by the segments or *corolla-lobes,* corresponding to the "petals" (Figs. 13, 15).

The different types of sympetalous corollas are described as *tubular, saccate, rotate, funnelform,* etc. (Figs. 13, 15, 16). Those characteristic of certain families are described by adjectives derived from the names of the respective families, as *labiate, solanaceous, papilionaceous, campanulate,* etc. The papilionaceous type is characteristic of the members of the subfamily *Papilionoideæ* of the *Le-*

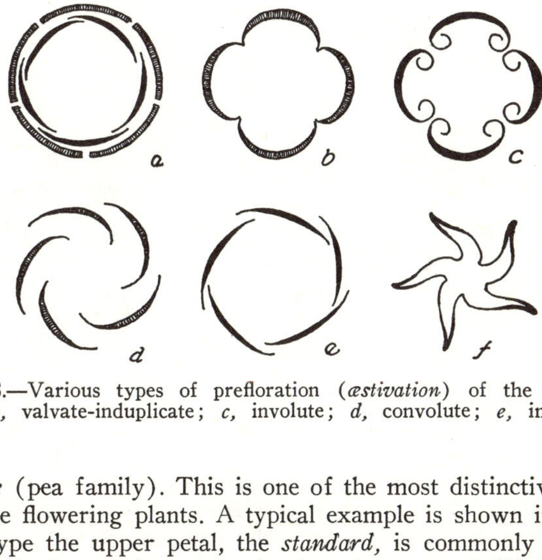

FIG. 18.—Various types of prefloration (*æstivation*) of the corolla: *a,* valvate; *b,* valvate-induplicate; *c,* involute; *d,* convolute; *e,* imbricate; *f,* plicate.

guminosæ (pea family). This is one of the most distinctive corollas among the flowering plants. A typical example is shown in Fig. 14. In this type the upper petal, the *standard,* is commonly large and broad, while the two lateral petals, the *wings,* are of different shape and usually smaller and narrower. The two lowest petals coalesce by their lower edge to form the distinctive *keel.* The labiate type is a two-lipped corolla, so called because the two upper lobes are usually so developed as to form a characteristic upper lip and the three lower lobes a lower lip (Fig. 15). This type of corolla is characteristic of the majority of the families included in the large order *Tubifloræ* of the *Metachlamydeæ.* Hence, any flower possessing this type of corolla, associated with hypogyny, can be at once referred to this order. However, in some of the lower families of the order the corolla is not of this type (see *Convolvulaceæ, Polemoniaceæ, Solanaceæ*). In these members the corolla is of the rotate, campanulate, or funnelform type. A characteristic campanulate corolla is seen in *Campanula*—*Campanula rotundifolia* (bluebell). A distinctive *ligulate* type of corolla occurs in the composites. Here it appears as if the corolla were drawn out on one side into a long strap-shaped limb, the *ligule* ("little tongue"). This type is characteristic of the "ray-flowers" of composites (*e.g., Helianthus,* sunflower), but in the chicories and their allies of the same family all the flowers are ligulate (*e.g.,* dandelion). A distinctive urn-shaped type of corolla is characteristic of some members of the heath family (*e.g., Vaccinium,* Fig. 16).

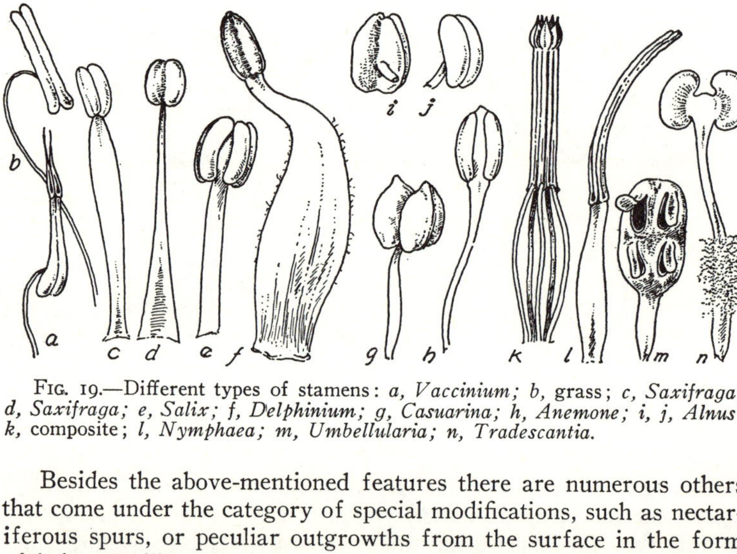

FIG. 19.—Different types of stamens: *a, Vaccinium; b,* grass; *c, Saxifraga; d, Saxifraga; e, Salix; f, Delphinium; g, Casuarina; h, Anemone; i, j, Alnus; k,* composite; *l, Nymphaea; m, Umbellularia; n, Tradescantia.*

Besides the above-mentioned features there are numerous others that come under the category of special modifications, such as nectariferous spurs, or peculiar outgrowths from the surface in the form of hairs, papillæ, glandular processes, etc., etc. (Fig. 17). All such characters are of relatively minor importance, but may serve a useful purpose in distinguishing genera, subgenera, or species. Thus one of the outstanding recognition marks of the ranunculaceous genus *Aquilegia* (columbine) is the prominent nectariferous spur on each petal. And many buttercups, *e.g., Ranunculus septentrionalis,* possess a characteristic scale at the base of each petal. Other forms may be seen in *Lychnis* and *Phacelia.* Then again such striking modifications as the "tongue" in *Strelitzia* (Fig. 485), an organ formed by the union of the two lower petals and presenting an almost reptilian appearance, and the saccate third petal in *Cypripedium*—these features at once set these genera off from their allies.

It is possible, within limits, to use the color of the "flower" or the corolla as a basis of a scheme of classification, but such a scheme, while it may subserve certain practical purposes, can only be more or less artificial. Within genera, however, a color key is often very convenient, and is frequently used in standard floras.

It has been pointed out in preceding discussions that in primitive flowers the floral organs are relatively numerous and that with evolutionary advance the tendency seems to be towards a reduced and definite number. It is not surprising, therefore, to find many instances in which stability of floral design has not yet been attained, so that the number of floral organs in any one set will be found to vary, often considerably. Thus with respect to the corolla in particular considerable variation will be found both in number and shape of the petals or segments. Good examples of such variations may be seen in the common buttercup, *Ranunculus glaberrimus,* and the mock-orange or syringa, *Philadelphus Lewisii* and other species of that genus, both of the Northwest. Whenever such variation occurs, it is obvious that to seize upon a definite number of petals, for example, as the basis for segregating species, would lead only to confusion. Better say that the species, in the large, is variable with respect to the corolla.

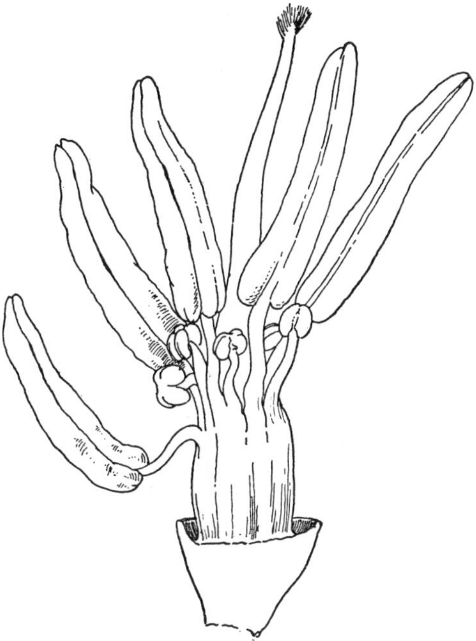

FIG. 20.—The two sets of stamens in a lupine (*Lupinus*).

Prefloration (*æstivation*). This term refers to the mode in which the petals or corolla, and sepals as well, are disposed in the bud. It is necessary to distinguish this term from *vernation,* which refers to the disposition of foliage leaves in a leaf-bud. While for ordinary purposes we need not concern ourselves with the problem of prefloration, still it has its importance in distinguishing certain genera and even families in which a certain type of prefloration is of constant occurrence. Specific instances, by way of illustration, are the gentian family (*Gentianaceæ*), the lilac (*Syringa*), *Saxifraga.* Prefloration can be studied to best advantage in flower-buds before they

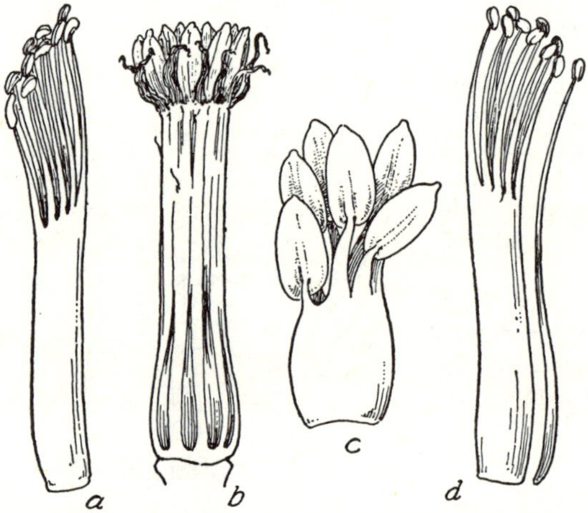

FIG. 21.—Various types of union of stamens: monadelphous stamens (*a*) in *Spartium; b,* in *Melia Azedarach; c,* in *Geranium; d,* diadelphous stamens in *Lathyrus.*

open. Thus in large flower-buds like those of the rose or the poppy, the way in which the petals are folded is clearly shown. Transverse sections of buds will also show the manner of folding. To describe the various ways in which the petals are folded in the bud a number of adjectival terms are in use. These are explained below.

When the floral leaves overlap like shingles or tiles of a roof, they are said to be *imbricate.* Several types of overlapping or imbrication occur, and these will be briefly explained. Thus if there are four segments present, the innermost pair may be overlapped by the outermost, a mode of overlapping that corresponds to the two-ranked or *distichous* arrangement of foliage leaves. If three petals

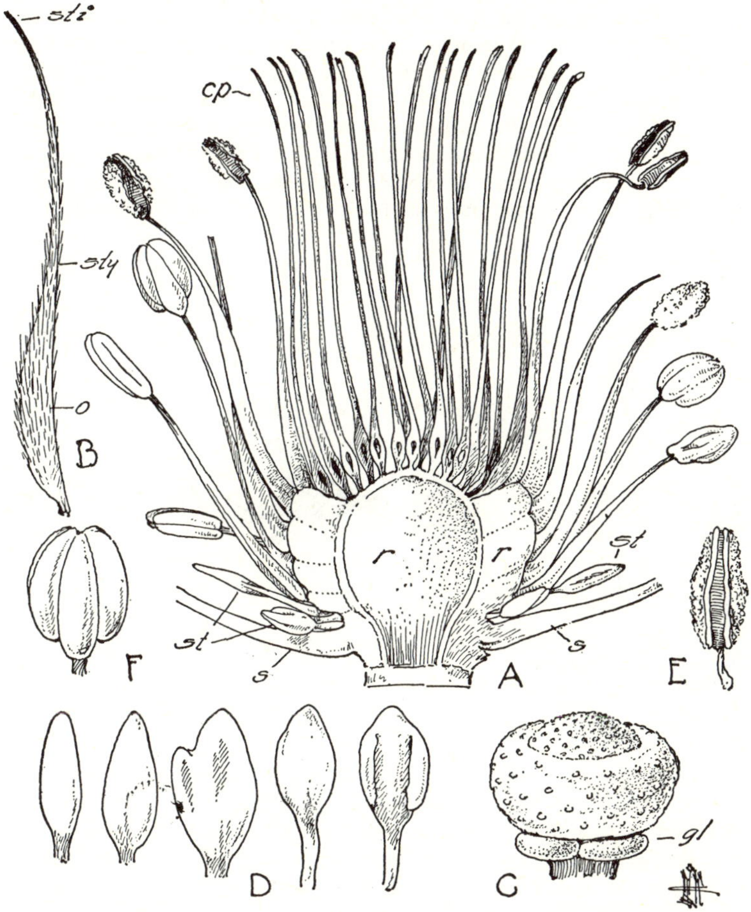

FIG. 22.—Staminody in *Anemone patens,* var. *Wolfgangiana: A,* median section of flower with perianth removed; *st,* staminodes; *s,* sepals; *cp,* carpels; *r,* receptacle. *B,* a carpel; *o,* ovary; *sty,* style; *sti,* stigma. *C,* the receptacle, after removing floral organs, showing surface occupied by the stamens and carpels; *gl,* glands. *D,* a series of staminodes, showing transition to stamen, and conversely. *E,* an anther after dehiscence; *F,* anther of normal stamen.

(or sepals) are present, the first or lowest will be outside, the third or upper will be inside, and the second or lateral will partly overlap the third, but will be overlapped by the first on the opposite side.

This conforms to the three-ranked or *tristichous* mode of leaf-arrangement. If five members are present in a whorl, two of the members are outside and two are inside, while the fifth is disposed in the manner of the second just described for the trimerous whorl. This has been called the *quincuncial* arrangement and conforms to the five-ranked or *pentastichous* arrangement of foliage leaves (Fig. 18).

When the segments of a whorl lie edge to edge without folding or wrinkling, their disposition is said to be *valvate*. A modification of the

FIG. 23.—Petalody and carpelody in a species of *Saxifraga: a,* supernumerary but abortive carpels, and petaloid stamens; *b,* an organ partly stamen *st* and partly carpel *cp; an,* anther; *c,* an organ partly sepal *s* and partly petal *p; d,* an organ partly stamen and partly petal in *Philadelphus; e,* petalody and fasciation of stamens in *Saxifraga aequidentata.*

valvate disposition, called the *induplicate,* occurs when the edges of the segments are turned inward. If the edges are inrolled, the disposition is *involute.* The *convolute* type occurs when the segments are twisted in a body as one would twist a roll of several sheets of paper. If the segments do not touch, the disposition is said to be *open* (Fig. 18).

In sympetalous corollas we frequently find a *plicate* or *plaited* disposition, in which the folding is lengthwise, somewhat on the order of a fan. In this case the edges of the folds either turn inward or outward. If the folds are twisted as in the convolute type, the disposition is *supervolute* (Fig. 18).

CHAPTER V

ANDRŒCIUM AND GYNŒCIUM

Andrœcium or stamens. Taxonomically the andrœcium is one of the most important parts of the flower. Whether the component parts or stamens are numerous or few, whether they are spiral or cyclic, their many peculiar and distinctive shapes and dispositions and other distinctive features all yield valuable criteria of relationships and besides serve as useful guides to the identity of species. Likewise for purposes of keying, the stamens often furnish clearly defined distinguishing characters.

In spite of the seemingly endless variety of stamens, nevertheless, the andrœcium is often of remarkably constant design even for large groups. Thus for the entire composite family the same type of andrœcium runs through all, and in the pea family (*Leguminosæ*) the subfamily *Papilionoideæ* is distinguished by a distinctive type. Many other instances of constancy of type could be mentioned, but it will suffice to suggest studies of the andrœcia of the grasses and sedges, the orchids, the mustards, and the heaths.

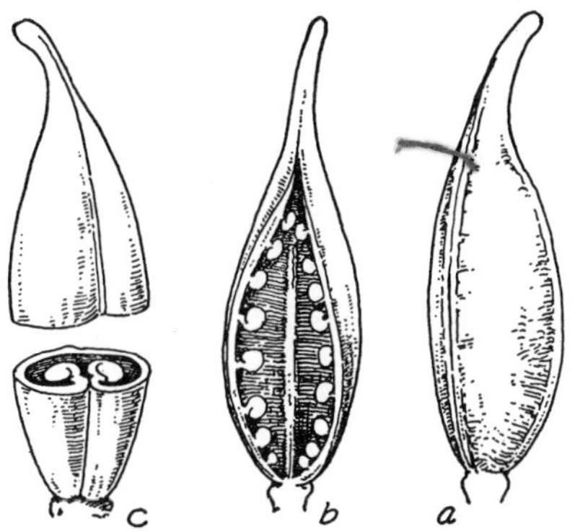

FIG. 24.—Illustrating the supposed foliar (sporophyllous) nature of a carpel: *a,* closed, *b,* open, and *c,* transverse section; *d,* dorsal suture; *v,* ventral suture. In *b* and *c* note that the ovules are borne on the infolded margins of the carpels, *i.e.* on marginal placentas.

In general a stamen consists of the pollen-sacs or *anther* and a stalk or *filament* (Fig. 19). The anther usually consists of two 2-lobed longitudinal halves (*thecæ*), each lobe forming a *pollen-sac* (or *microsporangium*) in which are produced the *pollen-grains* (*microspores* or *microgametophytes*). The two halves of the anther are connected by a tissue called the *connective,* which is often prolonged beyond the anther or is otherwise peculiarly modified, and thus becomes a distinctive feature (see *Salvia, Tradescantia*). Sometimes only two pollen-sacs are present in the anther ("two-celled") or the pollen-sacs of each half may form a "cell" by confluence. In other cases a multiplication of pollen-sacs may be brought about by the formation of dissepiments.

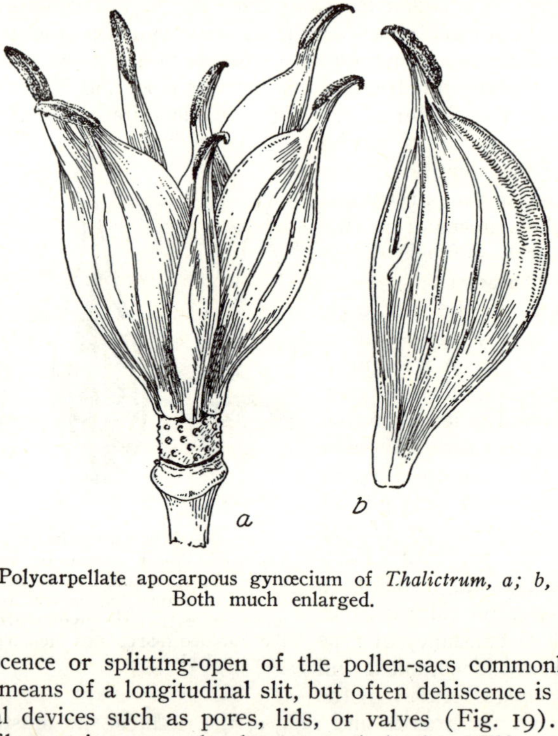

FIG. 25.—Polycarpellate apocarpous gynœcium of *Thalictrum, a; b,* a carpel. Both much enlarged.

Dehiscence or splitting-open of the pollen-sacs commonly takes place by means of a longitudinal slit, but often dehiscence is effected by special devices such as pores, lids, or valves (Fig. 19).

The filament is commonly elongate and slender (*filiform, terete*) but is often thickened or flattened or expanded above or below (*subulate* or *awl-shaped, club-shaped, spatulate*), and is often provided with hairs or appendages of various sorts (Fig. 19). The filaments elongate rapidly after the opening of the flower and become characteristically disposed to facilitate pollination. By studying the stamens in various stages of the flower-buds, the changes they undergo up to the fully expanded flower will be better understood.

In some cases the filaments are wanting, and the anthers are then *sessile.*

The anthers are attached to the filament in several different ways. If attached at a point on the connective, near the middle, so as to allow it to swing freely, the anther is *versatile.* If attached by the base, the anther is *innate* or *basifixed.* On the other hand, if the anther is affixed for its whole length to the filament, it is *adnate* (Fig. 19). The anthers, furthermore, show a definite orientation with respect to the gynœcium or the center of the flower; thus if they face the gynœcium, they are said to be *introrse* ("turned inwards"), while if they face the corolla, they are *extrorse* ("turned outwards"). Sometimes both of these orientations occur in the same flower (see genera of the *Lauraceæ* and Fig. 196).

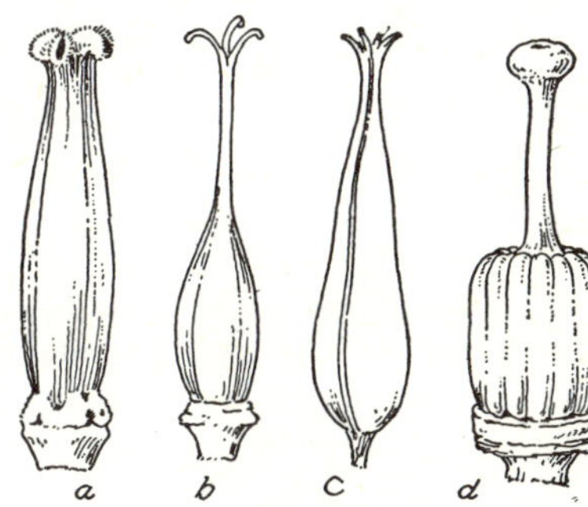

FIG. 26.—Polycarpellate syncarpous gynœcia in (*a*) *Tulipa,* (*b*) *Uvularia.* (*c*) *Salix,* (*d*) *Primula.*

While the stamens are mostly free or *distinct,* there are many cases in which they are united in various ways. The pea family (*Leguminosæ*) offers good examples of two characteristic modes of such union. Thus in some genera the filaments are all united more or less below to form a tube around the ovary. Such a union is said to be *monadelphous* ("in one brotherhood"). Moreover the filaments in these genera are often of unequal lengths—a short set and a long set (Fig. 20). In other genera of this family the stamens are *diadelphous,* that is, in two sets or brotherhoods—nine in one set and one in the other (Fig. 21). Various other methods of union will be met with, *viz. triadelphous, tetradelphous,* etc., *polyadelphous.* A different sort of union is to be seen in the *Lobeliaceæ* and the *Compositæ.* In these families the stamens are united in a characteristic way by their anthers, the filaments remaining free (Figs. 408, 415). This is an outstanding recognition-mark of these two families. Such stamens are said to be *syngenesious* or *synantherous.* In some forms the stamens unite to form a column or *synandrium,* in which the anthers may be free (*Araceæ—Colocasia, Caladium*).

It frequently happens that the anthers shed their pollen before the stigma is sufficiently matured to receive it: in such case the flowers are said to be *proterandrous* (or *protandrous*). On the other hand, if the stigma becomes receptive before the anthers are mature, the flowers are *proterogynous* (or *protogynous*). In any case where the stamens and the stigma reach maturity at different times, the condition is termed *dichogamy*.

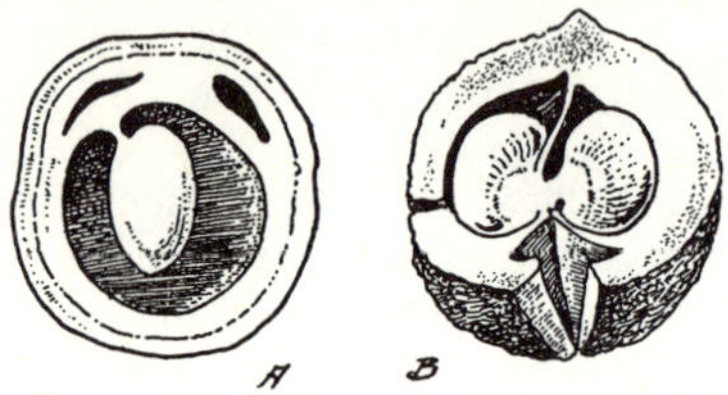

FIG. 27.—Suppression of carpels (locules) in *A, Quercus rubra, B, Æsculus hippocastanum.*

The occurrence of several stamens in a place ordinarily occupied by one stamen has been described as a case of doubling (*dédoublement*). However, it is not always easy to tell whether a particular set of stamens is the result of branching or of union, and for ordinary purposes the exact situation need not concern us (Fig. 262).

Stamens that have become modified in one way or another and have thereby lost their normal function are called *staminodes* (*staminodia*) (Fig. 22). Staminodes may be entirely functionless

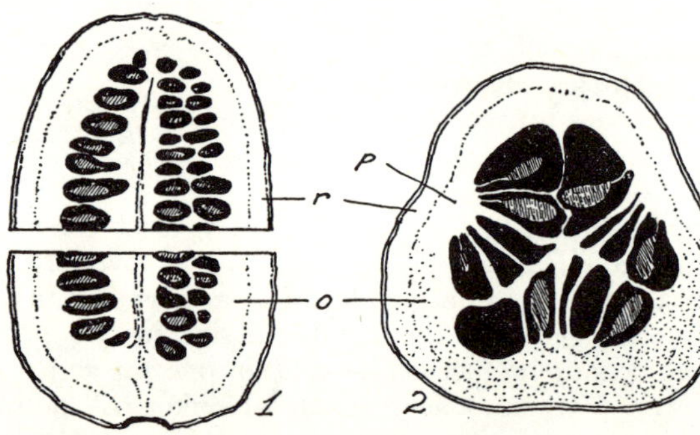

FIG. 28.—Median (*1*) and transverse (*2*) section of a cucumber: *o*, ovary; *p*, placenta with several ovules attached; *r*, receptacle.

(*apostaminodia*), or they may function as glands of one sort or another (*metastaminodia*). Sometimes stamens become transformed into petals, a condition known as *petalody;* such modified stamens may be seen in *Canna, Saxifraga* species, and *Philadelphus* (Figs. 22, 23).

A number of convenient terms are in use to describe the number of stamens present in a flower: thus if one stamen only is present, the flower, or the andrœcium, is *monandrous;* if two stamens are present, the flower, or the andrœcium, is *diandrous;* and so on through *tri-, tetra-,* etc., to *polyandrous,* meaning one, two, three, four, to many stamens respectively. When a flower has four stamens grouped in two pairs, one pair longer than the other, the stamens are said to be *didynamous,* as in *Agastache, Tecoma radicans,* and others. When six stamens are present, four of them longer than the other two, as in all mustards, the andrœcium is *tetradynamous.* It was on the number of stamens that Linnæus based his classic grouping of the flowering plants into *Monandria, Diandria, Triandria,* etc. In many cases the stamens will be found to vary in number even in the same species (*Saxifraga, Valeriana*); on the other hand the number five is constant throughout the *Compositæ,* for example (Fig. 408).

After pollination the stamens wither and fall separately or, when epipetalous, along with the corolla, or they may persist in the withered condition even in the fruit (strawberry, blackberries, apple).

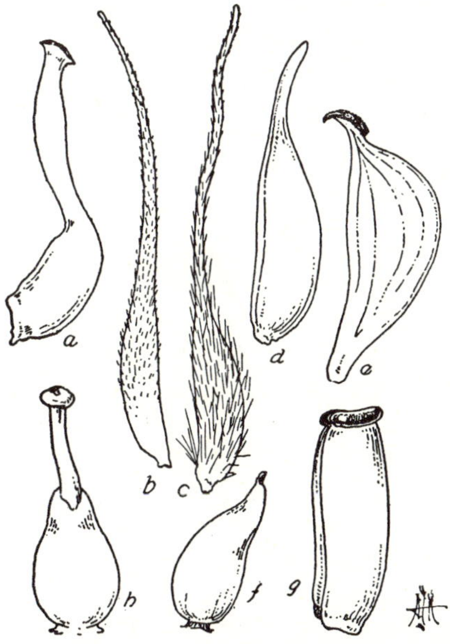

FIG. 29.—Carpels of apocarpous gynoecia of various dicotyledons: *a, Rubus; b, Anemone patens; c, Rosa; d, Delphinium; e, Thalictrum; f, Magnolia; g, Akebia; h, Prunus.*

To illustrate the composition and nature of the andrœcium, a few distinctive types selected from different families may be noted, as follows:

One stamen is present in some *Chenopodiaceæ, e.g., Aphanisma, Monolepis,* and *Allenrolfea;* in the grass *Cinna,* and in the willow *Salix Uva-Ursi.*

Two stamens are present in *Allenrolfea* (*Chenopodiaceæ*), in some willows (*Salix*), and in *Valerianella* (*Valerianaceæ*).

Monadelphous stamens are found in some *Leguminosæ* and in the *Malvaceæ.*

Diadelphous stamens occur in some *Leguminosæ.*

Pentadelphous stamens occur in the basswood (*Tilia*) and in the St. John's-worts (*Hypericum*).

. Epipetalous stamens are characteristic of the *Convolvulaceæ, Solanaceæ,* and allies (Figs. 347, 348, 362, 366, 368, 371).

Stamens of an especially distinctive sort occur in *Tradescantia, Orchidaceæ, Ericaceæ, Malvaceæ, Euphorbiaceæ, Compositæ, Lauraceæ, Berberidaceæ.*

The anthers are extrorse in *Liliaceæ* (*Veratrum, Zygadenus*) and *Lauraceæ* (*Persea*). The anthers are introrse in *Liliaceæ* (*Tofieldia, Lilium, Tulipa*) and *Lauraceæ* (*Sassafras*).

For anthers opening by pores, valves, etc., see *Lauraceæ, Berberidaceæ, Melastomataceæ, Ericaceæ.*

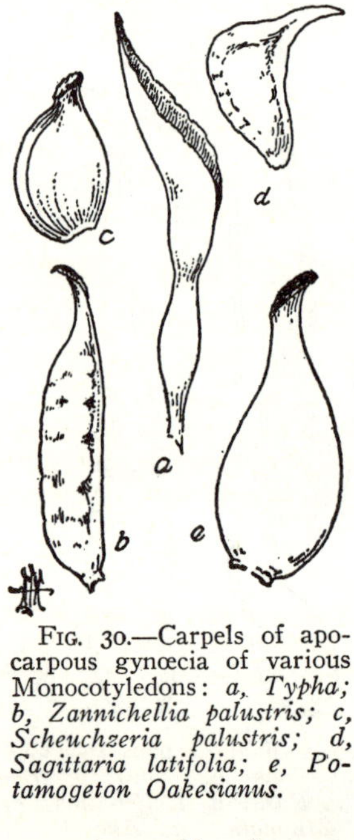

FIG. 30.—Carpels of apocarpous gynœcia of various Monocotyledons: *a, Typha; b, Zannichellia palustris; c, Scheuchzeria palustris; d, Sagittaria latifolia; e, Potamogeton Oakesianus.*

Pollen. Pollen has not been used to any great extent in classification, yet it often possesses certain distinctive markings or sculpturings that are more or less constant for one group or another. Thus Wodehouse, who has made extensive studies of pollen in relation to classification, has this to say about it: "The forms of pollen grains are as useful as any other characters in the classification of plants; as a general rule they serve best in distinguishing and showing relationships among higher groupings, as families, tribes, and genera, but sometimes they are useful among species; and frequently the pollen grains reveal relationships which are obscure and difficult to demonstrate by other means. Their usefulness is limited only by the number of characters observable; in some groups of plants these are so few that little can be told of their phylogeny without aid from other characters of the plants. In other groups, however, the structure of the grain is so clearly defined and remarkably various, offering so many characters of taxonomic value, that any attempt at classification of such groups is apt to lead to error unless account is taken of the pollen-grain structure." *

Numerous examples of peculiar sculpturings of the pollen-grains

* Wodehouse, R. P., in *Bulletin Torrey Botanical Club* (55:8:449, 1928).

will be found among the various genera of the *Acanthaceæ*. The pollen of wind-pollinated plants also shows some marked differences as contrasted with that of insect-pollinated kinds. Compare, for ex-

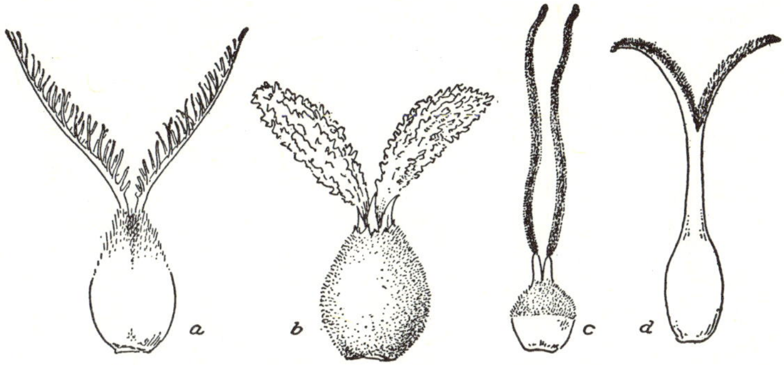

FIG. 31.—Stigmas in wind-pollinated (anemophilous) flowers: *a*, grass; *b*, walnut; *c*, hazelnut; *d*, sedge (*Scirpus*).

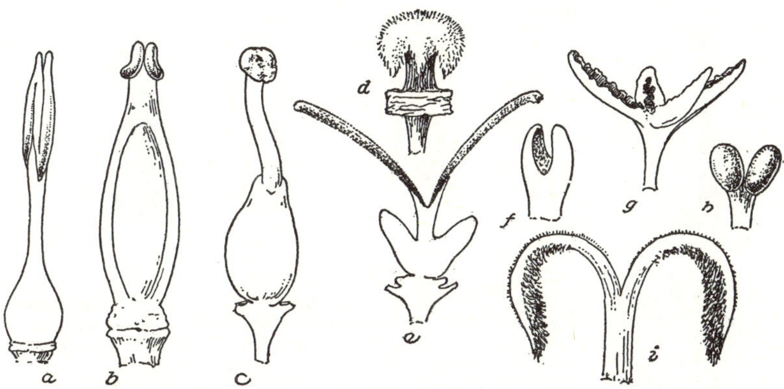

FIG. 32.—Stigmas in insect-pollinated (entomophilous) flowers: *a, Philadelphus; b, Sanguinaria; c, Prunus; d, Vinca; e, Acer Negundo; f, Monardella; g, Œnothera; h, Forsythia; i, Helianthus.*

ample, the pollen of a ragweed with that of a goldenrod or sunflower. In some groups of plants the pollen-grains cling together in groups of four (*tetrads*), while in the orchids and milkweeds the pollen-grains cling together in masses called *pollinia*—one of the distinctive features of these plants.

The gynœcium. The gynœcium is an aggregation of carpels (megasporophylls). In flowers of the more primitive type the carpels are usually numerous and spirally arranged, but the general evolutionary tendency is toward few carpels with a whorled or cyclic arrangement. Thus a three-carpellate gynœcium is characteristic of most Monocotyledons, while a two-carpellate gynœcium is common among Dicotyledons, *e.g., Saxifragaceæ, Umbellifloræ, Tubifloræ,*

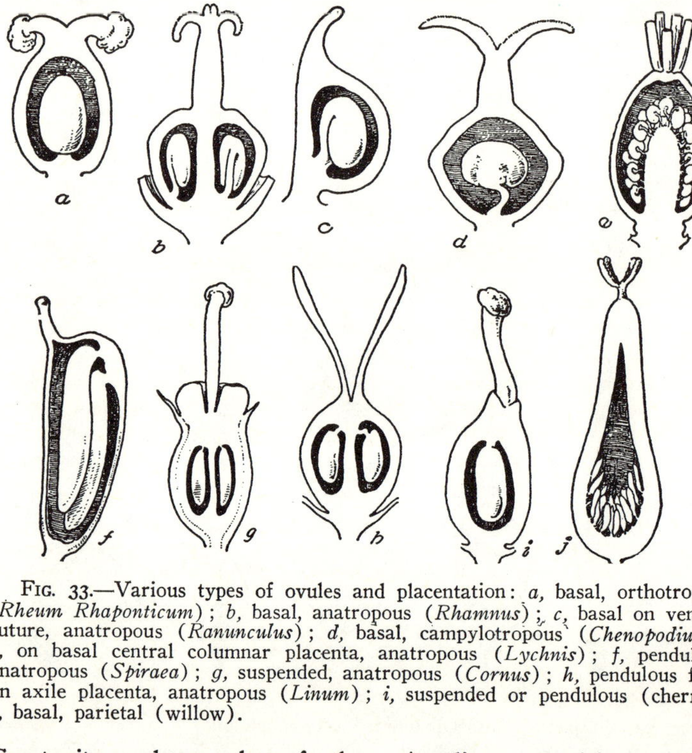

FIG. 33.—Various types of ovules and placentation: *a,* basal, orthotropous (*Rheum Rhaponticum*); *b,* basal, anatropous (*Rhamnus*); *c,* basal on ventral suture, anatropous (*Ranunculus*); *d,* basal, campylotropous (*Chenopodium*); *e,* on basal central columnar placenta, anatropous (*Lychnis*); *f,* pendulous, anatropous (*Spiraea*); *g,* suspended, anatropous (*Cornus*); *h,* pendulous from an axile placenta, anatropous (*Linum*); *i,* suspended or pendulous (cherry); *j,* basal, parietal (willow).

Compositæ, and a number of others. A solitary carpel is present in the barberries (*Berberis*) and some others. Often the number of carpels present will vary in the same genus or even species, especially in flowers of the primitive type (*Anemone, Magnolia, Saxifraga Lyalli, Sagittaria*), but the general tendency is toward a definite and constant number.

A simple carpel usually consists of an *ovary,* a *style,* and a *stigma*

(Figs. 1, 24, 29, 30). In the cavity or *locule* of the ovary are produced the *ovules* or *megasporangia.* If a carpel be regarded as a primitive spore-leaf or sporophyll that has become folded in the manner of a book, then the midrib of the spore-leaf becomes the *dorsal suture* and the fused edges the *ventral suture* of the carpel. The fused edges along the ventral suture are pushed more or less into the cavity of the ovary to form the placenta, on which the ovules are borne.

Union of the carpels: syncarpy. In flowers of the primitive type the carpels are free or *distinct,* and the gynœcium in all such cases is said to be *apocarpous* (Figs. 1a; 25). But as we advance to the higher cyclic forms, we shall find that the carpels tend to become

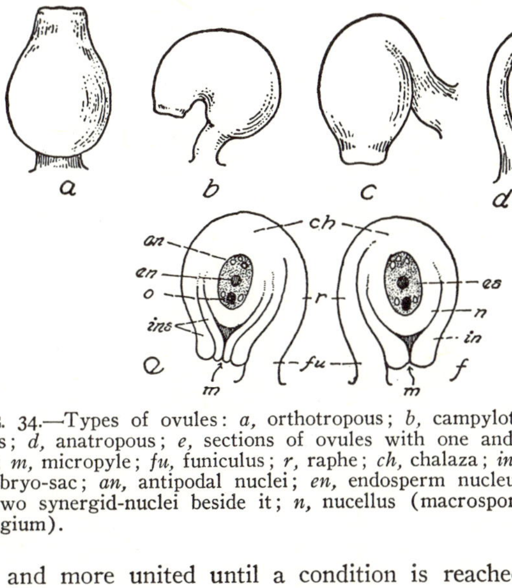

FIG. 34.—Types of ovules: *a,* orthotropous; *b,* campylotropous; *c,* amphitropous; *d,* anatropous; *e,* sections of ovules with one and with two integuments; *m,* micropyle; *fu,* funiculus; *r,* raphe; *ch,* chalaza; *in, ins,* integuments; *es,* embryo-sac; *an,* antipodal nuclei; *en,* endosperm nucleus; *o,* egg-nucleus, with two synergid-nuclei beside it; *n,* nucellus (macrosporangium, or megasporangium).

more and more united until a condition is reached of such complete union that little external trace of the component carpels remains. The result is a *compound* or *syncarpous* gynœcium. Such a syncarpous gynœcium may on the surface have all the appearances of a single or simple carpel, that is, one ovary, one style, and one stigma (see *Primula* and Figs. 26, 329).

The extent to which carpels are united may vary more or less in the same genus. Thus among the saxifrages the progressive steps from an apocarpous to a syncarpous gynœcium can be instructively shown (Fig. 208). Similar instances will be found in the *Crassulaceæ, Cratægus, Pyracantha, and other forms* (Figs. 207, 231, 233).

As the number of carpels present in a gynœcium is always of fundamental taxonomic importance, it is necessary that this point be determined when analyzing flowers, particularly when the gynœcium is syncarpous. In the case of apocarpous gynœcia, the number is easily determined if the carpels are relatively few, but it is worth while to count them even if numerous. For syncarpous gynœcia the number of carpels can usually be determined from the number of

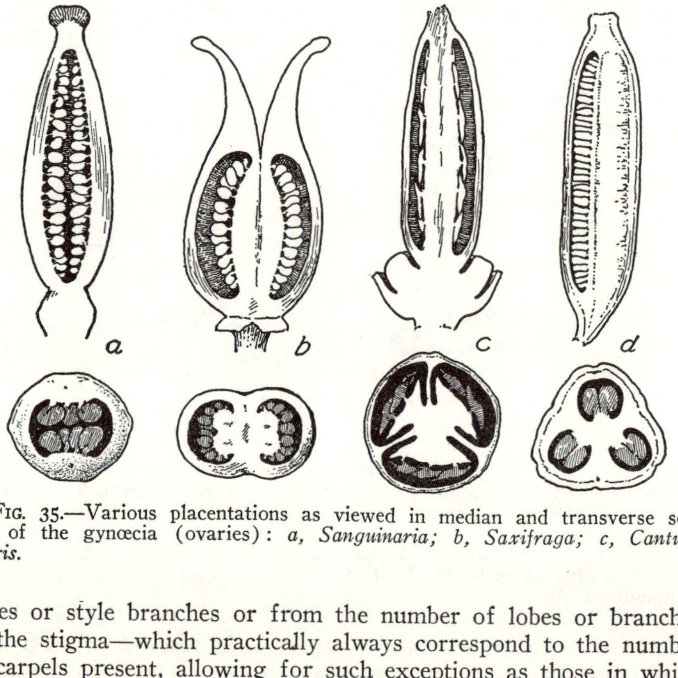

FIG. 35.—Various placentations as viewed in median and transverse section of the gynœcia (ovaries): *a, Sanguinaria; b, Saxifraga; c, Cantua; d, Iris.*

styles or style branches or from the number of lobes or branches of the stigma—which practically always correspond to the number of carpels present, allowing for such exceptions as those in which the stigmatic lobes are themselves divided or in which the stigma shows no definite lobing. But as a check on these clues, a transverse section of the ovary should be made, as the number of locules present will generally correspond to the number of carpels. Exceptions will be found quite frequently, however, such being syncarpous gynœcia in which false partitions multiply the number of locules, or the ovary may be unilocular (one-celled) even though the gynœcium is composed of several carpels (*Caryophyllaceæ, Primulaceæ, Compositæ*). The process of suppressing locules (or carpels) may be observed in the buckeyes (*Æsculus*), oaks (*Quercus*), loquat (*Eriobotrya japonica*), and others (Fig. 27). In the gourds (*Cucurbitaceæ*) a multiplication of the locules has been brought about by the outpushing of the placentas to meet the ovary-wall (Figs. 28, 402). Variations in the number of locules, or of carpels, may occur even in a species, as in the apple, in *Ribes malvaceum,* and in others. Among the Monocotyledons suppression of locules may be observed in the grasses, sedges, cocoanut, and others.

Progressive reduction in the number of carpels may be observed in the *Ranunculaceæ* and the *Rosaceæ,* for instance, the particular

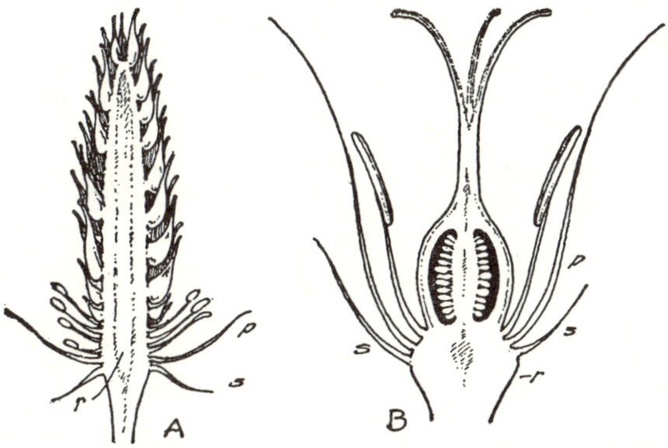

FIG. 36.—Sequence of organs in flowers: *A,* in mouse-tail (*Myosurus*); *B,* in bellwort (*Uvularia*); *p,* petals; *s,* sepals; *r,* receptacle.

types all being apocarpous. Thus in the *Ranunculaceæ* the situation may be illustrated by types from the genera *Anemone, Caltha, Aquilegia, Delphinium,* in the order given. Other genera may be used as well; and a similar progressive reduction-series could be shown even by species selected from the genus *Cimicifuga* alone. Similarly in the *Rosaceæ,* selecting for convenience types from *Physocarpus, Spiræa, Rhodotypus,* and *Prunus* (Fig. 237). The *Crassulaceæ, Saxifragaceæ, Leguminosæ,* and *Papaveraceæ* will show similar instances.

Since the characters of the gynœcium are more constant than those of the other floral organs, except the andrœcium, they are of fundamental significance in taxonomy, especially with respect to the solution of problems of relationships. Species, genera, families, orders—all rest fundamentally on definable characters of the gynœcium. Hence it is to the gynœcium that we must constantly refer in order to be able to place a plant in its proper niche in the classificatory scheme. And the more familiar we become with the different types of gynœcia, the quicker we shall learn to comprehend the meaning of phylogenetic classification and to recognize on sight the category in which a particular plant or species belongs. Some further details connected with the gynœcium are discussed in the next chapter.

In common terminology the gynœcium is usually spoken of as the *pistil*. An individual carpel will then be a *simple pistil,* and an apocarpous gynœcium will consist of as many such pistils as there are

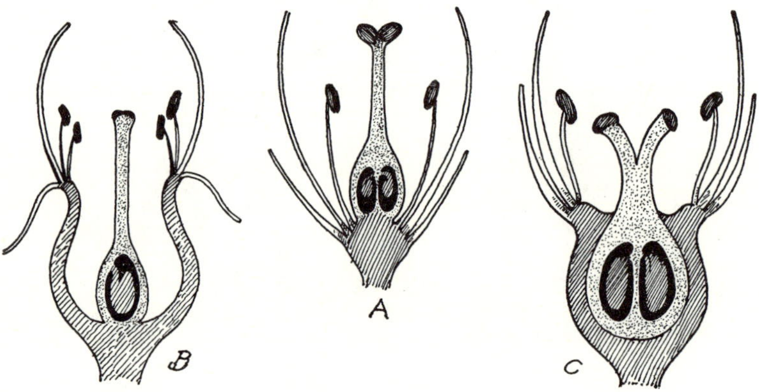

FIG. 37.—Diagrammatic sections of hypogynous (*A*), perigynous (*B*), and epigynous (*C*) flowers.

carpels present. On the other hand, a syncarpous gynœcium will be a *compound pistil;* and such a pistil will have a *compound ovary.* Often one sees the term *carpel* applied to the locules or "cells" of such a compound ovary, but such misuse of the term can lead only to confusion.

The style. The function of the style is to place the stigma in a favorable position for the reception of the pollen, and in correlation with this function it has undergone various adaptational modifications. Thus in many gynœcia of the primitive apocarpous type there is often only a slight demarcation of the style, the transition from ovary to style being very gradual, while on the other hand the style may be sharply defined in other apocarpous forms (Figs. 25, 29, 30). In syncarpous gynœcia the styles may be entirely distinct or they may show stages from partial to complete union (Figs. 26, 237, 329). There is also great diversity in length, very long slender styles being a conspicuous feature of the fuchsias, for instance, while in some forms such as *Primula sinensis* and *Lythrum Salicaria* three different lengths of styles are found, a condition known as *heterostyly*. Usually the style is terminal, but owing to inequalities of de-

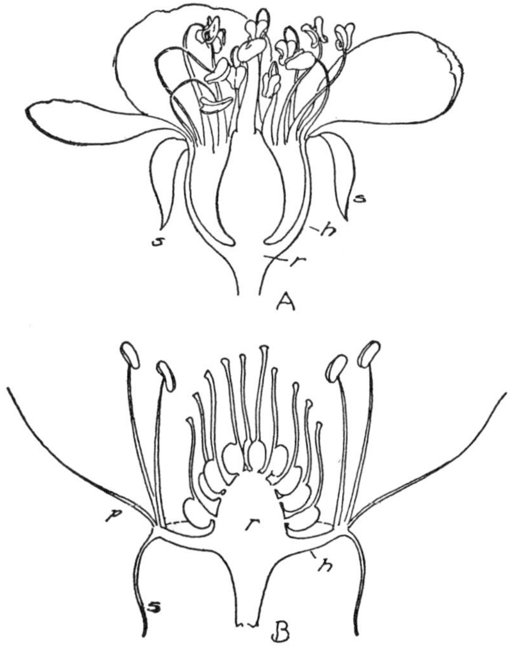

FIG. 38.—Types of perigynous flowers: *A*, of cherry; *B*, of raspberry; *h*, hypanthium of receptacle; *r*, receptacle; *s*, sepals; *p*, petals.

velopment of the sides of the ovary it may be dorsal, ventral, or basal (Figs. 29, 30, 50). In the *Umbelliferæ* there is a prominent swelling or *stylopodium* at the base of the style (Figs. 315, 316). Long, plumose, persistent styles, which serve as devices for effecting dispersal of the fruit, occur in *Clematis, Cercocarpus,* and *Geum;* in *Geum* the style is, generally, jointed near the middle in such a way as to cause the distal part to disarticulate, leaving the lower part as a persistent hooked appendage. In many cases the style persists as a small peg at the summit of the ovary, as in the water-plantains, various genera of the buttercup family, certain legumes, and many others. Often in types of this sort the orientation and special features of the style are of considerable taxonomic value in distinguishing species; this is particularly the case in the sedge genera *Eleocharis, Rhynchospora, Scirpus,* as well as in the water-plantains.

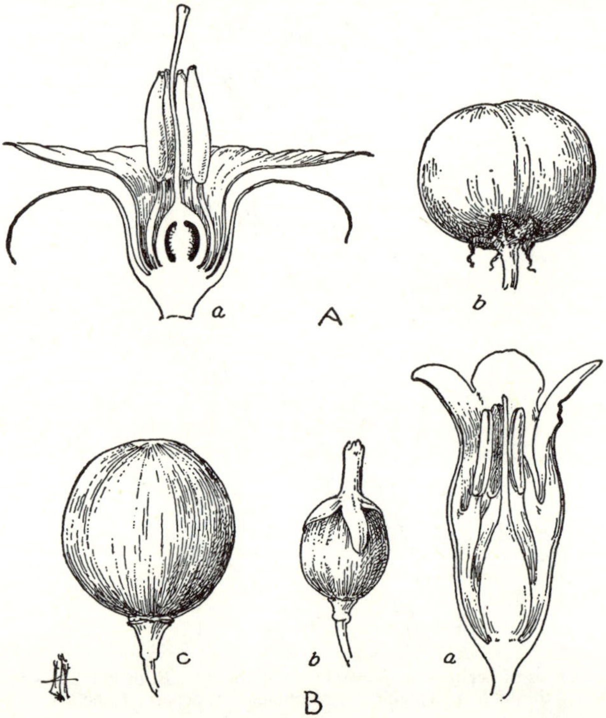

FIG. 39.—Fleshy fruits derived from superior ovaries, and median sections of the flowers from which they are derived: *A,* potato plant; *B,* Solomon's seal; *b,* immature stage with perianth clinging to summit.

Fleshy fruits such as cherries, plums, peaches, and tomatoes lose the style very early, only a scar remaining to mark its location. (Figs. 229, 440.)

The stigma. The stigma is the organ that receives the pollen. As a special part of the gynœcium it presents many diverse forms. Thus in wind-pollinated flowers it is commonly plumose (Fig. 31), while in insect-pollinated flowers it is often marked by a viscid or rough or papillate or even bristly surface, and may be club-shaped, discoid, capitate, or lobed (Fig. 32). As a rule each particular type is generically constant for a genus, or a certain type may be constant for an entire family, as in the *Compositæ*. In such cases the stigma is a reliable guide to the particular categories. Usually in syncarpous gynœcia the stigma is lobed or cleft or forked, the divisions as a general rule corresponding to the number of carpels of which the gynœcium is composed. But sometimes these lobes or divisions are themselves variously lobed or divided so that close scrutiny may be required to make out the exact connection with the corresponding carpels. Even in those syncarpous gynœcia in which one or more carpels have been suppressed their stigmas may persist and thus give the clue to the number of carpels originally present. Instances of this sort may be observed in the grasses, sedges, willows and poplars, oaks, walnuts, composites, and many others. In some cultivated walnuts, for example, a plurality of stigmas may be present. In the primulas, on the other hand, the stigma shows scarcely any trace of lobing. In some cases the style is absent so that the stigma rests on the summit of the ovary. Such a stigma is said to be

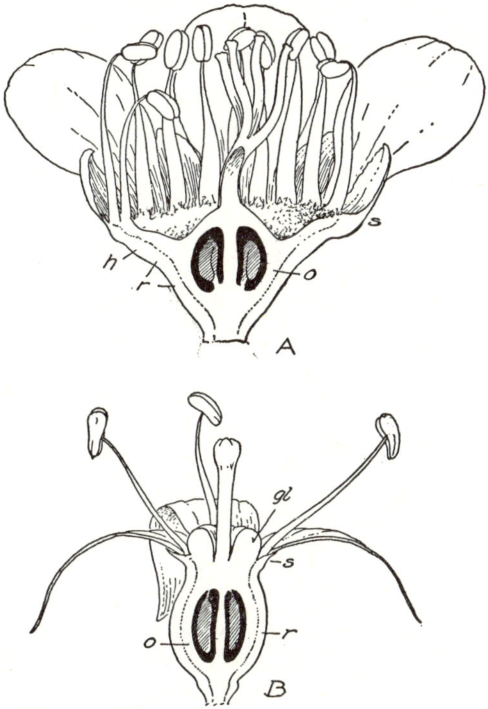

FIG. 40.—Types of epigynous flowers: *A, Aronia* (*Rosaceæ*); *B, Cornus; r,* receptacle; *h,* hypanthium of receptacle; *o,* ovary; *s,* calyx.

sessile; it is found in the barberries, poppies, holly, tulip, and the mariposa lily (*Calochortus*) (Figs. 188, 190, 268, 463).

Ovules. The ovules are the *megasporangia* (or *macrosporangia*). In the flowering plants they are produced in the ovary-chamber (*locule*) of the carpel, solitary, in pairs, several together, or numerous, on a special cushion of tissue called the *placenta.* The ovules may be borne on a short or elongate stalk (*funicle* or *funiculus*), or they may be stalkless (*sessile*). At the time fertilization takes place the ovule is an ovoid body consisting of an inner somewhat elliptical

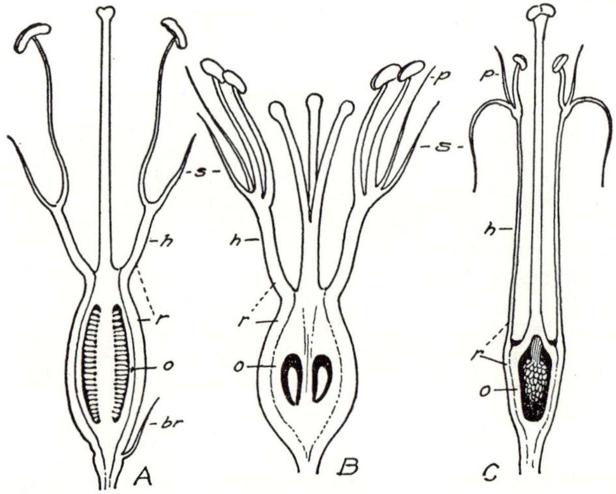

FIG. 41.—Types of epigynous flowers with prominently developed hypanthium: *A, Agave; B,* apple (*Malus*); *C,* golden currant (*Ribes odoratum*); *h,* hypanthium of receptacle; *br,* floral bract; *o,* ovary; *p,* petals; *s,* sepals; *r,* receptacle.

portion, the *nucellus* or macrosporangium proper, which is invested by one or two coats, the *integuments.* The basal portion of the ovule is called the *chalaza.* The integuments do not close completely over the nucellus but contract into a short neck through which an orifice, the *micropyle,* leads down to the nucellus. The micropyle usually serves as a passage-way for the pollen-tube in its course down to the *embryo-sac* (*female gametophyte*) or *egg-apparatus* within the nucellus. The point where the funicle is attached to the chalaza is called the *hilum.* The hilum is marked by a scar when the ovule later, as the *seed,* becomes detached from the funicle (Fig. 62).

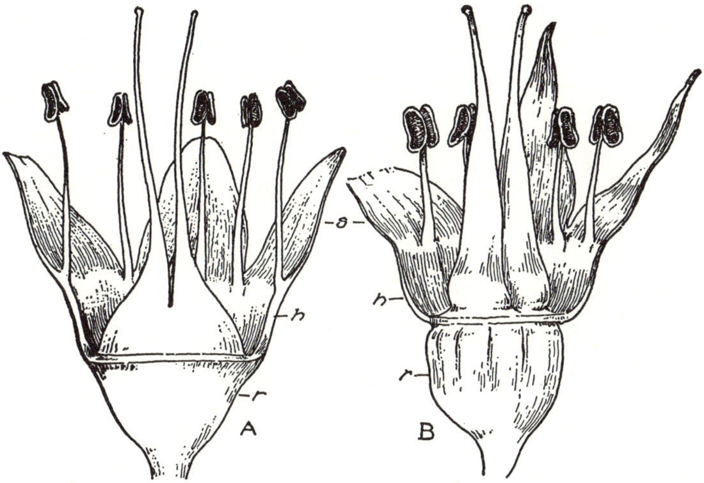

FIG. 42.—Special types of epigyny in saxifrages: *A, Heuchera americana; B, Boykinia major; h,* hypanthium; *r,* part of receptacle coalesced with ovary; *s,* sepals.

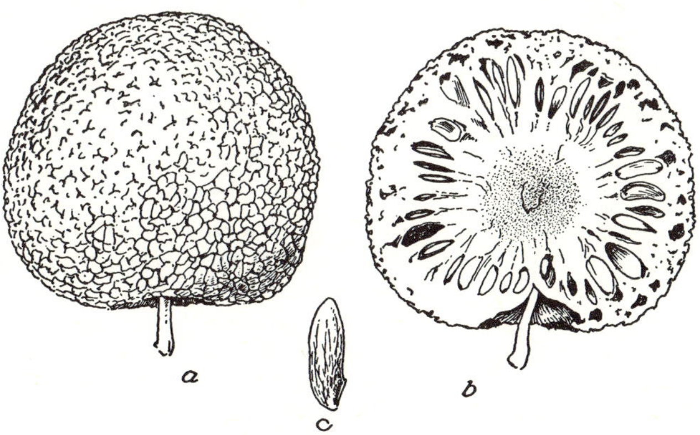

FIG. 43.—Fruit of the osage-orange (*Maclura pomifera*), *a; b,* in median section; *c,* seed.

As the ovules develop they may grow straight out from the placenta, horizontally or erectly, as the case may be, or they may turn upward (*ascending*), or they may be suspended (*pendulous*) from the upper part of the locule, or from the top (*suspended*) (Fig. 33). The particular mode of orientation of the ovules in their outgrowth from the placenta is used in Engler and Gilg's system of classification to distinguish the two orders *Geraniales* and *Sapindales* —a taxonomic point worth noting.

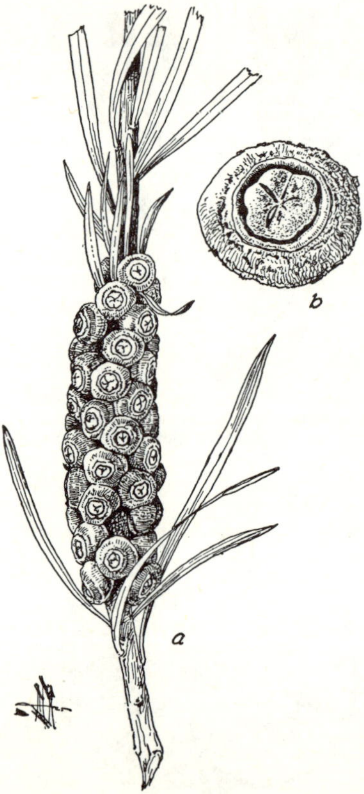

FIG. 44.—Fruit cluster of *Callistemon, a; b,* an individual fruit (spurious).

If the body of the ovule lies in a straight line with its funicle, the ovule is described as *orthotropous;* but if the body is bent or curved, the ovule is *campylotropous.* The ovule is *amphitropous* if the body is bent over at the chalazal end so far as to become fused for a part of its length with the funicle, and it is *anatropous* if it is bent completely over and fused with the funicle (Fig. 34).

In amphitropous and anatropous ovules the hilum is at the point where the free portion of the funicle joins the body of the ovule, and the adnate portion of the funicle appears as a ridge called the *raphe,* visible on the body of the ovule. In anatropous ovules the raphe extends from the chalaza to the micropyle, but in amphitropous ovules it is shorter (Fig. 34).

The characters of the ovules are taken into consideration in all critical taxonomic work. Special emphasis is placed on the number of integuments, the placentation, and the number of ovules present. Thus the number of integuments present in the ovule is a point of distinction between the families *Betulaceæ* and *Fagaceæ*. In general, two integuments are characteristic of the *Monocotyledones* and the *Archichlamydeæ,* and one integument of the *Metachlamydeæ*. But certain families vary in the number of integuments—*e.g.*, the *Salicaceæ* (willow family), in which the poplars possess two integuments and the willows one; and the same is true of the genera *Piper* and *Peperomia* of the *Piperaceæ* (pepper family), respectively. In a few cases the integuments are wanting (*Santalaceæ*).

Placentation. The different types of placentation are described as *axile, parietal, marginal, basal, pendulous, suspended* (Figs. 24, 33, 34, 35). The two last named have been sufficiently explained in a foregoing paragraph. The axile placentation occurs in syncarpous gynœcia of two or more carpels in which the ventral edges or sutures of the respective carpels form the central axis of the compound ovary; the placenta is thus at the inner or axial angle of each carpel. In the parietal type the placenta is on the wall of the ovary-chamber, in the marginal it is on the margins of the carpel (spore-leaf) (Fig. 24), while in the basal type the placenta arises from the base of the ovary-chamber (Figs. 33, 35). These different types of placentation are characteristic of families, genera, and even orders (*Geraniales, Sapindales, Tubifloræ, Compositæ, Saxifraga, Pyrus*). In most families of the *Tubifloræ,* for instance, the placentation is so distinctive that little difficulty will be found in referring plants to this order on this character alone.

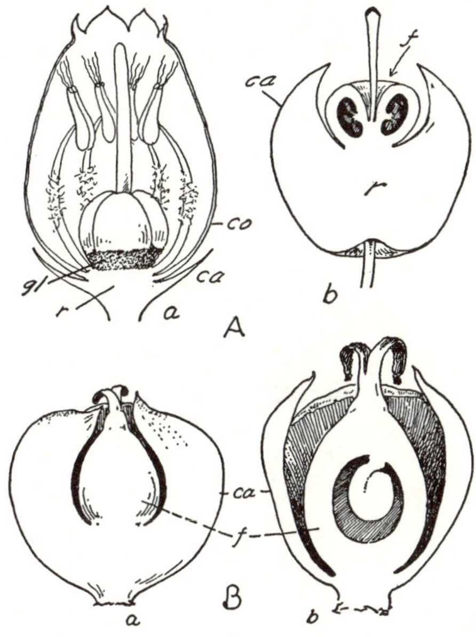

FIG. 45.—*A, a,* section of flower of *Gaultheria procumbens; b,* section of fruit; *ca,* calyx; *co,* corolla; *f,* true fruit; *gl,* gland; *r,* receptacle; *B,* mulberry (*Morus*); *a,* section of "fruit"; *b,* section of carpellate flower; *f,* the true fruit; *ca,* calyx.

CHAPTER VI

THE RECEPTACLE

The receptacle is the terminal portion of the flower-stalk or pedicel on which the floral organs are borne. In primitive flowers in which the floral organs are spirally arranged, the receptacle is more or less elongated, but in flowers of the advanced cyclic type it is much shortened and often more or less globose or flattened

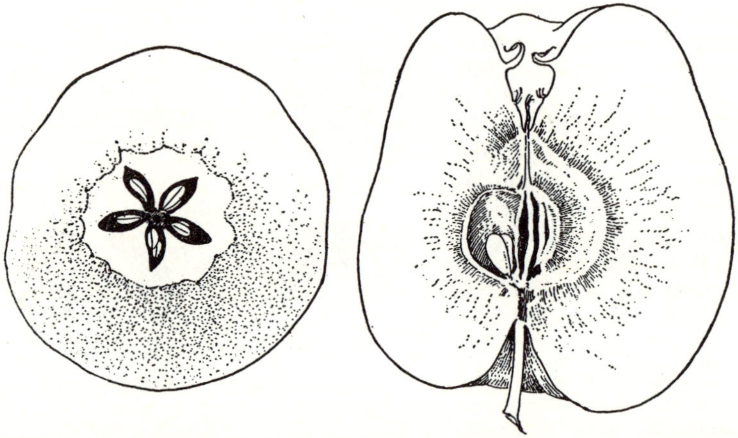

FIG. 46.—Pome of the apple, in transverse and median sections—the latter showing calyx segments and withered style and stamens in depression at top. Black strips in center (in right hand figure) are openings into the ovaries of adjoining carpels.

or variously modified (Fig. 36). Because of its bearing on classification and on phylogenetic grouping in particular, the configuration of the receptacle should always be carefully studied when analyzing flowers, especially when dealing with those advanced types in which modifications of the receptacle have brought about changes in position or elevation of the calyx, corolla, and andrœcium with respect to the ovary. The special cases discussed below are of fundamental importance to a clear understanding of classification.

Hypogyny, perigyny, epigyny, superior and inferior ovary (Fig. 37). It was pointed out on a preceding page that the different sets of floral organs follow one another in definite sequence on the receptacle. In no case is this sequence disturbed, whatever may be the apparent situation. Thus the sequence of the organs may be clearly shown in a median longitudinal section of the flower of a buttercup and of a lily (Figs. 1, 36). The differences in the shape of the receptacle should be noted—the buttercup representing a Dicotyledon and the lily a Monocotyledon; in the one the gynœcium is

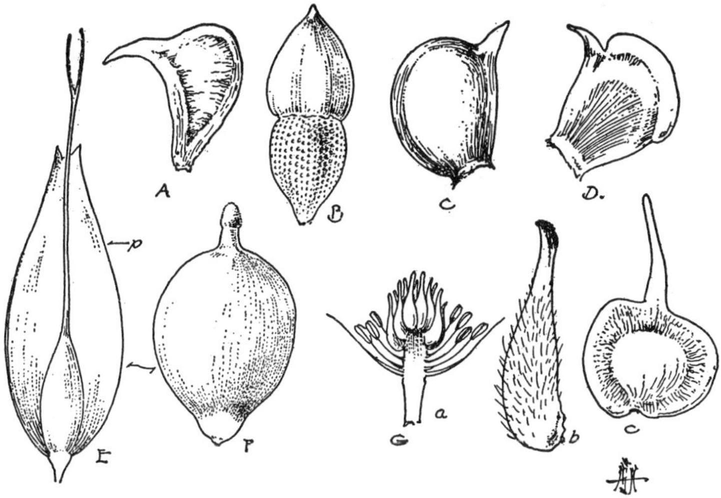

FIG. 47.—Various types of achenes: *A*, of *Sagittaria latifolia; B*, of *Eleocharis tuberculosa; C*, of a *Ranunculus; D*, of a magnolia; *E*, gynoecium, perigynium (*p*), and fruit (*F*) of a sedge (*Carex*); *G*, section of flower of *Anemone canadensis; b*, carpel, and *c*, achene of same.

apocarpous, in the other syncarpous (in this connection compare the monocotyledonous genera *Sagittaria* and *Lophotocarpus* with the buttercup). In both types the sepals, petals, and stamens stand *below*, and free from, the gynœcium—or ovary as we usually say when the gynœcium is syncarpous. Hence these organs are said to be *hypogynous* with respect to the gynœcium or ovary. And since the gynœcium or ovary stands *above* the other organs, it is said to be *superior*. A more advanced condition is found in *Rubus* and in *Prunus* (Fig. 38). Here the part of the receptacle below the gynœcium is expanded into a saucer-shaped or cup-shaped structure called the *hypanthium,* which bears the sepals, petals, and stamens on its margin; and as these organs now appear to stand in a circle *around* the gynœcium, they are said to be *perigynous.* It is to be observed that the gynœcium still occupies the morphological apex of the floral axis, that is, the gynœcium, or for convenience the ovary, is still superior—and is entirely free from the hypanthium. Many different degrees of perigyny will be found; that is, the receptacle may be only slightly expanded or it may form a very deep hypanthium. When, however, the receptacle is united or coalesced with the ovary more or less, the sepals, petals, and stamens will now often appear to stand on or on top of the ovary and are therefore said to be *epigynous.* As the ovary is now immersed in the receptacle, it is said to be *inferior,* but its actual morphological position at the apex of the floral axis has not been disturbed (Figs. 37C, 39). There are many different types and degrees of epigyny; these may be seen in irises, narcissi, orchids, walnuts and oaks, currants and gooseberries, apples and pears and quinces, fuchsias and evening primroses, aralias, carrots and parsnips and dogwoods, cranberries and blueberries, all madders and honeysuckles, harebells and lobelias, and all gourds and composites. Such genera as *Ribes, Pyrus,** *Fuchsia,* and *Œnothera* shows clearly the evolutionary advances in the development of the hypanthium (Figs. 40, 213, 235). These types should not be confused, however, with such forms as are found in the spiræas and the cherries and plums—forms in which there is a well-developed hypanthium but in which the ovary is superior. A case of epigyny of a special type is that found in *Heuchera* and other saxifragaceous

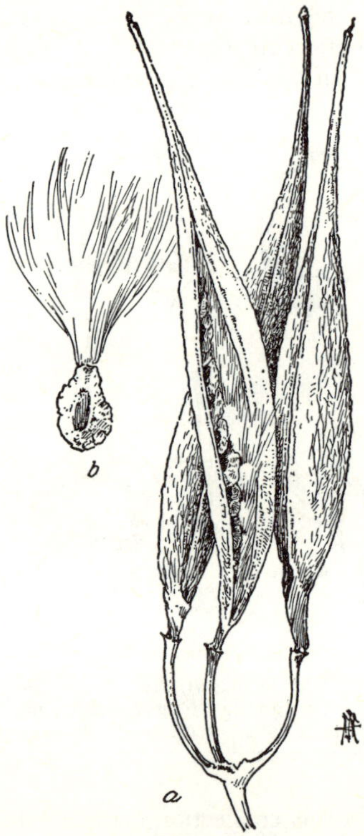

FIG. 48.—Follicles (*a*) and seed (*b*) of a milkweed (*Asclepias*).

* Including *Malus.*

genera (Fig. 41). In these forms the ovary is only partly inferior, yet a more or less prominent hypanthium has been developed on which the stamens arise at various levels instead of on the margin. In certain rosaceous genera (*Hirtella, Acioa*), an unusual modification of the floral axis has been developed; here the floral axis forms a deep cup-shaped or funnel-shaped tube below the floral organs, on the margin and one side of which the andrœcium and the gynœcium are seated.

The hypanthium of epigynous flowers is frequently colored, so that it is likely to be confused with a tubular sympetalous corolla.

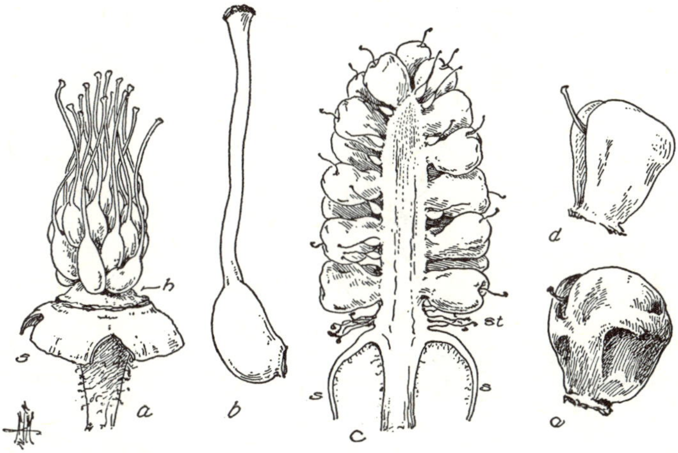

FIG. 49.—Blackberry (*Rubus*); *a*, the polycarpellate gynœcium; *b*, carpel; *c*, median section of the aggregate fruit; *d*, a young drupelet; *e*, mature drupelet; *h*, hypanthium; *s*, sepals; *st*, stamens.

But the position of the calyx (hypogynous or epigynous) and of the ovary (superior or inferior) will explain the situation. Compare *Ribes, Fuchsia, Œnothera,* with *Pyrus, Heuchera, Philadelphus, Cucurbita, Campanula, Helianthus, Iris, Canna.*

Epigynous flowers are characteristic of a number of important orders and families, all represented in our flora by many common and familiar plants. As these epigynous groups are relatively few and well defined, they are not difficult to remember once we are familiar with the floral type of each. The orders are listed in the chart presented on a succeeding page. As epigynous flowers are regarded as the most advanced of the three types just discussed, it

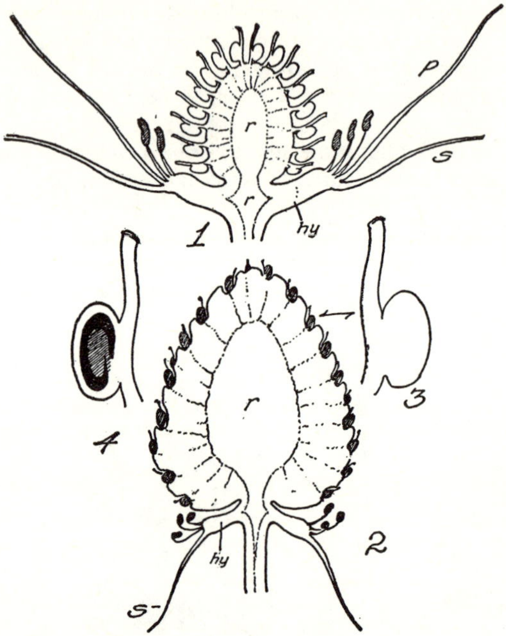

FIG. 50.—Strawberry (*Fragaria*): *1*, median section of flower; *2*, median section of the accessory fruit; *3*, a carpel; *4*, median section of carpel; *hy*, hypanthium, with stamens, petals (*p*) and sepals (*s*) on margin.

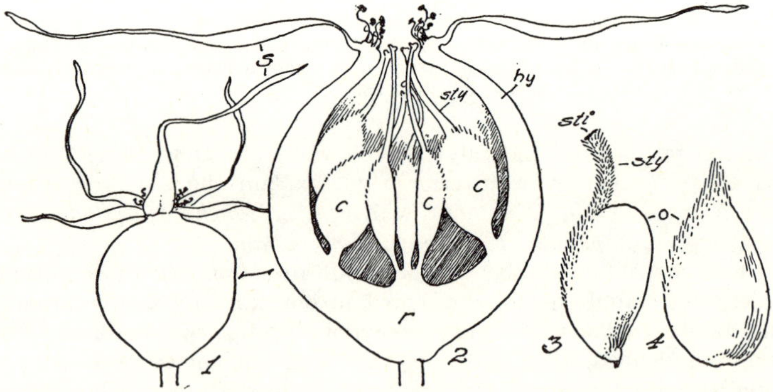

FIG. 51.—Rose: *1*, fruit (hip); *2*, median section of hip; *3*, a carpel; *4*, fruit (achene) derived from *3*; *c*, carpels; *r*, receptacle; *hy*, hypanthium of receptacle; *o*, ovary; *sti*, stigma; *sty*, style; *s*, sepals. Stamens in throat of hip.

follows that the groups possessing them will stand at the top, generally speaking, of a phylogenetic series.

Gynophore, androphore, androgynophore. The apical portion of the receptacle is sometimes more or less elongated so as to lift the gynœcium or the androecium or both above the parts below. The stalk thus formed, if it supports the gynœcium alone, is called a *gynophore.* If it supports the androecium alone, it is an *androphore.* If it supports both gynœcium and androecium, it is an *androgynophore.* (See *Silene, Cleome, Cassia, Euphorbia,* and some of their allies). In the *Umbelliferæ* and the *Geraniaceæ* the floral axis between the carpels forms a *carpophore,* to which the carpels (*mericarps*) cling on separation at maturity. All such structures as these serve as useful recognition-marks of the genera possessing them (Figs. 181, 182, 204).

CHAPTER VII

THE FRUIT

In common usage the term *fruit* connotes a multitude of different plant structures. That fruits spring from flowers in one way or another is generally understood, but what parts of the flower are immediately involved and whether or not structures outside of the flower may also be involved is not always clear. Botanically the term is difficult to define. In the simpler cases where the ovary is superior, as in the cherry, pea, or milkweed, the fruit is developed from the ovary alone. Such a fruit is a *true fruit* (Figs. 1, 22, 25, 29, 30, 38, 39). And such a fruit will always be formed from hypogynous or perigynous flowers. The situation is the same whether the gynœcium be apocarpous or syncarpous, and irrespective of the number of carpels, except that when there are several or numerous distinct carpels there will result as many true fruits as there are carpels, provided all the carpels developed (Figs. 1, 25). When the ovary is inferior, however, as in all epigynous flowers, the receptacle—which is here coalesced with the ovary, becomes a part of the fruit. Fruits of this type are called *spurious fruits* (Figs. 37, 40, 41, 42). This term may, however, be applied to any fruit which is not a true fruit, *e.g.*, a pineapple. But fruits like the pineapple, that is, composite fruits consisting of numerous individual fruitlets derived from as many separate flowers, are usually termed *multiple* or *collective* fruits. Thus an ear of maize, a mulberry, or the osage-orange (*Maclura pomifera*), and the fruits of the bottle-brushes, are similarly mul-

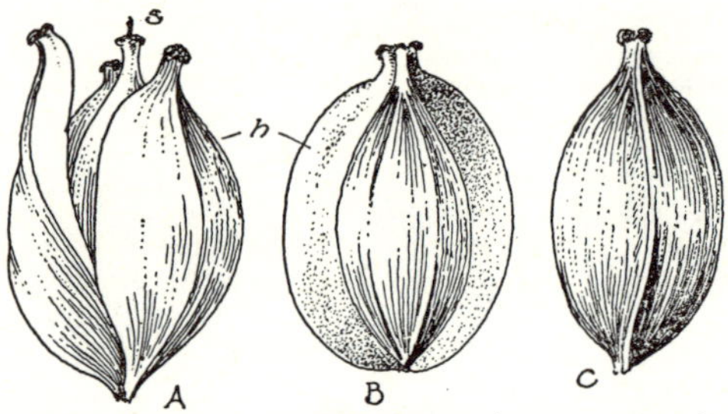

FIG. 52.—Fruit of a hickory (*Carya glabra*), dehiscing at *A*, in section at *B*; *C*, nut; *h*, exocarp —receptacular part of fruit; *s*, style.

tiple or collective fruits, except for this difference, that the individual fruitlets of the maize, the mulberry, and the osage-orange are true fruits, since they are derived from superior ovaries, while in the bottle-brushes the individual fruitlets are derived from inferior ovaries (Figs. 43, 44). In the mulberry there is a further feature to be observed, namely, that it is the persistent calyx of each flower that forms the succulent part of each fruitlet, while the fruit proper, a true fruit, is completely invested by the calyx and is of the nature of an achene (Fig. 45). Such fruits are termed *accessory* fruits, but they might as conveniently be called *calycular* fruits, just as we might with equal reason, by way of contrast, call all fruits derived from an inferior ovary *receptacular* fruits. Other forms

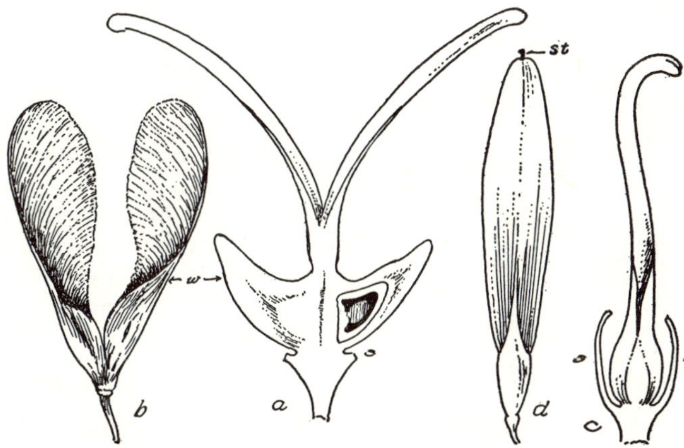

FIG. 53.—Samaras and flowers from which derived: *a,* gynoecium of box-elder (*Acer Negundo*) and (*b*) the fruit; *w,* wing (developed from edge of ovary); *c,* gynoecium and calyx (*s*) of an ash (*Fraxinus*); *d,* the samara; *st,* style.

similar to the mulberry may be seen in the common wintergreen (*Gaultheria procumbens*) and its relative, the Salal (*Gaultheria Shallon*) of the Pacific Coast, except that the separate fruitlets are not crowded together in a compact mass (Fig. 45).

The apple, pear, quince, currants and gooseberries, pomegranates, all gourds, blueberries and cranberries, acorns, beech-nuts, walnuts, the fruits of the eucalypts and of the entire orders *Umbellifloræ, Rubiales,* and *Campanulatæ* are all spurious fruits derived from inferior ovaries and adnate receptacle. To distinguish these from other spurious forms, it might be convenient to call them receptacular fruits, inasmuch as the receptacular portion often forms the bulk of the fruit (apple, pear, quince) (Figs. 28, 46, 391, 392, 408).

True fruits derived from polycarpellate apocarpous gynœcia are characteristic of the *Ranales,* many of the *Rosales,* and of such Monocotyledons as the water-plantains (*Alismataceæ*). As a rule these fruits are small and dry at maturity. The smaller one-seeded ones, which do not split open or *dehisce* at maturity, are called *achenes* (Figs. 25, 47, 424); the larger, with few to many seeds, which dehisce along one side at maturity, are called *follicles* (Fig. 48). Because of their small size achenes are often, but erroneously, called seeds. The entire mass of such achenes is called an *aggregate* fruit. An aggregate fruit thus is always formed from the gynœcium of a single flower (Figs. 1, 25, 187, 194).

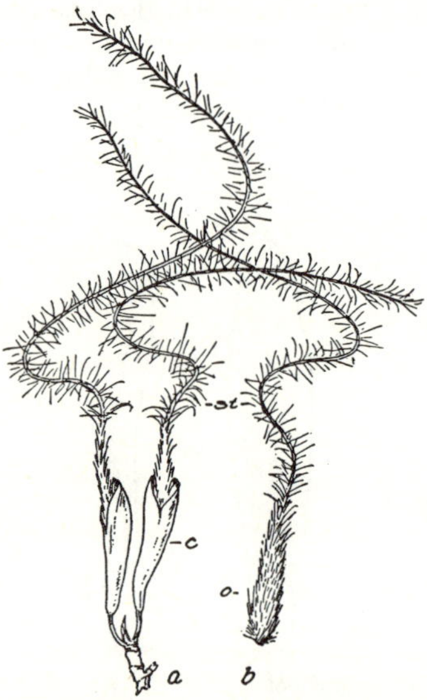

FIG. 54.—Fruits of *Cercocarpus* (*Rosaceae*): *a,* fruits seated in calyx (*c*); *b,* an individual fruit (carpel); *st,* styles.

The "fruits" of raspberries and blackberries are likewise aggregate fruits, but in these forms the individual fruitlets are succulent; in outward appearance they resemble the multiple fruit of the mulberries, however. Furthermore, the flowers of raspberries and blackberries are perigynous, and the fruitlets are borne on a columnar part of the receptable which rises from the shallow hypanthium (Fig. 49). A modification of this type is seen in the strawberry: here the column develops into a large succulent structure—the "strawberry"—on the surface or in small pits of which are borne the numerous minute *true fruits* or achenes (Figs. 50, 9). The flower of the strawberry is also perigynous like that of the raspberries and the blackberries. Thus the main differences are seen to reside in the features of the gynœcium-bearing column and of the true fruit—succulent drupelets on the one hand and dry achenes on the other.

A spurious fruit of a distinctive type, derived from a perigynous flower, is found in the roses. Here it is the receptacle (hypanthium) that becomes the "fruit" as generally known—botanically termed the *hip*. The true fruits or achenes are the so-called "seeds" borne inside, the styles of which may be seen projecting from the orifice (Fig. 51). A fruit of a similar sort, though not succulent, is that of the Carolina allspice (*Calycanthus,* Fig. 195). A transition from the rose type to the pomaceous type of spurious fruit may be ob-

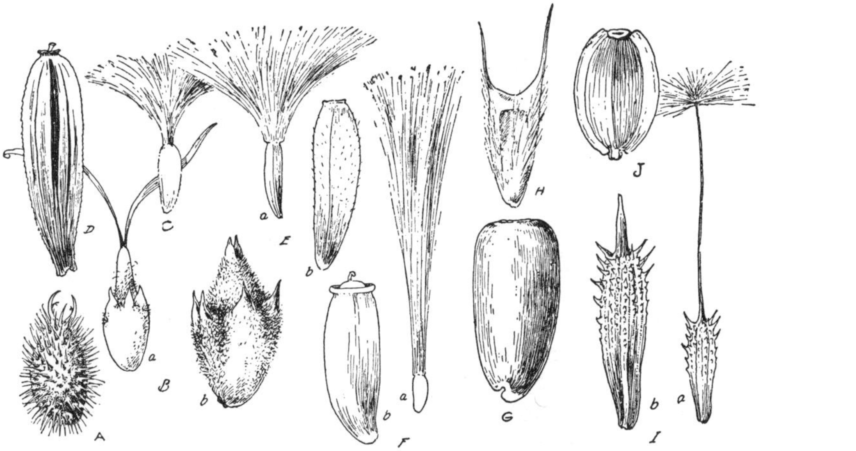

FIG. 55.—Fruits (spurious; cypselas) of composites: *A,* bur-fruit of *Xanthium*. *B,* bur-fruit of *Ambrosia; a,* early stage, with styles in place; *b,* mature stage. *C,* fruit of sow-thistle (*Sonchus*). *D,* same enlarged, with pappus removed. *E,* fruit of an aster enlarged (*a*), and with pappus removed at (*b*). *F* (*a*) fruit of Canada thistle, enlarged at (*b*). *G,* fruit of a sunflower (*Helianthus*). *H,* fruit of *Geraea canescens*. *I,* fruit of dandelion (*a*) body enlarged at (*b*).

served in the hawthorns, *Cratægus* and *Cotoneaster* (Figs. 231, 232, 233). It may be interpolated, in passing, that the various families of the orders *Ranales* and *Rosales* will yield the student bountiful sources for the study of different types of fruits.

Nut is another term having various applications both in botany and in popular usage. As commonly used the term may connote a fruit or a seed. Thus the acorn and hazel-nut, both spurious fruits derived from epigynous flowers, are called nuts. The walnut is the "stone" of a spurious drupaceous fruit, while the cocoanut of commerce is the stone of a true fruit. Brazil-nuts are seeds; shelled peanuts are seeds. *Nutlets* are very small, hard-shelled, seed-like fruits of various plants (Fig. 52).

An unusual modification of the flower-stalk to form a large pear-shaped "fruit" occurs in the cashew-tree (*Anacardium occidentale*)

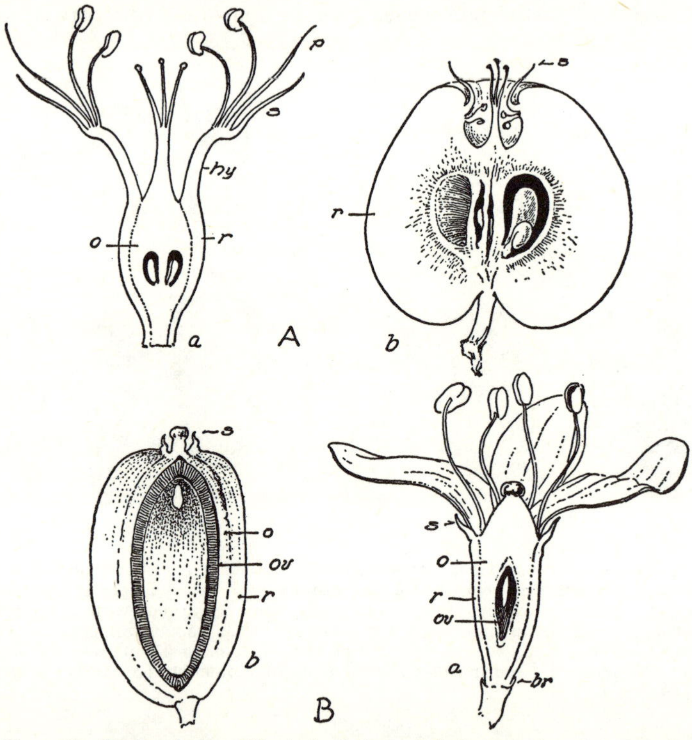

FIG. 56.—Fleshy fruits (*b*) derived from inferior ovaries (*a*): *A*, apple (*Malus*), *B*, a dogwood (*Cornus*); *hy*, hypanthium; *o*, ovary; *ov*, ovule; *r*, receptacle; *p*, petals; *s*, sepals. Apple shows style and stamens in depression at top.

of the Brazilian tropics; above this structure is the fruit proper, a true fruit derived from the ovary of the solitary carpel. In the custard-apple (*Anona*) the many carpels fuse with the receptacle to form a pulpy mass—an aggregate fruit. Another curious situation may be seen in the blue cohosh (*Caulophyllum thalictroides*), where the ovary-wall is ruptured by the developing ovules, these taking on the appearance of berries.

In many spurious fruits derived from epigynous flowers the calyx-segments persist at the top of the fruit, both of succulent and dry forms. Thus in the pomegranate the persisting calyx is very prominent and of a succulent texture (Fig. 302), while in the apple, pear, quince, cranberry, and blueberries it is quite small. In small dry spurious fruits it is often much reduced, and in

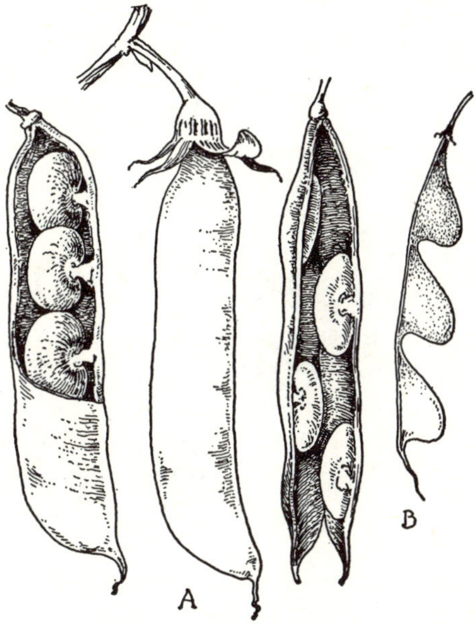

FIG. 57.—*A,* pods (legumes) of a bean; *B,* loment of a tick-trefoil (*Desmodium*).

many cases caducous (*Umbelliflorœ, Rubiales, Campanulales, Philadelphus;* Figs. 313, 316, 318, 388, 390, 404, 408).

Appendages and surface features of fruits. Many fruits develop wing-like outgrowths from thc ovary-wall which serve as devices for effecting dispersal. Such appendages are characteristic of the fruits of maples, elms, birches, hop-trees, and many others. Fruits of this sort are known as *samaras* or *key-fruits.* In numerous other cases spines, bristles, tubercles, or felt-like coatings of hairs of various sorts, often gland-bearing, are formed. A persistent style or calyx also becomes an appendage, serving mainly as an organ of dispersal, as noted under preceding headings (Figs. 53, 54).

Coming under the general category of fruits are the specially modified heads or capitula of certain composites, such as the burdock (*Arctium*), the cocklebur (*Xanthium*), and the ragweeds (*Ambrosia*). Here the entire head is usually called the fruit, although strictly it is an inflorescence, enclosing the fruits proper, which are spurious. In the burdocks the involucral bracts become

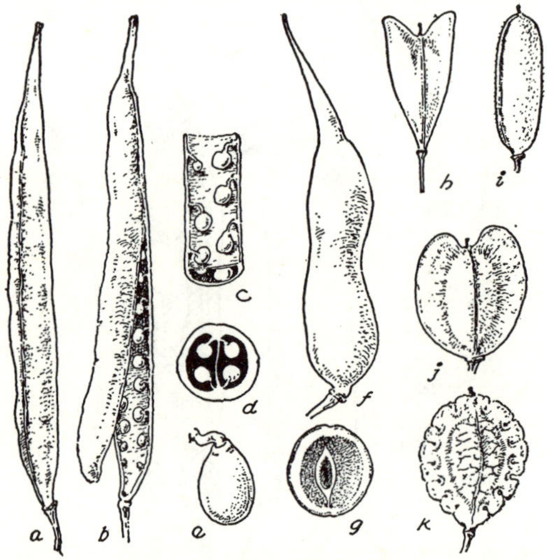

FIG. 58.—Fruits of various mustards: *a, b,* siliques of *Brassica; c,* section of a part of same, and *d,* transverse section of ovary; *e,* ovule of *Brassica; f,* silique of radish (*Raphanus sativus*); *g,* transverse section of same; *h,* silicle of shepherd's purse (*Capsella Bursa-pastoris*); *i,* silicle of *Draba verna; j.* silicle of "French-weed" (*Thlaspi arvense*); *k,* silicle of a *Thysanocarpus.*

barbs, but are not coalescent as in the cocklebur and the ragweeds (Fig. 55).

Parts of the fruit. In all true fruits the wall or *pericarp* corresponds to the ovary-wall, whereas in spurious or receptacular fruits the pericarp corresponds to the ovary wall and the adnate receptacle. In fleshy fruits the pericarp consists of three layers—an outer "rind," the *epicarp;* a middle succulent layer, the *mesocarp;* and an inner, often hard or stony layer, the *endocarp* (Figs. 56, 52A).

A knowledge of the fruit is important to a full understanding of the relationships and classification of the flowering plants. Often the fruit, and the seeds as well, are the last resort in differentiating species, genera, etc., and in trying to identify a particular plant.

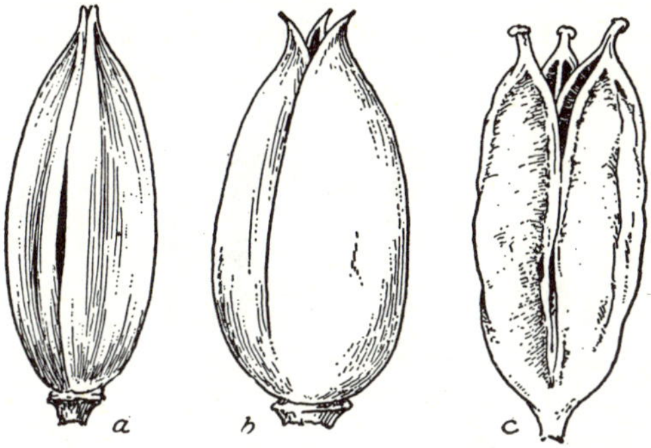

FIG. 59.—Capsules of some Monocotyledons: *a, Juncus; b, Veratrum; c, Iris.*

Generally speaking, fruits are stable organs, and so furnish reliable diagnostic characters. Thus one can recognize without difficulty an apple, a gourd, or a rose hip, in spite of the many species and varieties.

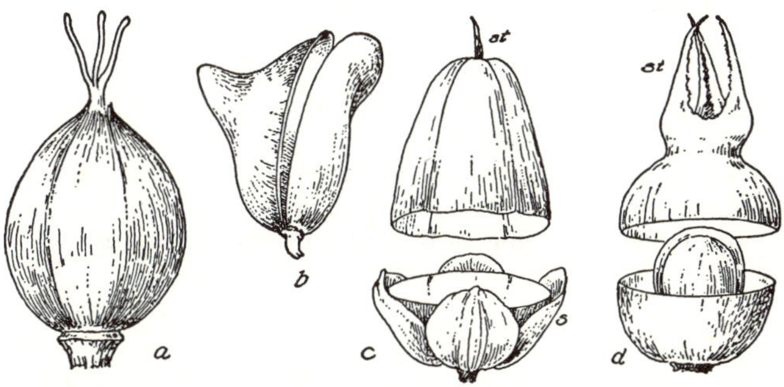

FIG. 60.—Capsules of some Dicotyledons: *a, Phlox; b, Scutellaria; c, Plantago; d, Amaranthus retroflexus; st,* style; *s,* calyx.

Classification of fruits. It was shown above that there exist several kinds of fruits according to their structure or organization. These were true fruits, spurious fruits, aggregate fruits, etc., etc. In other words, these appellations merely designate certain categories into which fruits may be grouped or classified, according to their morphological organization. Some of these different kinds, it was also noted, were dry fruits and others were succulent or fleshy, thus showing two different evolutionary trends with respect to texture. On the basis of texture, then, we may classify all fruits as either *dry* fruits or *fleshy* fruits. Dry fruits are of two sorts according to whether they split open to discharge the seeds or remain closed, namely:

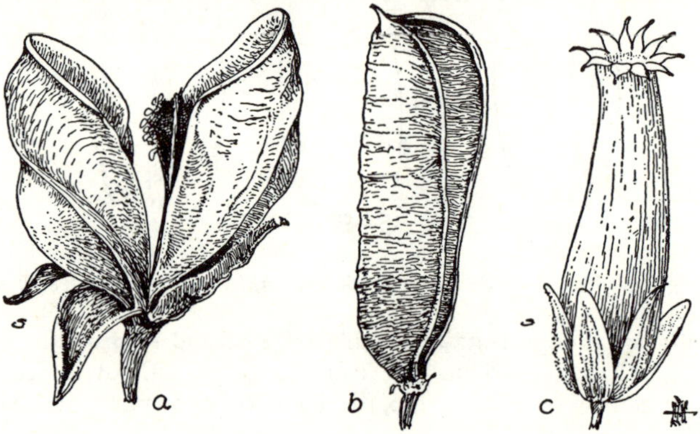

FIG. 61.—Various dry fruits: *a*, follicle of *Veronica; b*, follicle of *Delphinium; c*, capsule of *Stellaria; s*, sepals.

dehiscent and *indehiscent*, respectively. Fleshy fruits rarely dehisce. The dehiscence of dry fruits proceeds in various ways, and is often violent or "explosive," scattering the seeds over considerable areas. The various types of fruits coming under these different categories are classified below, and in addition there is appended a classification of the fruits according to their morphological origin.

Dry Fruits

I. DEHISCENT

Follicle. A pod-fruit arising from a single carpel with a superior ovary, which dehisces along one edge. Pæony, larkspur, milkweed (Figs. 48, 60, 186, 345, 346).

Legume. A pod-fruit arising from a single carpel with a superior ovary, which dehisces along both edges. Pea, bean, vetch, lupine (Fig. 57).

Loment. A legume that breaks up crossways, usually into one-seeded sections (Fig. 57).

Silique (*siliqua*). An elongate two-celled pod-fruit arising from a two-carpellate gynœcium with a superior ovary. A false dissepiment

FIG. 62.—Mode of dehiscence of capsules: *1*, septifragal; *2*, loculicidal; *3*, septicidal.

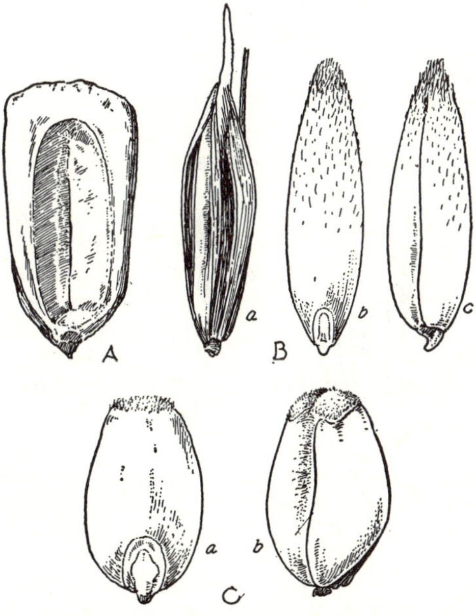

FIG. 63.—Fruits of some grasses (cereals): *A*, maize; *B*, oat: (*a*) with husk, (*b*) dorsal and (*c*) ventral aspects; *C*, wheat: (*a*) dorsal and (*b*) ventral aspects.

divides the ovary into two cells. When ripe the two halves (*valves*) of the fruit separate from the dissepiment from below upwards. Some siliques break up transversely into one-seeded segments in the manner of a loment. Many mustards (Fig. 58j).

Silicle (*silicula*). A short, broad, flat, pouch-like silique. Shepherd's purse, penny-cress (*Thlaspi;* Fig. 58).

Capsule. A pod-fruit arising from a syncarpous gynœcium with a superior or inferior ovary. Dehiscence various: by lengthwise splitting into sections, by pores, or by a transverse lid (*circumscissile*). Flax, morning-glory, *Philadelphus* (Figs. 59, 60, 61, 62).

Pyxidium or pyxis. A capsule which dehisces by a transverse lid. Henbane (*Hyoscyamos niger*), *Amaranthus, Plantago* (Fig. 60).

Diplotegium. A capsule arising from an inferior ovary (Fig. 59c). *Iris, Philadelphus.*

2. INDEHISCENT

Achene. A small one-seeded dry fruit arising from a single carpel with a superior ovary. In most cases the style persists as a peg or slender elongate appendage. Buttercup, anemones, strawberry, rose, water-plantains (Figs. 25, 47, 51, 424).

Cypsela. A small achene-like fruit arising from the inferior ovary of a syncarpous gynœcium. Sunflowers, asters, dandelions (Fig. 55).

Utricle. A small one-seeded dry fruit arising from the superior ovary of a syncarpous gynœcium, in which the pericarp is bladdery and loosely invests the seed. Lamb's-quarters, dock, rhubarb, buck-wheat (Figs. 167, 171).

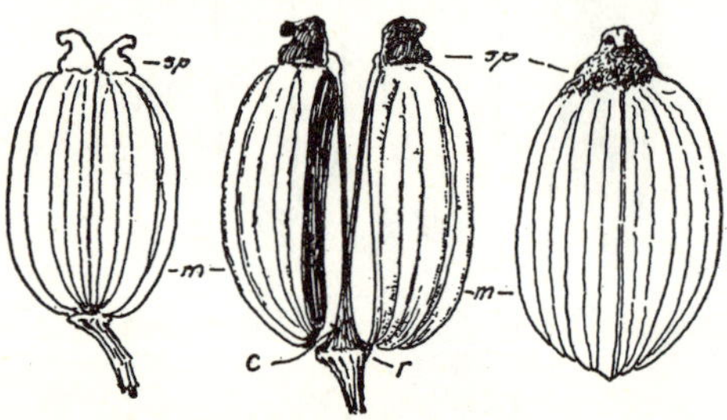

FIG. 64.—Fruit (schizocarp) of an umbellifer (*Foeniculum vulgare*): *sp,* stylopodium; *m,* mericarps; *r,* receptacle; *c,* carpophore.

Caryopsis or *grain.* A small, one-seeded, achene-like fruit arising from the superior ovary of a syncarpous gynœcium, in which the pericarp is fused with the seed-coat. Wheat, oats, barley, millet, Indian corn, and all other grasses (Fig. 63).

Samara or key-fruit. A small, achene-like, winged fruit arising from the superior ovary of a syncarpous gynœcium. Maple, box-elder, ash, elm (Fig. 53).

Schizocarp. A small dry fruit arising from a superior or inferior ovary of a syncarpous gynœcium, which splits into sections (*mericarps*) corresponding to the individual carpels or segments of carpels. The *nutlets* of the borages, mints, vervains, and some mallows (*Malva*) arise from superior ovaries; the fruitlets of the *Umbelliferæ* arise from inferior ovaries (Fig. 64).

Nut. A hard-shelled fruit arising from the inferior ovary of a syncarpous gynœcium, as the acorn, hazel-nut. Or the "stone" or bony endocarp of some drupaceous fruits arising from inferior ovaries, as the walnut, or of some true fruits, such as the cherry, plum, almond, peach; or a seed as applied to Brazil-nuts or shelled peanuts (Fig. 52).

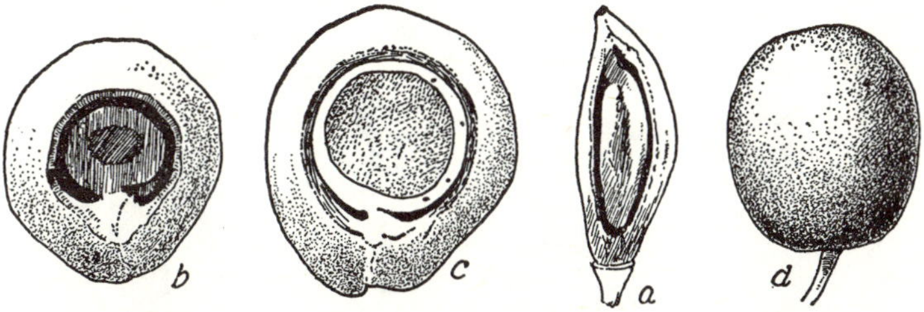

FIG. 65.—Stages in the development of the fruit of a cherry: *a,* median section of a young ovary; *b, c,* early and late stages,—*c,* showing well-advanced "stone"; *d,* mature fruit.

Nutlet. A small hard-shelled fruit arising from a superior ovary, including mericarps. Borages, mints, vervains (Figs. 64, 359).

Diclesium. An achene invested by the persistent base of the perianth. *Mirabilis Jalapa.*

Fleshy Fruits

Drupe or stone-fruit. A fleshy, usually one-seeded fruit arising from the superior ovary of a single carpel or of a syncarpous gynœcium. Cherry, plum, peach, apricot, almond, olive, palms (Figs. 65, 228).

Drupelet. A diminutive drupe. Fruitlets of raspberries and blackberries (Fig. 49).

Berry or bacca. A small or large fleshy fruit arising from the superior or inferior ovary of a syncarpous gynœcium. The fruits of the tomato and the grape arise from superior ovaries; those of the currants and gooseberries, cranberry, blueberries, strawberry, guava, and pomegranate arise from inferior ovaries (Figs. 39, 66, 67).

Hesperidium. A thick- but soft-rinded berry arising from the superior ovary of a syncarpous gynœcium. All citrus fruits (Fig. 67).

Pepo or gourd. The thick-rinded, often large berry of the gourd family (*Cucurbitaceæ*), arising from an inferior syncarpous ovary and adnate receptacle (Figs. 400, 401, 402).

Pome. A firm fleshy fruit arising from the inferior ovary of a syncarpous gynœcium and adnate receptacle. The carpels are some-

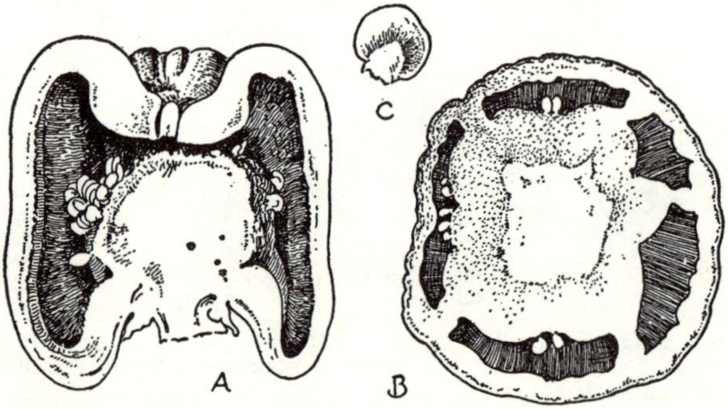

FIG. 66.—Fruit (berry) of red pepper (*Capsicum frutescens*): *A,* median; *B,* transverse, section; *C,* ovule.

times more or less loosely united by their inner edges. Apple, pear, quince, hawthorns (Figs. 41, 46, 56, 229, 230, 231, 233).

Hip. The fleshy fruit of roses, consisting of the succulent hypanthium of the receptacle, which encloses the several true fruits or achenes (Figs. 51, 226).

Summary of fruits according to their morphological nature:

I. Fruits arising from a superior ovary, *i.e.,* from hypogynous or perigynous flowers. *True fruits.*
  1. Dry true fruits.
    Dehiscent: Follicle, legume, loment, capsule, siliqua, silicle, pyxidium (pyxis).
    Indehiscent: Achene, diclesium, caryopsis, utricle, samara, schizocarp (mericarps, cremocarps), nutlets.
  2. Fleshy true fruits: Drupe, drupelet, berry, hesperidium.

II. Fruits arising from inferior ovaries and adnate receptacle, *i.e.,* from epigynous flowers. *Spurious or receptacular fruits.*
  1. Dry spurious fruits.
    Dehiscent: Diplotegium (capsular).
    Indehiscent: Cypsela (achene-like).
  2. Fleshy spurious fruits: Berry, pepo or gourd, pome.

Special types of fruits involving extra-floral parts:

Aggregate fruits. A mass of fruitlets arising from polycarpellate apocarpous gynœcia. Anemones, water-plantains, blackberries, and raspberries.

Multiple fruits. Fruits composed of numerous fruitlets of distinct flowers massed on the sometimes more or less succulent axis of the inflorescence. Also called *collective fruits.* Mulberry, pineapple.

Accessory fruits. Fruits composed not only of the gynœcium but of parts accessory to but not coalesced with it, such as calyx or receptacle. Wintergreen, strawberry.

Syconium. A multiple fruit, the fleshy part of which is the axis of the inflorescence. Fig (*Ficus*).

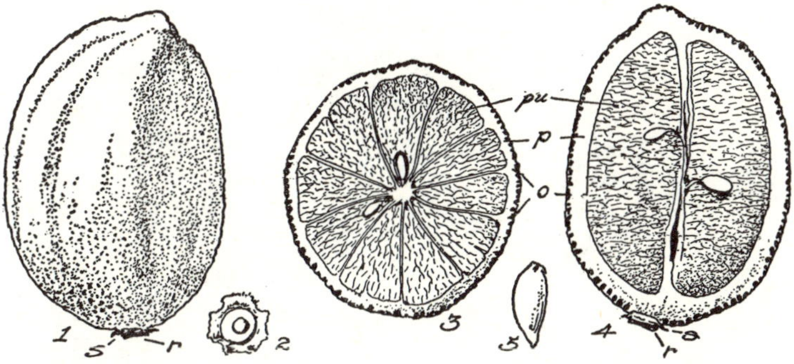

FIG. 67.—Fruit (hesperidium) of lemon (*Citrus Limonia*): *1*, mature fruit; *2*, remains of calyx; *3*, transverse section; *4*, median section; *5*, ovules; *o*, oil-glands; *p*, pericarp; *pu*, pulpy substance of ovary (locules); *r*, receptacle; *s*, calyx.

The seed. After fertilization the ovary and whatever accessory parts may be involved undergo certain notable changes that result in a fruit—which may be dry or fleshy, true or spurious, as the case may be. In the meantime the ovules within the ovary develop into seeds. In the course of this development the integuments of the ovule become the seed-coat or *testa;* the nucellus either shrinks to form a thin layer next to the testa or it becomes a storage organ of reserve foods called the *perisperm;* and in the space formerly occupied more or less by the embryo-sac, there is now present an immature plant or *embryo* surrounded by a food-storage tissue, the *endosperm*—which, however, is often absorbed by the developing embryo, but leaves remains in the form of a thin coating around

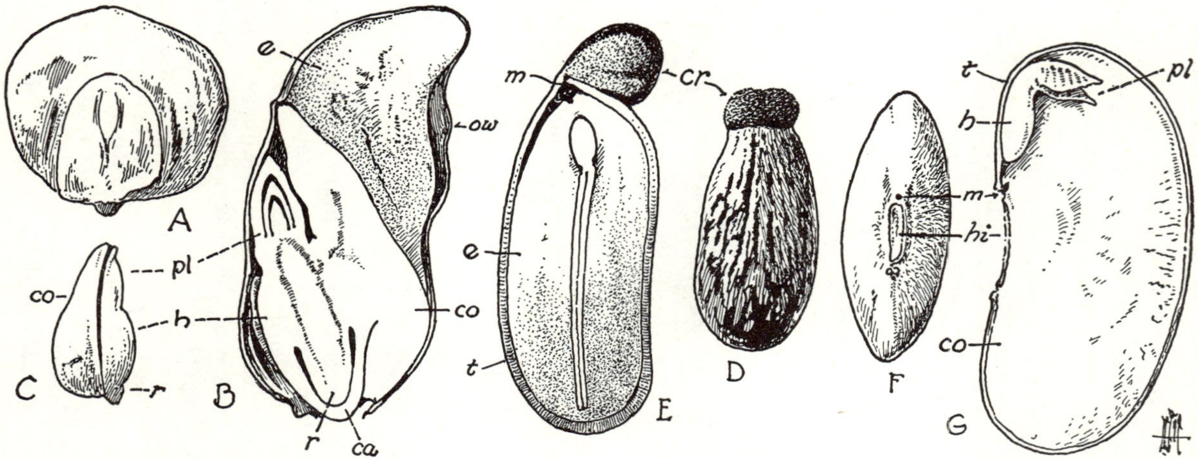

FIG. 68.—Albuminous (*A-D*) and exalbuminous (*F-G*) seeds. *A*, the "seed" (fruit) of maize, embryo side; *B*, median section of same; *e*, endosperm; *ow*, ovary-wall (pericarp,—coalesced with testa of seed); *pl*, plumule (epicotyl); *h*, hypocotyl; *r*, radicle. *C*, embryo; *co*, cotyledon (scutellum); *r*, radicle. *E*, section of seed of castor bean (*Ricinus communis*); *m*, micropyle; *e*, endosperm; *t*, testa; *cr*, caruncle; *D*, dorsal aspect of seed. *F*, basal aspect of bean; *G*, sectional view with one cotyledon removed; *pl*, plumule (epicotyl); *t*, testa; *h*, hypocotyl; *m*, micropyle; *hi*, hilum; *co*, cotyledon.

the embryo. Seeds in which an endosperm is present are called *albuminous* or endospermous seeds, those in which it is absent are called *exalbuminous* or exendospermous seeds. In some cases both endosperm and perisperm are present (*Piper;* Fig. 68).

The terms *hilum, micropyle, chalaza,* and *raphe,* as applied to the ovule on a preceding page (58), have a corresponding application to the seed. In some seeds a thick outgrowth, called the *caruncle* or *strophiole,* covers the micropyle (castor bean or *Ricinus* and other spurges; Fig. 68).

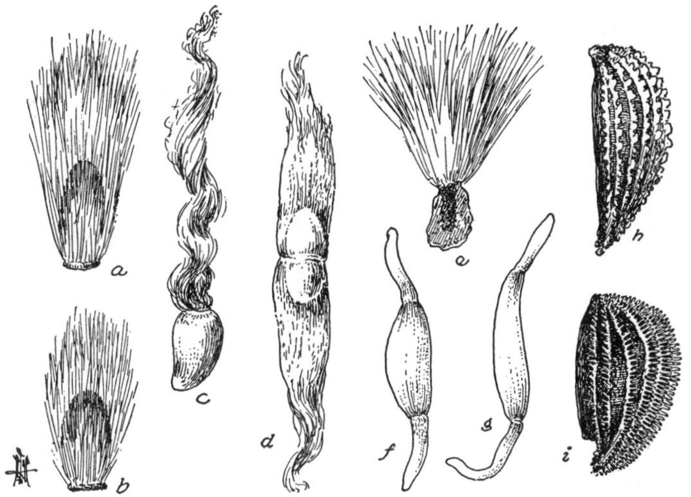

FIG. 69.—Various types of appendaged seeds: *a,* of willow; *b,* of cottonwood; *c,* of *Zauschneria* (*Œnotheraceæ*); *d,* of *Catalpa;* *e,* of a milkweed (*Asclepias*); *f,* of *Juncus canadensis;* *g,* of *Juncus Vaseyi;* *h,* of *Saxifraga virginiana;* *i,* of *Saxifraga saximontana.*

The characters of seeds are generally constant and therefore often of considerable taxonomic value. Not infrequently the seed characters are the last resort in defining the limits of a species. Many examples could be given of the value of seed characters in distinguishing species, but it will suffice to suggest that many such instances will be found in the genera *Juncus, Brassica, Saxifraga, Eucalyptus, Trifolium, Lupinus,* and *Phlox.* Most seeds when studied in quantities will show certain bulk or mass characters which may yield useful criteria for purposes of classification or identification. Characteristic winged seeds are produced by the catalpas, for instance, and the cottons are well known for their long-haired seeds, the hairs of which are known as the "staple." Seeds of willows and milkweeds bear a tuft of hairs at one end (Fig. 69). Many very small fruits are commonly called seeds, such as those of grasses, sedges, buckwheat, buttercups, roses, sunflowers and all other composites.

The embryo (Fig. 70). The embryo is the essential part of the seed; it is a product of the fertilized egg or *zygote* lying in the em-

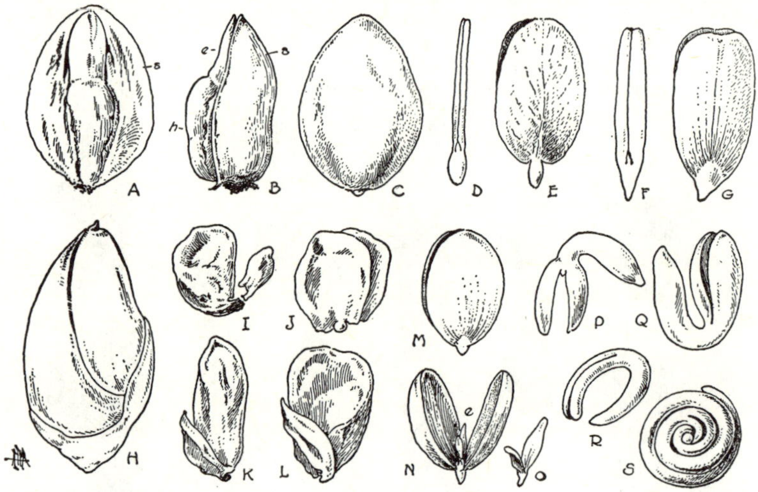

FIG. 70.—Various embryos: *A, B, C,* of maize—*A,* face view, *B,* side view, *C,* dorsal view of cotyledon (*scutellum*); *s,* cotyledon; *e,* epicotyl; *h,* hypocotyl. *D, E,* edge and side views of embryo of castor bean. *F, G,* edge and side views of embryo of sunflower. *H,* multiple embryo of an orange; *I-O,* seven smaller embryos found in *H; e,* minute embryo invested by larger embryo *N; I, J, K, L,* embryos with unequal cotyledons (irregular shape due to pressure of contiguous embryos); *M, N,* perfectly symmetrical embryos. *P,* embryo of sweet clover (*Melilotus alba*). *Q,* embryo of hemp (*Cannabis sativa*); *R,* embryo of lamb's quarters (*Chenopodium album*); *S,* embryo of Russian thistle (*Salsola Kali* var. *tenuifolia*).

bryo-sac. In "ripe" seeds the embryo is in a resting condition, and may or may not be invested by endosperm or perisperm or both. The shape of the embryo and its disposition with respect to the endosperm are always of significance in deciding questions of phylogenetic affinities. In many groups the embryo is straight, and if endosperm is present the embryo may lie in the center of it or at either end. In other groups the embryo is spirally coiled, or curved around the endosperm (Figs. 71, 171, 177). Numerous shapes intermediate between these are found both in endospermous and exendospermous seeds. In exendospermous seeds the embryo usually occupies all the interior space of the seed (Fig. 71).

The parts of the embryo are the *hypocotyl, cotyledons, epicotyl,* and *radicle.* The hypocotyl is the stem of the embryo; at its lower tip is the radicle, from which springs the primary root. The cotyledons or "seed-leaves" are attached at the upper end of the hypocotyl, at the *cotyledonary node.* Terminating the upper end of the hypocotyl is the *epicotyl* or terminal bud of the embryo, which is to form the shoot (Figs. 68, 72). The number of cotyledons in the embryo, namely, one or two, distinguishes the two major divisions of the flowering plants, the *Monocotyledones* and the *Dicotyledones.* In such Monocotyledons as the grasses, the embryo lies to one side of usually copious endosperm. In the gourds (*Cucurbitaceæ*) a peculiar outgrowth called the "peg" arises on the hypocotyl below the cotyledons and serves to dislodge the seed-coat as the cotyledons are lifted out of the ground after germination (Fig. 72). A tuberous hypocotyl is developed in certain geophilous peppers of South America. While two cotyledons are characteristically present in the Dicotyledons in general there are a

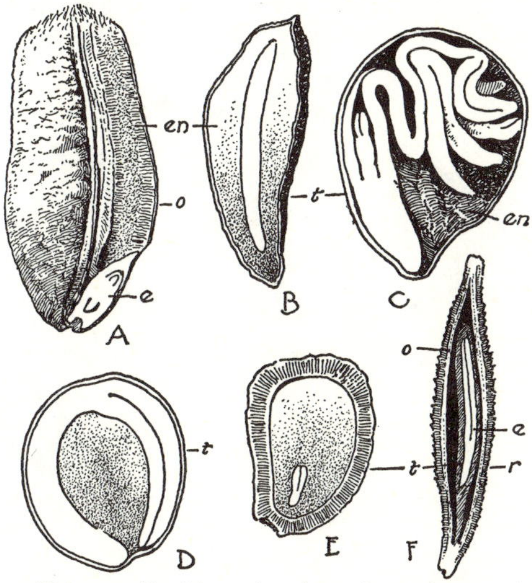

FIG. 71.—Position of embryo in various seeds: *A,* caryopsis (fruit !) of wheat with embryo (*e*) at base and to one side of endosperm (*en*); *o,* ovary-wall (pericarp), coalesced with testa. *B,* section of onion seed, with embryo in center of endosperm (*en*); *C,* section of seed of a morning-glory—embryo with folded cotyledons; *en,* portion of endosperm; *t,* testa. *D,* section of seed of a pig-weed (*Amaranthus retroflexus*); embryo coiled around endosperm. *E,* section of seed of Ribes, with small embryo at base of endosperm; *t,* testa. *F,* section of cypsela (achene) or fruit of oyster-plant (*Tragopogon porrifolius*); *e,* embryo within seed; *o,* ovary-wall, coalesced with receptacle, *r; t,* testa.

few exceptions to the rule; thus in *Bunium* (*Unbelliferæ*), the embryo has only one cotyledon, and in some *Proteaceæ* there may be as many as three to eight cotyledons. In contrast to these are such plants as the orchids, the dodders (*Cuscuta*), and the *Pyrolaceæ,* in which the cotyledons are absent. In the *Cruciferæ* the cotyledons are folded in various ways, which, along with correspondingly constant orientations of the embryo as a whole, afford useful bases for subdividing the family into minor groupings.

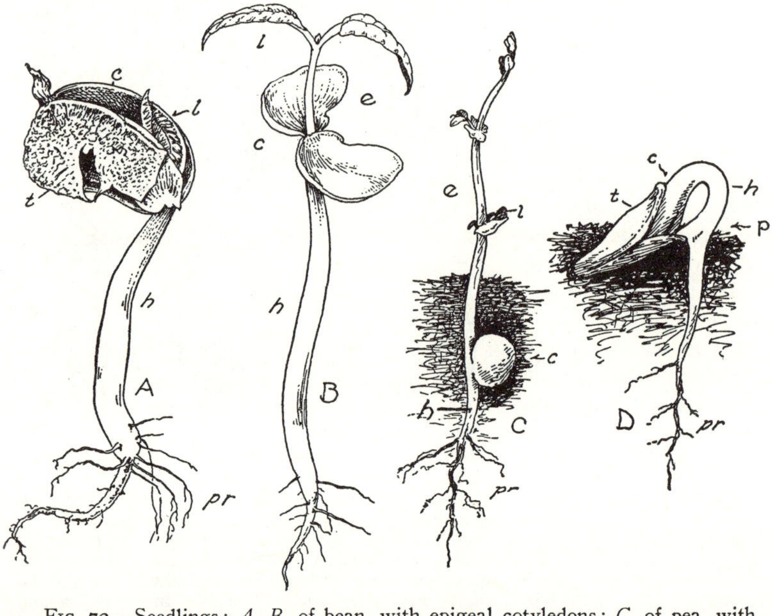

FIG. 72.—Seedlings: *A, B,* of bean, with epigeal cotyledons; *C,* of pea, with hypogeal cotyledons; *D,* of squash, with epigeal cotyledons; *t,* testa; *c,* cotyledons; *h,* hypocotyl; *e,* epicotyl; *l,* first leaves (of epicotyl); *p,* peg.

The seedling. In the development of the seedling the cotyledons are *hypogeal* or *epigeal* according to whether they remain in the ground or are lifted above the ground (Fig. 72). The cotyledons of the garden pea and that of the maize are of the former type, while those of the garden pea and the squash are of the latter type. In the gourds and the castor bean the cotyledons remain on the plant for a long time, functioning as leaves. Usually the cotyledons possess rather clearly marked generic characters that serve as useful recognition-marks. For instance, the basswoods, the maples, the gourds, the mustards, and other plants may be recognized quite readily in the seedling stage if we are acquainted with the shape of the cotyledons. Seedlings of alders, elms, clovers, and lupines will be found to bear bacterial nodules of more or less distinctive form according to the kind of plant. In the gourds the seedlings are often found developing inside the fruit, a condition known as vivipary; this is especially noticeable in the chayote (*Sechium edule*).

CHAPTER VIII

THE INFLORESCENCE

An inflorescence is a flower-cluster (Fig. 73). The simplest kind of inflorescence is one the stalk or *peduncle* of which bears a solitary flower at the summit; if the flower-stalk arises from the ground, such an inflorescence is called a *scape* (*Tulipa,* Fig. 74). *Scapose* or scape-like flower-clusters are those bearing two or more flowers, as in some lilies and orchids, and the dandelion. The flowers of an inflorescence may be stalked (*pedicellate*) or stalkless (*sessile*).

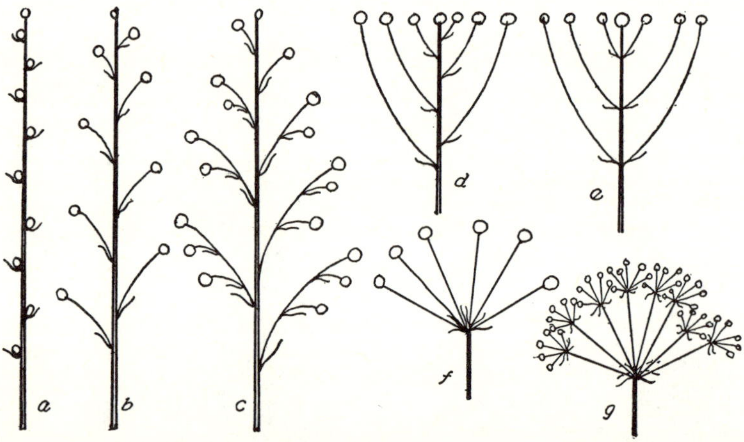

FIG. 73.—Inflorescences: *a,* spike; *b,* raceme; *c,* panicle; *d,* corymb; *e,* cyme; *f,* umbel; *g,* compound umbel.

The inflorescence may or may not bear leaves. When leaves are present they are often much reduced in size and are then called *bracts,* or there may be a gradual transition from foliage-leaves to bracts. The inflorescence always assumes some characteristic shape, such as conical, cylindrical, wand-like, umbrella-like, or capitate. Sometimes a certain type of inflorescence occurs as a constant feature of an entire order or family (*Umbelliflorœ, Compositœ*), or different forms may occur in the same genus (*Spiræa, Prunus*). The inflorescence may be open and loose or compact, and in the latter case the flowers are often crowded together into smaller compact masses called *glomerules*. An inflorescence of this latter type is said to be *glomerate*.

Morphologically an inflorescence is a flower-bearing shoot; it may be limited in its growth or it may grow indefinitely in length. The main axis of an inflorescence is called the *rachis*. According to the mode of growth, two general types of inflorescences are distinguished, namely, the *racemose* and the *cymose*. In the racemose or indeterminate type the shoot continues to grow more or less indefinitely until it sometimes attains considerable length (*Verbascum; Epilobium angustifolium,* Fig. 311). The lowest flowers of such an inflorescence, being the oldest in the order of succession, are the first to open, and the blossoming proceeds from below upwards, that is, in *acropetal* succession or, in other words, from the circumference inwards towards the center—in *centripetal* succession (Fig. 75a). In the cymose type the elongation of the shoot is stopped (determinate) by the terminal flower, but the branches may continue to grow until they reach the level of or overtop the terminal flower before their flowers open, or flowers may continue to be formed in the bract- or leaf-axils below the terminal flower. The terminal flower, being in this case the oldest, is now the first to open and is followed in succession downwards by those below. The order of flowering thus proceeds from the top downwards, that is in *basipetal* succession, or from the center outwards—in *centrifugal* succession (Fig. 75b).

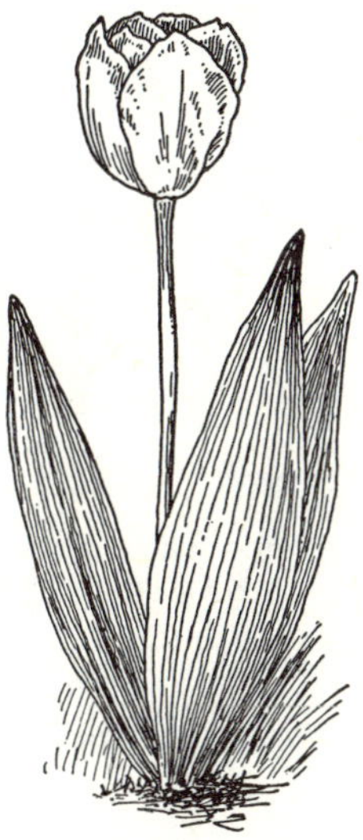

FIG. 74.—Scape of the tulip.

In many inflorescences the flowers arise from the axil of a small or sometimes large bract (*floral bract*), in addition to which there are often to be found on the pedicel two smaller bracts (*bracteoles*), sometimes in close proximity to the flower. The bracteole to the left of the median plane of the flower is designated by the Greek letter α, the one on the right by β. From this we observe that the flower together with its pedicel is in reality a shoot arising from the axil of a leaf (the floral bract) (Fig. 3). The bracteoles are ab-

FIG. 75.—Inflorescences: *a*, racemose inflorescence of a mustard (*Brassica*); *b*, cymose inflorescence of a sunflower (*Helianthus divaricatus*).

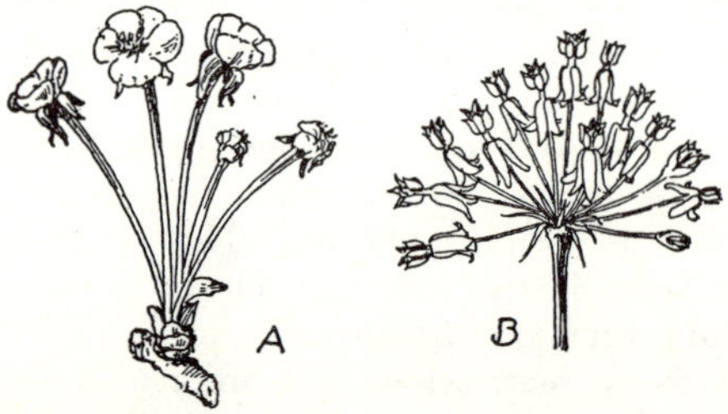

FIG. 76.—Umbellate inflorescence of a cherry, *A*; *B*, umbel of a milkweed (*Asclepias mexicana*).

sent, for instance, in the *Cruciferæ*, but they are present in the *Caryophyllaceæ*. *Supra-axillary* flowers, that is flowers that arise at some distance above the axil of the subtending leaf or floral bract, are present and characteristic of certain members of the *Solanaceæ*, of *Cuphea* (*Lythraceæ*), and certain species of *Sparganium*.

The following types of racemose and cymose inflorescences are generally met with:

Racemose Inflorescences

1. SIMPLE (UNBRANCHED)

Raceme. In which the flowers are pedicellate (Figs. 73, 75).

Spike. In which the flowers are sessile or very short-pedicellate. See *Plantago,* Fig. 73.

Spadix. A spike with a fleshy axis (rachis) surrounded by a large bract (*spathe*) (Figs. 418, 464).

Catkin or ament. A spike or raceme with a slender rachis, which falls as a whole after flowering (staminate) or when the fruit is mature (carpellate) (Figs. 138, 141, 143, 144, 149, 154).

Umbel. A shortened rachis bearing long-pedicellate flowers spreading umbrella-like (Figs. 73, 76).

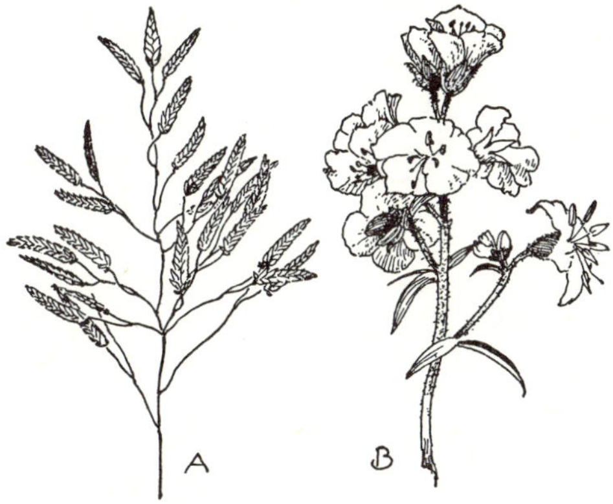

FIG. 77.—Inflorescences: *A,* panicle of a grass (*Eragrostis*); *B,* corymbose inflorescence of *Polemonium.*

Capitulum or head. In which numerous small flowers are crowded together on a very short rachis (*disk*) (Figs. 74b, 77). In rare cases it is compound (or branched) (Fig. 135).

Corymb. An umbel-like or cyme-like raceme (Fig. 73).

2. COMPOUND (BRANCHED)

Panicle. The main branches of which are racemes (Figs. 73, 77).

Compound umbel. An umbel the branches of which terminate in smaller umbels (umbellets) (Fig. 73).

Cymose Inflorescences

Cyme. A more or less flat-topped inflorescence, the flowers of which arise from terminal buds. *Cymule,* a diminutive cyme. (Figs. 73, 74b).

Dichasium. An inflorescence resulting from a dichotomous or forked mode of branching in which the branches are of equal rank (Fig. 78).

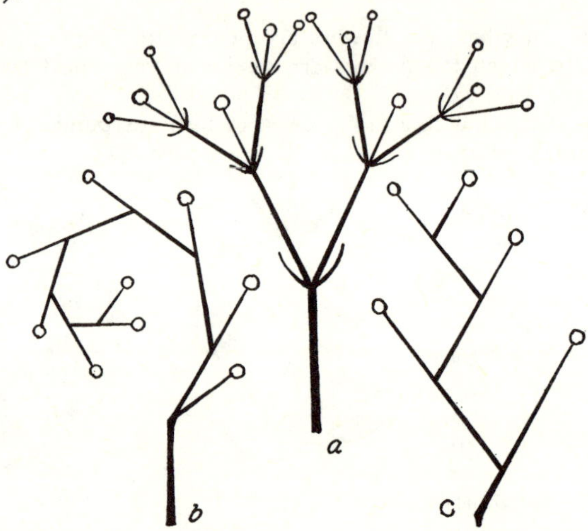

FIG. 78.—Types of inflorescences (diagrammatic) : *a,* dichesium (compound cyme) ; *b,* helicoid inflorescence ; *c,* scorpioid inflorescence.

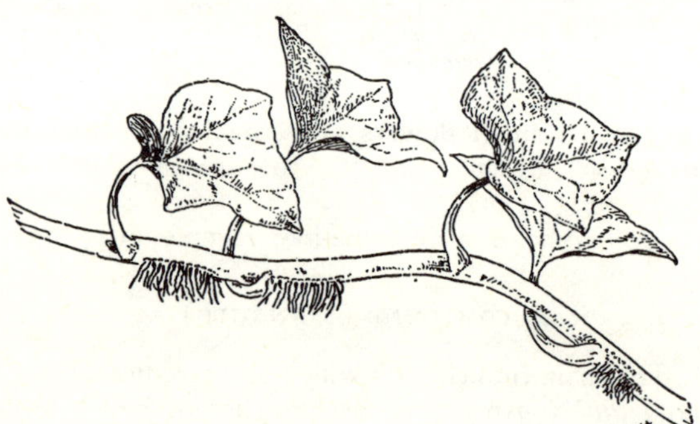

FIG. 79.—Aërial roots of the ivy (*Hedera helix*).

Helicoid cyme. An inflorescence in which the branches are on one side only of the rachis (Fig. 78).

Scorpioid cyme. An inflorescence in which the branches develop alternately on opposite sides of the rachis (Fig. 78).

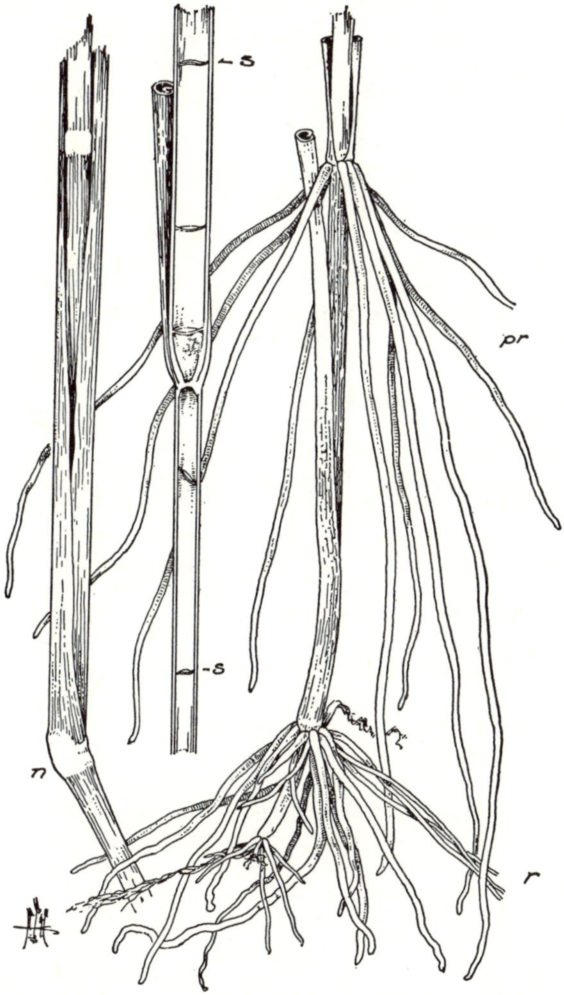

FIG. 80.—Prop-roots (*pr*) and soil-roots (*r*) of wild rice (*Zizania aquatica*); *s*, parchment-like septa in hollow culms; *n*, joint (node).

Epiphyllous inflorescences borne on leaf-like shoots (phylloclades) may be seen in *Ruscus* (*Liliaceæ*, Fig. 95c). An inflorescence of a similar sort is found also in the basswood (*Tilia;* Fig. 293).

Monœcism and diœcism. These conditions have been brought about by the segregation of the essential organs into separate flowers. Generally the andrœcium and the gynœcium are present in the same flower, which is then *bisexual*—or *perfect, monoclinous,* or *hermaphroditic;* but when these organs are borne in different flowers these flowers are *unisexual*—or *imperfect* or *diclinous.* The flower containing the stamens is *staminate,* while the one containing the carpels is *carpellate* or *pistillate* (Figs. 1, 2).

When staminate and carpellate flowers are borne on the same plant or individual, the flowers, or the plant, or the species, is said to be *monœcious,* but if they are borne on separate individuals these or the flowers, or the species, are said to be *diœcious.* It is clear that no fruit can be produced by the staminate flowers, but these flowers are not therefore "sterile" unless their pollen is functionless or fails to develop properly. Among unisexual flowers it is not unusual to find the organs of the opposite sex present in a vestigial condition (*Sagittaria, Lychnis, Rhamnus, Schinus molle, Æsculus;* (see under families, Figs. 181, 182, 265, 276, 278, 279, 424).

The presence of bisexual and unisexual flowers on the same individual is described as *polygamy* (flowers *polygamous*); in such cases both kinds of flowers are often found in the same flower-cluster (red maple, horse-chestnut). For a complete taxonomic description of such species it is obviously necessary to have both kinds of flowers at hand. A special case of polygamy results when an individual bears some flowers that are staminate along with others that are bisexual; such an individual or condition is *andromonœcious.* In contrast to this is the *gynomonœcious* condition, in which carpellate flowers are present along with the bisexual ones. In other words these individuals are monœcious with respect to the staminate or carpellate flowers only, as the case may be. An *androdiœcious* condition obtains when some individuals bear bisexual flowers and others of the same species bear staminate flowers (*Dryas octopetala*). On the other hand, if some individuals bear bisexual flowers and others of the same species bear carpellate flowers we have a *gynodiœcious* condition (*Thymus Serpyllum*).

When studying diœcious species it is of course necessary to have specimens of both individuals at hand. The two individuals are sometimes readily distinguishable in the field, as for example in the case of the box-elder (*Acer Negundo*). Even in the case of willows the distinction is not difficult to make and can often be detected at a distance.

CHAPTER IX

VEGETATIVE CHARACTERS

By vegetative characters are meant all peculiarities of form and structure shown by all organs of the plant other than the flower. These characters are exceedingly varied and numerous and for accurate description demand an extensive technical vocabulary. Study of the descriptions in standard floras will help one to acquire a knowledge of the use of the terms as well as a knowledge of the style of the descriptions. Conciseness, precision, orderly sequence, and clearness are aimed at always. First of all the characters of the plant we wish to describe should be carefully studied and systematically arranged and the outstanding or *diagnostic* characters should be sifted out from the rest. Minutiæ, unless of diagnostic significance, may be omitted except when a very complete description is required. What characters are important and what are not can only be determined, however, by carefully comparing our plant with other specimens of its kind and with other species of the same genus. The diagnostic characters are the ones that will be found most valuable when we come to making a key. The relative value of a character depends upon its constancy: variable characters offer no dependable criteria, usually, although variable tendencies are often characteristic of certain species as opposed to others of a genus. Too much reliance should not be placed on a single character in defining a species, but other characters should be considered if constant. From a single specimen, however, it is not possible to tell whether or not a species is variable: that can only be learned from adequate material collected from all parts of the range. Nor is it always safe to place too much reliance upon immature material.

The Root-System

Roots are characteristically subterranean organs. Exceptions are the *aërial* roots of certain epiphytes and climbing plants and floating aquatics. The prop roots of maize, sugar-cane, and barnyard-grass, the rope-like roots that drop to the ground from the stems of *Monstera deliciosa* of the American tropics and from the branches of the banyan of India, and the roots that spring from the bole of the mangrove trees of tropical tide-flats are all essentially aërial roots (Figs. 79, 80).

In old botanical books all the underground parts of plants were called roots, and outside of botanical laboratories this usage still persists. But precise botanical description requires that we distinguish between organs that are properly roots and those that are properly stems—such as root-stocks or rhizomes, runners, stolons, stem-tubers, and corms. Even Theophrastus, who wrote on botany in the fourth century B.C., recognized these distinctions.

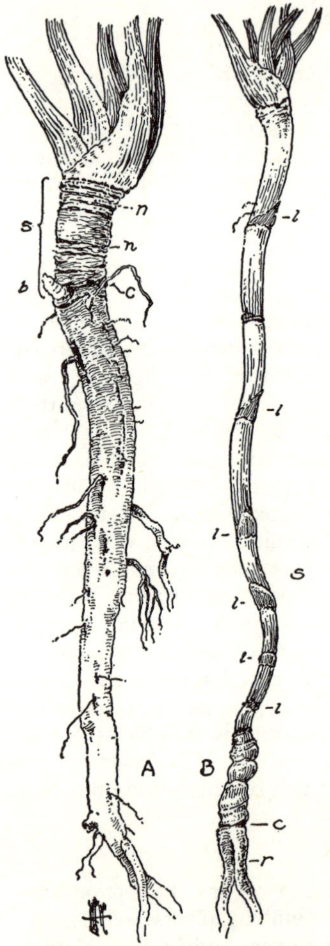

FIG. 81.—Sprouts developed from cut tap-roots of a dandelion: *A*, old tap-root cut off at (*c*); *s*, stem of new sprout; *n*, leaf-scars; *b*, a bud; *B*, young root (*r*) cut off at (*c*); *s*, deep underground stem developed from root at (*c*); *l*, scale-leaves.

Roots are ordinarily produced by stems, but sometimes they spring from leaves (*Bryophyllum, Begonia*). Roots differ from stems (1) in the anatomical organization of the primary tissues; (2) in the absence of nodes and internodes; (3) in the absence of leaves. Vegetative shoots or "root sprouts" do sometimes spring, however, from roots (Canada thistle, dandelion) (Fig. 81).

The descriptive terminology of roots based on external features involves the following terms. Roots that are fine and thread-like are termed *fibrous* (Fig. 82); those that are thickened as the result of storage of nutritive materials and water are termed *fleshy*, such as the carrot, parsnip, turnip, radish, and beet (Fig. 83). The roots of the dahlia, sweet potato, and the sunflower, *Helianthus giganteus*, are *tuberous* roots or *root tubers*. All fleshy roots are essentially tuberous. Because the roots of the dahlia are produced in a spreading cluster, they are termed *fascicled roots*. Similar roots occur in some buttercups and in *Syndes-*

FIG. 82.—Fibrous secondary (adventitious) root-system of a Verbena; *bd,* crown of basal buds; *st,* old stem.

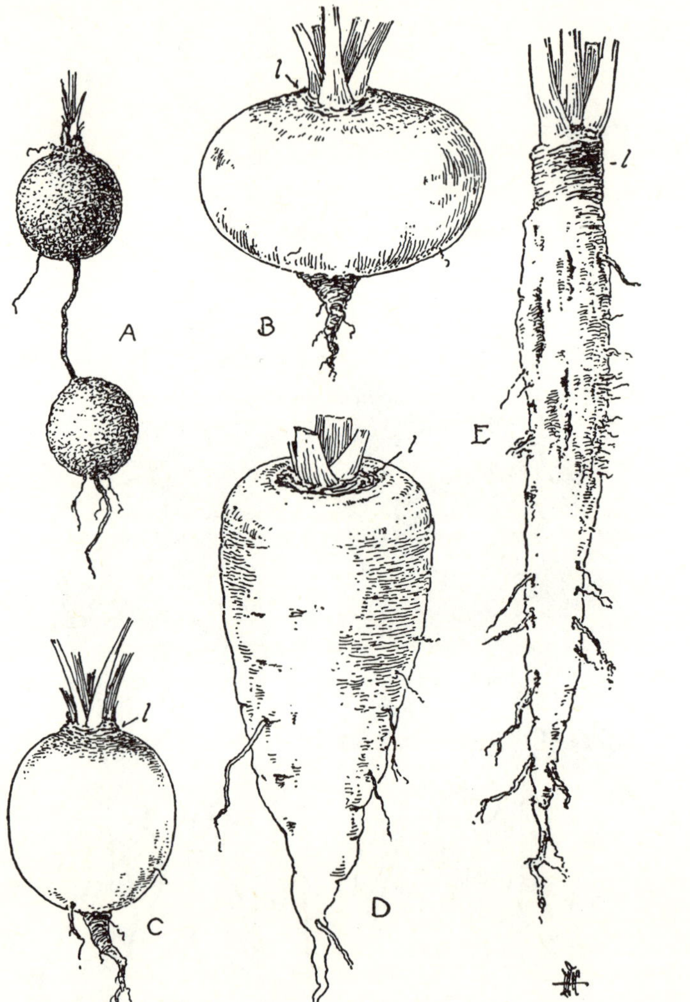

FIG. 83.—*A,* moniliform tap-root (primary root) of an umbellifer (*Lomatium Geyeri*). *B-E,* fleshy tap-roots (primary roots) of (*B*) turnip, (*C*) radish, (*D*) carrot, (*E*) salsify; *l,* leaf-scars on "root-crown."

mon (Fig. 84). Enormous fleshy roots weighing many pounds are produced by various Californian species of *Echinocystis* (man-root).

With respect to shape fleshy roots are described as *fusiform* if like those of the dahlia or sweet potato; *napiform* if like those of the turnip; *carrot-shaped* if like those of the carrot, etc., etc. A *monili-*

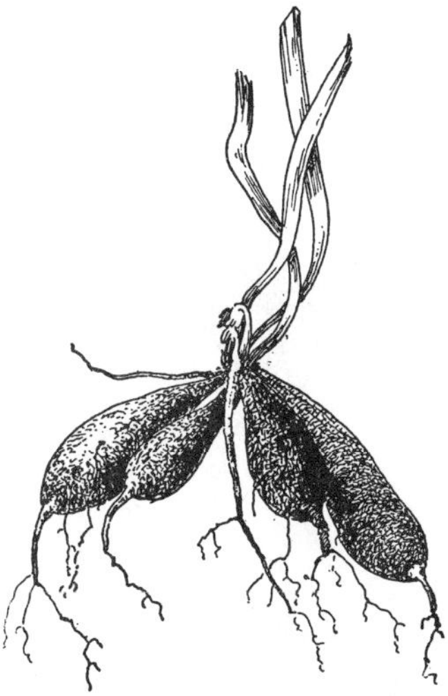

FIG. 84.—Fascicled tuberous secondary roots of rue-anemone (*Syndesmon thalictroides*).

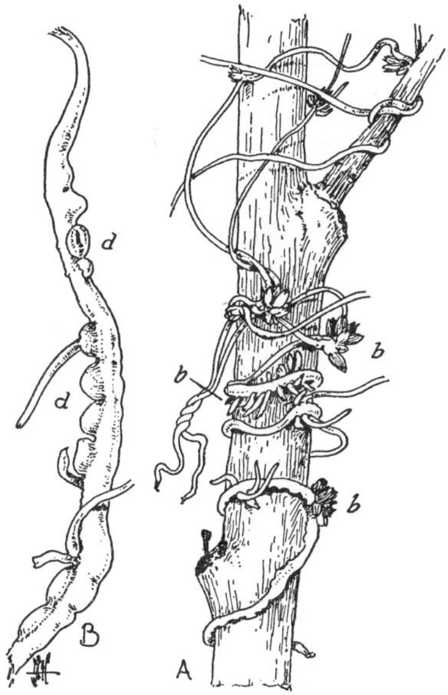

FIG. 85.—Twining stems (*A, B*) of dodder (*Cuscuta*) with haustorial disks (*d*); *b*, different stages of flower-buds.

form type of tap-root (one like a chain of beads) occurs in the Far Western umbellifer, *Lomatium Geyeri* (Fig. 83).

Fibrous roots are common to grasses, sedges, and other Monocotyledons, and to herbaceous Dicotyledons, and they form the ultimate branches of the root-sysems of trees and shrubs. Likewise, the roots that spring from carrots and beets, rootstocks, and other underground stems, are all fibrous. Root-like organs or *haustoria* are produced by some parasitic plants like the dodders (Fig. 85) and the mistletoes, while some parasitic or saprophytic plants such as the

Indian pipe (*Monotropa uniflora*) produce a mass of coral-like "roots" in the rich humus of the forest floor.

DURATION OF LIFE OF ROOTS

The duration of life of a plant is dependent upon the duration of life of its root-system. We speak of plants as *annuals, biennials,* or

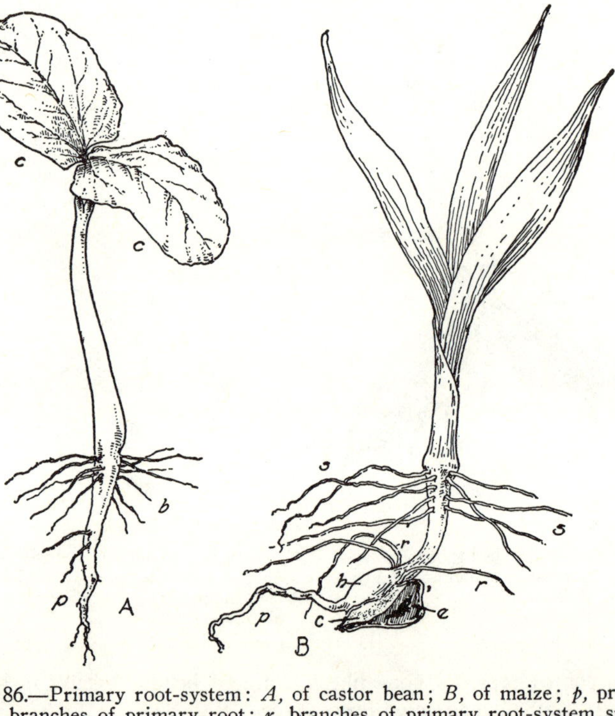

FIG. 86.—Primary root-system: *A,* of castor bean; *B,* of maize; *p,* primary root; *b,* branches of primary root; *r,* branches of primary root-system springing from hypocotyl; *s,* secondary roots springing from first internode of epicotyl; *h,* hypocotyl; *c,* cotyledon; *e,* endosperm. The cotyledon (scutellum), *c,* of maize remains within the seed in the ground.

perennials, by which we imply that the root-systems live for one, two, or many years, as the case may be. A *winter-annual,* such as the winter-wheat, may be sown in the fall and, with a good start, will survive

FIG. 87.—The primary root as a tap-root: *a,* in sweet clover (*Melilotus alba*); *b,* in the mallow (*Malva rotundifolia*); *c,* in the pigweed (*Amaranthus hybridus*), an annual; *a, b,* biennials; *bd,* buds on root-crown (base of stem); *s,* sprout developed from root-crown.

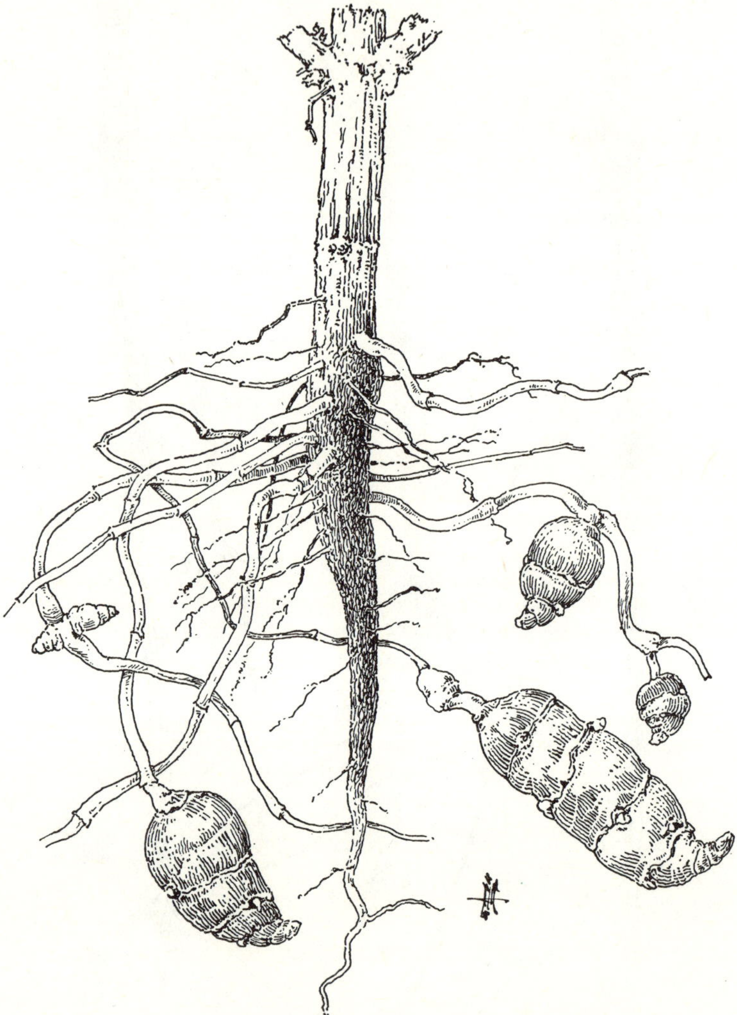

FIG. 88.—Stem tubers on rhizomes of a sunflower (*Helianthus tuberosus*).

the winter and renew its growth the following spring. Some species may be both annual and biennial according to the climatic conditions.

THE PRINCIPAL ROOT-SYSTEMS

The principal root-systems of plants are the *primary* and the *secondary* or *adventitious* (Fig. 86). The primary root-system develops from the hypocotyl of the embryo; it includes any root springing from the hypocotyl. The main part of the primary root-system is the *primary root,* which is formed from the tip of the hypocotyl as a deeply going root with radiating branches. The branches arise in more or less evident longitudinal rows and may be termed primary or secondary branches according to rank. If the primary root persists, it is called a *tap-root* (Fig. 87). Tap-roots are usually stout roots; they are prominent and deep-going in such plants as the sweet clover, alfalfa, and jack-pine. Carrots, beets, turnips, and similar roots are merely fleshy tap-roots (Fig. 83). The seedlings of some plants (pumpkin) produce instead of a tap-root a fascicle of primary roots (*multiple primary roots*) (Fig. 84).

In many cases the primary root-system is of short duration, being replaced by a secondary root-system which springs from the epicotylary portion of the seedling or from any vegetative shoot. In *Bryophyllum* and *Begonia,* secondary roots will spring from the leaves or fragments of leaves when placed in moist soil. The roots of all such vegetative organs as rootstocks, stem-tubers, runners, stolons, bulbs, and corms are secondary roots, and likewise all roots springing from cuttings or slips. Secondary roots are the characteristic permanent roots of Monocotyledons. As a rule secondary roots are fibrous, but in many cases they may be considerably thickened. The root-system of our most important crop plants, such as the grains or cereals, is the secondary root-system (Figs. 86, 89, 90, 91).

The Stem in General

The stem is the axis of the shoot and bears the leaves. All stems have *nodes* or "joints" and *internodes*—the intervals between the nodes. The leaves arise at the nodes. At the tip of the stem, or branch, or shoot, is the *terminal bud,* which continues the growth in length of the shoot, and in the axils of the leaves are the *axillary* or *lateral buds,* from which spring the branches. Axillary or lateral buds have the same arrangement on the stem as their subtending leaves. In our

Northern deciduous trees and shrubs only the shoots of the season bear leaves, these leaves falling in the autumn and leaving *leaf-scars* at the nodes below the buds. The following spring new leaves are

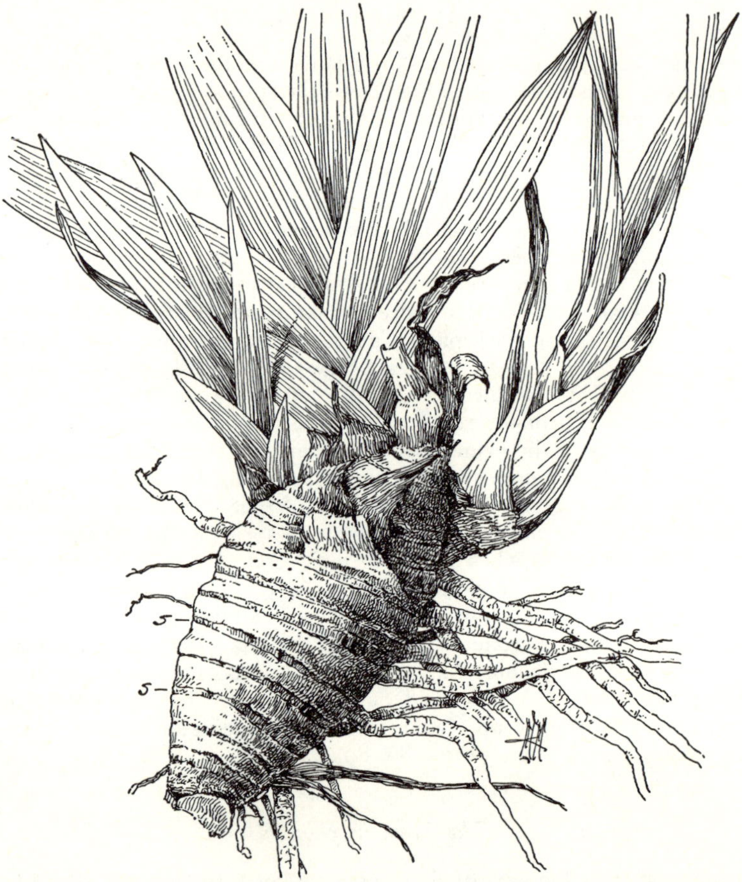

FIG. 89.—Rhizome of a garden iris, with stout secondary roots arising from the internodes; *s*, annular leaf-scars of scale-leaves, a few of which are seen above.

produced by the new shoots that spring from the terminal and lateral buds of the previous season's shoots. The winter state of our deciduous trees and shrubs is as distinctive for each species as is the summer state, and to the experienced woodsman these trees and shrubs are as familiar in winter as in summer. General appearance, special features of the bark, branches or twigs, and appearance of the buds or the leaf-scars should be studied closely in order to learn to distinguish the different species.

The distinctive features of stems in general are: (1) that they have nodes and internodes; (2) that they bear leaves. In addition they

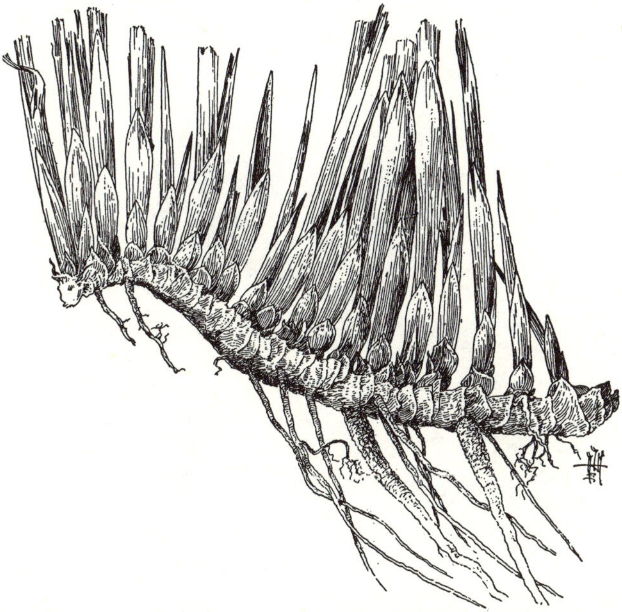

FIG. 90.—Rhizome of a *Juncus,* with stout secondary roots and closely overlapping scale-leaves. Numerous aërial shoots are seen to arise from the nodes of the rhizome.

bear buds and flowers. In old stems such as tree-trunks, the nodes become in time obscured, but often their location may be determined by the presence of traces of the leaf-scars and of course by stubs of branches (birches, poplars, hickories). In some herbaceous forms the stems are often more or less swollen at the nodes (smart-weeds, carnations) or the internodes may be more or less inflated (*Eriogonum inflatum* of Western deserts).

CLASSIFICATION OF STEMS

Stems may be classified in various ways according to what our immediate purpose may be. As the duration of life of the plant corresponds to the duration of life of its root-system, stems, like roots, may correspondingly be classified as annuals, biennials, or perennials. In the case of many "perennial" herbs, the "plant" lives but one season, that is, it grows to maturity—producing fruit and seed—

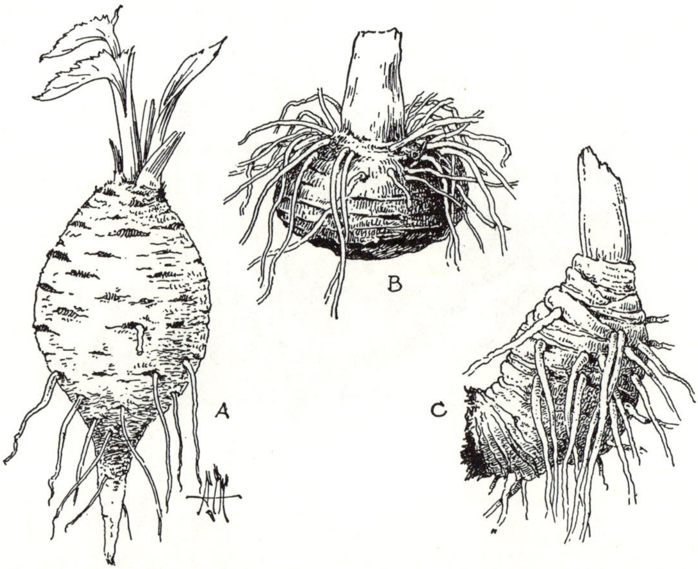

FIG. 91.—Types of tuberous underground stems: *A*, corm of celeriac, *Apium graveolens* var. *rapaceum* (*Umbelliferæ*); *B*, corm of Jack-in-the-pulpit (*Arisæma triphyllum*); *C*, corm of a *Trillium*.

in one growing season, and then dies. But such plants, while they are developing fruit and seed, are at the same time producing subterranean rootstocks or rhizomes, stolons or tubers, from which new plants of the same species are developed the following season. These plants thus possess two methods of propagation. Sunflowers, goldenrods, asters, potatoes, and irises are plants of this sort.

Below is given a series of classifications of stems from different points of view, together with the descriptive terminology appropriate thereto.

As to texture, stems are:

herbaceous, when soft or more or less succulent, and dying down to the ground at maturity; these plants develop little woody tissue or xylem;

woody, with abundantly developed woody tissue or xylem, such as trees and shrubs;

shrubby or *fruticose,* woody perennials of low growth;

frutescent or *suffrutescent,* when woody below and herbaceous above.

As to life-form (growth-form or habit), stems are:

arboreous, when tree-like, *i.e.,* having a definite trunk;

arborescent, when tree-like in appearance and size;

diffuse, when the branches spread loosely in all directions;

decumbent, when reclining on the ground;

declined, when bending over;

assurgent or *ascending,* when inclined (leaning);

prostrate or *procumbent,* when lying flat on the ground;

creeping, trailing, or *repent,* when running along on the surface of the ground;

climbing or *scandent,* when growing upward by clinging to other objects for support; here belong also the tropical *lianas;*

twining, when coiling around other plants for support; lianas also here included.

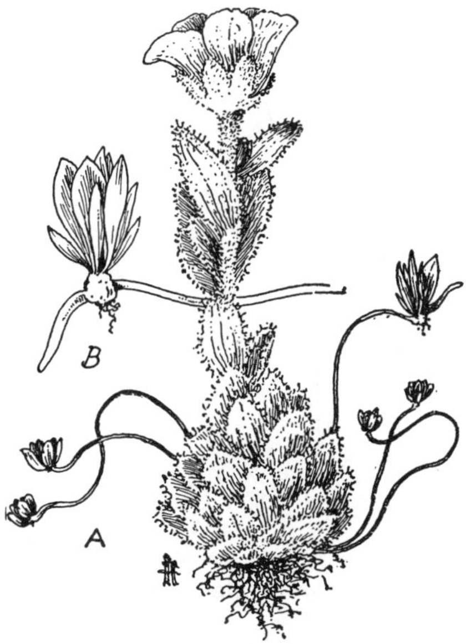

FIG. 92.—Stolons of *Saxifraga flagellaris.* *A,* with buds at ends; *B,* a bud enlarged, with continuation of stolon at left.

As to the medium in which they grow, stems are:

aërial or *epigeous,* when growing above the ground;

subterranean or *hypogeous,* when growing beneath the surface of the ground (Fig. 88);

aquatic, when growing in water, either on the surface (*floating*) or beneath the surface (*submersed*).

SPECIAL FORMS OF STEMS

In this category may be placed all forms generally known as *modified stems.* Most of these function as propagative organs; they may be subterranean or aërial.

Tuber. A thick subterranean stem or branch, the thickening being the result of storage of food materials and water. The shape of the tuber will depend upon the localization of the storage materials. Potato, Jerusalem artichoke. Leaves generally greatly reduced or vestigial (Fig. 88).

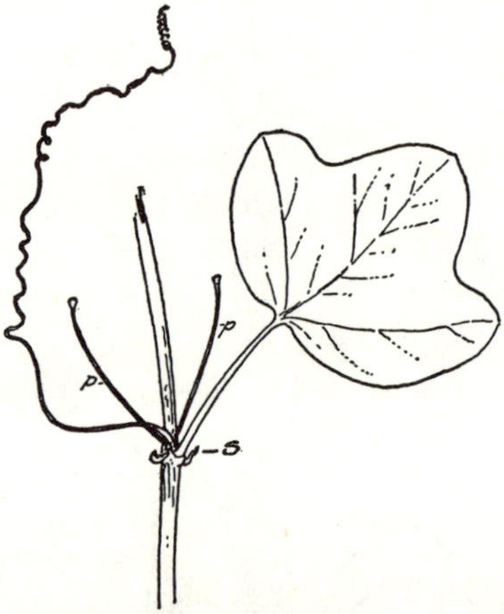

FIG. 93.—Tendril of *Passiflora lutea*; *p,* flower-stalks; *s,* stipules.

Rootstock or *rhizome.* A stout or slender subterranean stem, simple or branched. Internodes usually very short, and leaves reduced to scales (*scale-leaves*) which in time disintegrate, leaving only the fibrous framework (Figs. 88, 89, 90).

Corm. Usually a short, tuberous, more or less erect, subterranean stem of solid texture, sometimes broader than long. The older portion dies away from year to year, and new roots are produced from the nodes and internodes of the new portion. A corm differs from a bulb in that the thickened part is the stem and not the leaves. *Cyclamen, Trillium* (Fig. 91).

Caudex. A short or elongated rootstock which persists at or near the surface of the ground after the season's aërial shoot dies down. New shoots arise from buds produced on the upper portion, that is below the base of the old stem. Many plants arising from a caudex appear to be stemless (*acaulescent*) or, if the caudex rises above the ground to form a short stem, the plant is described as *caulescent. Heuchera, Geum,* strawberry, violets, plantain (Fig. 385).

Sucker. A subterranean branch that comes above the ground and continues as an aërial shoot. Currants, gooseberries, blackberries, some grasses and sedges.

Stolon. A slender stem (branch) that creeps over or slightly beneath the surface of the ground, taking root and sending up new shoots at the nodes. The internodes of a stolon may be more or less elongate, and as they disintegrate the new shoots become independent plants. Quack-grass (Fig. 92).

Runner. A slender, stolon-like branch, commonly with a single node near the middle between long internodes, running along the ground and taking root at the end, that is, at the terminal node. But a shoot may also develop at the middle node, especially if the terminal node or bud be injured or lost. Strawberry, *Potentilla anserina.*

Offset. A stolon-like or sucker-like branch (offshoot) bearing a cluster of leaves (bud) at the end. Houseleek.

Cane. A slender weak stem such as those of raspberries or blackberries.

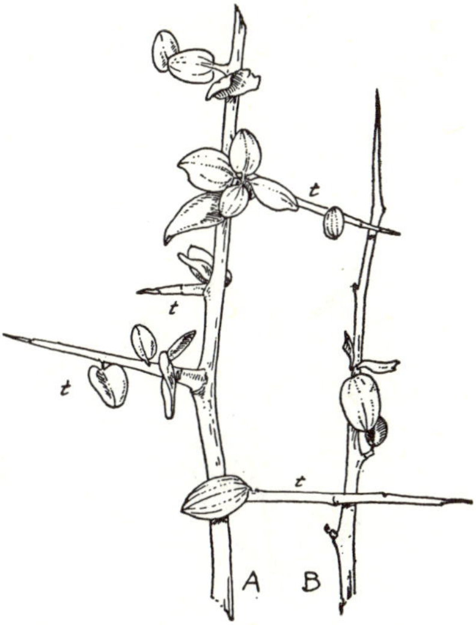

FIG. 94.—Thorns (*t*) of a *Ceanothus,* showing transformation from a leafy shoot. *A,* thorns developing from lateral branches. *B,* a terminal thorn.

Turions. New shoots or sprouts appearing from aërial or subterranean stems (suckers, etc.). Blackberries, currants, gooseberries, service-berries.

Tendrils. Long slender branches distitute of leaves as a rule. Their stem nature may be deduced from their axillary origin and the presence of more or less obvious nodes and internodes. Virginia creeper, Passion-flower (Fig. 93). (See *Sapindaceæ,* p. 363.)

Thorn. A woody branch whose growth has been arrested and which has become hardened and sharp-pointed. Hawthorns (Fig. 94).

Scape. A flowering stem, naked or with a few reduced leaves, arising from a caudex, rhizome, corm, or bulb. Tulip (Fig. 74).

Succulent green stems, showing extreme specialization and functioning as leaves, are characteristic of the cacti, and species of African spurges (*Euphorbia*), and of the African *Stapelia* and *Hoodia* (*Asclepiadaceæ*). In these forms the leaves are greatly reduced and modified. *Phylloclades* or leaf-like stems occur in *Asparagus* and *Ruscus* among the Monocotyledons and in *Muehlenbeckia platyclados* (*Polygonaceæ*), *Carmichaelia* (*Leguminosæ*), *Phyllanthus* (*Euphorbiaceæ*), *Opuntia* (*Cactaceæ*), and *Allenrolfea* and *Salicornia* (*Chenopodiaceæ*). A strong inflation of a basal internode occurs in *Eriogonum inflatum* (*Polygonaceæ*), and in the kohlrabi (*Brassica caulorapa*) the stem becomes tuberous above the ground (Figs. 95, 96).

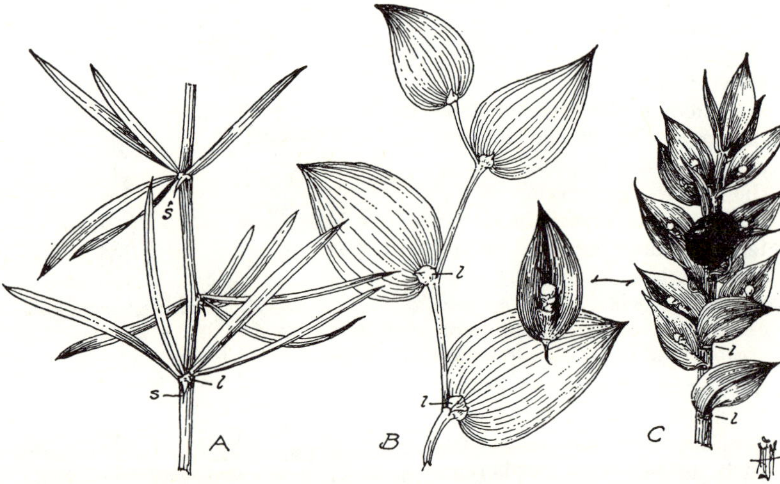

FIG. 95.—Leaf-like stems: *A*, phylloclades of *Asparagus Sprengleri*. *B*, phylloclades of *Asparagus asparagoides*. *C*, phylloclades of *Ruscus aculeatus*, bearing flower-buds on upper surface and one ripe fruit (black); at left a phylloclade with bud, enlarged; *l*, true leaves, reduced to spinescent-scales (*s*) in *A*.

Stems of aquatics. These stems are usually of a soft texture and are provided with spongy aërating tissues, but the rhizomes of the perennial forms, such as water-lilies, water-arum, cat-tails, bulrushes, sweet-flag, and others, are more or less stout and of firmer texture (Fig. 312).

Stems of desert shrubs and trees are usually of slow growth and develop very hard wood, while the branches are rigid and gnarled, and often spinous. "Broom" and "switch" forms are common among these plants.

BARK

It is common observation that trees and shrubs can be distinguished by their bark. The woodsman, from long experience, will recognize the log of a tree on sight; indeed, to him a knowledge of the features of the bark and wood of a tree is an indispensable asset. While the differences in the bark of our forest trees are often very obvious, there are many cases in which the differences are quite obscure, even as between species of unrelated genera. Thus certain maples and hickories, especially the younger trees, may prove puzzling.

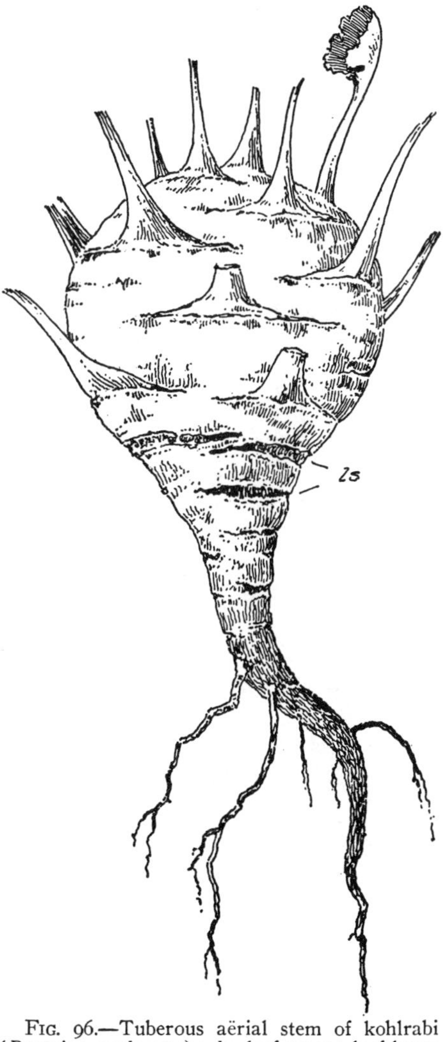

FIG. 96.—Tuberous aërial stem of kohlrabi (*Brassica caulorapa*); *ls*, leaf-scars; leaf-bases above.

Perhaps no characters in taxonomy are so difficult to describe as are those of the bark. Such descriptive terms as "smooth" and "rough" when applied to bark are merely relative. There is usually a considerable difference between the bark of the trunk and that of the branches, and between the barks of young and old trees of the same species. Among the oaks, poplars, willows, birches, and maples, for example, these

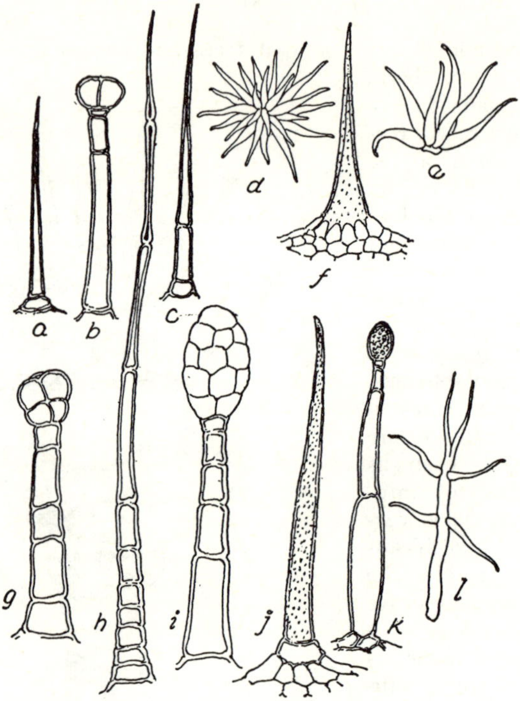

FIG. 97.—Cauline hairs of various plants: *a, b, c,* of a mint, —*b,* gland-bearing; *d, e,* branching hairs of *Malvastrum* (*Sphæralcea*); *f,* stout acicular hair of a radish; *g,* multicellular gland-bearing hair of involucral bract and *h,* multicellular acicular hair of a composite (*Heterotheca grandiflora*); *i,* multicellular hair of a saxifrage (*Saxifraga*); *j,* unicellular acicular hair of a crucifer (*Brassica*); *k,* multicellular gland-bearing hair of storksbill (*Erodium*); *l,* branching hair of a "sycamore" (*Platanus*).

differences are very marked. The twigs also may show certain marked differences, often useful as recognition-marks.

LENTICELS

These are special aërating organs found on the young branches and on the twigs of trees and shrubs. They have the appearance of small excrescences or eruptions dotting the surface. Usually they are of more or less distinctive shape according to the genus or species.

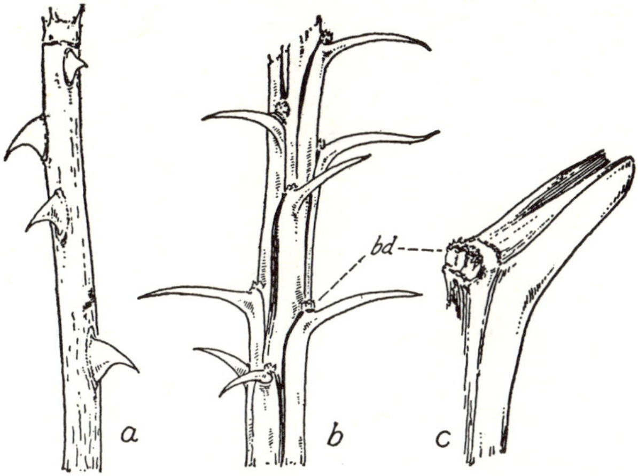

FIG. 98.—Spinous processes: *a,* spines of a rose; *b,* of the ocotillo (*Fouquiera splendens*); *c,* basal portion of spine of same; *bd,* axillary buds.

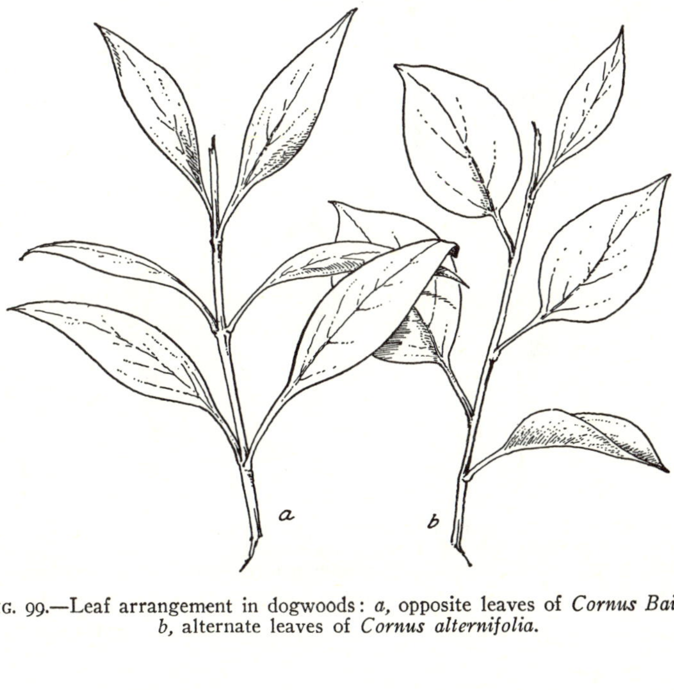

FIG. 99.—Leaf arrangement in dogwoods: *a,* opposite leaves of *Cornus Baileyi;* *b,* alternate leaves of *Cornus alternifolia.*

FIG. 100.—Leaf arrangement: *a*, whorled (verticillate) and alternate (spiral) leaves in the prairie tiger-lily (*Lilium umbellatum*); *b*, opposite (distichous) sheathing leaves of the fennel (*Fœniculum vulgare*); *s*, stipular appendages; *c*, diagram of leaf arrangement in fennel.

They are particularly prominent in the cherries, birches and alders, dogwoods, maples, and hazels (Fig. 128).

LATEX

Latex, as the milky sap of certain plants is called, is always a useful recognition-mark of these forms. In our flora the plants possessing latex belong to relatively few families and these by observation and association soon become familiar. The families are *Moraceæ, Papaveraceæ, Euphorbiaceæ, Apocynaceæ, Asclepiadaceæ, Campanulaceæ,* and the chicories of the *Compositæ.*

COATINGS AND OTHER SUPERFICIAL MARKINGS OF STEMS

The stems of many herbs and shrubs, and the twigs and young shoots or sprouts of many trees, are covered with hairs of various sorts or with prickles or resinous or mucilaginous coatings, all of which are of taxonomic significance. In many cases species may vary with respect to a character of this sort, and the relative taxonomic value of the character must consequently be appraised according to the problem in hand. Not infrequently the degree of pubescence, for instance, will vary with the age of the plant.

Prickles are superficial outgrowths of the stems of many plants, *e.g.,* roses, blackberries, gooseberries, and some nightshades. If the prickles are of considerable size they are referred to as spines. In the plants mentioned the prickles and spines are of taxonomic significance and useful in keying.

The hairs of plants show many peculiar forms, two or more forms occurring frequently on the same plant. They may be simple or branched, unicellular or multicellular, and may or may not bear a terminal unicellular or multicellular gland. Frequently they are of considerable taxonomic value, *e.g.,* in *Saxifraga* and other saxifrages, mulleins, mustards, hawkweeds, *Malpighiaceæ.* In some forms scaly coatings are present (*Elæagnaceæ*), and many plants are covered with a bluish or whitish waxy coating; these forms are described as *glaucous* (cabbage, some asters). Stinging hairs are characteristic of several nettles (Figs. 97, 98).

The Leaf

Leaves have been described as flattened lateral outgrowths of stems. No other organs of the plant display such diversities of form or are so variable or plastic. Many organs that morphologists tell us are leaves are hardly recognizable as such on superficial inspection. But leaves are not all flat nor are they always green, and neither are they always aërial, for some are borne on subterranean stems and others on submersed aquatic stems. (See *Mesembryanthemum, Asparagus,* rhizome of *Iris, Utricularia.*

The primary functions of all green leaves are (1) photosynthesis, or the manufacture of carbohydrate foods out of the raw materials, water and carbon dioxide; and (2) transpiration, or the exhalation of water-vapor. But in many cases leaves subserve other useful functions, such as storage of reserve food materials, as holdfasts or climbing organs, etc., etc., and ordinarily in such instances they are more or less modified or specially adapted for the particular function.

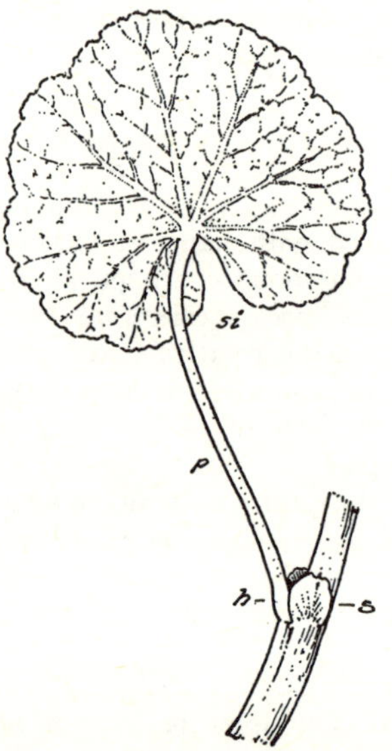

FIG. 101.—General type of leaf (Geranium): *si,* sinus of the blade (lamina); *p,* petiole; *h,* hypopodium; *s,* stipules.

No taxonomic description of a flowering plant would be complete without taking due account of its leaves. We know from experience that we can distinguish many plants by the shapes of their leaves, on sight, but we know, too, that many times we have been unable to do so with much assurance. Indeed, it happens that many plants of quite remote relationships have very similar leaves, while on the other hand many closely related plants have very dissimilar leaves. Hence it is seen that the value of leaves as furnishing trustworthy distinguishing characters has certain limitations, outside of the more minute and less obvious details. Repeated and careful field observations will guide the student here.

Since leaves are so varied in form, structure, etc., it necessarily follows that a rather extensive vocabulary will be required for purposes of proper description and classification. And what has been said on preceding pages appertaining to the bases of classification of other organs of plants applies with equal force here. A detailed account of the taxonomic features of leaves is given below.

ARRANGEMENT OF LEAVES (PHYLLOTAXY)

Leaves ordinarily have a definite arrangement on the stem. They may be *opposite, alternate* or *spiral, whorled* or *verticillate* (Fig. 99). *Alternate* and *spiral* are interchangeable terms, yet are convenient to

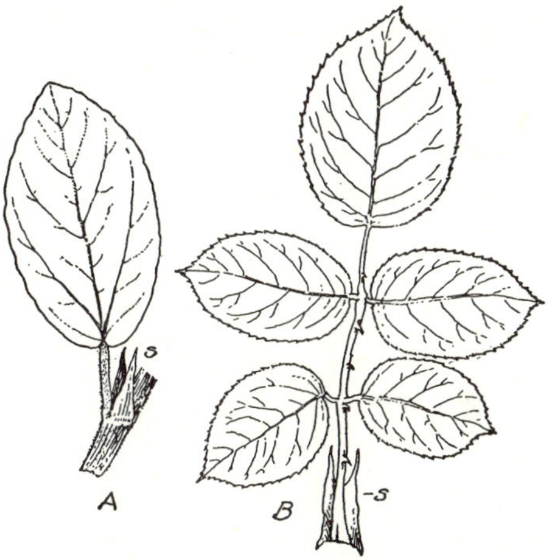

FIG. 102.—*A,* simple petiolate and stipulate leaf of *Ficus pumila; B,* compound pinnate and stipulate leaf of a rose; *s,* stipules.

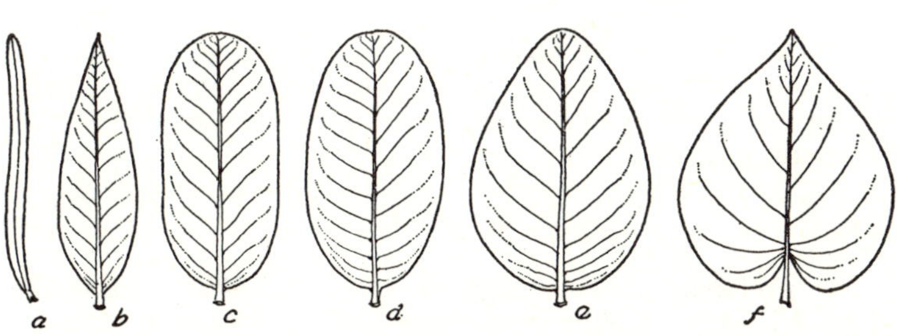

FIG. 103.—Leaf-forms: *a,* linear; *b,* lanceolate; *c,* oblong; *d,* elliptical; *e,* ovate; *f,* ovate-cordate.

distinguish special cases; strictly alternate leaves lie in the same plane, an arrangement sometimes termed *distichous* (Fig. 100). As a matter of fact, any type of leaf arrangement conforms to the spiral. Sometimes opposite and alternate leaves occur on the same plant. In the bedstraw (*Galium*), which possesses whorled "leaves," some of the "leaves" are in reality stipules. The technical term for leaf-arrangement is *phyllotaxy*.

THE PARTS OF A LEAF

The ordinary leaf, that of a Dicotyledon, for instance, consists of (1) a blade or *lamina;* (2) a leaf-stalk or *petiole;* (3) a leaf-base or *hypopodium,* which bears two lateral appendages called *stipules* (Fig. 101). One or the other of these parts is often more or less reduced or modified or wanting. If the blade is wanting, the leaf is described as *elaminate* (*Fouquiera splendens*); if the petiole is want-

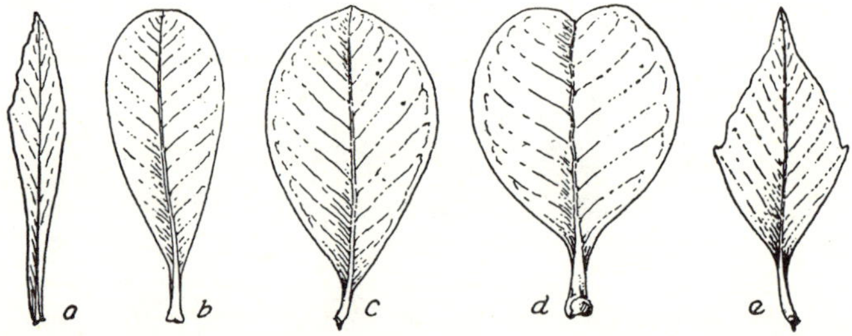

FIG. 104.—Leaf-forms: *a,* oblanceolate (*Myrica cerifera*); *b,* spatulate (*Pittosporum Tobira*); *c,* obovate (*Psidium Guajava*); *d,* obcordate (*Photinia* sp.); *e,* cuneate (*Pittosporum rhombifolium*).

ing, the leaf is *apetiolate,* and the leaf-blade is *sessile;* if the stipules are wanting, the leaf is *exstipulate* (Fig. 115). *Stipels* or stipule-like structures are found on some compound leaves (*Thalictrum, Phaseolus, Robinia*).

CLASSIFICATION OF LEAVES

With reference to the configuration of the leaf-blade, leaves are:

Simple, when the blade is not divided into separate parts or *leaflets* (Figs. 101, 102).

Compound, when what corresponds to the blade consists of separate leaflets. The leaflets or *pinnæ* of compound leaves have, in general, all the earmarks of simple leaves, except stipules and hypopodium. The axis or continuation of the petiole of a compound leaf is termed the *rachis* (Figs. 102, 112). Compound leaves may be once, twice, or thrice compound.

With reference to the configuration of the leaf-blade, simple leaves are:

cordate or *heart-shaped* when rounded and indented at the base, that is, with a *sinus* at the base, and tapering to a point at the tip (Fig. 103f);

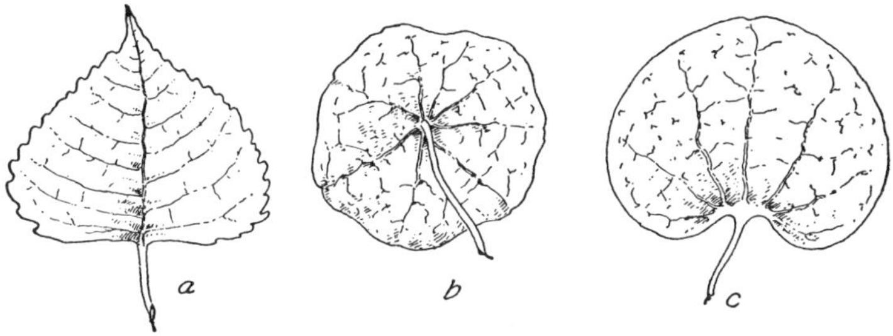

FIG. 105.—Leaf-forms: *a*, deltoid leaf of Populus deltoides; *b*, peltate leaf of *Tropæolum majus; c,* reniform leaf of *Aristolochia cymbifera.*

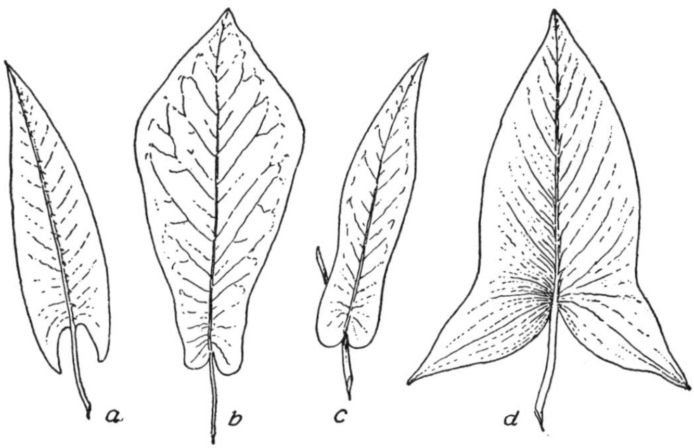

FIG. 106.—Leaf-forms: *a*, sagittate petiolate leaf of *Polygonum sagittatum; b,* broadly oblanceolate, auriculate, petiolate leaf of *Magnolia Fraseri; c,* lanceolate, auriculate, sessile leaf of *Aster phlogifolius; d,* hastate petiolate leaf of *Sagittaria latifolia.*

ovate when egg-shaped, with a rounded base and a blunt or rounded apex (Fig. 103c);

elliptical when like an ellipse in outline, commonly broadest at the middle and tapering slightly to a broadly rounded base and apex (Fig. 103d);

oval when intermediate between ovate and elliptical and broadest below the middle (Fig. 103d, e);

oblong when rectangular in outline but with rounded base and apex, the sides nearly straight (Fig. 103c);

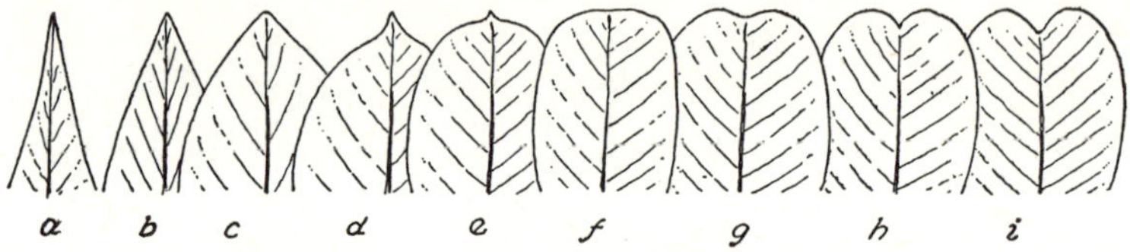

FIG. 107.—Leaf apices: *a*, acuminate; *b*, acute; *c*, obtuse; *d*, cuspidate; *e*, mucronate; *f*, truncate; *g*, retuse (shallowly notched); *h*, emarginate (notched); *i*, obcordate (deeply notched).

lanceolate when lance-shaped, mostly broadest below the middle and tapering gradually to a pointed apex, and more abruptly so towards the base (Fig. 103b);

linear when narrow and usually much longer than broad, with straight sides or tapering more or less towards the apex (Fig. 103a);

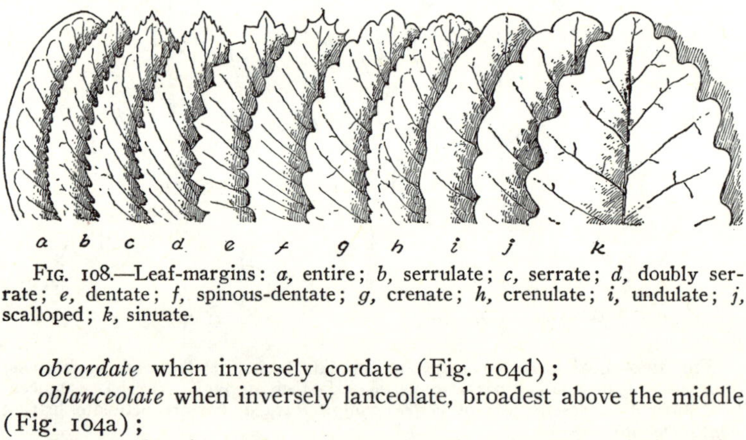

FIG. 108.—Leaf-margins: *a*, entire; *b*, serrulate; *c*, serrate; *d*, doubly serrate; *e*, dentate; *f*, spinous-dentate; *g*, crenate; *h*, crenulate; *i*, undulate; *j*, scalloped; *k*, sinuate.

obcordate when inversely cordate (Fig. 104d);

oblanceolate when inversely lanceolate, broadest above the middle (Fig. 104a);

obovate when inversely ovate and broadest above the middle (Fig. 104c);

palmate when having the form of a palm leaf (Fig. 112);

spatulate when spoon-shaped—broad and rounded above the middle and tapering more or less gradually towards the base (Fig. 104b);

cuneate or *cuneiform* when wedge-shaped, broad above the middle and tapering towards the base, the sides rather straight, the apex various (Fig. 104c);

orbicular when wheel-shaped or circular (Fig. 105b);

deltoid when like the Greek letter Δ (Fig. 105a);

reniform or *kidney-shaped* when broader than long and with a sinus at the base (Fig. 105c);

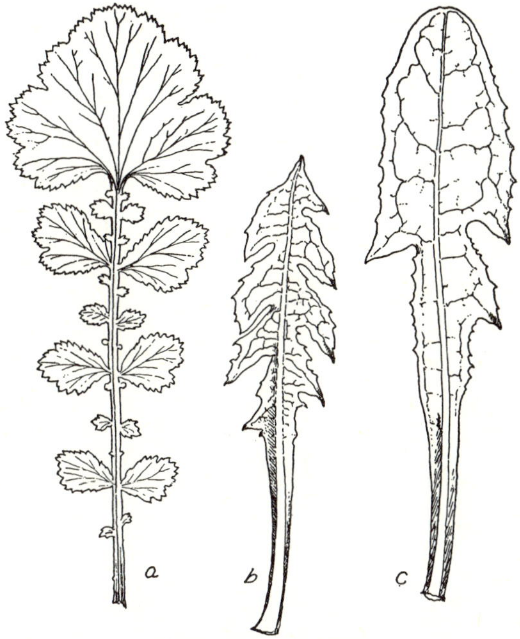

FIG. 109.—Leaf-forms: *a*, pinnately parted (*Geum*); *b*, *c*, sinuately lobed or incised (dandelion).

sagittate or *arrow-shaped* when having a deep sinus at the base and the auricles turned inwards (auricles—the lateral basal lobes of blade) (Fig. 106a);

hastate or *halberd-shaped* when spear-shaped, with the auricles turned outwards (Fig. 106d);

peltate or *shield-shaped* when the leaf-stalk is inserted at or near the center of the blade (Fig. 106b);

seculate or *sickle-shaped* (Fig. 117);

auriculate when the sides of the leaf project at the base in the form of prominent lobes (ear-flaps) (Fig. 106b, c).

With reference to the apex, leaf-blades are (Fig. 107):
acute when tapering to a sharp, narrow point;
acuminate when tapering more narrowly to a sharp point;
obtuse when tapering rather abruptly to a point;
truncate when cut off square or nearly so, transversely;
retuse when rounded and only slightly notched;

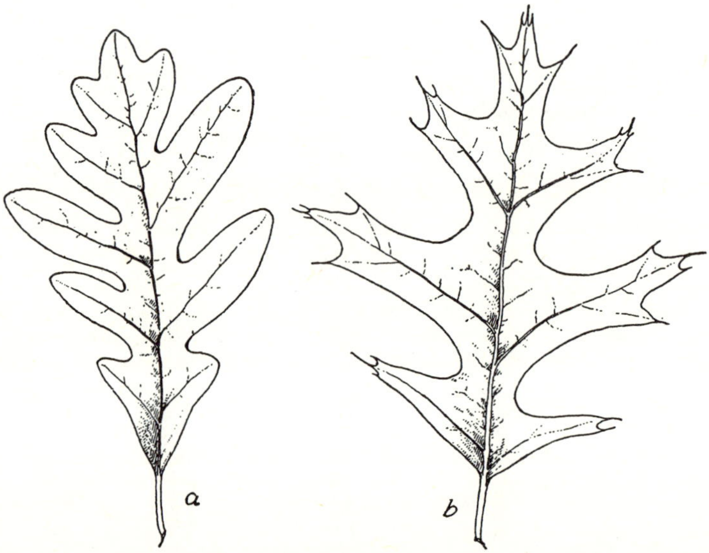

FIG. 110.—Leaf-forms: pinnately lobed or incised leaf of (*a*) *Quercus alba* and (*b*) of *Quercus velutina*.

emarginate when decidedly notched;
cuspidate when toothed—coming to a sharp, rigid point;
mucronate when coming abruptly to a sharp, rigid point;
mucronulate (diminutive of *mucronate*);
aristate when awl- or bristle-pointed.

With reference to the margin, leaf-blades are (Fig. 108):
entire when without cuttings of any kind;
serrate when saw-toothed, the teeth projecting forward;

serrulate when serrate, but with the teeth very small;

dentate when cut into teeth which project straight outwards;

crenate or *scalloped* when having broad, rounded, marginal teeth;

repand or *undulate* when having a wavy margin;

sinuate when sinuous, the indentations being deeper than in the crenate or repand type (Fig. 108k);

incised when deeply cut into jagged or irregular teeth (Fig. 109c);

lobed when deeply cut into lobes (Figs. 108, 109);

cleft when cut half way to the midrib or to the base into lobes (Figs. 110b, 111);

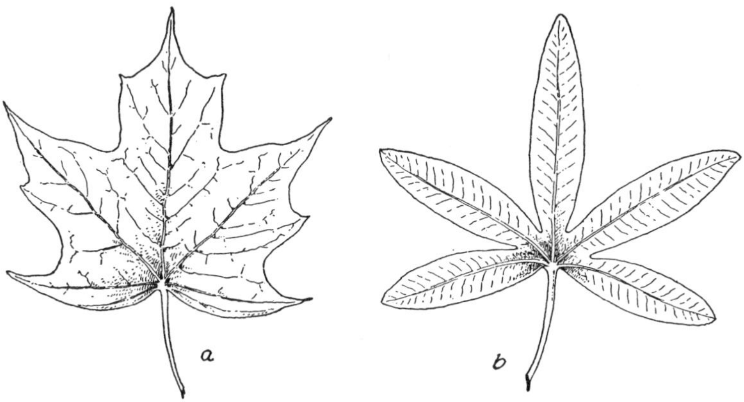

FIG. 111.—Leaf-forms: *a,* palmately lobed and palmately veined leaf of maple (*Acer saccharum*); *b,* digitate or deeply palmately lobed leaf of Passion-flower (*Passiflora cœrulea*).

parted when cut nearly to the midrib or base into lobes (Fig. 111b);

divided when cut quite to the midrib or base so as to simulate a compound leaf (Fig. 109a).

Various terms are used to designate the different forms of lobing of the leaf-blade, as: *pinnately lobed, pinnately cleft, pinnately parted* or *divided,* these cuttings corresponding to the pinnate type of venation. When the venation is of the palmate type we have, accordingly, the *palmately lobed, palmately cleft, palmately divided* forms, or the synonymous *digitately lobed, digitately cleft,* etc., etc. The number of lobes or divisions is indicated thus: *palmately three-lobed, palmately three-cleft, palmately three-parted,* etc. Other combinations may be made as required. A leaf cut into numerous fine segments is described as *dissected, pinnatifid,* or *multifid.* The segmentation of compound leaves is described as *twice* or *doubly compound, bipinnate* or *twice-pinnate, thrice-pinnate, pinnatifid* (Figs. 109-114).

With reference to their insertion, leaf-blades are:

sessile when the blade is attached directly to the node, the petiole being wanting (Fig. 115a);

clasping when sessile and partly investing the stem (Fig. 115b);

FIG. 112.—Leaf-forms (compound): *a,* odd-pinnate leaf of pepper-tree (*Schinus molle*); *b,* evenly pinnate leaf of karob (*Ceratonia Siliqua*); *c,* trifoliolate leaf of clover (*Trifolium*); *d,* palmately compound leaf of horse-chestnut (*Æsculus Hippocastanum*).

amplexicaule when completely investing the stem (Fig. 115c);

sheathing when the expanded base of the petiole invests the stem (Fig. 100b);

perfoliate when the stem appears to pass through the base of the blade (Fig. 116);

connate or *connate-perfoliate* when the bases of opposite leaves appear to have fused around the stem (Fig. 116b);

decurrent when the blade appears to continue down the sides of the stem (Fig. 115);

oblique leaves have the sides of blade unequal at the base (Fig. 117).

VENATION OF LEAVES

By *venation* is meant the frame-work of the vascular bundles or *veins* (*nerves*) as they course through the leaf-blade. With respect to venation leaf-blades are (Fig. 118):

feather-veined or *pinnately veined* when the branches of the *mid-rib* diverge towards the margins of the blade as do the pinnæ of a feather from the shaft (Figs. 103, 117, 118a);

palmately or *digitately veined* when the veins diverge from a common point (the point of insertion of the petiole) at the base of the blade (Figs. 112, 119);

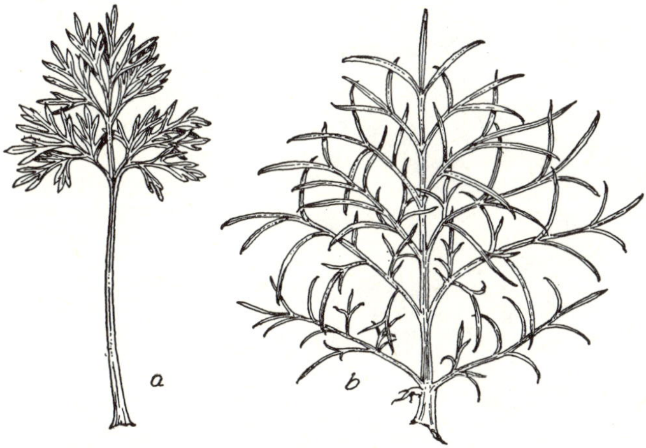

FIG. 113.—Compound leaves: pinnately dissected leaf (*a*) of carrot and (*b*) of *Cosmos.*

parallel veined when the veins run more or less closely together from the base to the apex of the blade. A pinnate venation is simulated when the veins diverge towards the margins of the blade, as in some Monocotyledons (Figs. 118, 125, 126).

THE PETIOLE

The petiole shows many variations in length and thickness. It is sometimes so short as to be hardly distinguishable, or it may be several times longer than the blade. Long and stout petioles are characteristic of the rhubarb, cow-parsnip, celery, water-lilies, palms, and numerous other plants. In trees the petioles may be seen to be progressively shorter upward on the shoot. Thus the length of the petiole may vary considerably in the same species (see rosette plants).

The petiole may be cylindrical (*terete*), or flattened dorsiventrally or laterally, and is often concave or channeled above. It is more or less dilated at the base to form the *hypopodium* (Fig. 101), the latter being clasping or sheathing and sometimes very large and prominent (*Umbelliferæ, Platanus;* Figs. 100, 119, 125). Often the petiole is winged on the sides, especially in cases where the blade is decurrent on the petiole (*Plantago*). When the petiole has a clasping base or hypopodium, it covers and conceals the axillary bud (*Platanus, Rhus;* Figs. 100, 101, 119, 125). In some species of *Aristolochia* the petiole serves as a climbing organ, while in some acacias it is transformed into a phyllode (Figs. 238, 239). In the ocotillo (*Fouquiera splendens*) the petiolar part of the leaf becomes a stout spine (Fig. 98). Sometimes the petiole is beset with glandular structures and various other excrescences. In "motile" leaves the base of the petiole is enlarged and functions as a tactile organ or *pulvinus* (*Mimosa pudica*). Inflated petioles are characteristic of *Utricularia* and *Eichhornia* (Fig. 125). In Monocotyledons, well-developed petioles are found among the palms, the bananas, the cannas, and others. A gradual transition from blade to petiole occurs in lilies, orchids, and other plants. In some acacias the petioles are modified as leaves (phyllodia; Figs. 238, 239).

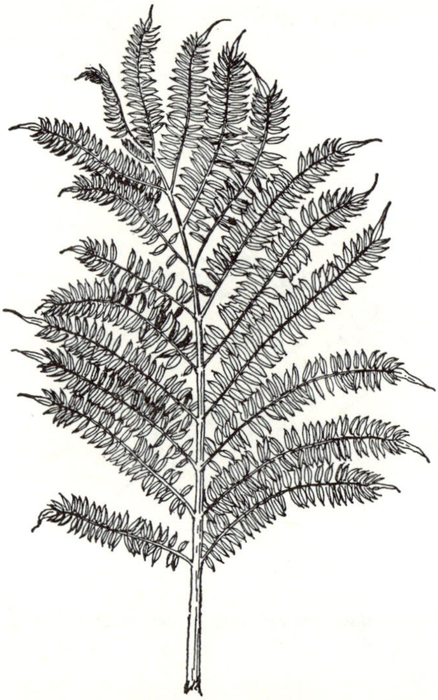

FIG. 114.—Leaf-forms: twice pinnate leaf of *Jacaranda ovalifolia.*

STIPULES

Stipules are lateral outgrowths of the base of the petiole or hypopodium (Fig. 101). They are absent in the *Anonaceæ, Lauraceæ, Cruciferæ,* and some other groups. When present, they show wide range in size and shape, being often very large and leaf-like (garden pea, pansy; Fig. 120). They may fall soon after the leaf unfolds (cherries, some willows), or they may persist permanently, especially when adnate to the petiolar base (roses) or when they constitute the

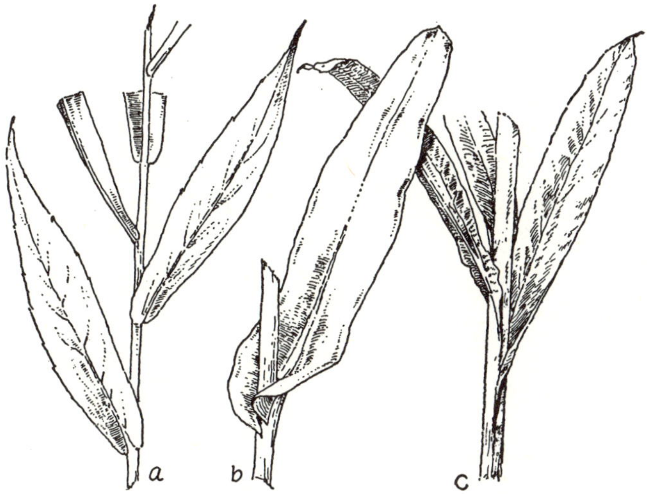

FIG. 115.—Leaf-forms: *a,* sessile leaves of an aster; *b,* sessile clasping leaf of a *Brassica; c,* decurrent leaves of *Gnaphalium decurrens.*

whole "leaf" as in *Lathyrus Aphaca* (Fig. 102). In *Cobœa scandens,* the lowest pair of leaflets of the compound leaf is so placed as to simulate stipules. Concrescence of stipules of the same or of opposite leaves occurs, *e.g.,* in *Dipterocarpus alatus, Humulus Lupulus, Urtica dioica.* Concrescence of stipules of adjacent leaves gives rise to *interpetiolar stipules,* as in *Onobrychis,* forming here a dry membranous structure. In *Galium* the stipules are hardly distinguishable from the leaves. Axillary stipules occur in *Caltha palustris, Gunnera, Potamogeton, Tetrapanax* (Fig. 121). The *ocreæ* of *Polygonum, Rumex,* and *Fagopyrum* are sheathing stipules (Fig. 169). Sheathing or *amplexicaule* stipules of another sort occur in *Ficus elastica,* where they cover the terminal bud as a hood (Fig. 121). Asymmetrical stipules occur in *Vicia, Pisum, Lathyrus, Spiræa, Ulmaria, Agrimonia.* In some species of *Ceanothus* the stipules are modified as warty cushions (Fig. 280). Some species of a genus may possess stipules, while others lack them (*Helianthemum*). In some genera (*Rosa, Salix, Populus*) the specific differences in the stipules are sufficiently marked to give them taxonomic value.

LEAF VARIATIONS

Variations in leaf-form in the same species are of common occurrence. The mulberries are familiar examples. Such *dimorphisms*

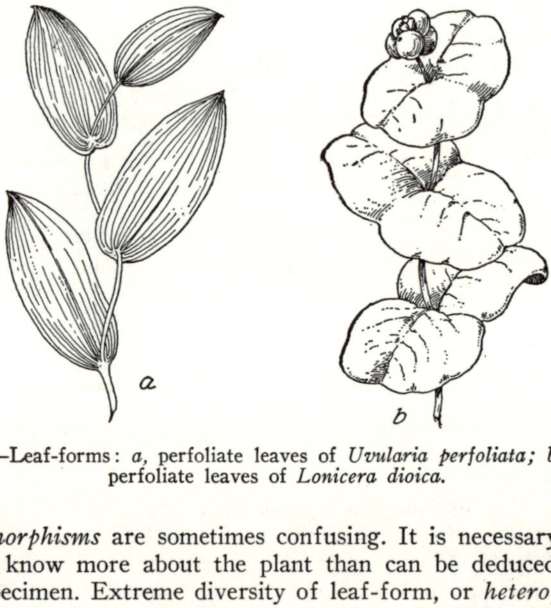

FIG. 116.—Leaf-forms: *a*, perfoliate leaves of *Uvularia perfoliata; b*, connate-perfoliate leaves of *Lonicera dioica.*

or *polymorphisms* are sometimes confusing. It is necessary in such cases to know more about the plant than can be deduced from a single specimen. Extreme diversity of leaf-form, or *heterophylly,* is characteristic of many aquatics (*Ranunculus aquatilis*), while certain *amphibious* plants develop a distinctly different type of leaf from the aquatic form when growing out of water (*Polygonum amphibium*). Sometimes the leaves on opposite sides of a shoot may differ both in size and shape, a condition (*anisophylly*) not uncommon among prostrate shoots. Many "rosette" plants show striking differences between the basal or rosette leaves and the later cauline leaves (*Cam-*

panula rotundifolia). The seedlings of *Phœnix canariensis* (Canary Island palm) and of many other plants show striking differences between the primary or *juvenile* leaves and the *adult* leaves (Fig. 462). A distinctive juvenile form of leaf is also characteristic of the sprouts of many trees and shrubs. Such differences are very noticeable in the blue gum (*Eucalyptus globulus*), in the big-tooth poplar (*Populus grandidentata*), in *Amelanchier, Symphoricarpos, Rubus,* and many others. Not only are the juvenile and adult leaves of different shapes but they may have a different phyllotaxy (*Eucalyptus globulus;* Figs. 142, 304).

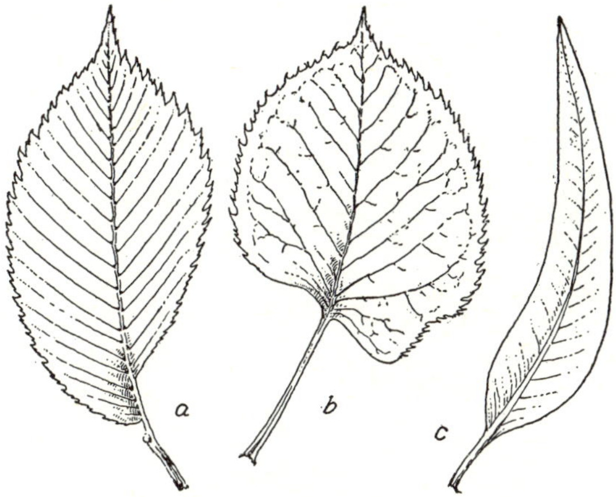

FIG. 117.—Leaf-forms: oblique leaf, *a,* of an elm; *b,* of a basswood; *c,* seculate leaf of *Eucalyptus globulus.*

SPECIAL FORMS OF LEAVES

In this category are *spines, leaf-tendrils, storage leaves, fly-trap leaves, pitcher-form leaves, scale-leaves, bracts, bracteoles.* These forms are usually spoken of as *modified leaves.* Except for the fly-trap and pitcher-form leaves, they are all quite common.

The spines of some cacti are supposed to be modified leaves. In the ocotillo (*Fouquiera splendens*) the spines are probably modified petioles (Figs. 98, 353).

Certain forms of *tendrils* are modified leaves or leaflets of compound leaves. Sometimes the petiole or the bracteoles (*prophylls*) below the flower may be modified as tendrils. Leaf tendrils occur in the garden pea, the sweet-pea, in *Smilax, Gloriosa, Littonia,* and other plants (Fig. 120a).

Storage leaves are characteristic of many succulent plants, including bulbiferous forms such as the onion and the hyacinth. They are prominent in *Mesembryanthemum, Cotyledon, Agave, Aloe,* and some cacti (Fig. 122).

Fly-trap leaves occur in *Drosera* and *Dionæa* (Fig. 206).

Pitcher-form leaves are characteristic of the common "pitcher plant," *Sarracenia purpurea,* of *Darlingtonia, Nepenthes,* and *Aristolochia* (Fig. 205).

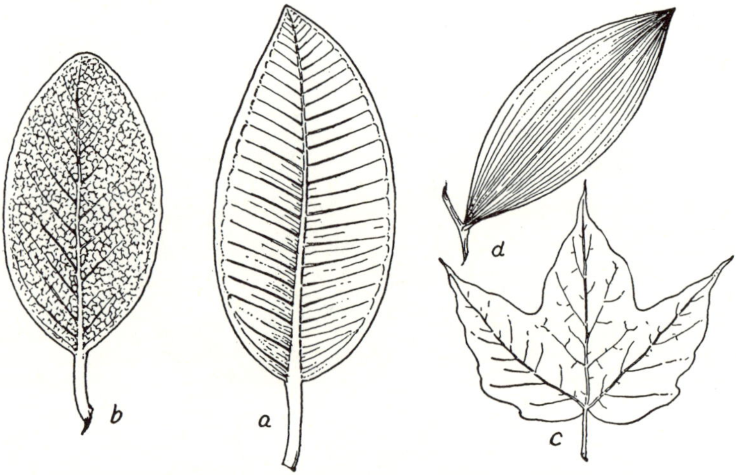

FIG. 118.—Leaf venation: *a,* pinnately veined leaf of rubber-plant (*Ficus elastica*); *b,* pinnately and net-veined leaf of *Ficus pumila; c,* palmately veined leaf of a maple; *d,* parallel-veined leaf of a Monocotyledon.

Bracts are much reduced leaves occurring in inflorescences or on the basal portion of stems (primary leaves) (Figs. 3, 11, 150-154, 408).

Bracteoles are minute leaves occurring on the pedicels of some flowers (Fig. 3).

Scale-leaves are reduced leaves lacking chlorophyll; they are common on subterranean stems, on bulbs, and on buds. In *Asparagus* the leaves are reduced to small scales or spines (Figs. 89, 444, 466, 470).

Transformations of parts of leaves into hooks for climbing purposes may be seen in *Stylidium scandens, Quisqualis indica,* and in certain members of the *Bignoniaceæ*. In some species of *Astragalus* and of *Caragana* the rachis of the compound leaf is transformed into a spine after the fall of the leaflets. Spinous leaflets are also found in some palms, *e.g., Phœnix canariensis.* The *ericaceous* type of leaf is a needle-like form common to certain plants of arid regions (*Erica, Acacia, Adenostoma*).

THE COATINGS OF LEAVES

What was said on a preceding page about the coatings of stems applies as well to leaves; in fact the leaves display the various kinds

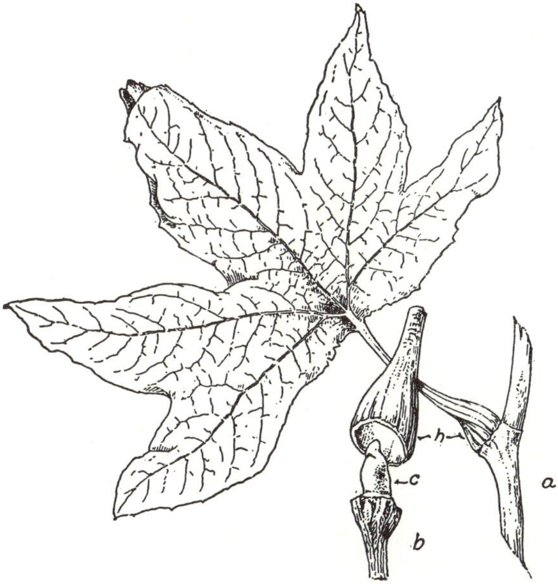

FIG. 119.—Leaf-forms: *a,* palmately lobed and palmately veined leaf of a sycamore (*Platanus racemosa*), with prominent hypopodium (*h*) which caps the axillary bud *b* (*c*).

of coatings more noticeably than do stems, sometimes. In leaves the coating, such as hairiness or other markings, does not always cover both surfaces, or it may be restricted merely to the veins. Thus in many plants the lower surface of the leaves may be coated with hairs (pubescent) while the upper surface is entirely bare (glabrous). In some species of *Ribes,* for instance, the lower surface of the leaves may be dotted with mucilaginous glands, while the upper surface is devoid of these markings. It is evident that all such peculiarities and differences in the markings of leaves serve a useful purpose in classification, at the same time being useful guides or recognition-marks when trying to identify an unfamiliar plant.

Below is given a list of the commoner terms used in describing the pubescence and other coatings or surface features of leaves.

Canescent. When covered with fine white hairs (grayish white, hoary).

FIG. 120.—Stipules: *a*, leaf (*l*) of garden pea with terminal tendril and large leaf-like stipules (*s*); *b*, leaf (*l*) of the pansy with very large and leaf-like, pinnately lobed, stipules (*s*); *c*, a bud.

Floccose. When covered with tufts of soft, woolly hairs.
Glabrate. When almost devoid of pubescence.
Glabrous. When devoid of hairy coatings.
Glutinous. When sticky or mucilaginous.
Glandular-punctate. When dotted with small resinous glands.
Glaucous. When covered with a whitish or bluish bloom (wax).
Hirsute. When covered with stiff hairs.
Hispid. When covered with bristly hairs.
Lanate. Woolly.
Pannose. When covered with a felt of woolly hairs.
Pilose. When covered with soft slender hairs.
Puberulent. When covered with very fine down.
Scabrous. When rough or harsh to the touch.
Sericeous. When covered with a silky pubescence.

Squamate, or *squamose.* When covered with minute scales.
Strigose. When beset with rigid bristles or bristly hairs.
Velutinous. When covered with a velvety pubescence.
Verrucose. When beset with minute projections; warty.
Villose. When covered with long shaggy hairs.

VERNATION OR PREFOLIATION

These terms denote the manner of folding or rolling-up of the leaves in the bud. When the leaves are not folded or rolled up in

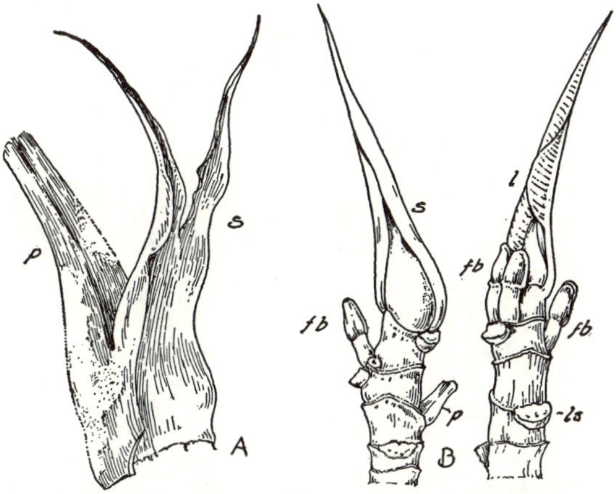

FIG. 121.—Stipules: *A,* connate axillary stipules (*s*) of Japanese rice-paper plant (*Tetrapanax papyriferum*); *B,* bud-investing stipules (*s*) of *Ficus elastica; l,* leaf; *fb,* flower-buds; *p,* base of a petiole; *ls,* leaf-scar.

any way, they are *straight.* By cutting transverse sections of the leaf-buds of trees and shrubs, for instance, the mode of folding may be seen. The principal types are (Fig. 123):

conduplicate, in which the leaf is folded like a book (oak, cherry);

convolute, when the leaf is rolled from one side to the other as one would roll a sheet of paper (plum);

circinate, when rolled from apex to base over the upper surface (ferns);

involute, when the sides are rolled towards the midrib over the upper surface (violet, water-lily);

revolute, when rolled similarly over the lower surface (*Ledum, Rhododendron*);

plicate or *plaited,* when folded like a fan.

ENVIRONMENTAL LEAF-FORMS

Leaves of aquatics. Heterophylly or the condition in which two or more forms of leaves are borne on the same plant is characteristic of many aquatics, especially those bearing both floating and submersed leaves. The differences between the floating and the submersed leaves are sometimes striking, and often in such cases, to be sure of the species, it is necessary to have the two forms of leaves

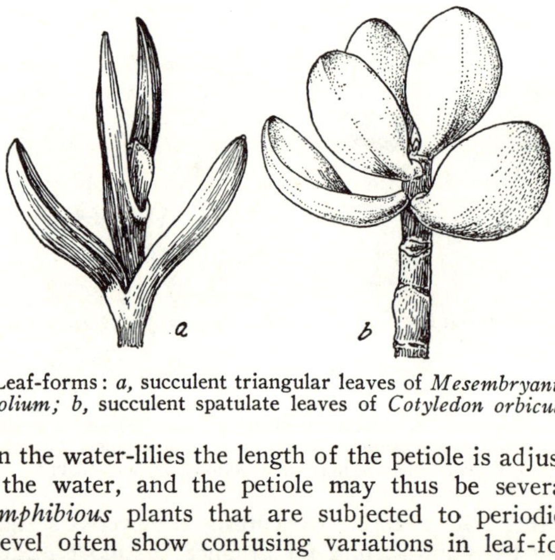

FIG. 122.—Leaf-forms: *a,* succulent triangular leaves of *Mesembryanthemum trianifolium; b,* succulent spatulate leaves of *Cotyledon orbiculata.*

at hand. In the water-lilies the length of the petiole is adjusted to the depth of the water, and the petiole may thus be several feet in length. *Amphibious* plants that are subjected to periodic changes in water-level often show confusing variations in leaf-form; such plants should be studied closely in their habitat in order to be certain of the species. Some of the commoner heterophyllous plants that will be met with frequently are *Sium cicutæfolium, Proserpinaca palustris, Sagittaria latifolia,* and *Polygonum amphibium.* The two last named and *Ranunculus delphinifolius* and *R. aquatilis* vary considerably as to leaf-form according to whether they grow in or out of water. Similar differences are to be seen strikingly displayed also in *Sagittaria arifolia* (Fig. 424).

Leaves of mesophytes or plants of moist or well-watered habitats are generally rather soft and thin as well as broad and flat.

They exhibit great diversity as to size, shape, cuttings, texture, and surface features, those of the shade-loving species being usually softer and thinner than those of species growing in the open. Plants of tropical rain-forests are remarkable for the great size of the leaves, in many cases, and these are usually of a leathery texture, rather than soft, a feature that seems strangely anomalous but which may be explained as an adaptation against loss of water during the rainless periods of intense heat.

Leaves of xerophytes or plants of arid regions are commonly of a firm and leathery texture (*sclerophyllous*) and generally much reduced in size or dissected; often they are more or less copiously clothed with hairs, especially on the lower surface. There are, however, many such sclerophyllous plants with broad leaves ("broad-sclerophylls"), these being often characterized by a glossy upper surface. The trees and shrubs of the "chaparral" vegetation of the Far West are typically sclerophyllous.

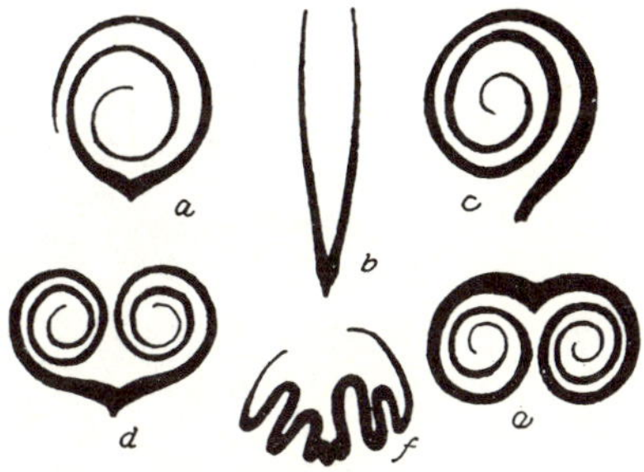

FIG. 123.—Leaf vernation: *a*, convolute; *b*, conduplicate; *c*, circinnate; *d*, involute; *e*, revolute; *f*, plicate.

Leaves of halophytes. Halophytes are xerophytic plants of saline or alkaline habitats. The leaves of these plants are more or less succulent and may be broad and flat or angular, or of the ericaceous type as to appearance. Plants of this sort are *Chenopodium album* (lamb's-quarters), *Salsola Kali* var. *tenuifolia* (Russian thistle), species of *Suæda* (sea blite), species of *Atriplex* (orach), and *Sarcobatus vermiculatus* (grease-wood).

Leaves of rosette plants. Rosette plants are generally biennials or perennials. The rosette is a crown of leaves borne close to or against the ground, and is produced during the first season of growth from the seed (in biennials). Actually such a rosette is merely a very short shoot, that is a shoot in which the internodes of the stem do not elongate, so that the leaves become closely crowded. Along with the rosette there is developed a stout tap-root which becomes charged with reserve food materials. In this condition the plant is carried over the winter season, and the following spring a tall, leafy, flowering and fruiting shoot is developed from the naked terminal bud of the rosette. The terminal bud is protected during the winter resting period by the leaves of the rosette, which usually remain green over the winter, although the outermost leaves may die and form a protective covering. In a sense a rosette is merely a large bud, the outermost leaves of which are spreading (Fig. 124).

Rosette plants are quite common, and as the rosette-forming habit is an intimate feature of the life-cycle of the particular plants, it is worth while to become familiar with them. They may be observed

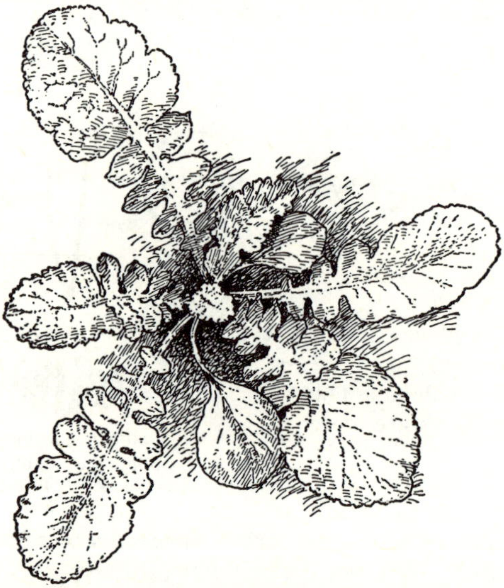

FIG. 124.—Rosette leaves of a crucifer.

among the evening primroses, in the dandelion, the hare-bell, mullein, burdock, sour dock, shepherd's purse and many other members of the mustard family. In the Western perennial, *Salvia apiana,* rosettes are formed at the ends of branches high above the ground, and similar rosettes, but borne close to the ground, are formed in *Sempervivum.*

Cushion plants are such as form dense hummocky growths (*Saxifraga Aizoön, Draba alpina*). In these the leaves are generally very small, succulent or firm, and leathery or spinescent. *Cæspitose* (*cespitose*) forms are those that grow in tufts or loose cushions (species of *Phlox* and *Saxifraga,* many grasses and sedges).

LEAVES OF MONOCOTYLEDONS

The leaves of Monocotyledons differ in general from those of Dicotyledons in being parallel-veined, in having a sheathing (*amplexicaule*) base, and often in showing little demarcation between blade and petiole. The venation in some forms resembles the pinnate and palmate types of the Dicotyledons and, as among the latter, the blade may be cut or lobed in various ways. Simple and compound leaves occur here also; but typical stipules are said to be unknown. Among

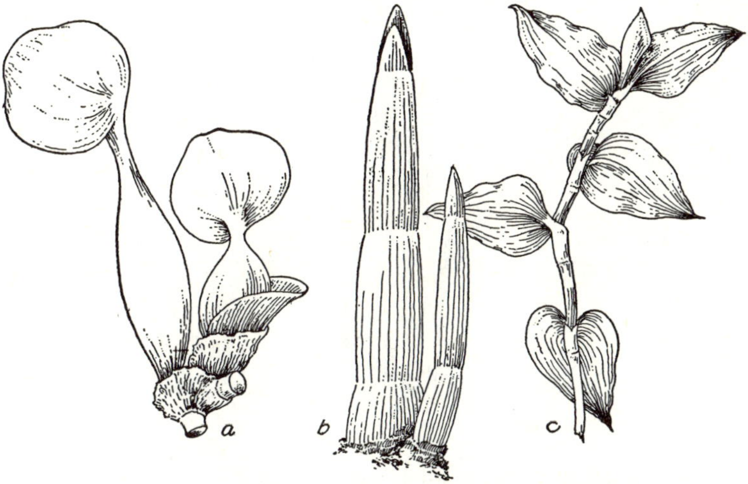

FIG. 125.—Leaf-forms of monocotyledons: *a*, leaves with inflated petioles in water-hyacinth (*Eichhornia*); *b*, sheathing leaves on sprouts of *Narcissus; c*, leaves with sheathing base in spiderwort (*Zebrina*).

the various forms of monocotyledonous leaves those mentioned below stand out rather prominently (Fig. 125).

Leaves of palms. Palm leaves are among the largest leaves known. They occur in two forms, the *palmate* or fan-shaped and the *pinnate* or feather-shaped. This difference is used to distinguish "fan-palms" from "feather-palms." The stately *Washingtonia filifera* of Southern California is a characteristic fan-palm (Fig. 463). At the base of the leaf-blade, in this palm, may be seen a prominent ligule (compare with grasses), while the margins of the petiole are beset with stout spinous teeth (Fig. 463). *Phœnix canariensis,* a common ave-

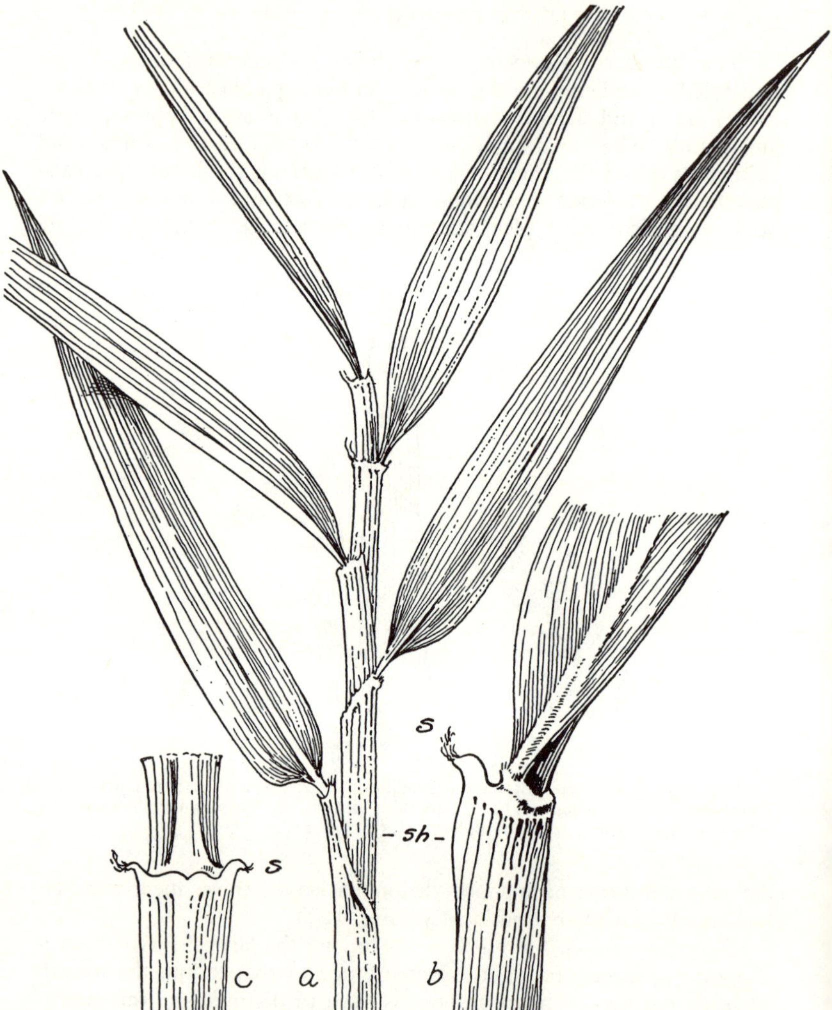

FIG. 126.—Leaf-forms in monocotyledons: *a*, branch of a bamboo, the broadly lanceolate leaf-blades contracted to a petiole-like base at junction with sheath *sh*; *b*, junction of leaf-blade with leaf-sheath, side view; *c*, dorsal view of *b*; *s*, stipule-like processes.

nue palm in California and other Southern States, is a feather-palm. Here the lower pinnæ of the leaf form stout spines. This species also shows a striking difference in shape between the juvenile and adult form of leaf (Fig. 453). Another prominent feature of palm leaves is the prominent sheathing base of the petiole (hypopodium).

Leaves of the banana. Here, also, blade and petiole are clearly differentiated. The venation resembles the pinnate type seen in the leaves of many Dicotyledons. Due to the type of venation, the leaves are easily lacerated by winds. As in the palms the petiole expands into a prominent sheathing base. Similar leaves are characteristic of *Canna, Ravenala,* and some large aroids.

Leaves of agaves. These are very striking rigid sword-shaped leaves crowded closely together in gigantic rosettes. The leaves of the aloes are of a similar type. Rigid, sword-shaped, rosette leaves terminating in a hard needle-point are characteristic of yuccas. A type somewhat intermediate between that of the agaves and the yuccas occurs in the sansevierias.

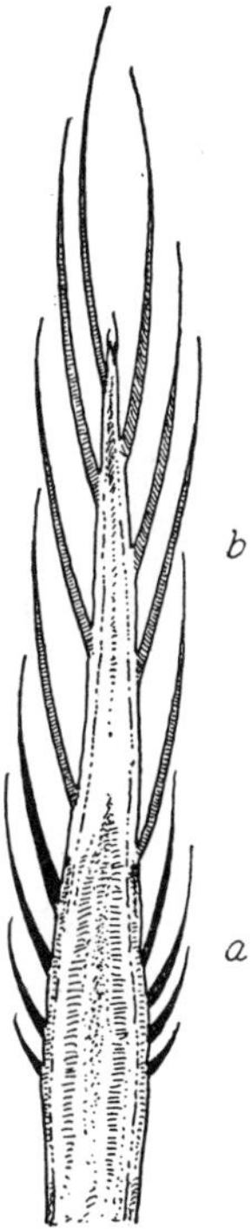

FIG. 127.—Median section of a leaf-bud of *Pittosporum Tobira; a,* bud-scales; *b,* leaves.

Leaves of lilies. The lily family shows a wide variety of leaf-form, one extreme being the leaves of the yuccas and the sansevierias just mentioned, while another extreme is to be seen in the grass-like leaves of onions and garlics. A form suggesting a Dicotyledon, with differentiated blade and petiole, occurs in *Smilax.* Whorled leaves occur in *Trillium, Medeola,* and some species of *Lilium.* In the onions and many broad-leaved forms, the base of the leaf is characteristically sheathing. In the bellworts, *Uvularia perfoliata* and *U. grandiflora,* the base of the leaf-blade is perfoliate (Fig. 124a).

Leaves of amaryllises and irises. These are generally of the sword-shaped type, although grass-like forms are characteristic of some members, such as *Hypoxis* of the *Amaryllidaceæ* and *Sisyrinchium* of the *Iridaceæ.* In most forms the leaf-base is characteristically sheathing (Figs. 95, 472, 473).

Leaves of orchids. These are in general very similar to the forms found among the lilies.

Leaves of aroids. The leaves of the aroids are among the largest and most striking known. Both simple and compound leaves occur, and these are commonly differentiated into blade and petiole. In some members the leaves resemble those of the water-plantains and the arrow-heads. Representative forms are *Symplocarpus, Lysichiton, Arisæma, Monstera, Amorphophallus, Alocasia.* An aberrant form

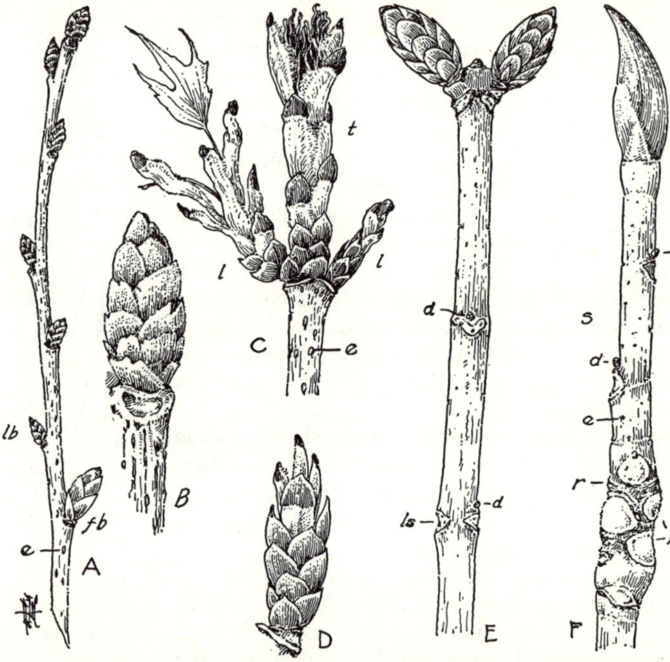

FIG. 128.—Buds: *A*, twig of elm, with leaf-buds (*lb*) and flower-bud (*fb*); *B*, a flower-bud of elm, enlarged; *C*, terminal bud (*t*) and lateral (axillary) buds (*l*) of sugar maple; *D*, lateral bud of sugar maple expanding; *E*, lateral buds of horse chestnut; *d*, latent buds; *F*, terminal bud of a magnolia; *d*, latent buds; *r*, scars of terminal bud-scales of previous season; *s*, previous season's growth of branch; *ls*, leaf-scars; *e*, lenticels.

occurs in the sweet-flag, *Acorus calamus,* in which the leaves are elongate and sword-shaped.

Long sword-shaped leaves are characteristic of some other Monocotyledons, also, such as the cat-tails (*Typha*) and the bur-reeds (*Sparganium*). Floating leaves with inflated petioles occur in *Eichhornia* (Figs. 135, 416, 417).

Leaves of grasses and sedges. The characteristic features of the leaves of grasses are: the *sheath,* which invests the stem or *culm;* the *blade,* which is more or less elongate and narrow or linear; and the *ligule,* a thin, annular, membranous upgrowth at the base of the blade (Figs 428, 438). Immediately above the ligule there is a pale cross-band of tissue through the activity of which the blade grows in length. Often the blade is more or less auricled at the base. The sheath in grasses is *open* or split down the side opposite the blade. The blade may be flat or trough-shaped, soft or wiry, and often keeled beneath. The different configurations of the blade, sheath, and ligule, are of taxonomic value in distinguishing genera and species. Compare wheat, oats, barley, rye, maize, sugar-cane, sorghum, millet, bamboo (Fig. 126).

The leaves of sedges are similar to those of the grasses in outward appearance, but their sheaths are *closed* and in some forms there is a more or less conspicuous thin, membranous, V-shaped tissue at the throat of the sheath, on the side opposite the blade (Fig. 438). Ordinarily the leaves are three-ranked, while those of the grasses are in two or several ranks.

Grass-like leaves are characteristic also of a number of other Monocotyledons, *e.g., Juncus, Luzula, Tradescantia, Hypoxis, Allium,* the similarities arising, however, chiefly from the appearance of the leaf-blade.

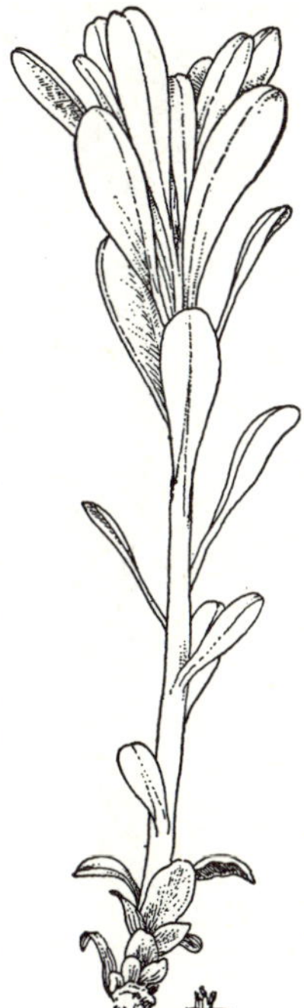

FIG. 129.—Developing shoot of *Pittosporum Tobira* showing transition from bud-scales to leaves.

BUDS

A bud is an incompletely developed shoot in a dormant or latent condition (Fig. 127). Many buds, especially the *winter buds* of Northern trees and shrubs, are invested by a series of modified leaves or *bud-scales* (Fig. 128). With reference to their location on the shoot buds are *terminal* or *lateral.* A terminal bud is one that occurs at the tip of a shoot, and from which the growth in length of the shoot proceeds. Lateral buds are *axillary* when they arise in the axils of leaves and *supra-axillary* or *extra-axillary* when they arise above the leaf-axil. Sometimes more than one bud will appear

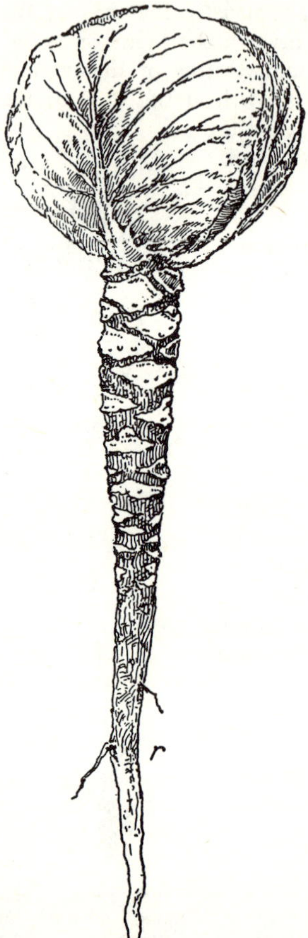

FIG. 130.—Head of cabbage (a bud) with large lower leaves removed, leaving large leaf-scars; *r*, tap-root (primary root).

in the axils, and these supernumerary ones, which are on the sides of (*collateral*) or above (*superposed*) the middle bud, are termed *accessory* or *supernumerary buds. Adventitious buds* are those that may appear elsewhere on the stem, as on the internodes; they usually appear following injury. But they are also produced on leaves in certain cases (*Bryophyllum*). *Naked buds* are those that have no protecting bud-scales; such buds may be seen in *Geranium, Alnus, Viburnum alnifolium, Kalmia, Platanus,* and *Rhus*. In *Geranium, Platanus,* and *Rhus glabra,* the buds are concealed beneath the expanded base (hypopodium) of the petiole (Figs. 101, 119).

Buds that give rise to flowers only are termed *flower-buds*. If a bud gives rise to a leafy shoot, it is a *leaf-bud* (Fig. 129). But if a bud produces a combination of a leafy shoot and a flowering shoot, it is a *mixed bud*. A transition from bud-scales to leaves is very well illustrated in *Pittosporum* (Figs. 129, 217).

Buds vary greatly in size and in superficial features according to the genus or species. A cabbage head, for instance, is in reality a huge leaf-bud (Fig. 130), and a cauliflower a huge mixed bud. In the category of buds are also all bulbs, such as those of the onion, hyacinth, and banana (Figs. 131, 132). The young ear of maize is similarly a bud. Large buds are produced on the subterranean parts of the rhubarb, the skunk-cabbage, and on the stout rhizomes of the tall reed-grass and on Solomon's seal, and large terminal buds may be seen in rosette plants like the burdock, the sour dock, and evening primroses. Small bulbs termed *bulblets* or *bulbils* are produced by some lilies (Fig. 133), onions, and other plants (*Cicuta bulbifera*).

The taxonomic characters of buds are found in their size, shape, coatings; color, arrangement of the bud-scales and their size, shape, and textures; inner parts of the bud and any other features that may be prominent or of diagnostic value. The importance of bud characters in distinguishing species may be observed in the oaks, hickories, elms, willows and poplars, maples, and ashes.

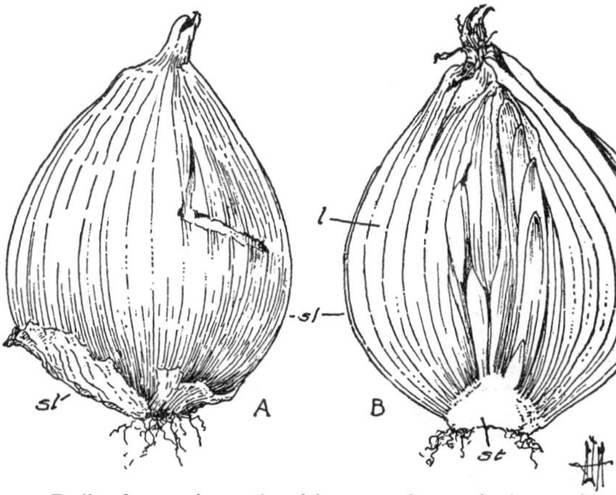

FIG. 131.—Bulb of an onion: *A*, with outer dry scaly leaves in place. *B*, median section of the bulb, showing short stem (*st*) with a few secondary (adventitious) roots, and succulent leaves (*l*); from the axils of the inner leaves spring two buds.

Teratological Phenomena

Under this heading are classified all sorts of malformations or monstrosities due to various causes. Malformations or abnormalities are of frequent occurrence and sometimes confusing in that they are likely to be mistaken for normal features. Thus the pine-cone gall of the willow, *Salix cordata,* caused by the gall-gnat, *Cecidomyia strobiloides,* is sometimes mistaken for the "fruit," and so are likewise the common "oak-apples" produced on various species of oaks by certain gall-flies of the genus *Amphibolips* (Fig. 134). A common form of monstrosity is that known as a *fasciation,* in which the organ or organs involved assume grotesque proportions. These are frequently met with among the chicories and among the thistles. The cock's-comb, *Celosia cristata,* may be explained as of the same category. Fasciated organs often have the appearance of several shoots grown together into one broad, flat body. Abnormalities in flowers are numerous and are not restricted to any particular set of organs (Fig. 135). But all floral abnormalities may be included under the caption of *peloria.* To mention particular instances, five-spurred

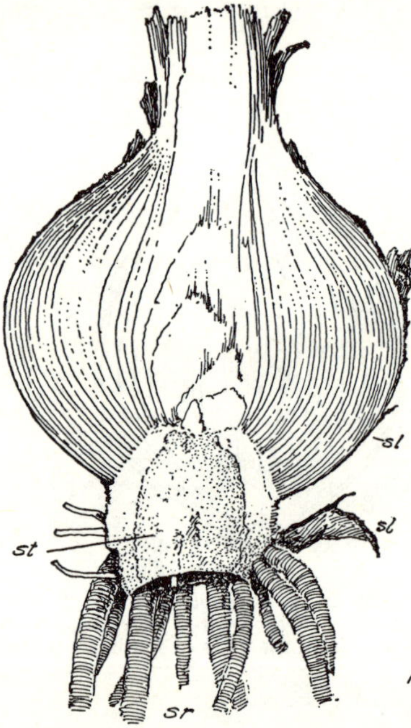

FIG. 132.—Median section of the bulb of *Hymenocallis; st,* stem; *sr,* secondary (adventitious) roots; *sl,* scale-leaves.

FIG. 133.—A, bulbils (*a*) of wild prairie tiger-lily (*Lilium umbellatum*); *B, b,* axillary bulbil of *Saxifraga bulbifera; c,* same with two outer scale-leaves; *d,* a bulbil enlarged, showing two succulent leaves.

corollas have been found in species of *Linaria* which normally produced flowers with a one-spurred corolla, and *petalody* of the stamens, that is a transformation of stamens to petals, is a common feature of the flowers of *Philadelphus Lewisii.* Even a transformation of the ovules into leaves, or *phyllody* of the ovules as it is called, has been found in *Alliaria officinalis* and other plants (Goebel). Transformation of the stamens into staminodes (*staminody* of the stamens) is of common occurrence among the *Lauraceæ* and among some of the *Ranunculaceæ, e.g. Anemone patens.* Abortion or atrophy of the stamens occurs in some species of *Saxifraga* (Fig. 23).

The reversion of an organ to an ancestral type, such as the occurrence of juvenile leaves in some plants, or the production of leafy buds in place of flowers, or of an inflorescence in place of a single flower, may all be regarded as teratological conditions. Often such manifestations result from injury, as in the case of the appearance of juvenile leaves on the sprouts arising from the stumps of the bigtooth poplar, *Populus grandidentata,* or the appearance of several flowers in the perigynia of some species of *Carex* that produce new shoots after being cut or mowed. Leafy winter buds are produced by the pond-weeds, *Potamogeton gemmiparus* and *P. Robbinsii,* and by the rush, *Juncus pelocarpus.* All forms of mycorrhiza may perhaps be considered, also, as being essentially of teratological nature (Figs. 142, 304).

FIG. 134.—Insect galls: *a,* of *Salix cordata; b,* "oak-apple" of *Quercus chrysolepis; c,* leaf of *Quercus chrysolepis.*

VESTIGIAL ORGANS

Vestigial organs are those which, through some cause or another, have been arrested or retarded in their development. "The more elaborate the differentiation a plant exhibits the more common is the appearance which we designate as arrest," says Goebel.* And, according to this authority, if "the primordium of the organ can be proved to have a developmental existence," that is if it exists as a "rudiment or vestige," it is *aborted;* if no trace of the organ exists it is *suppressed.* To take a particular case—that of the campion, *Lychnis alba,*

* *Organography of Plants,* I, 56.

we shall find in some cases a small rudiment of a gynœcium in the staminate flowers (Fig. 182), while in other cases no such rudiment is discernible (Fig. 181c). Likewise in the carpellate flowers of the same species a whorl of very small rudimentary stamens is to be seen on the annular disk surrounding the base of the ovary (Figs. 181, 182). Similarly in the buckeye, *Æsculus glabra,* a conspicuous rudimentary gynœcium will be found in many of the flowers (Fig.

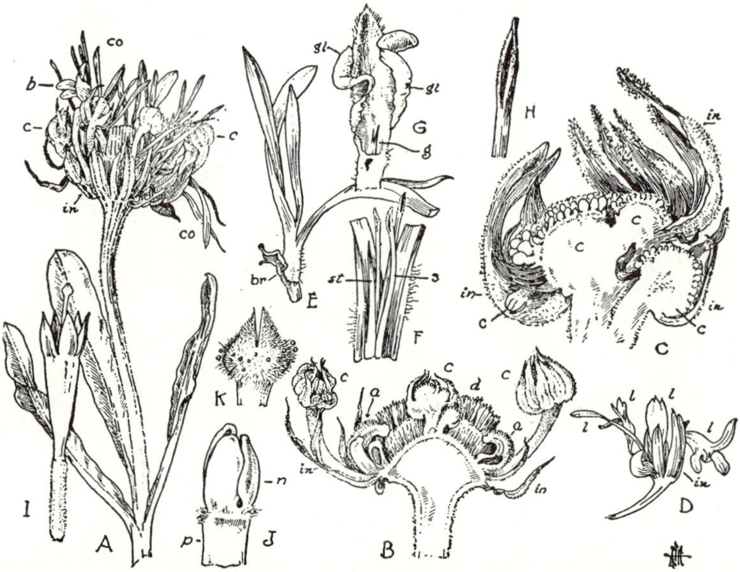

FIG. 135.—Teratalogical features in *Tagetes lucida* (*Compositæ*): *A,* a flowering shoot; *in,* involucral bracts of capitulum; *c,* small secondary capitula (two others towards center of capitulum); *co,* ligules of ray-flowers. *B,* median section of a capitulum; *a,* achenes; *c,* secondary capitula; *d,* disk-flowers; *in,* involucral bracts. *C,* section of a secondary capitulum; *c,* four smaller (tertiary) capitula, the surface of the three larger densely covered with numerous minute and curiously modified disk-flowers,—one shown enlarged in *J; in,* involucral bracts. *D,* modified small secondary capitulum shown at *b,* in *A,* with a few-bracted involucre (*in*) and florets with ligules of different shapes. *E,* anterior view of a floret of *A; br,* floral bract with glandular, orange-colored margin. *F,* posterior side of floret in *E,* with ligules cut off, and showing three slender styles each terminating in a stigma; *st,* a fourth style without stigma. *G,* inner face of bract (*br*) in *E,* with fleshy undulate, orange-colored, glandular margin (*gl*), and two peculiar bristle-like structures (modified florets ?) at base. *H,* stigma of style (*s*) in *F. I,* a normal disk-flower of *A. J,* modified floret of a small capitulum in *C,* with stout pedicel (*p*) and three succulent leaf-like organs (*n*). *K,* stigma of normal floret shown **at *I.***

274b). Again, in the pepper tree, *Schinus molle* (*Anacardiaceæ*), a diœcious species, a small rudimentary gynœcium is to be seen in the staminate flowers (Fig. 263). So too in *Rhamnus crocea* (Fig. 277) and other species of this genus, a rudimentary gynœcium appears in the staminate flowers. Many other similar instances could be enumerated, taken from different families. These instances may be regarded as showing the ultimate stages in the evolution of the unisexual flowers, the final step being the complete suppression of the particular organ, as in some flowers of the campion.

PART II

SYSTEMATICS

	Syllabic Chart of the Grouping of the Orders (Read up)				
Ovary inferior	*Cucurbitales*		*Rubiales*		*Campanulatæ*
Ovary superior	*Contortæ*		*Tubifloræ*		*Plantaginales*
			TETRACYCLICÆ		
Ovary inferior		*Ericales*	*Primulales*	*Ebenales*	
Ovary superior	*Diapensiales*	*Ericales*	*Primulales*	*Ebenales*	*Plumbaginales*
			PENTACYCLICÆ		
			METACHLAMYDEÆ		
Ovary inferior	*Opuntiales*	*Rosales*	*Parietales*	*Myrtifloræ*	*Umbellifloræ*
Ovary superior	*Geraniales* *Pandales* *Ranales*	*Rosales* *Sapindales*	*Parietales* *Rhamnales* *Rhœadales*	*Myrtifloræ* *Malvales* *Sarraceniales*	*Centrospermæ*
			DIALYPETALÆ		
Ovary inferior	*Julianiales*	*Santalales* (?)	*Juglandales*	*Aristolochiales*	*Fagales*
Ovary superior	*Urticales* *Batidales* *Verticillatæ*	*Myricales* *Salicales*	*Balanopsidales* *Leitneriales*	*Polygonales* *Garryales*	*Centrospermæ* *Proteales* *Piperales*
			MONOCHLAMYDEÆ		
			ARCHICHLAMYDEÆ		
	(Read up)		—*DICOTYLEDONES*—		

CHAPTER X

THE APPROACH TO CLASSIFICATION

Essentials. In order to understand any system of classification it is first of all necessary to familiarize ourselves with the principles upon which that system is based; and since these principles are derived mainly from the structure of the flower, we must approach the system with an understanding of that very important phase of the subject. Merely to "run down" a "flower" in a key to find its name may satisfy our curiosity to that extent if that be all we desire to know, but very likely in so doing we shall be inclined to make no more than a cursory examination of the flower, sufficient to satisfy the specifications of the key, and very likely we shall fail to apprehend the significance of the structure of our flower in relation to the general scheme. Very likely, too, if confronted with another flower of a similar type, we should again be compelled to resort to the key to place it in its proper category. But if on the other hand we have at our command a knowledge of a sufficient number of types of flowers characteristic of the principal categories, we shall find little difficulty in placing a flower in these categories, from which it is but a relatively simple step to the orders and families, and even genera. Know the types, and any strange flower we may find can be placed with relative ease. Nothing can be more futile than to attempt to approach the subject by memorizing the characters of families and genera, or learning floral formulas, or by having to resort continually to keys.

It was pointed out in Part I that one of the most important and useful guides to the proper classification of a plant is the median longitudinal section of the flower. This at once discloses whether the ovary is superior or inferior, that is, whether the flower is hypogynous, perigynous, or epigynous, and the degree of perigyny or epigyny—points which are essential to the placing of the flower in its proper category. Observations of the median longitudinal sections can be made with very little trouble in the field, where on many an occasion they may aid one in placing an unfamiliar flower. Indeed there is no more significant guide to the affinities of a flower than the median longitudinal section.

The approach to the system. In developing the approach to the present system, therefore, it is assumed that the reader is familiar with the various types of flowers, fruits, etc., discussed in Part I, and that he is familiar, further, with the primary distinctions between the two major groups of the flowering plants, namely, the Monocotyledons and the Dicotyledons. As the Monocotyledons are relatively simple, and the principal groups are quite clearly defined, they may be omitted from the present discussion.

We may turn, then, to the general classification of the Dicotyledons as presented in the chart opposite the beginning of this chapter. Here are shown the major categories of these plants and an arrangement of the orders under each major subdivision according to the relation of the ovary to the receptacle, *i.e.,* superior or inferior. This grouping of the orders is not to be construed as phylogenetic, but rather as a convenient way of facilitating study, although the orders with inferior ovary are to be understood as being more advanced than are those with superior ovary. The placing of the order names is also a matter of convenience.

It will be observed that any flower (exceptions aside) with a choripetalous corolla, *i.e.,* with a corolla of distinct petals—such as a magnolia, a buttercup, a stonecrop, a mustard, a rose, a pea, or a dogwood—or any flower lacking a corolla—such as a birch, an oak, an elm, or the buckwheat—will fall in the first major division, the *Archichlamydeæ;* and any flower (exceptions aside) with a sympetalous corolla—such as a heath, a morning-glory, a snapdragon, or a sunflower—will fall in the second major division, the *Metachlamydeæ.* The *Archichlamydeæ,* it will be noted, are subdivided into two large groups, the *Monochlamydeæ* and the *Dialypetalæ.* To the former belong those plants whose flowers possess neither calyx, nor corolla, nor perianth; or if a perianth is present, it is not differentiated into calyx and corolla. Such flower types are found among the peppers and lizard's-tails, the willows and poplars, oaks and beeches, walnuts and hickories, elms, hemp, mulberries, buckwheat, lamb's-quarters, and amaranths. We have only to observe which of these members possess superior ovaries and which inferior ovaries to segregate them according to the chart, with relatively few orders in each category to remember. If now we proceed to eliminate those that are already familiar, say the oaks, beeches, birches, hazels, and walnuts, we thereby reduce the number of forms—of those with inferior ovary, for example—to a relatively few unknowns which may be readily determined from the accompanying keys in the text. This method of eliminating the knowns will be found very useful, and as we become increasingly familiar with more and more types, we shall soon discover that we can place a "flower" in its proper order and even family with relative ease, and, as our experience grows, even the genera will become familiar. It is by such procedure that the characters of the orders, the families, and the genera will begin to dawn upon us. In this way we proceed from the familiar open highways into the unknown byways and trails, which can only become familiar by repeatedly retraversing them. Then, too, with constant comparison, contrasting, and association of one form with another, the accumulating mass of taxonomic data will be seen in their proper relations and the different categories will stand out in clearer perspective.

The members of the other group, the *Dialypetalæ,* all possess a corolla, the outstanding feature of which is that it is made up of distinct petals. Hence any flower (exceptions aside) with such a corolla belongs in this subdivision. Here belong a large proportion of the flowering plants. It is also one of the most difficult groups to deal with as regards the orders and families. But the problem is considerably simplified by segregating them on the basis of superior and inferior ovary. And as the chart shows, the orders with inferior ovary are thus reduced to a very few, at least two of which are so clearly defined that they will be recognized with little difficulty. These are the *Opuntiales* and the *Umbellifloræ,* the former comprising the cacti and the latter the aralias and their allies, the carrots and parsnips and their allies, and the dogwoods. With these eliminated there remain only three to differentiate, and these are difficult. As a rule the *Parietales* can be segregated by their parietal placentation (hence the name of the order), but there are a number of apparent exceptions; the *Myrtales* generally possess an exile placentation, although there are a number of exceptions here also. In the (epigynous) *Rosales* the placentation is axile or parietal. But if we take such pairs of flowers as an evening primrose and a eucalyptus, a begonia and a mentzelia, a gooseberry and an apple, we may obtain a general idea of the characters of these epigynous *Myrtales, Parietales,* and *Rosales,* respectively. (But see text on subsequent pages.)

Similarly the order characters of the superior-ovaried members are difficult to differentiate and recognize without familiarity with a number of representative floral types. There are a few, however, that stand out rather conspicuously and can thus be easily eliminated. These are the *Sarraceniales* which can be recognized at once by their fly-trap leaves, if not by the flowers; the mustards among the *Rhœadales,* and some of the mallows among the *Malvales.* Many of the

Ranales are easily recognizable, although some are likely to be confused with certain forms of the *Rosales,* and conversely. The one useful character that will separate the *Geraniales* from the *Sapindales,* although technical, is the position of the ovule (see text). The *Pandales* may be eliminated since they do not occur in our flora. And it may be said that while some of these orders contain a large number of families, yet the situation is considerably simplified by the fact that relatively few of these families are actually represented in our flora, so that a knowledge of some of the floral types of these families, using the more familiar forms as a foundation, will sufficiently explain the dominant characters of the respective orders and families. It may be pointed out that the presence of an hypogynous gland (disk) in many *Geraniales, Sapindales,* and *Rhamnales* will be found to be a significant guide to these orders. (But see families.)

In the *Metachlamydeæ* the most difficult group is the *Pentacyclicæ,* a group which, as the name indicates, possesses normally five whorls of floral organs; but as one whorl, the outer, is often suppressed, flowers of these forms are likely to be confused with the *Tetracyclicæ.* However, inspection of the andrœcium of these forms will show that the stamens are *opposite the petals,* that is *obdiplostemonous,* a character that will distinguish these members from the *Tetracyclicæ.* The obdiplostemony indicates that the outer whorl of stamens has been suppressed in these forms. And it should be remembered, too, that a choripetalous corolla will be found in some members of the first three families. Other features are the presence in many cases of four or three carpels instead of five, and a simple style—often with obscure stigmatic lobes (*Primulaceæ*). Two orders of the *Pentacyclicæ* can be distinguished by characters of the stamens. These are the *Ericales,* the anthers of which open by pores and are often furnished with horn-like appendanges; and the *Ebenales,* in which in some cases two or more whorls of stamens, or numerous stamens, are present. In several cases in the *Pentacyclicæ* the outer whorl of stamens is transformed into staminodes. As will be seen by the chart, some members of the *Pentacyclicæ* are epigynous, a feature which readily sets these apart. Attention should also be directed to the fruits, as these are always useful guides to the respective groups.

The *Tetracyclicæ,* with the possible exception of the *Plantaginales* and some of the *Contortæ,* are the most readily distinguishable of all the *Metachlamydeæ.* The outstanding feature is the universal presence of tetracyclic flowers, in which the stamens alternate with the lobes of the corolla. Here we have only six orders to remember, three with superior ovaries and three with inferior. Of the superior-ovaried members the *Contortæ* often show a twisted æstivation of the corolla in the bud, a feature from which the group derives its name and which is especially prominent in the gentians; and the leaves are usually opposite and exstipulate. Naked and unisexual flowers will be found in some forms (*Oleaceæ*). Trees and shrubs are frequent in the *Oleaceæ* and *Loganiaceæ*. In the *Apocynaceæ* and *Asclepiadaceæ* occur many forms with milky sap—always a significant recognition-mark—and free carpels and follicular fruits are common. But the *Asclepiadaceæ* are distinguished further by the presence of pollen-masses or pollinia, a feature found nowhere else except among the orchids. The *Plantaginales* are a small, relatively unimportant group which may be recognized by remembering the common dooryard plantain.

The outstanding order of the superior-ovaried *Tetracyclicæ* is the *Tubifloræ*. The superior ovary associated with the characteristic tubular actinomorphic or bilabiate corolla and the generally 2-carpellate gynœcium with the characteristic parietal or marginal (on margins of carpels), basal, or thick axile placentas and often numerous seeds are the outstanding recognition-marks of this order. In the *Convolvulaceæ*, the *Polemoniaceæ*, and in a few other cases more than two carpels are present, while in the *Boraginaceæ*, *Verbenaceæ*, and *Labiatæ* the ovary is often deeply cleft so as to form nutlets at maturity. Many labiates are further characterized by quadrangular stems; and the aromatic properties of many of these plants are well known. Useful guiding characters will be found in the number and position of the stamens in many families of the order.

The epigynous *Tetracyclicæ* are relatively easy to recognize. With few exceptions the *Cucurbitales* or gourds stand out clearly, for we have only to recall the common garden cucumber, or the melons and squashes and pumpkins to recognize their characters. But more specifically their peculiar stamens and distinctive fruits and placentation set them apart. As for the *Rubiales*, a constant character, aside from epigyny, is the opposite leaves, and associated with these characters are the frequent occurrence of trees and shrubs, and the usually small flowers aggregated in dense clusters. Generally, too, the calyx is very much reduced, often to mere bristles or teeth, and in some members the corolla is zygomorphic. The only puzzling members, perhaps, are the *Adoxaceæ* and the *Dipsacaceæ*. The former is of doubtful status and is restricted to a solitary herbaceous species (see family). The latter suggest the composites, from which they differ, however, mainly in the possession of only four stamens—which are distinct—and a cup-shaped involucel or epicalyx. The teasel and the scabiosa of the gardens are typical representatives. Of the remaining order, the *Campanulatæ,* the composites stand out so clearly by the compact capitulate inflorescence with its characteristic ray-flowers and disk-flowers that most of them are recognized on sight. The only other members of this order in our flora are the harebells and the lobelias and their allies, the former possessing actinomorphic, the latter zygomorphic corollas. In the lobelioid members the anthers coalesce to form a tube around the style.

The madders (*Rubiaceæ*) are rather poorly represented in our flora, but the common bedstraws (*Galium*) and the small bluets (*Houstonia*) belong here and will exemplify some of the more important floral characters. A marked character is the four-parted corolla. In the honeysuckles (*Caprifoliaceæ*) both actinomorphic and zygomorphic flowers occur. In our madders the fruits are fleshy or dry. Structurally the flowers are very similar to those of the dogwoods (*Cornaceæ*) among the *Archichlamydeæ*—indeed they may be regarded as the sympetalous derivatives of the dogwoods. (Compare Figs. 390, 391, 393, 394.) Stipules are lacking.

Applying the system. For the purpose of demonstrating the application of the system to specific cases, a few types of supposedly unfamiliar flowers will be selected and analyzed as far as necessary to show their connections with the groups in the chart. It is taken for granted that we have at hand the entire plant or as much of it as may be necessary for proper identification. Suppose, then, that we have before us a wild flower possessing five distinct petals, five sepals, and many distinct stamens and carpels. Obviously a corolla of distinct petals will place the flower in the *Dialypetalæ.* But the many distinct carpels interposes a doubt as to the minor categories. From the standpoint of our flora, however, there are but two possible dialypetalous orders to which it might be referred: it must be either to the *Rosales* or the *Ranales.* On general appearance the flower might be either a buttercup or an anemone, or even close to the strawberry. To settle this question we cut a median longitudinal section of the flower, and this shows, indeed, certain features of a buttercup or an anemone, but the flower is perigynous, and that at once throws it out of the buttercup group of the *Ranales* and into the *Rosales.* Here, as far as our flora is concerned, it could come nowhere else but in the rose family or the *Rosaceæ,* and here it properly belongs. This is as far as we can trace it in this volume. but with the aid of a "flora" and key we shall find the plant in the genus *Potentilla*—not so far removed, indeed, from the strawberry. It may be interpolated here that hypogyny does not occur in the rose family, and this fact alone will guide the student in distinguishing between ranunculaceous and rosaceous forms with numerous distinct carpels. Had our flower possessed only three distinct petals and three sepals, it would have been a Monocotyledon—a water-plantain.

As another example we may choose a small low plant with alternate leaves and small white flowers aggregated in an umbel-like inflorescence at the top of the stems. Examining a flower, we find but one whorl in the perianth, and this is petaloid and composed of distinct segments; evidently it is the calyx, and the corolla is wanting. The plant should fall, therefore, in the *Monochlamydeæ*. From the median longitudinal section we observe that the ovary is inferior. Hence there is but one place to which this flower can be assigned in the *Monochlamydeæ* and that is to the inferior-ovaried group. As the *Julianiales* do not occur in our flora, there remain but four orders to decide upon. The *Juglandales* and the *Fagales* are all trees and shrubs and perhaps so familiar that we can eliminate them from consideration, leaving only the *Santalales* and the *Aristolochiales,* both scantily represented in our flora—which again simplifies our problem. Among the former the mistletoes will probably be familiar, and among the latter we have the wild gingers, the birthworts, and the Dutchman's pipe, to neither of which, obviously, our flower belongs yet to which it may be allied. Familiarity with these forms, however, will enable us to eliminate them in using a key or a "flora" —in which our plant will be found in the genus *Comandra* of the *Santalaceæ* or sandalwoods. Reflecting for a moment on what was done in locating *Comandra,* it will be noted that nothing was said about the number of parts in the flower nor was any attempt made to demonstrate the floral formula. Just three significant points were sufficient to enable us to locate the affinities of the plant, and these were: the presence or absence of a corolla, the fact that the perianth-segments were distinct, and the fact that the ovary was inferior. The last-named feature, it will be observed, was settled by the median longitudinal section. The minor details have been intentionally omitted since they must be sought in a flora or in the text on a succeeding page of this volume.

For the next illustration we may choose a type that presents certain technical difficulties yet not of the obscure sort. The corolla is sympetalous, which puts the flower in the *Metachlamydeæ*. Furthermore, the corolla is actinomorphic and somewhat campanulate. This suggests several possible allocations. However, there is present only one whorl of stamens and these alternate with the corolla-lobes and are epipetalous. This points to the *Tetracyclicæ*. Resorting to the median longitudinal section, we discover a gynœcium of two apparently distinct carpels as well as two distinct styles arising between the ovaries; and, furthermore, the ovaries are superior. There is now no other choice but the *Tubifloræ*. But in this order there are several places in which a deeply cleft or divided ovary of this sort is present; however, the presence of some small, gland-like bodies at the base of the ovary points to the *Convolvulaceæ*—where, indeed, the plant actually belongs, in the genus *Dichondra*. (See family.)

One other type, also of a somewhat technical sort, will suffice for these examples. This is a tall shrub with slender stems and branches thickly covered with minute leaves, in every way suggestive of a heather (*Erica*). The flowers are small and are crowded at the ends of the branches, each terminating a small spur shoot. The corolla is actinomorphic and decidedly vasiform, with a slender tube which is dilated above and crowned by small flaring lobes. This would appear to strengthen our notion of its heath affinities, but the median section dispels the idea. The stamens are epipetalous and in one whorl, and alternate with the corolla-lobes. Furthermore, the ovary is superior and supports a long, slender style which terminates in a prominent discoid stigma. Undoubtedly the plant belongs to the *Tubifloræ,* but to substantiate this point we now resort to a test that has not heretofore been called into service, namely the transverse section of the ovary. At once we see the characteristic two-celled ovary and thick axile placentas of so many of the *Tubifloræ*. As for the family, the nature of the style and the characteristic fleshy gland at the base of the ovary place the plant in the *Solanaceæ,* in the genus *Fabiana*—a strange genus in comparison with the forms we commonly associate with this family.

By this method of procedure the student should acquire a working knowledge of the characteristic types of flowers of the major categories and orders, and with practice, even of the families and their representative genera. And having acquired facility in recognizing orders and families, he will be spared the laborious task of depending upon a key to place an unknown plant except in connection with genera and species. If we reflect for a moment and try to fix in our minds the outstanding features—or what appear to be outstanding features, judging by the material available—we shall be surprised to discover that relatively few points are necessary to place a flower in its proper category, as was shown in the preceding paragraphs. Not infrequently one single character will settle the problem; for example, we have before us a strange and strikingly rose-like flower, which leads us to suspect rosaceous affinities, but a look at the stamens reveals at once that the flower belongs in the *Melastomataceæ*. Of course this illustration implies some previous experience with melastomataceous forms, nevertheless it serves to show the significance of a single character as a guide in identification.

CHAPTER XI

PRACTICAL SUGGESTIONS

As a means of stimulating interest in the subject and injecting variety into the laboratory routine, a few suggestions may be offered of a sort that can be conveniently carried out at any time during the year, provided that material be collected and prepared for the purpose. The aim is to broaden the point of view and to develop a certain degree of skill and confidence in dealing with taxonomic problems, while at the same time awakening an appreciation of the problems that confront the taxonomist.

Species and specific distinctions. For the study of this subject certain genera offer excellent material. Flowering and fruiting specimens of as many species of the genus as can be found in the neighborhood should be available. Species of the following genera may be suggested: *Sagittaria, Panicum, Scirpus, Carex, Juncus, Lilium, Erythronium, Calochortus, Trillium, Salix, Populus, Betula, Alnus, Quercus, Ulmus, Polygonum, Rumex, Ranunculus, Anemone, Philadelphus, Ribes, Rubus, Spiræa, Prunus, Rosa, Amelanchier, Cratægus, Trifolium, Acer, Viola, Œnothera, Cornus, Viburnum, Loniçera, Ambrosia, Aster, Helianthus.*

Sagittaria usually may be found to show certain differences in growth-form due to conditions of the habitat, these being sometimes so marked as to suggest different specific affinities. But a study of the structure of the flowers at this point will show the more constant taxonomic characters.

The different species of *Panicum* are very useful for showing specific differences in fruit, glume, and vegetative characters. *Scirpus* illustrates instructive tendencies as regards growth-form and fruit; and similarly for *Carex,* in addition to showing unisexual flowers and monœcism. *Juncus* is excellent for showing specific differences in the seeds and the importance of seed characters in classification (*e.g.,* keying). Many other features may be studied to advantage, but these may be sufficient to indicate the possibilities.

Genera and generic characters and distinctions. Any one of the genera mentioned under the preceding heading are suited to the study of generic characters and the genus concept, provided an adequate number of the species are available. *Ribes, Cornus, Salix, Quercus, Ulmus, Helianthus, Aster, Panicum, Viola, Viburnum, Prunus, Amelanchier, Ranunculus,* and *Anemone* are particularly advantageous. For *Ribes* both currants and gooseberries should be available. This genus illustrates certain evolutionary tendencies with respect to the development of the hypanthium that are especially well marked (see text). In this respect it may be compared with *Œnothera.* The different species of *Prunus,* in like manner, illustrate similar tendencies in a perigynous type of flower. *Anemone* and *Ranunculus* will show tendencies in the development of the receptacle. The different species of *Acer* will show presence or absence of the disk, and differences in its configuration when present; it is also an excellent genus for studying specific differences in leaf characters and the tendencies toward diœcism. *Cornus* shows presence or absence of the involucre, and in those species in which the involucre is absent, we encounter a technical group in which species are rather difficult to distinguish, compelling us to consider not only floral characters but the characters of leaves and fruits as well. The genus should be compared with *Viburnum* and *Symphoricarpos.* In *Quercus,* the oaks, we have another rather technical group, in which not only the flowers but the leaves, buds, and acorns and their cupules must be considered. *Aster* will test one's ability to diagnose the specific characters, as the genus is a very technical one. *Helianthus* is somewhat less difficult although quite technical as regards flowers, fruit, and capitula, and even as to the leaves in several cases. It is a genus in which some interesting diagnostic features will be found in the underground parts. *Salix* offers opportunity to study the time of appearance and of flowering of the catkins in relation to the time of appearance of the leaves, and is excellent for studying taxonomic features in both staminate and carpellate flowers. In all cases, leaf characters and growth-form should be included in the study, and the underground parts also where possible. In connection with trees and shrubs, the buds and twigs in the winter state offer material for many a profitable exercise in taxonomic diagnosis.

The other genera above mentioned and many others besides are replete with material for studies of this nature, and in addition to what was said at the outset of these suggestions, we may in this way gain an idea of what is meant by taxonomic species and genera and an insight into evolutionary tendencies and bases of phylogenetic connections.

Fruit and seed characters. These are often neglected in courses in taxonomy in spite of the important and significant rôle they play in all taxonomic work. At all times the morphology and taxonomic characters of the fruit and the seed should be studied along with the flower, if possible. Only in this way can we learn to associate their type with the particular flower that gave rise to them. The genera already mentioned will furnish ample material for studies along this line.

Leaf characters. Leaf characters are important but sometimes confusing. But they are what the average person most generally relies upon in telling one plant from another. Some genera are especially suitable for studies of similarities and differences in the characters of the leaves of the various species; among these are *Ribes, Acer, Quercus, Populus, Salix, Betula, Fraxinus, Ulmus, Cornus, Helianthus, Trifolium, Rosa, Prunus, Rubus.* In some of these some strikin variations will be encountered. In *Eucalyptus* we find a rather baffling situation. The value of leaf characters will be brought out more forcibly when we come to the problem of keying.

Comparisons. A most important and essential part of good taxonomic work is the making of comparisons. Having carefully assembled all our data, there is no other method by which we can make safe deductions. Thus in studying a species we should compare its characters carefully with the characters of other species of the genus, and generic characters should be compared with the characters of other genera of the particular family. Profitable studies in this respect may be made of different species of *Salix, Populus, Quercus, Betula, Ulmus, Prunus, Rosa, Anemone, Ranunculus, Helianthus, Trifolium,* and various others.

It is also profitable to contrast various genera of the same or of different families, as in this way some of the distinctive characters will be brought out. For this purpose the following groups of genera may be suggested, although other genera may serve equally well. Thus we may contrast the floral and fruit characters of *Iris, Canna,* and *Amaryllis; Lilium, Erythronium, Uvularia, Tulipa, Calochortus,* and *Amaryllis; Allium, Brodiæa, Hyacinthus, Lilium, Erythronium; Ranunculus* and *Anemone* with *Sagittaria* and *Lophotocarpus; Cornus* with *Viburnum, Symphoricarpos,* and *Sambucus; Ribes* with *Fuchsia, Œnothera,* and *Chænomeles; Malus* or *Pyrus* with *Cratægus, Pyracantha, Photinia,* and *Chænomeles; Delphinium, Aconitum,* and *Aquilegia* with *Spiræa* and *Physocarpus; Spiræa* and *Physocarpus* with *Prunus* and *Rosa; Potentilla, Fragaria,* and *Rubus* with *Ranunculus* and *Anemone; Aralia, Daucus,* and *Cornus; Diervilla, Scabiosa,* and *Valeriana,* with *Campanula, Aster, Helianthus,* and *Taraxacum; Convolvulus* and *Solanum* with *Antirrhinum* and *Pentstemon; Sola-*

num, Lycopersicon, Physalis, and *Datura; Betula, Quercus, Fagus,* and *Juglans; Salix* and *Populus* with *Leitneria, Quercus, Betula; Mimosa* and *Cassia* with *Pisum* or *Phaseolus* or *Lupinus.*

Construction of keys. Practice in the making of keys will test one's ability to pick out diagnostic characters as few other practices will. The following problems along this line may be suggested:

1. Construct a key to the trees and shrubs of your neighborhood on the basis of leaf characters.
2. Construct a key to the trees and shrubs of your neighborhood on the basis of bud characters.
3. Using the same plants, construct a key based on fruit characters.
4. Construct keys to the common herbs or "flowering" plants of your neighborhood on the bases of (a) floral characters; (b) fruit characters; (c) seed characters; (d) leaf characters; (e) color of the flowers.

Suggestions as to field studies.

What plants of your locality produce rosettes?

What species are weeds? To what families do they belong? Which are weeds of yards? of gardens? of grain fields? Which are annuals? Which are biennials? Which are perennials? In what countries are they native?

What species of your locality are relatively rare?

What species are in danger of extermination?

What are the characteristic species of a tamarack swamp? of a cedar swamp? of a pine woods? of the river banks? of a maple-basswood or maple-beech woodland? of an oak knoll?

What species are shore plants? Which are aquatic?

What species of your locality are wind-pollinated?

What genera of your locality are represented by the most species? Are any of the genera monotypic? What families are best represented with respect to species and genera?

In what trees and shrubs of your locality do the flowers open before the leaves? with the leaves? after the leaves?

What species are the first to blossom in the spring? Which blossom during the winter months?

What species of your locality have a wide geographic distribution? Which have a relatively restricted distribution?

What species of your locality are introduced?

Suggestions concerning the herbarium. Collect and prepare herbarium specimens of

(1) the juvenile and adult shoots of trees and shrubs of your locality;
(2) species showing variations in the shapes of the leaves;
(3) the seedlings of the trees and shrubs and other plants of your locality (construct a key to these plants based on the characters of the seedlings);
(4) specimens showing the various types of inflorescences;
(5) specimens showing various forms of rhizomes, stolons, runners, stem-tubers, root-tubers, etc., etc.;
(6) specimens showing leaf and stem tendrils;
(7) specimens showing cleistogamous flowers;
(8) specimens of climbing plants, and of prostrate, carpet- or mat-forming plants;
(9) specimens showing various forms of spines and thorns;
(10) specimens of various rosette plants;
(11) specimens of fruits and seeds showing various devices for securing dispersal.

CHAPTER XII

THE ANGIOSPERMS

Subdivision Angiospermæ, the Angiosperms. The Angiosperms or flowering plants constitute the higher group of the seed plants or Spermatophytes (*Spermatophyta*), the lower group being the Gymnosperms. The Angiosperms differ from the Gymnosperms in bearing the ovules in a closed "spore-leaf" (*sporophyll*) or *carpel,* whereas in the Gymnosperms the ovules are borne on an open sporophyll. The characteristic organ of the Angiosperms is the *flower.* Pollination is effected through the agency of wind, insects or other animals. The pollen is received on the *stigma* of the carpel or *gynæcium* and fertilization is effected by the development of a *pollen-tube* which grows downward through the style to the *embryo-sac* in the ovule where a *sperm-cell* from the pollen-tube unites with the *ovum* or *egg-cell* lying at the micropylar end of the embryo-sac. A second sperm-cell from the pollen-tube unites with another cell (or nucleus) lying in the center of the embryo-sac to form a fusion-nucleus from which the *endosperm* is developed. This other nucleus with which the second sperm-cell united was itself a fusion-nucleus, having been formed by the fusion of a nucleus from the micropylar end of the embryo-sac with a nucleus from the opposite or antipodal end of the embryo-sac. From the fertilized egg, after a series of preliminary cell-divisions, is developed the *embryo.* The embryo with the investing endosperm, nucellus, and integuments of the ovule, forms the seed (Fig. 34).

The Angiosperms are divided into two classes, the *Dicotyledones* and the *Monocotyledones.* These classes are distinguished from each other primarily by the number of cotyledons in the embryo, the former having two and the latter one. There are, however, some exceptions to the rule among the Dicotyledons. Thus, among the *Proteales,* from three to eight cotyledons are to be found, while in the *Balanophoraceæ* and certain other groups the cotyledons are wanting. Cotyledons are also wanting in some Monocotyledons (orchids).

CHAPTER XIII

DICOTYLEDONES

(THE DICOTYLEDONS)

Embryo with two cotyledons. Flowers differing widely in the number of parts but in general pentamerous. Leaves usually pinnately veined but palmate venation common. Stems with open vascular bundles and a definite cambium through the activity of which growth in thickness takes place. Plants of diverse habits.

The subclasses of the Dicotyledons are the *Archichlamydeæ* and the *Metachlamydeæ*. These are distinguished on the characters of the corolla. In the *Archichlamydeæ,* the corolla when present consists of distinct petals. In the *Metachlamydeæ* the petals are "united," that is, the corolla is a one-piece, more or less tubular, organ which is lobed above. Each of these subclasses is conveniently subdivided into two lower categories, which are characterized in the text following.

THE ARCHICHLAMYDEÆ

Dicotyledons with a simple undifferentiated perianth or with calyx and corolla, in either case the segments distinct (*choripetalous, polypetalous*), rarely united. Perianth in some cases wanting or greatly reduced. Integuments two or one.

According to the nature of the perianth the *Archichlamydeæ* are conveniently subdivided into two groups, the *Monochlamydeæ* and the *Dialypetalæ*. It is to be understood, however, that the component orders of these two subdivisions do not in all cases represent phylogenetically related groups; but the grouping is a very convenient one for practical purposes. Many members of one subdivision show affinities with those of the other, and members of both show affinities with the *Metachlamydeæ*.

The Monochlamydeæ

Flowers with a predominantly undifferentiated perianth, calyx and corolla appearing only in the higher orders. Perianth wanting in some members. Stamens as many as the perianth-segments or more numerous, rarely fewer. Wind-pollination predominant in the lower orders. The lower orders are also characterized by the predominance of the woody habit (trees and shrubs) and in these members the "catkin" or *ament* type of inflorescence is a marked characteristic along with unisexual flowers, although unisexual flowers occur also in the higher forms. Epigyny is attained in some orders. The question as to whether the various orders of the *Monochlamydeæ* are primitive or specialized as regards their floral structures has been much in dispute. That the epigynous members represent a more advanced type than the hypogynous ones seems certain. Because members of certain different orders bear aments, they are not for this reason to be regarded as related.

ORDER VERTICILLATÆ (CASUARINALES)

Flowers minute, very simple, unisexual and monœcious or diœcious, each seated in the axil of a small bract. Staminate flowers in whorls in a long, slender, linear or club-shaped, terminal inflorescence, the subtending bracts united to form a protecting involucre. Each flower consists of 1 stamen and 2 minute scale-like perianth-segments below which are two bracteoles (Fig. 136). Carpellate flowers in subglobose or ovoid heads on short lateral spur-shoots, each flower subtended by a pair of minute bracteoles. Perianth wanting. Gynœcium 2-carpellate, with two long filiform styles which are stigmatic for the greater portion of their length. The carpels are medianly placed and only the anterior one develops. Ovary 1-celled by the suppression of the posterior carpel. Ovules 2-4, erect, orthotropous. Fruit dry and indehiscent, invested by the two bracteoles, which at this stage are woody. The entire cone-like cluster of these small fruits forms a multiple or collective fruit. Seeds at length without endosperm. Embryo straight, with two large flat cotyledons and an upwardly directed hypocotyl. A group remarkably different from other flowering plants as regards the development of the sexual or gametophytic generation.

One family: *CASUARINACEÆ,* with characters as above, and one genus, *Casuarina,* of about 40 species of trees of Australia and the Indo-Malayan region. In Australia the trees are called oaks, from the appearance of the wood. Several species are cultivated in California and the Southern States. In aspect the trees resemble pines, for which they are often mistaken. They are frequently planted as avenue trees in California, a purpose for which they are suitable where not exposed to strong winds. The striking feature of the trees

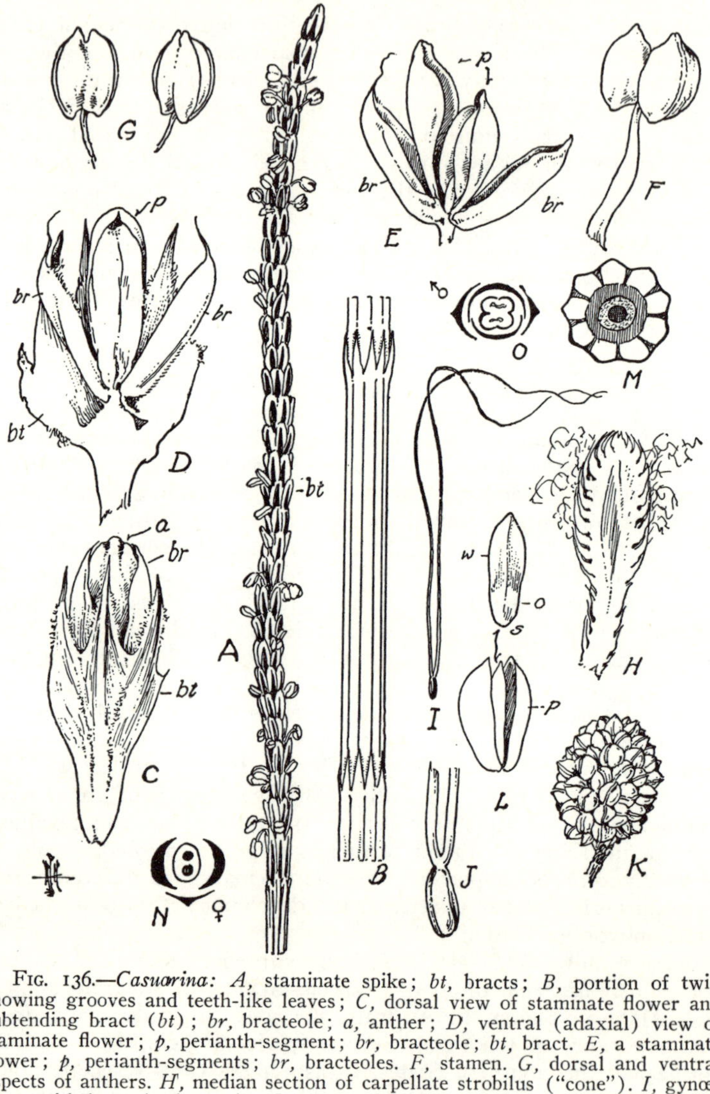

FIG. 136.—*Casuarina: A,* staminate spike; *bt,* bracts; *B,* portion of twig showing grooves and teeth-like leaves; *C,* dorsal view of staminate flower and subtending bract (*bt*); *br,* bracteole; *a,* anther; *D,* ventral (adaxial) view of staminate flower; *p,* perianth-segment; *br,* bracteole; *bt,* bract. *E,* a staminate flower; *p,* perianth-segments; *br,* bracteoles. *F,* stamen. *G,* dorsal and ventral aspects of anthers. *H,* median section of carpellate strobilus ("cone"). *I,* gynœcium, with long slender styles. *J,* ovary, with base of styles attached. *K,* mature strobilus. *L,* fruit, with investing bracteoles (*p*); *s,* fruit, —*o,* ovary, *w,* wing. *M,* transverse section of stem. *N,* diagram of carpellate flower. *O,* diagram of staminate flower.

is the grooved horsetail-like (*Equisetum*-like) green twigs which have the appearance of pine needles. The whorled leaves are reduced to minute scales (Fig. 136b). This is an isolated group, believed by some to show affinities with the Gymnosperms.

ORDER PIPERALES

Flowers minute, bisexual or unisexual, monœcious or diœcious, regular, hypogynous, without perianth, each borne in the axil of a bractlet and aggregated in spikes or catkins, or sometimes in racemes. Bracteoles wanting. Stamens 1-10. Gynœcium of 1-4 free or united carpels, usually with 1 orthotropous ovule in each carpel. Integuments 2 or 1. Embryo small, at apex of a copious endosperm. Small trees, shrubs, or herbs. By Engler regarded as a very primitive group of Dicotyledons, by Wettstein as of doubtful affinities, although the structure of the flower would appear to connect them with the *Polygonaceæ,* and serum diagnosis points to affinities with the *Myricaceæ* and the *Fagaceæ.*

Families: *Piperaceæ, Saururaceæ, Chloranthaceæ, Lacistemaceæ.*

KEY TO THE FAMILIES

Stamens 1-10.
  Herbs, shrubs, or trees, Carpels 1-4, when more than one then united. Ovary 1-celled, 1-ovuled. *Piperaceæ.*
  Herbs. Carpels 3-4, free or united; when united then with two or more ovules in each carpel. *Saururaceæ.*
Stamens 1-3.
  Carpels 1. Stamens 1-3. Aromatic trees with opposite stipulate leaves. *Chloranthaceæ.*
  Carpels 2-3. Stamens 1. Shrubs or trees with alternate exstipulate leaves. *Lacistemaceæ.*

PIPERACEÆ (*Pepper Family*). Flowers in dense fleshy spikes (catkins) or sometimes in racemes, bisexual or unisexual, regular. Herbs, woody climbers, shrubs or small trees. Leaves alternate, opposite, or whorled, with broad, sometimes very large, and often pellucidly dotted, blades. Species about 1,300, in about 9 genera, mostly of the tropics. The most important genera are *Piper* (Fig. 2) with 600—700 species, and *Peperomia,* with about 500 species.

Several species of *Piper* (Pepper) are cultivated, among which are *Piper nigrum,* the pepper of commerce; *P. Cubeba* (Cubeb) and *P. ornatum;* these are climbing plants. The fruits are small, somewhat fleshy, berries. The genus is rich in aromatic properties.

FIG. 137.—*Anemiopsis californica: A,* habit sketch. *B,* young plants from stolon (*s*). *C,* inflorescence; *p,* involucre. *D,* a cluster of flowers as seen in superficial view of inflorescence. *E,* median section of inflorescence (spike); dark spots ovaries immersed in axis of the spike. *F,* section of margin of spike showing stamens and styles. *G,* detail of *F,* showing stamens, two styles and corresponding immersed unilocular ovary with three placentas. *H,* detail of an ovary with one placenta,—the style to right above. *I,* detail of an open carpel, the ovary with two placentas but only one (compound ?) style. *J,* style, showing long stigmatic surface. *K,* dorsal aspect of an anther. *L,* an ovule.

Peperomia comprises mostly low, annual or perennial, somewhat succulent herbs that propagate by tubers or rhizomes. Several species are cultivated (*Peperomia Sandersii, P. nummularifolia, P. blanda*).

SAURURACEÆ (*Lizard's-Tail Family*). Flowers bisexual, in spikes. Gynœcium of 3—4 free or united carpels, each with 2 or more ovules borne on the ventral suture of the carpel when the carpels are free, but on parietal placentas when the carpels are united—these placentas corresponding to the fused margins of the adjacent carpels. Ovary superior, becoming a capsule or berry. Perennial herbs. The known species number only four.

Genera: *Saururus, Anemiopsis, Houttuynia.* The two last-named are monotypic.

Saururus (Lizard's-Tail) is represented in Eastern United States by *S. cernuus,* a slender but tall herb of swamps and shallow waters. The flowers are borne in a long slender peduncled spike which is drooping at the tip. Leaves broadly ovate-cordate and entire. Fruit somewhat succulent but becoming strongly wrinkled in drying. The only other species of this genus, namely *S. Loureirii,* occurs in the Philippines and Japan.

Anemiopsis californica (Yerba Mansa) is a vigorous, broad-leaved herb, sometimes growing to two feet in height, occurring in saline and low grounds from Southern California to Utah and Texas and Northern Mexico. The leaves are basal, ovate to oblong, with stout sheathing petioles, arising from a stout rhizome. The sap of the plants is aromatic and astringent. The small flowers are densely crowded in a conical spike which is subtended by a number of showy white involucral bracts, the whole inflorescence bearing a marked resemblance to the flower of some anemones (*e.g., Anemone canadensis*). Each flower, except the lowest ones, is subtended by a small white bract. The inflorescence is borne on a scapose stem bearing a spathe-like leaf near the middle. The plants propagate by means of stolons and as a consequence are usually in more or less dense colonies which are very conspicuous when the plants are in bloom. A distinctive feature of the flowers is the complete immersion of the ovary in the tissues of the axis of the spike, with which it is consolidated. The carpels may be open on the ventral side and thus incompletely united ventrally. Numerous minute ovules are suspended from thick placentas that project into the cavity of the ovary. The styles are thick and conical, with thick, elongate, fleshy stigmas on the ventral side (Fig. 137).

The geographical distribution of these genera brings to our attention at the very outset an instance among many others of some remarkable affinities existing between certain elements in the flora of North America and in that of eastern Asia. Throughout the pages of this volume attention is called here and there to a number of other similar instances. It will be observed that one of the genera under discussion above, namely *Saururus,* is represented by one species in eastern North America and by another in eastern Asia, two geographic regions distantly remote from each other. As for the other two genera, each of which contains but a single species, one, *Anemiopsis,* occurs in southwestern United States and adjacent Mexico while the other, *Houttuynia,* is Asiatic, being found from the Himalayas to Japan. The significance of these interrupted ranges has an immediate bearing on the problem of the evolutionary history and relationships of these forms. That these plants are closely related is unquestionable, as evidenced by the structure of the flowers and even as it appears from the gross vegetative features; in fact, *Anemiopsis* and *Houttuynia* are very similar as to the form of the inflorescence. But how did they come to be so widely separated? On the theory of spontaneous generation their presence in their respective regions might be easily explained, but the more likely explanation would appear to be that the areas which they now inhabit were at one time in the remote geological past continuous, and when the North American and Asiatic continents were separated the family was, so to speak, broken up. The close affinities of these plants are, therefore, not surprising; undoubtedly they have sprung from some common ancestral form that originated at some particular place from which the descendents spread and became differentiated until finally separated by changes in the land masses.

CHLORANTHACEÆ. This is a small family of tropical and subtropical herbs, shrubs, and trees, with opposite, simple, stipulate leaves and small bisexual or unisexual flowers in spikes or cymes. A rudimentary perianth is present in some forms. Stamens 1 or 3; when three they are united with each other (monadelphous) and with the ovary. The gynœcium consists of but 1 carpel. Ovary superior, with a single ovule suspended from the top of the ovary-cavity. Fruit drupaceous. Seed with copious oily endosperm at one end of which is a minute embryo.

Three genera: *Chloranthus,* with about 10 species in eastern Asia and the East Indies; *Ascarina,* with 3 species on certain islands of the Pacific; *Hedyosmum,* the largest genus, with about 20 species in tropical America.

LACISTEMACEÆ. Comprises only one genus, *Lacistema,* with about 16 species of shrubs and small trees with alternate exstipulate leaves, distributed over tropical America from Mexico and the West Indies to Brazil and Peru. Flowers small, bisexual, in spikes. Perianth present or absent. Stamens 1. Gynœcium 2—3-carpellate, syncarpous. Ovary superior, with 1—2 ovules suspended from the top of the ovary-cavity. A distinctive feature of the flowers is a cup-shaped annular disk which is conspicuously developed.

ORDER SALICALES

Flowers hypogynous, unisexual, diœcious (rarely monœcious), naked, in catkins (aments), each flower subtended by a small bract. Staminate flowers with 1 to several free or united stamens. Carpellate flowers with 1 to several united carpels. Fruit a 1-celled pod-like capsule with parietal placentation and numerous small seeds furnished with a silky down springing from an annular base. Integuments 2. Trees and shrubs mostly of the northern hemisphere. In a number of forms the catkins are androgynous. Stipules present but often fugaceous.

One family: *SALICACEÆ* (Willow Family) with 2 genera, *Salix* (Willow) of 200 or more species, and *Populus* (Poplar) of about 30 species.

Salix is widely distributed through north temperate and arctic climates, with a few in the southern hemisphere. The species range in size from a few inches high (arctic forms) to tall trees, with leaves of various shapes from short and broad to long and narrow. The wood is soft. The branches of some species are lithe, in others they are more rigid. The buds of all willows are covered by a single bud-scale. The unisexual flowers are borne in erect or ascending catkins which appear before, with, or after the leaves. Staminate flowers with 2—10 (rarely 1) free or more or less united stamens. The very simple carpellate flowers consist of a 2-carpellate, syncarpous gynœcium with 2 stigmatic lobes at the summit of a short or elongate style. The gynœcium is elevated on a stalk (gynophore) of various lengths. Both staminate and carpellate flowers bear one or more peculiar glandular bodies at the base of the stamens or the gynophore, and both are subtended by a small bract. All willows are insect-pollinated. The anthers give a bright yellow color to the expanded staminate catkins, while the carpellate catkins are greenish.

The willows are a technical group, and they easily hybridize, which adds to the difficulties in systematic studies. The leaves, how-

FIG. 138.—*Salix lasiolepis* (?) : *A*, staminate twig. *B*, *a*,—portion of a staminate branch at later stage; *b*, terminal portion of a carpellate branch. *C*, a carpellate catkin. *D*, a staminate flower with subtending bract (*br*) and posteriorly placed gland (*gl*). *E*, a carpellate flower (late stage) with subtending bract (*br*) and posteriorly placed gland (*gl*) ; *s*, bud-scale.

ever, are generally recognizable, but the most reliable characters reside in the flowers, fruits, and seeds.

The stipules of willows are more or less prominent and specifically distinctive but fugaceous, so that for accurate study specimens in

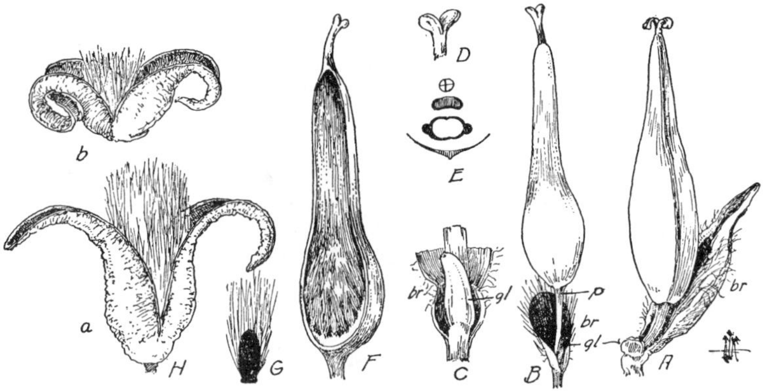

FIG. 139.—*Carpellate* flowers and fruits of two different species of willow (*Salix*), showing differences in structural details. *A*, gynœcium with short style and relatively short pedicel. *B*, gynœcium with long style and long pedicel. *C*, basal portion of *B*. *D*, stigma of *B*. *E*, floral diagram of *B*. *F*, a nearly mature fruit in section, with seeds below. *G*, a seed. *H*, *a*, *b*, fruits of the two species, dehiscing; *br*, bracts; *gl*, glands.

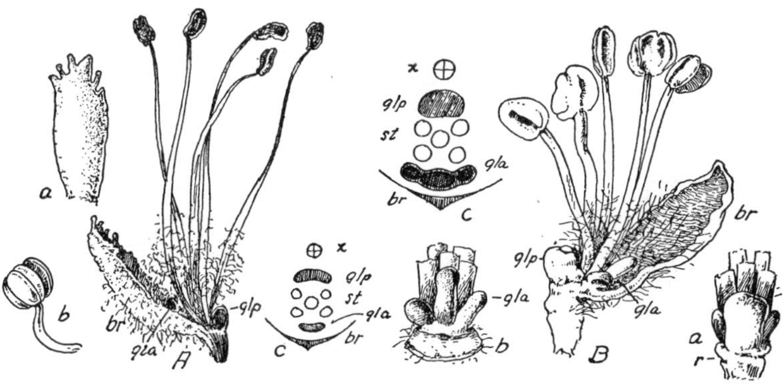

FIG. 140.—Staminate flowers of two different species of willow, showing differences in stamens, glands, and subtending bracts. *A*: *a*, bract; *b*, stamen, early stage; *c*, floral diagram; *br*, bract; *st*, stamens; *gla*, anterior gland; *glp*, posterior gland; *x*, axis of catkin. *B*: *br*, bract; *gla*, anterior gland; *glp*, posterior gland; *a*, enlarged view of posterior gland (three-lobed); *r*, receptacle; *b*, enlarged view of three anterior glands (*gla*); *c*, floral diagram of *B*; *br*, bract; *st*, stamens; *x*, axis of catkin; *gla*, anterior and *glp*, posterior glands.

various stages of development are desirable. In some cases the leaf-forms intergrade with those of poplars. Numerous willows are found in our flora. They are characteristic plants of river-bottoms, lake shores, marshes, and mountains. Several species are grown for various purposes, as *Salix viminalis,* the Osier or Basket Willow, of Eurasia,

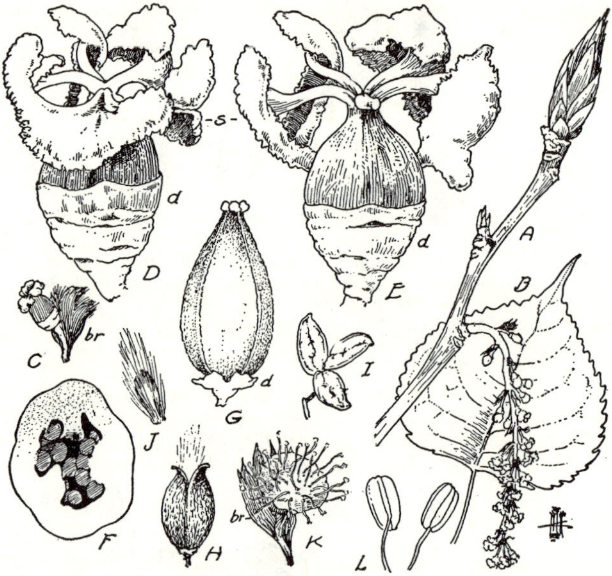

FIG. 141.—*Populus deltoides: A,* a twig with terminal bud and carpellate catkin. *B,* leaf. *C,* carpellate flower with subtending bract. *D,* carpellate flower, much enlarged. *E,* same with one style-branch removed; *s,* lobed stigmas; *d,* cup-like disk. *F,* transverse section of ovary, with three parietal placentas, corresponding to the three carpels. *G,* the three-carpellate gynœcium, late stage; *d,* disk. *H,* fruit dehiscing. *I,* fruit after shedding of seeds, showing three valves corresponding to the three carpels. *J,* a seed. *K,* cup-like disk of a staminate flower, with filaments of stamens, and subtending bract (*br*). *L,* dorsal and ventral aspects of stamens.

used in basketry. *S. pentandra,* another Eurasian species, was introduced in early times into New England, mainly for ornamental purposes, as were also *S. alba* (White Willow) of Europe, and the Chinese *S. babylonica* (Weeping Willow). Monœcious flowers occur in *S. Medemii* of Persia, and bisexual flowers in *S. Caprea* of southeastern Europe. One of the handsomest of our willows is *S. lucida,* the Shiny-leafed Willow of northern lake-shores (Figs. 138, 139, 140).

The poplars (*Populus*), including cottonwoods, are all trees, sometimes of large size. They differ from the willows in having more or less ovate, cordate, or deltoid leaves, drooping and more robust catkins, and buds covered with hairy or mucilaginous scales. The staminate flowers consist of many distinct stamens borne in a cup-shaped disk, and the carpellate flowers of a 2—4-carpellate gynœcium which is much stouter than that of willows, fits closely into a shallow disk, and has thick, fleshy, lobed stigmas on a short, thick style (Fig. 141).

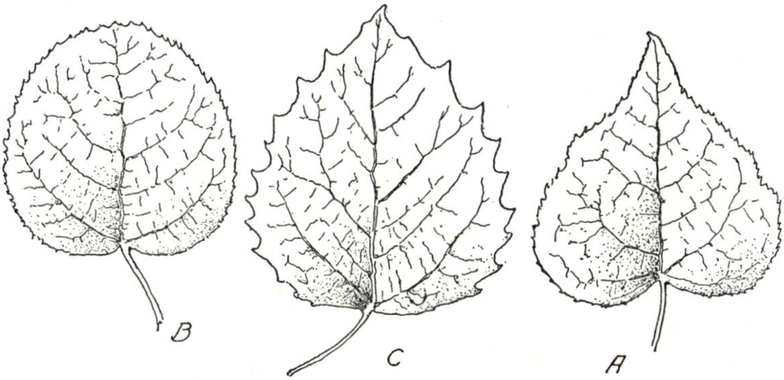

FIG. 142.—Leaves of *Populus grandidentata: A, B,* juvenile leaves of a sprout from the stump. *C,* adult form of leaf.

Two classes of poplars are distinguished, *viz.:* (1) Aspens and (2) Cottonwoods. In aspens the bud-scales are silky-hairy, the stigmatic lobes narrow or filiform, and the capsules thin-walled, whereas in the cottonwoods the bud-scales are glutinous, the stigmatic lobes broad and fleshy, and the capsules thick-walled. The 30 or more species of poplars are natives of the northern hemisphere, with 18 or 20 in our flora, some of which are introduced. The wood of poplars is soft and spongy and is at present used to a considerable extent in the pulp industry and for cheaper grades of lumber. A number are planted as avenue trees, such as *Populus alba* (White or Silver Poplar), a European tree, *P. nigra* var. *italica* (the stately Lombardy Poplar), *P. candicans* and *P. balsamifera* (the Balm of Gileads or Balsam Poplars), and the natives *P. deltoides* and *P. trichocarpa* (Western), the common cottonwoods. Our commonest and most widely distributed species is the American Quaking Aspen, *P. tremuloides,* easily recognized by its small, quaking, dark-green leaves and white-mealy smooth bark. Its wood is now extensively used in the pulp industry. This tree and the cottonwoods were an important source of fuel and timber to the pioneers of the prairies and plains, while the groves of these trees afforded shelter as well. The quaking aspen furthermore plays an important rôle in the repopulating of burnt-over forest lands. *P. grandidentata* (Big-Tooth Poplar) is found scattering or in small stands in the coniferous and mixed forest belts of the upper Mississippi and Great Lakes basins. This tree rarely attains large size.

Willows and poplars sprout readily from the root or stump, as may be seen wherever groves of these trees have been cut over. The sprouts produce a foliage quite unlike that of the adult tree, with leaves sometimes of enormous size (*P. grandidentata*) (Fig. 142). These sprouts (juvenile forms) also show the stipules to advantage. Both willows and poplars are rapid growers, sprouts of some cottonwoods growing to a height of five feet in one season.

The fact that willows and poplars bear catkins does not imply that they are therefore in any way allied to other catkin-bearing groups of the *Monochlamydeæ*. On the other hand their unique floral structures stamp them rather as an isolated group.

ORDER GARRYALES

A small order of evergreen shrubs or sometimes trees, with unisexual diœcious flowers in catkins, the staminate with a simple perianth of 4 (or 3) segments, often united at the tip and between which stand the 4 stamens. The carpellate flowers are naked, with a syncarpous 2—3-carpellate gynœcium with free, spreading styles. The ovary is superior and 1-celled, with 2 anatropous ovules suspended from the top of the cavity. Integuments 1, often incompletely developed. Fruit a berry with a thin pericarp and crowned by the remains of the styles.

One family, *GARRYACEÆ* with one genus, *Garrya,* of the arid parts of the southwestern United States, and Mexico, with one species in the West Indies. Four species occur in California (Fig. 143).

ORDER MYRICALES

Flowers hypogynous (?), unisexual, monoœcious or diœcious, without a perianth, in catkins, each flower in the axil of a small bract. Stamens mostly 4, sometimes 2—16. Gynœcium of 2 united carpels, with 2 elongate styles and stigmas. Fruit nut-like or drupa-

FIG. 143.—*Garrya elliptica: A,* shoot with carpellate catkins. *B,* a carpellate catkin. *C,* two carpellate flowers (gynœcia) invested by connate bracts (*br*). *D,* gynœcium, early stage. *E,* median section of gynœcium, the ovary containing two ovules. *F,* fruit stage.

ceous, with a single seed, and secreting waxy granules on the surface. Endosperm none. Integument 1.

One family, *MYRICACEÆ* (Sweet-Gale Family). Characters of the order. In the broadest sense one genus, *Myrica,* with about 60 species of wide distribution. The plants are shrubs or trees with simple or pinnately incised leaves containing aromatic resin-glands. Androgynous catkins are produced in some species, and sometimes bisexual flowers. *Myrica Gale* (Sweet-Gale), a low shrub, occurs in the Great Lakes and St. Lawrence basins and south to Virginia, as well as in Europe and northern Asia. *Myrica asplenifolia,* or *Comptonia asplenifolia* as it is preferred by some writers, ranges from the Atlantic to Saskatchewan. This is a low sweet-scented shrub with deeply cut fern-like leaves (hence the common name Sweet-Fern), found in dry sandy pine or mixed woods. The carpellate catkins develop as bur-like "fruits." This is the only species in which stipules occur, but these soon fall. The order is of doubtful affinities, perhaps an isolated one, although in some works it is treated as connected with the birches.

ORDER BALANOPSIDALES

Flowers unisexual, diœcious, the staminate in spikes, with a simple perianth, the carpellate solitary and naked but subtended by a number of bracts. Gynœcium 2-carpellate, syncarpous. Fruit a drupe subtended by a cupule of loosely imbricated bracts.

One family, *BALANOPSIDACEÆ,* with one genus, *Balanops* of about 7 species of trees and shrubs of New Caledonia.

ORDER LEITNERIALES

Flowers hypogynous, unisexual, diœcious, in catkins which appear before the leaves. Staminate flowers with 3—12 stamens and without perianth. Carpellate flowers with a single carpel and a much-reduced perianth of united segments. Ovule 1, on a dorsal placenta. Fruit drupaceous. Endosperm thin and fleshy. Integuments 2.

One family, *LEITNERIACEÆ* (Cork-Wood Family) with the monotypic genus *Leitneria.* The solitary species, *Leitneria floridana* (Cork-Wood), is a tree-like shrub inhabiting the swamps of southern parts of the United States from Missouri to Florida and Texas. The plant suggests a willow but is thought by some writers to be related to the *Myricaceæ,* while on the other hand it suggests a connection with the Sweet Gum (*Liquidambar*) and to the Plane-Trees (*Platanus*). (Fig. 144.)

ORDER JUGLANDALES

Flowers epigynous, unisexual, monœcious, naked or with a simple, usually undifferentiated, perianth. Staminate flowers in catkins; stamens 4—40. Carpellate flowers solitary or in small clusters or spikes. Gynœcium 2-carpellate, syncarpous, with conspicuous stigmas (in

FIG. 144.—*Leitneria floridana: A,* staminate shoot and catkins. *B,* carpellate catkin, with styles projecting above. *C,* staminate catkin enlarged; *br,* bract; *s,* staminiferous scales. *D,* staminiferous scale and stamens. *E,* carpelliferous scale and carpellate flower (consisting of gynœcium only). *F,* a cluster of mature fruits. *O,* ovule.

some cases several carpels are present). Ovary inferior, 1-celled enclosing a single basal ovule. Endosperm none. Integuments 1. Fruit drupaceous (spurious), with a hard crustaceous shell. Trees with alternate pinnately compound exstipulate leaves. Petals present in the carpellate flowers of some species.

One family, *JUGLANDACEÆ* (Walnut Family). Characters of the order. Over 30 species in 6 genera, chiefly of north temperate climates but extending south along the cordillera to Bolivia. Our genera are *Juglans* (Walnuts and Butternuts) and *Carya* (Hickories). These differ in certain characters of the fruit and inflorescence; in

Juglans the husk or shell of the fruit is indehiscent, and the staminate catkins simple, while in *Carya* the husk splits into 4 valves (Fig. 46), and the staminate catkins are branched. *Juglans cinerea,* the Butternut, is distinguished from walnuts by its ellipsoid viscid-glandular pointed fruit and deeply and irregularly grooved (black) nut. In *Carya* the fruit is globular or oblong. The cotyledons in both genera are more

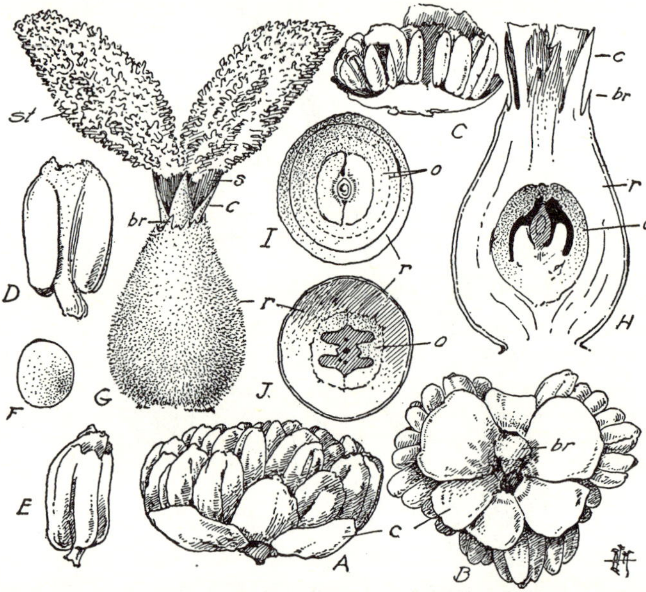

FIG. 145.—*Juglans: A,* side view of staminate flower; *B,* same viewed from below; *c,* perianth segments; *br,* subtending bract of flower. *C,* sectional view of staminate flower. *D, E,* stamens. *F,* pollen-grain. *G,* carpellate flower; *br,* bracts; *s,* style; *st,* coarsely papillate stigma; *c,* calyx. *H,* median section of carpellate flower; *c,* calyx; *br,* bract; *r,* receptacle, coalesced with ovary (*o*). *I,* early, and *J,* later stage of ovary in transverse section; *o,* ovary with ovule within; *r,* receptacle.

or less corrugated or convoluted and (in *Juglans* in particular) are separated by false partitions from the wall of the "stone"; they remain in the ground (hypogeal) on germination. (Figs. 145, 146.)

The butternut, walnuts, and hickories can all be distinguished by their leaves, fruits, and bark. Some species, *e.g.,* hickories, are sometimes confused with ashes, but the ashes have opposite leaves and winged fruits.

The outstanding characters of the order are the staminate catkins associated with monœcism, epigyny, union of the perianth and bracteoles with the subtending bract, and reduction of the number of carpels and ovules. The stigmas are usually large and colored, with coarsely papillate ridges or lamellæ. All are wind-pollinated. Walnuts of commerce are the product of *Juglans regia* or crosses. Pecans come from *Carya Pecan*.

FIG. 146.—Leaf-bud of *Carya ovata*: *s*, foliaceous bud-scales; *l*, leaf-scars.

ORDER BATIDALES

Flowers unisexual, diœcious, in catkins (spikes). Staminate flowers in a saucer-shaped, lobed perianth, with 4 stamens and 4 staminodes in the axils of 4-ranked scale-like bracts. Carpellate flowers naked, 2-carpellate, in 4 ranks in axils of deciduous bracts of a short spike but becoming consolidated into one mass. Ovary superior, 2-celled, but becoming 4-celled by development of false partitions. Ovules 4, one in each cell. Endosperm none. Embryo curved.

One monotypic family, *BATIDACEÆ*. Characters of the order. The solitary species, *Batis maritima*, is a low subshrub with linear-oblanceolate leaves, occurring on seashores of southern and Lower California, the Hawaiian Islands, and from Florida to Brazil. Apparently an isolated type.

ORDER JULIANIALES

Flowers unisexual, diœcious, small, green, and inconspicuous, the staminate in catkins or racemes, with a perianth of 3—9 segments and as many stamens, the carpellate naked, 3 or 4 together in a more or less fused mass in an involucre. Gynœcium 3—5-carpellate. Ovary superior, 1-celled, with a solitary ovule. Fruit aggregate, nut-like, invested by the persistent involucre and consolidated with the axis of the inflorescence. Endosperm none. Ovule with 1 integument and a peculiarly enlarged funicle.

One family, *JULIANIACEÆ*. Characters of the order. Small resinous trees and shrubs with pinnately compound leaves.

Genera two: *Juliania,* with 4 species in Mexico, and *Orthopterigium,* with 1 species in Peru. Apparently an isolated group.

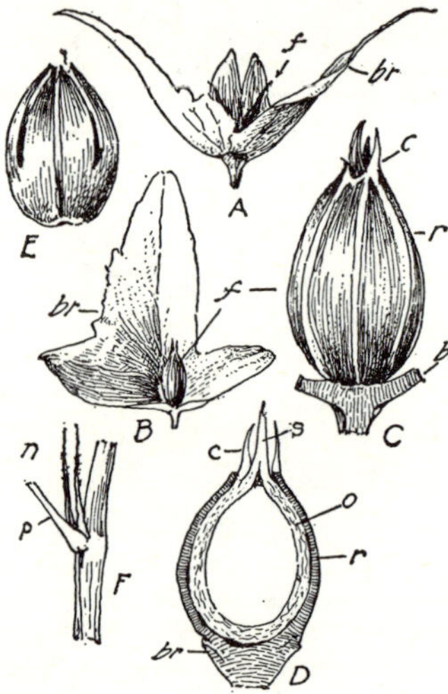

FIG. 147.—*Carpinus virginiana: A,* carpellate flowers (*f*) and subtending bracts (*br*). *B,* a carpellate flower (*f*) and subtending bract (*br*). *C,* gynœcium enlarged; *c,* calyx; *r,* receptacle; *br,* bract. *D,* sectional view of *C; s,* style; *c,* calyx; *o,* ovary; *r,* receptacle; *br,* bract. *E,* ovary after removing husk (receptacle). *F,* portion of twig showing stipules (*n*) of leaf; *p,* petiole.

ORDER FAGALES

Flowers epigynous (except the staminate), cyclic, unisexual and monœcious, rarely bisexual or diœcious; small, greenish, and inconspicuous, in compound inflorescences. Perianth of free or united segments, rarely wanting, that of the staminate flowers adnate to the subtending bract, the carpellate surrounded by the bract and bracteoles. Gynœcium of 2—6 united carpels with distinct styles. The inferior ovary with 1 or 2 pendulous ovules in each cell. Endosperm none. Integuments 1 or 2.

Only the families in the accompanying key.

Staminate and carpellate flowers both in catkins or the carpellate few together at the ends of shoots of the season; both kinds in the axils of scaly bracts. Integuments 1. *Betulaceæ.*

Staminate flowers only in catkins, the carpellate solitary or few together, subtended by an involucre of few to numerous scaly bracts, which in the fruit becomes a bur or cup. Integuments 2. *Fagaceæ.*

BETULACEÆ (*Birch Family*). Monœcious trees and shrubs with the flowers in compound inflorescences (catkins). Perianth when present of small, free or united, scale-like segments, in some forms greatly reduced, or obsolete in the carpellate flowers. Staminate flowers of 2—10 or 12 stamens, the flowers adnate to their subtending bract and bracteoles. Ovary 2-celled, containing a pendulous anatropous ovule. Fruit a 1-seeded nutlet. Embryo with large oil-containing cotyledons. About 75—80 species in 6 genera, widely distributed in the northern hemisphere, with a few in the southern hemisphere.

Two subfamilies, *Coryloideæ* and *Betuloideæ*, characterized as follows:

Coryloideæ. Staminate flowers without perianth, solitary in a simple spicate inflorescence (catkin). Stamens divided. A perianth is present in the carpellate flowers, and these are surrounded by a husk (cupule) formed of the united bracteoles and subtending bract. *Ostryopsis, Ostrya, Carpinus, Corylus*.

Betuloideæ. Staminate flowers 3 together in the axil of the cover-scale (fused bract and bracteoles). Perianth present in staminate flowers. Carpellate flowers without perianth, 2 or 3 together, subtended by bracteoles and bract. In fruiting stage, bracteoles and bract are fused to form the cover-scale. Bisexual flowers are said to occur. *Betula, Alnus*.

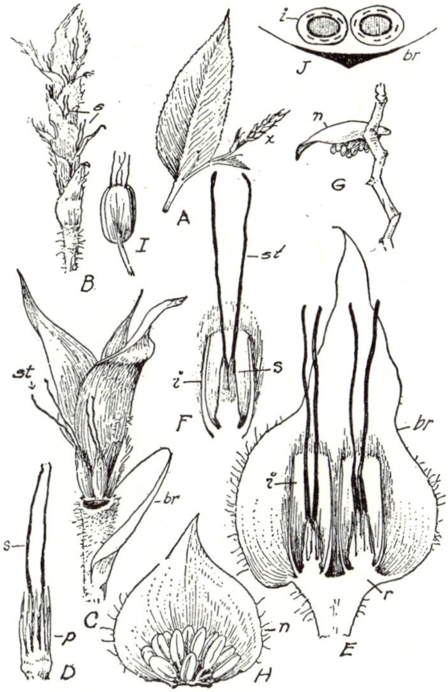

FIG. 148.—*Ostrya virginiana: A*, a twig with carpellate inflorescence at *x*. *B*, carpellate inflorescence, enlarged; *s*, projecting styles. *C*, portion of base of carpellate inflorescence, showing two carpellate flowers; *st*, stigmas; *br*, bract. *D*, a carpellate flower; *s*, stigmas; *p*, perianth. *E*, bract of inflorescence, investing two carpellate flowers; *i*, involucre; *r*, receptacle; *br*, bract. *F*, a carpellate flower, enlarged, with perianth removed; *i*, involucre; *s*, style. *G*, portion of staminate catkin; *n*, staminiferous scale, with stamens beneath. *H*, staminiferous scale (*n*) with stamens. *I*, stamen. *J*, diagram of carpellate flowers in axil of bract (scale) *br; i*, involucre, with ovary and surrounding small perianth-segments within.

The inflorescence (catkin) of the *Betuloideæ* is compound (cymose). The elements (simple inflorescences) of this compound inflorescence consist of 2 or 3 flowers with their bracteoles in the axil of the subtending bract (Fig. 153). The fusion of the bracteoles with the subtending bract produces the 3- or 5-lobed cover-scale.

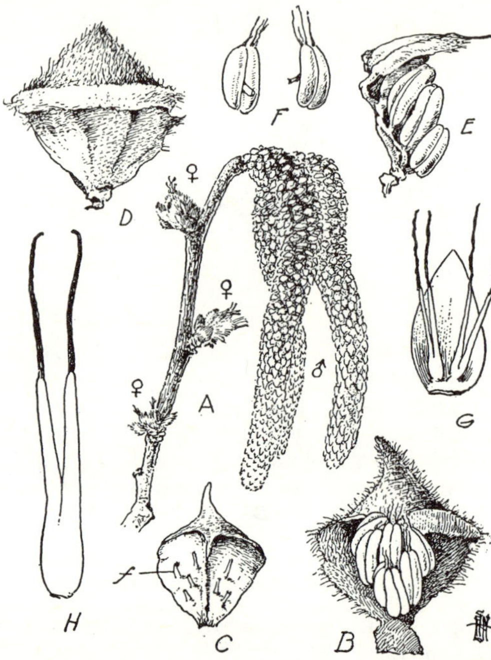

FIG. 149.—*Corylus: A,* twig with staminate (♂) and carpellate (♀) inflorescences. *B,* staminate scale; *C,* staminate scale showing insertion of filaments, with anthers removed. *D,* dorsal aspect of staminate scale. *E,* side view of staminate scale. *F,* anthers. *G,* carpellate scale with two carpellate flowers. *H,* a carpellate flower.

Coryloideæ. Ostryopsis is a monotypic genus of eastern Mongolia and nothern Siberia. The anthers are undivided and the husk consists of 3 segments which enclose the fruit.

Carpinus (Hornbeam) comprises 12 species of Europe and Asia, with one, *Carpinus caroliniana* (American Hornbeam, Water-Beech), in eastern North America. The plants are trees and shrubs with straight-veined leaves, and catkins that expand before the leaves.

The cylindrical staminate catkins are sessile at the ends of lateral branches of the preceding season. The stamens are forked, each fork bearing a solitary anther. Stamens 3—12. The carpellate catkins terminate leafy branches. The 2-carpellate gynœcium becomes an

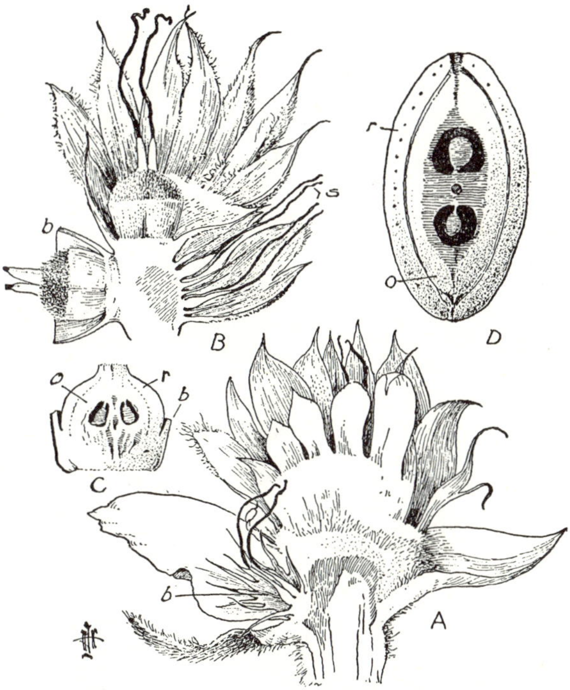

FIG. 150.—*Corylus: A,* section of carpellate inflorescence, with a carpellate flower at lower left and its subtending involucral bract (*b*). *B,* detail of carpellate cluster, with two young carpellate flowers at right, and two more advanced at left and above; *b,* inner bract (involucre); *s,* stigmas, *C,* median section of gynœcium (styles removed); *o,* ovary; *r,* receptacle; *b,* involucral bract. *D,* transverse section of "ovary"; *o,* ovary, coalesced with receptacle (*r*) and two locules within, each containing one ovule.

ovoid ribbed nutlet which adheres to a much-enlarged and variously lobed bracteole that serves as an organ of dispersal. The bracts are deciduous. (Fig. 147.)

Ostrya (Hop-Hornbeam, Ironwood). A genus of trees; 6 species according to some writers, native to Eurasia and North America.

Only one species, *O. virginiana,* occurs in eastern North America, from Canada to Texas. The staminate catkins are borne without protecting bud-scales at the ends of short branches of the preceding season. The filaments are forked, each fork tipped with an anther. The carpellate catkins are on short spur shoots of the season. Flowers in pairs at the base of the subtending bract. Fruit a nutlet enclosed in a conspicuous sac formed of the bracteoles. At maturity the carpellate catkin has the appearance of a cluster of hops (Fig. 148).

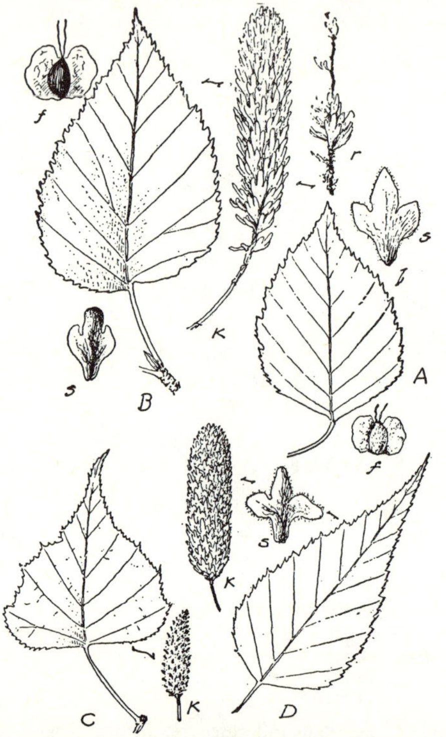

FIG. 151.—Various species of birch: *A, Betula papyrifera; B, Betula papyrifera* var. *cordifolia; C, Betula populifolia; D, Betula cœrulea-grandis. k,* carpellate catkin; *r,* axis of catkin after shedding of scales; *s,* ovuliferous scale; *f,* the winged fruit (spurious).

Corylus (Hazel, Filbert) comprises 7 species of shrubs of Eurasia and North America. The staminate catkins (spikes) are borne on short lateral shoots and pass the winter unprotected by bud-scales. They expand before the leaves in spring. The flowers are solitary in the axil of each bract. Filaments forked, each fork supporting an anther. The carpellate flowers are developed in clusters from scaly buds at the ends of short branches of the season. Two flowers are seated in the axil of each bract and are subtended by bracteoles. The 2-carpellate gynœcia terminate in elongate styles and stigmas. Fruit a nut enclosed in a husk formed of the two subtending bracteoles (Figs. 149, 150).

Two hazels, *Corylus americana* and *C. rostrata,* occur in our flora, the latter with a variety on the Pacific Coast. The two species are

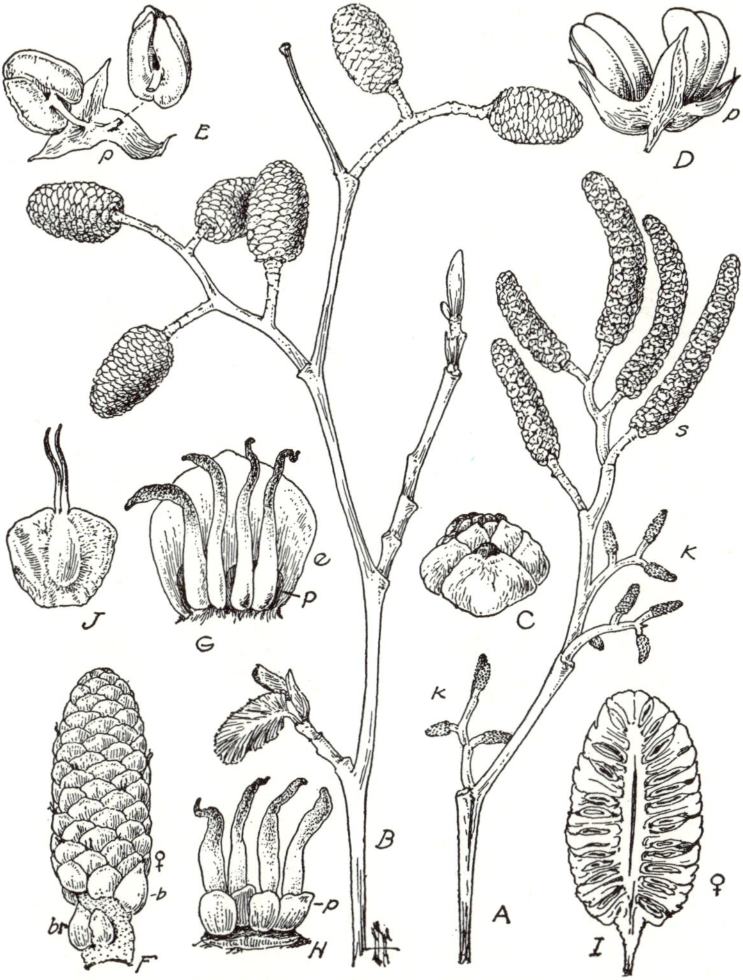

FIG. 152.—*Alnus rhombifolia: A,* twig with staminate (*s*) and carpellate (*k*) catkins. *B,* branch with mature "cones." *C,* staminate flower-cluster with subtending scales. *D,* staminate flower; *p,* perianth. *E,* staminate flower opened; *p,* perianth. *F,* carpellate strobilus ("cone"); *br,* bracts. *G,* ovuliferous scale (*e*) with carpellate flowers and inner small scales (*p*). *H,* dorsal view of carpellate flowers, showing position of small inner scales (*p*). *I,* median section of carpellate strobilus. *J,* "fruit," immature.

readily distinguished when in fruit by the husk, which in *C. americana* is short and broad, in *C. rostrata* elongate, narrow, and beak-like (hence *rostrata,* rostrate); and in addition the foliage of the latter forms a flat spray, while the leaves are trough-shaped in the former.

Betuloideæ. Betula (Birch). About 35 species of trees and shrubs of colder parts of the northern hemisphere. The buds are naked, protected by the stipules only, and the leaves are straight-veined. The catkins expand before or with the leaves. The staminate flowers are

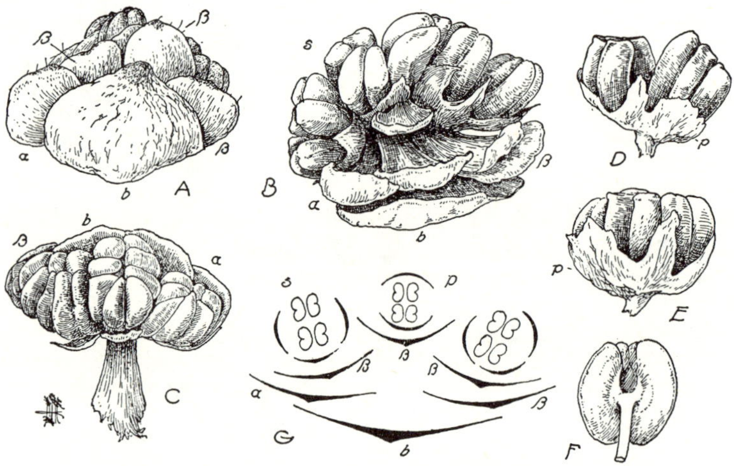

FIG. 153.—*Alnus rhombifolia:* Staminate flowers and subtending scales: *A,* dorsal aspect. *B,* the cluster in *A* laid open. *C,* inner (adaxial) view of staminate cluster. *D, E,* staminate flowers; *p,* perianth. *F,* dorsal view of a stamen. *G,* diagram of a staminate cluster shown in *A* and *B.*

borne 3 together at the base of a 3-lobed cover-scale which is formed of the subtending bract and the bracteoles α and β of the middle flower—in catkins which were developed the previous season. The perianth is membranous and irregularly lobed. There are 2 forked stamens. A like number of carpellate flowers are subtended by a cover-scale of similar origin, but the perianth is wanting. The 2-carpellate gynœcia terminate in elongate styles and stigmas. The fruit is a laterally winged nutlet (samara)—3 nutlets to each cover-scale—which, along with the cover-scale, fall away from the axis of the catkin at maturity (Fig. 151).

Our commonest birch is *Betula papyrifera,* the White, Canoe, or

Paper Birch, a tree closely resembling *B. alba* of Eurasia. The wood is durable and is utilized for various purposes commercially. In the Great Lakes and St. Lawrence basins the tree attains considerable size. *Betula lutea,* the Yellow Birch, is sometimes a larger tree, with darker bark which exfoliates in much thinner strips. The inner bark and twigs possess a marked wintergreen flavor. *Betula pumila,* a low species of bogs and prairies, is common from Newfoundland to Illinois and Minnesota.

Our birches fall into two well-defined classes, Black Birches and White Birches. In the Black Birches the scales of the fruiting catkins are more or less persistent and the bark aromatic. Here belong *Betula lutea, B. lenta* (Black Birch), *B. nigra* (Red or River Birch). In the White Birches the scales of the fruiting catkins are deciduous. that is the "cones" break up, and some differences in the venation of the leaves are to be noted. The species of this class are *Betula papyrifera, B. pendula, B. pumila,* and *B. glandulosa.*

Alnus (Alder). The alders differ from the birches in certain obvious gross features, such as coarser leaves and conspicuous woody persistent carpellate catkins ("cones"). The buds are naked and the catkins are developed the previous season. The staminate catkins are pendulous, with the flowers in clusters of 3 at the base of 5-lobed cover-scales which consist of the subtending bract fused with the α and β bracteoles of the middle flower and the same of the two lateral flowers; the α bracteoles of the lateral flowers are suppressed. A perianth is present. The carpellate flowers are in pairs in the axil of a similarly constructed cover-scale, but the middle flower is suppressed as is also the perianth. The gynœcium is similar to that in *Betula.* The cover-scales persist as the woody scales of the "cone." The 14 or more species of alders are mainly boreal, but two of them press down the cordillera into the Andes (Figs. 152, 153.)

Like the birches our alders are for the most part separable into two classes, *viz.,* Green Alders and Swamp Alders (or Black Alders). The Green Alders are a circumpolar group; their carpellate catkins are protected by scales during the winter, and the staminate appear with the leaves in spring. Their fruits are winged. In the Swamp Alders the carpellate catkins are naked, and the staminate catkins appear long before the leaves. The fruit is wingless or only slightly margined. The Green Alders are *Alnus crispa, A. mollis;* the Swamp Alders, *A. incana* and *A. rugosa,* and the Pacific Coast species *A. oregana* and *A. rhombifolia.* A third group is represented by the Eastern *A. maritima,* the flowers of which appear in autumn from shoots of the season. The Pacific Coast alders named are tall, handsome, smooth-barked trees. The wood of some species is used in the making of charcoal.

FAGACEÆ (*Beech Family*). These are monœcious trees and shrubs with alternate simple leaves. The staminate flowers are small,

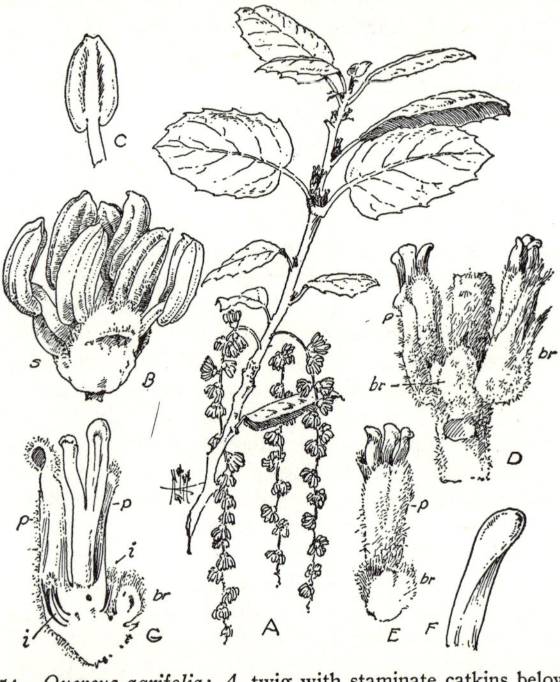

FIG. 154.—*Quercus agrifolia: A,* twig with staminate catkins below and carpellate flowers in axils of upper leaves. *B,* a staminate flower; *s,* perianth. *C,* stamen. *D,* carpellate flowers; *br,* bracts; *p,* perianth. *E,* carpellate flower; *br,* bracts; *p,* perianth. *F,* style and stigma. *G,* section of carpellate flower; *br,* subtending bract; *i,* involucral bracts (cupule); *p,* perianth segments, with two styles between,—united below.

in catkins, with a variously lobed perianth and few to many stamens. The carpellate flowers are solitary or in clusters surrounded by an involucre of bracts which in the fruit forms a cup (or cupule). There is a perianth of 6 segments. The gynœcium is 3-carpellate, with distinct stout styles. The inferior ovary bears 2 pendulous ovules in each cell, with 2 integuments. Endosperm none. The fruit is a 1-seeded nut (spurious fruit), seated in the cup. The embryo is large, with fleshy cotyledons which remain in the ground (hypogeal) in germination (except in the beech). The 1-seeded fruit results from suppression of two of the carpels and their ovules. Species about 350, in 6 genera —*Nothofagus, Fagus, Castanea, Castanopsis, Lithocarpus, Quercus.*

Nothofagus comprises trees and shrubs of antarctic South America and of Australia, Tasmania, and New Zealand. Species about 12; diœcious.

Fagus (Beech). The beeches are stately trees with smooth light-gray bark and horizontally spreading straight-veined leaves in two rows. The flowers appear with the leaves, the staminate in capitate clusters (catkins) subtended by deciduous bracts, the carpellate in pairs subtended by an involucre, in axils of leaves of the season. The staminate flowers consist of a campanulate perianth and 8—16 (20) stamens, the carpellate of a 3-carpellate gynœcium and a perianth of 6 narrow segments. The 3-celled ovary contains 2 ovules in each cell. The fruit is a 3-angled coriaceous-walled nut (spurious fruit) invested by a 4-valved husk (bur, involucre). In *Fagus* the simple inflorescence (dichasium) is 2-flowered, the middle flower suppressed. Only 4 species of beech are known, one in Europe, two in Japan, and one, *F. grandifolia,* in eastern North America. The European beech, *F. sylvatica,* was long ago introduced

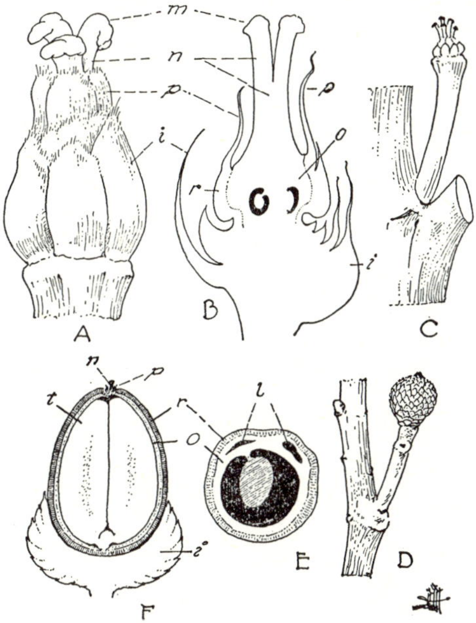

FIG. 155.—*Quercus rubra: A,* carpellate flower. *B,* same in sectional view; *i,* involucral bracts (cupule); *p,* perianth; *n,* styles (united below); *m,* stigmas; *o,* ovary; *r,* receptacle. *C,* axillary portion of a twig with long-stalked carpellate flower at summit. *D,* young acorn, as developed from *C. E,* transverse section of "ovary"; partly suppressed sterile locules at *l. F,* median section of immature acorn; *n,* style; *p,* perianth (remnants); *r,* receptacle; *o,* ovary; *t,* cotyledon of embryo; plumule (epicotyl) between cotyledons at base; *i,* involucre (cupule).

into the Atlantic States. It is said to attain its best development in Denmark.

Castanea (Chestnut). The chestnuts are trees and shrubs with lanceolate, straight-veined, sharply serrate leaves, staminate flowers in conspicuous elongate catkins and carpellate below in prickly involucres. The branches are without terminal buds. Species 4 or 5 to 10 according to the interpretation, distributed over temperate regions of the northern hemisphere. *Castanea dentata* (American Chestnut) and *C. pumila* (Chinquapin) are native in eastern United States. Two species of *Castanopsis*, trees very similar to *Castanea* but having entire leaves, occur in the Pacific States. Besides these, *Lithocarpus densiflora* (Tan-Bark Oak), a lone representative of an east-Asiatic genus, is found in Oregon and California. It has the inflorescence of a chestnut but the fruit of an oak. About 100 species are known.

FIG. 156.—Leaves and acorns of *Quercus alba* (White Oak). White oak group.

Quercus (Oaks) is the largest genus of the family, with 200 or more species, more than half the family, common to the northern hemisphere but with stragglers into the tropics (mountains). About one-third of the species are found in North America. Many are evergreen (live-oaks). The leaves are alternate, simple, and often lobed or incised. The very small greenish flowers appear before or with the leaves. The staminate flowers are in slender drooping catkins from the axils of leaves of the previous season, or from the axils of the inner bud-scales of the terminal bud, or from axils of leaves of the season. They are solitary and without bracteoles and bract, or with fugaceous bracts. Stamens 3—12. The very small carpellate flowers are solitary or few together in the upper axils of the season's shoots, subtended by 2 bracteoles and a deciduous bract. Gynœcium 3-, rarely 4- or 5-carpellate, with as many elongate styles as carpels. The inferior ovary is seated in an involucre of imbricated scales which at maturity forms the characteristic cup. The tip of the style is stigmatic above. There are 2 ovules in each cell of the ovary. The embryo is large, the cotyledons fleshy, and hypogeal in germination. The fruit (spurious) is an acorn which detaches from the cupule when ripe. Technical classificatory characters are found in the acorn and the cup (Figs. 154, 155).

The genus as a whole is very technical, certain species being difficult to differentiate in a completely satisfactory way. Intergrading forms appear to be common and hybridization is said to occur between certain species, and others appear to be exceedingly variable, particularly as to foliage. Of the three groups of oaks that have been recognized, two are represented by our species, *viz.,* White Oaks and Black Oaks. These are distinguished thus:

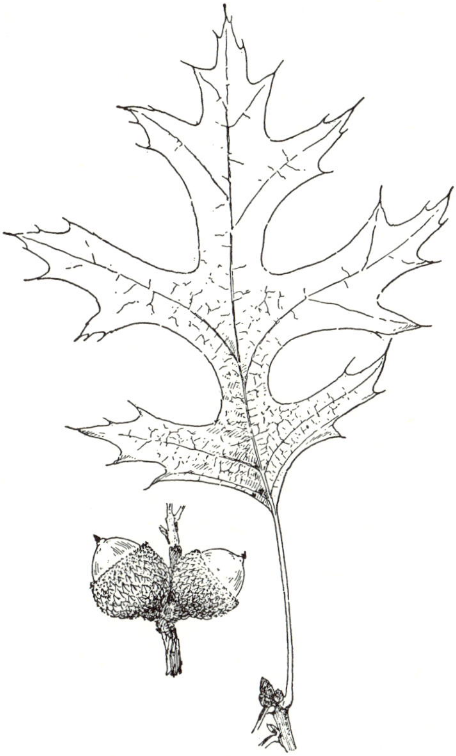

FIG. 157.—Leaf and acorns of *Quercus coccinea.* Black oak group.

White Oaks. Bark pale, and the lobes of the leaves not bristle-tipped. Stamens 6—8. Scales of the cupule woody at the base, and abortive ovules in the lower part of the acorn. Stigmas sessile or nearly so. Shell of acorn glabrous inside. Acorns ripening the first season. (Fig. 156.)

Here belong, among others, the eastern *Quercus alba, Q. macrocarpa, Q. Michauxii,* and *Q. Garryana,* and *Q. lobata* of the Pacific Coast.

Black Oaks. Bark dark, smooth on younger parts; lobes of leaves bristle-tipped. Stamens 4—6. Scales of the cupules soft and membranaceous. Styles long and spreading. Aborted ovules in upper part of acorn. Shell of acorn tomentose inside. Acorn ripening the second season. (Fig. 157.)

This group includes, among others, *Q. rubra, Q. coccinea, Q. velutina, Q. borealis, Q. marilandica* of eastern North America and *Q. agrifolia* and *Q. Kelloggii* of the Pacific Coast.

The wood of oaks is hard and durable and in some species makes valuable lumber. Some are planted as avenue trees, but their slow growth makes them less desirable for this purpose. *Quercus lobata* of California is the largest of all oaks.

ORDER URTICALES

Flowers hypogynous, cyclic, bisexual or unisexual, monœcious or diœcious, with a simple perianth of 4—5 segments or rarely naked. Stamens mostly as many as the segments of the perianth and opposite them. Carpels 2, united, or sometimes the style usually with 2 elongate stigmatic branches. Ovary generally 1-celled, with one ovule. Endosperm none or scanty. Integuments 2. Fruit nut-like, drupaceous, or an achene.

Families: *Ulmaceæ, Moraceæ, Urticaceæ, Cannabinaceæ.* Some writers have preferred to regard these families as tribes or subfamilies of one large family, *Urticaceæ.*

Key to the Families

Trees or shrubs with alternate leaves.
- Sap milky. Fruit multiple or spurious (syconium). *Moraceæ.*
- Sap not milky. Fruit a samara, nut, or drupe. *Ulmaceæ.*

Herbs (in ours).
- Plants aromatic, erect or twining; flowers diœcious; fruit an achene. *Cannabinaceæ.*
- Plants not aromatic, often armed with stinging hairs; flowers monœcious, diœcious, or polygamous; fruit an achene. *Urticaceæ.*

ULMACEÆ (*Elm Family*). Trees or shrubs with alternate, simple, often oblique, leaves and small, greenish, bisexual or unisexual, monœcious or diœcious, or polygamous flowers, the staminate in clusters, the carpellate solitary. Fruit a samara, drupe, or nutlet. About 13 genera and 140 species of temperate and tropical regions. Of the 13 genera 6 are monotypic.

There are two well-marked subfamilies, distinguished as follows:

Ulmoideæ. Flowers mostly polygamous, on branches of the preceding year. Anthers extrorse. Embryo straight, with flat cotyledons. Fruit a drupe. Represented in our flora by *Ulmus* and *Planera*.

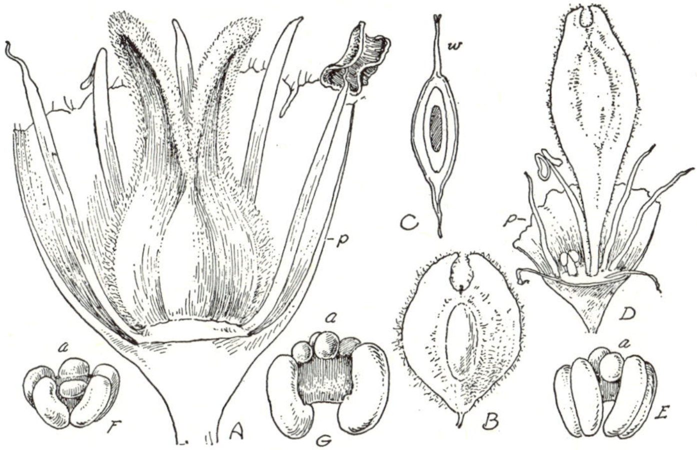

FIG. 158.—*Ulmus americana: A,* section of flower; *p,* perianth. *B,* fruit (samara). *C,* transverse section of ovary; *w,* wing. *D,* section of flower at a late stage; aborted stamen at left. *E, F, G,* supernumerary anthers on upper surface of thick connective.

Celtoideæ. Flowers monœcious-polygamous, on branches of the season. Anthers introrse. Embryo curved, with folded or rolled cotyledons. Fruit a drupe. Comprising 9 of the 13 genera. Represented in our flora by *Celtis*.

Ulmus (Elm). The elms have polygamous flowers with a cup-shaped perianth. The 2-carpellate gynœcium has 2 elongate spreading styles which are stigmatic on the edges. The edges of the ovary project and become the wings of the small 1-seeded nut-like samara. The trees bloom in the spring, except *Ulmus serotina* of the eastern United States, which blooms in autumn.

The leaves of elms are doubly serrate, straight-veined, and more or less oblique at the base, and two-ranked on graceful sweeping branches, forming a flat spray. The leaves of *Ulmus fulva* (Slippery

Elm) are rough above and on vigorous shoots large and commonly trough-shaped. The inner bark is mucilaginous. *Ulmus americana* (White or American Elm) and *U. racemosa* (Cork Elm) are quite similar, but the bark of the Cork Elm is generally formed into corky ridges on the branches and deep furrows and ridges on the trunk. The best marks of distinction between the species are found in the fruits.

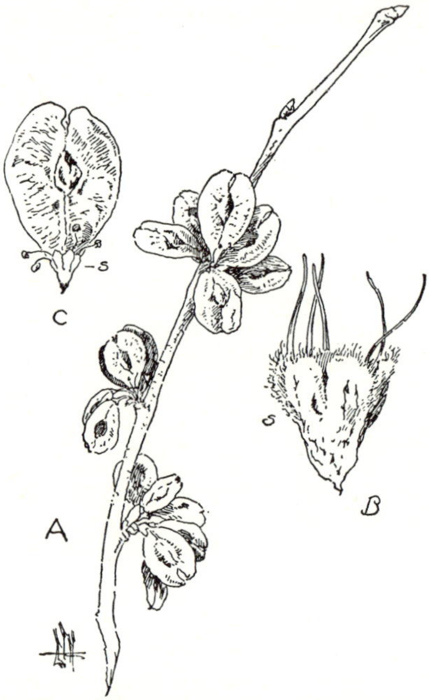

FIG. 159.—*Ulmus* sp.: *A,* twig with fruit clusters. *B,* flower (anthers removed from filaments); *s,* perianth. *C,* immature fruit, with persisting perianth and withered stamens.

The wood of elms is spongy and tough, especially when green or wet. It is utilized mostly as rough lumber. The trees are rapid growers and much used for avenue planting. In the natural state the trees appear to be rather selective of their habitat, occupying usually rich bottom lands, yet will generally do well when transplanted to drier and more barren situations. The European *Ulmus campestris* (English Elm) and several Asiatic species have been introduced. (Figs. 158, 159.)

Planera aquatica (Planer Tree or Water Elm) is a small elm-like tree of the southeastern United States, but differs in its coriaceous nutlet which is beset with irregular fleshy projections. The genus is monotypic.

Celtis (Hackberry) has thinner, paler leaves than the elms, with the lateral veins branched and not rigid and straight. An easy recognition-mark is the very irregularly corrugated or corky-ridged bark. Structurally the flowers resemble those of elms, but the fruit is a drupe. *Celtis occidentalis* and its varieties are widely distributed over eastern parts of the country, and in the West and South is found *C. mississippiensis.* The 60 species are widely distributed in temperate and tropical regions.

MORACEÆ (*Mulberry Family*). This is a large family of trees and shrubs with milky sap. The leaves are of various shapes, with fugaceous stipules. The small unisexual flowers are monœcious or diœcious, clustered or solitary, with a 4—5-lobed perianth which

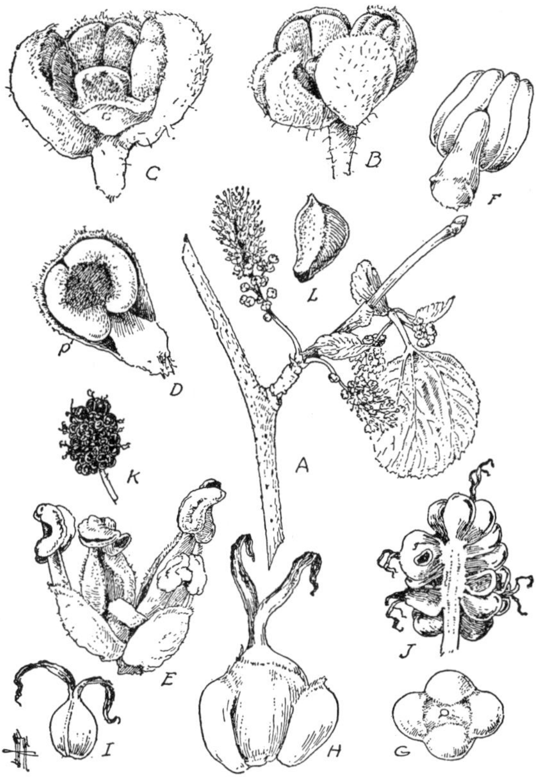

FIG. 160.—*Morus* sp.: *A*, twig with staminate catkins. *B*, staminate flower. *C*, staminate flower with a perianth-segment removed to expose anthers; in center a vestigial gynœcium. *D*, a stamen invested by perianth-segment (*p*). *E*, an older staminate flower; in center a vestigial gynœcium. *F*, dorsal view of a stamen. *G*, bud of carpellate flower. *H*, carpellate flower. *I*, gynœcium. *J*, young fruit cluster. *K*, mature collective fruit. *L*, seed.

in the carpellate flowers becomes fleshy in the fruit. Fruit multiple (aggregate). Endosperm present or none. Embryo curved or spiral. Genera about 55, species 900—1,000, of wide distribution in temperate and tropical climates.

Three subfamilies, two of which, *Moroideæ* and *Artocarpoideæ,* are represented by native, cultivated, or hot-house plants.

In the *Moroideæ* the stamens are curved inwards and the leaves folded in the bud, with small stipules. It is represented here by *Morus, Maclura,* and *Broussonetia.*

The *Artocarpoideæ* have straight stamens, leaves inrolled in the bud, and intrapetiolar stipules which unite as a cap over the bud (Fig. 121B). Represented in hot-houses and in cultivation by *Ficus* and *Artocarpus.*

Morus (Mulberry). The mulberries are trees or shrubs with milky sap and alternate leaves of diverse shapes and lobings. The small monœcious or diœcious flowers are borne in axillary spikes (catkins). There is a perianth of 4 segments, which persist and become succulent in the fruit, completely investing the achene (true fruit), the whole being a sort of spurious fruit. The whole cluster of these spurious fruits forms an aggregate or multiple fruit, resembling a blackberry (Fig. 160).

In *Maclura pomifera* (Osage-Orange) the aggregate fruit is of spherical shape and commonly as large as an orange. The true fruit is an achene (Fig. 38). *Broussonetia papyrifera* (Paper Mulberry) ranges from New York to Minnesota and Georgia as an escape from cultivation.

URTICACEÆ (*Nettle Family*). Trees, shrubs, or herbs with watery sap, and simple, alternate, or opposite leaves. The small greenish flowers are in cymes, often axillary, monœcious, diœcious, or polygamous. Perianth of 2—5 distinct or united segments. Filaments curved inwards in the bud. Gynœcium of 1 carpel. Stigma various. Ovary 1-celled, 1-seeded. Fruit an achene, invested by the perianth. Endosperm present and oily. Embryo straight. Species about 550, in 40 genera; widely distributed.

Genera in our flora: *Urtica, Laportea, Pilea, Bœhmeria, Parietaria.*

Urtica (Nettle). The nettles are simple or branched annuals or perennials with stinging hairs, opposite leaves with distinct or connate stipules, and small, greenish, unisexual, monœcious or diœcious or androgynous flowers in axillary cymes or glomerate cymules. Perianth-segments 4. Stamens 4. Gynœcium with a single erect, orthotropous ovule and sessile or nearly sessile stigma. Fruit an achene. Embryo with broad cotyledons. Species about 30, widely distributed.

Urtica gracilis and *U. Lyallii* are common and widely distributed

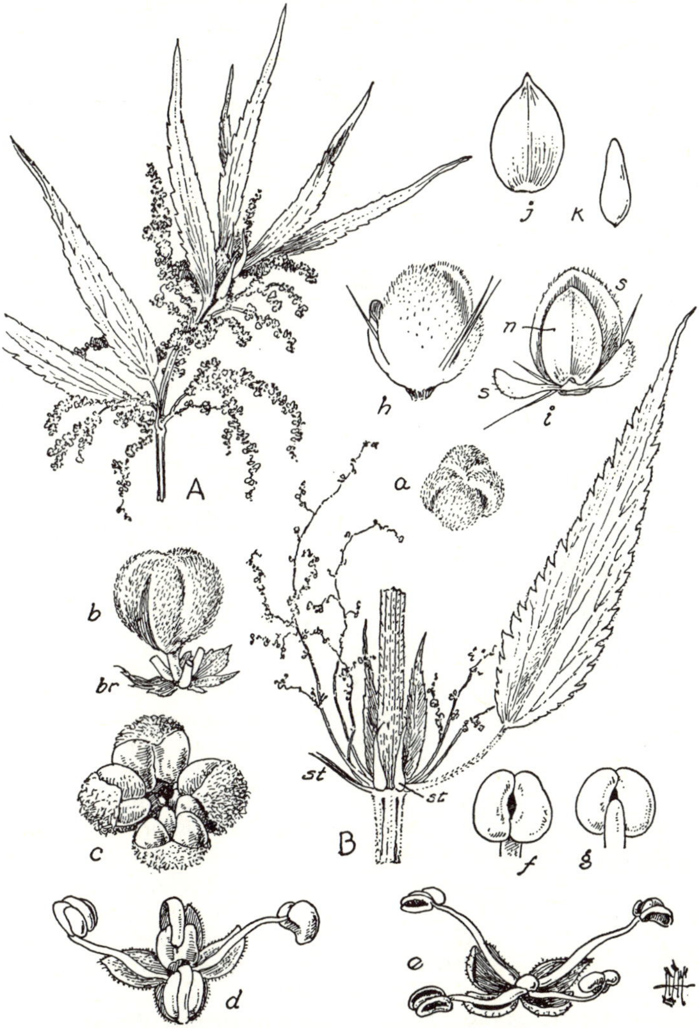

FIG. 161.—*Urtica gracilis: A,* carpellate shoot; *B,* staminate shoot; *st,* stipules; *a,* a staminate flower-bud; *b,* cluster of staminate flower-buds with subtending bracts,—three of the buds removed from pedicels; *c,* staminate flower-bud opening; *d,* staminate flower open; *e,* later stage,—vestigial gynœcium in center; *f,* ventral, and *g,* dorsal, aspect of anther; *h,* carpellate flower, with bristles in axils of perianth-segments; *i,* same with one perianth-segment removed; *s,* perianth-segments; *n,* fruit; *j,* ripe fruit; *k,* ovule, early stage.

northward and westward. The exotics *U. dioica* and *U. urens* are common as weeds (Fig. 161). *Laportea canadensis* (Wood Nettle) forms dense societies in deep, rich, deciduous Eastern woods. The stems are furnished with very tough fibers (Albany Hemp). About 25 species in all. *Pilea* is a large tropical genus of about 200 species, one, *P. pumila,* pressing north as far as Ontario and New Brunswick. It is an annual of cool, swampy or boggy situations, and lacks stinging hairs. *Bœhmeria,* another large tropical genus (50 spp.), is represented in our eastern flora by *B. cylindrica.* This, too, lacks stinging hairs, and the achene is invested by the persistent withered perianth. *B. nivea* is cultivated in parts of the Southwest for its fiber (ramie). *Parietaria pennsylvanica* is a low Eastern annual.

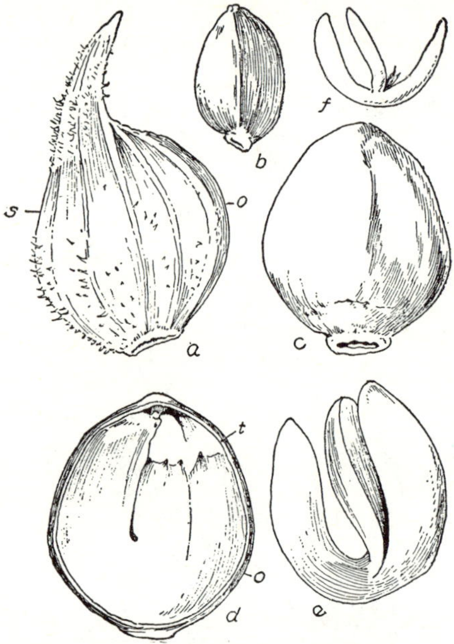

FIG. 162.—*Cannabis sativa: a,* fruit (*o*) and investing bract (*s*); *b,* fruit, edge view; *c,* fruit, dorsal view; *d,* section of fruit, with curved embryo; *t,* testa; *o,* ovary-wall (pericarp); *e,* embryo; *f,* embryo with cotyledons parted to show plumule.

CANNABINACEÆ (*Hemp Family*). In this small family the flowers are diœcious, the staminate in paniculate-racemose inflorescences, the carpellate in spikes. Perianth segments and stamens 5—3, respectively. The style forks into 2 long stigmas. Fruit an achene with crustaceous pericarp. The single seed contains fleshy endosperm and curved or coiled embryo.

Genera 2: *Humulus* (Hop), *Cannabis* (Hemp). *Humulus Lupulus* (Hop) is a twining vine of wide distribution. The leaves are palmately lobed, the staminate flowers in paniculate inflorescences, the carpellate in pairs in the axil of each bract of a spike. The bracteoles of the flowers become enlarged and membranous and enclose the achene. Only two species known. *Cannabis* is a monotypic genus of eastern and central Asia. The solitary species, *Cannabis sativa* (Hemp), a tall aromatic annual, is cultivated in warm countries for its fiber, but occurs as a weed in many parts of the country. The leaves are palmately divided into narrow segments, and the flowers are small, greenish, and diœcious, the staminate in cymes, the carpellate in axillary spikes. The perianth invests the ovary, the fruit is an achene, and the seed contains a curved embryo and an oil from which, in the Orient, is made the strongly intoxicating drink hashish. (Fig. 162.)

ORDER PROTEALES

Flowers bisexual or unisexual by abortion, regular or irregular, cyclic, hypogynous, with a simple petaloid perianth. Stamens opposite the segments of the perianth and united with them, but the anthers free. Ovary superior, with 1 to many ovules on a ventral placenta. Integuments 2. Endosperm none. Cotyledons sometimes 3—8. Fruit follicular or indehiscent.

One family: *PROTEACEÆ,* with about 1,100 species in 50 genera of shrubs and trees of Australia and South Africa. There are two subfamilies:

Proteoideæ, in which the flowers are solitary in the axils of bracts, and the fruits 1-seeded. Here belong *Protea* with 80 species in South Africa, and *Leucodendron* with 70 species in the same region;

Grevilleoideæ, in which the flowers are in pairs in the axils of bracts and the fruits indehiscent, with 1 to many seeds. Including *Grevillea* with 160 species, and *Hakea* with 100 species, both Australian.

Several species of *Grevillea* and *Hakea* are cultivated in the warmer parts of the United States. *Grevillea robusta,* a large tree with silvery compound leaves and showy clusters of orange-colored flowers which are produced in profusion, is a handsome avenue tree in southern California. More common in California is the shrub, *G. Thelemanniana,* with small dissected leaves and reddish flowers, which is planted as an ornamental. The flowers are richly nectariferous. (Fig. 163.)

This is a curious and anomalous group, apparently without any close affinities among living flowering plants.

FIG. 163.—*Grevillea robusta: a,* inflorescences; *b,* section of flower; *p,* perianth; *s,* anther; *st,* style; *o,* ovary; *gl,* gland; *c,* a flower-bud; *r,* pedicel; *d,* median section of base of flower; *o,* ovary; *n,* ovule; *gl,* gland; *f, g,* anthers; *e,* fruit.

ORDER SANTALALES

Flowers epigynous, rarely hypogynous, cyclic, regular, bisexual or sometimes unisexual, diœcious or polygamous. Stamens mostly as many as the segments of the perianth and opposite them. Carpels mostly 2—3, united, rarely 1. Endosperm oily. Integuments 2.

Families: *Myzodendraceæ, Santalaceæ, Opiliaceæ, Grubbiaceæ, Olacaceæ, Octoknemataceæ, Loranthaceæ, Balanophoraceæ.*

KEY TO THE FAMILIES

Evergreen, often small and inconspicuous, dichotomously branched parasites on branches of trees. Flowers greenish, minute. *Loranthaceæ.*

Green, herbaceous or shrubby, root parasites (on tropical trees). Petals none. *Santalaceæ.*

MYZODENDRACEÆ. A small family of semiparasitic plants of antarctic regions of South America, growing on species of *Nothofagus* ("Beech"). Flowers unisexual, the staminate without perianth, the carpellate with a 3-merous perianth. The 3-angled fruit is provided with 3 feathery brushes which function as organs of dispersal.

SANTALACEÆ (*Sandalwood Family*). Herbs, shrubs, or trees with alternate or opposite exstipulate leaves and bisexual or unisexual, monœcious or diœcious flowers, solitary or in axillary or terminal clusters. Calyx often petaloid; petals none. Ovary inferior, 1-celled, with 2—4 ovules suspended from the top of the central placenta. Fruit (spurious) a drupe or nut. Seed 1, without testa. Endosperm copious and fleshy. Embryo small. About 250 species, in 26 genera, mostly tropical, with a few reaching north to Canada.

Only the following genera with us: *Comandra, Pyrularia, Nestronia.*

Comandra (Bastard Toadflax) is represented by 4 or 5 species of low perennial, somewhat woody, rhizomatous herbs with more or less elliptical blunt-tipped alternate leaves and numerous small, white or greenish-white flowers in axillary or terminal cymose clusters. The flowers are bisexual, with a bell-shaped perianth (calyx). The anthers are connected to the calyx by tufts of hairs, and a 5-lobed disk lines the hypanthium above the ovary. The fruit is drupaceous (spurious) or a nutlet, crowned by the persistent perianth. The plants are parasitic on roots of trees and shrubs (Fig. 164).

Pyrularia is represented by *P. pubera* in rich woods from Pennsyl-

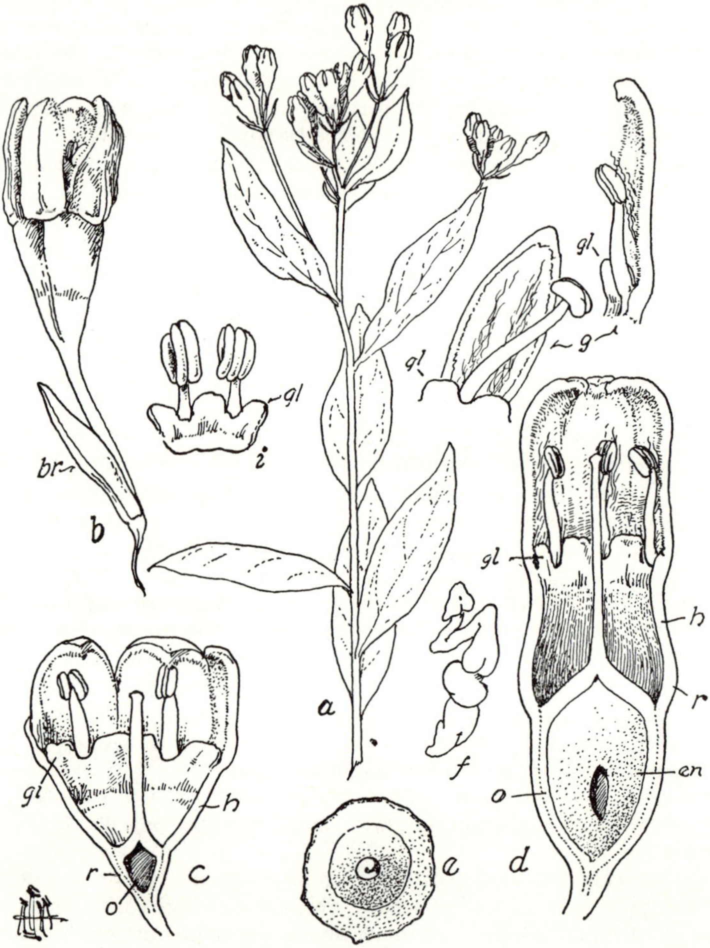

FIG. 164.—*Comandra Richardsiana: a,* flowering shoot; *b,* flower; *br,* bract; *c,* median section of marginal flower; *h,* hypanthium of receptacle; *r,* receptacle; *o,* ovary; *gl,* gland; *d,* median section of central flower of cymule; *o,* ovary; *r,* receptacle; *h,* hypanthium; *gl,* gland; *en,* endosperm,—embryo-cavity in center; *e,* transverse section of ovary; *f,* embryo; *g,* inner face and side view of perianth-segment with stamen and gland (*gl*),—tufts of hairs behind stamens.

vania to Florida. Two others occur in Asia. Ours is a straggling shrub up to 15 feet high, with small flowers in terminal racemes. *Nestronia* is a monotypic genus of eastern United States, with the solitary *N. umbellata* occurring as a parasite on roots of trees along streams.

OPILIACEÆ is a small family of the tropics, with bisexual flowers and without integuments in the ovules.

GRUBBIACEÆ comprises the solitary genus *Grubbia,* with 3 species in South Africa. The flowers are 4-merous, with a perianth of 4 undifferentiated segments and 8 stamens (4+4) with the posterior cells of the anthers rudimentary. Gynœcium 2-carpellate, syncarpous, with one style. The inferior ovary is at first 2-celled, later 1-celled, with 2 ovules suspended from the central placenta. Integuments 1.

OLACACEÆ is a small family (150 spp.) of trees, shrubs, climbers, and some hemiparasites, of South America and Africa. Flowers regular and mostly bisexual, with one style. Fruit drupaceous or nutlike, 1-seeded.

OCTOKNEMATACEÆ is a monotypic family of woody plants of West Africa. Flowers bisexual, with a perianth of 2+3 segments and 5 (2+3) stamens. *Octoknema.*

LORANTHACEÆ (Mistletoe Family) includes hemiparasites on trees, from which they derive nourishment by means of specialized absorbing organs (haustoria). The plants are yellowish or olive-green, with dichotomous branching and generally well-developed broad to narrow opposite leaves (reduced to scales in some forms). The flowers are epigynous, bisexual or unisexual, monœcious or diœcious, and regular, with an undifferentiated perianth or with calyx and corolla (of the formula 2+2 or 3+3). Gynœcium of a varying number of carpels. Style 1 or wanting. Stamens as many as the perianth-segments and opposite them, free or more or less united. Fruit (spurious) a berry containing a single seed with seed-coat merged with the usually copious endosperm.

About 900 species, in over 20 genera, mainly tropical. This and the following family are further distinguished by certain technical features of the ovule and embryo.

Two subfamilies are generally recognized, *Loranthoideæ* and

Viscoideæ, distinguished by the presence in the one (*Loranthoideæ*) of a *calyculus* (diminutive of calyx)—consisting of the bracteoles fused with the perianth, or an enlargement of the pedicel—and the absence of this structure in the other (*Viscoideæ*). Our plants all belong to the *Viscoideæ,* in the genera *Phoradendron* and *Arceuthobium.* Here belongs also the classical European mistletoe, *Viscum album. Phoradendron flavescens* and *Arceuthobium pusillum* occur in eastern United States, and several other species in the West. *Arceuthobium pusillum* causes the "witches' brooms" on certain conifers (spruce, larch). In *Phoradendron* leaves are well-developed, the stems much branched and swollen at the nodes, the flowers imbedded in the joints of the inflorescence, the anthers 2-celled, and the fruit is a pulpy sessile berry. In *Arceuthobium* the stems are angled and brittle-jointed, the leaves reduced to scales, the anthers 1-celled, and the berry pediceled and flattened. The flowers of both are unisexual and diœcious. Ripe berries of *Arceuthobium* are explosively dehiscent. Some Phoradendrons are parasitic on deciduous trees, others on conifers, but the arceuthobiums are restricted to conifers. (Fig. 165.)

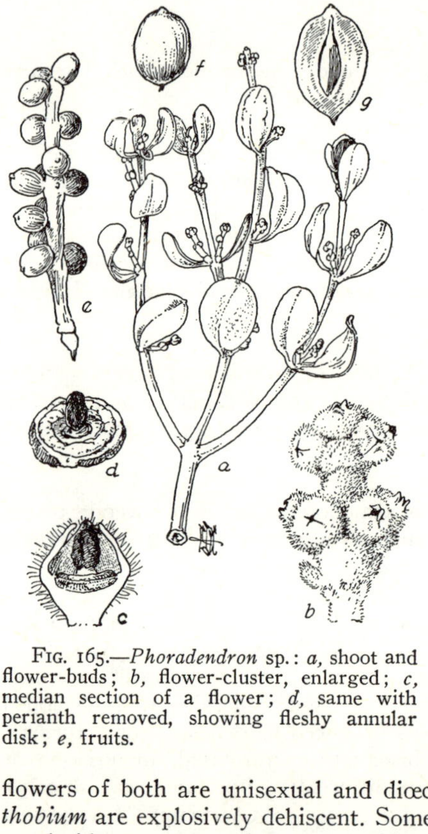

FIG. 165.—*Phoradendron* sp.: *a,* shoot and flower-buds; *b,* flower-cluster, enlarged; *c,* median section of a flower; *d,* same with perianth removed, showing fleshy annular disk; *e,* fruits.

BALANOPHORACEÆ. Flowers naked or with a simple undifferentiated perianth, and unisexual, monœcious or diœcious. Staminate flowers with a perianth of 3 or 4 segments, but sometimes with 2 or as many as 8, the segments more or less united (gamophyllous) at the base. Anthers 1-celled to several-celled, the cells opening by pores or slits. Carpellate flowers generally naked, and consisting of 1 to 5 carpels with a corresponding number of styles. Ovary superior, 1-celled, containing 1 to 3 ovules. The ovules are peculiar in being without integuments and often undeveloped, in which case the embryo-sac lies imbedded in the tissues of the free placenta or in the tissues of the ovary wall. The fruit is a 1-seeded nut, containing a small undifferentiated embryo lying in the apical region of the oily endosperm.

The plants of this strange and rather isolated family are fleshy, yellowish or reddish root-parasites of trees and shrubs. The stems are tuberous branching rhizomes bearing small, at first fleshy but later scale-like, deciduous leaves, attaching themselves to the roots of the host-plants by means of haustorial suckers. Numerous small, sessile or short-pedicellate flowers are produced in dense heads or spikes, or, as in the South African *Sarcophyte,* in a panicle. The inflorescence is said to develop in many cases in the inner tissue of the rhizome, from which it breaks forth. Most of the members, of which there are known some 40 species, in 14 genera, are confined to tropical woodlands. About six subfamilies have been described, these being differentiated chiefly on characters of the flowers and of the embryo-sac. Starch is present in the rhizome of some members, while a chemical principle, balanophorin, is present in others.

The largest genus is *Balanophora,* with 11 or 12 species in tropical Asia and in Australia. *Lophotophytum, Scybalium, Helosis,* and *Rhopalocnemis* are found in the American tropics. With respect to the development of the embryo-sac, endosperm, and embryo, the family presents some anomalous features.

ORDER ARISTOLOCHIALES

Flowers epigynous, cyclic, regular or irregular, mostly bisexual, with a petaloid perianth (in some cases fleshy). Corolla wanting. Stamens few to many, free, or in some cases united with the style to form a column (gynostemium). Gynœcium of 4—6 carpels, syncarpous. Seeds numerous, with endosperm or sometimes perisperm. Integuments 1—2.

Families: *Aristolochiaceæ, Rafflesiaceæ, Hydnoraceæ.*

ARISTOLOCHIACEÆ (*Birthwort Family*). A small family of herbs or woody climbers, chiefly South American. Leaves alternate, cauline or basal, petiolate, exstipulate, cordate, or reniform. Flowers epigynous, bisexual, with petaloid calyx, and petals usually none.

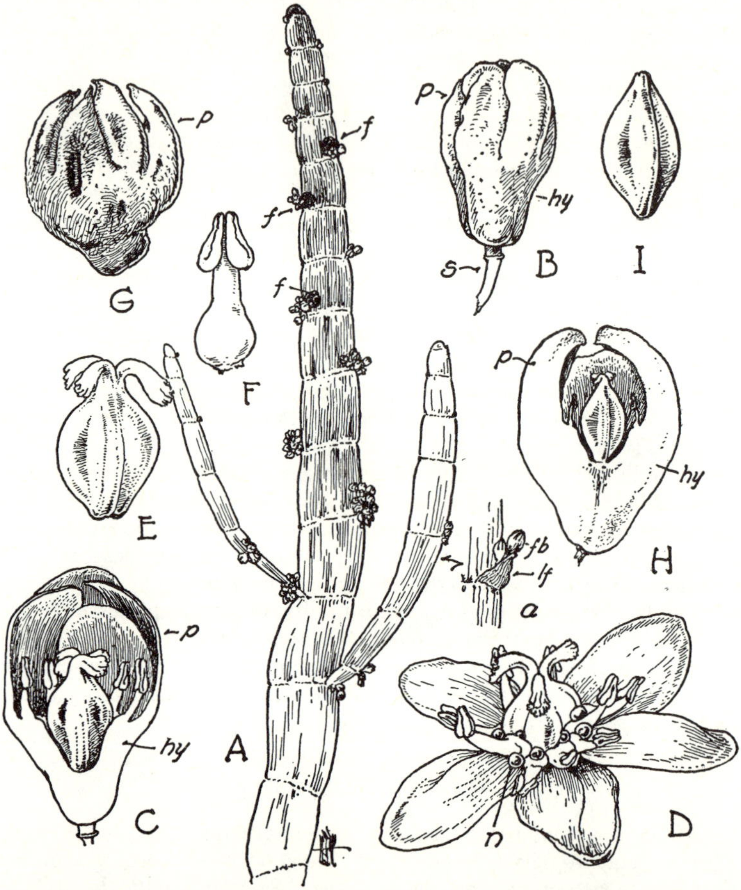

FIG. 166.—*Muehlenbeckia platyclados: A,* a portion of a branch showing flattened character, with flower-clusters on alternate sides at the nodes, *f;* ripe "fruits" (blood red); at (*a*) a portion of a node showing reduced scale-like leaf (*lf*) with flower-buds (*fb*) in axil. *B,* a flower-bud (enlarged) on jointed pedicel (*s*). *C,* median section of flower-bud; *hy,* hypanthium; *p,* perianth (Compare with *B, D, G,* and *H*). *D,* a flower; *n,* droplets of nectar at base of stamens. *E,* gynœcium. *F,* a stamen (ventral aspect). *G,* a "fruit," with succulent perianth ("calyx") (*p*) and hypanthium (*hy*). *H,* median section of "fruit." *I,* achene—the true fruit.

Stamens 5 or 6 to numerous, free or united with style. Ovary 4—6-celled, inferior, rarely superior, with many ovules. Fruit a capsule (spurious). About 200 species, in 5 or 6 genera.

Only *Asarum* (Wild Ginger) and *Aristolochia* (Birthwort) represented in our flora. The wild gingers are plants of rich woods, creeping by slender branching rootstocks, the branches ending in a pair of long-petioled cordate or reniform leaves between which is borne a solitary long-pedicelled flower. The regular perianth consists of 3 broad or narrow segments. Petals none or 3, and 12 distinct stamens. Fruit a loculicidal or irregularly dehiscent capsule. The rootstocks possess a pronounced gingery juice. The birthworts are woody climbers with ovate or cordate leaves and irregular flowers without petals, and 6 or fewer stamens united with the style. The fruit is a septicidal, 6-valved capsule containing flat seeds. A number of species are horticultural (*Aristolochia macrophylla,* Dutchman's Pipe). *Asarum,* about 13 species; *Aristolochia,* 180.

RAFFLESIACEÆ. This small family of scarcely over 20 species of the tropics is remarkable on account of the parasitic habit of its members and the peculiar structure and sometimes enormous size of the flowers. In the last particular *Rafflesia* is especially noteworthy, as one of its species, *Rafflesia Arnoldi,* produces a flower with a diameter of over 3 feet, and is the largest flower known.

HYDNORACEÆ is a still smaller family of parasitic plants. Genera 2: *Hydnora* with about 8 species in Africa, and *Prosopanche* with 2 in Argentina.

ORDER POLYGONALES

Flowers hypogynous, cyclic, regular, mostly bisexual, with undifferentiated perianth. Stamens 4—12, mostly 6—9. Gynœcium of 3 or 2, rarely 4, united carpels. Ovary 1-celled, with one ovule. Fruit a nutlet or achene. Endosperm mealy. Integuments 2.

One family, *POLYGONACEÆ* (Buckwheat Family). Characters of the order, but in addition distinguished by jointed stems and sheathing stipules (ocreæ) and polygamous, monœcious or diœcious flowers borne in inflorescences of various forms. The style is 2—3-cleft or -parted, with relatively large capitate, fringed, or tufted stigmas. The achene is lenticular when developed from a 2-carpellate gynœcium and 3-angled when from a 3-carpellate gynœcium. The embryo is straight or curved, with flat cotyledons. With respect to the perianth, which is often petaloid, the flowers are 3-merous or 5-

merous. In some cases the perianth invests the achene. Diversity in habit, even in the same species, is notable; some forms are climbing or trailing, and some have inflated stems; in others, such as *Muehlenbeckia platyclados,* the stems are modified as cladodes (Fig. 166). These diverse characters are generally of taxonomic value. Ocreæ are wanting only in a few. In many cases an involucre subtends the flower-clusters. Species 800, in about 40 genera, widely distributed, and including a number that are useful to man.

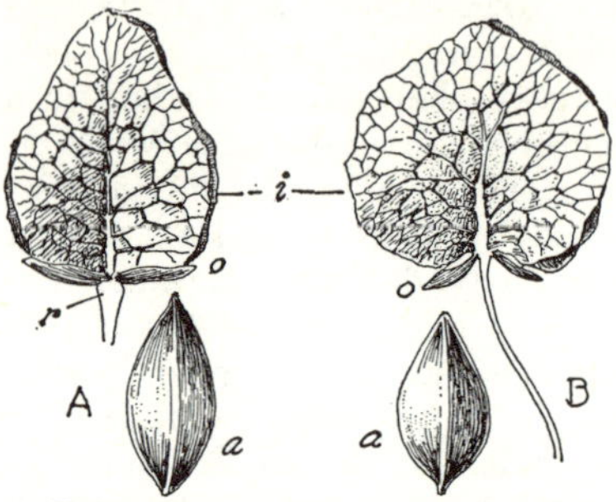

FIG. 167.—*A,* Flowers and fruits (*a*) of (*A*) *Rumex occidentalis* and (*B*) *Rumex patientia; i,* inner perianth-segments; *o,* outer perianth-segments; *r,* receptacle.

The following subfamilies are distinguished:

Eriogonoideæ, in which the flowers are 3-merous and cyclic, ocreæ wanting, and endosperm not cleft. Including *Pterostegia, Chorizanthe, Eriogonum, Oxytheca.*

Polygonoideæ, in which the flowers are 3-merous or "pseudopentamerous" (3+2) and acyclic, and the endosperm is not cleft. Including *Rumex, Oxyria, Rheum, Fagopyrum, Kœnigia, Polygonum.*

Coccoloboideæ, in which the flowers are 3-merous or pseudopentamerous, bisexual or unisexual (diœcious), ocreæ are present or much reduced or wanting, and endosperm is cleft. Including *Antigonon, Muehlenbeckia, Coccoloba.*

The *Eriogonoideæ* are native to western North America, including Mexico. In the monotypic *Pterostegia,* with *P. drymarioides* from Oregon to Lower California, the flowers are 3-merous—a character rarely occurring outside the Monocotyledons. *Chorizanthe* ranges from California to Chile, with 28 of the 34 species in California. The plants are low annuals with dichotomous branching and the leaves mostly in a basal rosette. The flowers are subtended, and sometimes invested, by an involucre. The 6-parted calyx is petaloid, and there are 3—9 stamens. The species are inhabitants of deserts and arid foothills.

Eriogonum, with 120 species (next to the largest genus of the family), is abundantly represented in western North America. The bisexual flowers are clustered and subtended by a more or less conspicuous, often calyx-like, involucre. There is a 3-carpellate gynœcium with 3 styles and minute capitate stigmas, developing into an achene which is invested by the persistent petaloid calyx. The species are annual or perennial herbs or subshrubs with the leaves alternate or whorled or in basal rosettes. *Oxytheca* is restricted mainly to California, but has one species in Chile. The plants are annuals with the

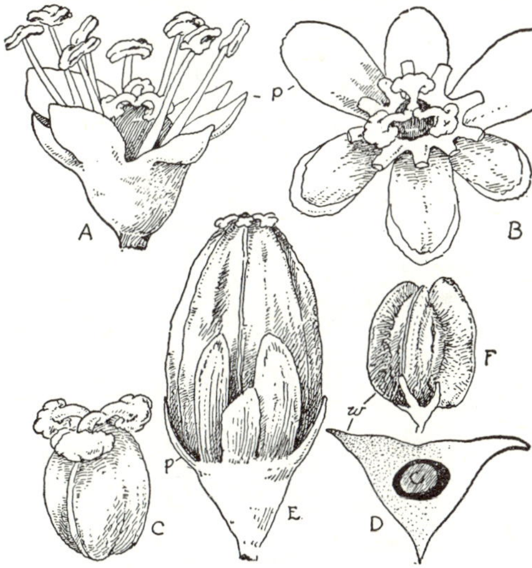

FIG. 168—*Rheum Rhaponticum: A,* flower; *B,* flower seen from above, with stamens removed; *p,* perianth-segments; *C,* gynœcium; *D,* transverse section of ovary; *E,* older stage of gynœsium; *p,* perianth; *F,* fruit, with persistent calyx below; *w,* wing.

leaves in basal rosettes, and the internodes of the stem are glandular-pubescent. An involucre of bristle-tipped lobes subtends the flowers. Stamens 9. Achene mostly lenticular.

Polygonoideæ. Rumex comprises over 100 species of annual or perennial small to large and robust herbs of wide distribution. The stem is usually more or less grooved, in some cases decumbent or creeping, and bears lanceolate or hastate leaves with the margins frequently wavy or crisped, and with more or less prominent cylindrical ocreæ. The numerous small greenish flowers are crowded (often in whorls) in paniculate inflorescences. There are 6 perianth segments, the 3 outer of which are small and inconspicuous, the 3 inner much larger and more or less veiny and often furnished with a more or less prominent tubercle over the basal portion of the midrib (Fig. 167). The 3 inner segments are persistent as a covering of the

FIG. 169.—*Polygonum amphibium: a,* flowering shoot; *b,* flower; *c,* median section of flower; *gl,* gland; *h,* hypanthium; *d,* portion of hypanthium showing insertion of stamens and lobes of gland (*gl*); *e,* gynœcium; *f,* ventral, and *g,* dorsal, views of stamens; *o,* ocrea.

achene. The flowers are bisexual, polygamo-diœcious or diœcious, with noticeably developed hypanthium (perigynous). Some species possess an acrid juice. One of the commonest species is *Rumex*

Acetosella (Sheep Sorrel), which occurs as a weed in acid soils, of which it is a sure indicator. Another species common as a weed is *Rumex crispus* (Yellow Dock). The solitary *Oxyria digyna* is noteworthy because of its 3-merous or 2-merous flowers and circumpolar and alpine distribution.

One of the most important genera of the family is the Eurasian *Rheum,* with about 20 species ranging from Siberia to Palestine and the Balkans. The plants are stout perennial herbs with large lobed leaves and prominent stipules (ocreæ), and numerous small flowers borne in massive inflorescences. The flowers are 3-merous, with 6 perianth-segments in two whorls, the outer smaller than the inner, and 6 stamens. The 3-carpellate gynœcium ends in a stout style bearing prominent stigmas. Perigyny has been brought about by the development of a prominent hypanthium. The most important species is *Rheum Rhaponticum,* the Rhubarb, long cultivated for the pleasantly acrid petioles of the leaves. Another species, *R. palmatum,* is grown as an ornamental. (Fig. 168.)

Polygonum (Knotweed), the largest genus (over 200 spp.), comprises annual or perennial herbs or subshrubs of wide geographic distribution. The species are of diverse habits—erect, decumbent, or climbing—while some are aquatics with dimorphic leaves. Conspicuous features are the swollen nodes and prominent ocreæ. The small and numerous flowers are in axillary clusters, racemes, spikes, or panicles. The perianth is mostly 5-merous and there are 3—9 stamens. Fruit a 3-angled or lenticular achene or nutlet. *Polygonum aviculare* and *P. erectum,* the knotweeds, are common plants of dooryards, roadsides, and waste grounds. *P. lapathifolium* is common in wet places, and the polymorphic *P. amphibium* is ubiquitous along the borders of shallow lakes. Sharp prickles on the angles of the stems and on the midrib of the leaves readily distinguish the reclining *P. sagittatum* of low grounds. A common and sometimes noxious weed in the twining *P. Convolvulus.*

The most important genus is the Eurasian *Fagopyrum,* the 6 species of which comprise annuals or perennials with striate or grooved stems, hastate leaves, and cylindrical ocreæ. The small white or greenish flowers are bisexual, in axillary or terminal racemose-paniculate clusters. There is a 5-merous petaloid perianth, and there are 8 stamens. The 3-carpellate, 3-styled gynœcium becomes a 3-angled achene. The most important species is *Fagopyrum esculentum,* the Buckwheat.

In *Kœnigia* we meet again with a monotypic genus, founded on *K. islandica* which, like *Oxyria digyna,* is a circumpolar species, being found outside of its range only in the high Himalayas. *Antigonon,* with 3 or 4 species of Mexico and Central America, is represented in cultivation by the robust vine, *A. leptopus,* which is characterized by hastate leaves and bisexual pink flowers borne in slender racemes which terminate in a tendril. The flowers vary from acyclic to cyclic, associated with variation in the number of stamens (7—9). But the gynœcium is definitely 3-carpellate.

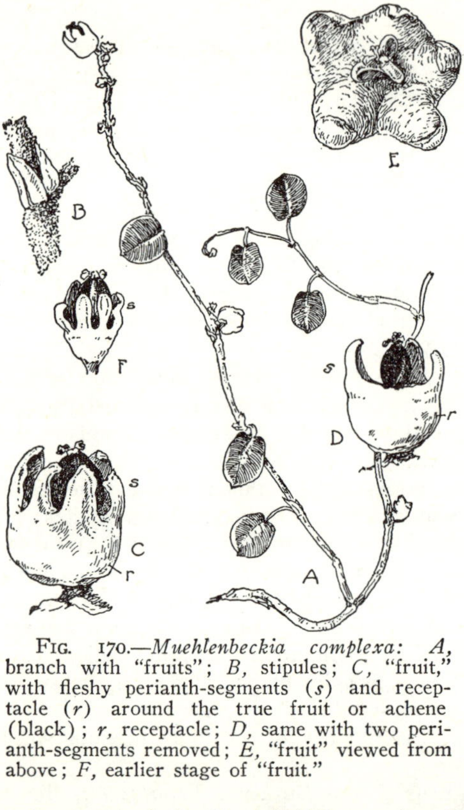

FIG. 170.—*Muehlenbeckia complexa: A,* branch with "fruits"; *B,* stipules; *C,* "fruit," with fleshy perianth-segments (*s*) and receptacle (*r*) around the true fruit or achene (black); *r,* receptacle; *D,* same with two perianth-segments removed; *E,* "fruit" viewed from above; *F,* earlier stage of "fruit."

Muehlenbeckia, with about 15 species, is restricted to the southern hemisphere. The plants are of very diverse habits, sometimes with the stems modified into cladodes, as in *M. platyclados* (Fig. 166), and the leaves often reduced to scales. The perianth consists of 5 segments, and there are 8—9 stamens and a 3-carpellate gynœcium with distinct styles. In *Muehlenbeckia complexa* the perianth and receptacle are modified as the succulent part of the "fruit" (Fig. 170). *Coccoloba,* one of the larger genera, belongs in the American tropics. The genus is noteworthy as comprising trees and shrubs, some of which are planted for their massive foliage. Structurally the flowers are similar to those of *Muehlenbeckia.*

ORDER CENTROSPERMÆ (CHENOPODIALES)

Flowers hypogynous or in some cases perigynous or epigynous, spiral or cyclic, unisexual or bisexual, mostly regular. Perianth differentiated or undifferentiated. Stamens as many as the segments of the perianth and opposite them, (sometimes) numerous, or one. Carpels several or one, mostly united. Ovary 1-celled, with many ovules or one. Integuments mostly 2. Embryo curved or coiled.

An order of rather distinctive plants, fairly easily recognizable.

Families: *Chenopodiaceæ* (incl. *Illecebraceæ*), *Amaranthaceæ, Nyctaginaceæ, Cynocrambaceæ, Phytolaccaceæ, Aizoaceæ, Portulacaceæ, Basellaceæ, Caryophyllaceæ.*

Key to the Families

A. Perianth not differentiated into calyx and corolla.

a. Carpels 1.

Perianth usually petaloid, persistent around the achene. Flower-clusters subtended by an involucre (often calyx-like) of distinct or united bracts. *Nyctaginaceæ.*

Perianth not petaloid, persistent. Fruit a berry or dry. *Phytolaccaceæ.*

b. Carpels 2 or more, often many.

1. Fruit a thin-walled achene closely invested by the persistent perianth. *Nyctaginaceæ.*
2. Fruit a utricle.

Perianth membranaceous or scarious. *Amaranthaceæ.*

Perianth mostly herbaceous. Utricle with a thin or coriaceous coat, sometimes winged, and sometimes closely invested by a pair of foliaceous bracts. *Chenopodiaceæ.*

3. Fruit a berry (in some exotics a capsule or samara). *Phytolaccaceæ.*
4. Fruit a capsule, variously dehiscent. Succulent or woody plants. The numerous stamens often transformed into strap-shaped petaloid staminodes simulating the "rays" of Composites.

Carpels 2-many. Very succulent plants in *Mesembryanthemum.* *Aizoaceæ.*

Carpels 3—5. Plants very succulent, and ovary inferior in *Portulaca.* *Portulacaceæ.*

B. Perianth differentiated into calyx and corolla. *Caryophyllaceæ.*

CHENOPODIACEÆ (*Goosefoot Family*). These are annual or perennial, more or less succulent herbs or in some cases shrubs, rarely tree-like, with alternate or opposite exstipulate leaves and numerous small, generally greenish, unisexual or bisexual, polygamous, monœcious or diœcious flowers borne in dense clusters. Petals none, the perianth of distinct or united segments, in some cases wanting in the carpellate flowers, in rare cases reduced to a single segment. Anthers 2-celled. Ovary 1-celled, with a solitary amphitropous ovule. Styles and stigmas various. Fruit a utricle, enclosed in the persistent perianth. Because of their small size the fruits are often mistaken for seeds. Over 500 species, in about 75 genera, widely distributed. Chiefly plants of saline soils, some occurring as weeds. To this family belong a number of very useful plants.

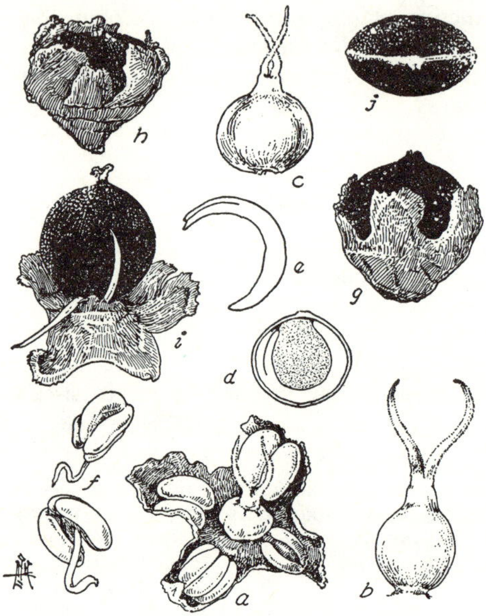

FIG. 171.—*Chenopodium californicum: a,* flower viewed from above; *b,* gynœcium; *c,* older stage of gynœcium; *d,* section of seed, showing embryo curved around endosperm; *e,* embryo; *f,* stamens; *g,* ovary and investing calyx; *h,* a younger stage of same; *i,* gynœcium (ovary) with perianth pushed back (two filaments are shown); *j,* ovary laid on side and viewed from above.

Our native and cultivated species belong in the following subdivisions and genera:

Cyclolobeæ, in which the more or less annular embryo surrounds the endosperm. Includes *Beta, Chenopodium, Atriplex, Spinacia, Kochia, Corispermum, Salicornia, Suæda, Monolepis.*

Spirolobeæ, in which the embryo is spirally coiled, dividing the endosperm into two parts, or endosperm is wanting. Includes *Sarcobatus, Salsola.*

In *Beta* (Beet) the flowers are bisexual, and the ovary is inferior with 1—3 ovules in the one cell. The inflorescence is terminal and spicate. The sugar-beet, the garden beet, and mangels are all varieties of *Beta vulgaris.* Swiss chard and spinach-beet are varieties of *Beta cycla.* In *Spinacia* the flowers are unisexual and diœcious, in axillary clusters. *Spinacia oleracea* (Spinach) is native in southwestern Asia. Unisexual flowers occur also in the close ally, *Atriplex,* along with a development of both monœcism and diœcism, and the carpellate flowers are subtended by two characteristic bractlets which persist in the fruit; but in the staminate flowers these bracts are wanting. *Atriplex* numbers about 130 species of wide distribution, of which about 50 occur in arid parts of the West and in

Mexico. The plants are annual or perennial herbs but some are shrubs. Technically this is a difficult genus, but very useful taxonomic characters are found in the bracts of the carpellate flowers. Several species of *Chenopodium* are familiar as weeds, such as *C. album* (Lamb's-Quarters, Pigweed), *C. ambrosioides* (Mexican Tea), *C. murale, C. Botrys* (Jerusalem Oak), and the native *C. capitatum* (Strawberry Blite) and *C. hybridum* (Maple-leaved Goosefoot). The herbage of *C. album* is sometimes used as greens. Species in all about 60. Flowers bisexual, bractless (Figs. 171, 172).

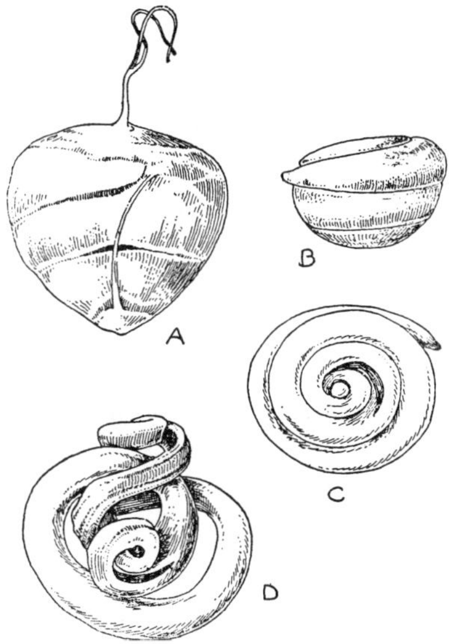

FIG. 172.—Fruit (*A*) and embryo (*B*) of *Salsola Kali* var. *tenuifolia; C, D,* embryo uncoiling.

Kochia, with 30—40 species, has bisexual flowers but without subtending bracts. *K. scoparia* and its variety *trichophylla* are ornamentals of flower gardens, under the name Summer Cypress or Belvedere (Fig. 173). The monotypic *Sarcobatus* is Western, where *S. vermiculatus* occurs as the common "greasewood" of the plains. The flowers are monœcious or diœcious, the staminate without a perianth and in spikes, the carpellate with perianth and solitary in the axils of the leaves. The fruit is bordered by a membranous wing which is an appendage of the perianth, and distinguishes the plant at once. The embryo is green and endosperm is wanting. *Suæda,* with over 40 species, comprises annuals and perennials (shrubs), more or less succulent plants with linear leaves. They are restricted mainly to salt-marshes and saline areas. The flowers are bisexual or unisexual (monœcious) and bracteolate, with a succulent perianth which encloses the keeled or crested utricle. *Suæda depressa* is found in alkaline areas in the northern part of the Mississippi basin and westward across the plains to coastal southern California.

Monolepis, with about 5 species, ranges over western United States and northern Asia. Its distinctive features are the solitary perianth-segment and solitary stamen. *M. Nuttalliana* inhabits alkaline soil from Manitoba and Minnesota westward. *Salsola* (59 species) comprises annuals and perennials with spinescent leaves, and the segments of the perianth are furnished with wing-like appendages which persist in the fruit. Embryo spiral and endosperm none. *Salsola Kali* and its varieties are pernicious weeds (Russian Thistle). The 10 species of *Salicornia* are succulent herbs with the leaves reduced to scales. The bisexual and staminate flowers are imbedded in the axils of succulent scales in elongate terminal spikes. The plants inhabit salt-marshes and saline areas. Very similar is the stouter *Allenrolfea occidentalis* of deserts of the Southwest, but it differs in its shrubby habit and in the alternate instead of opposite scales of the spikes. The solitary *Cycloloma atriplicifolia* of sandy shores and banks of the North Central States is an easily recognized low herb with characteristic winged fruit as in *Atriplex* and *Salsola.*

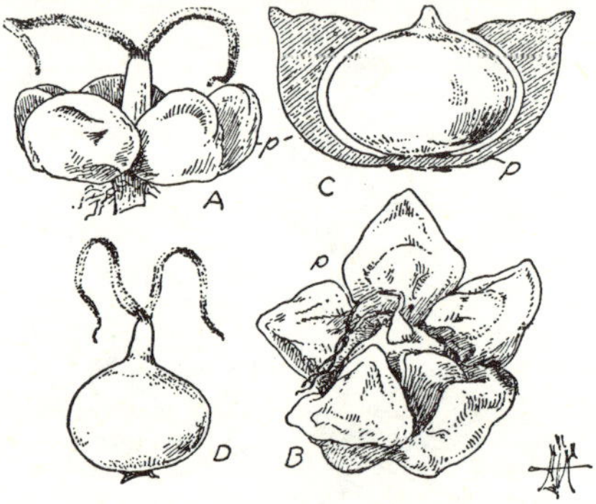

FIG. 173.—*Kochia* sp.: *A,* flower; *B,* flower viewed from above; *C,* ovary and perianth-segments (*p*); *D,* gynœcium.

The minuteness of the flowers of the *Chenopodiaceæ* make the plants a rather difficult and technical group to work with, yet the simplicity of the flowers and the often very distinctive features otherwise, compensate for the difficulties. The bent or coiled embryo which is a characteristic feature of the plants (Fig. 172) is best dissected out from fruits that are not quite mature; however, this type of embryo is not restricted to this family alone, but is found in the succeeding families as well. Close study of the fruits will materially facilitate recognition of the different genera. Another feature that is well marked in many cases is the pronounced succulence, which may be interpreted as an adaptation to conditions in saline habitats.

AMARANTHACEÆ (*Amaranth Family*). These are mostly rough and coarse herbs, or in some cases shrubs, with alternate leaves, and flowers essentially like those of the *Chenopodiaceæ*—bisexual or unisexual, monœcious or diœcious, or polygamous, and always subtended by 2—3 small bracts. The fruit is a small utricle which is circumscissily dehiscent in some genera and indehiscent (achene-like) in others. Species about 475 in 40 genera, chiefly of the warmer parts of the world.

The family is rather poorly represented in our flora, mainly by

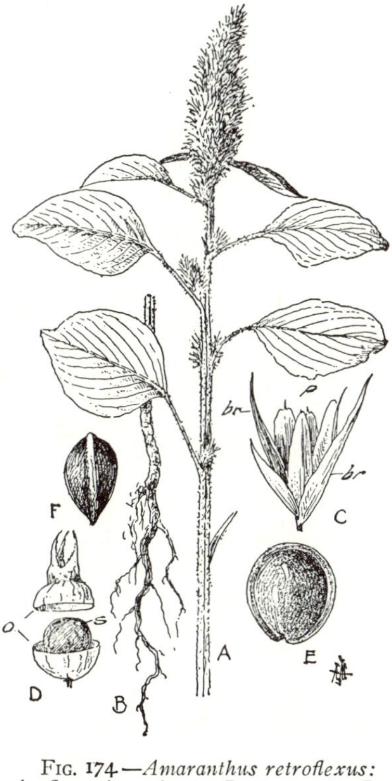

FIG. 174.—*Amaranthus retroflexus: A,* flowering shoot. *B,* tap-root. *C,* flower; *br,* bracts; *p,* perianth. *D,* fruit dehiscing,—*o,* ovary; *s,* seed. *E,* side, and *F,* edge view, of seed.

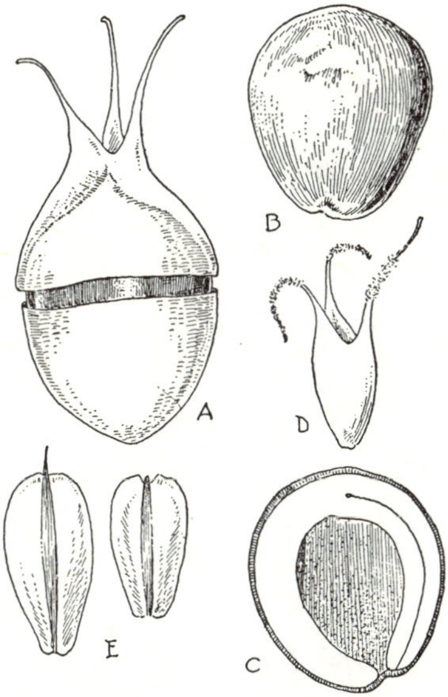

FIG. 175.—*Amaranthus retroflexus: A,* fruit dehiscing. *B,* seed. *C,* section of seed showing embryo curved around endosperm. *D,* gynœcium. *E,* perianth-segments.

species of *Amaranthus, Acnida, Celosia, Iresine, Froelichia, Gomphrena.*

In *Amaranthus, Acnida,* and *Celosia* the anthers are 2-celled and the leaves alternate; in *Celosia* the flowers are bisexual, the filaments united at the base, and a primitive condition is retained in the

FIG. 176.—*Acnida tuberculosa: A,* carpellate, and *B,* staminate shoot. *C,* staminate twig and flower. *D,* anther. *E,* staminate flower with bracts below. *F,* carpellate flower. *G,* gynœcium. *H,* fruit. *I, J,* side and edge views of seed. *K,* diagram of glomerule.

more numerous ovules (2—-8), while in *Amaranthus* and *Acnida* the ovules have been reduced to one and an advance is seen in the unisexual flowers, which are monœcious or polygamous in *Amaranthus,* diœcious in *Acnida,* with loss of the perianth in the carpellate flowers; the filaments in both, however, remain distinct. *Iresine, Froelichia,* and *Gomphrena* are distinguished by 1-celled anthers and opposite leaves, with advance from bisexual flowers in *Gomphrena* and *Froelichia* to unisexual, polygamous, or diœcious ones in *Iresine;* in all three the filaments are united—into a well-developed tube in *Froelichia* and *Gomphrena.* There are, besides, some differences in the inflorescences. *Cladothrix* and *Tidestroemia* are represented in the West. Species of *Amaranthus* (Amaranth), *Iresine, Telanthera, Gomphrena, Deeringia,* and *Celosia* are grown in flower gardens. *Celosia argentea* var. *cristata* is the Cockscomb, *Gomphrena* the Globe Amaranth (Figs. 176, 177, 178).

NYCTAGINACEÆ (*Four o'clock Family*). Flowers regular, bisexual or unisexual, hypogynous, in terminal or axillary clusters, in some cases subtended by an involucre of distinct or united bracts which often simulates a calyx. Perianth undifferentiated, usually petaloid, tubular or funnel-form, of 4—5 lobes or teeth, usually persisting and investing the fruit, the whole forming an anthocarp. Stamens 1—5 or up to 30, the mostly unequal filiform filaments generally united below into an hypogynous ring. Gynœcium always 1-carpellate, the ovary containing a solitary campylotropous or anatropous basal ovule. Embryo straight or curved, or coiled around the mealy endosperm and perisperm. Fruit an achene (anthocarp). Needle-like crystals of calcium oxalate usually present.

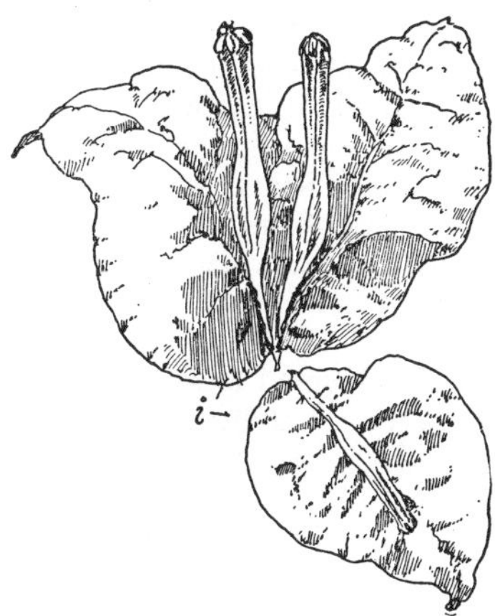

FIG. 177.—Two flowers and large subtending foliaceous involucral bracts (*i*) of *Bougainvillea glabra.*

Herbs, shrubs, or trees, with simple, alternate, or opposite exstipulate leaves, widely distributed but best developed in southwestern and southern United States, and in Chile and Argentine. The herbaceous forms are usually characterized by swollen nodes. Species about 350, in about 25 genera. Characteristic genera are *Mirabilis, Oxybaphus* (*Allionia*), *Bougainvillea, Abronia, Pisonia, Boerhavia.*

A common representative is *Mirabilis Jalapa* of flower gardens, a plant with thick tuberous roots, petiolate lower leaves and sessile upper ones. An involucre of 5 sepal-like bracts subtends a solitary flower. The calyx is trumpet-shaped, with a tube 1 to 2 inches long. In *Boerhavia* there is an involucre of minute deciduous bracts, a many-flowered inflorescence, and a calyx-tube of various shapes and lengths. Several species are native in the arid West. *Oxybaphus* is abundantly represented on the Western plains. The plants are perennial herbs with opposite leaves, from thick roots. The calyx is campanulate, with a short tube and the fruit a ribbed or angled achene. About a half-dozen species occur in the eastern United States.

Bougainvillæa is a South American genus represented in the Southern and Southwestern States by the favorite porch climbers *B. glabra* and *B. spectabilis.* These are favored because of their showy "flowers"—involucres consisting usually of 3 bracts enclosing 1—3 small flowers attached to the midrib of the bracts (Fig. 174). The ovary is stipitate, with a lateral style. The cotyledons are unlike. *Abronia,* with about 45 species, is restricted mainly to the southwestern United States and Mexico. The plants are viscid, with the leaves of opposite pairs unequal. The flowers are usually showy, and solitary or in numerous-flowered clusters (heads) subtended by more or less conspicuous involucres of distinct bracts. The perianth-tube is elongate, with 5 broad spreading lobes, the whole very suggestive of a phlox or verbena, whence the common name "Sand Verbena." The 1—5-winged or ridged achene contains a cylindrical shiny seed. A notable feature is the suppression of one of the cotyledons. The plants are erect or prostrate annuals or perennials. Some species form vivid masses of bloom on the deserts of the Southwest. When collecting specimens of these plants one will usually be disappointed to find that the perianth collapses very rapidly, and to aggravate the situation they are usually covered with wind-blown sand owing to their viscidness. A few species are grown in flower gardens. *Pisonia* comprises 40 or more species of tropical American trees and shrubs, except for a solitary species occurring in Africa. The flowers are mostly unisexual.

CYNOCRAMBACEÆ. Limited to *Cynocrambe prostrata* of the Mediterranean and the Canaries, and *C. macrantha* of central Asia, both prostrate herbs with unisexual, monœcious flowers.

PHYTOLACCACEÆ (*Pokeweed Family*). These are trees, shrubs, or herbs, of wide distribution in temperate and tropical countries, especially America and Africa. The leaves are simple, entire, glabrous, exstipulate or with minute stipules. The small flowers are hypogynous, regular, bisexual or unisexual, with an undifferentiated perianth of 4—5 or occasionally 8—10 segments. Stamens 4—5—8, with the filaments in some cases united. Carpels 1 to several, free or united, with styles corresponding. Ovary inferior in rare cases (*Ag-*

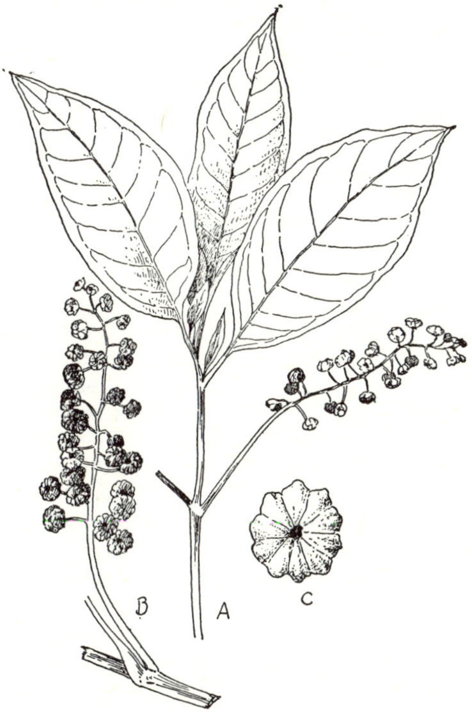

FIG. 178.—*Phytolacca decandra: A,* shoot with inflorescence. *B,* inflorescence enlarged, arising at side of petiole. *C,* fruit viewed from above.

destis, Mexico), with 1 campylotropous ovule in each carpel. Fruit capsular, samaroid, or a berry. Embryo curved around the mealy perisperm. The 2-carpellate fruits form schizocarps, while the many-celled apocarpous kinds form nutlets. (Fig. 175.)

The family shows considerable variation in the structure of the flower, in a number of cases even in the same species. The unisexual flowers are the result of abortion. The gynœcium is particularly notable in this respect, showing reduction from a pluricarpellate to a unicarpellate condition. In all cases each carpel contains but one ovule. There may be one or two whorls of stamens. The inflorescence is a simple or compound raceme or cyme. Species 110—115, in 22—24 genera. The family is represented in our Eastern flora by *Phytolacca decandra* (Pokeweed), a tall large-leaved herb from a large root, bearing numerous small flowers in ample terminal racemes, which with growth become lateral and opposite the leaves. The calyx is white and the berries are dark purple. The root is poisonous. *Rivina humilis* is grown as an ornamental chiefly for its red berries. The gynœcium consists of but 1 carpel. The juice of the berries of these two plants is used as a pigment, and the young sprouts of *Phytolacca decandra* have been used as greens.

AIZOACEÆ (*Carpet-Weed Family*). This is an assemblage of erect or prostrate herbs or subshrubs of diverse habits, inhabiting chiefly deserts and sandy seashores. The leaves are opposite and alternate or whorled, flat, linear, very succulent and angular, or reduced to scales. Stipules scarious or none. The flowers are hypogynous, perigynous, or epigynous, regular, and usually with a simple perianth of 4—5 commonly more or less united segments. Stamens 5, 3, or numerous; when numerous the outer ones transformed into petaloid staminodes, causing the flower to simulate strongly a portulaca or a composite. The fruit is capsular, nut-like, or drupaceous (spurious in all epigynous forms), and many-seeded from a 2- to several-celled ovary. Ovule amphitropous, with the embryo curved around mealy endosperm (perisperm-endosperm).

There are two subfamilies, distinguished on the following characters:

Molluginoideæ, in which the perianth-segments are but slightly united, petaloid staminodes usually wanting, and the ovary superior. *Mollugo.*

Ficoideæ, in which the perianth is tubular, staminodes present or wanting, and ovary superior or inferior. The ovary is superior in *Sesuvium,* inferior in *Mesembryanthemum, Tetragonia.*

The *Ficoideæ* as expressed in *Mesembryanthemum* and *Tetragonia, e.g.,* appear to be rather anomalous members of this family; it is difficult to connect them with any other group of flowering plants. On the other hand, the *Molluginoideæ* tie up in certain respects with the *Portulacaceæ, Phytolaccaceæ,* and *Caryophyllaceæ.*

The *Molluginoideæ* are represented here by the introduced *Mollugo verticillata* (Carpet-Weed), a prostrate mat-forming herb of waste grounds, with small whorled leaves and small flowers on elongate axillary pedicels. There are 5 perianth-segments, 3 stamens, and a 3-carpellate, 3-styled gynœcium with a 3-celled ovary which develops into a 3-valved capsule.

Among the *Ficoideæ* similar decumbent or prostrate plants with opposite leaves, dilated and connate at the base of the petiole, are

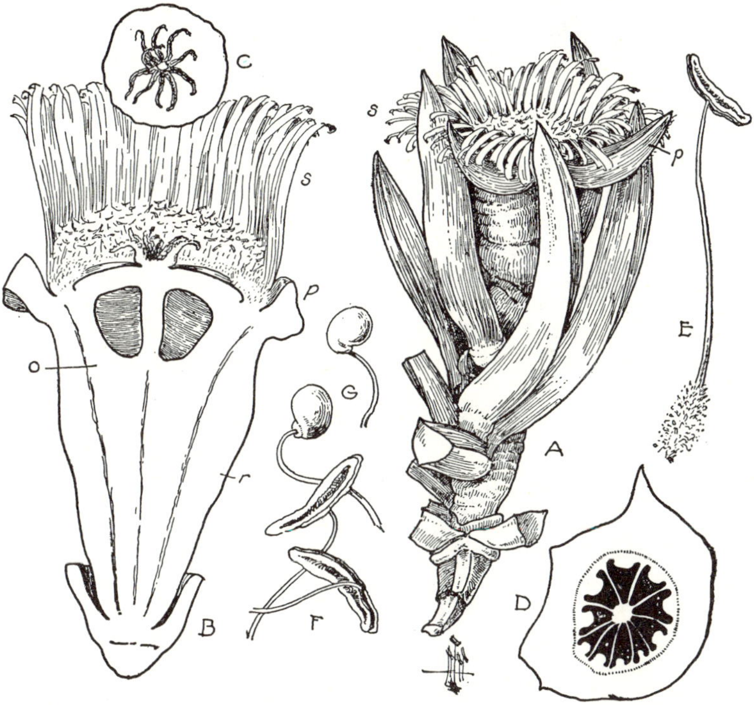

FIG. 179.—*Mesembryanthemum* sp.: *A,* flowering stem. *B,* median section of flower; *s,* petaloid staminodes surrounding numerous stamens; *p,* perianth; *o,* inferior ovary, coalesced with receptacle (*r*). *C,* ovary viewed from above, showing styles. *D,* transverse section of ovary. *E,* stamen. *F,* dehisced anthers. *G,* ovules, with long stalks.

found in *Sesuvium,* and the flowers are solitary or several together in the axils of the leaves. There is a pinkish or purplish perianth, 5-many stamens with filiform filaments, sometimes united at the base, and a 3—5-celled ovary, usually with numerous ovules in each cell. The fruit is a loculicidal or circumscissile capsule. *Sesuvium maritimum,* an annual, is native along the Atlantic Coast from Long

Island to Florida. Two others occur in the West and South. *Mesembryanthemum* comprises a unique group of prostrate herbs which in addition to an inferior ovary have developed extreme succulence as well as extreme modification of the stamens into petaloid staminodes of brilliant hues, the flower as a result strongly simulating a composite. The ovary in this genus is 5—12-celled (carpellate) with as many styles, contains numerous very small seeds, and forms a spurious capsular fruit which dehisces at the top, or the fruit is in some cases of fleshy texture. The widely distributed *Mesembryanthemum crystallinum* (Ice-Plant), with ovate to spatulate leaves and herbage densely coated with glistening watery vesicles, and the stouter *M. edule* (Sea Fig), a profusely spreading plant with very succulent 3-angled leaves and large brilliant yellow or purple flowers, are typical. These plants form extensive mats on the dunes and bluffs of the southern California coast. The genus in the broad sense numbers from 300 to 400 species and is the largest in the family. The Australian *Tetragonia expansa* (New Zealand Spinach) has been introduced in coastal California. It is the second largest genus of the family (60 spp.). The herbage may be used as greens. Other representatives of the family are also found in California, *viz., Glinus lotoides* (*Molluginoideæ*), and *Cypselea humifusa* (*Ficoideæ*) (Fig. 179).

PORTULACACEÆ (*Purslane Family*). The purslanes are succulent herbs, rarely shrubby, with hypogynous or epigynous, cyclic, regular, bisexual flowers, and perianth differentiated into calyx and corolla of 2 (or 4—8) and 4—5 segments respectively. Stamens 5 to numerous. Gynœcium syncarpous, 3—5-carpellate, with as many distinct styles. Seeds 2 to numerous, with embryo curved around the endosperm. A small family of about 180 species, in 17 genera, chiefly American.

The family is represented in our flora by *Talinum, Calyptridium, Calandrinia, Claytonia, Montia, Portulaca,* and *Lewisia.*

Of these genera all but *Lewisia* have 2 sepals and all have superior ovaries except *Portulaca.* In *Talinum* and *Lewisia* the stamens are 5 to numerous, in *Calandrinia* 5—14, or in some cases 3. The seeds are numerous in *Talinum* and *Calandrinia,* few to many in the other genera. Corms, rootstocks, and bulbiferous stolons occur in *Claytonia,* and succulence is well marked in *Montia, Portulaca,* and *Lewisia.* In *Lewisia* the leaves are in basal rosettes from a fleshy root; the roots were formerly a source of food among Indians. A characteristic species is *Lewisia rediviva,* the Bitter-Root, the State flower of Idaho. In this species the leaves appear very early in the spring and are gone before the flower appears, much later, the flower thus appearing to spring right out of the ground. *Montia perfoliata* (Miners' Lettuce), a succulent herb widely distributed in the West, is easily recognized by the connate-perfoliate leaf borne below the inflorescence. *Portu-*

FIG. 180.—*Stellaria longifolia: A,* flowering and fruiting shoot. *B,* flower. *C,* petal. *D,* calyx. *E,* ventral (*a*) and (*b*) dorsal view of stamen. *F,* gynœcium. *G,* capsule, with calyx. *H,* capsule with calyx removed. *I,* median section of capsule. *J,* floral diagram.

laca, Montia, and *Calandrinia* are represented in flower-gardens. *Portulaca oleracea* (Common Purslane) is a widespread weed, often abundant in gardens. The numerous stamens of *Portulaca* and *Lewisia* suggest affinities with the *Aizoaceæ*.

BASELLACEÆ. A very small family (15 species, in 5 genera) of succulent perennial climbing herbs from tuberous rootstocks, all

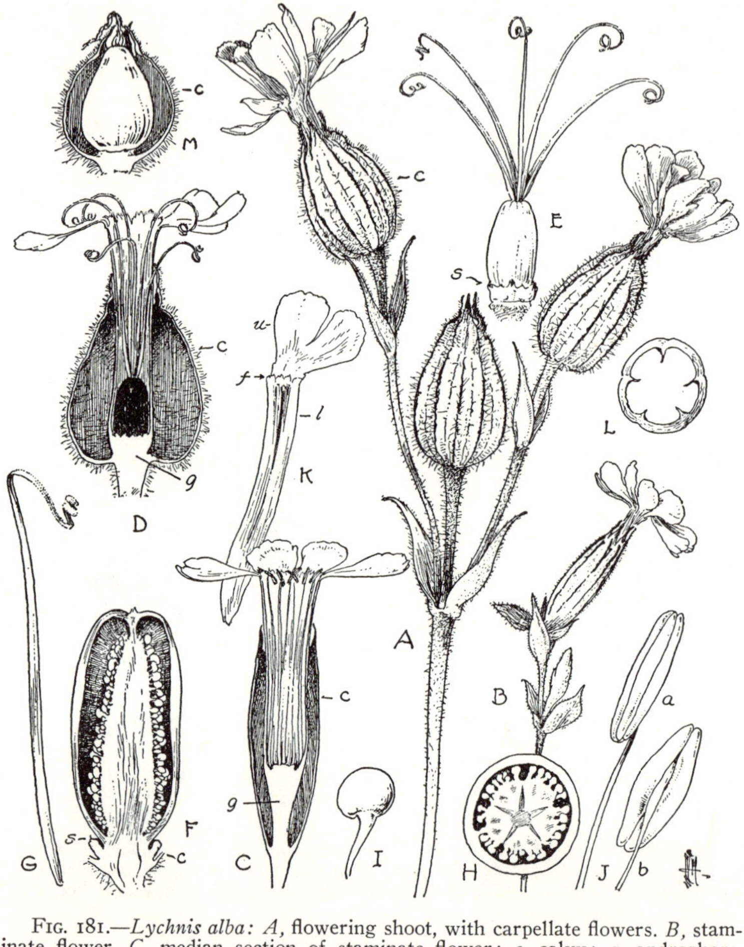

FIG. 181.—*Lychnis alba: A,* flowering shoot, with carpellate flowers. *B,* staminate flower. *C,* median section of staminate flower; *c,* calyx; *g,* androphore. *D,* median section of carpellate flower; *c,* calyx; *g,* gynophore. *E,* gynœcium; *s,* vestigial stamens, on margin of gynophore. *F,* median section of ovary, with central placenta; *s,* vestigial stamen, inside insertion of petals on margin of cup-shaped gynophore. *G,* style. *H,* transverse section of ovary. *I,* ovule. *K,* petal, —*u,* limb; *l,* claw; *f,* corona. *L,* transverse section of ovary showing incomplete septa. *M,* fruit (capsule) invested by gamosepalous calyx.

except one Asiatic genus native to tropical America. The small, regular, bisexual flowers, borne in racemose inflorescences, are subtended by 2 small bracts. Species of *Basella* and *Boussinggaultia* are horticultural.

CARYOPHYLLACEÆ (*Pink Family*). The pinks are herbs, or in some cases subshrubs, with terete stems swollen at the nodes, opposite, entire, commonly narrow leaves, and hypogynous or perigynous, regular, 5-merous, bisexual or unisexual flowers in cymose

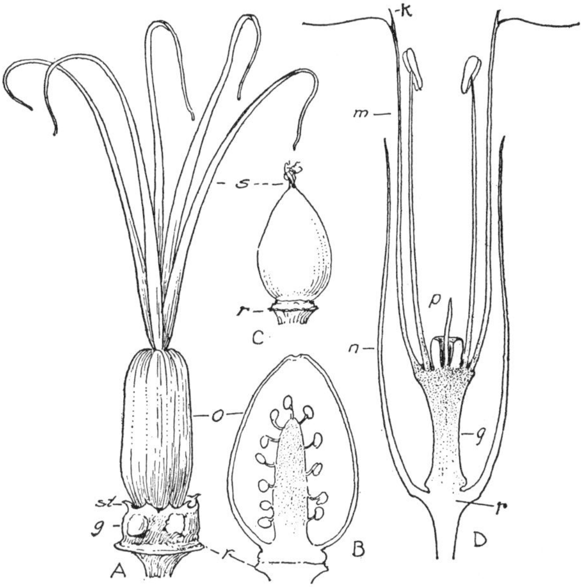

FIG. 182.—*Lychnis alba: A,* gynœcium, with vestigial stamens (*st*) on margin of cup-shaped discoid gynophore (*g*). *B,* median section of ovary (*o*); *r,* receptacle. *C,* immature capsule, with remnants of styles at summit; *r,* receptacle. *D,* median section of staminate flower, with vestigial gynœcium (*p*) seated in a cup-shaped disk; *m,* petal; *n,* calyx; *g,* androphore; *r,* receptacle; *k,* corona.

inflorescences; and generally with calyx and corolla. Gynœcium syncarpous, 5—2-carpellate, with as many styles; ovary usually 1-celled, with numerous (rarely 1) anatropous or campylotropous ovules on a central columnar, or basal, placenta. Fruit a capsule or berry. Embryo curved around the endosperm. About 1,450 species in 70 or more genera, widely distributed, mostly in temperate climates. Compare with *Primulaceæ*.

There are two subfamilies, distinguished by the following characters:

Alsinoideæ, the more generalized group, has a calyx of distinct sepals, but is specialized in the direction of perigyny in some forms.

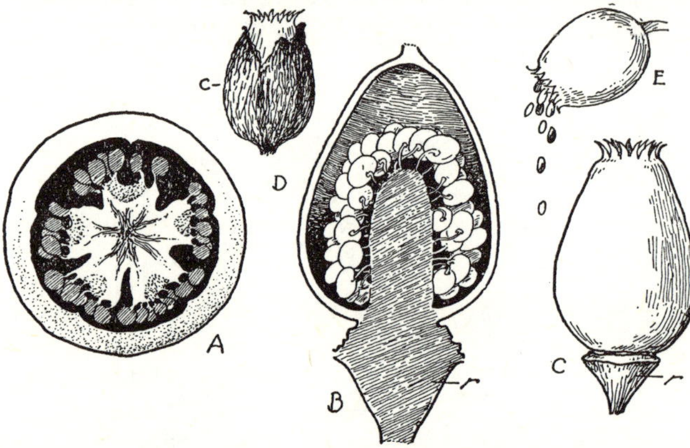

FIG. 183.—Transverse section of ovary (*A*), median section (*B*), and capsule (*C*) of *Lychnis alba: A,* showing five-lobed placenta and vestigial dissepiments on inner edge of ovary-wall. *B,* showing free central columnar placenta; *r,* receptacle. *D,* capsule with investing calyx of *Lychnis dioica. E,* capsule of same discharging seeds.

Includes *Spergularia, Sagina, Arenaria* (incl. *Moehringia*), *Stellaria, Cerastium, Holosteum.* (Fig. 180).

Silenoideæ, the more specialized group, has a gamosepalous calyx, and a stipitate gynœcium, *i.e.,* elevated on a stalk (gynophore). Includes *Agrostemma, Lychnis, Silene, Saponaria, Gypsophila, Dianthus.* (Figs. 181, 182, 183.)

The members of this family are fairly readily recognizable as a whole in the field, but the different genera are not so easily distinguished, especially those of the *Alsinoideæ.* The technical characters to be particularly noted, aside from those mentioned in the above key, are the presence of stipules, shape of petals, shape and nervation of the calyx, number of styles, and shape and mode of dehiscence of the capsule. Thus in *Spergularia, e.g.,* stipules are present and there are three styles; but a like number of styles is also present in *Arenaria,* which, however, lacks stipules. In the same way the closely similar *Lychnis* and *Silene* are distinguished mainly on the number of styles—5 in *Lychnis,* 3 in *Silene.*

A number of species are common weeds, such as *Cerastium vulgatum* (Common Mouse-Ear), *Agrostemma Githago* (Corn Cockle), frequent in grain fields, and *Saponaria offinalis* (Soapwort or Bouncing Bet). *Dianthus* (Carnation), *Gypsophila* (including Baby's Breath), *Lychnis* (Campion), and *Silene* (Catch-Fly, Campion) are represented in flower-gardens.

The *Caryophyllaceæ* by their clearly differentiated calyx and corolla mark the culmination of the *Monochlamydeæ* and a transition to the *Dialypetalæ.*

CHAPTER XIV

THE DIALYPETALÆ

Perianth differentiated into calyx and corolla. Calyx in some forms greatly reduced. Stamens as many as the segments of the corolla or more, often numerous. Sympetalous corolla rare. Insect pollination predominant. Herbs, shrubs, and trees. Herbaceous forms predominant.

Besides these features there is a general evolutionary tendency from polycarpellate apocarpous gynœcia to 2-carpellate syncarpous gynœcia, from many-seeded to few-seeded forms, and from hypogynous to perigynous and epigynous flowers, with corresponding superior and inferior ovary.

Some members show phylogenetic affinities with the *Metachlamydeæ.*

ORDER RANALES

Flowers bisexual, sometimes unisexual; hypogynous or in some cases perigynous, epigynous; cyclic, spirocyclic, or acyclic; regular or irregular. Perianth undifferentiated or differentiated into calyx and corolla, sometimes wanting. Stamens mostly numerous. Carpels mostly numerous, sometimes 1, free or sometimes united. A large order of herbaceous or woody plants, the lower families of which are characterized in particular by a generalized or primitive type of floral structure.

Families: *Nymphæaceæ, Ceratophyllaceæ, Trochodendraceæ, Cercidiphyllaceæ, Ranunculaceæ, Lardizabalaceæ, Berberidaceæ, Menispermaceæ, Magnoliaceæ, Himantandraceæ, Calycanthaceæ, Lactoridaceæ, Anonaceæ, Eupomatiaceæ, Myristicaceæ, Gomortegaceæ, Monimiaceæ, Lauraceæ, Hernandiaceæ.*

Key to Families Represented by Native or Cultivated Plants

A. Aquatics.

Sepals and petals wanting. Carpels 1. Leaves finely dissected and whorled. *Ceratophyllaceæ.*

Sepals and petals present, with transition from one to the other. Stamens numerous. Stout plants with large shield-shaped or cordate long-stalked leaves and long-flowering stems from stout rootstocks. Water-lilies. *Nymphæaceæ.*

B. Terrestrials, few aquatics.
  1. Herbs.
    Carpels 1. Anthers dehiscing by valves. *Berberidaceæ.*
    Carpels numerous to few, all distinct. Anthers opening by lateral slits. *Ranunculaceæ.*
  2. Vines. Carpels few to many.
    a. Flowers hypogynous.
      Sepals 4-many. Petals 6. *Menispermaceæ.*
      Sepals 4-5, small to very large, petaloid. Petals none. *Ranunculaceæ (Clematis).*
    b. Flowers perigynous. Receptacle hollow, urn-shaped. *Monimiaceæ.*
  3. Trees and shrubs.
    a. Carpels 1.
      Flowers perigynous, the hypanthium bearing numerous stamens inside.
        Flowers bisexual. *Calycanthaceæ.*
        Flowers mostly unisexual, monœcious or diœcious. *Monimiaceæ.*
      Flowers hypogynous.
        Corolla none.
          Stamens monadelphous. Fruit fleshy. *Myristicaceæ.*
          Stamens distinct, some often transformed into staminodes. Fruit a berry or drupe. *Lauraceæ.*
        Corolla present; sepals and petals often in two or more whorls. Fruit a berry or capsule. *Berberidaceæ.*
    b. Carpels 2 to 5 or numerous, rarely 1.
      Calyx and corolla both wanting. Stamens numerous. Carpels 2 or more, in one whorl. *Trochodendraceæ.*
      Corolla wanting. Stamens 6. Carpels 3-9. *Lardizabalaceæ.*
      Calyx and corolla present. Carpels when numerous often on an elongate conical or cylindrical axis. Flowers large and showy.
        Sepals 3, petals commonly 6. Fruit fleshy. *Anonaceæ.*
        Sepals and petals in three or more 3-merous whorls. Fruits dry. *Magnoliaceæ.*

NYMPHÆACEÆ (*Water-Lily Family*). Cosmopolitan aquatic herbs with submersed or floating leaves and bisexual, regular, hypogynous or perigynous flowers. Floral organs few to numerous, the sepals, petals, and stamens commonly passing into one another. Ovules few to numerous on the inner wall of the carpels. Fruits achenes or berries, often with arillate seeds. Perisperm and endosperm present. Cotyledons fleshy, hypogeal. Rootstocks stout.

Three subfamilies are recognized, *viz.:*

I. *Nelumbonoideæ.* Aquatics with long-stalked peltate leaves and showy flowers raised above the water on long pedicels. Flowers spiral, with numerous perianth-segments and stamens, the distinct carpels borne in cavities on the upper surface of the greatly enlarged obconical receptacle. Endosperm none. One genus, *Nelumbo* (Sacred Bean), with one species, *Nelumbo lutea* (Yellow Nelumbo) in eastern North America, and another, *N. nucifera,* in tropical and eastern Asia. (Fig. 184.)

II. *Cabomboideæ*. Aquatics with dissected submersed leaves and long-stalked, peltate, floating leaves, and mucilaginous stems, from thick rootstocks. Flowers spirocyclic or cyclic, usually with 3 sepals and 3 petals. Stamens 3—18, and 3—18 free carpels. Fruits coriaceous, 3-seeded. Seeds with perisperm and endosperm. Two genera.

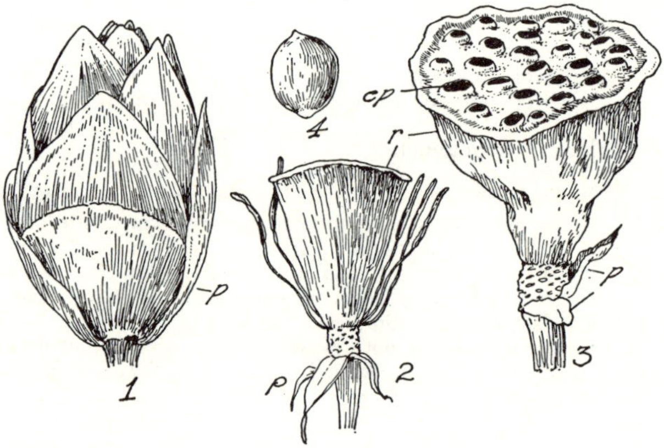

FIG. 184.—*Nelumbo lutea: 1,* flower; *2,* same with most of perianth-segments and stamens removed; *3,* older stage of same; *4,* fruit (nutlet); *p,* perianth-segments; *r,* obconical upper carpel-bearing portion of receptacle; *cp,* cavities, in each of which a carpel is seated.

Cabomba with 6 species in tropical and subtropical America, and *Brasenia* with *B. purpurea* in America, Asia, Africa, and Australia, and *B. Schreberi* in California.

III. *Nymphæoideæ*. Aquatics with submersed or floating long-stalked leaves with narrow to broad or orbicular blades cut into a deep sinus at the base, growing from stout rootstocks. Flowers bisexual, raised above the water on long peduncles. Perianth differentiated into 4—6 sepals and numerous petals, the inner passing into staminodes or stamens. Carpels several to many, united, the more or less sessile stigma discoid, with radiating stigmatic lines. Ovules numerous, on the inner wall of the ovary. Fruit a leathery many-seeded berry. Five genera and about 45 species, widely distributed in north temperate regions. Only *Nymphæa* (*Nymphozanthus*) and *Castalia* represented in our flora. *Nymphæa advena,* the Large Yellow Water-Lily, is common in shallow fresh waters from Labrador and Nova Scotia to Utah and Texas, and the almost as com-

FIG. 185.—*Castalia odorata* (White Water-Lily): *a,* a flower-bud; *b,* median section of flower-bud before opening, showing stamens and innermost petals arising from ovary-wall; *r,* receptacle. *c,* section of gynœcium, with radiating styles capped by fleshy decurrent stigmas; in center a fleshy projection from axis of ovary; *d,* a style and stigma; *e,* a series of stamens; *f,* an inner stamen; *g,* leaf.

mon *Castalia odorata* (White Water-Lily) from Newfoundland to Manitoba and Kansas, south to Louisiana and Florida. (Fig. 185.)

CERATOPHYLLACEÆ (*Hornwort Family*). In the broadest sense a monotypic family represented by the solitary *Ceratophyllum demersum* (Hornwort), a submersed plant of lakes and streams, with long slender stem, whorled filiform-dissected leaves, and unisexual, monœcious or diœcious flowers solitary in the axils of the leaves. Stamens numerous. Gynœcium a single carpel producing a single seed. Ovary superior. Perianth none, but the flowers subtended by an involucre which persists in the fruit—a nutlet or achene beaked by the persistent style. Endosperm none. A peculiar feature of the embryo is the whorl of 4 ovoid cotyledons and a well-developed epicotyl of several nodes and leaves.

TROCHODENDRACEÆ is a small family of about 6 species of trees of Asia and Japan. Genera two, *Trochodendron* (Japan) and *Euptelea* (eastern Asia). The flowers are cyclic, bisexual or unisexual, and monœcious.

CERCIDIPHYLLACEÆ is another very small family, comprising but two species of trees of the genus *Cercidiphyllum* of Japan. The flowers are unisexual and diœcious, the gynœcium 2—5-carpellate, the stamens 16—35. Endosperm copious.

RANUNCULACEÆ (*Crowfoot Family*). Flowers spiral, spirocyclic or cyclic, mostly bisexual, hypogynous, regular or irregular. Perianth of varying number of segments differentiated into calyx and corolla, the latter sometimes wanting and the sepals then petaloid. Stamens indefinite, free. Carpels numerous to few or 1, usually free, with many to few (1) ovules. Fruits achenes, follicles, or berries. Endosperm present. Integuments 1 or 2. About 35 genera and 1,200 species of annual or perennial herbs and some woody plants, with acrid juice and alternate (except in *Clematis*), clasping or sheathing, exstipulate leaves; widely distributed, with numerous representatives in America.

The flowers are solitary or in cymose inflorescences, and staminodes and nectariferous structures of various sorts are frequently present. Petaloid sepals are present in some forms having both calyx and corolla. The leaf-blades are of diverse shapes, from lanceolate to broad, cordate, palmately lobed or divided or palmately digitate, or pinnate. Dissected leaves are characteristic of a number of aquatics, associated with heterophylly. The sheathing petiole is in some cases furnished with stipular lobes. In *Clematis* the leaves act as climbing organs. The cotyledons are usually epigeal but remain hypogeal in *Pæonia* and a species of *Clematis*. Rhizomes and tuberous roots are characteristic of perennials.

In certain respects the family is suggestive of the Monocotyledons. Thus the cotyledons are sometimes connate by their bases, forming a tube through which the epicotyl grows out laterally. And in the arrangement of the vascular tissues a somewhat monocotyledonous type is approached.

Engler recognizes four tribes, *viz.:*

I. *Hydrastideæ,* in which the flowers are solitary, with an undifferentiated perianth, and nectariferous organs are wanting. In the ovule the outer integument is strongly developed. Fruit a berry. Represented by *Hydrastis* and *Glaucidium.*

II. *Pæonieæ,* in which the flowers (and ovules) are as in the *Hydrastideæ,* but with calyx and corolla. The fruit is follicular. *Pæonia.*

III. *Helleboreæ,* in which the flowers are solitary or several, the integuments equally developed, and the fruit a follicle or berry. Includes *Caltha, Trollius, Helleborus, Eranthis, Isopyrum, Coptis, Zanthorrhiza, Aquilegia, Actæa, Delphinium, Aconitum, Nigella.*

IV. *Anemoneæ,* in which the flowers and the ovule are as in the *Helleboreæ,* but the fruit is an achene. *Anemone, Clematis, Myosurus, Ranunculus, Thalictrum, Adonis, Hepatica.*

In the first three tribes, a varying number of ovules (rarely 1) spring from each side of the ventral suture of the carpel, but in the *Anemoneæ* the solitary ovule arises at the base of the dorsal suture. Rudimentary ovules are sometimes present. The general situation may be illustrated by the following representatives of the different tribes.

Hydrastideæ. Hydrastis canadensis (Orange-Root), a perennial with solitary greenish-white flowers and palmately lobed leaves, from a thick yellow rhizome. There are 3 fugacious sepals, and petals are wanting, but stamens and carpels are numerous and distinct, and 2 ovules are produced in each carpel. At maturity the carpels become a head of crimson berries. Only two species.

Pæonieæ. Pæonia Brownii, the wild Pæony of the Pacific States, a perennial herb with ternately compound leaves and solitary flowers. The sepals are foliaceous and persistent, the thick purplish petals rather fugacious, falling soon after the opening of the flower. Stamens numerous, and carpels 2—5, distinct, stout, erect, with very short thick styles. Below the ovary is an annular angular-lobed disk. Fruits follicles, enclosing several seeds. At maturity the follicles are brought into contact with the ground by the bending over of the plant (Jepson). About 15 species of Pæonies are known. (Fig. 186.) *Pæonia suffruticosa* includes the shrubby "tree" peonies.

Helleboreæ. Caltha palustris (Marsh-Marigold), a common yellow-flowered marsh-plant with reniform-cordate leaves. Petals are wanting and the petaloid sepals fall soon after the opening of the flower. There are numerous stamens, and carpels numerous to few each with several ovules on each side of the ventral suture. The fruits are small follicles. The 15 or more species of the genus are widely distributed through temperate and arctic regions.

Delphinium (Larkspur, cultivated). This is a tall herb with palmately divided leaves and dark-blue irregular flowers in a racemose-paniculate inflorescence. Both sepals and petals are present, the sepals petaloid and the posterior one produced into a spur. The petals are small and the two posterior ones are spurred. The numerous stamens are spirally arranged. The 1 to 5 carpels have elongate ovaries contracted above into slender styles with small apical stigmas. The fruits are follicles. The flowers are bilaterally symmetrical. The genus comprises about 125 species, widely distributed. *Aconitum* is very similar. In *Nigella* we have an exceptional case in this family of the development of a syncarpous gynœcium, the carpels being united by the ovaries. Spurred, cornucopia-like petals are developed in *Aquilegia,* but the flower is regular.

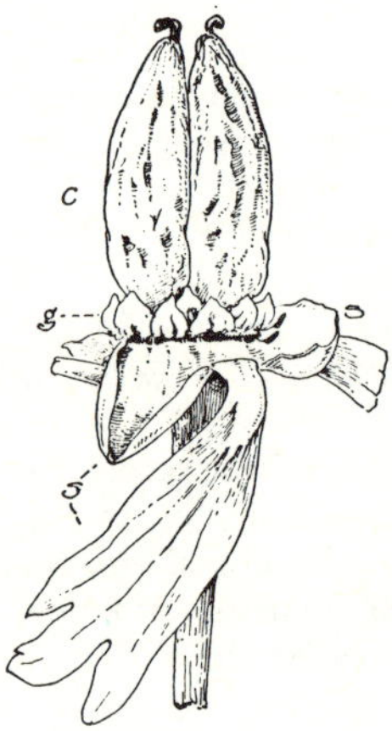

FIG. 186. — *Pæonia Brownii* (Western Wild Peony): a 2-carpellate gynœcium (*c*) with hypogynous lobed staminodial (?) disk (*a*); *s,* sepals, one foliaceous.

Anemoneæ. A common species of this tribe is *Anemone canadensis* (Wind-Flower), inhabiting sunny banks and prairies from New England to the Rockies. It is a stout hairy species with showy white flowers and palmately lobed leaves, the basal ones long-petioled, the stem-leaves sessile and forming involucres (primary and secondary). Petals are wanting, these being replaced by petaloid sepals. Stamens numerous. The usually numerous carpels develop into flattened achenes beaked by the slender persistent style. The whole globular head of achenes is a sort of collective or multiple fruit of which the individual achenes are the true fruits. In some species the carpels are arranged densely in cylindrical heads (*A. cylindrica*). The genus comprises over 80 species. *Anemone patens* var. *Wolfgangiana* (*Pulsatilla hirsutissima* of Britton), the Pasque Flower, has large blue sepals, and numerous stamens, the outer of which are modified into staminodes. The numerous carpels are provided with long, slender, hairy styles as in *Clematis.* The re-

FIG. 187.—*Anemone canadensis: a,* habit sketch; *b,* polycarpellate apocarpous gynœcium; *c,* gynœcium in sectional view, showing elongate receptacle; *d,* a young carpel; *e,* more mature carpel, edge view; *f,* mature carpel (achene); *g,* dorsal and ventral view of stamens.

ceptacle is much shortened to a sub-globose shape. (Figs. 187, 22.)

In *Clematis* the flowers are solitary or in cymose-paniculate inflorescences, with petals wanting or these replaced by petaloid staminodes. In some species the flowers are very small and diœcious (in ours), in others there are large and showy sepals. The plants are erect herbs or woody vines with compound leaves, climbing by means of the leaf-stalks. About 170 species in all. In *Myosurus* (Mouse-Ear) the leaves are linear, in basal rosettes, and the receptacle is greatly elongated. The plants inhabit moist open places.

The largest genus of the family is *Ranunculus* (Buttercup, Crowfoot), comprising about 175 species widely distributed over the cooler regions of the northern hemisphere. The plants are annual or perennial herbs of wet or dry habitats, bearing simple lanceolate or palmately divided or dissected or compound leaves and solitary or clustered, white or yellow, flowers. There are usually 5 sepals (fugacious) and as many or more petals, the number varying even in the same species. Carpels numerous, in fruit becoming achenes, spirally arranged in heads or spikes on the more or less elongated receptacle. The style persists as a beak on the achene and nectar glands are commonly present at the base of the petals. Dissected leaves with filiform segments are characteristic of some aquatic species and these and others of muddy shores commonly spread by prostrate rooting stems or runners. Some species develop fleshy fascicled roots. The genus is well represented in American vegetation.

In *Thalictrum* (Meadow Rue) there is advance from bisexual to unisexual or polygamous flowers and diœcism, associated with shortening of the receptacle and reduction in the gynœcium to a few carpels. Numerous petals persist in *Adonis* along with five sepals, while the receptacle remains elongate and the carpels numerous. In *Hepatica* there is a loss of the corolla, compensated for by the petaloid calyx, and in addition the involucre is crowded against the calyx, which it closely simulates. The receptacle is shortened but the gynœcium remains relatively pluricarpellate.

To summarize the general situation in the *Ranunculaceæ,* we find the following significant features: (1) Flowers of a primitive or generalized type, in which the floral organs are numerous and spirally arranged on an elongate floral axis (receptacle); (2) presence of certain anatomical features in the stems that are suggestive of the Monocotyledons; (3) dominance of the herbaceous habit, with persistence of aquatics. Significant evolutionary steps are seen in the suppression of the corolla and the substitution of a petaloid calyx;

reduction of the number of floral organs, associated with tendencies in the direction of shortening of the floral axis, and cyclic arrangement of the floral organs; development of a woody habit in some forms, and a development of unisexual flowers and diœcism.

LARDIZABALACEÆ (*Lardizabala Family*). A small family of woody climbers of Asia, Japan, and Chile, with regular unisexual or polygamous flowers in racemes or solitary. Rudiments of the other

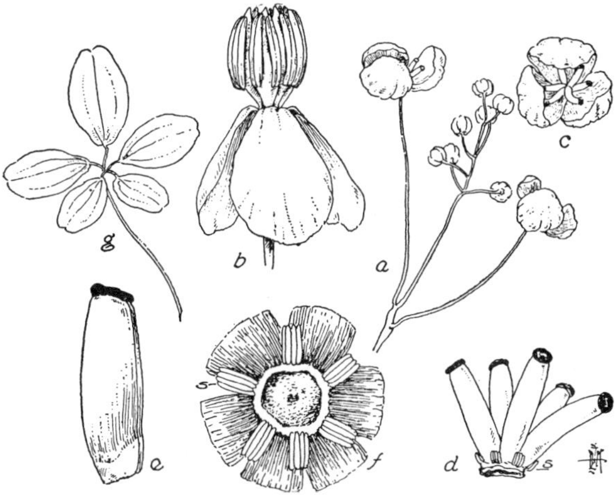

FIG. 188.—*Akebia quinata: a,* inflorescence with carpellate (large) and staminate (small) flowers; *b,* staminate flower; *c,* carpellate flower; *d,* the five-carpellate gynœcium, with rudimentary stamens at base of carpels and alternating with them; *e,* a carpel, enlarged; *f,* part of a carpellate flower with carpels removed to show rudimentary stamens (*s*); *g,* leaf.

essential organs are present in the respective flowers. The perianth is undifferentiated, in two (or one) 3-merous whorls. Two whorls of nectaries are often present. Gynœcium of 3 to 9 (or more) distinct carpels, each producing many ovules. Fruit commonly a berry. Seeds with endosperm. Eight genera and about 20 species. The leaves are digitately compound and the fragrant purple flowers monœcious, in racemes. *Akebia quinata* and *A. lobata* are horticultural. (Fig. 188.)

BERBERIDACEÆ (*Barberry Family*). The barberries are characterized by regular, bisexual, hypogynous, 3-merous or 2-merous flowers with an undifferentiated perianth or with calyx and corolla, borne solitary or in racemes. Stamens as many as the petals and opposite them, and distinctive in that the extrorse anthers open by means of lids (hinged above). The gynœcium is reduced to a single carpel with a large, more or less sessile, stigma, and containing 2 ovules. Fruit a berry or capsule. The plants are shrubs or herbs with alternate, cauline or basal, simple or compound leaves. Ten genera and about 130 species mostly of the northern hemisphere, with a few in the tropics. There are two subfamilies, distinguished mainly by the leaves and the presence or absence of nectaries (staminodes), *viz.:*

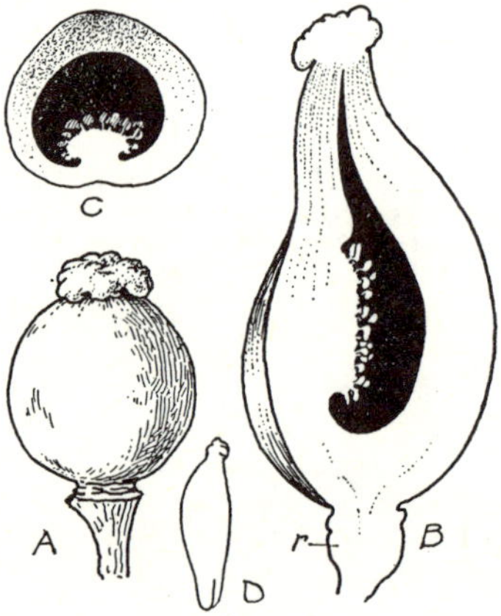

FIG. 189.—*Podophyllum peltatum: A,* gynœcium after flowering. *B,* median section of gynœcium, with placenta at left; *r,* receptacle. *C,* transverse section of ovary. *D,* ovule.

Podophylloideæ, in which nectaries are wanting and the leaves are not pinnate. Includes *Podophyllum, Diphylleia, Jeffersonia, Achlys, Vancouveria.*

Berberidoideæ, in which nectaries are present and the leaves are compound or reduced to a single leaflet. Includes *Berberis, Mahonia, Caulophyllum, Leontice* (in Orient).

Podophylloideæ. Among the representatives of this subfamily in eastern North America is the peculiar *Podophyllum peltatum* (May-Apple), a low perennial herb with peltate, palmately lobed basal leaves from a thick rootstock (poisonous). The flowering stems, which spring from separate rootstocks, bear two palmately divided leaves between which is borne a solitary large flower on a long peduncle. The flower buds are subtended by 3 fugacious bracteoles. The 3-merous condition of the lower appears in the 6 petaloid sepals, the 6—9 petals, and stamens—which are usually twice as many as the petals. The stamens are exceptional in the family in that they dehisce lengthwise, perhaps a significant feature in connection with the problematical affinities of this genus. The gynœcium is reduced to a solitary carpel which becomes a many-seeded berry, edible and of the flavor of a banana. An exceptional feature, further, are the connate cotyledons. Similar cotyledons occur also in *Leontice.* Possibly there is some connection here with the *Ranuncu-*

laceæ. Furthermore the trimerous condition of the flower is very suggestive of the Monocotyledons. Similar suggestive features appear in the next following genus, below. The few other species of *Podophyllum* are found in eastern Asia. (Fig. 189.)

Of *Diphylleia* only two species are known, one, *D. cymosa* (Umbrella-Leaf) of the Appalachians, and the other, *D. Grayi,* of Japan and Sakhalin Island. These are perennial herbs of the aspect of *Podophyllum* but springing from horizontal rootstocks and bearing a cluster of flowers. As in *Podophyllum,* the floral organs are in 3-merous whorls and the gynœcium consists of a solitary carpel which becomes a few-seeded berry, but the anthers open by lids. In *Jeffersonia* we have again a genus of but two species, remarkably widely separated geographically, one, *Jeffersonia diphylla* (Twin-Leaf) in Atlantic North America, the other, *J. dubia,* in far-off Manchuria. Here, however, the floral organs are in 4-merous (or 2-merous?) whorls, conforming to the formula *S*4 *P*8 *A*8 *G*1. The plants are perennial herbs with palmately lobed leaves and solitary scapose flowers, growing from a short fibrous-rooted caudex.

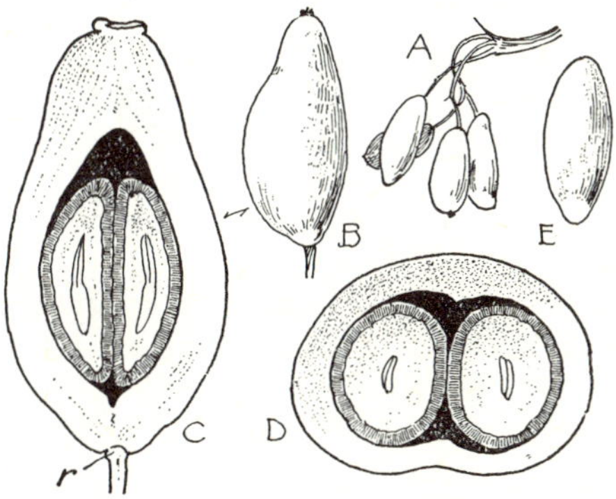

FIG. 190.—*Berberis Thunbergii: A,* cluster of fruits; *B,* fruit; *C,* median, and *D,* transverse, section of fruit, containing two ovules with stony seed-coat and small embryo in center of copious endosperm; *r,* receptacle. *E,* seed.

Again in *Achlys* we have a genus of but two species, similarly widely apart geographically, with *Achlys triphylla* (Vanilla-Leaf) ranging through the moist coniferous forests from California to British Columbia, and its solitary congener, *A. japonica,* in Japan. In this case the perianth is wanting and there is a marked instability in the number of stamens (6—12—13). But as in the preceding genera there is a solitary carpel, containing, however, only one ovule. The fruit is fleshy at first, but dry and indehiscent at maturity. The plants spring from a creeping rootstock. The presence of cumarin in the leaves gives them a vanilla-like odor when dry. Two species of *Vancouveria* occur from California to Washington.

From the standpoint of plant geography, the preceding genera are of special interest, among a large number of others, as illustrating the close affinities of the flora of North America with that of eastern Asia.

Berberidoideæ. The characteristic features of this subfamily are well illustrated in *Berberis* (Barberry). The barberries are shrubs or subshrubs with yellow wood and simple (unifoliolate) leaves, those

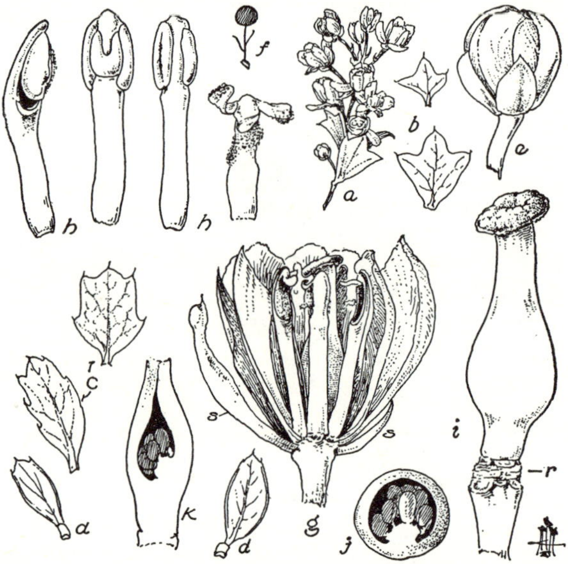

FIG. 191.—*Berberis Darwinii: a,* inflorescence; *b,* upper foliage leaves; *c,* lower foliage leaves; *d,* leaves at base of inflorescence; *e,* flower-bud; *f,* flower and bracteoles (diagrammatic); *g,* flower with petal and sepal removed; *s,* sepals; *h,* stamens; *i,* gynœcium; *r,* receptacle; *j,* transverse section of ovary, with single parietal placenta; *k,* median section of ovary.

of the primary shoots reduced to spines. The flowers are yellow, in racemose inflorescences. Sepals and petals are present and, with the stamens (6), are arranged in 3-merous whorls. The anthers open by lids (valves). The gynœcium is reduced to a solitary carpel which is capped by a broad stigma and becomes a few- to 1-seeded, mostly red, berry. Species about 80, of Eurasia and the Americas. Many as ornamentals. *Berberis vulgaris* (Common Barberry), a native of Europe, which has spread widely from cultivation in this country, is the host-plant of the æcidial stage of the destructive wheat-rust, for which reason a national campaign has been waged for a number of years to effect its extermination. *Berberis Thunbergii,* also cultivated, is immune to this rust. (Fig. 190.)

Mahonia is by some writers incorporated in *Berberis*. The plants differ from *Berberis* in having pinnately compound leaves (with spinescent leaflets) and no spines on the branches. The flowers are yellow, in more or less dense racemose inflorescences, and of the same structural plan as in *Berberis*. Species about 80, of Asia and Pacific North America; they are evergreen shrubs of ornamental habit. *Mahonia repens, M. aquifolia,* and *M. nervosa,* together known as Oregon Grape, are widespread on the Pacific Coast (Fig. 191).

The distinctive feature, in particular, of *Caulophyllum* is the peculiar development of the "fruit," namely, as the two ovules develop they burst the wall of the ovary and continue their development into a pair of blue stalked "berries," in reality the seeds. Only one other species is known, and that is found in Japan and in the Amur region of eastern Asia (!).

MENISPERMACEÆ (*Moonseed Family*). Chiefly a tropical family, but with representatives reaching into north temperate regions. The plants are woody climbers with alternate exstipulate, simple or palmately lobed or palmately veined leaves, and small diœcious flowers in cymes, panicles, or racemes. The calyx shows a variation from 4 to 12 sepals, but the petals are definitely 6; there is also an inconstant number of carpels, from several to 3, and distinct, sometimes 1. The ovary is superior and contains a single ovule. The fruit is drupaceous. There is no endosperm, or it is scanty when present. Species 250—300 in 70 genera, of which *Menispermum, Cocculus,* and *Calycocarpum* are represented in Atlantic North America (Fig. 192).

Menispermum shows considerable variation in the number of floral organs, the sepals ranging from 4 to 8, petals 6—8, and stamens 12—24, with 4-celled anthers, with variation from 4 to 2 in the number of carpels, the stigmas of which are broad. Each carpel becomes a drupe which in ripening bends so as to bring the stigma into a basal position, and causes at the same time the development of a curved or crescent-shaped stone, and a similarly shaped embryo, the cotyledons of which are filiform. There are two species, one, *Menispermum canadense* (Moonseed) in eastern North America, the other in Asia. *M. canadense* is a climbing vine with broadly heart-shaped or palmately lobed leaves, and small white flowers in axillary clusters, inhabiting wooded stream banks. The fruits are black drupes covered with a bloom, in grape-like clusters. In *Cocculus* the carpels vary from 3 to 6, but the other floral organs are definitely 6. The anthers are 4-celled, and the fruit is a 1-seeded drupe. Our only species, *C. carolinus* (Carolina Moonseed) is a slender vine of river banks from Virginia to Kansas and south to the Gulf. The other 8 or 10 species are African and Asiatic. In *Calycocarpum* petals are wanting. The gynœcium is constantly 3-carpellate, with laciniate stigmas. The anthers are 2-celled. The genus is monotypic, the solitary *Calycocarpum Lyoni* (Cup-Seed), a climbing vine with palmately lobed leaves, occurring in rich woods of central and southern United States, west to Texas.

FIG. 192.—*Cocculus carolinus:* different leaf-forms, inflorescence (*i*), in fruiting stage, and fruit (*f*).

MAGNOLIACEÆ (*Magnolia Family*). In the *Magnoliaceæ,* the flowers are of the primitive or generalized type (spiral or spirocyclic), associated with a woody habit. In addition to the primitive phyllotaxy of the organs, the flowers are regular, hypogynous, bisexual or unisexual, with perianth mostly differentiated into calyx and corolla, and numerous stamens and carpels (distinct, in rare cases united). The receptacle is usually quite elongate. Fruit a follicle, a winged nut, or a berry, bearing the seeds on parietal placentas. Endosperm copious and embryo small. The plants are trees or shrubs with evergreen or deciduous, entire, often stipulate leaves. About 100 species in 9 genera, distributed through Atlantic America and from tropical Asia to the Himalayas, with a few in Australasia. Three subfamilies, based on characters of the receptacle, presence or absence of stipules, and habit, *viz.:*

Magnolioideæ, in which the receptacle is elongate and the leaves are furnished with stipules or stipular sheaths. Flowers bisexual. *Magnolia, Liriodendron, Michelia.*

Illicioideæ, in which the receptacle is shortened and stipules are wanting. Flowers bisexual or unisexual. A peculiar and distinctive feature is the absence of vessels in the secondary wood of some members, in which respect these forms resemble the conifers. The plants are evergreen trees or shrubs. *Illicium, Drimys, Zygogonum.*

Schizandroideæ, in which the receptacle is short and convex, but often becoming elongate with age. Stipules and stipular sheaths wanting. Flowers unisexual. Climbing shrubs. *Schizandra, Kadsura.*

Magnolioideæ. (*Magnolia*). In the magnolias the flowers conform very closely to the primitive type, especially with respect to the numerous carpels and stamens spirally arranged on an elongate cylindrical axis, but a cyclic condition is approached in the perianth, where the sepals and petals are in 3-merous whorls. The anthers are linear and introrse, and the fruits are 2-seeded follicles from which the seeds become suspended by slender threads. At maturity the entire mass of carpels has a cone-like (strobiloid) appearance. The 30 or more species are divided between North America (including the West Indies and Mexico) and Asia. (Figs. 193, 194.) The flowers of *Liriodendron* are structurally similar to those of the magnolias, but the anthers are extrorse and the carpels more elongate, in fruit becoming samaras, and enclosing 1 or 2 pendulous seeds. Only two species are known, one, *Liriodendron Tulipifera* (Tulip-

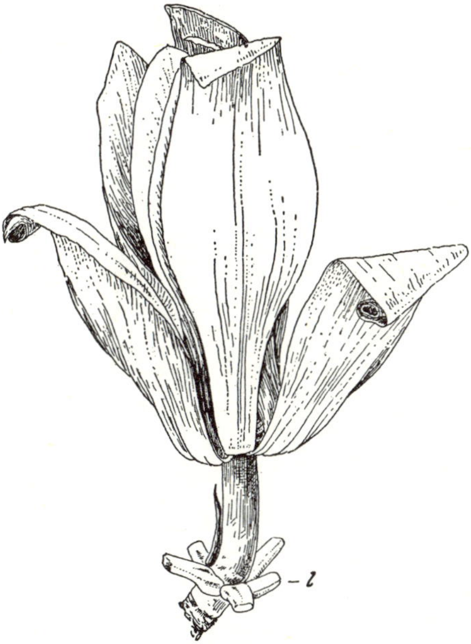

FIG. 193.—Flower of a magnolia; *l,* bases of petioles.

Tree), one of the largest and handsomest of our forest trees, restricted to Atlantic North America, the other, scarcely distinct, restricted to China.

Michelia differs technically from *Magnolia* and *Liriodendron* in having the carpel-bearing portion of the receptacle separated from the stamen-bearing portion by a prominent internode. A further distinction appears in the mostly axillary inflorescence. About a dozen species of trees and shrubs of tropical Asia, China, and the Himalayas.

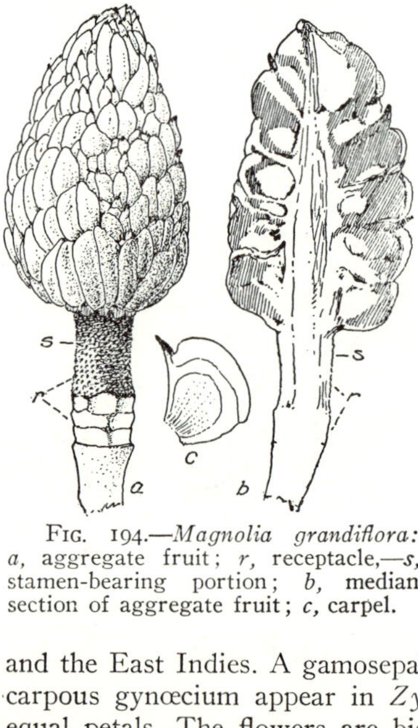

FIG. 194.—*Magnolia grandiflora:* *a*, aggregate fruit; *r*, receptacle,—*s*, stamen-bearing portion; *b*, median section of aggregate fruit; *c*, carpel.

Illicioideæ. In *Illicium* the flowers are bisexual, with the outer perianth segments sepaloid. Represented in Atlantic North America by two shrubs, in Asia and Japan by five shrubs and trees. Absence of vessels in the secondary wood is peculiar to *Drimys* and *Zygogonum,* in this respect simulating the confers. This aside, *Drimys* has bisexual or unisexual, polygamous flowers, a calyx of 2—4 connate sepals which at length become disconnected, numerous to few petals, and a single whorl of 2 or more carpels. The 10 species are inhabitants of Mexico, South America, and the East Indies. A gamosepalous calyx (perigyny ?) and a syncarpous gynœcium appear in *Zygogonum,* associated with few unequal petals. The flowers are bisexual. It is a monotypic genus of New Caledonia.

Schizandroideæ. Schizandra (about 6 spp.) is Asiatic, with one species in southern United States. *Kadsura* has 7 species in tropical Asia, and one in Japan.

HIMANTANDRACEÆ. A small family of the Moluccas, Papua, and East Australia. They are woody plants with naked, spiral, bisexual flowers. Stamens perigynous, and above and below them are staminodes. A beginning of syncarpy is shown in the basal union of the carpels. The plants are coated with scale-like hairs. *Himantandra,* 2 species.

CALYCANTHACEÆ (*Calycanthus Family*). A small family of shrubs with aromatic bark, opposite, entire, exstipulate leaves, and large solitary perigynous flowers on leafy lateral branches. Associated with numerous carpels and stamens is an undifferentiated perianth. Anthers extrorse. The perigyny here results from a peculiar development of the receptacle as a hollow urn-shaped structure comparable to the condition in the rose, on the inner surface of which are borne the carpels, while the outer surface is formed of a number of spiral adnate bracts. The carpels become achenes ("seeds"), enclosed in the dry urn-shaped receptacle, the whole forming a sort of spurious fruit similar to the hip of roses. In the broad sense *Calycanthus* is the only genus, with *C. floridus* (Carolina Allspice) and *C. fertilis* in the Alleghenian region, and *C. occidentalis* in northern California. Three others in Asia are by some writers separated as *Chimonanthus*. The plants are often used as ornamentals. Florally they suggest the *Rosoideæ* (Fig. 195).

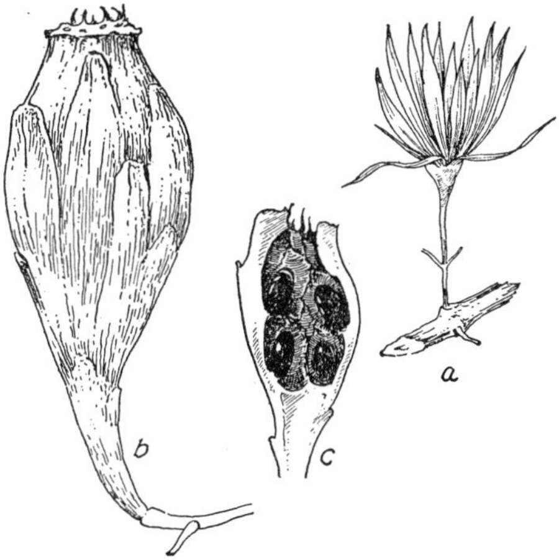

FIG. 195.—*Calycanthus fertilis: a,* flower; *b,* receptacular fruit, with adnate bracts on outside; *c,* median section of fruit, with four achenes (black), the true fruits.

LACTORIDACEÆ. A monotypic family, with *Lactoris fernandeziana* on the island Juan Fernandez; a species with polygamo-monœcious flowers with an undifferentiated perianth, 6 stamens in 2 whorls, and 3 slightly united carpels.

ANONACEÆ (*Custard-Apple Family*). Mainly a tropical family of aromatic trees and shrubs with alternate exstipulate leaves and hypogynous, spirocyclic, mostly bisexual, regular, solitary and axillary flowers. A 3-merous condition obtains, but with petals in 2 whorls, while there is a persistence of numerous spiral stamens and numerous distinct carpels, the latter, however, tending to be reduced (to 1 in some cases) and united. Species about 800, in 62 genera, widely distributed in the tropics, with an extratropical ramification into Atlantic North America.

Principal genera: *Asimina, Cananga, Polyalthia, Xylopia, Anona, Artabotrys.*

Asimina alone of the genera reaches beyond the tropics, being represented in eastern North America by a half-dozen species, of which *Asimina triloba* (N. Am. Papaw) has penetrated to the Canadian border. At its best the species may attain a height of 40 feet. It is characterized by alternate obovate-lanceolate leaves, naked buds, and nodding axillary flowers showing a 3-merous condition with respect to the perianth, which consists of one whorl of sepals and two of petals; but stamens and carpels vary from several to few. The carpels become a cluster of oblong pulpy fruits containing several dark-brown, fleshy-arillate seeds. The plants commonly grow in patches in low, rich grounds. The flowers appear before the leaves.

Cananga includes two or three trees of the Malay region. Their special features are the pointed connectives, two whorls of carpels, and the presence of staminodes. The fruits are fleshy and many-seeded. The species are a source of perfume extracts. *Cananga odorata* (Ylang-Ylang) of the East Indies and the Philippines is grown in Florida. A perfume is extracted from its flowers. *Polyalthia,* one of the largest genera of the family (70 or more spp.), comprises trees and shrubs of Africa, Madagascar, and tropical Australia. *Xylopia,* with about 60 species of trees and shrubs, is also tropical. The plants produce elongate fleshy fruits. *Anona* (Custard-Apple), with 50 species of trees and shrubs, is restricted to tropical America. The genus yields a number of important edible fruits, such as Cherimoya (*Anona Cherimoya*), and the common Custard-Apple (*A. reticulata*). The plants are evergreen or deciduous, with solitary or clustered *supra*-axillary flowers (compare with *Solanaceæ* and *Sparganiaceæ*). The floral structure is that of *Asimina,* but the calyx is tubular and the inner series of petals is wanting or represented by scales. In addition the carpels are consolidated with the receptacle to form a large, fleshy, distinctive type of fruit—a syncarp. A climbing habit, by means of specialized peduncles, has been developed in *Artabotrys,* besides a constriction of the corolla below the middle, a crowding of the stamens, and reduction of the gynœcium to relatively few carpels. These ripen into 2-seeded drupes of the size of plums. In all there are about 40 species, some of which are cultivated.

Special interest attaches to the *Anonaceæ* because of their often remarkably constructed flowers and fruits. In some respects the characters point to affinities with the *Magnoliaceæ*. Compare.

EUPOMATIACEÆ. A small family of New Guinea and Australia. The flowers have no perianth. *Eupomatia,* 2 species.

MYRISTICACEÆ (*Nutmeg Family*). Evergreen trees with alternate exstipulate leaves and small, diœcious, regular, acyclic flowers in axillary clusters. There is no corolla, instead a 3-lobed calyx, and many to few monadelphous stamens, the filaments forming a tube. The gynœcium is reduced to a solitary carpel which becomes a fleshy, 1-seeded, dehiscent fruit. The seed is furnished with a fleshy laciniate aril and contains a convoluted or channeled endosperm and a small embryo with foliaceous cotyledons. Species about 100, in 8 genera, mostly of tropical Asia. The great majority of the species belong in *Myristica,* of which *M. Fragrans* (Nutmeg) is the most important.

GOMORTEGACEÆ. A monotypic family based on *Gomortega nitida* of Chile. Flowers spirocyclic, with an undifferentiated perianth of 7 segments, and with 2—3 stamens and carpels. Fruit drupaceous. Endosperm copious and embryo large.

MONIMIACEÆ. Flowers bisexual or unisexual, regular or irregular, with differentiated or undifferentiated perianth, perigynous or epigynous. Stamens and carpels numerous; ovules 1. Placentation various. Filaments appendaged. The true fruits are dry and indehiscent and borne within a hollow urn-shaped receptacle very much as in *Calycanthus,* in many cases in a depression of the inner surface. Embryo small, in copious endosperm. Species about 350, in 32 genera; trees and shrubs and some climbers, of tropical and subtropical climates, having their best development in the southern hemisphere. In their relatively numerous stamens and carpels and in the pronounced perigyny the plants strongly suggest the roses or the *Calycanthaceæ,* but on the other hand in the valvate dehiscence of the anthers of some members there is a suggestion of the *Berberidaceæ* or the *Lauraceæ*. Rosaceous and lauraceous types are also expressed in the epigynous forms. *Siparuma, Mollinedia.*

LAURACEÆ (*Laurel Family*). Flowers bisexual or unisexual, monœcious, regular, cyclic, 3-merous, perigynous, yellowish or greenish, with undifferentiated perianth. Stamens in 3 to 4 3-merous

FIG. 196.—*Persea americana* (Avocado). *A,* branch, with inflorescences above. *B,* flower, the black organs staminodes. *C,* median section of flower; *s,* staminodes. *D,* a 2-carpellate (apocarpous) gynœcium. *E,* a stamen with six anthers. *F,* dorsal and ventral aspects of a 2-anthered small inner staminode. *G,* a 2-anthered larger staminode. *H,* a 4-anthered stamen. *I,* lower leaf of a branch. *J,* flower-bud.

whorls, often transformed into staminodes; anthers opening by 2 to 4 lids or valves, mostly introrse, but may be both introrse and extrorse in the same flower, often with glandular structures on the filaments. Gynœcium usually 1-carpellate, with superior, 1-ovuled ovary. Fruit a berry or in some cases a drupe, commonly seated in the persistent cup-shaped hypanthium or enveloped by the persistent base of the perianth. Endosperm wanting, the seed completely filled by the large fleshy cotyledons of the embryo. Integuments 2. Flowers sometimes solitary, otherwise commonly in compound inflorescences of various sorts. Species about 1,000 in 40 genera.

One of the most important families of the order, comprising aromatic trees and shrubs with coriaceous, mostly evergreen, simple alternate leaves, widely distributed in the tropics, with outposts in the cooler belts. Represented in eastern North America by species of *Persea, Sassafras, Litsea,* and *Benzoin.*

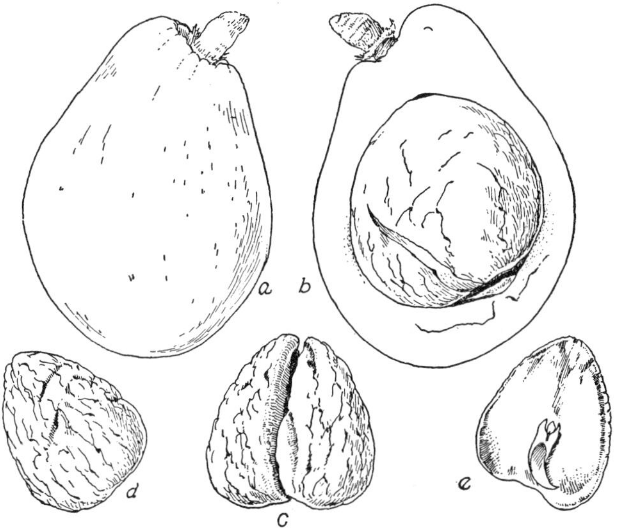

FIG. 197.—Avocado: *a,* fruit; *b,* section of fruit, with large central seed ("stone"); *d,* face view, and *c,* side view, of embryo,—only the two fleshy cotyledons visible; *e,* embryo with one cotyledon removed, showing hypocotyl and plumule.

The species fall into two subfamilies, distinguished chiefly on characters of the anthers, *viz.:*

Perseoideæ, in which the anthers open by 4 valves. Here belong *Cinnamomum, Persea, Nectandra, Sassafras, Litsea, Umbellularia.*

Lauroideæ, in which the anthers open by 2 valves. Includes *Laurus* and *Benzoin,* and the parasitic tropical *Cassytha* (Dodder-like plants).

Perseoidæ. Cinnamomum comprises about 54 species of evergreen trees and shrubs of tropical Asia and Australia. The leaves are alternate or opposite and 3-nerved from the base; the flowers are bisexual or polygamous, in panicles. There are 9 or fewer functional stamens in 3 whorls, followed by a fourth whorl of staminodes. The fruit is a berry, seated in the persistent hypanthium. *Cinnamomum Camphora* (Camphor Tree) of China and Japan, and *C. zeylanicum* (Cinnamon Tree), and one or two others, are grown in the Southwest. The first-named is the source of camphor. In *Persea* the small, greenish, bisexual flowers show a perianth of 6 segments, 12 stamens in four 3-merous whorls, the inner whorl transformed into gland-like staminodes: the anthers of the third whorl are extrorse, the others introrse. The ovary of the solitary carpel becomes a succulent, globular or oblong or inversely pear-shaped berry containing a large "stone." *Persea Borbonia* (Red Bay) and *P. pubescens* are native in Atlantic North America. *P. americana* (Avocado) of tropical America and *P. drymifolia* (Mexican Avocado) are cultivated more or less extensively in the Southwest. Some species of *Nectandra,* a large genus (90 spp.) of tropical America, are useful in the arts. (Figs. 196, 197).

Sassafras has one species, *S. Sassafras* (Sassafras or Ague Tree), in eastern North America and another in Asia. The American plant is a deciduous shrub or tree (to 125 feet high) with palmately lobed leaves and yellow diœcious flowers in umbellate-racemose clusters, with an involucre of small scaly bracts. The perianth consists of 6 segments (two 3-merous whorls), persistent in the carpellate flowers. There are 9 stamens in three 3-merous whorls, with the filaments of the inner whorl provided with stalked glands at the base. The anthers are introrse. In the carpellate flowers there are 6 staminodes. Fruit an oblong or globular blue drupe. The characteristic spicy bark is used in medicine and in making "sassafras tea." *Litsea* is represented by *L. denticulata* (Pond Spice) in swamps of the Atlantic coastal plain. A hundred or more species occur in tropical Asia and Australia. A notable advance is shown by the diœcious flowers and 2-carpellate syncarpous gynœcium. Staminodes are present in the carpellate flower. Fruit a drupe.

In the monotypic *Umbellularia* there is a perianth of 6 segments (two whorls), and 9 stamens in 3 whorls, the inner whorl with stipitate glands at the base alternating with 3 staminodes. The outer anthers are introrse, the inner extrorse. Fruit a globular drupe. The genus is Californian, where the solitary *Umbellularia californica* (California Bay Tree) is common in the canyons of the coastal belt. It is an evergreen tree or shrub, unpleasantly sweetish-aromatic, with lanceolate leaves and small greenish-yellow flowers in umbellate clusters. The wood is prized for its hardness and beautiful grain. (Fig. 198.)

Lauroideæ. Laurus (Laurel) has diœcious or polygamous 4-merous flowers. Four staminodes are usually present in the carpel-

FIG. 198.—*Umbellularia californica: a,* flowering branch; *b,* umbel-bud opening; *c,* umbel opened; *d,* flower; *e,* flower with some parts removed, showing gynœcium and stamens; *f,* ventral and *g,* side view of stamen; *h,* stamen with two large glandular appendages (*i*); *j,* staminode; *k,* gynœcium; *l,* fruit; *m,* section of fruit showing single seed (stone); *n,* seed.

late flowers. The fruit is a small berry. *Laurus nobilis* of the Mediterranean region is the Laurel of the classics. The only other species, *L. canariensis,* occurs in the Canaries and in Madeira. *Benzoin* is represented in Atlantic North America by *B. æstivale* (Spice-Bush) and *B. melissæfolium;* the remaining four or five species are found in Asia. The floral structure is essentially as in the preceding.

HERNANDIACEÆ. Probably less than 40 species of trees and shrubs of the tropics. Flowers cyclic, bisexual or unisexual, regular, epigynous. Perianth undifferentiated, of 4—10 segments, and one whorl of stamens. *Hernandia, Gyrocarpus.*

ORDER RHŒADALES

Flowers hypogynous or occasionally perigynous, cyclic—except the stamens, which are spiral in some members—bisexual, regular or irregular, with calyx and corolla. Gynœcium syncarpous, of 2 to several carpels usually with many ovules. Integuments 2. Endosperm present or wanting. Mostly herbs.

Families: *Papaveraceæ, Fumariaceæ, Capparidaceæ, Cruciferæ, Tovariaceæ, Resedaceæ, Moringaceæ, Bretschneideraceæ.*

Key to the More Important Families

A. Flowers regular.
  a. Gynœcium elevated on an elongate stipe — *Capparidaceæ.*
  b. Gynœcium not so elevated.
    1. Stamens numerous. Sepals 2 (rarely 3 or 4), caducous. Carpels 2 to many. — *Papaveraceæ.*
    2. Stamens 6. Sepals 4. Carpels 2. — *Cruciferæ.*
B. Flowers irregular.
  a. Flowers hypogynous.
    1. Gynœcium and stamens elevated on a short or elongate stalk (androgynophore).
      Androgynophore short, expanded into a disk on one side under the stamens. — *Resedaceæ.*
      Androgynophore long and slender, not forming a disk. — *Capparidaceæ.*
    2. Gynœcium and stamens not elevated on a stalk. Sepals 2. — *Fumariaceæ.*
  b. Flowers perigynous. Gynœcium 3-carpellate, on a short gynophore. — *Moringaceæ.*

PAPAVERACEÆ (*Poppy Family*). The poppies are herbs with milky or colored sap and bisexual, regular, 2-merous (occasionally 3-merous) flowers. Stamens numerous, with 2-celled anthers. Carpels 2 to many, the ovary 1- to several-celled, with parietal placentas. Fruit a capsule, opening by pores or valves. Seeds numerous to few, with fleshy or oily endosperm at the base of which lies the small embryo. About 24 genera and 450 species, widely distributed.

In some works the *Fumariaceæ* is included in the *Papaveraceæ.* As here treated the *Papaveraceæ* consists of two subfamilies, *Hypecoöideæ* of central Asia and the Mediterranean region, and *Papaveroideæ* of subtropical and temperate regions, to which our species belong. The *Hypecoöideæ* differ from the *Papaveroideæ* mainly in the absence of milky sap and in having constantly 2-merous flowers. The principal genus is *Hypecoum* of the Mediterranean region and central Asia. With respect to our flora, then, we have to deal only with the *Papaveroideæ,* as represented by species of the following genera: *Sanguinaria, Stylophorum, Chelidonium, Papaver, Argemone, Platystemon, Romneya, Eschscholtzia, Meconella, Dendromecon, Glaucium.*

The solitary *Sanguinaria canadensis* (Blood-Root) is a common herb of rich woodlands of eastern North America. The plant sends up in early spring a single, palmately lobed, long-petioled leaf and a one-flowered scape from a thick, horizontal, scaly-leaved rootstock. The two green sepals fall with the opening of the flower, and the 8—12 pure-white petals are almost as fugacious. The petals are not crumpled in the bud. The numerous stamens are of different lengths. The gynœcium consists of a 1-celled ovary contracted into a short thick style with a thick 2-lobed stigma, and contains many ovules on parietal placentas. The fruit is an elongate valvate pod. The whole plant contains a sticky blood-red sap to which it owes its generic name. (Figs. 32, 35.)

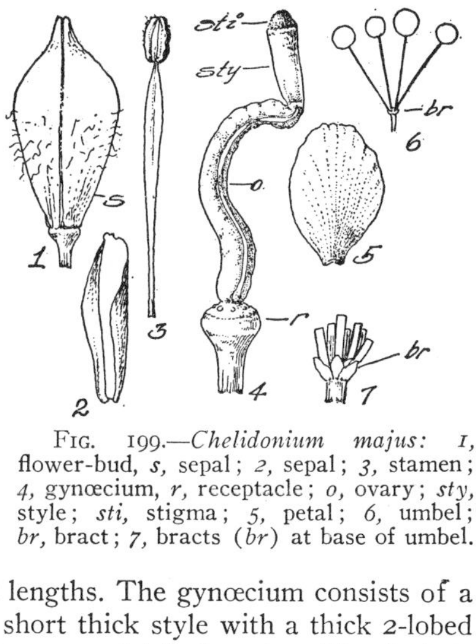

FIG. 199.—*Chelidonium majus:* *1,* flower-bud, *s,* sepal; *2,* sepal; *3,* stamen; *4,* gynœcium, *r,* receptacle; *o,* ovary; *sty,* style; *sti,* stigma; *5,* petal; *6,* umbel; *br,* bract; *7,* bracts (*br*) at base of umbel.

Chelidonium (Celandine) differs from *Sanguinaria* in having pinnately lobed leaves and a corolla of 4 petals which are crumpled in the bud, besides a more slender and nearly styleless 2-carpellate gynœcium with a small stigma. *Chelidonium majus,* a native of Europe, has become naturalized in the Atlantic States. The petals and sap are yellow. (Fig. 199.) Like *Sanguinaria,* the genus is monotypic. *Stylophorum* is similar to *Chelidonium* as regards color, number, and crumpling of the petals, but differs in having a 3—4-carpellate gynœcium with as many styles and placentas, and a bristly pod. One

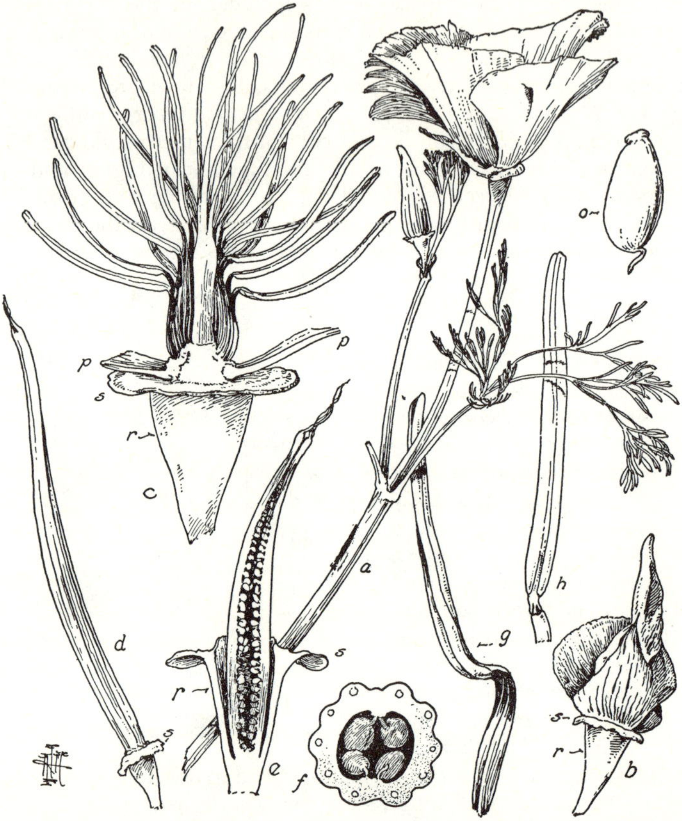

FIG. 200.—*Eschscholtzia californica: a,* flowering stem; *b,* unfolding flower-bud, —*s,* calyx; *r,* hypanthium; *c,* flower, partly in sectional view, —*p,* petals, *s,* calyx, *r,* hypanthium; *d,* advanced stage of gynœcium; *e,* median section of gynœcium and hypanthium (*r*); *f,* transverse section of ovary; *g,* stamen; *h,* anther; *o,* ovule.

species, *Stylophorum diphyllum* (Celandine Poppy), inhabits damp Eastern woods; two or three others occur in Asia. *Glaucium* has yellow flowers and sap, alternate lobed or dissected leaves, 4 petals, numerous stamens, and a 2- or occasionally 3-carpellate gynœcium and lattice-veined seeds. The half-dozen species are native in Mediterranean countries. *Glaucium flavum* (Horn Poppy) is adventive in the North Atlantic States.

Papaver, with 40 to 100 species according to different writers, is an Old World genus but with a few representatives in the Far West. The plants are herbs with milky sap, alternate lobed or dissected leaves, and showy flowers from drooping buds. The sepals vary, sometimes numbering 3, and petals often 4 to 6, but the stamens are numerous, with extrorse anthers. The gynœcium is pluricarpellate and completely syncarpous, with an ovary of as many cells and crowned by the sessile radiating stigmas, at maturity forming a capsule which dehisces by pores at the top. The numerous seeds are minutely pitted. The Mediterranean *Papaver somniferum* is the source of opium. Garden poppies include the arctic *P. nudicaule* (Iceland Poppy), *P. glaucum* (Tulip Poppy), *P. orientale* (Oriental Poppy), and *P. Rhœas* (Corn Poppy).

Argemone (Prickly Poppy), with about 10 species, is mainly west American. The species are herbs with yellow sap and alternate, thistle-like leaves, and large yellow flowers from erect buds. The floral plan is similar to that of *Papaver* but the gynœcium has fewer carpels and the capsule is oblong and bristly and dehisces at the top by valves instead of pores. A number of species are horticultural.

Romneya and *Dendromecon* are shrubs, both monotypic, as broadly interpreted. *Romneya Coulteri* (Matilija Poppy), which sometimes grows to a height of eight feet, ranges from the Santa Barbara region of California into Lower California. It bears large, pure-white flowers with 3 sepals, 6 petals in two whorls, numerous stamens, and a gynœcium of several united carpels capped by an annular stigma. The capsule is bristly and contracted above, and dehisces to the middle by 7—12 valves. The higher-growing *Dendromecon rigida* (10 ft.), also Californian, lacks milky sap; its leaves are entire and rigid, and the flowers are yellow and showy, with 2 sepals, 4 petals, many stamens, and a 2-carpellate gynœcium which develops into an elongate, slender, somewhat curved capsule, dehiscing to the base by two valves.

Eschscholtzia digresses from the preceding genera in a number of important features. There is an advance to a decided perigyny and the sepals are united into a hood which is sloughed off over the petals as the flower opens. The stamens, however, remain numerous, and there are 4 petals, but the gynœcium shows variation from several to 2 carpels, with elongate slender styles and small stigmas. The elongate slender capsule dehisces by valves from the base. The genus is west American, in the broadest sense consisting of relatively few species. These are annual or perennial herbs with dissected leaves, showy flowers, and watery sap. Concerning the dehiscence of the capsule Dr. Jepson writes: "Dehiscence of the capsule commonly occurs in flight, after the capsule parts from the receptacle and before it reaches the ground, usually beginning at the moment that the base of the capsule is released from the vase-like hollowed receptacle, this action allowing the valves, which are elastically dehiscent from base to apex, to separate." *Eschscholtzia californica* (California Poppy) is a choice horticultural. (Fig. 200.)

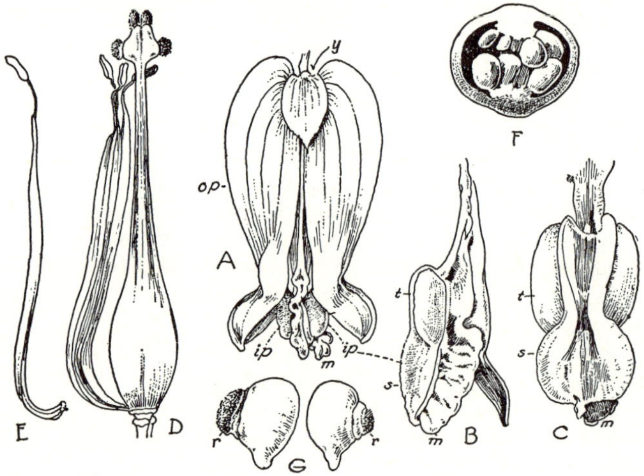

FIG. 201.—*Dicentra spectabilis: A,* side view of flower; *p,* gibbous petal; *y,* sepal. *B,* side view, and *C,* inner face, of inner petal (*ip*); *m, s, t,* marking corresponding parts. *D,* gynœcium and stamens (those on right removed). *E,* stamen. *F,* transverse section of ovary; placentation parietal. *G,* ovules, with prominent capillate crests (*r*).

In *Platystemon californicus* (Cream Cups) of the Far Southwest, the gynœcium consists of many slightly united carpels with filiform styles. The carpels become moniliform with age and at maturity separate and break up into 1-seeded, indehiscent joints (Jepson). One or two other genera are represented in the Far West.

FUMARIACEÆ (*Fumitory Family*). The fumitories are smooth, weak, annual (biennial or perennial) herbs with watery sap, alternate dissected leaves, and bilaterally symmetrical, hypogynous flowers in terminal racemes. The calyx is reduced to two small scale-like sepals. The 4 petals are in two 2-merous whorls, the two inner petals more or less spurred at the base and forming a hood over the stigma. There are 6 stamens in two 3-merous whorls, diadelphous, the anther of the middle stamens of each whorl 2-celled, the lateral 1-celled. Gynœcium 2-carpellate, syncarpous, with a superior ovary which develops into a 2-valved, several-seeded capsule, or a 1-seeded utricle. Species about 170, in 5 genera, mostly of northern latitudes and South Africa.

Adlumia, Dicentra, Corydalis, Fumaria are represented. Prominent distinctions appear in the corolla; this is 2-spurred in *Adlumia* and *Dicentra,* 1-spurred in *Corydalis* and *Fumaria,* the last being further distinguished by its fruit—a 1-seeded nutlet. Useful generic distinctions reside also in the seeds, in the shape of wings or crests. Union of the petals appears in *Adlumia* and *Dicentra,* and in the former they persist as an investment of the pod, while in the latter they are either persistent or deciduous. (Fig. 201.)

Adlumia is monotypic, with *A. fungosa* (Climbing Fumitory) in eastern North America. It is a delicate climbing vine with thrice-compound leaves and drooping pinkish flowers in ample panicles. The 6 stamens are monadelphous below and diadelphous above, and the fruit is an oblong capsule enclosed in the persistent corolla. The plant is horticultural. The seeds are without crests. *Dicentra* includes erect or sometimes climbing rhizomatous herbs with dissected leaves and racemose flowers, and winged or crested seeds from elongate pods. The 16 or more species are native to North America and Asia. *Dicentra spectabilis* is the Bleeding Heart of the gardens. (Fig. 201.) *Corydalis,* with over 100 species, is a genus of Europe, Asia, and North America; the seeds are crested or winged. *Fumaria* is an Old World genus of about 40 species.

CAPPARIDACEÆ (*Caper Family*) (including *Koeberliniaceæ*). The capers are herbs or shrubs, chiefly of the tropics and subtropics, with watery sap and simple or palmately compound leaves, and cruciform flowers borne solitary or in racemes. Sepals 4—5, free or united; petals as many or none and in some cases clawed. Stamens 6 to many, not tetradynamous. Gynœcium 2(or more)-carpellate, completely syncarpous, sessile or elevated above the stamens on a gynophore, in some forms with a prominent internode between stamens and corolla. Ovules numerous on parietal placentas. Fruit a capsule, berry, or drupe. Seeds with endosperm and coiled embryo; cotyledons folded or twisted. About 450 species in 35 or 40 genera. Some forms are clothed with stellate hairs or scales. Engler recognizes five subfamilies, of which only two are represented in our flora; these are distinguished mainly by their fruits, *viz.:*

Capparidoideæ, in which the fruit is fleshy (berry). Includes *Capparis, Koeberlinia.*

Cleomoideæ, in which the fruit is dry (pod). Includes *Cleome, Cristatella, Cleomella, Polanisia, Isomeris, Wislizenia, Oxystylis.*

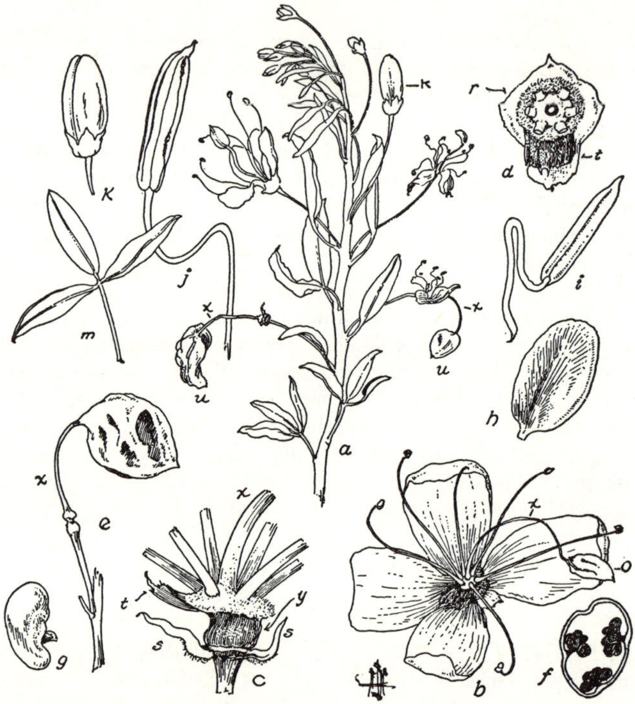

FIG. 202.—*Isomeris arborea: a,* flowering stem; *u,* young fruit; *x* gynophore; *b,* flower; *o,* ovary; *c,* basal part of flower, showing sepals (*s*), excentric disk (*t*), forming a fleshy androgynophore (*y*) and gynophore (*x*) in center of whorl of filaments; *d,* disk (*r*) with eccentric lobe (*t*); *e,* ovary on gynophore (*x*); *f,* transverse section of ovary with three parietal placentas; *g,* ovule; *h,* petal; *i,* dorsal, and *j,* ventral, views of stamen; *k,* flower-bud; *m,* leaf.

Cleome is native to the American tropics and Africa but has several species in colder latitudes. The plants are herbs or low shrubs with simple palmately lobed leaves in bracteate racemes. Sepals 4, and petals 4 and clawed, but mostly 6 stamens. The 2-carpellate gynœcium is elongate and stipitate, without a partition in the ovary. A prominent internode separates the stamens and petals. Fruit an elongate many-seeded capsule. *Cleome spinosa* (Spider Flower) is a common garden plant; it is widely distributed in the eastern and southern United States. *Cristatella* differs mainly in its laciniate petals. The two species are restricted to the Western parts of the United States. In *Cleomella* the ovary is short and dilated laterally, with a gland at the base. There are about 8 species in the Southwest and Mexico.

Isomeris arborea (Burro-Fat), the only species, is a shrub (4 ft.) with 3-foliolate leaves and a large, inflated, long-stipitate, few-seeded pod. The species is restricted to the desert region of southern California (Fig. 202).

Polanisia comprises mostly viscid-glandular annuals with 3—5-foliolate leaves, and flowers in bracteate terminal racemes. It is especially distinguished by its comparatively numerous stamens (8 to many) and well-developed perigyny. The elongate, nearly sessile pod dehisces at the summit. The seeds are rugose or reticulate. Of the 30 or more species ranging from the tropics into cooler climates, only two have spread over the United States—*Polanisia graveolens* over the Eastern parts and *P. trachysperma* over the Western. Two lone species of *Wislizenia* range from southern California to New Mexico. These are distinguished by their 2-carpellate, 2-seeded, didynamous fruits and the presence of stipules. The solitary *Oxystylis lutea* is restricted to the Death Valley region. It differs from *Wislizenia* chiefly in the short-stipitate fruit and absence of stipules.

CRUCIFERÆ (*Mustard Family*). A large and technically well-defined family of herbs with pungent sap, alternate leaves of diverse shapes, and cruciform racemose or corymbose flowers with 4 early deciduous sepals, 4 equal clawed petals with spreading limb, mostly 6 stamens in two sets—2 (short) in the lower and 4 (long) in the upper; and a 2-carpellate completely syncarpous gynœcium with a more or less developed style and 2-lobed stigma. Ovary superior, 2-celled (by a longitudinal partition), forming a short or elongate dehiscent or indehiscent pod (silique, silicle) of diverse shapes, or a nutlet. Seeds campylotropous, without endosperm, attached along the inner edges on both sides of the partition (septum). Embryo straight but with the cotyledons folded in three distinct ways,—incumbent, accumbent, or conduplicate. About 1800-2000 species in 200 or more genera, widely distributed in the Old and the New World, and largely represented in our flora. The family is especially well developed in the Mediterranean region. Irregular flowers occur in some forms.

Because of the uniformly similar floral structures the *Cruciferæ* are a difficult and technical group of plants, but the characteristic cruciform corolla, and the tetradynamous stamens in addition, make the family connections readily recognizable. The more technical characters are to be sought in the fruits and the cotyledons. The fruit (pod) of mustards is generally termed a silique if longer than it is broad; if short, it is termed a silicle. In some cases the fruit is a small globular nutlet. The manner of folding of the cotyledons may be worked out from sections of immature seeds, or mature seeds softened in water, and will appear as in the following symbols adapted from Warming (the circle indicating the section of the hypocotyl):

(1) ○ =, cotyledons accumbent; (2) ○ ||, cotyledons incumbent; (3) ○ > >, cotyledons conduplicate; (4) ○ || ||, ○ || || ||; cotyledons cut several times.

The cotyledons are epigeal and often notched at the apex.

Several schemes of classification of the mustards have been proposed, but the following, adapted from various sources, will suffice for our genera, only a few of which are given.

I. Fruit a silique.
- Tribe I. *Arabideæ.* Cotyledons accumbent. Pod short and globose to long and slender, cylindrical or flattened. *Arabis, Radicula.*
- Tribe II. *Sisymbrieæ.* Cotyledons incumbent. Pods elongate, dehiscent. *Sisymbrium, Erysimum.*
- Tribe III. *Brassiceæ.* Cotyledons conduplicate. Fruit elongate. *Brassica.*

II. Fruit a silique with a broad septum.
- Tribe IV. *Alysseæ.* Cotyledons mostly accumbent. Pods orbicular to linear, flattened parallel to the septum and dehiscent. *Alyssum, Draba.*
- Tribe V. *Camelineæ.* Cotyledons incumbent. Pods short, about as long as broad. *Camelina, Subularia.*

III. Fruit a silique flattened perpendicularly to the narrow septum.
- Tribe VI. *Lepideæ.* Cotyledons incumbent. Pods strongly flattened, short. *Lepidium, Capsella.*
- Tribe VII. *Thlaspideæ.* Cotyledons accumbent. Pods short, strongly flattened, obcordate. *Thlaspi.*

IV. Fruit constricted between the seeds or divided by transverse walls into 8-seeded segments.
- Tribe VIII. *Raphaneæ.* Cotyledons conduplicate. Pods more or less constricted or 1-seeded. *Raphanus* (pods beaked, indehiscent).
- Tribe IX. *Cakileæ.* Cotyledons accumbent. Pod constricted into two segments; indehiscent. *Cakile.*

Arabideæ. In *Arabis* the fruit is elongate, slender, and linear or linear-oblong. The plants are annuals, biennials, or perennials, glabrous or pubescent, with simple or branched hairs, and with simple or pinnatifid leaves and white or purplish flowers. Species about 120, Northern, and well represented in northern and western United States. *Arabis Holboellii* ranges from Greenland to British Columbia and Alaska. *Radicula* has a short, ovoid or oblong, terete fruit, from small yellow or white flowers. The plants are low annuals, biennials, or perennials, with simple lobed or dissected leaves, commonly inhabiting low moist or wet grounds. The 50 or more species are widely distributed, with about 20 in North America. *Radicula Nasturtium-aquaticum* is the common Water-Cress.

Sisymbrieæ. Sisymbrium has elongate terete or somewhat flattened fruits from small whitish or yellow flowers. In the broad sense the genus is a large one, comprising annuals, biennials, or perennials, some of which are tall, with pinnatifid or dissected leaves and clothed with simple or branched hairs. Included here are *Sisymbrium officinale* (Common Hedge Mustard), *S. altissimum* (Tumble or Jim Hill Mustard), *S. Sophia,* and *S. Thalianum* (Mouse-Ear Mustard); these are common weeds. *S. canescens* and *S. incisum* are native. In *Erysimum* (Treacle Mustard) the fruit is elongate, tapering upwards. As broadly interpreted the genus comprises many species of Eurasia and America. These are annuals or biennials with simple or pinnatifid leaves and yellow to orange flowers. *Erysimum asperum* (Western Wall-Flower) is common Northward and Westward *E. Perofskianum* is horticultural.

Brassiceæ. The important genus here is *Brassica,* with probably over 100 species of Eurasia and northern Africa; it includes the common mustards, turnip, rutabagas, cabbages, cauliflowers, etc. The fruit is elongate, generally slender and terete, somewhat constricted between the seeds, and dehiscent. The leaves are more or less cut or pinnatifid-lyrate. Some bear coarse hairs (Fig. 203).

Alysseæ. In *Alyssum* the fruit is compressed, orbicular or oblong, from yellow or white flowers. There are about 100 species, of the Old World; low branching annuals or perennials, with variously shaped leaves and clothed with a stellate pubescence. *Alyssum alyssoides* occurs as a weed and *A. saxatile* is horticultural. *Draba* comprises low branching annuals or biennials producing elongate ellipsoid flattened fruits from small flowers. There are about 175 species mostly of northern to arctic regions. Cleistogamous flowers occur in some species (*Draba verna,* an introduced European species).

Camelineæ. Camelina comprises erect annuals with simple toothed

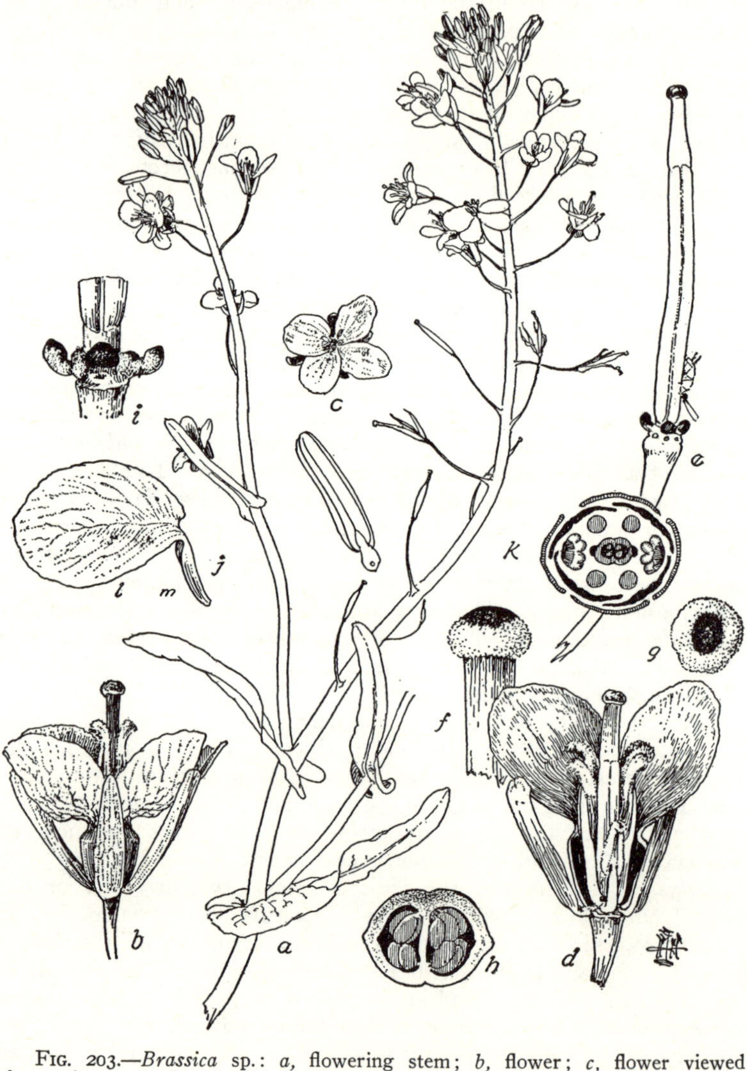

FIG. 203.—*Brassica* sp.: *a,* flowering stem; *b,* flower; *c,* flower viewed from above; *d,* flower with two petals and a sepal removed, showing one short outer and four long inner stamens; *e,* gynœcium, with glandular organs at base; *f,* stigma; *g,* stigma seen from above; *h,* transverse section of ovary; *i,* glands at base of gynœcium; *j,* petal; *l,* limb; *m,* claw.

or pinnatifid leaves, producing turgid obovoid fruits from small yellow flowers. Only about 5 species, native in Europe and eastern Asia. *Camelina sativa* is frequent in fields and waste places. *Subularia aquatica,* a dwarf aquatic which is widely distributed over North America, Europe, and Siberia, develops ovoid or globular fruits from minute small flowers. It has the distinction of being the only member of the family that has developed a perigynous flower.

Lepideæ. Lepidium (Pepperwort, Peppergrass) produces very flattened orbicular or obovoid fruits which are notched or beaked at the apex and sometimes more or less winged. The leaves range from linear to pinnatifid. The genus comprises plants of diverse habits, ranging from annual and perennial herbs to subshrubs. Those in our limits are mostly introduced from Europe and are frequent as weeds in waste grounds. *L. virginicum* (Wild Peppergrass) is native. The flowers are very small, greenish or whitish; some are apetalous. *Capsella Bursa-pastoris,* the common Shepherd's Purse, is very similar, but produces a prominently obcordate fruit. It is a very variable species, both as to leaves and as to fruit.

Thlaspideæ. Thlaspi (Penny-Cress) resembles *Lepidium* somewhat as to fruit, except that here the fruit is more conspicuously winged above. The species are low plants with white or purplish flowers. *Thlaspi arvense* (Field Penny-Cress or French Weed) and *T. perfoliatum,* which has a smaller fruit, are both introduced; the former is sometimes a pernicious weed in grain fields.

Raphaneæ. Raphanus (Radish) is distinguished by its usually large fruits which are more or less constricted between the seeds, especially in *Raphanus Raphanistrum* (Charlock), but much less so in *R. sativus* (Radish). The fruits are also prominently beaked. The plants are stout annuals or biennials with incised-lyrate leaves, and medium-sized whitish or purplish flowers.

Cakileæ. Cakile (Sea Rocket) comprises fleshy annuals of seashores, with purplish flowers producing a 2-jointed fruit. *Cakile edentula* occurs on the Atlantic Coast and shores of the Great Lakes.

TOVARIACEÆ. A very small family based on the single genus *Tovaria,* comprising one or two species of annual herbs of the West Indies and tropical South America. Flowers regular, bisexual, cyclic, conforming to the floral formula *S*3 *P*8 *A*8 *G*6-8. Fruit a berry with numerous seeds.

RESEDACEÆ (*Mignonette Family*). Flowers irregular, bisexual, spirocyclic, hypogynous or perigynous, with considerable variation in the number of the floral organs, these conforming to the formula *S*4-8 *P*8-0 *A*3-40 *G*2-6. Carpels not united by their inner edges so that the gynœcium is open above; in other cases distinct and the gynœcium apocarpous. Gynœcium and andrœcium raised on a prominent androgynophore which is expanded on one side into a prominent

FIG. 204.—*Reseda* sp.: *a,* flowering stem; *b,* flower viewed from beneath; *c,* section of flower; *p,* gynœcium; *s,* sepals; on left side of flower the disk (*r*); *d,* part of flower showing disk (*r*), gynœcium (*p*) and five of the stamens; *e,* petal; *f,* capsule, with bases of some filaments below; *g,* disk viewed from above, with bases of filaments and transverse section of ovary; *h,* transverse section of ovary; *i,* three stamens; *j,* dorsal view of a stamen.

disk. Ovary superior, 1-celled, with numerous ovules on parietal placentas, corresponding to the number of carpels when the gynœcium is syncarpous. Fruit a capsule, except in *Ochradenus,* where it is a berry. Seeds devoid of endosperm. Embryo curved.

The plants are herbs with simple, lobed or pinnate, stipulate, leaves, and small flowers in racemose inflorescences. Mostly of Mediterranean countries, including North Africa, but also in South Africa and western Asia, with one species in southwestern United States. Genera 6, species about 65.

The great majority of the species (50 or more) belong in the genus *Reseda.* Of these the most familiar is *Reseda odorata,* the mignonette of the gardens, which is believed to be a native of North Africa. The much coarser and taller *R. alba,* a Mediterranean species, is also horticultural, and often escapes to waste grounds along with the introduced *R. luteola* and *R. lutea.* (Fig. 204.)

Oligomeris, with 5 or 6 species, is represented in the Far Southwest by *O. linifolia,* a low branching plant with linear leaves and white flowers in terminal spikes. The other species inhabit South Africa and the region from North Africa to India. This species thus brings to our attention the connection of our Southwestern flora with that of the Mediterranean region and western Asia.

In *Caylusea* and *Astrocarpus* the carpels are distinct, marking the persistence of a primitive feature. Thus in passing to the other genera, in which the carpels are united, but not completely, we may observe the progressive evolutionary steps leading to syncarpy—which is attained in *Ochradenus.* A step in advance in another direction may be observed in *Randonia,* which is perigynous. Compare with the saxifrages, the *Rosaceæ,* and the gynœcium in *Xanthoxylum* (*Rutaceæ*) and *Nigella* (*Ranunculaceæ*).

MORINGACEÆ. A small family based on *Moringa,* a North African and west Asiatic genus of about three species of large trees with alternate twice- or thrice-compound exstipulate leaves and bearing white or red flowers in axillary racemes. The flowers are irregular, 5-merous, and perigynous—with a pronounced hypanthium. Stamens 5, and an outer whorl of 5 staminodes; anthers 1-celled, introrse. Gynœcium 3-carpellate, borne on a gynophore. Ovary 1-celled; ovules numerous on three parietal placentas; style elongate. Fruit an elongate capsule, dehiscing loculicidally by three valves. Seeds large, winged or wingless. Endosperm none. Embryo with large, thick cotyledons. A family of doubtful affinities.

BRETSCHNEIDERACEÆ. Based on the monotypic genus *Bretschneidera,* a tree found in the province of Yunnan, China. Leaves odd-pinnate. Flowers irregular, with perigynous corolla and andrœcium. Calyx and corolla 5-merous. Stamens 8. Gynœcium 3-carpellate, sessile, with 2 suspended ovules in each cell.

ORDER SARRACENIALES

Flowers regular, hypogynous, spirocyclic or cyclic, bisexual or unisexual, with calyx and corolla. Stamens few to numerous. Gynœcium 3- to 5-carpellate, syncarpous, with numerous ovules on parietal or central (axile) placentas. Endosperm present. Insectivorous herbs with peculiarly specialized leaves.

Families: *Sarraceniaceæ, Nepenthaceæ, Droseraceæ.* Except for the common insectivorous habit it is doubtful whether these families are sufficiently closely related to warrant the present grouping in the same order. Wettstein places the *Droseraceæ* in the order *Parietales,* and the two other families in the order *Polycarpicæ.*

Key to the Families

Leaves modified as "pitchers."
  Low herbs of bogs and swamps. Leaves in rosettes. *Sarraceniaceæ.*
  Climbing shrubs or subshrubs. *Nepenthaceæ.*
Leaves not modified as pitchers, but the leaf-blades clothed with gland-bearing hairs or the margins fringed with sensitive hairs. *Droseraceæ.*

SARRACENIACEÆ (*Pitcher-Plant Family*). A small family of insectivorous marsh herbs with pitcher-form radical leaves and large scapose-solitary or racemose flowers. Sepals 8 to 5, but petals constantly 5, followed by an indefinite number of stamens. Gynœcium 5- to 3-carpellate, with 1 style, the ovary with as many cells as carpels and bearing numerous ovules on axile placentas. Integuments 1. Fruit a loculicidal capsule. Endosperm fleshy and embryo straight.

Genera: *Sarracenia,* with 7 species in Atlantic North America; *Darlingtonia,* with 1 species in California; *Heliamphora,* with 1 species in Guiana.

In *Sarracenia* the style is unusually developed as a 5-lobed umbrella-shaped structure with a stigma under the apex of each lobe. The leaves are spreading and dilated upward from the slender petiole into the pitcher, with a rigid flange along the upper side, and terminating in an expanded apical flap. *Sarracenia purpurea,* the common Pitcher-Plant, occurs in peat-bogs from Florida to Labrador, west across the Great Lakes region and thence northwestward to the Mackenzie basin. The purplish flower is solitary on a long scape (Fig. 205).

In *Darlingtonia californica,* the only species, the leaves are elongate, and curved at the apex, with the terminal flap forked. The

FIG. 205.—*Sarracenia purpurea: a,* habit sketch; *b,* flower viewed from above and (*c*) from below; *d,* median section of flower; *p,* petal; *s,* sepals; *st,* stigma; *r,* umbrella-shaped expansion of style (stigma); *o,* ovule.

FIG. 206.—*Drosera rotundifolia: a,* habit sketch; *b,* median section of flower after pollination; *c,* section of ovary, with six styles at summit; *d, e,* five-styled (5-carpellate) gynœcia, with 1-, 2-, and 3-lobed stigmas; *f,* transverse section of ovary; *g,* ventral and dorsal views of stamen; *h,* flower-bud; *i,* petals; *j,* gland-tipped hair of leaf; *k,* leaf, upper surface.

scapes are 1-flowered. The plant is restricted to marshes from southwestern Oregon to northwestern California.

NEPENTHACEÆ. One genus, *Nepenthes,* of about 60 species of tropical Asia, the East Indies, Madagascar, and north Australia. The plants are prostrate, erect, or climbing shrubs or subshrubs, with alternate leaves dilated upward as a pitcher with an expanded lid, the lower part of the leaf tendril-like and broadening toward the base. The flowers are small, regular, and diœcious, with a perianth of two 2-merous whorls and 4 to many stamens united into a tubular column. Gynœcium 4- to 3-carpellate. Fruit a capsule. Seeds with a straight embryo surrounded by endosperm.

DROSERACEÆ (*Sundew Family*). Herbaceous plants with variously shaped insectivorous leaves. Flowers regular, bisexual, hypogynous or sometimes perigynous, 5-merous or 4-merous, solitary or in racemose or cymose inflorescences. Anthers 1-celled. Ovary 1-celled, with numerous to few ovules on parietal or basal placentas. Styles distinct. Fruit a 1-celled, loculicidal capsule. Seeds with fleshy endosperm and a straight basal embryo. Four genera and about 87 species, 84 of which belong in *Drosera*. The other genera are *Dionæa, Drosophyllum, Aldrovanda.*

Drosera has its greatest development in Australia, but a few species occur mainly in the northern United States and adjacent Canada. Of our American species *D. rotundifolia* is the most common; it is widely distributed over the continent, being found in peat-bogs from Labrador to the Great Lakes region and Alaska, and southward in the mountains to Alabama and to Montana and California in the West. The plant grows in deep peat-moss, up through which it pushes to form a rosette of leaves with orbicular blades covered with glandular hairs. The inflorescence is a one-sided raceme of small white flowers. In developing from the seed, no primary root is formed, but instead the hypocotyl is swollen into a cormose structure from which rises the rhizomatous epicotyl to produce the rosette of leaves. In our other species the leaves range from spatulate to oblanceloate and linear (Fig. 206).

Dionæa muscipula is the Venus's fly-trap of the Carolinas. *Drosophyllum lusitanicum* inhabits dry situations in Spain, Portugal, and Morocco. *Aldrovanda vesicula,* an aquatic, is found in Europe, Asia, the upper Nile region of Africa, and in Australia. It has a fly-trap leaf comparable to that of *Dionæa.*

ORDER ROSALES

Flowers hypogynous, perigynous, or epigynous, cyclic or sometimes spirocyclic, regular or irregular, usually bisexual, with few to numerous stamens, and with calyx and corolla. Carpels few to several, free or united, with few to numerous ovules. Endosperm present or only slightly developed or wanting. A large and important order of flowering plants.

Families: *Podostemonaceæ, Hydrostachyaceæ, Crassulaceæ, Cephalotaceæ, Saxifragaceæ, Pittosporaceæ, Byblidaceæ, Brunelliaceæ, Cunoniaceæ, Myrothamnaceæ, Bruniaceæ, Hamamelidaceæ, Roridulaceæ, Eucommiaceæ, Platanaceæ, Crossosomataceæ, Rosaceæ, Connaraceæ, Leguminosæ.*

Key to Families Represented by Native or Cultivated Species

I. Ovary superior.
  A. Flowers regular.
    a. Flowers hypogynous.
      1. Plants succulent.
        Aquatics of thalloid, alga-like habit. Carpels 2-5. Atlantic Coast. *Podostemonaceæ.*
        Terrestrial, with thick fleshy leaves. Carpels several. *Crassulaceæ.*
      2. Plants not as above.
        Leaves simple.
          Calyx and corolla obscure or apparently none. Leaves large, palmately lobed, with large clasping stipules. Large trees. *Platanaceæ.*
          Calyx and corolla (mostly) present.
            Herbs. Leaves various. *Saxifragaceæ.*
            Trees or shrubs. Leaves leathery. *Pittosporaceæ.*
        Leaves pinnately compound. Flowers often small, in small globose heads, spikes, or racemes. *Leguminosæ (Mimosoideæ).*
    b. Flowers perigynous.
      a. Fruit a legume. Stamens numerous. Leaves once- or twice-pinnate. Carpels 1 or more. *Leguminosæ (Mimosoideæ).*
      b. Fruit a capsule, achene, or drupe.
        Stamens numerous, sometimes 5, 4, or 1. Petals sometimes wanting. Carpels sometimes 1. *Rosaceæ.*
        Stamens 5 or 10. Fruit a capsule. *Saxifragaceæ.*
  B. Flowers irregular.
    a. Succulent aquatics of thalloid habit. *Podostemnoaceæ.*
    b. Terrestrial, rarely succulent.
      Carpels 2 or sometimes more, united. Fruit a capsule. *Saxifragaceæ.*
      Carpels 1. Fruit a legume. *Leguminosæ.*
II. Ovary inferior.
  a. Corolla none. Flowers bisexual or unisexual, polygamous or monœcious, in heads or spikes. *Hamamelidaceæ.*
  b. Corolla present.
    1. Receptacle produced beyond its connection with the ovary as a shallow to deep, saucer-shaped, cup-shaped, or cylindrical tube (hypanthium).
      Stamens numerous. Fruit a pome. *Rosaceæ.*
      Stamens few. Fruit a berry. *Saxifragaceæ.*

2. Receptacle not developed as an hypanthium.
   Petals long, narrow, and ribbon-like. *Hamamelidaceæ* (*Hamamelis*).
   Petals large and broad to small and narrow.
      Stamens mostly numerous. Shrubs or vines. *Saxifragaceæ*.
      Stamens 5. Shrubs or small trees. *Saxifragaceæ* (*Itea*).

PODOSTEMONACEÆ (*River-Weed Family*). A family of about 175 species (21 genera) of mainly tropical aquatic herbs represented in Atlantic North America by *Podostemum ceratophyllum*. They are mostly inhabitants of running water, in habit suggesting algæ. The gynœcium is 2- to 3-carpellate, with as many styles. The superior ovary contains numerous minute seeds with a straight embryo but no endosperm. The fruit is a capsule. Flowers regular or irregular.

HYDROSTACHYACEÆ. These are aquatics with tuberous stems, pinnate leaves, and naked diœcious flowers in spikes. The staminate flowers are reduced to 1 stamen, while the gynœcium is 2-carpellate, with free styles. One genus, *Hydrostachys*, of about a dozen species in Madagascar and South Africa.

CRASSULACEÆ (*Orpine Family*). Succulent herbs with hypogynous or sometimes perigynous, regular, cyclic, 3-merous or polymerous, bisexual flowers. Stamens as many or twice as many as the petals. Carpels as many as or fewer than the petals, free or slightly united, each carpel usually with a small scale at the base. The fruits are follicles, each producing many seeds. Endosperm scanty. Integuments 2. The family differs from the *Saxifragaceæ* chiefly in having constantly symmetrical flowers, more carpels, and in the absence of the gland (disk) associated with the carpels, and scanty endosperm. About 600 species in 30 genera, widely distributed in temperate and warm regions.

Principal genera: *Crassula, Sedum, Sempervivum, Echeveria, Bryophyllum, Cotyledon.*

In all but *Crassula* the stamens outnumber the petals, being generally twice as many, while in *Crassula* they are as many as the petals, and the petals are essentially distinct. *Sempervivum* differs from *Sedum* in its more numerous floral organs (more than 5 in a set) and in the dense basal rosettes of leaves. In *Bryophyllum, Cotyledon,* and *Echeveria* the petals are more or less united (sympetalous), with the special distinction in *Bryophyllum* that the flowers are 4-

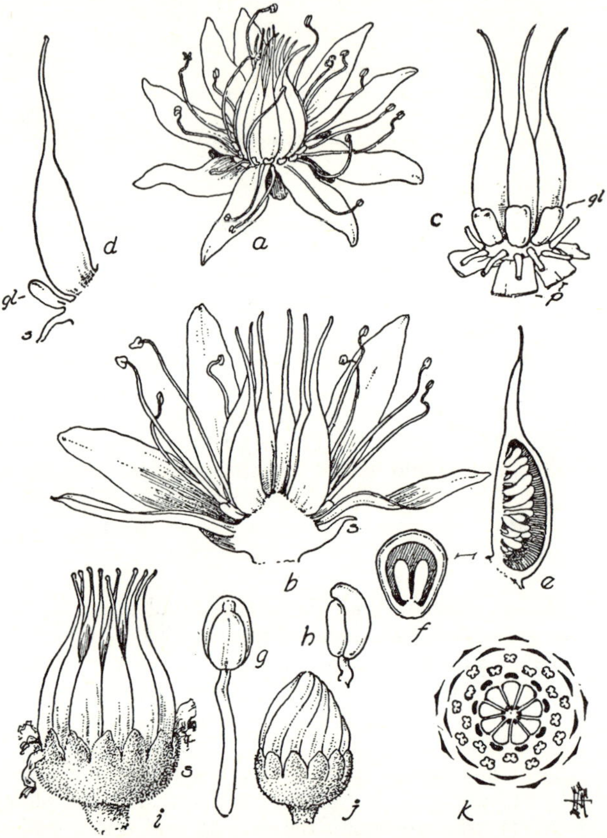

FIG. 207.—*Sedum: a,* flower; *b,* section of flower; *s,* sepal; *c,* part of flower showing glandular bodies (*gl*); *p,* petals; *d,* a carpel with subtending gland; *e,* median section of carpel; *f,* transverse section of ovary; *g,* ventral and *h,* lateral, view of stamen; *i,* late stage of carpels; *s,* calyx; *j,* flower-bud; *k,* floral diagram.

merous instead of 5-merous as in the latter two. Except for the corolla-tube and certain foliar characters, there is little to distinguish *Echeveria* from *Cotyledon.* The calyx of *Bryophyllum* is inflated and the leaves are often pinnate; the plants are easily propagated by means of the leaves or leaf-fragments. Compared with the others *Bryophyllum* is a very small genus—only four species, all tropical.

The largest genus of the family is *Sedum,* with about 300 species; *Crassula* has about 150; *Echeveria,* 100 or more; and *Sempervivum,* 60—70. All are widely distributed. (Fig. 207.)

Besides the above, two species of *Tillæa* occur in the Atlantic States, one of which is found also on the Pacific Coast, *Sedelia, Congdonia, Hasseanthus,* and *Rhodiola* are represented in California by one species each.

The solitary *Penthorum sedoides,* which is distinguished from all other members of the family by its syncarpous gynœcium and non-succulent habit, has been separated by various authorities as the basis of the family, *Penthoraceæ.* The species is an inhabitant of ditches and swamps of the Atlantic States to New Brunswick.

Besides herbaceous forms, a shrubby habit is characteristic of some members. The stems are usually fleshy, with simple, exstipulate, alternate or whorled leaves, in the whorled types either in basal or terminal rosettes. These characteristic rosettes associated with the conspicuous succulence afford easy recognition-marks of the family connections. The inflorescence is generally cymose, and the flowers are characteristically radially symmetrical.

CEPHALOTĂCEÆ. A monotypic family based on *Cephalotus follicularis,* a pitcher-plant of west Australian bogs. The leaves are basal, from a rhizome. Flowers bisexual, in scapose-paniculate inflorescences. Carpels 6, free.

SAXIFRAGACEÆ (Saxifrage Family). The saxifrages are herbs or shrubs with hypogynous, perigynous, or epigynous, regular or sometimes irregular, mostly bisexual flowers, in some cases with undifferentiated perianth. Stamens as many as the sepals or sometimes more, and carpels as many as the sepals or fewer, rarely more, free or more or less united, the styles distinct. Ovary usually with as many cells as carpels. Placentas mostly axile and fleshy, bearing numerous ovules. Integuments 2. Embryo small, in more or less copious endosperm. Fruits various (spurious where the ovary is inferior)—follicles, capsules, or berries. Leaves opposite or alternate, sometimes with stipules. Stamens often obdiplostemonous.

In certain respects a very technical family, with characters in common with the *Crassulaceæ* and the *Rosaceæ.* Many herbaceous forms are strikingly similar as to foliar characters although generically distinct, and on the other hand many congeneric forms are as strikingly dissimilar in foliage. Vegetative and floral characters are often so variable that generic and specific lines are difficult to draw. Species 650—700, in 70 or more genera. Widely distributed.

Engler recognizes the following subfamilies:

I. *Saxifragoideæ,* in which there are commonly 2 carpels, rarely 3—7, free or more or less united below. Ovules with 2 integuments. Herbs.

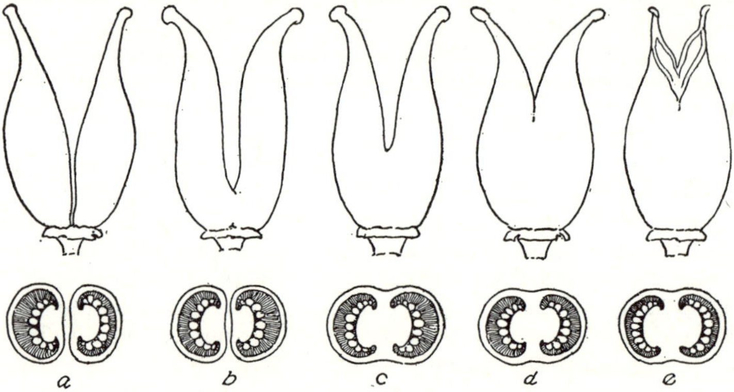

FIG. 208.—Left to right, stages in union of carpels in *Saxifraga; a-e* below, transverse sections of ovaries of corresponding gynœcia.

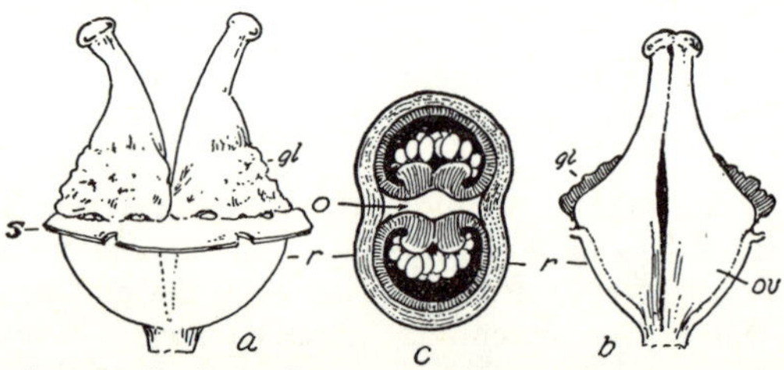

FIG. 209.—A type of epigyny in some saxifrages (*Saxifraga pennsylvanica*): *a,* the 2-carpellate gynœcium and adnate receptacle (*r*); *s,* sepal bases; *b,* section of (*a*) with one of the carpels removed; *c,* transverse section of (*a*) below calyx; *r,* receptacle; *ov,* ovary; *o,* free space between the two carpels; *gl,* gland.

Including: *Astilbe, Saxifraga, Bergenia, Peltiphyllum, Boykinia, Bolandra, Sullivantia, Tiarella, Lithophragma, Tellima, Mitella, Tolmiea, Heuchera, Chrysosplenium, Parnassia.*

II. *Francooideæ,* with tetramerous flowers and superior ovary. Herbs. Two genera: *Francoa, Tetilla.* Chile.

III. *Hydrangeoideæ,* with more or less inferior ovary and 5-merous perianth. Ovules with 1 integument. Shrubs and trees. Includes *Philadelphus, Deutzia, Carpenteria, Hydrangea.*

IV. *Pterostemonoideæ,* with 5-carpellate gynœcium and superior ovary. Stamens 10. One genus, *Pterostemon,* of two species. Shrubs. Mexico.

V. *Escallonioideæ,* ovary superior or inferior. Stamens 5. Integuments 1. Shrubs and trees. Species about 100 in 21 genera, of which *Escallonia,* the largest (50 species), is representative.

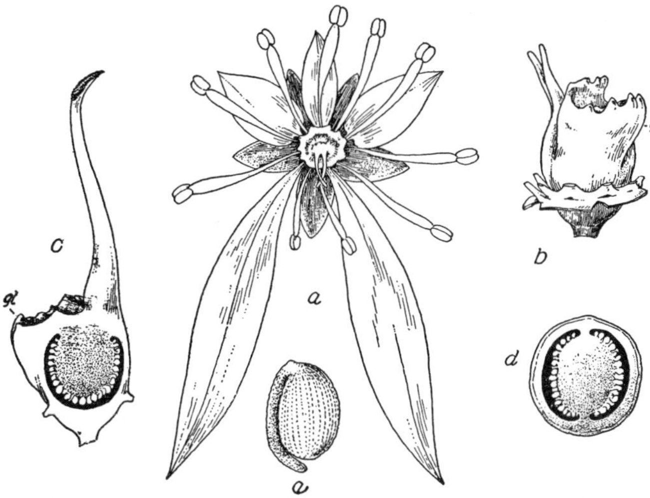

FIG. 210.—Zygomorphic flower (*a*) of *Saxifraga sarmentosa; b,* gynœcium, with dorsal gland (*gl*); *c,* gynœcium, the ovary in sectional view; *gl,* gland; *d,* transverse section of ovary; *e,* ovule.

VI. *Ribesioideæ,* with 2-carpellate gynœcium, inferior ovary, and parietal placentation. Stamens 5. Shrubs. One genus (in broad sense), *Ribes,* with about 150 species.

VII. *Baueroideæ,* ovary inferior, with parietal placentation. Fruit a capsule. One genus, *Bauera,* with 3 species. Australia and Tasmania.

Saxifragoideæ. Astilbe (False Goat's Beard) comprises tall herbs with ternately compound leaves and polygamous flowers in terminal panicles. The gynœcium is 2—3-carpellate, the carpels free nearly to the base but the basal portion fused with the receptacle to form a partly inferior ovary. Species about 7, of Atlantic North America and Asia. *Astilbe biternata* occurs in the Appalachians from Virginia to Georgia and Tennessee.

Saxifraga (Saxifrage) in the broadest interpretation is a large genus, comprising about half the species of the family. In this broad sense it includes a heterogeneous aggregation of forms, differing both in habit and in floral structure. The flowers range from regular

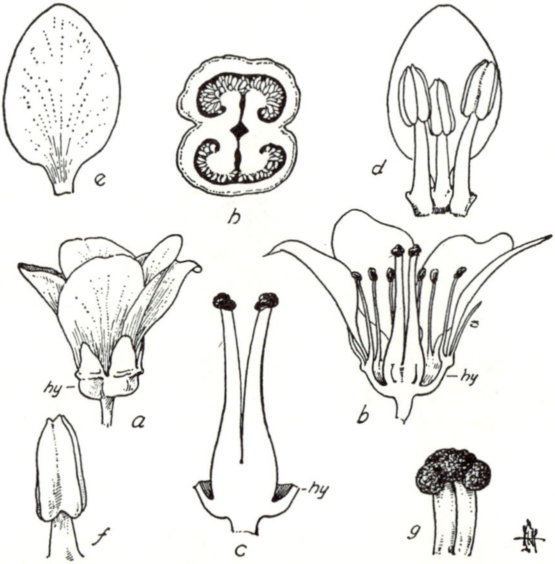

FIG. 211.—*Bergenia crassifolia: a,* flower; *b,* median section of flower; *hy,* hypanthium; *s,* sepal; *c,* gynœcium; *d,* stamens and petal; *f,* young stamen; *g,* stigma; *h,* transverse section of ovary.

to bilaterally symmetrical, but are uniformly 5-merous—except the gynœcium, which is commonly 2-carpellate, although as many as 7 carpels sometimes occur, these free or more or less united and with superior or inferior ovary. The ovary (or ovaries) is commonly girdled by an annular disk. The different groups thus distinguished by their distinctive floral characters are by some writers given the rank of genera. Some of the different types of fruits are illustrated in Fig. 208. Figures 209-212 and 72 will give an idea of the floral structures and fruits of various other members of this subfamily. The plants are widely distributed.

Hydrangeoideæ: Philadelphus. In this genus the flowers are epigynous, sepals and petals sometimes 4, stamens numerous, and the gynœcium 3—5-carpellate, with a deeply inferior ovary and styles as many as carpels. Ovules numerous, on thick axile placentas. The spurious fruit is a capsule, dehiscing by valves at the top and commonly crowned by the persistent calyx-segments. About 50 species of the United States, Mexico, Asia, and Central Europe. Ours are

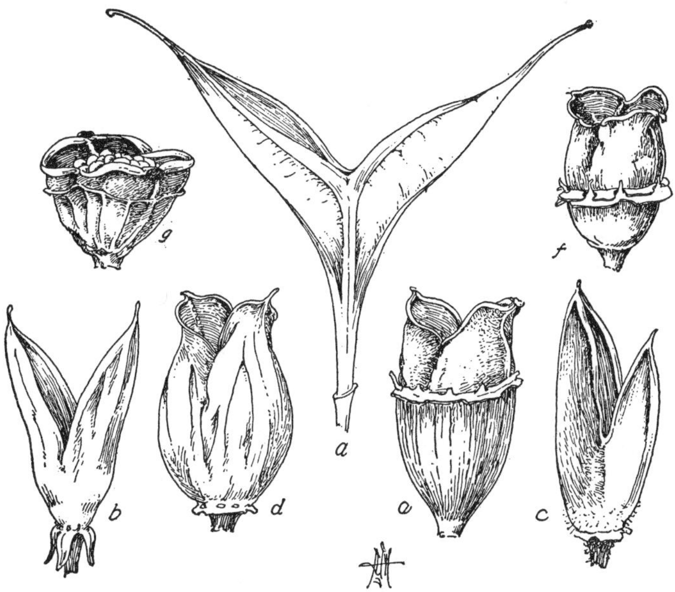

FIG. 212.—Capsules of various saxifrages: *a, Tolmiea Menziesii; b, Saxifraga micranthidifolia; c, Tiarella cordifolia; d, Saxifraga leucanthemifolia; e, Saxifraga debilis; f, Chrysosplenium americanum; g, Mitella pentandra.*

generally known as "Syringa" or Mock-Orange. The leaves are opposite and the flowers showy and fragrant. Petalody of the stamens is of frequent occurrence. The plants are common horticulturals in the North. *Deutzia* is very similar in floral structure except for the uniform pentamery outside of the 3—4-merous gynœcium (Fig. 213).

The solitary *Carpenteria californica* is a shrub of the Sierras. The leaves are opposite and entire, and the flowers very variable in the number of parts—sepals and petals 6—8, and stamens up to 200; carpels 2—8, free above but united below. Ovary inferior. In *Hydran-*

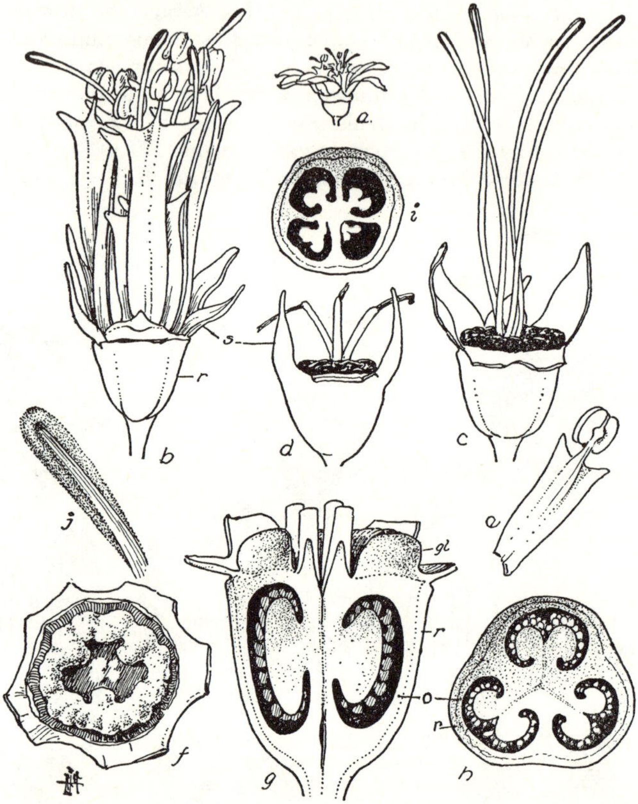

FIG. 213.—*Deutzia: a,* flower; *b,* flower with petals removed; *c,* same with parts removed to show styles and annular disk (black); *d,* later stage of *c; e,* stamen; *f,* disk viewed from above (other parts removed); *g,* median section of "ovary,"—*r,* receptacle, *o,* ovary; *gl,* gland; *h,* transverse section of a 3-carpellate ovary; *i,* transverse section of a 4-carpellate ovary,—ovules removed; *j,* stigma; *s,* sepals.

gea the outer flowers of the inflorescence are sterile, with enlarged and conspicuous calyx-segments. Carpels 2—4, with inferior ovary and distinct styles. Fruit (spurious) a capsule. About 35 species of North America, South America, and Asia. Several are horticultural, among which is the root-climber, *Hydrangea petiolaris* of Japan.

Escallonioideæ. Escallonia has the flowers 5-merous with respect to sepals, petals, and stamens, but the gynœcium 2- to 5-merous, with a single style and lobed stigma. The ovary is inferior, producing numerous ovules on divided axile placentas. The spurious fruit is a capsule crowned by the calyx-segments. South American shrubs and

FIG. 214.—*Escallonia montevidensis: a,* flowering branch; *b,* branch of inflorescence with terminal flower and small buds in axils of bracts (*br*) below; *c,* enlarged view of bract (*br*) of *b,* with axillary bud; above, diagram of same; *d,* flower-bud; *f,* median section of flower; *g,* median section of flower, enlarged, with parts removed; *h,* transverse section of "ovary"; *s,* sepals; *r,* receptacle; *o,* ovary; *gl,* gland (disk); *i,* dorsal, and *j,* ventral view of anther; *k,* leaf, with marginal teeth at left, enlarged.

trees, several of which are grown as ornamentals in the South and Southwest (Fig. 214).

Ribesioideæ: Ribes (Currants and Gooseberries). Flowers racemose, commonly 5-merous, epigynous, with a shallow to deep hypanthium, the stamens in one whorl, and the gynœcium 2-carpellate, rarely 3-carpellate, with short or elongate style more or less lobed or forked at the top. Ovary inferior, with numerous ovules on parietal placentas. The spurious fruit is a smooth or prickly berry. About 175 species of wide distribution. Some Asiastics are polygamous or diœcious (*e.g., Ribes alpina*). The plants are usually distinguished as currants and gooseberries, especially in horticulture, but the char-

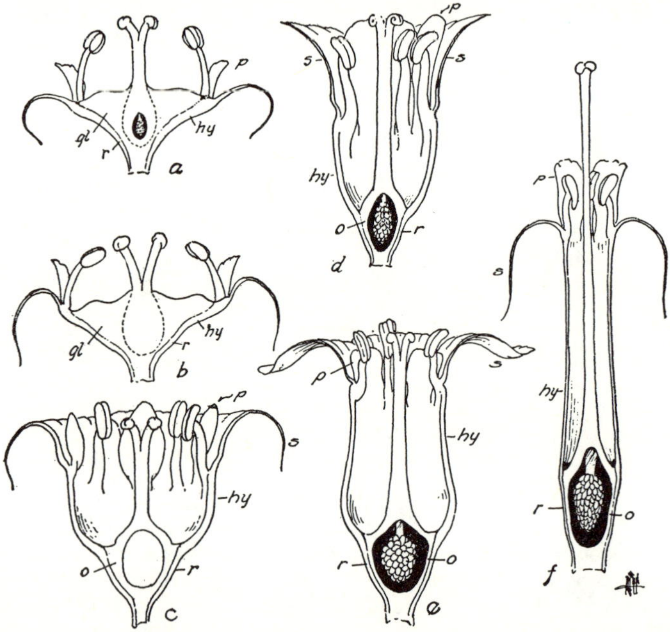

FIG. 215.—Median sections of flowers of different species of *Ribes: a, Ribes hirtellum; b, R. vulgaris; c, R. hudsonianum; d, R. floridum; e, R. malvaceum; f, R, odoratum* (*R. aureum*)*; gl,* disk; *p,* petal; *s,* sepal; *hy,* hypanthium; *o,* ovary; *r,* receptacle. Note increasing depth of hypanthium from (*a*) to (*f*).

acters on which the popular distinction is based are frequently common to both groups. Thus smooth and prickly stems as well as smooth and bristly berries are common to both currants and gooseberries. Technically the currants have many-flowered racemes and the pedicels are jointed at the base of the flower, whereas in the gooseberries the racemes are few-flowered and the pedicels are jointless. In either case certain species are rather similar in habit and easily confused in the field, *e.g., Ribes americanum, R. nigrum,* and *R. hudsonianum,* among

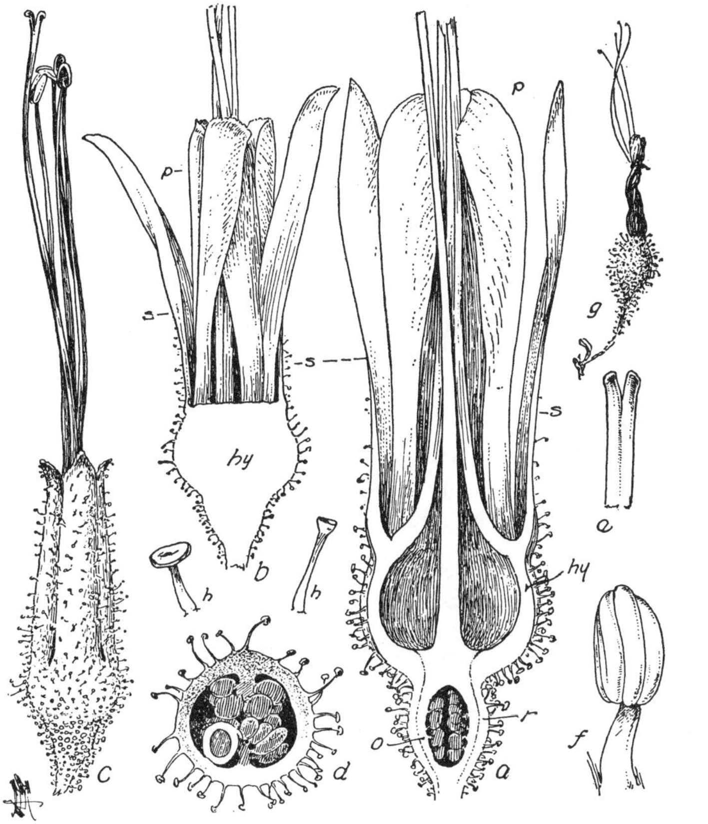

FIG. 216.—*Ribes speciosum: a,* median section of flower, *b,* section of flower with two sepals removed; *c,* flower, entire; *d,* transverse section of ovary; *e,* stigmas; *f,* stamen; *g,* flower after pollination; *h,* gland-bearing hairs of flowers; *s,* sepals; *p,* petals; *hy,* hypanthium; *o,* ovary; *r,* receptacle.

the currants, and *R. Cynosbati* and *R. oxyacanthoides* or *R. hirtellum,* among the gooseberries. On the other hand some species are easily recognized, *e.g., Ribes speciosum, R. viscosissimum,* and *R. odoratum.* Some writers prefer to regard the gooseberries as distinct from the currants both as to genus and family, *viz.,* as *Grossularia, Grossulariaceæ.*

Ribes like *Saxifraga* illustrates certain striking evolutionary tendencies in the structure of the flower, which may be readily traced step by step in median sections. A selected series is shown in Fig. 215. The particular feature to be noted is the progressive deepening of the hypanthium from a shallow saucer-shaped type in *Ribes vulgare* to a deep narrow tube in *R. odoratum*. Other species may be used as well for the same purpose.

Several species of currants and gooseberries are cultivated either for their berries or as ornamentals, the commoner ones being *R. vulgare* (Red Currant) of western Europe, *R. nigrum* (Black Currant) of Eurasia, and *R. Grossularia* (Prickly Gooseberry) of Eurasia and North Africa. *R. americanum* (Wild Black Currant), a native woodland species, is somewhat similar to *R. nigrum* and is frequently cultivated. Another black-berried species, *R. hudsonianum,* is frequent in cold Northern swamps; the ovary is somewhat glandular. *R. lacustre* (Swamp Black Currant) is found in similar situations and can be distinguished by its prickly stems and glandular, bristly fruit. Along with these there occur also two red-fruited species, *R. triste* (Swamp Red Currant), and *R. prostratum* (Skunk Currant), both reclined and straggling, the former with smooth fruit, the latter with rather fœtid smelling foliage and glandular bristly fruit. The tall clammy *R. viscosissimum* is common in deep woods in the West. *R. odoratum* (Golden Currant), a tall Western species with firm, smooth leaves, is a common ornamental. The common native gooseberries are *R. Cynosbati* (Prickly Gooseberry), *R. missouriense* (Missouri Gooseberry), and *R. oxyacanthoides* (Northern Gooseberry), and in coastal California there is found a tall, elegant, evergreen, and very prickly species, *R. speciosum* (Fuchsia-Flowered Gooseberry), with bright crimson flowers. (Fig. 216.)

PITTOSPORACEÆ (*Pittosporum Family*). Flowers hypogynous, cyclic, regular or somewhat irregular, bisexual, rarely unisexual, 5-merous, the gynœcium 2-carpellate, rarely 3- or 5-carpellate. Ovary superior, 1-celled, or with as many cells as carpels. Style 1. Ovules numerous, on axile or parietal placentas. Integuments 1. Fruit a berry or capsule. Endosperm copious. Embryo small.

About 150 species of trees and shrubs or woody climbers, in 9 genera, with alternate, commonly leathery exstipulate leaves of diverse shapes, and medium-sized, often very fragrant, clustered flowers. The genera are Australian, except *Pittosporum,* the largest genus, which is more widely distributed through the tropics of the Old World.

Several species of *Pittosporum* are cultivated in southern California and other parts of the Southwest. These are evergreen trees or

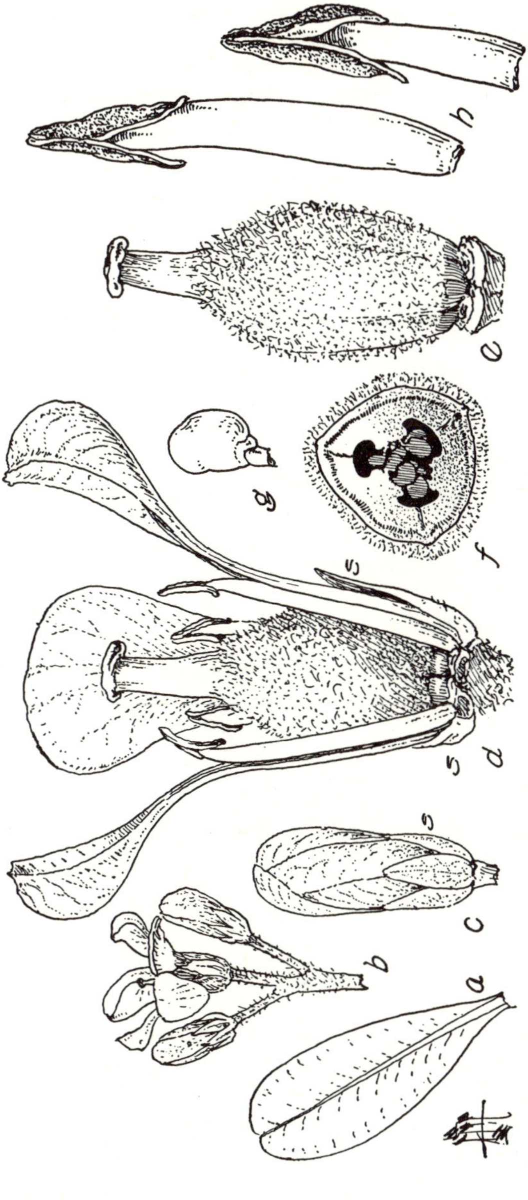

FIG. 217.—*Pittosporum Tobira: a*, leaf; *b*, cymule; *c*, flower-bud; *d*, flower with parts removed; *e*, gynœcium; *f*, transverse section of ovary; *h*, stamens after anthesis; *g*, ovule; *s*, sepals.

shrubs with more or less leathery lanceolate, oblancolate-spatulate, or rhomboidal leaves, and clustered creamy-white or purplish flowers. The sepals and petals are free or somewhat united at the base. The stamens have stout filaments and large anthers narrowed upward. The gynœcium is commonly 2- or 3-carpellate and 1-celled, with

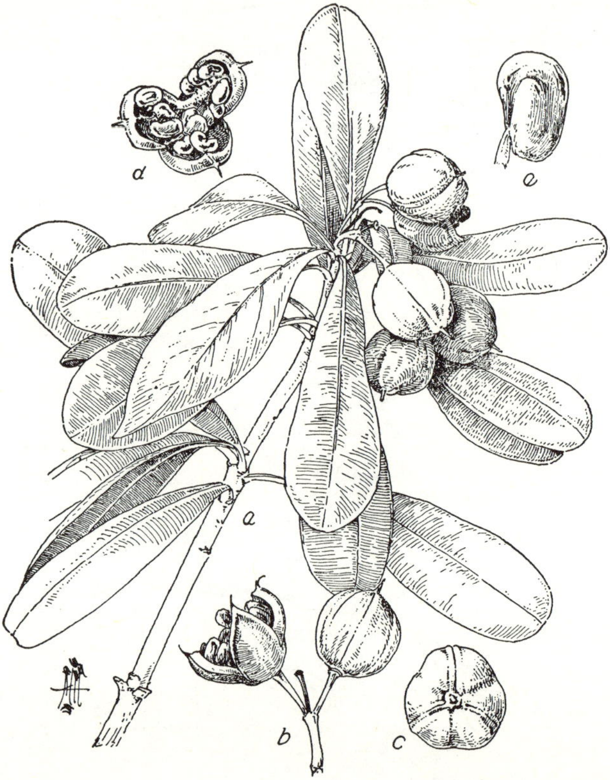

FIG. 218.—*Pittosporum Tobira: a,* branch with cluster of fruits; *b,* fruits (capsules); *c,* top view of capsule; *d,* dehisced capsule; *e,* seed.

numerous ovules on parietal placentas. The fruits are globose-ovoid, somewhat succulent at first, drier and greenish or bright orange with age, and ripen into loculicidal capsules. The species are very ornamental, especially those with clusters of orange-colored fruits. In all there are about 100 species. (Figs. 217, 218, 219.)

BYBLIDACEÆ. One genus, *Byblis,* of 2 species in Australia. These are glandular-pubescent herbs with bisexual 5-merous flowers and capsular fruits.

BRUNELLIACEÆ. Flowers unisexual, 5-merous or 7-merous. Carpels 5, free. Fruits 1- or 2-seeded follicles. Endosperm mealy. Woody plants with opposite or whorled leaves. *Brunellia,* 10 species from Mexico to Peru.

CUNONIACEÆ. Flowers bisexual, or unisexual by abortion, in racemose or paniculate inflorescences, perigynous or epigynous, calyx and corolla 4-merous or 5-merous, petals often wanting. Stamens 4—5 or numerous, inserted beneath an hypogynous disk. Carpels commonly 2 (or 4—5), 2- or many-seeded. Fruit capsular, follicular, or drupaceous. Endosperm fleshy. Embryo small. Trees or shrubs with opposite or whorled stipulate leaves. Over 100 species in about 20 genera, several of which are monotypic. *Weinmannia* the largest, with 70 or more species. Chiefly of the southern hemisphere.

FIG. 219.—Unfolding leaf-buds of *Pittosporum undulatum,* with evident transition from bud-scales to leaves.

MYROTHAMNACEÆ. Flowers regular, naked, unisexual, diœcious. Stamens 4—8; carpels 4—3, with as many styles with broad elongate stigmas. Ovary superior, with numerous ovules in two rows on the dorsal suture. Fruit a septicidal capsule. Shrubs with opposite leaves and quadrangular twigs. One genus, *Myrothamnus,* of 2 species in South Africa.

BRUNIACEÆ. Flowers regular or irregular, bisexual, perigynous or sometimes epigynous, 5-merous. Carpels 3—2, or sometimes

1, with as many styles. Fruit a 2-seeded capsule or nutlet. Seeds arillate, with copious endosperm. Integuments 1. Shrubby plants often of aspect of heaths or myrtles. Twelve genera and about 50 species of South Africa. *Myrothamnus,* 2 species.

HAMAMELIDACEÆ (*Witch-Hazel Family*). Flowers regular or irregular, bisexual, or unisexual and diœcious, sometimes naked, hypogynous or epigynous, commonly 4-merous or 5-merous. Gynœcium 2-carpellate, with 1 to several ovules in each carpel. Fruit a loculicidal or septicidal capsule dehiscing above, with 1—2 seeds in each cell. About 50 species of trees and shrubs in 20 genera, natives of North America, Asia, and South Africa. Two subfamilies;

Bucklandioideæ, with numerous ovules in each carpel. Includes *Liquidambar, Altingia.*

Hamamelidoideæ, with 1 ovule in each carpel. Includes *Hamamelis, Fothergilla, Corylopsis.*

Bucklandioideæ. Liquidambar is represented in our flora by *L. Styraciflua* (Sweet Gum), a large forest tree of Atlantic United States, Mexico, and Central America. It is distinguished by large palmately lobed leaves and monœcious flowers, the staminate borne in racemose clusters, the carpellate in solitary globular heads. The staminate flowers have no perianth but numerous stamens. In the carpellate flowers the corolla is wanting, and the calyces and 2-carpellate gynœcia become consolidated into a globular head. The ovary is inferior, with several to numerous ovules. The individual fruits are capsular, dehiscing by valves at the top. The species is often planted as an avenue tree. *L. macrophylla* of central Asia differs in the larger leaves and in the shape of the cotyledons. In *L. orientalis* of Asia Minor the flowers and leaves are smaller. The plants contain a balsamic juice. *Altingia* includes two Asiatic trees which differ from *Liquidambar* in having ovate leaves, stamens with short and thick filaments, and a deciduous style. The genus has been made the basis of the family *Altingiaceæ* in some systematic works.

Hamamelidoideæ. Hamamelis (Witch-Hazel) comprises shrubs or small trees with alternate straight-veined leaves and small, yellow, bisexual or polygamous, 4-merous flowers in axillary clusters or heads subtended by a 3-leaved involucre. The anthers open by valves and the inner whorl of stamens is transformed into staminodes. The gynœcium is 2-carpellate, with as many styles and a partly inferior 2-celled ovary containing 1 pendulous ovule in each cell. The spurious fruit is a capsule, dehiscing loculicidally at the top. The seeds are black, shiny, and bony. *Hamamelis virginiana* is native in the eastern

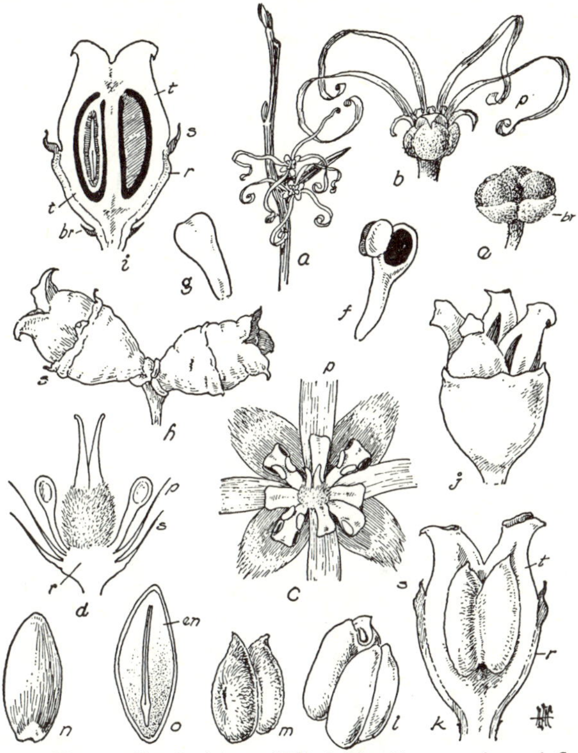

FIG. 220.—*Hamamelis virginiana* (Witch-Hazel): *a,* twig and flowers; *b,* flower; *c,* flower viewed from above, petals cut off; in center the 2-carpellate gynœcium; between stamens are alternating staminodes; *d,* median section of flower; *e,* a cluster of three flower-buds; *f,* stamen; *g,* staminode; *h,* fruits; *i,* median section of fruit; *j,* fruit at dehiscence; *k,* section of fruit showing "stone"; *l,* earlier and *m,* later stages in development of stone; *n,* seed; *o,* median section of seed; *en,* endosperm, with embryo in center; *t,* ovary; *r,* receptacle; *s,* sepals; *p,* petals; *br,* bracts.

United States. The flowers appear in late autumn and have long, narrow, ribbon-like petals. The plants sometimes attain a height of 25 feet; they are commonly cultivated as ornamentals (Fig. 220). Two other species occur in Japan. *Fothergilla* is very similar but differs in its apetalous flowers borne in spicate inflorescences. The filaments are thickened above. The ovary is inferior, with 2 slender styles. The fruit is a cartilaginous capsule. One species, *F. Gardeni,* occurs as a small shrub of low grounds from Virginia to Georgia.

FIG. 221.—*Platanus* sp.: *a,* carpellate, *b,* staminate shoot; *c,* transverse section of carpellate inflorescence (head); *d,* carpellate flower; *e,* carpels; *f,* staminodes; *g,* stamen; *h,* stamen viewed from above; *br,* bracts; *i,* branched hair of pubescence.

RORIDULACEÆ. Roridula, with two shrubs in the mountains of the Cape region of South Africa. Flowers 5-merous. Anthers opening by valves. Gynœcium 3-carpellate, with superior ovary. Fruit a loculicidal capsule.

EUCOMMIACEÆ. A monotypic family based on *Eucommia ulmoides* of China, a tree with alternate exstipulate leaves and unisexual flowers. Stamens 6—10. Carpels 2, one commonly aborting. Fruit a 1-seeded samara. Seed with endosperm. Latex in young growth. Bark prized in parts of China for its medicinal properties.

PLATANACEÆ (*Plane-Tree Family*). A small family of large trees with alternate palmately lobed leaves and numerous small, inconspicuous, monœcious flowers in dense globular heads. Stipules sheathing. Sepals and petals greatly reduced or obsolete, 5-merous. Stamens 5 or 3—8, with short filaments and elongate anthers crowned above by the expanded connective. Gynœcium commonly of 1 slender carpel gradually narrowed upward into the elongate curved style which is stigmatic on the inner side. The whole carpellate inflorescence forms a globular multiple fruit composed of the numerous nutlets (the true fruits) derived from the ovaries of the individual carpellate flowers. The nutlets are usually subtended by a ring of bristles. Integuments 2. (Figs. 221-222.) The name Plane-Tree owes its origin to the fact that the bark comes off in planes, a characteristic feature of these trees which at the same time enhances their beauty. The trees are also called Sycamore and Buttonwood. The eastern species, *Platanus occidentalis,* which sometimes attains a height of over one hundred feet and a trunk diameter of 10—14 feet, has its counterpart in the beautiful Californian *P. racemosa.* The Eurasian *P. orientalis* and its varieties are commonly planted as avenue trees.

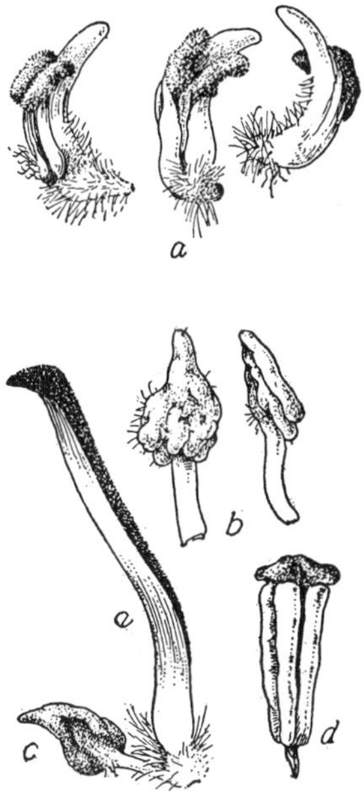

FIG. 222.—*Platanus* sp.: *a,* staminodes; *b,* staminodes of another shape; *c,* staminode; *d,* normal stamen; *e,* carpel, with long stigmatic surface.

CROSSOSOMATEÆ comprises but one genus, *Crossosoma,* with two species of shrubs, *C. Bigelowii,* with fascicled leaves, of the desert region of the Southwest and Lower California, and *C. californica,* with scattered leaves, of Catalina and Guadalupe islands. The flowers are white, perigynous, 5-merous, except the 2—9 free, several-seeded carpels. The 15—20 stamens are inserted on a disk lining the hypanthium. Fruits stipitate follicles.

ROSACEÆ (*Rose Family*). Flowers regular, perigynous or epigynous, bisexual, rarely polygamo-diœcious, 5-merous with respect to calyx and corolla (rarely wanting), stamens generally numerous and in several whorls, the anthers small and 2-celled, and the gynœcium of 1-several distinct or more or less united carpels with as many distinct styles or these more or less united, terminal or lateral. Ovaries superior or more or less inferior, with 1, 2 or several ovules on axile placentas. Integuments 2. Fruits derived from superior ovaries are true fruits—achenes, follicles, or berries; those derived from inferior ovaries are spurious—capsules, berries, or pomes. Endosperm scanty or none. Embryo mostly with thick cotyledons.

One of the most important families of flowering plants, numbering between 2,000 and 2,500 species of herbs, shrubs, and trees, generally with alternate, stipulate leaves, in 90 or more genera, widely distributed over the world. To this family belong a large number of useful plants that have long been cultivated for their luscious fruits, hard wood, or ornamental qualities. The cultivated varieties of some members, *e.g.*, roses and apples, run into the thousands.

Six subfamilies are recognized, which may be conveniently grouped as follows:

THE PERIGYNOUS ROSACEÆ: OVARY SUPERIOR (FIG. 237)

I. *Spireoideæ*. Receptacle forming a shallow cup (hypanthium). Carpels several to 1, commonly 5—2. Fruits follicles or in some cases achenaceous. Shrubs or herbs. Includes *Physocarpus, Spiræa, Aruncus, Exochorda, Holodiscus.*

II. *Rosoideæ*. Receptacle (hypanthium) shallow, with a more or less developed central column, or deep and urn-shaped. Carpels commonly numerous, in spirals, on the enlarged central column, or borne on the inner surface of the urn-shaped hypanthium. An epicalyx is present in some cases. Shrubs and herbs. Includes *Rhodotypus, Rubus, Fragaria, Potentilla, Geum, Alchemilla, Agrimonia, Sanguisorba, Rosa, Cercocarpus, Adenostoma, Chamæbatia, Coleogyne, Purshia.*

III. *Prunoideæ*. Receptacle (hypanthium) cup-shaped. Carpels commonly 1, sometimes as many as 5, distinct, with a terminal style. Fruit a drupe. Shrubs and trees. Includes *Prunus* (in broad sense), *Nuttallia, Osmaronia.*

IV. *Chrysobalanoideæ*. Hypanthium as in *Prunoideæ,* and carpels

5—1 but with basal style. Fruit a drupe. Evergreen trees and climbing shrubs of Mexico, Central America, Africa, and the East Indies. Includes *Chrysobalanus, Acioa.*

THE EPIGYNOUS ROSACEÆ: OVARY INFERIOR (FIG. 237)

V. *Neuradoideæ.* Receptacle cup-shaped, adnate to the ovary of the 5—10 united carpels. A small group of herbs of the deserts of South Africa and of North Africa to India. Two genera: *Neurada* (2 species) and *Grielum* (6 species).

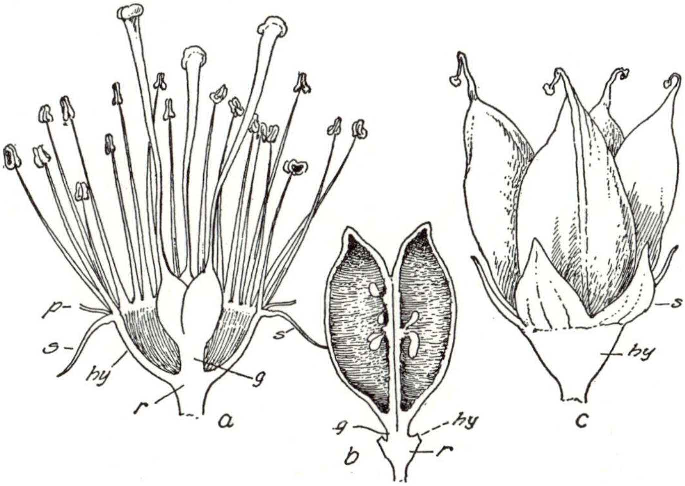

FIG. 223.—*Physocarpus opulifolius: a,* median section of flower; *b,* median section of ovary; *c,* nearly mature fruits; *p,* petals; *s,* sepals; *hy,* hypanthium; *r,* receptacle; *g,* gynophore.

VI. *Pomoideæ.* Receptacle adnate to and commonly prolonged beyond the ovary as a shallow to deep hypanthium. Carpels 5 (or more) to 2, often loosely united, with as many free or more or less united styles. Seeds numerous or commonly 2 in each cell. Fruit (spurious) a pome. Trees and shrubs with stipulate leaves, widely distributed in the northern hemisphere. A group of great economic importance as the source of a number of edible fruits.

Includes *Pyrus, Malus, Amelanchier, Cydonia, Cratægus, Cotoneaster, Pyracantha, Raphiolepis, Photinia, Eriobotrya, Osteomeles, Aronia, Sorbus, Chænomeles.*

Spirœoideœ. *Physocarpus* (*Opulaster*, Nine-Bark) comprises about a dozen species of chiefly American shrubs (one species in Manchuria) with palmately lobed leaves and corymbose clusters of white flowers. The bark usually strips off in thin layers, whence the common name Nine-Bark. The sepals, petals, and numerous stamens are arranged on the margin of the bowl-shaped hypanthium, and the 1—5 carpels, with ovoid ovaries and elongate slender styles terminating in broad stigmas, are elevated on a prominent stalk (gynophore). The fruits are 2— to several-seeded inflated follicles that split into 2 valves. *Physocarpus opulifolius* occurs on rocky hillsides and banks of streams in Atlantic North America, and a scarcely distinguishable form in the Rocky Mountain region and the Pacific Northwest. (Fig. 223.)

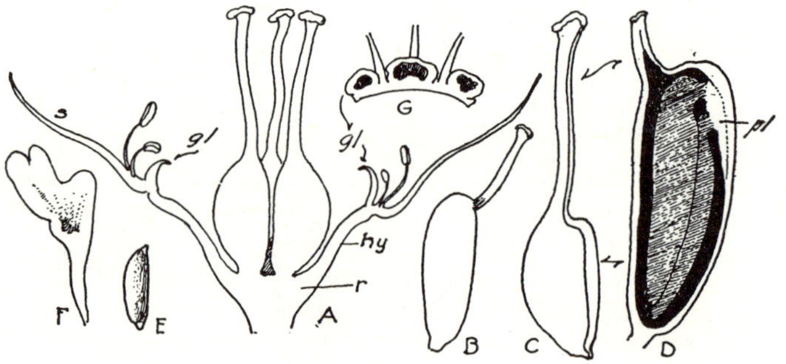

FIG 224.—*Spirœa Vanhoutii: A,* median section of flower; *r,* receptacle; *hy,* hypanthium. *B,* young and *C,* older carpel. *D,* median section of a carpel; *pl,* placenta. *E,* seed. *F,* petaloid stamen. *G,* part of margin of hypanthium, with filaments, and lobes of disk (*gl*).

In *Spirœa* the leaves are without lobings, the inflorescences are paniculate or corymbose, the stamens are sometimes more numerous, the carpels are sessile or stipitate and not inflated in the fruit, and a more or less prominently lobed disk lines the margin of the hypanthium. The 70 or more species are shrubs or subshrubs, widely distributed over the north temperate zone, about a third of the species occurring in North America. The inflorescence is corymbose in *Spirœa corymbosa* and *S. virginiana* of Atlantic North America and *S. densiflora* of the Northwest, while in *S. alba* (*S. salicifolia*) of eastern North America and *S. Douglasii* and *S. Menziesii* of the Pacific Northwest it is paniculate. Among the many cultivated species are *S. Vanhouttei, S. Thunbergii, S. alba, S. Douglasii,* and *S. tomentosa.* (Figs. 224, 225.)

Aruncus is diœcious. The plants are herbs with twice- or thrice-pinnate leaves and small white flowers in racemose-paniculate inflorescences. The 3—4 carpels, seated in a shallow hypanthium, ripen into follicles. *Aruncus sylvester* (Goat's Beard) of eastern North America, Europe, and Asia, is a tall herb very similar in habit to *Astilbe biternata* (*Saxifragaceæ*). The species is cultivated under the name Spiræa. Another species occurs in Japan.

In *Exochorda* the flowers are large and polygamo-diœcious. The 5 carpels are connate by their ovaries, with the styles distinct. The inner surface of the hypanthium is discoid. The three or four species are Asiatic shrubs and are commonly cultivated. In *Exochorda racemosa* the stamens are in clusters but apparently in two whorls, alternating with the sepals (Fig. 226). *Holodiscus discolor* (Ocean Spray)

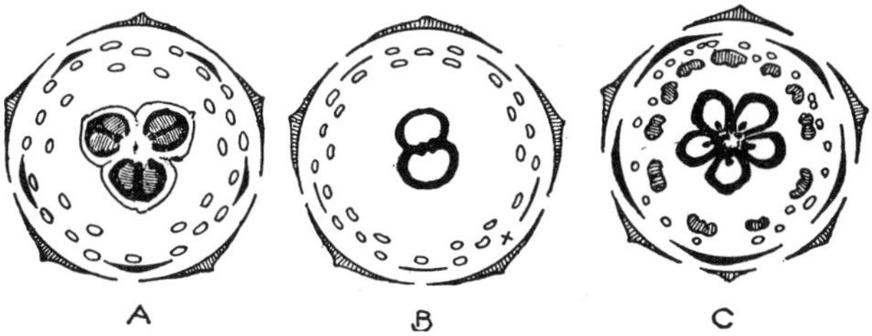

FIG. 225.—Empirical floral diagrams, *A, B,* of flower of *Physocarpus; C,* of a *Spiræa. A,* with tricarpellate, *B,* with bicarpellate gynœcium.

is a tall shrub of western North America to Guatemala, producing small white flowers in dense panicles. The 5 carpels become follicles; and of the approximately 20 stamens the 15 outer ones are united at the base. The hypanthium is shallow and discoid on the inner surface. One other species occurs in South America.

Rosoideæ. The solitary *Rhodotypus kerrioides,* native in Japan, has from 2 to 6 distinct carpels with ovoid ovaries and long slender styles, and the margin of the disk rising as a lobed ring over the ovaries. (Compare with *Spiræa.*) An epicalyx is present, and also endosperm in the seeds. The genus connects the spiræas and the potentillas and, according to Focke, is a relict from Tertiary times.

Rubus (Blackberries, Raspberries, Dewberries, Brambles). In this genus the hypanthial portion of the receptacle is shallow, while the axial portion, which bears the numerous spirally arranged carpels, forms a somewhat fleshy column (gynophore). Each carpel consists of an ovoid or oblong, somewhat curved ovary and a long, slender, nearly terminal style with a globose or discoid stigma, becoming a juicy drupelet when ripe, with the style on the inner edge. The whole cluster of drupelets forms an aggregate fruit popularly called the berry. In the blackberries and dewberries the "berry" is firmly attached to the gynophore but in the raspberries it detaches easily at maturity. The sepals are persistent. Some of the blackberries possess prickly canes. *Rubus* is a very technical genus of probably several hundred species, many of which are perplexingly variable and apparently hybridize freely. The taxonomic characters are especially difficult in the blackberries. For accurate studies it is necessary to have at hand not only specimens of the older canes but also the turions of the identical species. Numerous species and varieties are in cultivation. Among the more distinctive and easily recognizable wild species are *Rubus parviflorus* (Thimble-Berry), a soft shrub with broad palmately lobed leaves, inhabiting deep forests from the Lake Superior region to the Pacific Northwest and Alaska; it is similar in habit to *R. odoratus* of Atlantic North America. Another is the yellow- or red-berried *R. spectabilis,* a tall, elegant species of low, deep forests from Idaho to Alaska. The two dewberries *R. arcticus* and *R. pubescens* (*R. triflorus*) are restricted generally to cold Northern swamps. They are easily distinguished by their low, weak to trailing stems and large, lustrous, wine-colored berries. The former is distinguished by its purplish or rose-colored flowers. Our cultivated raspberries are

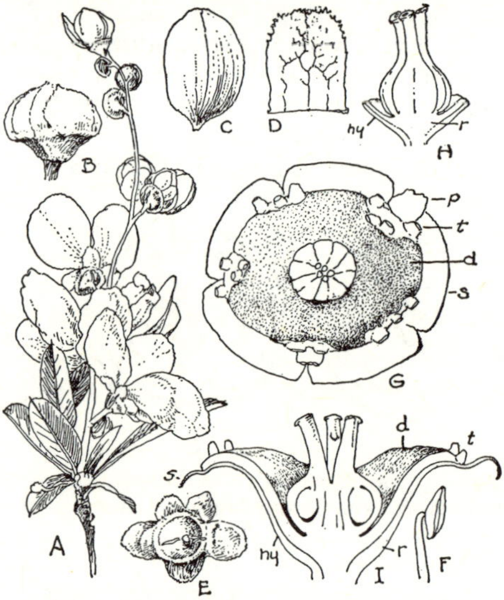

FIG. 226.—*Exochorda racemosa: A,* flowering shoot. *B,* flower-bud. *C,* petal. *D,* sepal (enlarged). *E,* calyx, viewed from below. *F,* stamen. *G,* center of flower with parts removed; *p,* petal; *t,* stamens; *d,* glandular disk; *s,* sepal. *H,* gynœcium; *hy,* hypanthium; *r,* receptacle. *I,* median section of flower with parts removed; *s,* sepal; *t,* stamens; *d,* disk; *hy,* hypanthium of receptacle; *r,* receptacle.

derived from *R. idæus* and its variety *strigosus,* or hybrids between these and other species. (Fig. 227.)

Fragaria (Strawberry). In *Fragaria* the gynophore of the receptacle becomes greatly enlarged in the fruit to form the familiar "strawberry," on which the numerous small, seed-like achenes (the true fruits—derived from the ovaries of the carpels) are spirally arranged, superficially or in pits. Owing to the peculiar development of the ovary, the style becomes basal in the achene. (Compare with the drupelets of the blackberries and raspberries.) Alternating with the persistent sepals are 5 segments of the epicalyx. The strawberries are low herbs with trifoliolate leaves, and white flowers in racemose or cymose inflorescences, arising from a subterranean caudex. The plants spread vegetatively by runners which are furnished with a node near the middle. Probably two dozen species of strawberries are known, these being confined generally to Northern climates, but with a representation down into the Andes of South America. One of our most distinctive species is *Fragaria chilensis* of the Pacific Coast and the coast of Chile. Numerous varieties and hybrids of strawberries are cultivated (Figs. 9, 44). *Duchesnea indica,* the Indian strawberry, a native of Asia but introduced in the Eastern States, is very similar to the strawberries, but differs in having yellow petals and an insipid fruit.

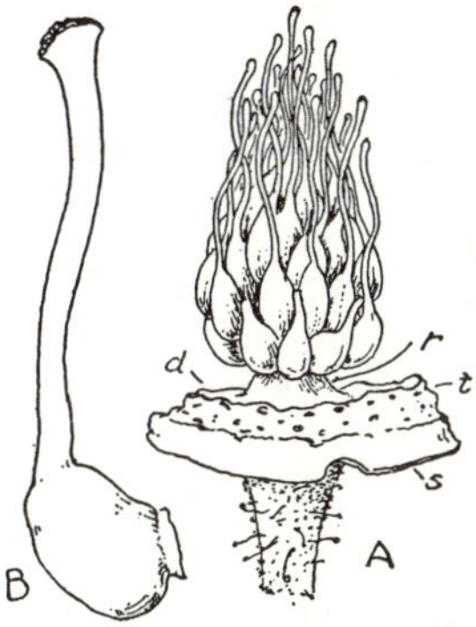

FIG. 227.—*Rubus setosus: A,* polycarpellate gynœcium on columnar part (gynophore) of receptacle (*r*); *s,* sepal; *d,* disk portion of receptacle (hypanthium); *t,* stamen-bearing margin. *B,* carpel.

Potentilla (Cinquefoil). In the cinquefoils the flowers are essentially like those of the strawberries, but the gynophore dries up at maturity and is mostly more or less hairy. The carpels develop into achenes with subterminal deciduous styles. The genus comprises several hundred species of herbs with palmately or digitately compound leaves and yellow or whitish flowers; chiefly of Northern regions, with a large representation in western North America especially. Several species and their varieties are horticultural. (Compare with *Fragaria, Rubus, Ranunculus, Anemone, Sagittaria, Lophotocarpus.*)

Geum (Avens). The gynœcium in *Geum* is dry and sessile or stalked in the fruit, with the numerous carpels becoming achenes, on which the style persists as a long slender "tail" or organ of dispersal.

In some species the styles are hairy, in others jointed in the middle and disarticulating at the joint so as to leave a hooked basal portion, either type serving as an efficient organ of dispersal. The large fruiting-heads of the species with long-hairy styles are conspicuous in the field and at once distinguish the plants. The species are herbs with pinnate or pinnatifid leaves, the terminal lobe much the largest, all growing from thick subterranean caudices. Species about 40, of the northern hemisphere, with a scanty representation in South America and one species in South Africa.

In *Alchemilla* the hypanthium is contracted in the throat so as to enclose the usually solitary carpel. The style is basal, petals are wanting, and the stamens are only 4, or sometimes reduced to 1. Flowers very small. About 30 species, mostly of northern regions. Herbs, mostly with palmately lobed leaves. *Agrimonia* is distinguished by an obconical hypanthium beset with hooked prickles at the throat or sometimes over the outer surface, and often grooved, and closing over the 2 carpels, which become achenes at maturity. The 5 sepals are connivent and there are 5—15 stamens. The plants are herbs with pinnate leaves and in some cases with tuberous roots. Species about 15, mostly Northern but extending in some cases into Mexico and the Andes. *Sanguisorba* is distinguished by a top-shaped or obconical hypanthium which is constricted at the throat. Petals are wanting, but there are 4 sepals and 4 stamens with long-exserted slender filaments. About 10 species of herbs with odd-pinnate leaves; all Northern.

Rosa (Rose). The hypanthium of the roses is deeply urn-shaped, enclosing several to numerous, commonly hairy, long-styled, sessile or stipitate carpels which become bony achenes in the fruit. At maturity the hypanthium is fleshy and commonly bright red and is known as the *hip*. The hip with the enclosed achenes is a spurious fruit. The 5 sepals are generally elongate and slender, simple or variously lobed, and persistent or deciduous. When persistent, they are spreading or erect—a feature which is of some taxonomic value. Petals 5 or sometimes more. The numerous stamens are inserted on the margin of the disk which lines the throat of the hypanthium. A very technical genus of numerous species, widely distributed over the northern hemisphere, and represented in cultivation by several thousand varieties. The plants are erect, decumbent, or climbing, often prickly shrubs with alternate, stipulate, odd-pinnate leaves, growing from rootstocks. The technical characters are found chiefly in the fruit, sepals, and stipules, and in the presence or absence of prickles and infrastipular spines. Of our American

FIG. 228.—*Rosa* sp.: *a,* flowering branch; *b,* base of *a; c,* stamens, dorsal and ventral aspects; *d,* median section of flower; *e,* carpel; *f,* portion of stem; *g,* twig with flower-bud and flower after blossoming; *s,* stipules.

species *Rosa setigera* is distinguished by its coherent styles (Figs. 228, 45).

Cercocarpus, Adenostema, Chamæbatia, Coleogyne, and *Purshia* are genera of western North America. *Cercocarpus* comprises half a dozen tall shrubs with hard wood, alternate, simple, straight-veined leaves and solitary or fascicled apetalous flowers. The hypanthium is a deep narrow tube with numerous stamens on the flaring throat. There is a solitary hairy carpel with a long, slender style which persists as an organ of dispersal in the fruit (achene). About 6 species of the arid belt, two species inhabiting the islands off the coast of California (Fig. 229). *Adenostoma fasciculatum* (Chamise) and *A. sparsifolium* (Ribbon-Wood, Yerba del Pasmo), the only species, are characteristic of the chaparral belt of California. The first-named has fascicled leaves and sessile flowers, the last-named scattered leaves and pediceled flowers. In each there is but 1 carpel seated in an obconical hypanthium. Petals 5, and stamens 10—15 in groups of 2 or 3, alternating with the petals.

FIG. 229.—*Cercocarpus betuloides*: *A*, twig with fruits. *B*, carpel, with filiform bristly style. *C*, persisting calyx.

Coleogyne ramosissima, the only species, is an intricately branched thorny shrub of the deserts from Colorado to California. The flowers are solitary and terminal on spur shoots and subtended by several 3-lobed bracts. There are 4 persistent sepals but no petals, and in the shallow hypanthium stands a solitary carpel with a bent lateral style and enclosed in an elongate tube formed by the disk, upon the outer surface of which the stamens are inserted. Fruit a glabrous achene. In *Purshia* the flowers are similarly terminal on short spur shoots, but there are 5 clawed and spreading petals present and 18—25 stamens in one series. There are 2 species, *Purshia tridentata* (Antelope Brush), a grayish diffusely branched shrub of the Sierras, and *P. glandulosa*, a dark-green shrub of the Mojave region. The leaves of both are cuneate, with a 3-lobed or 3-toothed apex.

Prunoideæ. The shrub *Osmaronia cerasiformis* (Oso-Berry), with alternate, lanceolate, deciduous leaves, caducous stipules, and fragrant white diœcious flowers in racemes terminating leafy shoots, is found in northern California. In the staminate flowers the 15 stamens are placed in three whorls on the discoid throat of the cup-shaped hypan-

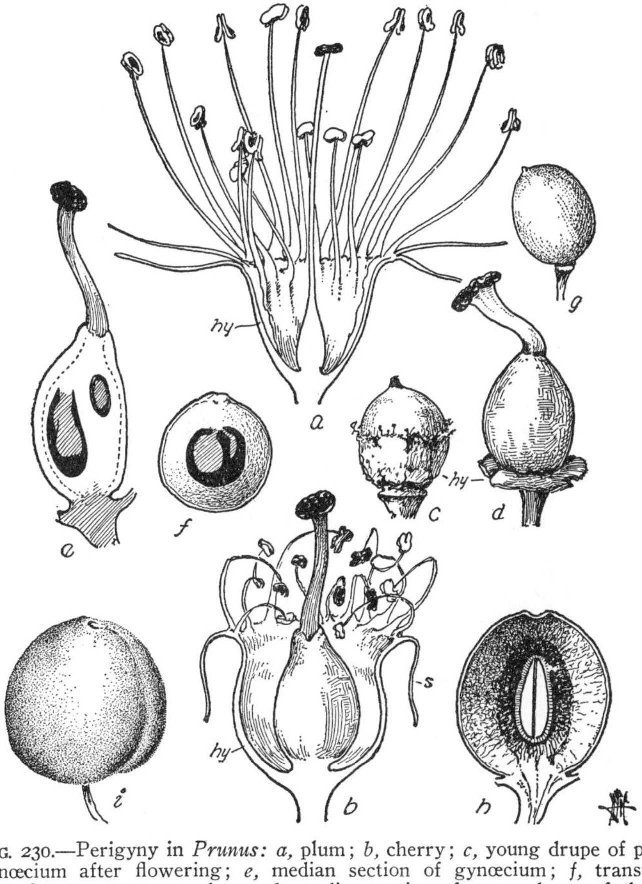

FIG. 230.—Perigyny in *Prunus: a,* plum; *b,* cherry; *c,* young drupe of plum; *d,* gynœcium after flowering; *e,* median section of gynœcium; *f,* transverse section of ovary; *g,* young drupe; *h,* median section of young drupe of cherry; *i,* ripe drupe; *hy,* hypanthium; *s,* sepal.

thium. In the hypanthium of the carpellate flowers there is a whorl of 5 distinct carpels with elongate slender styles and ovoid ovaries which become bitter drupes. Abortive stamens are also present in the carpellate flowers.

Prunus in the broadest interpretation includes the cherries, plums, peaches, apricots, and almonds. The plants are trees and shrubs with alternate leaves and white to rose-colored flowers in racemose or umbellate clusters. The stamens are numerous, and in the shallow or deep cup-shaped hypanthium is a solitary carpel with an elongat

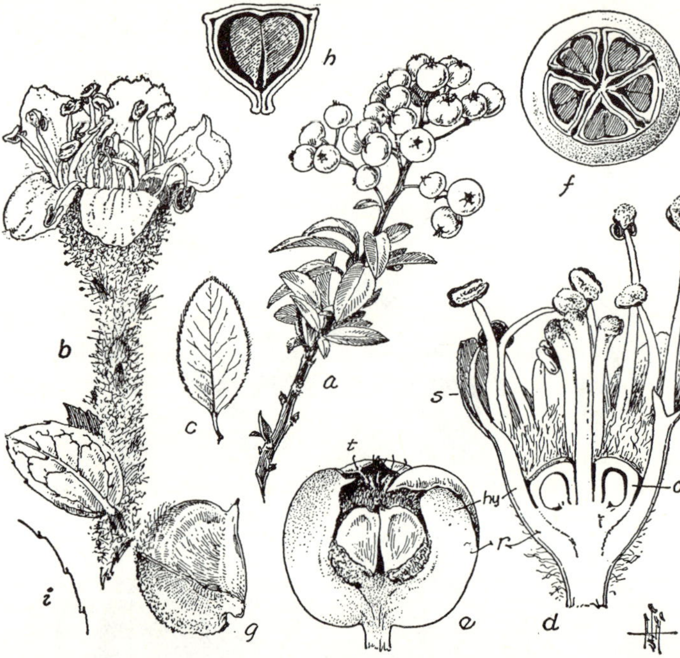

FIG. 231.—*Cotoneaster pannosa: a,* branch with fruit-cluster (fruit red); *b,* flower; *c,* leaf; *d,* median section of flower, showing ovaries (*o*) adnate basally to receptacle (*r*); *hy,* hypanthium; *e,* median section of fruit, exposing two stony "seeds" (ovaries); *t,* styles; *f,* transverse section of fruit, showing five stony "seeds"; *g,* carpel (stone "seed"), with style right and, below, the rough portion adnate to receptacle; *h,* transverse section of "stone," showing two seeds.

deciduous style and an ovoid ovary which ripens into a drupe. As the ovary develops the hypanthium is ruptured and sloughed off. The pericarp of the fruit is usually succulent, but in the almonds it becomes dry and detaches, leaving the wrinkled flattened stone. In the peaches the fruit is velvety and the stone is wrinkled as in the almonds; in the Crawford peach, *e.g.,* the thick juicy pericarp detaches easily from the stone when fully ripe. The plums are more or less thorny shrubs with umbellate inflorescences and often tart fruit with a flattened stone. In the cherries the stone is commonly smooth and not much flattened while the fruits range from sweet to tart and bitter. In the broadest sense the genus numbers around 175 species,

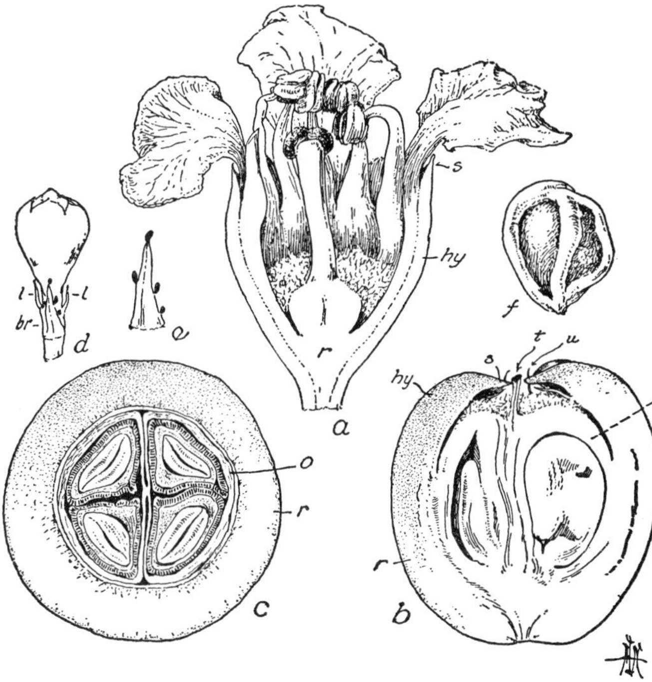

FIG. 232.—*Photinia arbutifolia: a,* median section of flower with one of the two carpels in place; *b,* median section of fruit, showing ovaries (*o*) adnate for more than half their length to receptacle (*r*); *c,* transverse section of fruit; *d,* flower-bud, with subtending bract (*br*) and bracteoles (*l*); *e,* bract (*br*) in *d; f,* seed; *r,* receptacle; *hy,* hypanthium; *s,* sepals; *t,* styles; *u,* filament.

differing in certain foliar and other vegetative characters and in the characters of the fruit, but all essentially similar in structure of the flower. A large number of species and varieties are cultivated for their fruits and in many cases as ornamentals. The genus is abundantly represented over the northern hemisphere, with an especially rich development in eastern Asia. (Fig. 230.)

Our cultivated forms are derived from various sources, among which are *Prunus avium* (Sweet Cherry), *P. cerasus* (Sour Cherry), the native plums *P. americana* and *P. nigra, P. Persica* (Peach), and *P. communis* (Almond), and *P. Armeniaca* (Apricot), besides many others. Some species are evergreen (*P. lusitanica*).

Pomoideæ. The genera of this subfamily may be conveniently grouped according to certain characters of the fruit. Thus in *Cotoneaster, Pyracantha, Osteomeles,* and *Cratægus,* the core of the fruit is stony, due to the hardening of the inner wall of the ovaries. In others the core is leathery or parchment-like, as in *Malus* and *Pyrus.* In *Cotoneaster, Pyracantha,* and *Osteomeles,* and more or less in *Cratægus, Malus,* and *Pyrus* (Pear), the carpels are free along their inner edges and separate easily at maturity, and the styles are distinct—at least above. With respect to vegetative features, *Osteomeles, Pyracantha,* and some species of *Cotoneaster* and *Cratægus* are evergreen, while *Osteomeles,* in addition, has odd-pinnate leaves, and pinnatifid leaves occur in some rare species of *Cratægus.* The flowers are small and numerous, in cymose or umbellate clusters. Stamens 15—20 or more, but reduced to 5 in some species of *Cratægus.* (Figs. 231, 232, 233.)

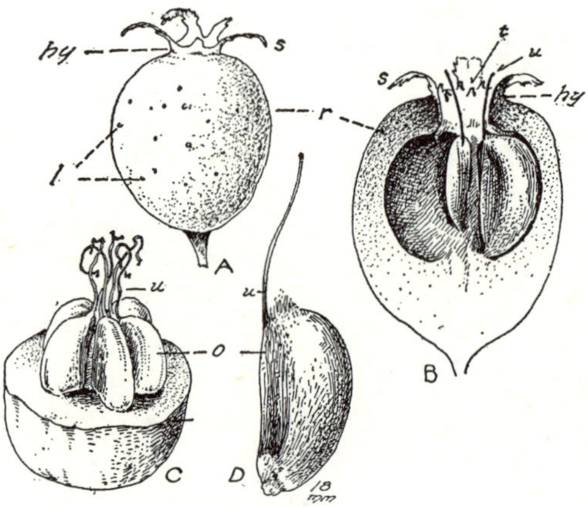

FIG. 233.—*Cratægus* sp.: *A,* fruit; *B,* median section of fruit; *C,* fruit with upper portion of receptacle removed; *D,* carpel; *hy,* hypanthium; *r,* receptacle; *l,* lenticels; *s,* sepals; *t,* bases of stamens, on margin of hypanthium; *u,* styles; *o,* ovary. Carpels loosely adnate to receptacle, but free above (*B*).

The genera with stony-cored fruits comprise hard-wooded shrubs and trees, many of which are cultivated as ornamentals, of which species of *Cotoneaster* and *Pyracantha* are especially attractive with their dense masses of bright orange-red or crimson fruits. *Cotoneaster* comprises about 40 species of the Old World; *Pyracantha,* a few species of eastern Asia and Europe; *Osteomeles,* three species of eastern Asia and Pacific islands. *Cratægus* (Hawthorn), a most difficult genus, is variously estimated to comprise from 300 to 1,000 species, largely represented in North America. The species are with difficulty distinguished on characters of leaves and fruits. (Figs. 232-237.)

The following genera of the *Pomoideæ* have leathery-cored fruits: *Cydonia, Chænomeles, Photinia, Amelanchier, Malus, Pyrus, Sorbus.*

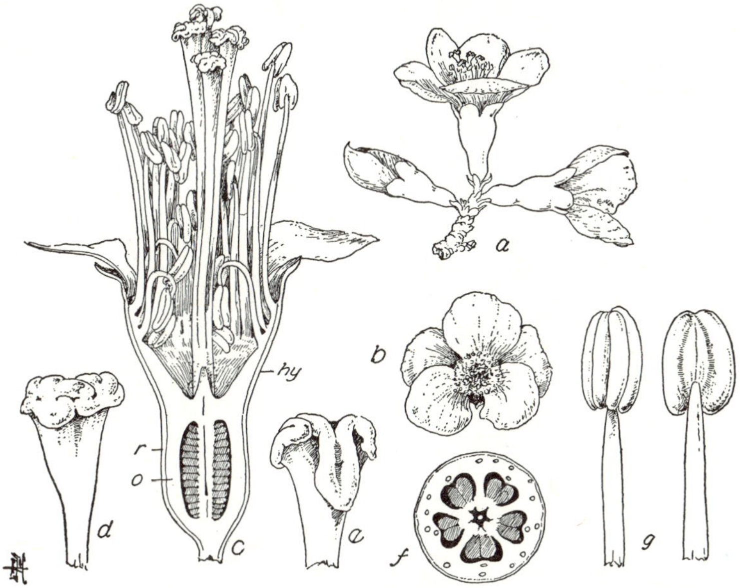

FIG. 234.—*Chænomeles* sp.: *a,* cyme of three flowers; *b,* flower viewed from above; *c,* median section of flower; *hy,* hypanthium; *r,* receptacle; *o,* ovary; *d,* tip of a style with fleshy convoluted stigma; *e,* another view of stigma; *f,* transverse section of ovary; *g,* ventral and dorsal aspects of stamens.

Of these *Cydonia* (Quince) and *Chænomeles* (Japanese Quince) have many-seeded fruits, the others having only 2 seeds in each cell. In *Cydonia japonica* (Japanese Quince) the flowers appear two or three in close proximity on branches of the previous season. The petals are rose-colored and rather large, the stamens numerous, with long filaments in three whorls. The 5-carpellate gynœcium terminates in slender elongate styles which are united only at the base. The receptacle is prolonged beyond the ovary as a cup-shaped hypanthium.

The numerous ovules are borne in two rows in each cell on axile placentas. The fruit (spurious) is a small pomaceous berry. Species about 3, of China and Japan. *Photinia arbutifolia* (Toyon), an evergreen shrub of California, has small flowers in cymose-paniculate clusters. The gynœcium is 2—3-carpellate, with a stout style parted

FIG. 235.—*Raphiolepis umbellata: A,* flowering shoot; *B,* flower-bud: *br,* bracteoles; *s,* sepals; *C,* petal; *D,* dorsal and lateral aspects of stamens; *E,* median section of flower: *s,* sepals; *hy,* hypanthium; *o,* ovary; *r,* receptacle; *F,* transverse section of "ovary"; *G,* ovule; *H,* fruit, with persistent calyx above.

above and the ovary only slightly adnate below to the fleshy receptacle. The hypanthium is shallow. The small fruits (pomes) form showy masses when ripe, and contain one or two seeds in each cell. Two dozen or more other species occur in China and Japan. (Figs. 232, 234, 235, 237.)

In *Amelanchier* (Service-Berry) there are 3—5 carpels with

styles distinct down to the middle, and a shallow, saucer-shaped hypanthium with a prominent annular, lobed disk, beneath the outer margin of which the numerous stamens are inserted. The carpels are distinct or nearly so along their inner edges. The fruit is a many-seeded berry-like pome. Species 25—30, widely distributed over the north temperate zone, over half of them in North America. Technically a very difficult genus, with many intergrading forms. *Aronia* is essentially similar to *Amelanchier* but has relatively stouter stamens, and the free portion of the ovary is very woolly (so also in some species of *Amelanchier*). (Fig. 236.)

Malus (Apple) has 5 carpels, although 6 or 7 occasionally occur in cultivated forms. The styles are distinct down to the middle. The compound ovary is wholly inferior but the individual ovaries are distinct or only slightly connate by their inner (ventral) edges and often separate completely at maturity. The receptacle is more or less constricted above the ovary and prolonged beyond it as an hypanthial tube which is discoid on the inner surface. The inner wall of the separate ovaries becomes strongly cartilaginous, and in each cell there are two seeds. The spurious fruit is a small to large pome on which the calyx-segments and the withered remains of the stamens persist in the hypanthial hollow of the deeply depressed apex. The genus includes the apples and the crab-apples to the number of 75 or more species, and a host of cultivated forms. Several species are native in the United States, including *Malus ioensis* (Western Crab-Apple) and *M. coronaria* (American Crab-Apple). Except for the shape of the fruit, more distinct carpels, and leaf-form, *Pyrus communis* (Pear) is very similar. (Figs. 37A; 61.)

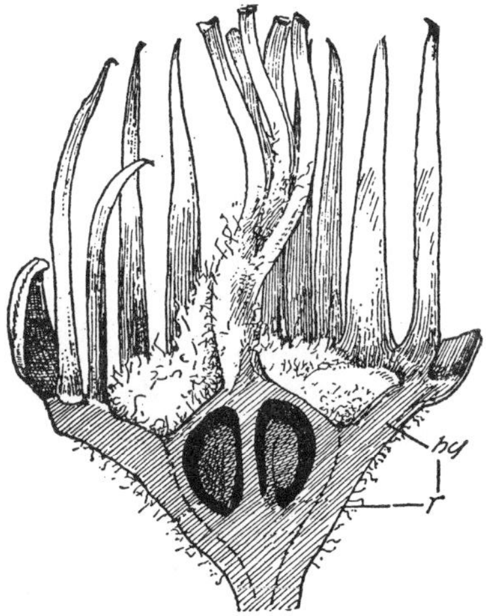

FIG. 236.—*Aronia* (*Pyrus*) *arbutifolia:* median section of flower, with petals and anthers removed, showing shallow hypanthium (*hy*) and styles united below; *r*, receptacle.

Sorbus is distinguished by its pinnate leaves. The flowers are small and numerous and borne in dense, terminal, cymose clusters. Carpels 3, with as many distinct styles, partially inferior ovary, and

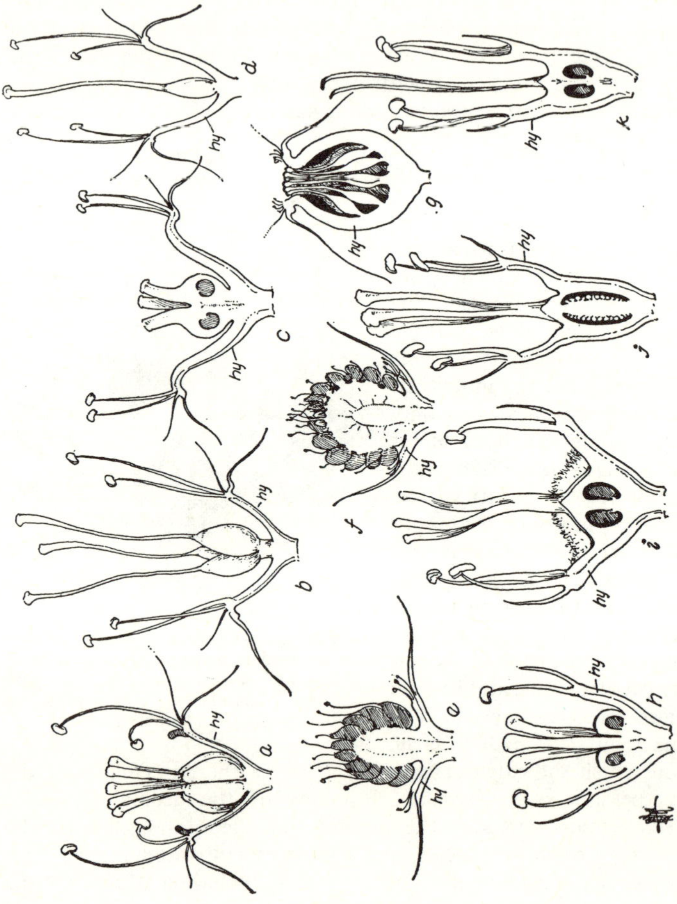

FIG. 237.—Perigyny and epigyny in the rose family (*Rosaceæ*). Perigynous forms: *a*, *Spiræa*; *b*, *Physocarpus*; *c*, *Exochorda*; *d*, *Prunus*; *e*, *Rubus*; *f*, *Fragaria*; *g*, *Rosa*; *hy*, hypanthium. Epigynous forms: *h*, *Cotoneaster*; *i*, *Aronia*; *j*, *Chænomeles*; *k*, *Raphiolepis*.

a prominent hypanthium. The plants are shrubs or trees of the northern hemisphere. Species about 10, three or four of them in the United States, including *Sorbus americana* (American Mountain Ash) of Atlantic North America and *S. sitchensis* (Western Mountain Ash) of the West. *S. Aucuparia* of Europe is commonly cultivated.

CONNARACEÆ. A small family of mostly tropical climbers and some trees and shrubs, with alternate pinnate leaves and small, regular, pentamerous, bisexual flowers in racemes or panicles. Carpels 5 or sometimes fewer, distinct. Stamens mostly 10. Integuments 2. About 250 species in a dozen or more genera. Largest genus *Connara,* with 50 or more species in tropical America, Asia, and Africa.

LEGUMINOSÆ (*Pea Family*). Flowers regular or irregular, hypogynous or perigynous, cyclic, pentamerous, mostly bisexual. Carpels 1, rarely more; ovary superior, sessile or stipitate, with many to few (sometimes 1) ovules alternating in two rows—one row on each side of the ventral suture. Style terminal, straight or curved. Stamens numerous, twice as many as the petals, 10, or fewer, free or more or less united (monadelphous or diadelphous). Petals free or sometimes united. Fruit a legume, sometimes follicular or indehiscent, and sometimes succulent. Integuments 2. Endosperm present or wanting. Cotyledons commonly large and fleshy. Herbs, shrubs, or trees, mostly with compound stipulate leaves.

The second largest family of flowering plants, comprising over 12,000 species in about 550 genera, widely distributed over the world. Here belong many important crop and forage plants, including peas, beans, clovers, lentils, alfalfa, sweet-clover, vetch, vetchling, peanut, and lupines. There are three well-marked subfamilies, namely:

I. *Mimosoideæ,* the Mimosa Group, in which the flowers are regular, and the stamens numerous to few (10 to 4) and distinct or united into a tube or united with the petals. Fruit mostly stipitate. Trees and shrubs, rarely herbs, mostly with twice-pinnate leaves, and small flowers in dense heads or spikes. About 40 genera, among which are *Albizzia, Acacia, Mimosa.* (Figs. 238, 239.)

II. *Cæsalpinioideæ,* the Cæsalpinia Group, in which the flowers are irregular but not distinctly papilionaceous, perigynous, with the two uppermost of the 5 sepals sometimes united, the petals 5 or fewer by abortion, the stamens mostly 10, sometimes more or fewer, free or united into a tube. Carpel subtended by a disk. Fruit dehiscent or indehiscent, stipitate or the stipe fused with the hypanthium. Funiculus arillate. Trees or shrubs, or sometimes herbs, with once- or twice-pinnate leaves, in rare cases the leaves reduced to a single leaflet. Flowers large to small, in terminal or axillary racemes or spikes.

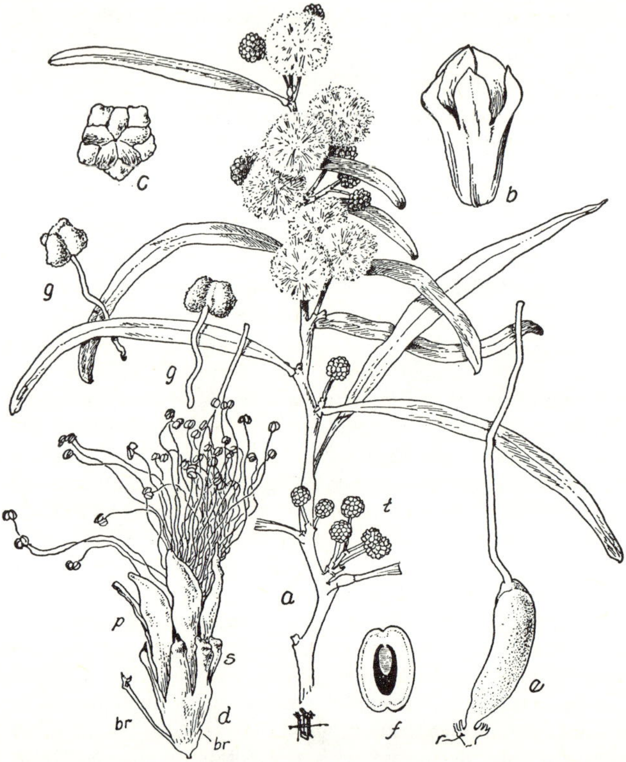

FIG. 238.—*Acacia,* with phyllodial "leaves": *a,* flowering branch; *t,* glomerules of minute flower-buds; *b,* flower-bud; *c,* flower-bud viewed from above; *d,* flower; p, petals; *s,* sepals; *br,* bracteoles; *e,* gynœcium; *r,* receptacle; *f,* transverse section of ovary; *g,* ventral and dorsal aspects of stamens.

Including, among others, *Cassia, Cercis, Ceratonia, Krameria, Gleditsia, Gymnocladus, Cercidium, Hoffmanseggia, Parkinsonia, Bauhinia, Tamarindus, Cæsalpinia.* (Figs. 240, 241.)

III. *Papilionoideæ,* the Pea Group, in which the flowers are irregular (papilionaceous, or bilaterally symmetrical), rarely regular; hypogynous or perigynous. Calyx regular or irregular, often campanulate or tubular, the 2 uppermost of the 5 segments often united, the 3 lower forming a lip, sometimes much reduced. Petals 5 or fewer, rarely wanting, the uppermost petal commonly the largest (standard), the two lateral smaller and usually clawed (forming the wings), the two lowermost similarly clawed and commonly united, at least by the lower edge, to form the *keel,* which is upcurved. Stamens 10, rarely 9 by abortion of the odd one, or rarely 5—free or more or less united into a tube (diadelphous, —$9+1$, or monadel-

FIG. 239.—*Acacia* with phyllodes (petioles); glomerules of flowers above; *b,* a leaf, showing transition to phyllode.

phous). The carpel with an elongate more or less upcurved style and with the ventral suture above, sessile or stipitate and often subtended by an annular disk. Trees, shrubs, and herbs, with digitate or once-pinnate leaves. Seeds several to few or 1. Endosperm scanty when present. (Figs. 242, 243, 244, 245.)

Here belong the great majority of genera and species of the family. Next to the Grass family and the Rose family, perhaps, no other family yields so many useful plants, important not only as sources of food for man and beast but also as restorers of nitrogen to the soil. The following tribes and characteristic genera, among many others, are represented in our flora.

Tribe *Sophoreæ,* in which the stamens are distinct and the leaves pinnate. *Sophora, Cladrastis.*

Tribe *Podalurieæ,* in which the 10 stamens are similarly distinct, but the leaves are simple or trifoliolate. *Thermopsis, Baptisia.*

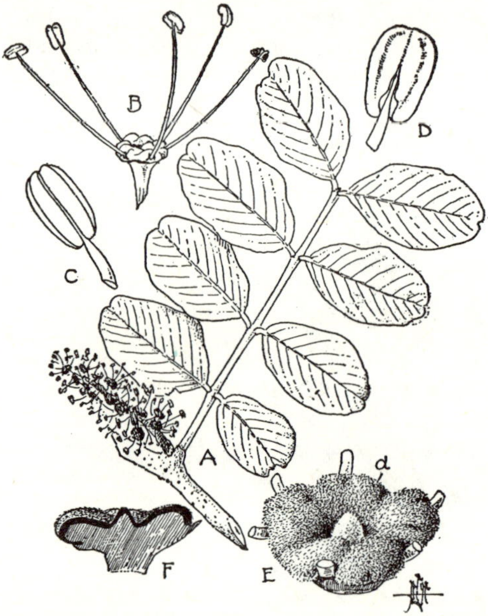

FIG. 240.—*Ceratonia Siliqua* (Carob): *A,* twig with leaf and staminate inflorescence; *B,* staminate flower; *C,* ventral, and *D,* dorsal aspects of stamens; *E,* flower with stamens removed, showing disk (*d*) with vestigial gynœcium in center; *F,* median section of disk,—conical projection in center the vestigial gynœcium.

In the following tribes the stamens are diadelphous or monadelphous.

Tribe *Hedysareæ,* in which the pod is a loment. Herbs.

Stamens diadelphous in *Hedysarum,* monadelphous in *Stylosanthes,* both monadelphous and diadelphous in *Meibomia.*

Tribe *Gelageæ,* stamens monadelphous or diadelphous; pod dehiscing by 2 valves. Herbs, shrubs (erect or climbing), or trees, with pinnate leaves.

Stamens monadelphous in *Indigofera, Amorpha;* diadelphous in *Robinia, Sesbania, Astragalus, Olneya;* monadelphous or diadelphous in *Cracca, Psoralea, Glycyrrhiza;* stamens 5 in *Petalostemum* (*Kuhnistera*).

Tribe *Genisteæ.* Stamens monadelphous. Leaves 3-foliolate with leaflets entire. Pod various, elongate and flat to ovoid. Herbs. *Crotalaria* (Fig. 242), *Lupinus, Genista, Cytisus, Ulex, Spartium* (Figs. 243, 245).

Tribe *Loteæ,* stamens monadelphous or diadelphous; leaflets 5; pod linear or ovoid. Ours herbs or subshrubs. *Lotus, Hosackia.*

Tribe *Trifolieæ,* stamens diadelphous (9 + 1), anthers all alike; leaflets 3, serrulate. Herbs. *Trifolium, Melilotus, Medicago.* (Fig. 244.)

Tribe *Vicieæ,* stamens diadelphous (9 + 1), or monadelphous below; leaves pinnate, tendriled. *Vicia, Lathyrus.* Herbs.

Tribe *Phaseoleæ,* stamens mostly diadelphous (9 + 1); leaves (in ours) pinnately 5- to 3-foliolate or simple, without tendrils. Herbs. *Amphicarpa, Apios, Phaseolus.*

The characters of the subfamilies will be sufficiently clear from a description of a representative of each.

Mimosoideæ: Acacia. The acacias are mostly trees or shrubs with twice-pinnate leaves or with phyllodes. The small, numerous, and commonly yellow flowers are borne in dense heads or cylindrical spikes or axillary racemes. The calyx is campanulate, toothed or lobed, the petals sometimes united or wanting, and the stamens numerous and free or united at the base. The carpel is sessile or stipitate and commonly dehiscent by two valves. Over 400 species are known, mostly of Africa and Australia, with several in the southern parts of the United States. Many are grown as ornamentals or as avenue trees in California and other parts of the Southwest and in the South. (Figs. 238, 239.)

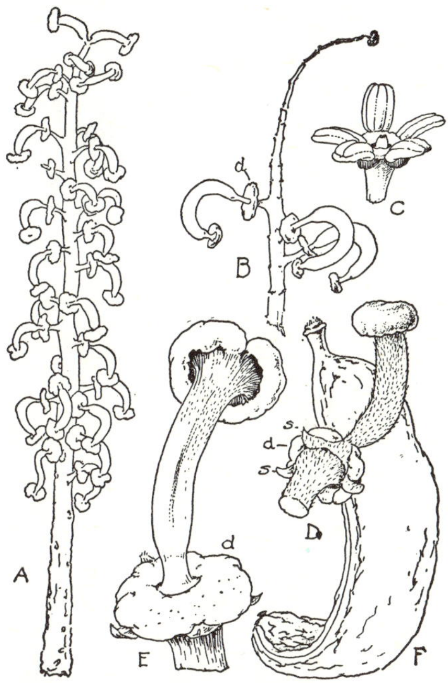

FIG. 241.—*Ceratonia Siliqua: A,* carpellate inflorescence (raceme); *B,* apical portion of inflorescence at later stage, showing persistence of only a few flowers; *C,* a flower at early stage, with rudimentary stamens; *D,* developing fruit; *E,* gynœcium after flowering; *F,* fruit (legume); *s,* sepals; *d,* disk.

Cæsalpinioideæ: Ceratonia Siliqua, the Carob. This is a handsome evergreen tree of the Mediterranean region, extensively planted as an avenue tree in southern California. The leaves are odd-pinnate,

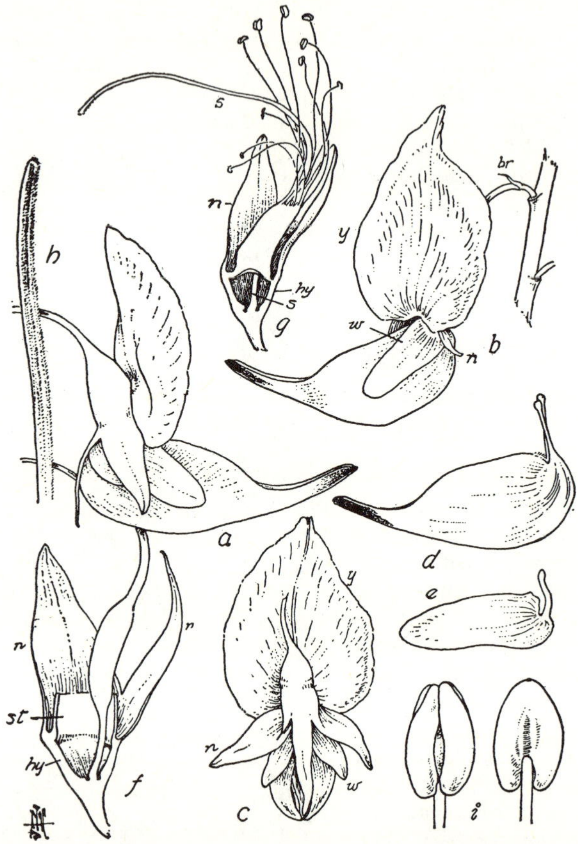

FIG. 242.—*Crotalaria agalliflora: a,* side view of flower; *b,* flower, with face view of standard (*y*), *w,* wing, *n,* sepal, *br,* bracteole; *c,* flower viewed from below,—*w,* wing, *n,* sepal; *d,* keel petal; *e,* wing petal; *f,* median section of flower, petals removed,—*st,* staminal tube, *hy,* hypanthium, *n,* sepals; *g,* median section of flower with staminal tube in place,—*s,* style, *n,* sepals, *hy,* hypanthium; *h,* stigma; *i,* stamens.

commonly with 4 or 5 pairs of thick entire leaflets. The flowers are small and polygamo-diœcious, in racemes. A thick, fleshy, lobed disk is present in both staminate and carpellate flowers. The carpel is somewhat curved and short-stipitate and ends in a broad, thick discoid stigma. A rudimentary carpel is present in the staminate flowers.

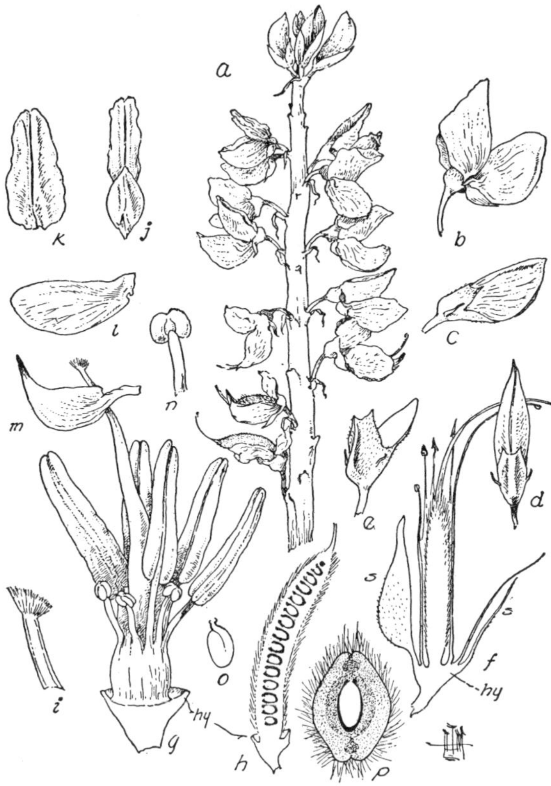

FIG. 243.—*Lupinus* sp.: *a,* inflorescence (raceme); *b,* side view of flower; *c,* side view of flower-bud; *d,* flower-bud viewed from above, showing united upper sepals; *e,* calyx; *f,* median section of flower,—*s,* sepals; *hy,* hypanthium; *g,* stamens and gynœcium; *h,* median section of ovary; *i,* stigma; *j,* rear view and *k,* face view of standard; *l,* wing petal; *m,* keel petal; *n,* small stamen; *o,* ovule.

The pod is large, and flattened and curved laterally, and is edible. Between the seeds there is a pulpy substance. (Figs. 240, 241.)

Papilionoideae. Tribe *Vicieæ. Lathyrus odoratus* (Sweet-Pea). Here the flowers, which are borne 1 to 3 or 4 together in racemes, are typically papilionaceous, with diadelphous stamens, the dorsal of which is free and the other nine more or less united into a tube which encloses the ovary. The style is curved upward and terminates in a club-shaped stigma. The fruit is a flat, dehiscent, pubescent pod in which are several seeds. The stem is winged and the compound leaves consist of a single pair of leaflets and a terminal branched tendril, while at the base there is a pair of small stipules.

The Pea Family is abundantly represented in our flora, both in genera and species, being especially well represented in the West, where herbaceous and woody forms constitute a conspicuous element of the vegetation. Here are numerous species of such large and technical genera as *Lotus, Lupinus* (herbs and shrubs), *Trifolium,* and *Astragalus,* and in the desert regions are many shrubs and tree-like forms of kramerias (*Krameria*), sennas (*Cassia*), paloverdes (*Cercidium Torreyanum* and *Parkinsonia microphylla*), mesquites (*Prosopis*), acacias, (*Acacia*), psoraleas (*Psoralea*), and the desert iron-wood (*Olneya tesota*). On the Atlantic side of the continent are found the honey-locust (*Gleditsia triacanthos*), the Kentucky coffee-tree (*Gymnocladus dioica*), the locust (*Robinia Pseudo-Acacia*), the stately yellow-wood (*Cladrastis lutea*), and the Judas-tree or red-bud (*Cercis canadensis*), which is matched on the Pacific Coast by the Western red-bud, *C. occidentalis.* From the standpoint of distribution and adaptability to a wide range of ecological conditions the common wild licorice, *Glycyrrhiza lepidota,* is unique. This species ranges from Ontario and the Hudson Bay region west to the Pacific and south into Mexico. The plant growing on the seaward bluffs of southern California and blossoming there in January is apparently identical with the one that clings to the edges of aspen groves in northern Minnesota and in Manitoba and which blossoms there in July.

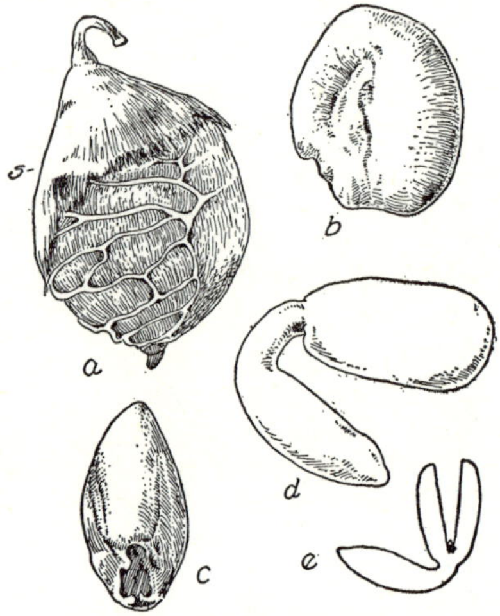

FIG. 244.—*Melilotus alba* (Sweet Clover): *a,* fruit, with remains (*s*) of outer wall of ovary; *b,* side view of seed; *c,* base of seed showing hilum; *d, e,* embryo.

For the identification of species, the fruit of leguminous plants is a valuable aid. While it is ordinarily a characteristic legume or loment, it is often very small and achene-like and may even be succulent, and in such cases it is a good guide to the genus. The characters of the andrœcium also serve as useful guides to the subfamilies, tribes, and genera.

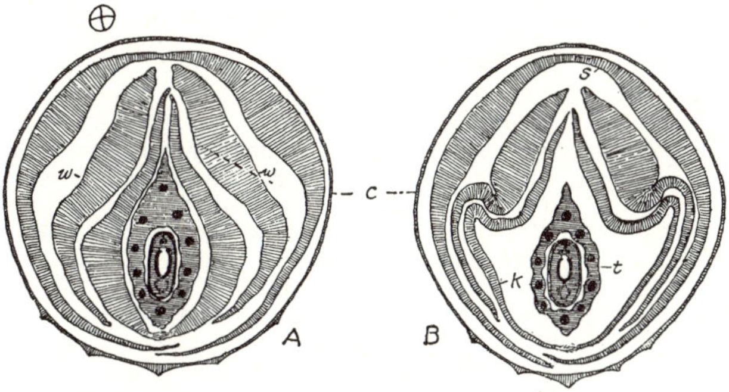

FIG. 245.—Floral diagrams of flower of *Spartium junceum* at two different levels—*A*, lower, *B*, upper; *c*, calyx; *s*, standard; *w*, wing petals; *k*, keel petals; *t*, staminal tube investing ovary (in center).

ORDER PANDALES

A small order comprising a single family, *Pandaceæ*, based on the genus *Panda* with one species, *P. oleosa*, in tropical West Africa. The flowers are diœcious and have a differentiated perianth, the carpellate with a 3-carpellate gynœcium. Fruit a drupe.

ORDER GERANIALES

Flowers hypogynous, cyclic, regular, mostly 5-merous, bisexual or unisexual (monœcious or diœcious), with calyx and corolla. Stamens mostly 10—5, sometimes more or fewer. Gynœcium generally of 5—2 carpels, these united but separating at maturity. Ovary usually with as many cells as carpels. Seeds mostly 2—1 in each cell, with or without endosperm. Integuments 2 or 1. According to Engler and Gilg (*Syllabus*), all members of this order have the ovule bent upward and inward. (Fig. 246.)

FAMILIES

Geraniaceæ
Oxalidaceæ
Tropæolaceæ
Linaceæ
Humiriaceæ
Erythroxylaceæ
Zygophyllaceæ
Cneoraceæ
Rutaceæ
Simarubaceæ
Burseraceæ
Meliaceæ
Akaniaceæ
Malphigiaceæ
Trigoniaceæ
Vochysiaceæ
Tremandraceæ
Polygalaceæ
Dichapetalaceæ
Euphorbiaceæ
Callitrichaceæ

KEY TO THE PRINCIPAL FAMILIES

lowers bisexual.
a. Herbage not covered with glandular dots, but often viscid.
1. Flowers regular.
(°). Carpels at maturity splitting apart as separate fruitlets.
Carpels splitting apart elastically from the central axis —from the base upward. Herbs. *Geraniaceæ.*
Carpels (in ours) splitting apart as spiny or tuberculate nutlets. Ours herbs. *Zygophyllaceæ.*
Carpels splitting into 3 commonly winged nut-like parts, but the fruit in some cases a drupe or nut. Stamens often united basally, some of them often staminodial. Flowers sometimes unisexual. Trees, shrubs, or vines, with opposite leaves. *Malphigiaceæ.*
(°°). Carpels at maturity not splitting apart as above.
Leaves compound, in ours 3-foliolate; capsule elongate; sap sour. *Oxalidaceæ.*
Leaves compound. Stamens commonly united into a tube. Flowers sometimes unisexual. Trees and shrubs. *Meliaceæ.*
Leaves simple, alternate or opposite, in ours narrowly lanceolate to linear. Capsule globose. *Linaceæ.*
2. Flowers irregular.
One sepal produced into a spur. Petals distinct, the two upper the smaller and unlike the others. Carpels 3, separating at maturity. *Tropæolaceæ.*
The two lateral sepals the largest, and colored. Petals more or less united, the lower often crested. *Polygalaceæ.*
b. Herbage covered with glandular dots. Aromatic trees and shrubs with simple or pinnate leaves. Flowers sometimes unisexual. *Rutaceæ.*

B. Flowers unisexual, monœcious or diœcious.
a. Land plants.
Plants with milky or acrid sap. Petals present or wanting. Carpels 3. Herbs, shrubs, or trees. *Euphorbiaceæ.*
Trees and shrubs with bitter bark and usually pinnate leaves. Carpels 2-5. *Simarubaceæ.*
b. Aquatics, rarely terrestrial, with linear or spatulate leaves. Flowers minute, monœcious. Sepals and petals none. *Callitrichaceæ.*

GERANIACEÆ (*Geranium Family*). The geraniums and their allies are mostly herbs with regular or sometimes irregular flowers, 10—15 stamens, or in some cases only 5 functional ones, and carpels with long slender styles, loosely adnate to the central axis—from which they separate as schizocarps at maturity. Each ovary mostly 1—2-seeded. Endosperm present. The plants have lobed or divided, stipulate or exstipulate leaves. About 11 genera and 650 species, mostly extratropical.

Principal genera: *Geranium, Erodium, Pelargonium. Geranium* alone is represented in our flora by native species.

Geranium. The geraniums are herbs with orbicular or reniform, variously lobed or dissected, stipulate leaves, naked buds, and regular 5-merous flowers variously clustered. Of the 10 stamens 5 are usually short. The ovary of each carpel is 2-ovuled, and the carpels at maturity separate from the central axis. The style is divided into 5 stigmatic limbs at the summit. In the common garden geraniums the flowers are somewhat bilaterally symmetrical, the calyx somewhat oblique, and the stamens united at the base (Fig. 247).

Authorities differ widely as to the number of species (190—260). Several species are cultivated. Probably over 70 species occur in North America. *Geranium maculatum* (Wild Cranesbill) is widely distributed in eastern North America.

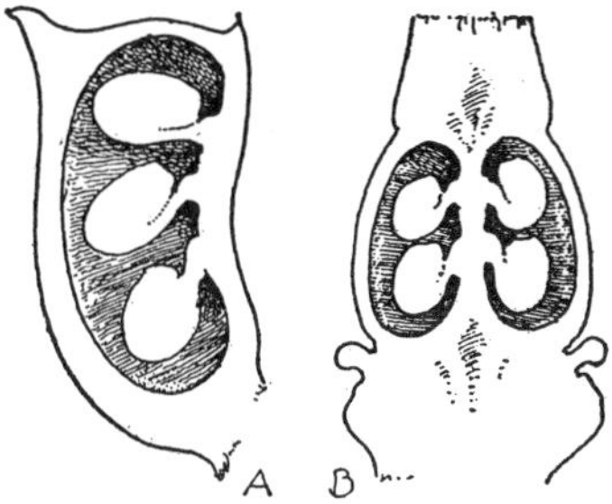

FIG. 246.—Mode of suspension of ovules in the *Geraniales: A, Dictamnus albus; B, Melia Azedarach.*

Erodium (Cranesbill) differs from *Geranium* technically in having but 5 functional stamens, and in having the tails of the carpels bearded on the inside and becoming spirally twisted in the fruit. *Erodium cicutarium,* a native of Europe, is widely distributed as a weed in waste grounds.

Pelargonium (Storksbill) differs from the preceding in having on the calyx a nectariferous spur which is joined to the flower-stalk. The flowers are irregular (bilaterally symmetrical), the two upper petals being usually larger and more strongly colored. The plants are herbs or subshrubs. Species about 250, mostly African, with a small representation in Australia and in Syria. Many are grown as potherbs and garden plants under the name geranium. (Fig. 248.)

OXALIDACEÆ (*Wood-Sorrel Family*). Flowers regular, 5-merous, the stamens 10—15, the gynœcium 5-carpellate, with 2-several ovules in each carpel. Styles united or distinct. Fruit commonly a capsule, dehiscing loculicidally. The plants are annual or perennial, caulescent or acaulescent herbs, or in some cases shrubs, with palmately 3-foliolate, pinnate, or peltate leaves and sour sap. Species about 300, in 15 genera; mostly tropical.

Oxalis, a genus of 170—180 species, is represented in our flora by probably less than a dozen species, among which is the common Wood-Sorrel, *Oxalis Acetosella,* which occurs up through the St. Lawrence basin to Saskatchewan. This plant sends up radical leaves and 1-flowered scapes from a creeping rootstock. *O. violacea,* a species of rocky woods, has a bulbous base, and violet flowers in an umbellate

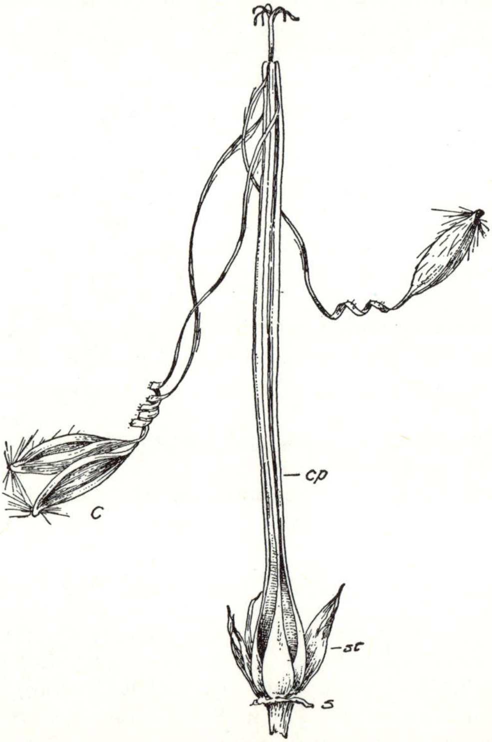

FIG. 247.—Dehiscing fruit of a geranium, the carpels elastically dehiscent; *cp,* carpophore; *s,* sepals; *st,* stamens.

cluster at the summit of the scape. *O. corniculata,* a yellow-flowered species, is a common weed in waste grounds and yards. (Fig. 249.)

TROPÆOLACEÆ (*Tropæolum Family*) comprises but one genus, *Tropæolum* (Tropæolum, "Nasturtium"), of about 50 species, ranging from Mexico to Chile. The plants are annual or perennial, more or less succulent herbs, with alternate, peltate or digitately lobed or dissected leaves. The flowers differ from those of the two preceding families in being markedly irregular (bilaterally symmetrical) and in being 5-merous only with respect to the calyx and corolla, and furthermore in having 8 stamens and a 3-carpellate gynœcium which, in the fruit, dehisces into three 1-seeded schizocarps. There is one style ending in three short stigmatic branches. The ovary is

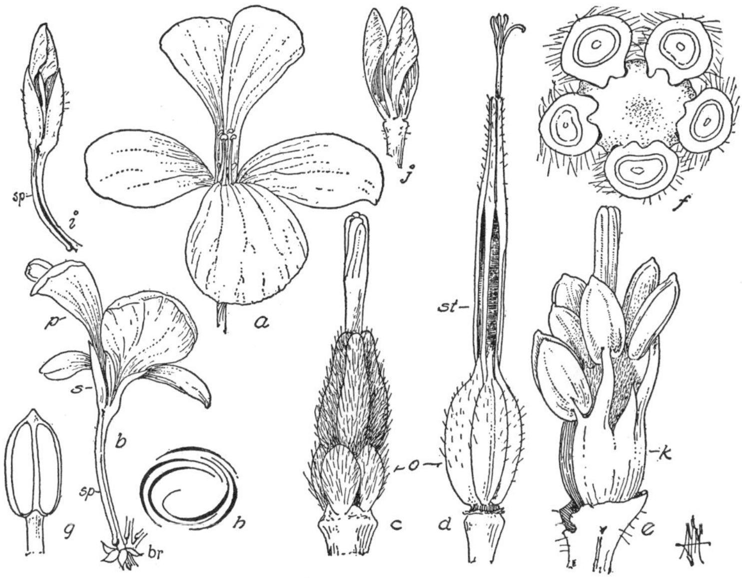

FIG. 248.—*Pelargonium peltatum: a,* flower (irregular); *b,* side view of flower; *c,* gynœcium from a bud; *d,* gynœcium from open flower; *e,* stamens, with stigmas protruding (form bud); *f,* transverse section of ovarian part of gynœcium; *i,* bud opening; *j,* corolla of bud in *i,* showing overlapping of petals; *g,* anther; *h,* diagram of arrangement of petals in the bud; *k,* staminal tube; *o,* ovary of individual carpel; *p,* petals; *s,* sepals; *st,* styles; *sp,* spur, adnate to pedicel; *br,* bracts.

globular. In addition the posterior sepal is developed downwards as a long spur with a nectariferous tip. Several species are cultivated, the commonest being *Tropæolum majus,* under the name "Garden Nasturtium"—not to be confused with the *Nasturtium* of the *Cruciferæ.* (Fig. 250.)

LINACEÆ (Flax Family) is a small family of herbs, shrubs, or trees, with alternate or opposite, stipulate or exstipulate leaves, and regular 5-merous or 4-merous flowers in mostly cymose or racemose inflorescences. The 5—20 stamens are united at the base into a ring which may or may not be provided with nectariferous glands. The gynœcium is commonly 5-carpellate, with distinct styles, but the ovary is sometimes many-celled by false partitions. Carpels in some cases fewer than 5. The fruit is a capsule, or sometimes drupaceous, with 1—2 seeds in each cell. Endosperm present. Between

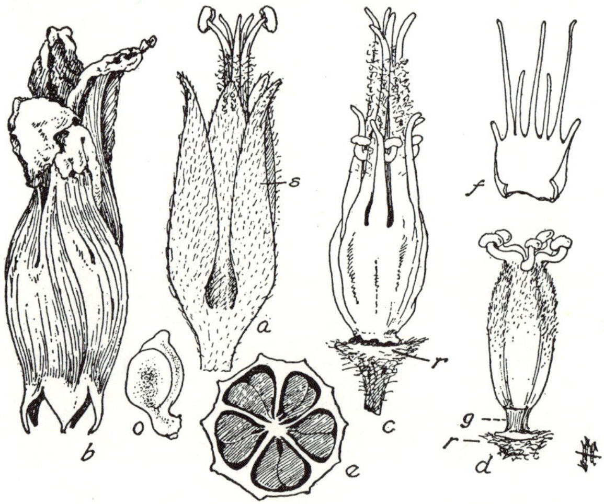

FIG. 249.—*Oxalis:* *a,* flower with corolla removed, stamens protruding; *b,* corolla; *c,* andrœcium, filaments free above, the styles protruding between them; *d,* gynœcium; *e,* transverse section of ovary; *f,* andrœcium laid open, anthers removed; *g,* gynophore; *r,* receptacle; *s,* calyx; *o,* ovule.

150—160 species in 14 genera, the largest of which is *Linum,* with about 90 species.

Linum usitatissimum, the common Flax, has for centuries been cultivated for its valuable fiber, which is made into linens, while its seeds yield a valuable oil. It is an annual with typically 5-merous flowers, fugacious petals, and the 5 stamens united at the base. False partitions make the otherwise 5-celled ovary several-celled. The fruit is a septicidally dehiscent, globular capsule containing flat, ovoid, glossy-brown seeds. The plant is a native of Europe. It is extensively cultivated in the prairie and plains States, and frequently appears as an escape from fields. Among other cultivated species are *L. grandiflorum* (Flowering Flax), a native of Africa, and the

European *L. flavum* and *L. perenne*. *L. Lewisii, L. rigidum,* and *L. sulcatum* are widespread native species. *Reinwardtia indica,* a shrubby plant of India, is sometimes seen in greenhouses. (Fig. 251.)

HUMIRIACEÆ is a small family of tropical American shrubs and trees with simple alternate leaves, regular bisexual 5-merous flowers, and drupaceous fruit. The inner stamens are sometimes

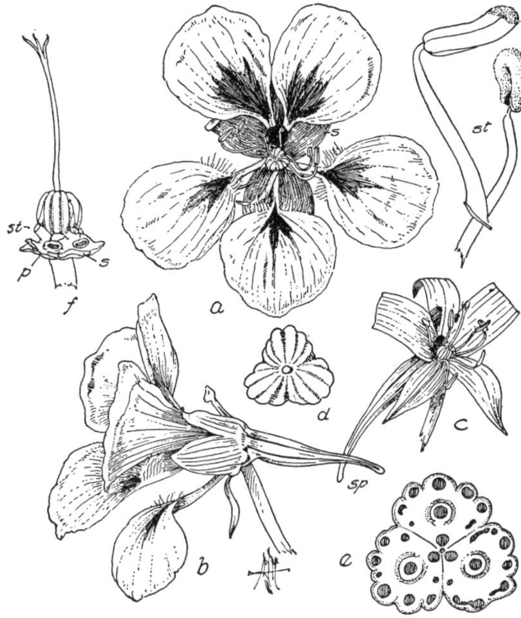

FIG. 250.—*Tropæolum majus: a,* flower, face view; *b,* side view; *c,* flower with petals removed; *d,* ovary, viewed from above; *e,* transverse section of ovary; *f,* gynœcium; *p,* petal; *s,* sepal; *st,* stamen; *sp,* spur of calyx.

reduced to staminodes. In one genus (*Vantanea*) the stamens are very numerous (50—180). The connective of the functional stamens is in some cases conspicuously prolonged (*Humiria*), and but one anther is present. Largest genus: *Saccoglottis,* with about half of the two dozen or more species, the others in *Vantanea* and *Humiria.*

ERYTHROXYLACEÆ. Shrubs or trees with alternate, stipulate leaves, inhabiting the tropics of both hemispheres. Flowers small, regular, bisexual, 5-merous (except gynœcium). Stamens 10, in two whorls, united at the base. Gynœcium 3- to 4-carpellate, all but one of the carpels usually aborting. Fruit drupaceous. All but one of the 194 species belong in the genus *Erythroxylon*. *Erythroxylon coca* is the Coca-plant of Peru and Bolivia. *E. novogranatense* of Colombia yields cocaine.

ZYGOPHYLLACEÆ. Mostly shrubs or subshrubs, rarely herbs or trees, with mostly opposite, pinnate, stipulate leaves, and solitary

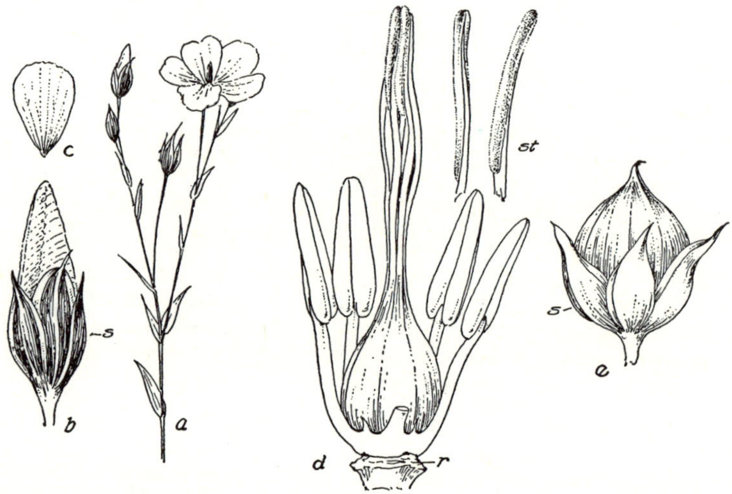

FIG. 251.—*Linum usitatissimum* (Flax); *a*, flowering shoot; *b*, flower-bud; *c*, petal; *d*, gynœcium and andrœcium; *e*, capsule and calyx (*s*); *st*, stigmas.

or clustered flowers which are regular, bisexual, 5-merous or sometimes 4-merous. Calyx sometimes gamosepalous at the base, and the petals in some cases wanting. Stamens twice or sometimes thrice as many as the petals, and obdiplostemonous, often with stipular appendages at the base. Gynœcium commonly 4- to 5-carpellate, with a single style, ripening into a loculicidal or septicidal capsule, or in some cases a berry or drupe, often strongly ridged or winged. Seeds with or without endosperm. The plants are inhabitants mostly of arid regions of the tropics and subtropics. About 160 species in 25 genera—of which only the following are represented in our flora: *Tribulus, Kallstroemia, Fagonia, Larrea.*

Tribulus terrestris, the Caltrop or Puncture Weed, occurs as an adventive here and there from the Mississippi Valley to California. It is a prostrate, spreading, pubescent annual with pinnate leaves. Where established it is a pernicious weed. The 5-carpellate gynœcium ripens into a stout-spiny dry fruit which separates into five nutlets. These nutlets are strong enough to puncture automobile tires.

Of the 20 or more species of *Kallstroemia,* 8 or 9 occur in the United States. The plants resemble *Tribulus,* except that the fruit is lobed or tuberculate instead of spiny. *Fagonia californica* is a subshrub of the desert region of Southern and of Lower California. The fruit breaks up into 5 separate, 1-seeded carpels which separate from the central axis. The 18 or more other species inhabit the desert regions of the Mediterranean, Southwest Africa, and Chile. There are 3 or 4 species of *Larrea,* of which *L. tridentata* (Creosote-Bush), a dark evergreen, strongly resinous, and rank-smelling shrub, forms often extensive zones in the desert regions of the Southwest. The ovoid hairy fruit is formed of 5 carpels which eventually separate from the central axis as 1-seeded achenes or schizocarps. The rather small yellow flowers terminate short lateral shoots.

CNEORACEÆ. One genus, *Cneorum,* of a few species of the Mediterranean region and the Canaries. The plants are shrubs with alternate, narrow, leathery leaves, and 3-merous or 4-merous, regular, bisexual flowers. Gynœcium subtended by a prominent disk. Fruit drupaceous.

RUTACEÆ (*Rue Family*). Flowers regular or irregular, hypogynous, bisexual or in some cases unisexual, 5-merous or 4-merous. Stamens as many or twice as many as the carpels or fewer by abortion, but rarely more. Gynœcium of 5 to 4 carpels—sometimes fewer (3, 1) or several (indefinite), often more or less loosely united, or distinct. Ovary surrounded at the base by a prominent annular disk, or supported on a gynophore. Fruit capsular, follicular, or a berry or drupe. Seeds with or without endosperm. Trees, shrubs, subshrubs, or herbs, with alternate or opposite, simple or compound leaves dotted with pellucid glands. The plants contain a strongly aromatic oil, and not only the herbage but parts of the flower, and the fruit, are often conspicuously gland-bearing. Economically the family is a very important one, yielding a large number of edible fruits, essential oils, and ornamentals. The number of known species is variously estimated to be between 900 and 1,000, distributed among 100 or more genera, and inhabiting tropical and subtropical climates although in some cases reaching well into colder belts.

There are five subfamilies, of which only the following need be considered:

Rutoideæ, in which the gynœcium consists of 4 or 5 (rarely 3 or 1) carpels that are loosely adnate by their styles and separate at maturity. Among the many genera are *Xanthoxylum, Ruta, Choisya, Dictamnus.*

Toddalioideæ, in which the gynœcium consists of 5 to 2 carpels (or sometimes only 1) which ripen into 2- or 1-seeded drupaceous or dry, winged samaras. *Ptelea* and several other genera.

Aurantioideæ, in which the gynœcium consists of 2 to 5 or more carpels, and the fruit is a berry, the cells of which are mostly filled with a juicy pulp. The most important group economically, yielding all our citrus fruits. Important genera: *Citrus, Fortunella, Poncirus.*

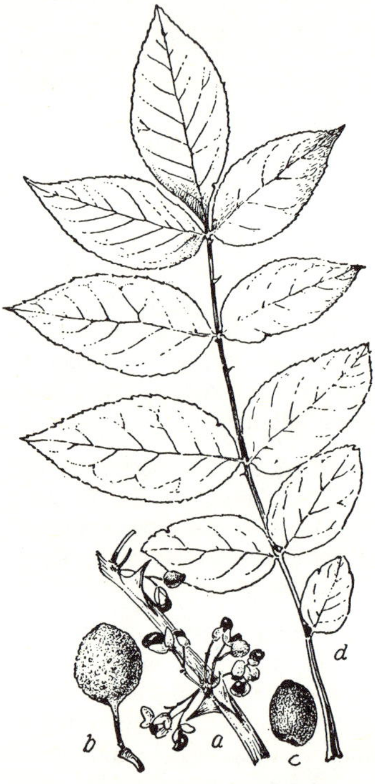

FIG. 252.—Fruit clusters (*a*), fruit (*b*), seed (*c*), and leaf (*d*) of prickly ash, *Xanthoxylum americanum.*

Rutoideæ. In *Xanthoxylum* the flowers are diœcious, and 5-merous except the gynœcium—in which the carpels vary from 5 to 2 and are distinct or but slightly connate by their styles. Each carpel is raised on a gynophore. Vestigial carpels are present in the staminate flowers. Each ovary is 2-seeded. The carpels often dehisce before maturity. Fruit follicular, the endocarp separating from the exocarp, and the seeds suspended, one above the other, from the loosened placenta. The plants are shrubs or trees with pinnate leaves, often beset with stout spines. Species about 150, of which less than a half-dozen exist in the United States. *Xanthoxylum americanum* (Prickly-Ash or Tooth-Ache Tree) is common in rich deciduous woods from western Quebec to Kansas. The small greenish flowers are borne in axillary cymes and have no calyx. *X. Clava-Herculis,* a much larger species occurring in the South, has much stouter spines and larger flowers in terminal cymes, and in addition possesses a calyx. (Figs. 252, 253.)

The solitary *Dictamnus albus* is a subshrubby perennial with odd-pinnate, pellucid-punctate leaves and large somewhat bilaterally symmetrical, 5-merous flowers, varying from rose-colored to white, in terminal racemes on bracteate pedicels. Stamens 10. The gynœcium is supported on a gynophore and its 5 carpels are distinct except at the base. On account of the basal union of the carpels the fruit is of capsular nature, but the individual carpels are follicular and flattened, and truncated above, the dorsal edge terminating in a spinous process and the ventral (inner) edge terminating in the short style. Each ovary is 2—3-seeded. The plant is native from southern Europe to the Amur region of Asia. It is commonly grown in gardens under the name Gas-Plant, from the fact that an inflammable gas escapes from the flowers (Fig. 254).

Ruta graveolens, the common Rue, also a common garden plant, is a glaucous strong-smelling herb with twice-pinnate leaves and yellow flowers. The inflorescence is distinctive in that the terminal flower of the cyme is 5-merous while the others are 4-merous. The ovary is 5-carpellate at the base and is subtended by a conspicuous annular disk. The plant is a native of southern Europe. Species about 40 in all. *Choisya ternata* is a handsome cultivated shrub with fragrant flowers.

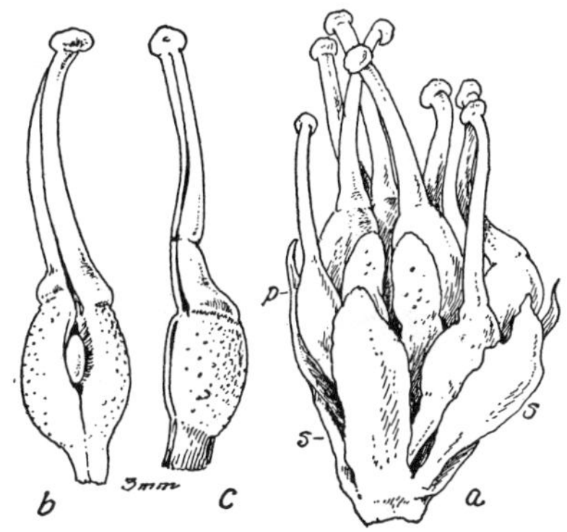

FIG. 253.—*Xanthoxylum americanum:* *a,* polycarpellate flower; *b,* ventral, and *c,* lateral view of a carpel (open); *p,* petal; *s,* sepal. In this case the carpels are not closed.

Toddalioideæ. Ptelea comprises about 3 species of shrubs or small trees of the United States and Mexico, among which is *P. trifoliata,* the Hop Tree, of Eastern woodlands (Fig. 255). The flowers are small, greenish, corymbose-paniculate and 5-merous or 4-merous, but the gynœcium is 2-merous. The fruit is an orbicular, flat, broadly winged samara, very similar in appearance to the fruits of some elms. *P. Baldwinii* var. *crenulata* occurs in California.

Aurantioideæ. The most important genus of the family is *Citrus,* which yields our citrus fruits. There are probably less than a dozen species, all natives of tropical and subtropical Asia. The leaves are evergreen, mostly shiny, unifoliolate, and glandular-punctate, with more or less spatulately wing-margined petioles. Some species are spiny. The flowers are white or cream-colored and fragrant, solitary or in cymes or panicles, and generally 5-merous. The 15 or more stamens have stout filaments and are in some cases arranged in clusters (polyadelphous). The gynœcium is several- to many-carpellate and syncarpous, with an oblong ovary and a single stout style terminating

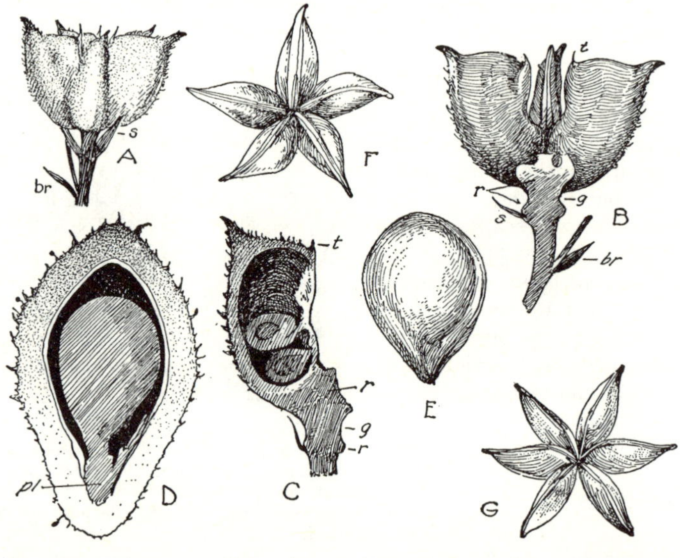

FIG. 254.—*Dictamnus albus: A,* gynœcium; *B,* gynœcium with two carpels removed; *C,* median section of a carpel; *D,* transverse section of *C; E,* seed; *F,* a 5-carpellate, *G,* a 6-carpellate gynœcium; *g,* gynophore; *pl,* placenta; *br,* bracts; *r,* receptacle; *s,* sepals; *t,* style.

in a usually large fleshy stigma. It is supported on a short, thick gynophore in the form of an annular hypogynous disk. The fruit is a large, globular or ellipsoid berry (*hesperidium*) the cells (*locules*) of which are filled with a juicy pulp, each cell containing 1 to several rather large seeds. The rind of the fruit is densely glandular, and glandular dots are conspicuous on the flower-buds (Fig. 256).

The common oranges are the fruits of *Citrus sinensis. Citrus nobilis* var. *deliciosa* yields the tangerines, *C. Limonia* the lemon

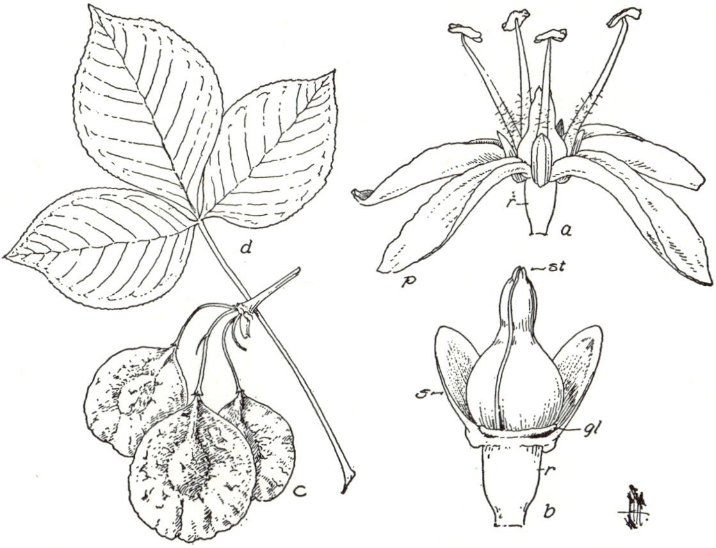

FIG. 255.—*Ptelea trifoliata: a,* flower; *b,* gynœcium; *c,* fruits (samaras); *d,* leaf; *gl,* gland; *p,* petals; *s,* sepal; *r,* receptacle; *st,* stigma.

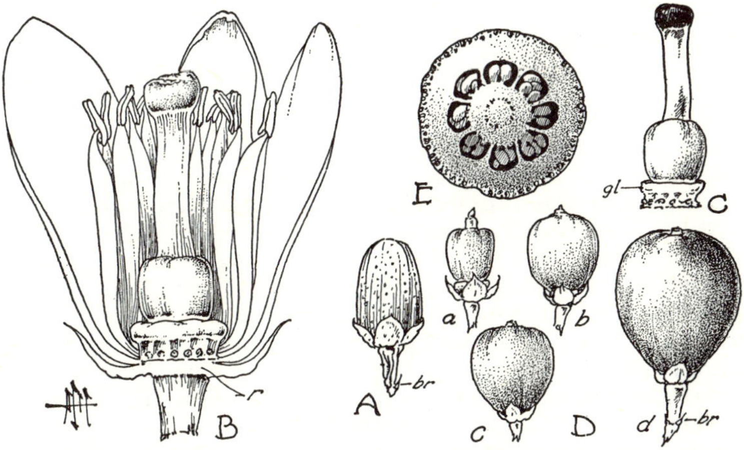

FIG. 256.—Orange (*Citrus sinensis*): *A,* a bud; *br,* bract; *B,* section of flower; *r,* receptacle; *C,* gynœcium; *gl,* gland; *D* (*a, b, c, d*), stages in the development of the fruit.

(Fig. 60), *C. maxima* the grapefruit, and *C. medica* the citron. Except the tangerines and the common orange, the trees are all spiny. In *C. medica* var. *sarcodactylus* the carpels develop as nearly distinct fleshy finger-like rays (Fig. 257). In the "navel orange" a secondary orange is formed at the top by the further development of the floral axis. *Fortunella* (Kumquat) and *Poncirus* (Trifoliate Orange) are native to China. In the deserts of Southern California are found *Thamnosma montana* (Turpentine Broom) and *Cneoridium dumosum.*

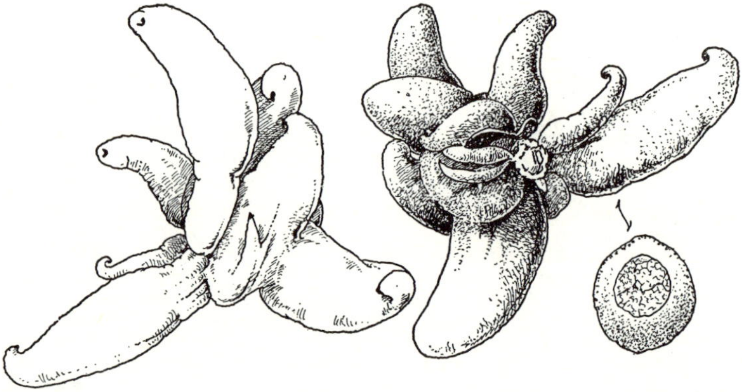

Fig. 257.—Fruits of *Citrus medica* var. *sarcodactylus.* At right a transverse section of a digit.

SIMARUBACEÆ (*Ailanthus Family*). Flowers regular, polygamous or diœcious, 3-merous to 5-merous, the gynœcium in some cases 2-merous, and the stamens sometimes twice as many as the petals. Carpels distinct or merely connate by the styles, or sometimes completely united. Ovary sessile, with or without an hypogynous disk, or elevated on a gynophore—which is discoid above. Scale-like appendages are in some cases present on the filaments. Fruit various, mostly 1-seeded. Endosperm present or wanting. Tropical and warm-temperate trees and shrubs, with alternate or sometimes opposite, usually pinnate (but not punctate) leaves, and bitter bark. Stipules present or wanting. About 150 species in 30 genera. The following Asiatic species is common in cultivation.

Ailanthus glandulosa (Tree of Heaven) is a tall sumac-like shrub or large tree (90 feet) with large odd-pinnate leaves which give off a disagreeable odor when bruised, and small, greenish, polygamo-diœcious flowers in terminal panicles. In the staminate flowers the 10 stamens arise from beneath a disk, and vestigial carpels are sometimes present. Gynœcium of 5—6 more or less distinct carpels, and each ovary containing 1 suspended ovule. Below the ovary is a prominent disk. The fruit is a samara. There are several species in East India and eastern Asia.

BURSERACEÆ. A small family of about 13 genera and 400 species of shrubs and often tall trees of the American tropics. Leaves compound, pinnate or trifoliolate, glandular-punctate. Flowers mostly small, unisexual, with 3-merous to 5-merous calyx and corolla, twice as many stamens as petals, these obdiplostemonous, placed beneath or on the margin of an annular hypogynous disk. Gynœcium 5—2-carpellate, syncarpous. Fruit various, usually 1-seeded. Endosperm wanting. Many species are valuable for their aromatic resins and balsams—*e.g.*, the Incense Tree (*Boswellia*) and the Myrrh Tree (*Commiphera*).

MELIACEÆ (*Mahogany Family*). Forty or more genera and over 600 species of hard-wooded tropical trees and shrubs with alternate or opposite, simple or pinnate, but not glandular-punctate, exstipulate leaves, and regular, bisexual or polygamous, often perigynous flowers with 4-merous or 5-merous calyx and corolla. Stamens mostly twice as many as the petals, the filaments united into a tube, rarely free, the anthers often inserted below the fringed or lobed margin of the tube. Disk when present various, sometimes forming a more or less pronounced gynophore. Gynœcium commonly 2—5-carpellate, sometimes without a style. Ovary free or sometimes adnate, more or less, to the disk. Fruit a capsule, berry, or drupe, or sometimes indehiscent.

Melia Azedarach, the China-Tree, and its variety *umbraculiformis,* the Texas Umbrella-Tree, are cultivated in the Southwest. The latter has pinnately compound leaves and produces masses of flowers and drupaceous fruits, while the branches form a dome-shaped or umbrella-shaped crown. These trees also yield valuable wood. *Swietenia Mahogani* yields the true mahogany of commerce. (Fig. 258.)

AKANIACEÆ. Limited to a solitary woody Australian (Queensland) species with odd-pinnate leaves and regular bisexual flowers.

MALPIGHIACEÆ (*Malpighia Family*). Flowers regular or obliquely irregular, bisexual or unisexual, 5-merous (except gynœcium commonly), the sepals free or united, the petals clawed, the stamens 10, free or the filaments forming a tube and the connective of the anthers often enlarged or variously extended. Carpels commonly 3, sometimes 2, 4, or 5, distinct or united, each producing 1 ovule. Fruit a schizocarp separating into winged nutlets, or sometimes a drupe or nut. Trees, shrubs, or woody vines, with simple, alternate or opposite, stipulate, and often gland-bearing leaves, inhabiting the tropics of both hemispheres, some occurring in South Africa. The family has

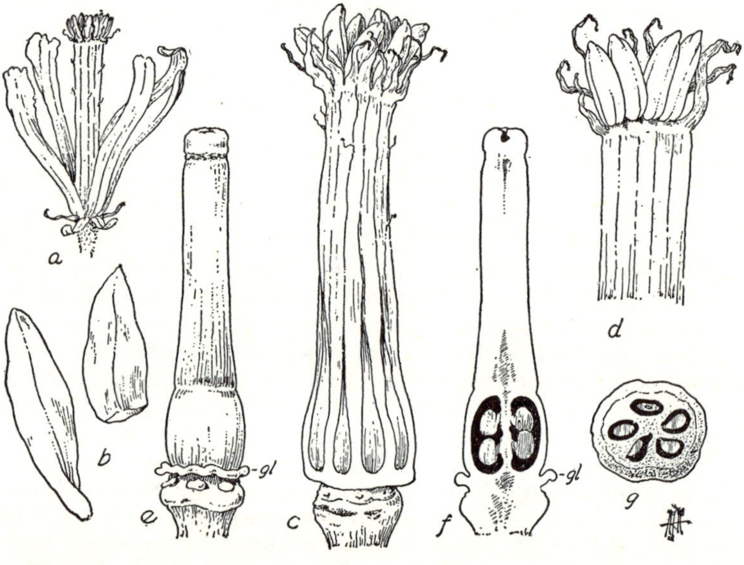

FIG. 258.—*Melia Azedarach: a,* flower; *b,* sepals; *c,* andrœcium; *d,* section of summit of andrœcium showing anthers and marginal linear lobes; *e,* gynœcium; *f,* median section of gynœcium; *g,* transverse section of ovary; *gl,* disk.

given name to a peculiar kind of unicellular branching hairs which make up the pubescence. Species of *Malpighia, Thryallis, Stigmaphyllon,* and others are cultivated. The peculiar glandular structures of the calyx, petiole, and base of the leaf-blades are useful recognition-marks. About 650 species in 55 genera.

TRIGONIACEÆ. Probably 30 species and 2 to 5 genera of tropical American trees and woody climbers, with simple opposite leaves, often interpetiolar stipules, and obliquely zygomorphic, bisexual, 5-merous flowers. Largest genus: *Trigonia,* with 26 or more species.

VOCHYSIACEÆ. Probably a half-dozen genera and 160 species of tropical South American trees and shrubs, with simple, opposite or whorled, stipulate or exstipulate leaves, and bisexual, obliquely zygomorphic, 5-merous flowers. Gynœcium 3-carpellate. Largest genera: *Vochysia* (over 50 species) and *Qualea* (over 30 species).

TREMANDRACEÆ. About 3 genera and 24 species of Australian herbs and shrubs with alternate, opposite or whorled leaves and regular, bisexual, 3-merous to 5-merous flowers. Fruit a capsule. The great majority of the species are in the genus *Tetratheca*.

POLYGALACEÆ (*Milkwort Family*). Flowers irregular, bisexual, solitary or in racemes or spikes, with considerable variation in the number of floral organs. The 5 sepals are free or more or less united, the two lateral ones larger and colored. Petals commonly 3, more or less united into a tube or united with the stamens, the lower petal commonly crested. Stamens in two whorls (5 + 5) but often reduced to 8 or fewer and commonly united into a tube which is open on the dorsal side. Gynœcium 5—2-carpellate, with a 5- to 1-celled ovary producing 1 seed in each cell. Fruit a capsule, drupe, or nut. Seeds with or without endosperm. About 10 genera and 1,000 species of herbs and a few trees and shrubs, chiefly of the tropics but also reaching into temperate climates. Leaves opposite or verticillate, exstipulate. The enlarged lateral petals suggest a papilionaceous flower, but the characters of the fruit or of the gynœcium will distinguish the plants from the *Papilionoideæ*.

Nearly half of all the species belong to *Polygala*, of which close to 60 species occur in our flora. One of the most widespread of the American species is *Polygala Senega* (Snake-Root) which ranges from the Atlantic States to the Rocky Mountains. In the early days the roots were extensively gathered by the Chippewa Indians of Minnesota and sold to drug dealers.

DICHAPETALACEÆ. About 3 genera and 136 or more species of tropical trees and climbing shrubs, with alternate, leathery, stipulate leaves and regular or irregular, bisexual or unisexual, 5-merous flowers in simple or branched inflorescences. Calyx often gamosepalous and the corolla often sympetalous. Gynœcium 2—3-carpellate. The great majority of the species in *Dichapetalum*—mostly of Africa and Madagascar.

EUPHORBIACEÆ (*Spurge Family*). Flowers regular or irregular, unisexual, monœcious or diœcious, often much reduced, naked or with calyx or corolla, borne in compound inflorescences of various sorts. In the staminate flowers the stamens are as many or twice as many as the perianth-segments, or sometimes numerous, or reduced to 1 (*Euphorbia*), with distinct or united filaments and 2-celled anthers. In *Euphorbia* the staminate flowers are naked and sometimes show a rudimentary gynœcium. In the carpellate flowers the gynœcium is commonly 3-carpellate and syncarpous, with the style 3-branched above. Ovules 1 or 2 in each cell, pendulous and anatropous, the micropyle covered by a *caruncle* (Fig. 62.) Staminodes may or may not be present. Endosperm abundant and oily,

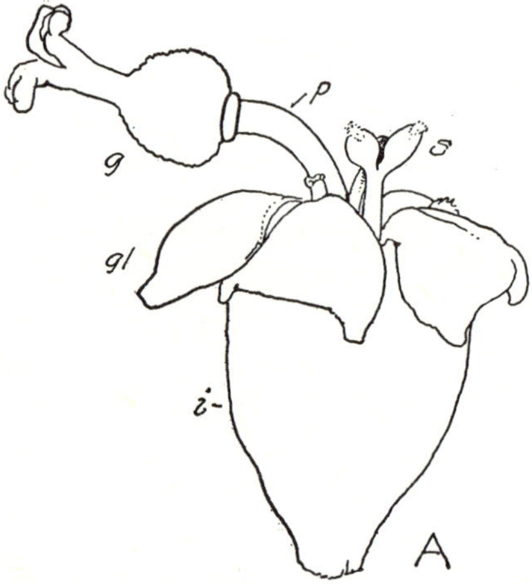

FIG. 259.—*Euphorbia Esula: A,* inflorescence (cyathium); *g,* gynœcium; *gl,* glands of involucre (*i*); *p.* pedicel; *s,* anthers opening. *B,* inflorescence with involucre removed; *r,* common receptacle of inflorescence; *sf,* staminate flowers. *C,* a staminate flower; *f,* filament; *p,* pedicel.

investing a large centrally placed embryo with large, flat, or variously folded cotyledons. Fruit a capsule, berry, or drupe. Capsule splitting into three partial fruits (*cocci*).

This is one of the largest families of flowering plants, comprising about 250 genera and 4,000 species of herbs, shrubs, or trees, with an acrid and often milky sap, and alternate, opposite, or verticillate leaves. Members of this family often show a remarkable diversity in habit (growth-form), which is especially noticeable in the Euphorbias, some of which—those inhabiting African deserts—are strikingly cactus-like. Several genera are represented in cultivation by many useful as well as ornamental species, a number of which are mentioned below.

There are four subfamilies, two of which are Australian (omitted here). The others are the *Phyllanthoideæ* and the *Crotonoideæ,* which are distinguished from the Australian subfamilies by the broad cotyledons of the embryo. (Fig. 62.)

Phyllanthoideæ. Distinguished by the absence of milky sap and in having two ovules in each cell of the ovary.

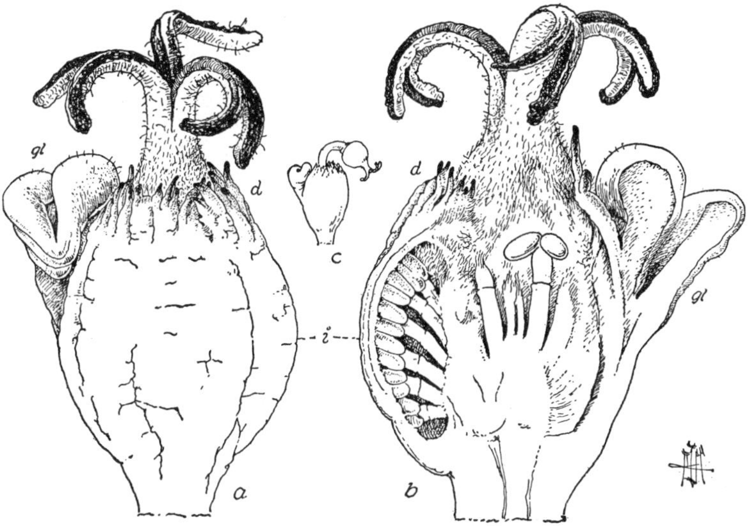

FIG. 260.—*Euphorbia pulcherrima* (Poinsettia): *a,* inflorescence with investing involucre (*i*); *b,* median section of inflorescence,—at left several stamens, in center three staminodes and a stamen; above, protruding style and spreading forked stigmas; *d,* succulent dactyliferous processes of involucre; *gl,* fleshy funnel-shaped gland; *c,* inflorescence after pollination.

The following genera are fairly representative: *Andrachne, Phyllanthus, Tetracoccus.* All three are represented with us by at least one species.

In these genera there is no involucre, but a calyx is present in both staminate and carpellate flowers. The corolla is wanting in *Phyllanthus* and *Tetracoccus,* and sometimes in *Andrachne.* All are monœcious (*Tetracoccus?*). A shrubby habit and a 4-merous gynœcium distinguish *Tetracoccus* from the other two.

Crotonoideæ. Distinguished by the presence or absence of milky sap and in having but 1 ovule in each cell of the ovary.

Among the numerous genera the following are represented by native or introduced species: *Croton, Crotonopsis, Eremocarpus, Ditaxis, Argythamnia, Mercurialis, Bernardia, Acalypha, Tragia, Ricinus, Jatropha, Hevea, Manihot, Stillingia, Euphorbia.*

The most abundantly represented genus is *Euphorbia,* with many native and introduced species, the latter often common as weeds. The floral characters may be illustrated by *Euphorbia pulcherrima,* the beautiful "Poinsettia" of the gardens and hothouses, a native of tropi-

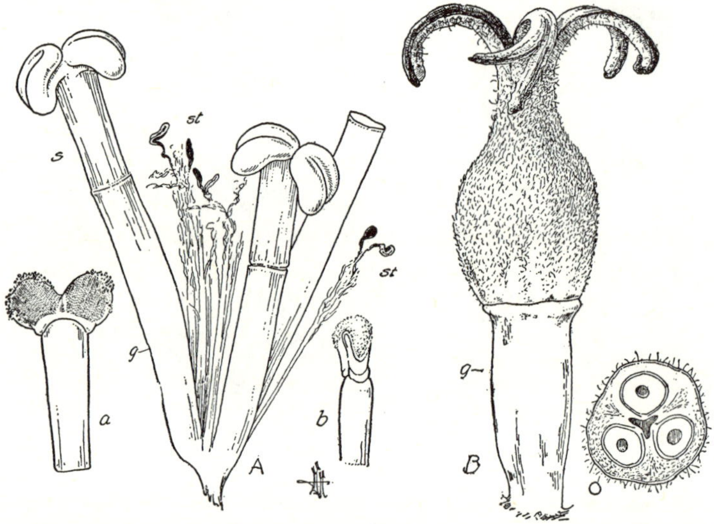

FIG. 261.—*A,* a cluster of staminate flowers and *B,* a carpellate flower of *Euphorbia pulcherrima; a, b,* stamens with anthers dehiscing; *s,* stamen; *g,* pedicel; *st,* staminodes; *o,* transverse section of ovary.

cal Mexico and Central America. The flowers are borne in cyathia, and these in terminal cymes which are subtended by a number of showy scarlet upper leaves. Each cyathium consists of an ovoid involucre with a conspicuous, fleshy, bright-yellow gland on one side, while the margin of the involucre is lined with a fringe of fleshy finger-like lobes. The carpellate flower rises on a more or less prominent stalk (pedicel) from the bottom of the cyathium. In many cases only a rudimentary carpellate flower is present, and the cyathium is consequently staminate only. The gynœcium is characteristically 3-carpellate, with a short style and three prominent forked stigmatic

FIG. 262.—*Ricinus communis* (Castor Bean): *A,* carpellate flower; *B,* section of carpellate flower; *C,* succulent spiniferous process from surface of ovary in *B; D,* staminate flower; *E,* branched "stamen" (*a*) and ultimate antheriferous branchlets (*s*) or (*a*); *p,* perianth.

lobes (branches). After pollination the gynœcium is pushed out of the cyathium by the elongation of the pedicel. Surrounding the carpellate flower, on fleshy outgrowths from the base of the cyathium, are the numerous staminate flowers. Each staminate flower consists of but a solitary stamen which is supported on an elongate pedicel comparable to the pedicel of the carpellate flower. Thus the cyathium as a whole is a rather complicated sort of inflorescence. Usually there are also numerous slender fringed staminodes crowded in among the staminate flowers. (Figs. 259, 260, 261.)

The Poinsettia is a tall perennial with a woody base, which, in California, blooms throughout most of the winter. The internodes are hollow and filled with an abundance of milky sap, as is also the foliage and other organs. The leaves are large and coarsely sinuate-toothed.

Among the other members *Ricinus communis,* the Castor Bean, is a common tall, large-leaved ornamental, grown out of doors the year round in the South and Southwest, where it also thrives as an escape and grows to tree-like proportions. It is the only species of its genus and comes originally, it is thought, from Africa. It is a monœcious species bearing numerous small flowers in terminal racemes, the carpellate above the staminate in the same inflorescence. A calyx is present in both staminate and carpellate flowers, but in the latter it soon drops off. The staminate flowers bear numerous profusely branched stamens, and there is no vestige of a gynœcium present. The carpellate flowers produce, besides the caducous calyx, a syncarpous 3-carpellate gynœcium with an ovoid ovary densely clothed with thick, fleshy, bristle-tipped processes. The short style terminates in three bright-purplish (when in full bloom), bifid, coarsely papillate, stigmatic branches. The fruit is a bristly or smooth septicidally dehiscent capsule, with one large seed in each part (coccus). The seeds are oblong-ovoid, smooth, and variously mottled, with a prominent caruncle over the micropylar end; they are very poisonous. There is an abundance of very oily endosperm closely investing the large embryo which has large, flat, and thin

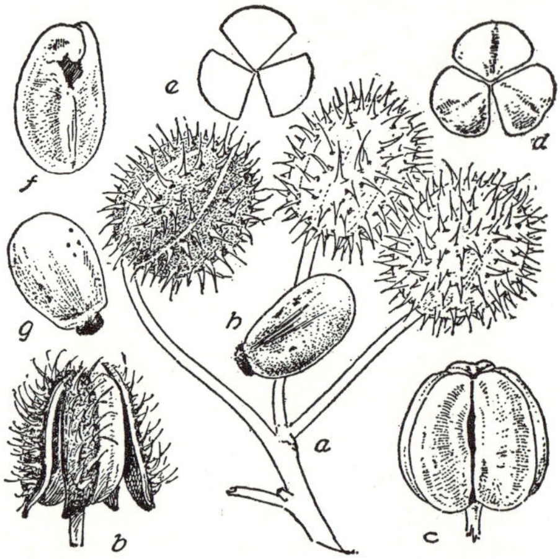

FIG. 263.—*Ricinus communis: a,* cluster of fruits (capsules); *b,* capsule dehiscing; *c,* cocci or nuts (endocarps) of capsule, each containing a seed; *d,* cocci separating, viewed from above; *e,* diagram of dehiscence of capsule and cocci; *f,* immature seed; *g,* dorsal and *h,* ventral aspect of seed.

cotyledons closely appressed to the endosperm (Figs. 262, 263).

Of the other genera, *Eremocarpus* (*setigera*), a low, matted annual, occurs from central California to Washington. *Bernardia myricæfolia,* a grayish shrub, inhabits the deserts of southern California and adjacent Mexico. *Jatropha* (*Cnidosolus*) *stimulosa,* an herb with digitately lobed leaves and armed with stinging hairs, is found in dry sandy soil from Virginia to Florida and Texas. Species of *Croton* are distributed from eastern and southern United States to Kansas and Texas and westward. *Ditaxis* has a couple of species in the Central and Southern States. *Acalypha* and *Tragia* are represented by several species each southward across the continent. *Stillingia sylvatica,* a monœcious herb, occurs throughout the South. *Mercurialis annua* (Mercury), a diœcious or monœcious annual introduced from Europe and Africa, is a weed in waste places in the East and South. *Hevea brasiliensis,* a tree of the Amazon region, is the source of Para rubber. *Manihot utilisissima,* the Manioc or Tapioca Plant of tropical Brazil, is a shrub with a tuberous root rich in starch, which yields the Brazilian arrow-root. Many species of the family are poisonous.

Summary of the outstanding characters of this family: (1) Remarkable diversity in habit; (2) presence of milky sap in many forms; (3) the constant occurrence of unisexual flowers; (4) suppression of the perianth (calyx or corolla, or both) in many forms and the development of a special apparatus (dichasium and involucre) as an "adaptation" to insect pollination.

CALLITRICHACEÆ (*Water Starwort Family*). A small family of about 20 species of widely distributed, slender, mostly aquatic herbs with small opposite, exstipulate leaves of various shapes and very small bisexual or unisexual (monœcious) flowers, which are without a perianth. The staminate flowers consist of a solitary terminal stamen, the carpellate of a 2-carpellate gynœcium with 2 long, slender styles. Ovary 4-celled by false partitions. Fruit indehiscent. Seeds with fleshy endosperm. The thick bracteoles subtending the flowers may be easily mistaken for a perianth. Some species are found in cold running waters. *Callitriche,* about 26 species, cosmopolitan.

ORDER SAPINDALES

Flowers regular or irregular, bisexual or unisexual (diœcious in some), hypogynous, perigynous, or epigynous, cyclic, mostly 5-merous, with calyx and corolla (sympetalous in some). Stamens 2 to many. Carpels 1 to several, in a syncarpous gynœcium, but sometimes distinct. In some families there is present a fleshy glandular disk. Seeds with or without endosperm. Integuments 2 or 1. A large order, mostly of trees and shrubs.

In all members of the order the ovule is bent *downwards* in such a way as to bring one side against the stalk (funiculus)—which may assume a raised or lowered position. This is the chief technical distinction between this order and the *Geraniales* (Fig. 264).

FAMILIES

Buxaceæ
Empetraceæ
Coriariaceæ
Limnanthaceæ
Anacardiaceæ
Cyrillaceæ
Pentaphylacaceæ
Corynocarpaceæ
Aquifoliaceæ
Celastraceæ
Hippocrateaceæ
Salvadoraceæ
Stackhousiaceæ
Staphyleaceæ
Didiereaceæ
Balsaminaceæ
Icacinaceæ
Aceraceæ
Æxtoxicaceæ
Hippocastanaceæ
Sapindaceæ
Sabiaceæ
Melianthaceæ

KEY TO THE PRINCIPAL FAMILIES

A. Flowers regular.
  Petals none. Flowers unisexual. *Buxaceæ.*
  Petals present.
    Low evergreen heath-like shrubs of subshrubs. *Empetraceæ.*
    Plants not as above.
      Herbs. Leaves pinnate. *Limnanthaceæ.*
      Vines. Disk well developed.
        Ovary often immersed in the disk. *Celastraceæ.*
        Ovary more or less elevated on a gynophore. *Sapindaceæ.*
      Trees or shrubs.
        Flowers unisexual or bisexual, polygamous, monœcious or diœcious.
          Leaves alternate, rarely opposite.
            Leaves simple. Trees and shrubs with resinous bark. *Anacardiaceæ.*
            Leaves pinnately compound.
              Stamens as many as the petals or twice as many, or more.
                Carpels mostly 3-1, rarely 5. *Anacardiaceæ.*
              Stamens commonly 8. Carpels 2-3. Disk mostly on one side. *Sapindaceæ.*
          Leaves opposite, simple or compound. *Aceraceæ.*
        Flowers bisexual or usually so.
          Leaves simple.
            Disk none. Leaves sometimes spiny-toothed. *Aquifoliaceæ.*

Disk prominent. Flowers sometimes unisexual. *Celastraceæ.*
Leaves compound (odd-pinnate). Disk prominent. *Staphyleaceæ.*
B. Flowers irregular.
Succulent herbs. *Balsaminaceæ.*
Trees or shrubs.
Leaves opposite, digitately 3—9-foliolate. *Hippocastanaceæ.*
Leaves alternate.
Flowers inverted by twisting of pedicels. *Melianthaceæ.*
Flowers not inverted, often with well developed gynophore. *Sapindaceæ.*

BUXACEÆ (*Box Family*). Flowers unisexual, rarely bisexual, monœcious or diœcious, in spicate or racemose, bracteate inflorescences. Sepals usually 4; petals wanting. Stamens 4—6 or sometimes more; when 4, opposite the sepals; when 6, opposite the inner sepals. Anthers large, 2-celled. Gynœcium 3-carpellate or sometimes 2- or 4-carpellate, with distinct styles, and 2—1 ovules in each cell. Fruit a capsule or drupe. Seeds with fleshy or scanty endosperm. Embryo straight. Integuments 2, About 6 genera and 35 species of mostly evergreen trees and shrubs and a few herbs of the East Indies, Asia, Japan, the West Indies, and a scanty number in South America, Natal, and the United States.

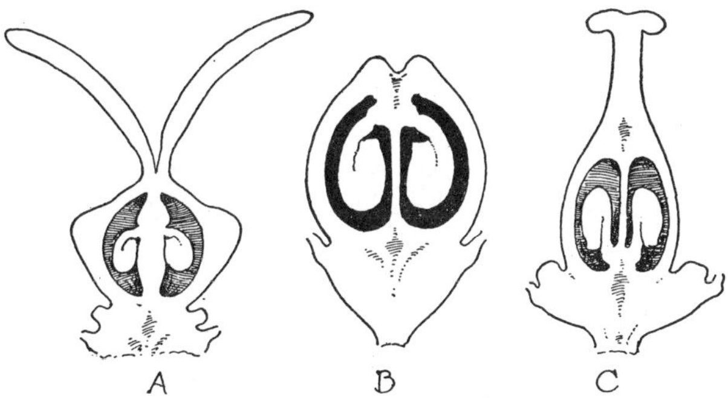

FIG. 264.—Mode of suspension of ovules in the *Sapindales: A, Acer macrophyllum; B, Ilex Aquifolium; C, Celastrus scandens.*

The following genera are representative: *Pachysandra, Buxus, Simmondsia.*

In *Pachysandra* and *Buxus* there are 2 seeds in each cell of the ovary, the staminate flowers have 4 stamens and usually a rudimentary gynœcium.

Pachysandra procumbens, a procumbent perennial monœcious herb, with alternate leaves and matted rootstocks, is found in woods from West Virginia southward and west to the Mississippi. Another species occurs in Japan. The fruit is a capsule. *Buxus sempervirens,* the common Box, is an evergreen shrub or small tree with quadrangular or slightly winged branches, opposite, leathery, shining, dark-green leaves, and pale-greenish flowers with yellow anthers. The carpellate flowers are terminal and the staminate lateral in the inflorescence. The fruit is an ovoid, loculicidally dehiscent capsule crowned by the persistent styles. Seeds shining black. This common hedge plant is a native of the Mediterranean countries. There are about 35 species in all, occurring in both the eastern and the western hemispheres.

The solitary *Simmondsia californica* (Jacoba or Goat-nut) is a much-branched monœcious shrub of arid hills of southern California, Arizona, and New Mexico. The leaves are opposite and the flowers unisexual. The staminate flowers are borne in capitate clusters and have a calyx of 5 distinct sepals and 10—12 stamens. The carpellate flowers are solitary and larger. The gynœcium is 3-carpellate, with a 3-celled ovary and distinct styles, but the ovary becoming a 2—1-celled and 1-seeded capsule by abortion; it is of the size and shape of an acorn.

EMPETRACEÆ (*Crowberry Family*). Flowers regular and unisexual, monœcious or diœcious or polygamous, with rudiment of the other sex present; commonly 2-merous or 3-merous, but the gynœcium 2- to 9-merous, with as many style-branches as carpels. Fruit drupaceous, with 6—9 stone seeds. Embryo straight, enclosed in copious endosperm.

Three genera and less than a half-dozen species of low evergreen heath-like shrubs, with small, linear, exstipulate, revolute leaves and small flowers in axillary or terminal heads. Genera: *Corema, Empetrum, Ceratiola.*

In the North Atlantic region *Corema Conradii* (Broom Crowberry) occurs as a low, densely leafy shrub in patches in rocky or sandy soil. The branches are scarred by the persistent pulvini of the leaf-bases. The plant bears small diœcious or polygamous flowers in terminal heads. The corolla is wanting. The drupaceous fruit produces 3 nutlet-like seeds. *Corema album,* the only other species, occurs in Spain, the Azores, and the Canaries. These species are of interest as tying up the floras of western Europe with Atlantic North America.

Empetrum nigrum (Black Crowberry) is a low, spreading shrub with a remarkably wide distribution over the boreal regions of North America and Eurasia. Isolated stations of it are found in the Lake Superior region and elsewhere beyond its northern range in western North America and in the North Atlantic region. Thousands of miles away, in Antarctic America, is another species, *E. rubrum,* which bears red fruit. How explain this strange isolation?

The small purplish flowers of *Empetrum nigrum* are solitary in the upper axils and are usually 3-merous, but the gynœcium has from 6 to 9 carpels. The fruit is black, and contains as many nutlets as carpels.

Ceratiola ericoides, the only species of this genus, is a heath-like diœcious shrub of the South Atlantic States. The flowers are dimerous and the fruit is a 2-seeded drupe surrounded at the base by the persistent calyx.

CORIARIACEÆ. This family comprises less than a dozen species of shrubs of Mediterranean countries, the Himalayan region, Japan, and New Zealand, and ranging in America from Mexico to Peru. The flowers are regular, bisexual or unisexual, and 5-merous, with calyx and corolla. The carpels vary from 6 to 8, separating at maturity into nutlets. Seeds with scanty endosperm. Leaves entire, opposite, or whorled. *Coriaria* is the only genus.

LIMNANTHACEÆ (*False Mermaid Family*). A very small family of annual North American herbs with alternate, pinnate, exstipulate leaves, and regular, bisexual, axillary, ebracteolate, commonly 3-merous or 5-merous, perigynous flowers. The carpels are almost distinct, the slender forked styles rising between the ovaries. Fruit of the schizocarpous type, the carpels separating into 1-seeded fruitlets (mericarps). The family is by some authorities referred to the *Geraniales.*

Genera: *Limnanthes, Floerkea.*

Three species of *Limnanthes* occur in California. These have solitary and axillary, mostly 5-merous flowers with 10 stamens, and petals with gland-like swellings at the base and a *U*-shaped band of hairs on the claw. *Limnanthes Douglasii* (Meadow Foam), a plant of wet grounds of the coastal region, sends up several flowering stems and pinnately dissected leaves from a short caudex. *Floerkea* comprises but one species, *F. proserpinacoides* (False Mermaid), a low annual with pinnate leaves and white flowers, growing in marshes and along streams in eastern parts of the continent. The calyx and corolla are 3-merous, but the stamens vary from 4 to 6, and the carpels from 2 to 3 (sometimes only 1). The carpels become 1-seeded achene-like fruits.

ANACARDIACEÆ (*Cashew Family*). Flowers regular, hypogynous, perigynous, or epigynous, bisexual or unisexual, polygamo-diœcious, 3-merous to 7-merous as to calyx and corolla, the stamens as many or twice as many as the petals, the gynœcium mostly 3- to 1-carpellate, in some cases 5-carpellate, with distinct styles; in some forms the carpels are distinct. Seeds with bony or crustaceous seed-coat and scanty or no endosperm. About 60 genera and 500 species of trees and shrubs with acrid or milky sap and simple or compound, alternate, rarely opposite, exstipulate leaves, inhabiting tropical and temperate climates. A number of species yield useful chemical principles, others yield edible fruits or are grown for their ornamental qualities, while a number are very poisonous.

Characteristic genera: *Mangifera, Anacardium, Spondias, Schinus, Cotinus, Rhus, Pistacia.*

Mangifera. The Mangiferas are large Asiatic trees with alternate, entire coriaceous leaves and numerous small, polygamous flowers in terminal compound panicles. The calyx and corolla are 4-merous or 5-merous, but the stamens vary from 1 to 5, and the gynœcium is 1-carpellate. When mature the fruit is a large fleshy drupe with a flattened fibrous stone. Species about 30. One of the important species is *Mangifera indica,* the Mango, a large evergreen tree with leaves up to 16 inches long, and flowers in large dense panicles. The fruit is from 2 to 6 inches long and edible. The tree is found from northern India to Burma.

Anacardium comprises about 8 species of tropical-American trees and shrubs with simple, alternate, entire, leathery leaves and numerous small, polygamous flowers in terminal panicles. The calyx and corolla are 5-merous, but the stamens vary from 7 to 10, and the gynœcium is 1-carpellate. The fruit as a whole is a composite structure consisting of a kidney-shaped fruit proper which is borne on a strikingly enlarged pear-shaped receptacle 3 or more inches in length and edible. *Anacardium occidentale,* the Cashew of tropical America, is cultivated in all tropical countries.

Spondias numbers about 12 species of trees of the tropics. The leaves are pinnate and alternate and clustered at the ends of the branches. The flowers are small and numerous, 4-merous or 5-merous, in racemes or panicles. The fruit is a fleshy, edible drupe. Several species are cultivated, including *Spondias purpurea* (Spanish Plum or Red Mombin), *S. Mombin* (Hog Plum or Yellow Mombin).

Schinus is a genus of 15 or more South American resinous trees with alternate odd-pinnate leaves and numerous small unisexual and diœcious flowers in terminal or axillary panicles. Stamens 10. Gynœcium 3-carpellate, with 3 styles and a 1-celled ovary which becomes

FIG. 265.—*Schinus Molle: A,* staminate inflorescence; *B,* bud of staminate flowers; *C,* ventral and dorsal aspects of stamen; *D,* staminate flower with prominent annular disk, and rudimentary gynœcium in center; *E,* gynœcium from *D; F,* fruit from carpellate tree; *G,* leaf.

a small globular drupe. An annular disk is present. *Schinus Molle,* the Pepper Tree, is extensively planted as an avenue tree in the Southwest and as such attains large size. Less common is the Christmas-Berry Tree, *S. terebinthifolius,* which is distinguished superficially by a more rigid habit and by smaller and brighter red fruits. (Fig. 265.)

The 20 or more species of *Pistacia* are trees or shrubs with alternate, persistent or deciduous, oddly or evenly pinnate or trifoliolate leaves and small, apetalous, diœcious flowers in axillary racemes or panicles. The staminate flowers are 5-merous, but the carpellate have a 3-merous or 4-merous calyx and a 3-carpellate gynœcium with a

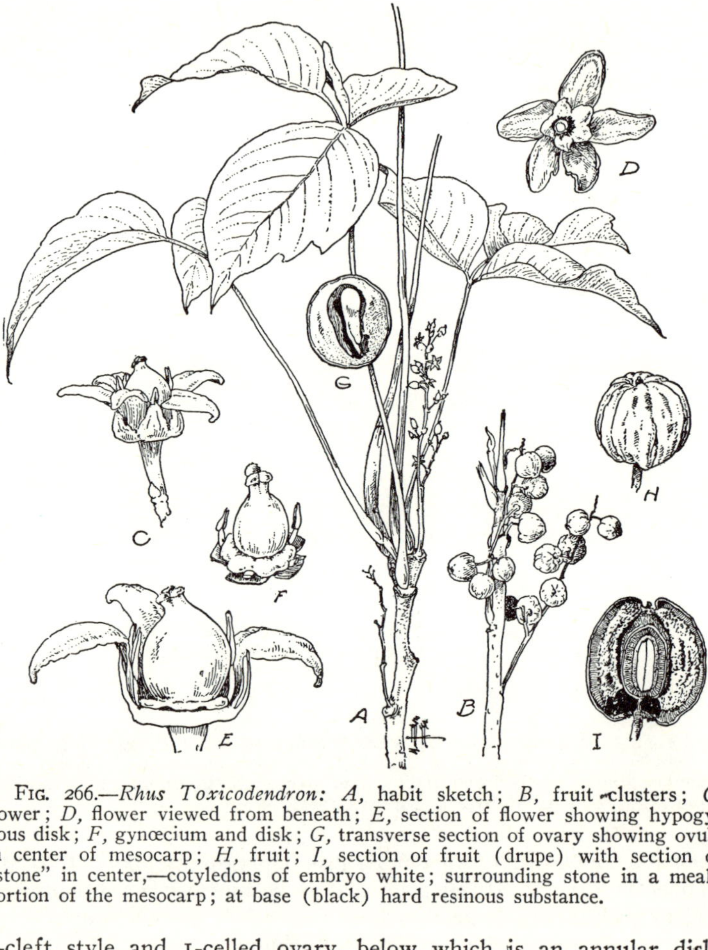

FIG. 266.—*Rhus Toxicodendron: A,* habit sketch; *B,* fruit-clusters; *C,* flower; *D,* flower viewed from beneath; *E,* section of flower showing hypogynous disk; *F,* gynœcium and disk; *G,* transverse section of ovary showing ovule in center of mesocarp; *H,* fruit; *I,* section of fruit (drupe) with section of "stone" in center,—cotyledons of embryo white; surrounding stone in a mealy portion of the mesocarp; at base (black) hard resinous substance.

3-cleft style and 1-celled ovary, below which is an annular disk. The fruit is a dryish drupe. The trees are confined mainly to the Mediterranean region, ranging eastward into Asia, with 1 species in Mexico and 1 in the Canaries. *Pistacia vera* (Pistachio) of the Orient and the Mediterranean region yields the Pistachio nuts of the confectionery.

Cotinus americanus (Smoke Tree) is a small tree of the South Central United States, bearing simple alternate leaves and small flowers in terminal panicles. The flowers are 5-merous except for a 3-carpellate gynœcium with 3 lateral styles. The fruit is a gibbous 1-seeded drupe. The only other species, *C. coccygria,* is Asiatic. The genus is by some writers incorporated in *Rhus.*

Rhus. This is by far the largest genus of the family, comprising 150 or more species of trees and shrubs of wide distribution over the tropical and temperate belts both of the Old and the New World. The species of the several subdivisions of the genus differ widely in habit and foliage, features which are beautifully illustrated in our American representatives. The flowers are polygamous and typically 5-merous—except the gynœcium which is 3-carpellate, with as many nearly distinct styles. A conspicuous, fleshy, lobed disk surrounds the base of the ovary. The fruit is drupaceous, dry, and smooth or hairy. The plants yield tannins, wax, dyes, varnishes, and medicinal compounds, and the fruits of some species contain a pleasant, acrid juice. On the other hand some species are very poisonous. A number are handsome ornamentals. Our American species are by some writers separated into distinct genera. The habit differences of our American species may be illustrated by the following.

Rhus hirta, the Staghorn Sumac, one of the commonest of our Eastern sumacs, is a shrub or tree with large, odd-pinnate leaves, densely velvety-hairy twigs and dense terminal panicles of small, greenish, polygamous flowers. In autumn the panicles are a solid mass of crimson, hairy drupes. *R. glabra,* which ranges far westward, is very similar, except that the twigs and leaves are smooth and glaucous. In both the leaves are showy crimson in autumn. Another species, also with pinnate leaves, is the very poisonous Swamp Sumac, *R. Vernix,* which, perhaps fortunately, is found only in unfrequented tamarack swamps. It is a shrub or small tree with smooth leaves and small grayish drupes in grape-like clusters.

Of very different habit is the common Poison Ivy, *Rhus Toxicodendron* (Fig. 266), which has trifoliolate leaves and produces a cluster of grayish or greenish yellow dryish drupes. The plant is very poisonous to the touch, at least early in the season. When growing in sandy dry soil it is usually a low plant with an upright stem, but under more favorable conditions in rich woods it becomes often a high-climbing vine. Somewhat similar are *R. canadensis* of our Eastern flora and *R. diversiloba* of the Pacific slope. *R. trilobata,* the Squaw-bush, is a straggling, rank-smelling species of the Pacific States. Its stems and branches are long and slender and wiry, with a velvety pubescence.

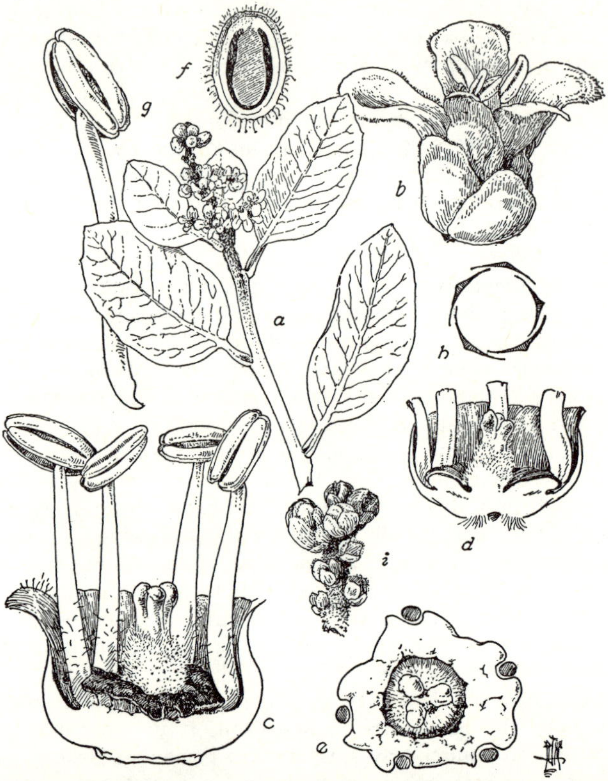

FIG. 267.—*Rhus integrifolia: a,* flowering twig; *b,* flower; *c,* median section of flower,—in center the small 3-carpellate gynœcium surrounded by fleshy annular disk (black); same in sectional view in (*d*); *e,* disk and gynœcium viewed from above, with sections of filaments in notches of disk lobes; *f,* transverse section of ovary, containing one ovule; *g,* stamen (side view); *h,* æstivation of calyx; *i,* a cluster of flower-buds.

More striking in their differences are the simple-leaved sumacs of Southern California. Occupying mainly a narrow coastal strip, they appear strangely isolated from most of their American relatives, and until they are in flower and fruit no one familiar with the commoner Eastern forms would suspect them of being sumacs. Their thick leathery leaves and their habit of growth make them pleasing ornamentals. The genus is widely distributed over tropical and temperate to cold temperate regions of the world. (Fig. 267.)

CYRILLACEÆ. Flowers regular, bisexual, with sepals and petals 4—8 and stamens 4—10 in two whorls. Gynœcium 1- to 4-carpellate, the superior ovary with as many 1—4-seeded cells. Fruit an ovoid spongy berry. Embryo in the center of a fleshy endosperm. Three genera and about 12 species of trees and shrubs with simple, alternate, exstipulate, evergreen leaves and small flowers in racemes. *Cyrilla racemiflora* (Southern Leatherwood) is found in swamps from Virginia to Florida and Texas, and *Cliftonia ligustrina* in similar habitats from Georgia to Florida.

PENTAPHYLACACEÆ. A monotypic (?) family based on *Pentaphylax euryoides,* a tree of the Hongkong region of China. The flowers are regular, bisexual, and 5-merous. Fruit a capsule.

CORYNOCARPACEÆ. Also a monotypic family, based on *Corynocarpus lævigatus,* a tree of New Zealand and Chatham Island.

AQUIFOLIACEÆ (Holly Family). Flowers regular, polygamodiœcious, small, solitary or in clusters, with 3—6 calyx-segments, 4—6 petals which are sometimes united at the base, and stamens as many as or more than the petals. Integuments 1. Disk none. Fruit a small berry. Three or more genera and 300 species of trees and shrubs with simple, alternate, stipulate or exstipulate, often evergreen leaves, inhabiting all the continents, with the center of distribution in Central and South America.

Ilex (Holly), by far the largest genus (280 species), is chiefly American, but has a few outliers in Asia, Africa, and Australia. The plants are trees or shrubs with entire or spiny-toothed leaves and solitary or cymose, bisexual or polygamous, axillary flowers. There are 4—5 calyx-segments, commonly 4—5 petals (sometimes united at the base), and stamens as many as or more than the petals and adnate to their base. The fruit is a small berry with 4—8 bony seeds. About 15 species are found in our flora, among which are *Ilex opaca* (American Holly), a small tree of Eastern and Southern woods; and *I. verticillata,* a shrub or small tree of Eastern swamps. *I. Aquifolium,* a Eurasian species, is the common English Holly. In Eastern swamps there is found a shrub or small tree, *Nemopanthus mucronata,* known as the Mountain Holly. (Fig. 268.)

FIG. 268.—*Ilex Aquifolium: a,* branch, with flower-clusters below; *b,* cluster of fruits; *c,* flower; *d,* median section of a bisexual (?) flower; *e,* a carpellate flower, petals removed; *f,* gynœcium of carpellate flower; *g,* summit of gynœcium (*f*); *h,* bud of carpellate flower; *i, j,* seeds; *k,* staminate flower.

CELASTRACEÆ (*Staff-Tree Family*). Flowers regular, bisexual or unisexual, perigynous or epigynous, mostly 4-merous or 5-merous, the gynœcium 2- to 5-merous and the base of the ovary coalesced with a prominent disk. Ovules 2 (or 1) or more in each cell. Fruit a capsule, samara, berry, or drupe. Seeds often arillate. About 45 genera and 375—400 species of trees and mostly climbing shrubs, occurring in most parts of the world outside of the frigid zones. Leaves simple, alternate or opposite, stipulate or exstipulate.

Flowers small, solitary or in terminal or axillary cymes or racemes.

Our genera: *Celastrus, Euonymus, Pachystima.*

In rich woods of eastern North America, *Celastrus scandens,* the Climbing Bittersweet, is a rather common high-climbing woody vine which twines itself tightly around the trunks of trees. The leaves are alternate and the flowers small, greenish, and diœcious or polygamo-diœcious, in terminal racemes. They are 5-merous except for the 2—4-carpellate gynœcium. The fruit is a 2—4-celled, loculicidally dehiscent, smooth capsule, the clusters of which make a brilliant display in autumn when the capsules burst and release the scarlet-arillate seeds. Twenty-five or 40 other species are found in Asia, Australia, Madagascar, and tropical America. (Fig. 269.)

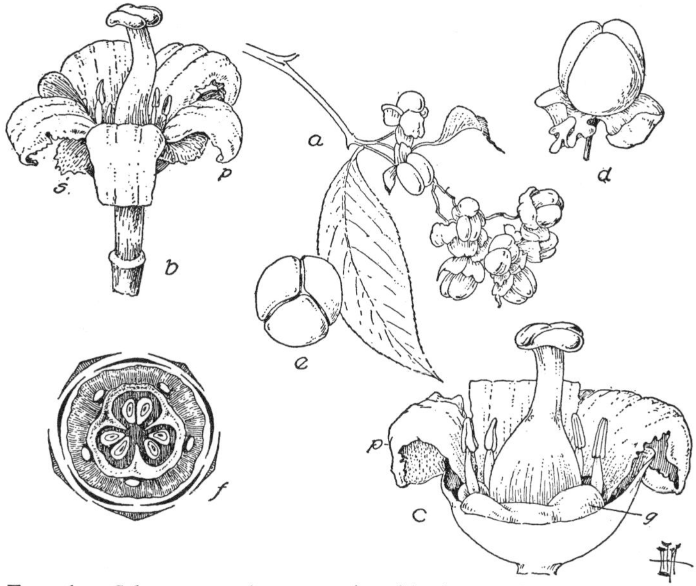

FIG. 269.—*Celastrus scandens: a,* twig with cluster of fruits; *b,* carpellate flower with rudimentary stamens; *s,* sepals; *p,* petals; *c,* section of carpellate flower showing glandular annular disk (*g*); *d,* fruit, with calyx below; *e,* summit of fruit; *f,* floral diagram.

In *Euonymus* the leaves are opposite and the flowers bisexual and 4-merous or 5-merous, the gynœcium 3- or 5-merous, and the capsule smooth or tuberculate. The plants are trees and erect shrubs, widely distributed in the north temperate zone, with probably fewer than a half-dozen in our range. A familiar species is *Euonymus atropurpureus,* the Burning Bush or Wahoo, of our Eastern woodlands. This species also makes a brilliant display in autumn when the fruits open and expose the scarlet seeds. There are about 65 species in all. (Fig. 270.)

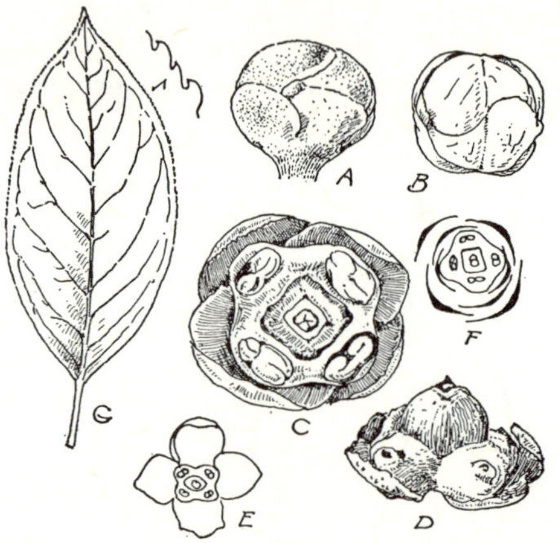

FIG. 270. — *Euonymus atropurpureus: A,* flower-bud; *B,* top of flower-bud; *C,* flower viewed from above; *D,* later stage of flower viewed from the side; *E,* flower with calyx spread out; *F,* floral diagram.

Pachystima is a strictly North American genus, but limited to only two species, *P. Canbyi,* a low shrub of the Appalachians of the Virginias, and *P. Myrsinitis,* also a low shrub, of the mountains of western North America. The plants bear small, leathery, opposite leaves and inconspicuous, brownish, solitary or clustered flowers. *Maytenus Boaria* (Mayten), a small to large and graceful evergreen tree of Chile, is planted as an ornamental. The seeds are white, black, or red, and are enclosed each in an orange-colored aril, which makes the fruit-clusters very showy. In Arabia the leaves of *Catha edulis* are used in the making of a tea (Khat-Tea).

HIPPOCRATEACEÆ. The plants of this family are mostly climbing shrubs with simple, opposite or alternate, stipulate or exstipulate, leaves and small to large flowers in axillary or terminal clusters. The calyx and corolla are 5-merous, the stamens commonly 3, or when 5 then 2 or 3 sterile, all arising from a conspicuous hypogynous disk or from beneath the ovary. Gynœcium 3-carpellate, with the ovary mostly free from the disk. Ovules several to 2 in each cell. Integuments 2. Fruit a berry or samara. Seeds with endosperm, and with or without aril. Cotyledons often large. Three genera and over 200 species, the great majority of the species belonging in *Hippocratea* and *Salacia.*

SALVADORACEÆ. A small family of trees and shrubs of Asia and Africa. Flowers regular, bisexual, or unisexual and diœcious, 4-merous or 5-merous, the corolla sometimes sympetalous. Gynœcium 2-carpellate, with 1—2 basel ovules in each cell of the ovary. Integuments 2. Fruit a berry or drupe, with cartilaginous endocarp, and mostly 1-seeded. Endosperm none. Cotyledons thick. The leaves are opposite, entire, and stipulate; the flowers small, in racemes or spikes. Three genera with less than a dozen species. *Salvadora*.

STACKHOUSIACEÆ. Annual or perennial herbs with narrow, simple, alternate stipulate or exstipulate, leaves, and small, bisexual, hypogynous or perigynous, 5-merous, bracteolate flowers in racemes or spikes. Gynœcium 2- to 5-carpellate, each cell with 1 ovule. Hypanthium discoid in perigynous flowers. Fruit a schizocarp which breaks up into indehiscent 1-seeded mericarps (cocci) Seeds with endosperm. Two genera: *Stackhousia*, containing most of the 20 or more species; and *Macgregoria*, a monotypic genus of Australia, New Zealand, and the Philippines.

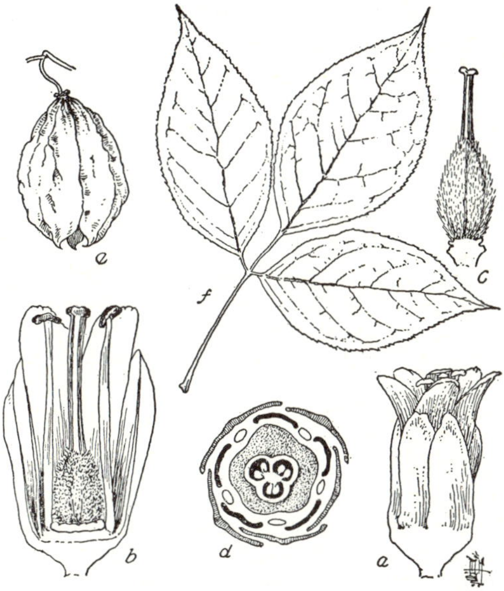

FIG. 271.—*Staphylea trifolia: a,* flower; *b,* section of flower, showing annular hypogynous disk; *c,* gynœcium, with disk beneath ovary; *d,* floral diagram; *e,* the inflated fruit (capsule); *f,* leaf.

STAPHYLEACEÆ (*Bladder-Nut Family*). Flowers regular, bisexual, hypogynous or perigynous, 5-merous, in axillary or terminal clusters. Gynœcium 2- to 3-carpellate with an hypogynous disk at the base. Ovules several to few in each cell, with 2 integuments. Fruit indehiscent, or a dehiscent bladdery capsule, with 1 to several seeds in each cell. Seeds with fleshy endosperm and a straight embryo. About 5 genera and 22 species of trees and shrubs with opposite, odd-pinnate or 3-foliolate, stipulate leaves, distributed over tropical and extra-tropical regions of the eastern and the western hemispheres.

In the woodlands of eastern North America *Staphylea trifolia,* the Bladder-Nut, occurs as a handsome shrub or small tree, mostly in deep ravines. The leaves are 3-foliolate and the rather small creamy-white flowers are borne on jointed bracteolate pedicels in axillary racemes. Below the ovary there is an annular disk. The fruit is a conspicuous bladdery capsule which dehisces at the top. In the Sierras of California is found another species, *S. Bolanderi,* the fruit of which opens by spreading valves at the summit. Another species is found in Mexico, one in Japan, and a few others in Asia and Europe—these being frequently cultivated. (Fig. 271.)

ICACINACEÆ. Flowers regular, mostly hypogynous, bisexual or unisexual, 5-merous, and mostly with a 3-carpellate gynœcium. Disk more or less prominent when present. Style 1. Fruit a 1-seeded drupe or sometimes a samara. Trees and shrubs (often climbing), and a few herbs, with simple, mostly alternate, entire or variously cut or lobed, often leathery leaves, and small flowers in cymose or paniculate clusters. About 38 genera and 150 or more species of the tropics and subtropics of the Old and the New World. Many of the genera apparently monotypic.

Villaresia congonha yields the maté of the Brazilians and the Paraguayans, and in Chile a tea is made from the leaves of *V. mucronata.* The stems of the water-lianas, *Phytocrene* (tropical Africa) and *Chlamydocarya* (tropical Asia) yield, when cut, an abundance of fresh drinking water.

ÆXTOXICACEÆ. A monotypic family of Chile, based on *Aextoxicon punctatum,* a scaly-coated tree with leathery leaves and small unisexual, diœcious, 5-merous flowers, and a fleshy fruit.

ACERACEÆ (*Maple Family*). Flowers regular, bisexual or unisexual, andromonœcious, androdiœcious, or diœcious, often with rudiment of the other sex, hypogynous or perigynous, the sepals, petals (rarely apetalous), and stamens varying from 4 to 10. Gynœcium 2-carpellate (occasionally more), with styles more or less united or nearly free, the branches long-stigmatic on the inner side. Ovary 2-celled, with 2 ovules in each cell. Fruit a schizocarp, the 2 carpels separating as large-winged samaras (key-fruits). The wings are developed as lateral outgrowths of the ovary-wall. Endosperm none. Trees and shrubs with opposite, simple, mostly palmately lobed, or occasionally odd-pinnate leaves, and small flowers in axillary, cymose or racemose, inflorescences. In some species the sap is very sugary. The cotyledons are flat or variously folded. The species are widely distributed over the northern hemisphere. Two genera: *Acer, Dipteronia.*

Acer (Maple). This large genus (about 100 species) is distributed widely over Europe, Asia, Japan, and North America. The plants are trees and shrubs with opposite, simple and commonly palmately

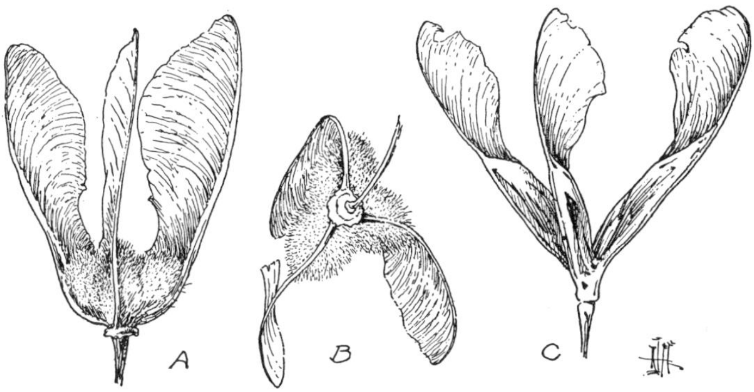

FIG. 272.—Tricarpellate gynœcia in (*A, B*) *Acer macrophyllum* and (*C*) *Acer Negundo.*

lobed, or compound odd-pinnate leaves and small, regular, polygamous or diœcious flowers in axillary or terminal, cymose or racemose inflorescences. Sepals generally 5, but may vary within the same species; petals as many or wanting. Stamens varying from 4 to 12, with filiform filaments. The gynœcium is 2-carpellate (occasionally more), with a 2-lobed, 2-celled, superior ovary terminating in 2 usually elongate slender styles which are stigmatic on the inner side for a considerable distance, or the style sometimes 1 and forked. Ovules 2 in each cell. There is often present a conspicuous lobed, hypogynous disk, inside of or upon which the stamens are often inserted. The fruit is a double samara, the wings of which are developed from the edges of the ovary. The samaras generally separate at maturity. Fifteen to twenty maples occur in the United States and Canada. The following are common.

Acer saccharum (Sugar or Hard Maple). This is a dominant

tree of Eastern deciduous forests and is widely planted as an avenue tree. Its flowers appear with the leaves. It yields a valuable wood and its sap yields the maple-sugar of commerce. For centuries it has been a source of sugar among American Indians. The disk is thin-lobed and one of the carpels is sometimes aborted.

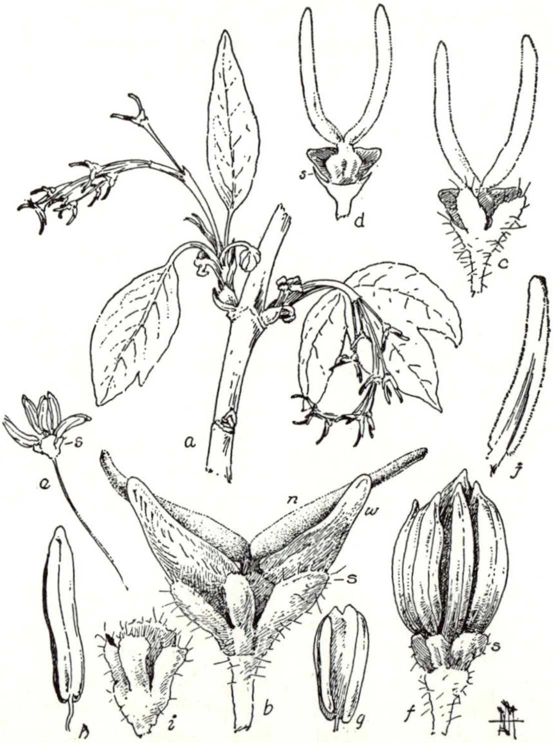

FIG. 273.—*Acer Negundo: a,* twig of carpellate tree with developing fruits; *b,* carpellate flower; *n,* stigma, *w,* developing wing of ovary; *c,* early stage of carpellate flower, with calyx below; *d,* gynœcium; e, staminate flower; *f,* early stage of staminate flower; *g,* dorsal, and *h,* lateral view of stamen; *i, s,* calyx; *j,* stigma.

Acer saccharinum (Soft Maple) is common in rich bottom-lands along streams in Central and Eastern parts of the continent. It sometimes attains considerable size and is often planted as an avenue tree, but as its branches are weak it is not resistant to severe storms.

The leaves are deeply palmately lobed and the bark is somewhat flaky. The sap is sugary but is not much used. The flowers are borne in dense clusters and appear before the leaves. *Acer macrophyllum* is a large tree of the low rain-forest of the Pacific Coast. It sometimes produces a 3-carpellate or 4-carpellate gynœcium (Fig. 272). *Acer rubrum,* the Red Maple, is a beautiful tree of deep forests of eastern North America. It forms a grayish bark, except on old

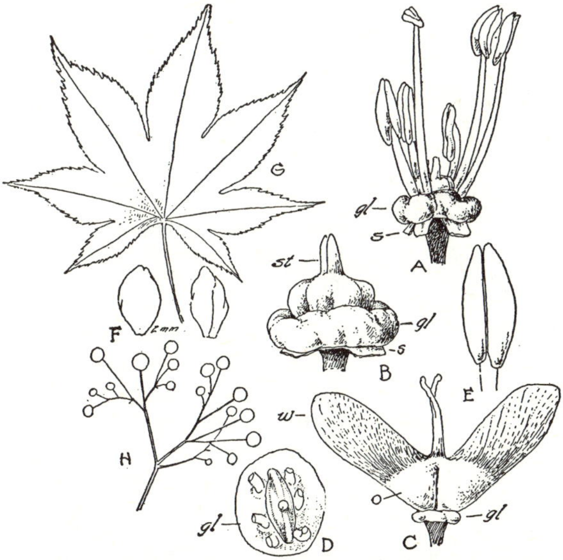

FIG. 274.—*Acer palmatum* var. *atropurpurem: A,* flower with floral envelop removed; *B,* gynœcium with thick fleshy annular disk below; *C,* developing fruit; *D,* gynœcium and disk viewed from above, and bases of filaments; *E,* anther; *F,* petals (2 mm. long); *G,* leaf; *H,* corymb; *gl,* disk; *s,* sepals; *o,* ovary; *w,* wing of ovary.

trunks, and the twigs are reddish. The leaves turn bright red, crimson, or yellow in autumn. The polygamous flowers appear before the leaves in the spring and are as conspicuously colored as are the leaves in autumn. A disk is present both in the staminate and the 2-sexual flowers. Associated with this species, especially in the Great Lakes region, one will generally find the smaller and often shrubby *A. spicatum* or Mountain Maple. This species bears soft leaves and numerous small flowers in terminal erect racemes. Somewhat similar is the Striped Maple, *A. pennsylvanicum,* of nearly the same range, but distinguished by its drooping, instead of erect, racemes.

Acer Negundo, the Boxelder, differs from all other maples in having compound, odd-pinnate, leaves. It is also more widely distributed over the continent than any of the other species, occurring as large trees throughout most of its range. Furthermore its flowers are entirely diœcious; they are borne in long drooping racemes. In the prairie States the tree has been extensively planted in groves for shelter and wind-breaks, a purpose for which it is well suited, as it is a rapid grower and very hardy. With age, however, it tends to become scraggly and has the habit of shedding its leaves continually during the growing season. The wood is of little value, except for fuel. Petals and disk are wanting, and the sepals vary in number. There is an evident perigyny here as well as in *Acer spicatum* and *A. pennsylvanicum.* (Fig. 273.)

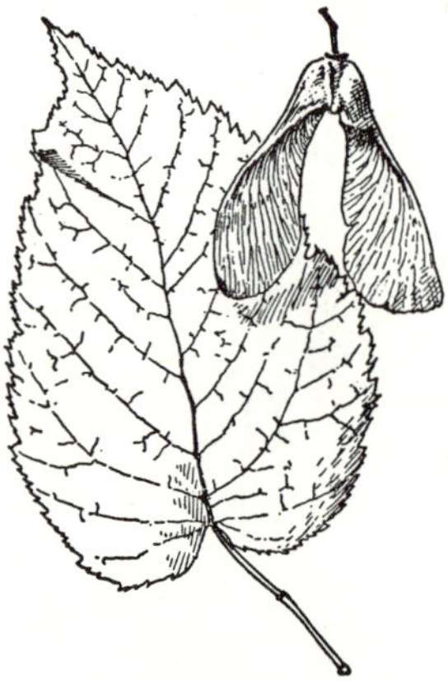

FIG. 275.—Samaras and leaf of *Acer tataricum.*

A number of handsome foreign species of maples are generally planted as avenue trees and ornamentals. The more common ones of these are *Acer Pseudo-Platanus* (Sycamore Maple), *A. platanoides* (Norway Maple)—both Eurasian, and *A. japonicum* and *A. palmatum,* two low-growing species of Japan. In the two last named the leaves have more lobes than in our maples, while in some Asiatic ones the leaves are not at all lobed—*e.g., A. tataricum* and *A. carpinifolium* (leaves similar to those of *Carpinus*). *Acer platanoides* produces an abundance of greenish-yellow flowers in conspicuous, corymbose clusters. (Figs. 274, 275.)

Dipteronia is a monotypic genus of central China, where *D. sinensis* occurs as a tree with odd-pinnate leaves. It differs further from *Acer* in having a fruit similar to that of *Ptelea,* that is a fruit winged all around.

HIPPOCASTANACEÆ (*Horse-Chestnut Family*). Flowers irregular (obliquely), bisexual or unisexual, hypogynous, with a 5-lobed, tubular or campanulate calyx, 4—5 unequal petals, and 5—8 stamens with elongate filaments. Gynœcium 3-carpellate, with 2 ovules in each cell. Style elongate. Fruit a leathery, smooth or prickly capsule which is 3-celled, or 1—2-celled by abortion. Endosperm none, cotyledons large. Integuments 2. A prominent disk is present.

Two genera, *Æsculus* with about 18 species, and *Billia* with 2 species, all trees and shrubs with opposite, digitately 3—9-foliolate

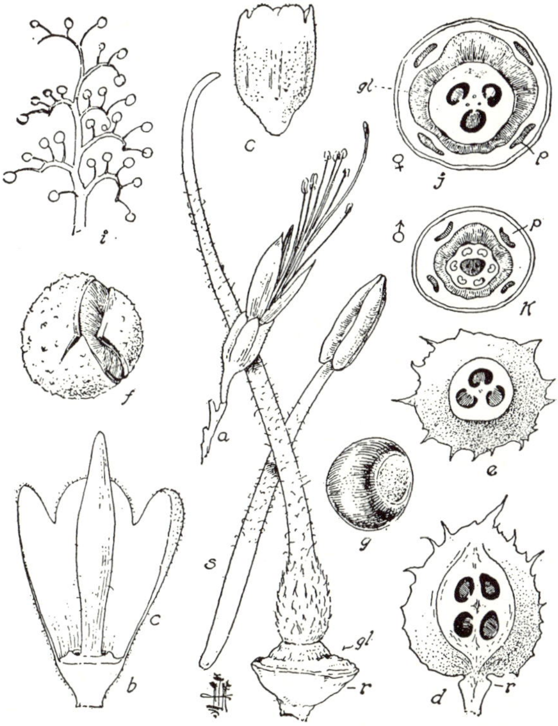

FIG. 276.—*Æsculus glabra: a,* bisexual flower; *b,* flower with rudimentary gynœcium, petals and stamens removed; *c,* calyx; *d,* median section of young fruit; *e,* transverse section of ovary at early stage; *f,* dehiscing fruit; *g,* nut; *i,* inflorescence; *j,* floral diagram of carpellate flower, and *k,* of staminate flower; *p,* petals; *gl,* disk; *s,* stamen.

leaves and more or less showy polygamous flowers in large terminal panicles.

The common avenue tree and ornamental, *Æsculus Hippocastanum* (Horse-Chestnut), is native from Greece to the Caucasus. The native American species are *A. glabra* (Fetid H.), *A. octandra*

(Sweet Buckeye), and *A. Pavia* (Red Buckeye), of our Eastern flora, and *A. californica* of California. In *A. Hippocastanum* and *A. glabra* the fruit is prickly and the petals are 5; in the others the fruit is smooth and the petals are 4. In *A. californica* the calyx is 2-lobed and the gynœcium is 2-carpellate (?). (Fig. 276.)

The flowers of the horse-chestnuts are produced in large cymose-paniculate inflorescences. They are andromonœcious, but with a rudimentary gynœcium in the staminate flowers. In all the disk is rather prominent, with the stamens arising from its inner margin. Only one cell of the ovary produces a seed, the other two being crowded out; the crowding-out process can be clearly seen in sections of young developing ovaries. The single seed developed is large and smooth, with a prominent hilum. The fruit splits into valves (Fig. 276).

SAPINDACEÆ (*Soapberry Family*). Flowers mostly obliquely irregular, bisexual or unisexual, polygamo-diœcious. Sepals and petals 3, 4, or 5, or petals sometimes wanting. Stamens mostly 8, sometimes many, or 5—10. Gynœcium 2—3-carpellate, mostly with 1 ovule in each cell of the ovary. Fruit various. Seeds often provided with large arils. Endosperm none. Embryo curved. A fleshy, annular disk is present, inside of which the stamens are inserted.

Trees, shrubs, and in some cases vines, with watery sap and mostly alternate, pinnate or palmate, stipulate or exstipulate leaves. Axillary tendrils sometimes present and said to be the modified axes of inflorescences.

About 125—130 genera and 1,000 of tropical and extratropical regions. The only genus represented in our flora is *Sapindus,* two or three species of which occur mainly in the Southern States. According to Bailey the following genera, besides *Sapindus,* are represented in cultivation: *Melicocca, Litchi, Euphoria, Dodonæa, Koelreuteria, Xanthoceras, Blighia, Cupania, Cardiospermum, Paullinia.*

In river valleys and on hillsides from Missouri and Kansas to New Mexico and southward, *Sapindus Drummondii* (Soapberry) is found as a small tree with odd-pinnate leaves and small polygamo-diœcious flowers in terminal or axillary panicles. The fruit is a globular berry. Another species, *S. Saponaria,* occurs in Florida. Seven or eight other species are native in the tropics of America and Asia.

Cardiospermum Halicacabum (Balloon Vine) of tropical America frequently occurs as an escape from cultivation in the Eastern States. It is a climbing vine with alternate, pinnate leaves and small, irregular, polygamo-diœcious flowers in axillary tendril-bearing corymbs.

Below the ovary is a one-sided undulate disk. The fruit is a conspicuously inflated 3-lobed capsule bearing several arillate seeds. *Litchi chinensis,* the Litchi of China, is cultivated for its edible fruit. *Koelreuteria paniculata* of eastern Asia and Japan is an ornamental with a 3-lobed bladdery pod.

The *Sapindaceæ* are especially notable for the many structural peculiarities of the flowers and fruits, these organs in many instances simulating in striking manner those of many other unrelated families. The disk in particular shows many diverse shapes, and in the lianas some rather anomalous anatomical structures appear. A large percentage of the genera are monotypic. Many species yield drugs and edible fruits, while some are poisonous. In nearly all cases the disk is outside of the stamen whorl.

SABIACEÆ. Flowers mostly regular, bisexual or polygamo-diœcious, hypogynous, mostly 4-merous or 5-merous (except gynœcium). Stamens opposite the petals, in exceptional cases 10 (some of them often sterile), free or epipetalous. Gynœcium 2—3-carpellate, with the styles more or less united; ovary 2—3-celled with 2 (rarely 1) ovules in each cell. An annular disk is present. Fruit dry and indehiscent or drupaceous, mostly 1-celled and 1-seeded. Seeds without endosperm. Embryo large, with a curved hypocotyl and contorted cotyledons.

Four genera and about 70 species of trees and shrubs with alternate, simple or pinnate, exstipulate leaves, and small flowers in panicles, inhabiting the tropics of both hemispheres. Of special morphological and taxonomic interest are the peculiar modifications of the petals and the stamens, and the configuration of the disk. In some cases a collar-like appendage subtends the anthers, and the outer stamens are transformed into pouch-like appendages of the petals. The family is of slight economic importance. The wood is generally soft. According to Warburg (*Pflanzenfamilien,* III, 5), species of *Mekiosma* are attacked by termites, and the fruit of a species of *Ophiocaryon* is used by the natives of the countries in which it grows as an antidote for snake-bite, apparently for no other reason than that the shape of the embryo suggests a snake.

Genera: *Sabia, Meliosma, Phoxanthus, Ophiocaryon.* The great majority of the species belong to *Sabia* and *Meliosma.*

MELIANTHACEÆ. Flowers regular or irregular, bisexual, rarely unisexual, 5-merous, but with variation in the number of stamens and carpels, and generally hypogynous but with a tendency toward perigyny. Between the stamens and the petals there is a completely or incompletely annular disk. Stamens free or more or less connate. Style 1, elongate and slender or stout. Ovules 1—4 in each cell of the ovary. Fruit a papery or woody capsule, loculicidally dehiscent by valves or opening only at the top. Seeds with endosperm and with or without an aril.

Three genera and about 30 species of shrubs and small trees of tropical and subtropical Africa. Leaves simple, alternate or pinnate, often with large intrapetiolar stipules. Flowers in racemes, small, and becoming inverted (resupinate) by the twisting of the pedicel.

Genera: *Bersama, Melianthus, Greyia.* Species of *Melianthus* and *Greyia* are grown as ornamentals. *Greyia* bears simple exstipulate leaves and perigynous flowers, and the disk has 10 gland-bearing projections. In *Melianthus* the leaves are odd-pinnate and the stamens (4) are didynamous. The fruit of both is a more or less papery inflated capsule.

DIDIEREACEÆ. Flowers regular, unisexual, diœcious, hypogynous. Sepals 2, petals 4, stamens 8—10 (united at base). Between the ovary and the stamens there is an annular disk. Gynœcium 3-carpellate but only 1 carpel fertile, and 1-ovuled. Style 1. Sterile stamens sometimes present in the carpellate flowers. Seeds with endosperm and a folded embryo with fleshy cotyledons. One genus, *Didierea,* of about 6 species of tall, spiny, cactus-like trees of Madagascar. Leaves simple.

BALSAMINACEÆ (*Jewel-Weed Family*). Flowers irregular, hypogynous, bisexual. Sepals 3 or sometimes 5, the 2 lateral small and greenish, the posterior one large, petaloid, saccate, and spurred. Petals 5 or sometimes 3, the 2 lateral ones united. Stamens 5, the short filaments connate above and the anthers connate around the ovary. Gynœcium 5-carpellate, with a short style and an oblong-ovoid ovary containing several ovules in each cell. Fruit an elastically dehiscent capsule, or berry-like. Two genera, *Impatiens* (220 species) and *Hydrocera* (1 species) of succulent herbs with opposite, alternate, or whorled leaves, and showy flowers borne solitary or in umbellate clusters.

Two species of *Impatiens,* namely *I. biflora* (Spotted Touch-me-not) and *I. pallida* (Pale Touch-me-not), are common in low, wet grounds in our Eastern flora. Technically the two species are distinguished mainly by the shape of the spur—which is incurved in the first named and spreading in the last named. The color of the flowers will also distinguish them (pale in *I. pallida*). The fruits of these plants are explosively dehiscent at the slightest touch. The great majority of the species are natives of Asia and Africa, with a very few in North America and Europe, and none in South America and Australia. Several species are cultivated as garden plants. (Fig. 277.)

Hydrocera is monotypic, the only species, *H. triflora,* occurring in swamps in India and Ceylon.

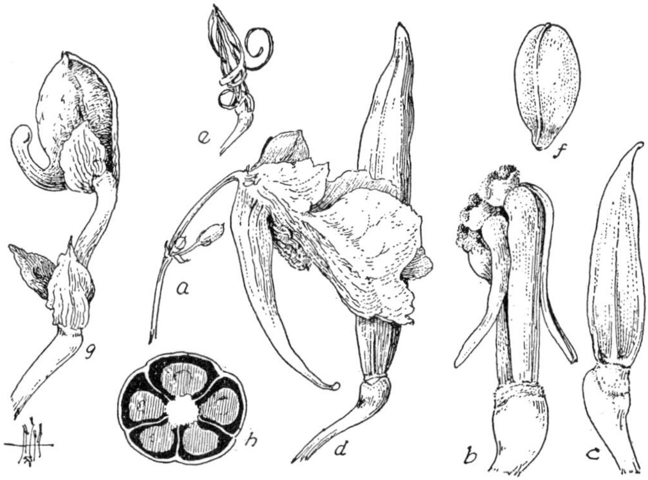

FIG. 277.—*Impatiens biflora: a,* lateral view of flower, with two flower-buds in axils of bracts below; *b,* flower with sepals and petals removed, showing stamens, two of which are spread out below, the stamens being coherent above; *c,* gynœcium; *d,* later stage of gynœcium (nearly mature fruit); *e,* dehisced fruit; *f,* seed; *g,* flower-bud, with subtending bracts; *h,* transverse section of ovary.

ORDER RHAMNALES

Flowers regular, cyclic, bisexual or unisexual, polygamous or diœcious, hypogynous, perigynous, or epigynous, 4-merous or 5-merous, with calyx and corolla, or the corolla sometimes wanting, but when present usually inconspicuous. Stamens 4—5, opposite the petals (obdiplostemonous). Gynœcium 5—2-carpellate. Seeds 2 or 1 in each cell, mostly with endosperm. Integuments 2. Fruit a drupe, berry, or capsule. Shrubs, small trees, or vines.

Families: *Rhamnaceæ,* comprising trees and shrubs, and *Vitaceæ,* comprising vines (in ours).

RHAMNACEÆ (*Buckthorn Family*). Flowers small, 4-merous or 5-merous, perigynous or epigynous, with a more or less lobed annular hypogynous disk. Fruit a drupe or capsule. Erect or climbing shrubs or small trees with alternate or opposite, stipulate, often 3—5-nerved leaves. Some forms thorny. Seeds solitary in each cell, usually with endosperm and a large embryo with flat cotyledons. Fifty genera and about 600 species, widely distributed in both hemispheres. The following genera are represented by native or cultivated species.

Fruit a drupe: *Berchemia, Rhamnus, Zizyphus, Condalia.*

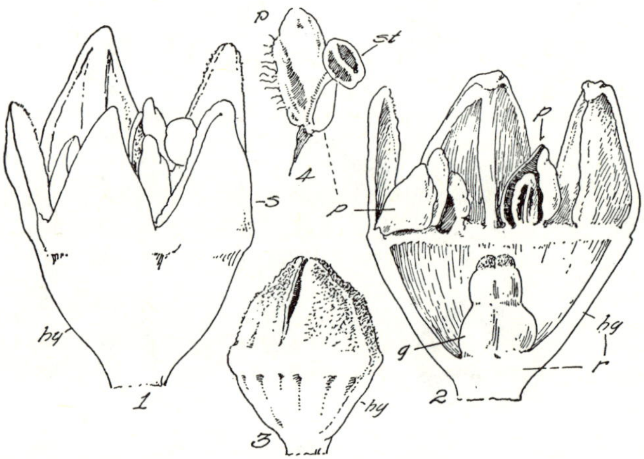

FIG. 278.—*Rhamnus* sp.: *1,* flower; *2,* median section of flower; *3,* a flower-bud; *4,* stamen (*st*) and petal (*p*); *hy,* hypanthium; *p,* petals; *s,* sepals; *r,* receptacle; *g,* rudimentary gynœcium.

Fruit a capsule: *Ceanothus, Paliurus, Pomaderis, Hovenia, Colletia, Adolphia.*

Berchemia scandens (Supple-Jack) occurs as a high-climbing shrub of low woods from Virginia to Florida and west to Missouri and Texas. It bears somewhat elliptic-oblong leaves, and small, typically pentamerous, perigynous flowers in terminal or axillary panicles. The disk forms an inner lining of the hypanthium. The drupaceous fruit contains a 2-celled stone. Eight or nine other species occur in Asia and tropical Africa.

Rhamnus is a large genus (about 90 species) of shrubs and small trees, mainly of north-temperate regions, with a scanty representation in the tropics and in south-temperate belts. About 15 species are found in the United States and Canada together. Some of these, as well as a few exotic species, are grown as ornamentals. The flowers are bisexual or unisexual, polygamous or diœcious, often with rudiments of the other sex present, and more or less perigynous, and 4-merous or 5-merous, except the gynœcium which is usually 3—4-carpellate. The disk is prominent, and there is a single 3—4-cleft style. The fruit is a drupaceous berry containing 2—4 small stones.

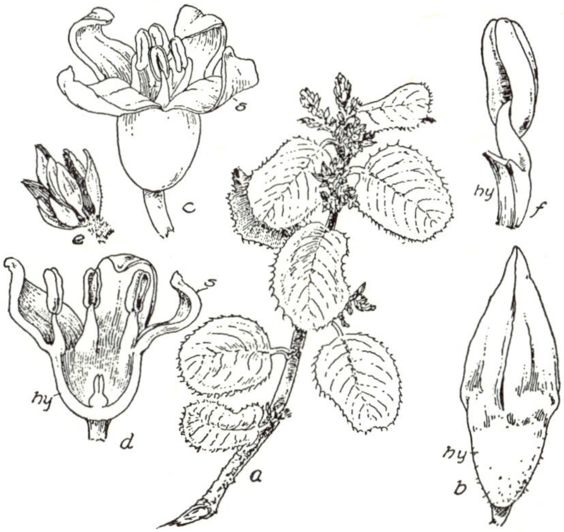

FIG. 279.—*Rhamnus crocea: a,* twig with flower-clusters; *b,* flower-bud; *c,* flower; *d,* median section of flower, with rudimentary gynœcium; *e,* cluster of flower-buds; *f,* stamen; *hy,* hypanthium; *s,* sepals.

The Eurasian *Rhamnus cathartica* (Buckthorn) is commonly used as a hedge-plant, but when growing freely will reach a height of 20 feet or more. It is generally thorny. The drupe yields a dye, and both the drupe and the bark possess powerful medicinal properties. *R. Frangula* (Alder Buckthorn) of Eurasia and Africa is also commonly planted. Its fruit is red and the bark medicinal. (Fig. 278.) *R. Purshiana* of the Pacific Coast is a handsome shrub which in many places has been practically exterminated for its medicinal bark (Cascara Sagrada); somewhat similar but smaller is *R. californica,* the Coffee-Berry or Yerba del Oso, of California. Another beautiful Californian species is the common *R. crocea* or Red-Berry, which bears shiny, coriaceous, and spiny-edged leaves; it is often strikingly similar in habit to *Prunus ilicifolia* of the same region (Fig. 279).

In the Colorado Desert there are found three species of *Condalia* which grow as spreading shrubs with spiny twigs and alternate leaves. The flowers are small, perigynous, and in axillary umbels. The

FIG. 280.—*Ceanothus macrocarpus:* *a,* flowering branch; *b,* flower, with prominent annular disk (black); *c,* gynœcium and disk (*g*) with bases of filaments on margin; *d,* median section of flower; *e,* stamens; *f,* capsule; *g,* disk,—style in center; *h,* flower-bud; *i,* cushion-like stipules; *p,* petals; *s,* sepals; *hy,* hypanthium.

gynœcium is 1-carpellate (?) (or at least possesses only 1 placenta). Petals are wanting and the calyx is persistent in *C. spathulata,* while in the other two species, *C. lycioides* and *C. Parryi,* the calyx is deciduous and the gynœcium 2-carpellate. Several other species occur in Mexico, Chile, Argentina, and Brazil.

Ceanothus, another large genus (about 55 species), comprises shrubs and small trees, in some cases thorny, ranging from Canada to Mexico, more than half of the species occurring in the southwestern United States, especially in California. Many of these species are handsome plants and are much planted as ornamentals. One of the handsomest is *C. arboreus* which is found on the islands off the coast of California. *C. americanus* (New Jersey Tea or Red-Root) and *C. ovatus* (Smaller Red-Root) are strictly Eastern. In the Rocky Mountains and northwestward, *C. sanguineus* (Buckbrush) is a common low, spreading shrub, growing in extensive patches. It is said to be a fattening food for sheep because of its oily seeds. The numerous species of California form a characteristic and conspicuous feature of the chaparral and montane vegetations.

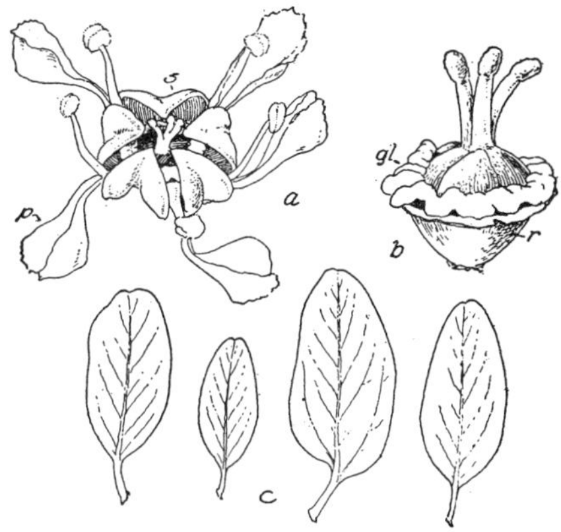

FIG. 281.—*Ceanothus integerrimus* (?) : *a*, flower; *b*, gynœcium; *c*, leaves; *gl*, disk; *p*, petals; *s*, sepals; *r*, receptacle.

The flowers of *Ceanothus* are typically 5-merous except for the 3-carpellate gynœcium. The style is 3-cleft. The prominent disk is often dark-colored and conspicuous. The upper part of each petal forms a hood which envelops a stamen. All the petals are long-clawed below. The carpels separate in the fruit, and each dehisces elastically to discharge the seeds. The genus is a very technical one and is divided by Weberbauer (*Pflanzenfamilien,* III, 5) into sub-genera according to the characters of the leaves, the inflorescences, and the fruits. In one subgenus (*Euceanothus*) the leaves are alternate, the branches flexible or rigid, the stipules thin and fugacious, the flowers in umbellate-racemose clusters, and the capsules ridged or crested on the lobes but never horned. Here belong all the above-named species and many others. In the second subgenus (*Cerastes*) the leaves are opposite (rarely alternate) and usually thick and small and often provided with sharp teeth, and the bases of the stipules persist as corky excrescences, while the flowers are borne in sessile umbels, and the capsules are both horned and crested. Here belong *C. cuneatus, C. rigidus,* and several others. (Figs. 280, 281, 282.)

In far southern California and Lower California *Adolphia californica* occurs as a low, divaricately branched, thorny shrub with opposite, stipulate, early-deciduous leaves, and a few small perigynous flowers in axillary clusters. The petals are strongly hooded as in *Ceanothus,* and the gynœcium is 3-carpellate. The disk forms the inner lining of the hypanthium. On dehiscence the valves of the fruit roll up spirally. The only other species occurs in Mexico.

The other genera above named are represented by cultivated plants. Of these *Colletia* is perhaps of most interest because of its peculiar and often grotesquely thorny habit. *Paliurus Spina-Christi* is the Jerusalem or Christmas Thorn; it is of Eurasian origin. *Zizyphus Jujuba,* the Common Jujube, is a spiny or unarmed tree which produces a large, oblong, dark-red fruit. In *Paliurus* the base of the style develops into a large, broad rim as of a hat.

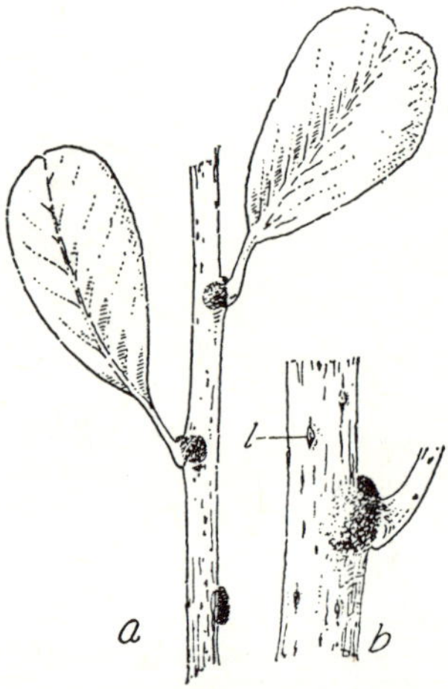

FIG. 282.—*Ceanothus macrocarpus: a,* portion of twig showing cushion-like stipules; *b,* part of same enlarged; *l,* lenticels.

VITACEÆ (*Grape Family*). Flowers small, regular, bisexual or unisexual, polygamo-monœcious or polygamo-diœcious, 4-merous or 5-merous but the gynœcium 2- to 8-carpellate. Ovary with as many cells as carpels and with 1—2 ovules in each cell. Disk when present prominent and variously lobed, with the stamens sometimes arising between its lobes. Fruit a berry. Seeds with a cartilaginous endosperm and small embryo. Ten genera and about 500 species of mostly tropical woody vines or erect shrubs with swollen nodes, watery sap, and simple or compound, alternate, stipulate leaves and small greenish flowers in cymes, panicles, or racemes.

Two subfamilies are recognized, one the *Vitoideæ,* in which the stamens are free and the petals cohere and fall together as a hood or remain separate, and the gynœcium is 2-carpellate with 2 ovules in each cell; the other, the *Leeoideæ,* in which the stamens are united into a tube and are epipetalous, and the gynœcium is 3—6-carpellate, with only 1 ovule in each cell.

Our genera all belong to the *Vitoideæ;* they are: *Vitis, Ampelopsis, Cissus, Parthenocissus.*

Vitis (Grape). In the grape the petals fall away as a hood, and there is present a prominent hypogynous disk. The leaves are more or less deeply lobed and coarsely serrate and the flowers mostly diœcious or polygamo-diœcious. The style is very short. Species about 50, natives of the warmer parts of the world. Our species number two dozen or more. *Vitis vinifera,* native from southern Europe to India, is the grape of history. A few native American species are also cultivated and have yielded a number of pomological varieties. Among these native species *Vitis labrusca* and *V. vulpina* are common in the Atlantic region. *V. californica* and *V. Girdiana* are restricted to California. Economically *Vitis* is the important genus of the family; the plants can be propagated from stem cuttings. The tendrils of grapes are regarded as modified stems (branches); sometimes the tendrils bear leaves and flowers.

In *Ampelopsis* and *Cissus* the petals do not form a hood. Species of the latter are always vines, but in the former there are also a number of shrubby forms. The flowers of *Ampelopsis* are polygamo-monœcious or polygamo-diœcious and 5-merous, while in *Cissus* they are bisexual and 4-merous. In both the disk is cup-shaped. About 225 species of *Cissus* have been described, while *Ampelopsis* is limited to 12 or 15. Both are represented in eastern United States.

Parthenocissus differs from the other three genera in having (in ours) digitately compound leaves and the disk much reduced or wanting. The genus, which comprises about 10 species, is mainly Asiatic, but with *P. quinquefolia* (Virginia Creeper) and *P. vitacea* rather common over the eastern half of this continent; both are climbing or trailing, woody, tendril-bearing vines, the tendrils usually terminating in adhesive disks. The flowers are polygamo-diœcious and the corolla expands before falling. The berries are blue and 1—4-seeded, and are borne in large clusters.

ORDER MALVALES

Flowers usually regular, bisexual, hypogynous, cyclic, 5-merous. Stamens mostly numerous, free or united in a column. Gynœcium 2— to many-carpellate, syncarpous, with a 2—to many-celled ovary. Ovules many—1 in each cell. Integuments 2. Seeds with or without endosperm. Fruit a nut, capsule, or schizocarp. Herbs, trees, or shrubs.

Families: *Elæocarpaceæ, Chlænaceæ, Gonystilaceæ, Tiliaceæ, Malvaceæ, Bombacaceæ, Sterculiaceæ, Scytopetalaceæ.*

KEY TO THE PRINCIPAL FAMILIES

Gynœcium and andrœcium supported on an androgynophore. Anthers 2-celled. *Sterculiaceæ.*
Gynœcium and andrœcium not as above.
  Stamens mostly numerous, free or in separate bundles. *Tiliaceæ.*
  Stamens with filaments united into a column.
    Gynœcium five- to several-carpellate. *Malvaceæ.*
    Gynœcium two- to five-carpellate. Tropical trees. *Bombacaceæ.*

ELÆOCARPACEÆ. Flowers bisexual, with 4-merous or 5-merous calyx and corolla, the calyx gamosepalous or the sepals distinct. Stamens free; anthers commonly opening by apical pores. Gynœcium with a single style and a 2—many-celled ovary; ovules many in each cell. Stamens and gynœcium elevated on an androgynophore. Fruit a loculicidal or septicidal capsule. Epicalyx none.

Seven genera and about 120 species of trees and shrubs with alternate or opposite, deciduous or evergreen, stipulate leaves, and flowers in racemose or cymose inflorescences. Mostly of the tropics of both worlds. The family appears to be most closely related to the *Tiliaceæ*. Largest genera: *Elæocarpus, Sloanea.*

CHLAÆNACEÆ. This is a small family of trees and shrubs, or in some cases lianas, restricted to Madagascar. The 22 or more species are distributed among 7 genera. The flowers are bisexual, regular, with sepals constantly 5 but the other floral organs in varying numbers. Gynœcium 3-carpellate. Fruit a capsule. The largest genera appear to be *Schizochlæna, Leptochlæna,* and *Sarcochlæna,* of about a half-dozen species each.

GONYSTILACEÆ. Restricted to the single genus *Gonystilus,* which comprises about 7 species of the Indo-Malayan region. The plants are trees with alternate, leathery, evergreen leaves, and regular, bisexual flowers in cymose inflorescences. The stamens are numerous, and the gynœcium 3—5-carpellate.

TILIACEÆ (*Linden Family*). Flowers regular, bisexual, rarely unisexual, the sepals and petals of varying number but commonly 5. Stamens mostly numerous and borne in separate bundles; anthers 2-celled, opening by a slit or pore. Staminodes sometimes present. Gynœcium of 2—several carpels, with a single style and a 2—several-

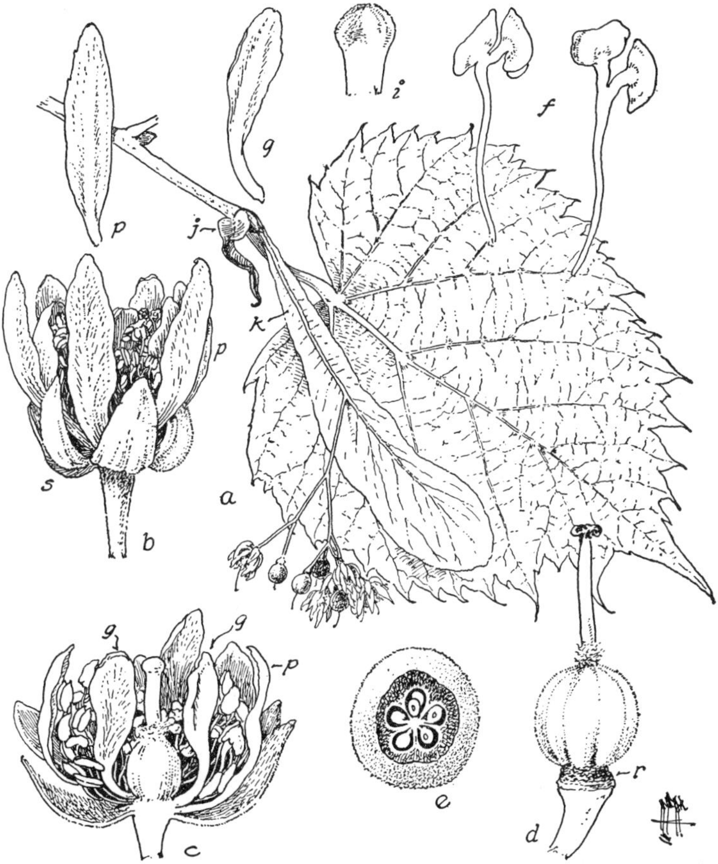

FIG. 283.—*Tilia americana: a,* twig with leaf and inflorescence; *b,* flower; *c,* section of flower; *d,* gynœcium; *e,* transverse section of ovary; *f,* stamens; *g,* above and in *c,* staminodes; *i,* stigma; *j,* bract; *k,* floriferous bract (*wing*); *p,* petals; *s,* sepals; *r,* receptacle.

celled ovary, or the ovary 1-celled by abortion. Ovules 1—several in each cell. Seeds with or without endosperm. Embryo with foliaceous cotyledons. Fruit drupaceous or baccate.

Trees or shrubs and some herbs. Leaves simple, mostly alternate, sometimes lobed, with caducous stipules. Flowers small to large, in terminal or axillary cymes or panicles. Between 35—41 genera and 350—370 species, widely distributed over tropical and temperate regions. Mucilage cells are scattered through the cortex and the pith. The bark of the woody forms is fibrous. Several genera are diœcious (*Vasicœa, Heliocarpus*). An androgynophore is present in *Grewina.* Fruit characters are very diverse, and hence useful in classification.

The following genera are represented by native or cultivated species: *Tilia, Corchorus, Sparmannia.*

Several species of *Tilia* (Linden, Basswood) are native in eastern parts of the continent. These are handsome trees with broad, obliquely cordate leaves and small, yellowish, fragrant flowers. The parts of the flowers are 5-merous, the condition being also characteristic of the andrœcium, in which the numerous stamens are arranged in five clusters. The 5-carpellate gynœcium has a 5-celled ovary and one style. Fruit a globular, dry, 1—2-seeded drupe. The flower-clusters are adnate for a considerable part of their length to a conspicuous membranous bract which serves efficiently to disseminate the fruit. The branches of the basswoods are brittle and their bark gritty, due to the presence in the cortical cells of crystals of calcium oxalate. Wood spongy and tough when green but makes serviceable lumber. The tough inner bark is used for making mats, cords, etc. The trees are often planted for their shade and ornamental qualities. The flowers yield abundant nectar.

Basswoods frequently grow to large size, with trunks several feet in diameter. Several species have been described for the United States, some of these being distinguishable, however, only on rather technical characters. Among our more distinctive species are *Tilia americana, T. heterophylla, T. Michauxii,* and *T. pubescens.* The European linden, *Tilia europœa* (*T. platyphyllos*) is a common avenue tree. There are probably less than two dozen species in the genus. The young seedlings of basswoods are easily recognized by their large, leaf-like, deeply five-lobed cotyledons. (Fig. 283.)

Corchorus comprises about 35 species of herbs or subshrubs of the tropics. *C. capsularis* is the jute of India. The small yellow flowers are mostly 5-merous, with 10—many stamens and a 2—5-carpellate gynœcium. The fruit is an elongate or subglobose loculicidal capsule. *Sparmannia africana,* one of five African species of shrubs and trees, is cultivated.

MALVACEÆ (*Mallow Family*). In the mallows the flowers are bisexual or in rare cases unisexual (polygamous or diœcious) and generally large and showy. The sepals vary from 3 to 5 and are often gamosepalous and frequently subtended by an epicalyx. Petals 5, convolute or contorted in the bud. Stamens numerous, the filaments united into a conspicuous column which encloses the gynœcium; anthers 1-celled, producing distinctive bristly pollen-grains. Gynœcium 5—several-carpellate, mostly with as many styles, which are united below but free above. Each cell of the ovary contains 1—several reniform seeds with copious or scanty endosperm. Fruit a loculicidal capsule, or sometimes baccate, or a schizocarp. Embryo curved, with the cotyledons variously folded.

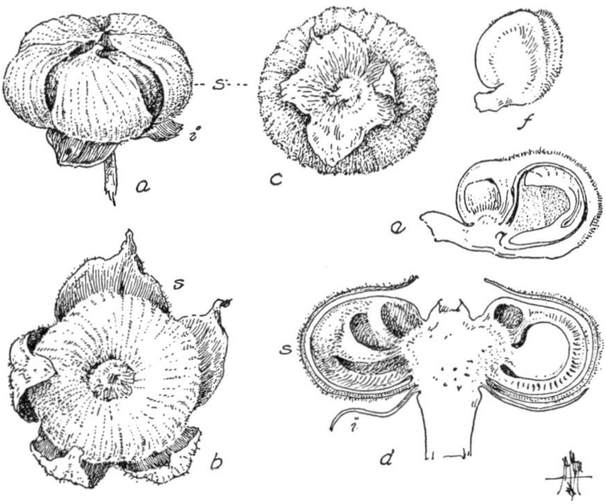

FIG. 284.—*Althæa rosea: a,* fruit invested by calyx (*s*), below which is the involucel (*i*); *b,* fruit viewed from above and (*c*) from beneath (showing involucel); *d,* median section of fruit; *e,* a mericarp (carpel) partly in section, with embryo at right; *f,* seed.

The mallows are herbs, shrubs, or trees (in tropics), with alternate, stipulate, mostly palmately veined and lobed leaves clothed with a stellate pubescence and containing mucilage sacs in the tissues. The flowers are solitary or disposed in inflorescences of various sorts. The family is a rather large one, in the neighborhood of 1,000 species in 45 genera, ranging from the tropics (especially American) to temperate regions. The most important members of the family are the cotton-plants, of which there are several species, besides a considerable number of highly prized ornamentals and some forms that are sought for their medicinal properties and mucilage. The large capsular fruits of many of the mallows are striking objects, and useful in identification and classification. The following genera are represented by native or cultivated species.

1. Fruit a schizocarp—the carpels separating from the central axis—and the stamineal column bearing anthers only at the top.

Althæa, Lavatera, Malva, Callirhoë, Sidalcea, Sida, Malvastrum, Sphæralcea, Abutilon, Napæa, Malvaviscus, Pavonia.

2. Fruit a loculicidal capsule. *Gossypium, Lagunaria, Hibiscus.*

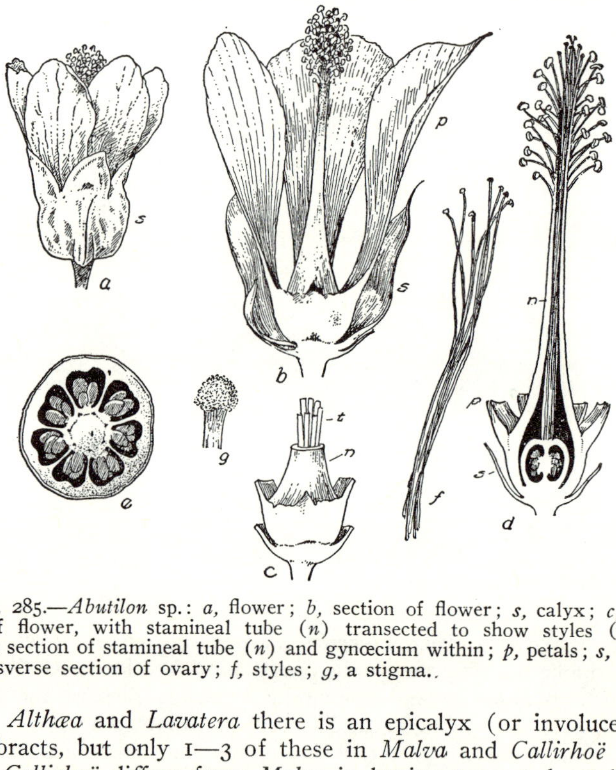

FIG. 285.—*Abutilon* sp.: *a,* flower; *b,* section of flower; *s,* calyx; *c,* basal part of flower, with stamineal tube (*n*) transected to show styles (*t*); *d,* median section of stamineal tube (*n*) and gynœcium within; *p,* petals; *s,* calyx; *e,* transverse section of ovary; *f,* styles; *g,* a stigma.

In *Althæa* and *Lavatera* there is an epicalyx (or involucel) of 3—9 bracts, but only 1—3 of these in *Malva* and *Callirhoë* (Fig. 284). *Callirhoë* differs from *Malva* in having truncated petals and beaked carpels, while *Lavatera* differs technically from *Althæa* in having the axis of the flower prolonged beyond the carpels. The stamineal column of these four genera consists of one series of stamens, but in *Sidalcea* the stamineal column bears two whorls of stamens—an outer whorl of five separate clusters of stamens and an inner continuous whorl, with the filaments free above. Capitate stigmas occur in *Malvastrum, Sida,* and *Napæa,* and besides, there is in *Malvastrum* an involucel of 1—3 bracts, which is wanting, however, in *Sida* and *Napæa;* they are all herbs. *Sphæralcea* and *Abutilon* produce two or more seeds in each cell of the ovary, while in all the other genera of the family there is but one seed in each cell. *Abutilon* may be distinguished from *Sphæralcea* by the absence of an involucel below its calyx. (Fig. 285.)

FIG. 286.—*A, Hibiscus Moscheutus: a,* median section of stamineal tube and gynœcium within; *b,* fruit after dehiscence; *c,* transverse section of fruit, showing loculicidal dehiscence; *B, d,*—section of stamineal tube of *Malva rotundifolia,* exposing gynœcium; *e,* top of fruit; *o,* mericarp or carpel of fruit.

Malva is an Old World genus of about 30 species, several of which have become established in this country and occur abundantly in some localities as weeds. The plants are herbs with palmately lobed, divided, or dissected leaves. *Malva rotundifolia* and *M. parviflora* are common in waste grounds and along roadsides (Fig. 286). *Althæa,* with about 15 species, is also an Old World genus. *A. officinalis* (Marsh-Mallow) occurs in salt-marshes of the coasts of

Massachusetts and Connecticut, and in saline places inland. *Althæa rosea* is the common hollyhock of the gardens. *Lavatera,* which comprises the tree mallows, ranges from the Mediterranean region to the Canary Islands, Asia, Australia, the Californias and the islands off their coasts. There are about 25 species, several of which are horticulturals. *Callirhoë* is exclusively American, its half-dozen or more

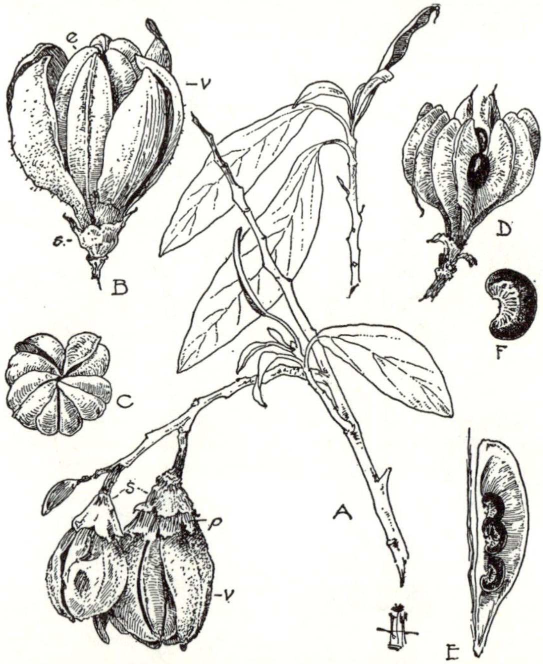

FIG. 287.—*Lagunaria Pattersonii: A,* branch with fruits, and leafy twig (above); *B,* dehisced fruit; *v,* valves (pericarp) of fruit; *c,* endocarps (inner part of ovaries of the several carpels); *C,* endocarps viewed from above; *D,* endocarps split open; *E,* endocarp in lateral aspect; *F,* seed; *p,* remnants of petals; *s,* calyx.

species ranging from Minnesota to North Carolina and Mexico. Of these, *C. triangulata* (Poppy Mallow) has perhaps the widest range—from Minnesota to North Carolina and Texas. The leaves of this and other species are strikingly similar in shape to those of some violets. *Sida,* with 70 or more species, is one of the largest genera of the family. It is chiefly American, but is well represented in Australia, and especially in the tropics. The plants are herbs or subshrubs. *Sida hederacea,* the Alkali Mallow, is abundant on the Pacific Coast, especially in Calfornia, where it is a pernicious weed in orchards. *Malvastrum,* another large genus—with 70—80 species—includes herbs and subshrubs of America and South Africa. A few species are cultivated for their showy flowers (*e.g., M. capense, M. Thurberi*). *Sphæralcea* has about 65 species chiefly in North America and South America, with a few in South Africa. Nineteen species occur in California (Jepson). Relatively few are horticultural. *Abutilon* includes about 100 species; these are the "flowering maples" of the gardener, so called because their leaves suggest those of maples. *Napæa* is a monotypic genus of east-central United States, where *N. dioica* occurs as a tall, stout herb with maple-like leaves and small flowers in a terminal panicle. The plant is diœcious, and is apparently the only mallow possessing this character.

Malvaviscus consists of ten or more species of stout herbs and shrubs of tropical America. Several are grown in hot-houses or out of doors in the South. A distinctive feature is the berry-like fruit. *Pavonia,* with about 100 species, is tropical. The plants are herbs or shrubs with lobed leaves and stellate pubescence. *Pavonia hastata* is grown in California.

About 29 species of *Gossypium* (Cotton) have been described, some of which are native in India, Australia, the Sandwich Islands, Africa, and Brazil. The cultivated ones are probably all hybrids. The plants are stout, annual or perennial herbs with large, palmately lobed leaves and large, showy flowers. An involucel of several large distinct or united bracts subtends the flower. Calyx and corolla are 5-merous, and the 3—5-carpellate gynœcium produces 2 or more ovules in each cell of the ovary. The fruit or "boll" is a loculicidal capsule. The globular seeds are clothed with a dense wool or fleece of varying length, besides a fine fuzz, although this is lacking in some species. Among the several cultivated cottons are *Gossypium barbadense* (Sea Island Cotton), *G. hirsutum* (Upland Cotton), *G. mexicanum* (Mexican Cotton), and *G. arboreum* (Tree Cotton). The fiber of the cotton seed is known as staple, and is sold in the markets as "long-staple" or "short-staple" according to its length. There may be both short-staple and long-staple varieties of the same species. The plants are also grown as ornamentals.

Lagunaria Pattersonii, an Australian tree, is planted as an ornamental in California. It bears 5-merous flowers, subtended by an involucel of 3 bracts. The 5-carpellate gynœcium develops into a loculicidal capsule which is covered with a copious indumentum and is hairy within. The pericarp of the capsule splits away as five valves, leaving the five endocarps free, and these split septicidally to expose the dark-brown or nearly black, shiny, kidney-shaped seeds (Fig. 287).

Hibiscus (Rose Mallow), in the broadest sense, comprises about 200 species of herbs, shrubs, and trees of tropical and cool climates. Many are prized ornamentals. The leaves are palmately lobed, and the flowers are large and showy and 5-merous. Each flower is subtended by an involucel of several bracts. The style-branches are slender and spreading. The fruit is usually a large, elongate, loculicidal capsule, the cells of which are sometimes septate by cross-partitions. *H. esculentus* (Okra) and *H. Sabdariffa* (Roselle) are cultivated for their edible fruits. *H. Rosa-sinensis* (Rose-of-China) and *H. syriacus* (Rose-of-Sharon) are common horticulturals. Among the several native Eastern species is *H. Moscheutos* (Swamp Rose-Mallow) which is found in brackish marshes along the Atlantic coast and occurs in swamps and on lake-shores inland. *Modiola caroliniana* occurs in the East and in California, while the Colorado Desert supports the tall, woolly *Horsfordia Newberryi*. (Fig. 288.)

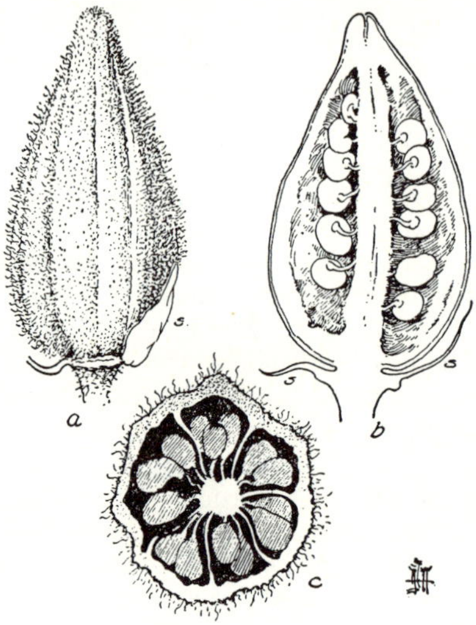

FIG. 288.—Capsule of *Hibiscus* (*a*) and in median section (*b*); *c*, transverse section; *s*, calyx.

BOMBACACEÆ (*Bombax Family*). The bombaxes are very similar to the mallows. The flowers are bisexual, regular, and 5-merous. Stamens often many, distinct or monadelphous, more or less in a column, and with 1—2-celled anthers which are more or less adnate and often curiously curved. Pollen smooth. Staminodes are present in some cases. The gynœcium is 2—5-carpellate, with a single style ending in a stigma which is divided into as many rays or lobes as there are carpels. Ovary 2—5-celled, with 2 or more ovules in each cell. Fruit capsular or baccate, commonly large.

Twenty genera and about 150 species of tropical trees with alternate, simple or digitately compound leaves clothed with a stellate or scurfy indumentum. Fruits in some cases covered with short, thick, spinous processes, or variously winged. The fruits are often large and striking objects. Seeds in some members arillate. Some of the trees grow to enormous size and often develop barrel-shaped trunks. Very distinctive also are the curious stamens. In some species the seeds are clothed with long, but brittle, hairs.

Some of the important genera are: *Adansonia, Bombax, Ceiba, Chorisia, Cavanillesia, Durio.*

The Adansonias are natives of Africa, Madagascar, and Australia. *Adansonia digitata,* the baobab of Africa, which is cultivated in India and South America, possesses the most massive trunk of any known tree. Its leaves are digitate, and the fruit is a large, oblong berry with a hard shell, which is borne suspended from the branches by a long, slender pedicel. The wood is light and soft and yields a bast which is made into paper or is utilized for fiber. The trunks serve the natives for dugouts.

The 50 or more species of *Bombax* are native chiefly to the American tropics. They are shrubs or tall trees with digitate leaves. The gynœcium is 5-carpellate and develops into a large, loculicidal capsule, the inside of which is lined with a dense coating of silky hairs. In the Antilles the seeds of *Bombax insignis* are roasted and used as is cocoa. *Ceiba pentandra,* the Silk-Cotton Tree, is sometimes seen in cultivation. It is a native of the tropics of America, Asia, and Africa. Its wide-spreading horizontal branches are peculiarly buttressed by prominent flanges. The flowers are borne in large, dense clusters and the fruit is a large, oblong capsule. About a half-dozen other species occur in the tropics. Three or more species of *Chorisia* are found in the South American tropics. These are huge trees with digitately compound leaves and large flowers and fruits. The stamens are peculiar in being much elongated and forking at the tip, and more or less curved. Of *Cavanillesia* there are two species in South America. Both are huge trees with barrel-shaped trunks, often fifteen feet in diameter. The 5-carpellate gynœcium ripens into a 1-seeded dry fruit which is broadly winged on the sides. *Durio zibethinus,* the Durian, is a large tree of southern Asia. It yields edible but unpleasantly smelling fruits.

STERCULIACEÆ (*Sterculia Family*). Flowers regular or in some cases irregular, bisexual or unisexual. Calyx gamosepalous, 3—5-lobed; petals 5 or none. Stamens 5 or more, in 2 whorls, the fila-

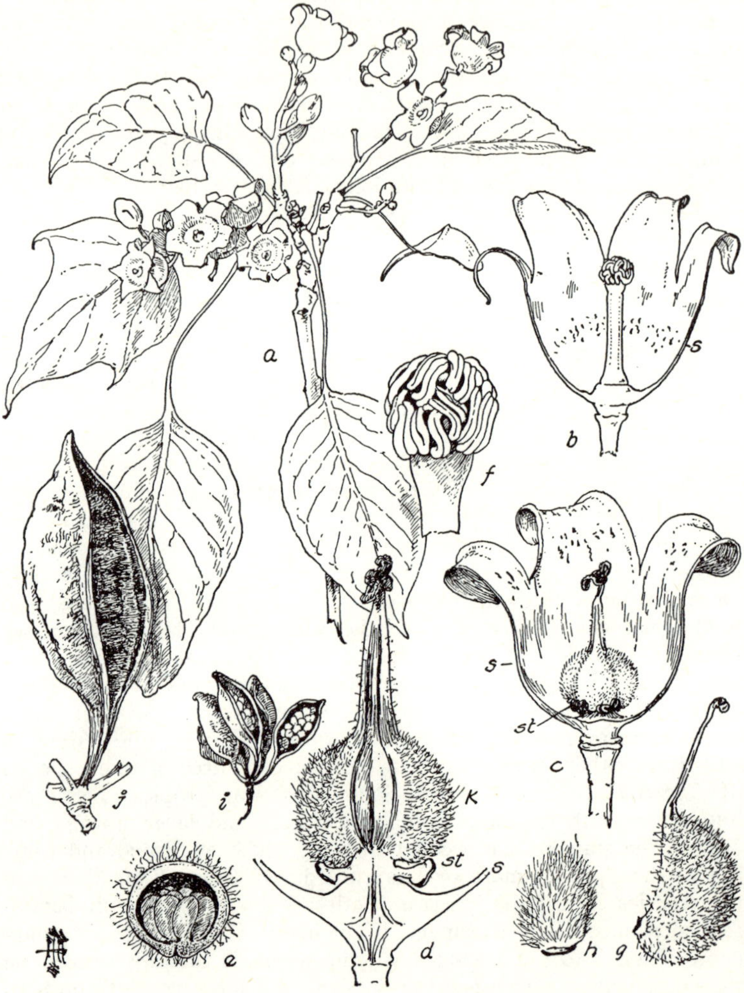

FIG. 289.—*Brachychiton* (*Sterculia*) *populneum: a,* flowering twig, with staminate flowers; *b,* median section of staminate flower; *s,* calyx; *c,* median section of a carpellate flower; *st,* staminodes; *d,* gynœcium, enlarged, and basal part of flower; *k,* carpel; *e,* transverse section of ovary of a carpel; *f,* anthers; *g,* carpel; *h,* seed; *i,* cluster of follicles; *j,* follicle.

ments commonly more or less united into a tube; the outer whorl of stamens often staminodial and curiously modified. Gynœcium mostly 5-carpellate (sometimes 10—12-carpellate), with simple, divided or distinct styles. An androgynophore supports the andrœcium and the gynœcium. Ovules 2—several in each cell of the ovary. Fruit a capsule, follicle, schizocarp, or berry. Seeds with or without endosperm. Embryo straight or curved.

Trees or shrubs, or rarely herbs, with alternate or opposite, simple or digitately compound leaves. Stipules caducous. Inflorescences compound, racemose or cymose. Over 50 genera and 700—750 species of the tropics and subtropics. According to Bailey, species of the following genera are planted: *Brachychiton* (*Sterculia*), *Firmiana, Cola, Theobroma, Dombeya, Mahernia. Fremontia* and *Ayenia* are represented in California.

Species of *Brachychiton* are planted as avenue trees in California. These bear pale foliage, and in their aspect suggest the camphor-tree (*Cinnamomum Camphora,* p. 256) except that the bark is smoother and the bole is prominently enlarged at the base. The flowers are unisexual or polygamous, in axillary racemes or panicles. There are no petals. The follicular fruit is hairy on the inside and yields many pea-sized and often hairy-coated seeds. The 10 or 12 species are Australian. (Fig. 289.)

Cola acuminata, one of a hundred or more species of tropical African trees, yields the cola-nut (seed) which is used in medicine and for beverages. The tree is cultivated in the West Indies. About 20 species of *Theobroma* are found in tropical America. Of these, *Theobroma Cacao* yields the chocolate of commerce.

Fremontia californica, the only species of this genus, is a tall, rigidly branched shrub of California. The flowers lack a corolla, but possess instead a showy, petaloid calyx. The filaments of the 5 stamens form a tube around the style. The fruit is a 4—5-celled loculicidal capsule. The plant is named for Colonel John C. Fremont, who discovered it (Jepson). *Ayenia californicia* is a low shrub of the Colorado Desert region. The unusually long-clawed petals are adnate to the stamineal column and form a hood over the stamens. The 5-carpellate gynœcium is supported on a gynophore and when ripe forms a schizocarp of 1-seeded mericarps. The genus comprises about 15 species ranging from the southwestern United States to the Argentine.

SCYTOPETALACEÆ. A small family of about 10 species of tropical west African trees. The leaves are simple, alternate, and without stipules. Flowers regular, bisexual, and perigynous, with a campanulate calyx, 3 or more petals, and numerous free or united stamens borne on a disk. The fruit is woody and drupaceous. Inflorescence a panicle or raceme.

Genera: *Scytopetalum, Brazzeia, Rhaptopetalum, Oubanguia.*

ORDER PARIETALES

Flowers regular or irregular, hypogynous, perigynous, or epigynous, cyclic or spirocyclic, mostly with 5-merous calyx and corolla. Stamens and carpels often numerous. Gynœcium of distinct carpels or syncarpous, with parietal or axile, rarely basal, placentation. Endosperm present or wanting. Integuments 2 or 1. Herbs, shrubs, or trees. A large order, comprising plants of diverse habit, and having a rather limited representation in our flora.

Hypogynous and perigynous families: *Dilleniaceæ, Actinidiaceæ, Eucryphiaceæ, Medusaginaceæ, Ochnaceæ, Caryocaraceæ, Marcgraviaceæ, Quiinaceæ, Theaceæ* (incl. *Camelliaceæ, Ternstroemiaceæ*), *Strasburgeriaceæ, Guttiferæ* (*Hypericaceæ*), *Dipterocarpaceæ, Elatinaceæ, Frankeniaceæ, Tamaricaceæ, Cistaceæ, Bixaceæ, Cochlospermaceæ, Lacistemaceæ, Canellaceæ* (*Winteranaceæ*), *Violaceæ, Flacourtiaceæ, Stachyuraceæ, Turneraceæ, Malesherbiaceæ, Passifloraceæ, Achariaceæ, Caricaceæ.*

Epigynous families: *Loasaceæ, Datiscaceæ, Begoniaceæ, Ancistrocladaceæ, Flacourtiaceæ.*

Key to the Principal Families

A. Ovary superior.
  a. Flowers regular.
    1. Styles distinct.
      Leaves opposite or whorled.
        Placentas axile. Low marsh or shore plants. *Elatinaceæ.*
        Placentas parietal or axile. Stamens often in bundles. *Guttiferæ.*
      Leaves alternate.
        Herbaceous or woody vines climbing by tendrils. *Passifloraceæ.*
        Trees or shrubs.
          Leaves small, scale-like or heath-like. *Tamaricaceæ.*
          Leaves leathery. *Theaceæ.*
          Leaves large, digitately lobed, in a crown at the summit of a straight trunk. Sap milky. *Caricaceæ.*
      Leaves alternate or opposite.
        Gynœcium mostly of distinct carpels, or carpels sometimes 1 and then styles 1. *Dilleniaceæ.*

Gynœcium syncarpous. Styles sometimes 1. *Flacourtiaceæ.*
2. Styles 1.
  Carpels 2. *Bixaceæ.*
  Carpels 3. *Violaceæ.*
  Carpels 3—5. *Cistaceæ.*
b. Flowers irregular.
  Styles distinct, sometimes 1. *Dilleniaceæ.*
  Styles 1. *Violaceæ.*
B. Ovary inferior.
  Flowers bisexual. Plants clothed with rough, stinging or barbed hairs. *Loasaceæ.*
  Flowers unisexual or bisexual. Trees and shrubs. *Flacourtiaceæ.*
  Flowers unisexual, monœcious. Succulent herbs. *Begoniaceæ.*

The Hypogynous and Perigynous Families. Ovary Superior

DILLENIACEÆ (*Dillenia Family*). Flowers regular or irregular, bisexual or in some cases unisexual, often spirocyclic. Sepals 3 or indefinite; petals commonly 5, sometimes fewer. Stamens generally numerous, some of them frequently staminodial. Carpels indefinite or sometimes reduced to 1, free or more or less united, with 1 to several ovules. Fruit a septicidal capsule, or berry. Endosperm more or less copious. Embryo straight. Trees and shrubs, or sometimes herbs and lianas, inhabiting the tropics around the world and especially well represented in Australia. About 12 genera and 300 species.

The largest genus, *Hibbertia,* with 100 or more species, is confined almost exclusively to Australia. Other large genera are *Dillenia* and *Loliocarpus.* The flowers of some members are quite rose-like in appearance (*e.g., Hibbertia*).

ACTINIDIACEÆ. Flowers regular, bisexual or unisexual, polygamous or diœcious, with 5-merous calyx and corolla, numerous to few stamens with versatile anthers, and 3 to many carpels in a syncarpous gynœcium. Styles united or distinct. Ovules 10 or more in each cell. Fruit a capsule or berry. Seeds with copious endosperm and a straight embryo. Placentation axile. Comprising the two genera *Actinidia* with 10 species of climbing vines of eastern Asia, and *Sauravia* with 250 species of trees and shrubs of eastern Asia and tropical America. Several species of *Actinidia* are planted for their edible fruits or as ornamentals. The sauravias are clothed with a distinctive pubescence of bristly hairs. Leaves simple, alternate, often leathery.

EUCRYPHIACEÆ. Flowers regular, spirocyclic, bisexual, with a 4-merous calyx and corolla, numerous stamens and 5—18 carpels in a syncarpous gynœcium with as many distinct styles. Ovules many in each cell, in two rows. Fruit a septicidal capsule, the style of each carpel persisting as a slender beak. Seeds with endosperm surrounding the embryo. One genus, *Eucryphia,* of about 4 species of resinous, glabrous, or tomentose trees, with opposite, simple or compound (pinnate) leaves and large, showy, rose-like, solitary, and axillary flowers. Two species in Chile and two in Australia and Tasmania.

MEDUSAGYNACEÆ. Based on the monotypic genus *Medusagyne* of the Seyschelles Islands (N.E. of Madagascar), a shrub with simple, opposite, leathery leaves and regular, bisexual flowers with 5-merous calyx and corolla, numerous free stamens, and a syncarpous gynœcium of many carpels. Fruit a septicidal capsule with 2 winged seeds in each cell.

OCHNACEÆ. Flowers regular or irregular, bisexual, sometimes spirocyclic, with a varying number (4—10) of sepals and petals, the stamens 10 to numerous, often borne on an elongate floral axis, and in some cases pushed to one side of the flower and thus bringing about irregularity—some of the stamens in certain cases modified into staminodes which may be connate into a tube; anthers opening by pores or slits. Gynœcium of 2—15 carpels which often have free ovaries but connate styles. Ovules 1—many in each ovary. Carpels often separating at maturity and then drupaceous or septicidal capsules. The elongate conical receptacle often becomes fleshy in the fruit, supporting the carpels on its surface, a condition similar to that seen in the strawberry, but the carpels much larger. Ovules with 1 or 2 integuments. Seeds with or without endosperm. The plants are trees or shrubs, and in some cases herbs, with simple or sometimes compound, alternate, leathery, smooth and shining, stipulate leaves and small to large flowers in axillary racemes or panicles. Species about 370, in 20 or more genera, of tropical South America (chiefly Brazil), Asia, and Africa.

Principal genera: *Ochna,* with 90 species in tropical Africa and Asia; *Ouratea,* with 190 species in tropical America and Africa; *Lophira,* with 2 species in tropical Africa; *Luxemburgia,* with 12 species in Brazil. In *Lophira* the fruit is samaroid by the unusual elongation of two of the persistent sepals. The spirocyclic floral structure of *Ochna* and *Ouratea* suggests an affinity with the *Ranales.*

CARYOCARACEÆ. A small family of tropical American trees or shrubs with alternate or opposite, digitately 3—5-foliolate, stipulate leaves and regular bisexual, sometimes more or less perigynous, flowers with 5-merous or 6-merous calyx and corolla, the segments of which are free or sometimes more or less united; numerous stamens united into a tube or into bundles, and a syncarpous gynœcium of 1—20 (rarely 1—3) carpels with distinct styles. Fruit drupaceous, the carpels separating at maturity, each 1-seeded. Seeds with scanty endosperm and an embryo with a spirally coiled or thick, fleshy, and curved hypocotyl.

Two genera, *Caryocar* with 10 species, and *Anthrodiscus* with 3 species, both of tropical America. In *Caryocar* the filaments are very slender and exceedingly elongated but equal in length; in *Anthrodiscus,* on the other hand, they are unequal and shorter; in both genera they are variously bent in the bud. *Caryocar nuciferum, C. glabrum,* and *C. amygdaliferum* yield edible seeds (Souari nuts).

MARCGRAVIACEÆ. A small family of tropical American shrubs or small trees and climbing epiphytic vines, with simple, alternate, and sometimes dimorphic, exstipulate leaves. Flowers regular, bisexual, hypogynous, in racemose or umbellate, terminal clusters, 4-merous or 5-merous, the stamens indefinite to 4, free or united, or epipetalous, the gynœcium syncarpous, commonly of 5 but sometimes of 2 or many carpels. Ovary at first 1-celled but eventually several-celled by the pushing-in of the parietal placentas. Fruit a loculicidal capsule or indehiscent. Seeds with starchy endosperm investing a straight or curved embryo.

Species 100 or more in about 5 genera, the largest of which are *Marcgravia* (42 species) and *Norantea* (40 species). The smaller genera are *Souroubea, Ruyschia,* and *Caracasia.* The Marcgravias are epiphytic vines with dimorphic branches—sterile ones with sessile orbicular leaves in two ranks—and flower-bearing ones with larger, petiolate, spirally arranged leaves. A peculiar feature of the sterile flowers is the curious pitcher-like development of the pedicellar bracts (bracteoles). The calyx and corolla are 5-merous. In *Souroubea* the calyx and the corolla are 4-merous, and the bracteoles are saccate. In *Norantea* and *Ruyschia* the calyx and corolla are 5-merous, but the gynœcium of the latter is 2-carpellate. In *Caracasia* there is likewise a 2-carpellate gynœcium, and the stamens are reduced to 3.

QUIINACEÆ. About 19 species of tropical American trees and shrubs and in some cases climbing vines, with simple or compound, opposite or whorled, stipulate leaves, and regular, bisexual or unisexual, often polygamous, hypogynous flowers in racemes or panicles. Calyx and corolla 4-merous or 5-merous; stamens numerous; gynœcium 2-carpellate, syncarpous, with distinct styles: ovules 2 in each cell. Fruit baccate or at length capsular. Seeds tomentose, without endosperm.

Genera: *Quiina, Touroulia.*

THEACEÆ (incl. *CAMELLIACEÆ,* and *TERNSTROEMIACEÆ* in part). Flowers regular, often more or less spirocyclic, bisexual, rarely unisexual, hypogynous, with sepals and petals varying but mostly 5-merous. Stamens numerous or 5, often united in bundles, sometimes connate below; anthers opening by pores. Gynœcium syncarpous, of 2 to several carpels, with distinct or connate styles. Ovules 1—2 or more in each cell, on axile placentas. Fruit a loculicidal or septicidal capsule, the carpels often separating from the central column, or indehiscent. Seeds with or without endosperm. Integuments 2.

Tropical and warm-temperate trees and shrubs with simple, alternate, mostly evergreen, exstipulate leaves, and often showy, solitary, paniculate or racemose flowers. About 375 or more species in 16 or more genera.

The following genera include some useful or ornamental plants: *Ternstroemia, Eurya, Camellia* (*Thea*), *Stewartia, Gordonia. Ternstroemia* and *Eurya* are large genera—about 80 species each—the former chiefly of tropical Asia and America, the latter of South America, Central America, and eastern Asia. The camellias (including *Thea*) are native in India, China, and Japan—about 40 species in all. *Stewartia,* with about 6 species, is represented in eastern and southern United States by the two shrubs *S. Malachodendron* and *S. pentagyna;* both are planted for their showy flowers. The others occur in Japan. The sepals and petals vary from 5 to 6. The stamens are monadelphous. The 5-carpellate gynœcium becomes a woody, loculicidal capsule. *Gordonia Lasianthus,* the Loblolly Bay, is found in low woods from Virginia to Florida, often becoming a large tree, with evergreen, leathery leaves and large, white flowers in axillary clusters. Calyx and corolla are 5-merous, while the stamens are bunched in 5 clusters. The 3—5-carpellate gynœcium develops into a woody, loculicidal capsule, the carpels of which separate from the central axis. Flat, winged seeds are produced. The remaining 15 or more species occur in eastern Asia and in Mexico. Species of *Camellia* yield the teas of commerce.

According to Engler these are *C. japonica* and *C. theifera* (= *Thea sinensis, T. bohea, T. viridis*).

STRASBURGERIACEÆ. Based on the solitary *Strasburgeria callicantha,* a tree of New Caledonia, bearing leathery leaves with connate stipules, and solitary, axillary, bisexual, and spirocyclic flowers. Sepals varying, but the petals 5. Stamens 10, with versatile anthers. Ovary surrounded by an annular disk. Fruit woody, indehiscent.

GUTTIFERÆ (*HYPERICACEÆ*) (*St. John's-Wort Family*). Flowers regular, hypogynous, more or less spirocyclic, bisexual or unisexual (polygamous), sepals and petals varying from 2 to 6, the stamens numerous and often united into a varying number of distinct bundles, some of the stamens in certain cases modified into staminodes. Gynœcium of 1 to several (15 ?) carpels, syncarpous, with free or partially united styles and often large stigmas. Ovules 1 to several in each cell. Fruit mostly a capsule, so dehiscing as to leave the septa, with the seeds between them, attached to the central column.

Trees, shrubs, and a few herbs, sometimes lianas, with simple, opposite or whorled, often evergreen, leaves, and showy, solitary or cymosely clustered flowers.

Forty-five genera and between 800 and 900 species of the warm regions of the world. The tissues generally contain resin or oil and the leaves are dotted with pellucid glands. By some writers the genera are grouped in two separate families, the *Guttiferæ* and the *Hypericaceæ. Ascyrum* and *Hypericum* are represented in our flora, the latter also in flower gardens. *Garcinia, Mammea,* and other genera likewise yield horticulturals. A number of trees yield useful woods, resins, oils, and edible fruits and seeds.

The St. Peter's-worts (*Ascyrum*) are low shrubs with opposite to oblong leaves, and bright-yellow flowers. Calyx and corolla are 4-merous, and the numerous stamens distinct or slightly clustered. The ovary of the 3—4-carpellate gynœcium is 1-celled, or several-celled by the intrusion of the placentas. The styles are distinct, and the fruit a capsule. *Ascyrum stans* and *A. hypericoides* inhabit sandy soils of the eastern United States. The St. John's-worts (*Hypericum*), of which there are over 200 species, are widely distributed over the world; about 30 species occur in our flora. The plants are herbs or shrubs with opposite yellowish- or black-dotted leaves and bright-yellow, pinkish, or purplish, mostly cymose flowers. Calyx and corolla are 5-merous, the stamens numerous and free or united in clusters. The 3—5-carpellate gynœcium has a 1-celled ovary or this becomes 3—5-celled by the intrusion of the placentas. The styles

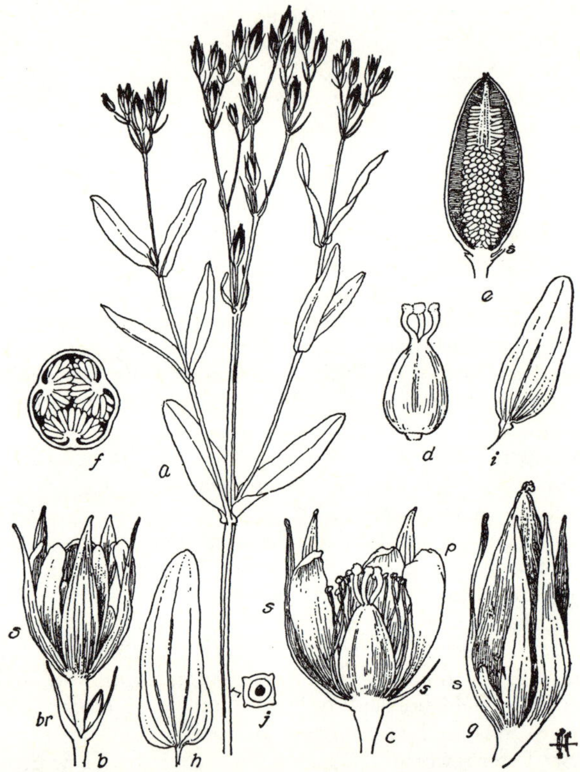

FIG. 290.—*Hypericum majus: a,* habit sketch with fruits; *b,* flower; *c,* flower with parts removed; *d,* gynœcium; *e,* median section of ovary; *f,* transverse section of ovary; *g,* immature fruit, with investing calyx (*s*); *h,* upper, and *i,* lower, leaf; *br,* bracts; *p,* petals.

are distinct. Fruit a capsule. Our species show preference for a wide range of habitats. Thus *Hypericum graveolens* is a mountain species of the Appalachians; *H. pseudomaculatum* inhabits woods or prairies; *H. dolabriforme,* dry rocky hills in Kentucky and Tennessee; while *H. ellipticum, H. boreale, H. majus,* and *H. virginicum* seek wet or boggy or swampy grounds. Linear leaves are characteristic of *H. canadense* and *H. Drummondii,* while in *H. gentianoides* the leaves are reduced to scales. *H. virginicum* is a stoloniferous species and its herbage is often conspicuously pinkish or purplish in color. Some species (*H. canadense, H. majus*) are annuals. (Figs. 290, 291.)

FIG. 291.—*Hypericum Moserianum: a,* flowering twig; *b,* flower-bud; *c,* flower; *d,* cluster of stamens; *e,* stamens; *f,* gynœcium and calyx; *g,* transverse section of ovary; *i,* floral diagram; *h,* ovule.

DIPTEROCARPACEÆ. Flowers regular, bisexual, perigynous or epigynous; calyx and corolla 5-merous; stamens numerous, 15, 10, or 5, often with the connective greatly prolonged into a filamentous appendage, or broad and short. Gynœcium commonly 3-carpellate, syncarpous, sometimes 1-carpellate, with 2—many ovules in each cell on intruded axile placentas. Fruit a 1-seeded nut, but in some cases dehiscing on germination into as many valves or carpels. Integuments 2. Seeds without endosperm and the embryo often with twisted or contorted cotyledons. In some cases bilateral symmetry (irregularity) of the flower has resulted from the unequal development of the anterior and posterior sepals. The calyx generally persists, the segments developing as large wings, and the base investing the fruit.

The family constitutes an important element of the flora of tropical Asia, the East Indies, the Philippines, and to some extent of Africa. There are about 330 species in 16 genera—the more important of which are *Dryobalanops, Dipterocarpus, Shorea, Vatica, Vateria, Anisoptera, Hopea, Pentacme, Doona.* Outside of a few shrubs, the plants are mostly trees, often very large, which yield valuable gums, resins, volatile oils, and timber. The most outstanding floral features are the remarkably developed calyx, which persists as a winged appendage of the fruit, and the peculiarly elongated connectives. Definite epigyny appears in *Pentacme,* but is less pronounced in *Vatica* and *Vateria. Shorea* appears to be hypogynous.

ELATINACEÆ (*Waterwort Family*). A small, unimportant family of about 40 species of aquatic or marsh herbs with opposite or whorled, stipulate leaves and very small, axillary, regular, hypogynous, 5-merous, bisexual flowers. Stamens as many or twice as many as the petals, the inner whorl sometimes aborted; anthers versatile. Gynœcium of 2—5 united carpels, with as many free styles. Ovules 2—many in each cell, on axile placentas. Fruit a septicidal capsule. Two genera, *Elatine* (Waterwort), with about 20 species of ponds and stagnant pools of America and Europe, and *Bergia,* with about 20 species of the tropics and temperate regions, —one species, *B. texana,* ranging from Illinois to Texas and California. *Elatine americana* inhabits margins of ponds across the continent. *E. triandra* and *E. brachysperma* range from Illinois westward and southwestward.

FRANKENIACEÆ. Flowers regular, bisexual, hypogynous, with a 4—7-lobed gamosepalous calyx. Petals as many as the calyx-lobes and appendaged at the base. Stamens numerous, in 2 or 3 whorls, free or united at the base, with versatile anthers. Gynœcium 2—4-carpellate, syncarpous, with styles connate at the top; ovary 1-celled, with many ovules on parietal placentas. Integuments 2. Seeds with endosperm. Fruit a capsule.

Four genera and about 64 species of herbs or subshrubs, with small, opposite, exstipulate leaves, inhabiting mainly saline or alkaline areas of the tropics and subtropics. The largest genus, *Frankenia* (60 species), has two species in California.

TAMARICACEÆ (*Tamarisk Family*). Flowers regular, mostly bisexual, hypogynous. Calyx and corolla mostly 4-merous or 5-

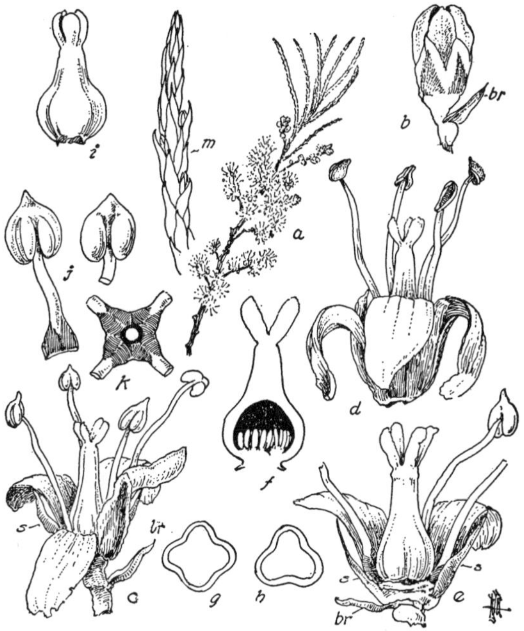

FIG. 292.—*Tamarix parviflora: a,* flowering branch; *b,* flower-bud; *c,* flower; *d,* flower after anthesis; *e,* section of flower exposing gynœcium and base of andrœcium; *f,* median section of gynœcium; *g,* transverse section of ovary of 4-carpellate gynœcium and (*h*) of a 3-carpellate gynœcium; *i,* gynœcium from a bud; *j,* ventral aspects of stamens; *k,* base of andrœcium; *br,* bracts; *s,* sepals; *m,* twig with flower-buds in axils of three of the leaves.

merous, the stamens as many or twice as many. Corolla and stamens inserted on a disk. Gynœcium 3—4-carpellate, with styles free or united at the base. Ovary 1-celled, with numerous ovules on parietal or basal placentas. Fruit a capsule. Seeds with or without endosperm, clothed with hairs, or these only in an apical tuft.

Shrubs or trees with small, alternate, cedar- or juniper-like leaves, closely appressed to mostly long, slender branches. Flowers very small, in lateral spikes or racemes. About 100 or more species in 4 genera—*Tamarix, Reaumeria, Hololachne, Myricaria*—chiefly of the northern hemisphere. *Tamarix,* by far the largest genus (75 species), is represented by a considerable number of horticulturals. The genus ranges from the Mediterranian region to the East Indies and Japan. (Fig. 292.)

CISTACEÆ (*Rock-Rose Family*). Flowers regular, or the calyx irregular, bisexual or sometimes unisexual, often cleistogamous. Sepals and petals varying from 3 to 5. Stamens numerous, unequal, often arising from an elongated zone of the receptacle. Gynœcium 3—10-carpellate, 1-celled or several-celled by the intrusion of the parietal placentas. Ovules 2—many in each cell, mostly orthotropous, on elongate stalks (funicles). Integuments 2. Fruit a capsule, dehiscing between the placentas. Seeds with endosperm and curved embryo. Sepals and petals usually folded in opposite directions in the bud.

Herbs or shrubs with simple, opposite or in some cases alternate, stipulate, leaves, the stipules often adnate to the petiole. Seven or 8 genera and 160—170 species, mostly of the northern hemisphere —where they are best developed in the Mediterranean region.

The important genera are *Helianthemum, Hudsonia, Lechea, Cistus.*

Helianthemum, with its 80 species, is the largest genus. Aside from a few American species, the genus is mainly Mediterranean. In our species two kinds of flowers (dimorphic) will be found —an early set with large, yellow petals, numerous stamens, and many-seeded capsules, and a later set of much smaller cleistogamous flowers with only a few stamens (3—10), the petals often wanting, and the capsules few-seeded. In both kinds the stigma is sessile or nearly so. *Helianthemum majus* and *H. canadense* are the common Eastern forms; both are hoary plants. They are known as "frost-weeds" from the curious fact that in late autumn crystals of ice are found to issue from the cracked bark at the base of the stems. The plants are also known as rock-roses. *H. Chamæcistus* is a horticultural form. A bilateral symmetry of the flowers results in some cases from the reduction, suppression, or increase in size of the two outer sepals. In the bud the petals are usually crumpled. The cleistogamous flowers, because of the absence of petals, may be easily mistaken for flower-buds. The embryo is spirally coiled.

The Hudsonias, *Hudsonia ericoides, H. tomentosa,* and *H. mon-*

tana, are distributed over eastern North America. *H. montana* is found only on the high mountain tops of North Carolina, while *H. ericoides* occurs in sandy soil along the coast from Newfoundland to Virginia. The most widespread is *H. tomentosa,* which occurs in sandy soils from New England to North Dakota and northwestward to the Mackenzie basin. These are all low, heath-like shrubs

FIG. 293.—*A, Lechea* sp.: *a,* habit sketch (fruiting); *b,* capsule, with persistent calyx; *s,* small outer sepal; *c,* capsule; *d,* median section of capsule, with calyx below; *e,* embryo. *B, Helianthemum* sp.: *a,* branch with fruits; *b,* calyx, investing capsule; *s,* small outer sepal; *c,* capsule; *d,* capsule from top of twig (*a*); *e,* transverse section of a 2-seeded capsule, *f,* of a 1-seeded capsule; *n,* embryo.

with small, awl-shaped leaves and numerous yellow flowers. Calyx and corolla are 5-merous, and the stamens are usually numerous. The 3-carpellate gynœcium becomes a 3-valved capsule.

Lechea (Pinweed) is also an exclusively American genus, with one species in Cuba, one in Mexico, and the remaining eight or nine ranging from Texas eastward. They are rather rigid, branching, perennial herbs with a somewhat woody base, small, linear or lanceolate leaves, and small, greenish or purplish flowers in panicles. The calyx is irregular, the two outer sepals being smaller than the others, and there are but 3 petals—which are persistent. The fruit is a 3-celled, 3-valved capsule. *Lechea tenuifolia* ranges from New Hampshire to Wisconsin and Nebraska and south to Florida and Texas; *L. stricta* from New York to Minnesota and Nebraska. Both are plants of dry, sandy soils. (Fig. 293.)

A half-dozen or more species of *Cistus* (Rock-Rose) are grown as ornamentals. These are shrubs with opposite, exstipulate leaves and showy flowers in simple or compound terminal cymes. There are 3—5 sepals, 5 petals, numerous stamens, a 5-carpellate gynœcium, and a capsular fruit loculicidally dehiscent by 5, or sometimes 10, valves. When five sepals are present, the two outer are smaller than the other three. The 20 or more species are natives of the Mediterranean countries.

BIXACEÆ (*Bixa Family*). Flowers regular or irregular (bilaterally symmetrical), bisexual, hypogynous, with 5-merous calyx and corolla—the petals large. Stamens numerous, free or slightly united at the base, with 2-celled, often horseshoe-shaped anthers. Gynœcium 2—3-carpellate, with an elongate curving style. Ovary 1-celled, or 2—3-celled by the intrusion of the placentas. Fruit a densely bristly or smooth loculicidal capsule. Seeds obovoid or kidney-shaped, with a fleshy testa or clothed with a dense coat of soft hairs. Embryo large and curved, with large foliaceous cotyledons, incurved at the top, and in copious endosperm.

Trees and shrubs with colored sap and large, alternate, simple or sometimes deeply digitately lobed leaves, and showy paniculate flowers. The plants yield gums, dyes, and various food products. Species about 20, in 4 genera—*Bixa, Maximilianea, Amoreuxia, Sphærosepalum,* mainly tropical, with a few in Australia. The solitary *Bixa orellana* (Annotto) of tropical America is cultivated; it is a small tree with large ovate leaves and large, showy, rose-colored flowers. Some species of *Maximilianea* develop thick, tuberous, underground stems. Bilateral symmetry appears in the corolla and stamens of *Amoreuxia,* three of the petals being broad and two narrow, while the stamens are in two sets—a short one opposite the narrow petals and a longer set opposite the broad petals. *Sphærosepalum* is restricted to Madagascar.

COCHLOSPERMACEÆ. A small family of less than 20 species of woody tropical xerophytes, belonging mainly to the genus *Cochlospermum.* The flowers are regular or irregular, bisexual, with 4-

merous or 5-merous calyx and corolla, numerous stamens, and a 3—5-carpellate gynœcium in which are produced numerous seeds with copious endosperm and a coiled embryo. Integuments 1. Fruit a capsule.

LACISTEMACEÆ. Comprising a single genus, *Lacistema,* of about 16 species of shrubs or small trees with alternate, simple, exstipulate leaves, and very small bisexual flowers composed of a two-whorled trimerous perianth, a single stamen, and a 2—3-carpellate gynœcium with 1 style. Fruit a capsule. Integuments 2. Seeds with endosperm. Receptacle modified into a cup-shaped disk. An anomalous family, apparently more closely related to the *Piperales* than to the *Parietales.*

CANELLACEÆ (WINTERANACEÆ). Flowers regular, bisexual, hypogynous to perigynous, mostly with 4-merous or 5-merous calyx and corolla, or the corolla in some cases wanting. Stamens many, monadelphous in a complete tube—to the outside of which the anthers are adnate. Gynœcium syncarpous, of 2—5 carpels and a thick style ending in a 2—5-lobed stigma. Ovary 1-celled, with numerous ovules on 2—5 parietal placentas. Integuments 2. Fruit a berry. Seeds with endosperm and a straight embryo. Trees with aromatic sap and simple, alternate, leathery, gland-dotted, exstipulate leaves.

Genera: *Canella (Winterana),* with about 7 species ranging from southern Florida through the West Indies to Colombia. *Cinnamodendron,* with about 3 species in the Antilles and Brazil, and one in tropical Africa. *Cinnamosma,* with 1 species in Madagascar. The plants yield gums, resins, and oils.

VIOLACEÆ (Violet Family). Flowers regular or irregular (mostly), perigynous, rarely unisexual and polygamous, occasionally diœcious, often cleistogamous, and 5-merous—except the gynœcium, which is usually 3-merous. Sepals free or somewhat connate, often unequal. Petals mostly unequal, the lowermost often large and saccate or spurred. Stamens connivent in a ring around the ovary, the 2 lower often spurred or appendaged; filaments in some cases connate in a tube, or sometimes very broad and thin. Ovary 1-celled, with few to many anatropous ovules on 3—5 parietal placentas. Style simple, straight or curving, usually with an oblique, dilated, disk-like or globular stigma. Integuments 2. Fruit a capsule with elastic loculicidal dehiscence, or a berry. The plants are herbs or

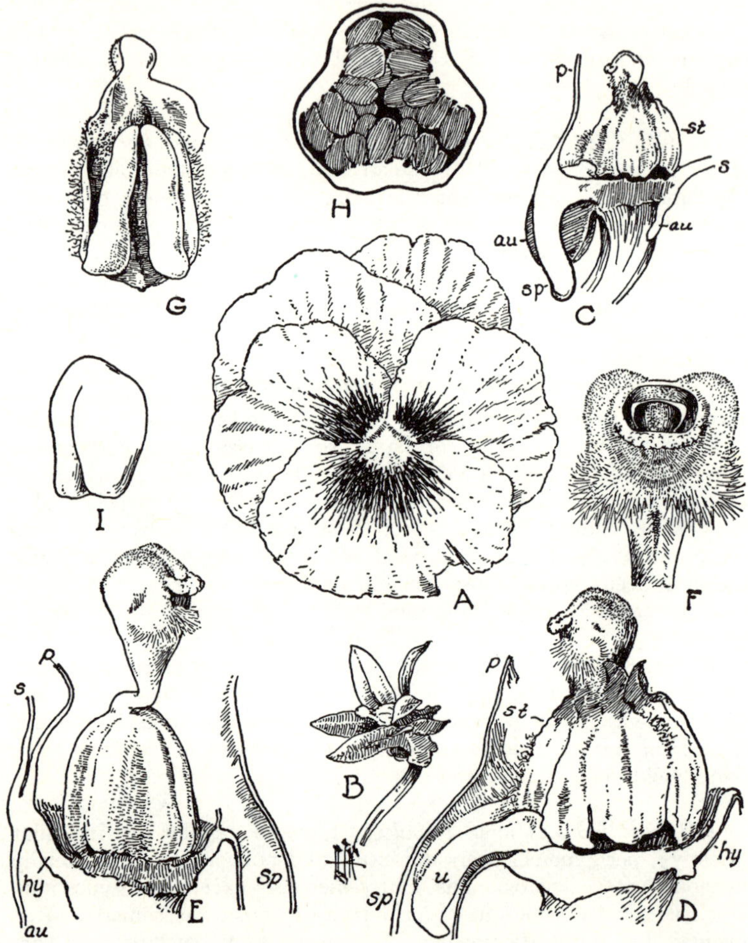

FIG. 294.—Pansy (*Viola tricolor* var. *hortensis*): *A,* face of corolla; *B,* calyx with gynœcium in center; *C,* flower with parts removed: *st,* stamens; *p,* petal; *s,* sepal; *au,* auricles of calyx; *sp,* spur of lowest petal in (*A*); *D,* same section as (*C*) enlarged, with spur of petal in section to show spurs (*u*) of anterior stamens (*st*): *hy,* hypanthium (flower perigynous); *E,* gynœcium: *p,* petal; *s,* sepal; *hy,* hypanthium; *au,* auricle of calyx; *sp,* spur of lowest petal; *F,* face view of stigma; *G,* ventral aspect of a stamen; *H,* transverse section of ovary; *I,* ovule.

shrubs with simple, alternate, rarely opposite, stipulate leaves which are often basal and variously lobed or laciniate. Stipules often large and leaf-like. About 15 genera and 300—400 species, mainly of tropical and temperate regions.

Principal genera: *Viola, Rinorea, Hybanthus.*

Three hundred or more species of violets (*Viola*) are distributed over the temperate regions of the world. These are annual or perennial herbs or in some cases subshrubs. The herbaceous forms are leafy-stemmed (caulescent) or the leaves are radical from an underground rootstock (acaulescent forms). The leaves are of exceedingly diverse shapes, especially in hybrids. Petaliferous flowers are usually produced in early spring and these are followed by cleistogamous flowers which usually lack petals and never open, yet produce seeds in abundance. The cleistogamous flowers are generally borne close to the ground, where they may be found covered by the leaf-mold; they may be easily mistaken for pods. In some cases, as in the pansy, the stipules are often large and leaf-like (Fig. 120). The flowers are bilaterally symmetrical, variously colored, and often nodding.

FIG, 295.—*Viola arenaria: a,* immature capsule, and calyx; *s,* sepals; *r,* auricles of calyx; *b,* transverse section of ovary; *c,* capsule dehiscing; *d,* seed; *e,* fruit, with bracts at base of pedicel; *o,* ovule.

The sepals, which are commonly auricled, persist in the fruit (capsule). The two lowest petals are more or less spurred, and into these spurs project the appendages of the lowest stamens. The filaments are usually broad and thin. Useful technical characters reside in the stigma—which in this genus shows many curious modifications. The capsule dehisces elastically when ripe, scattering the seeds for a considerable distance. Horticultural varieties are numerous (*Viola tricolor,* Pansy), and many wild species occur in our woodlands and on the prairies. One of these, *Viola pedata,* the Crowfoot Violet, does not produce cleistogamous flowers. (Figs. 294, 295.)

Hybanthus, with 50 or more species, is tropical, but *H. concolor* (or *Cubelium concolor*), if here included, ranges from New York to North Carolina and west to Kansas. *Rinorea* comprises 40 or more species of tropical trees.

FLACOURTIACEÆ (*Flacourtia Family*). Flowers regular, hypogynous, perigynous, or epigynous, bisexual or unisexual, polygamous or diœcious, sometimes spirocyclic. Sepals vary from 2 to 15; petals 10 or more (or none) in some cases with a basal scale. In some cases the corolla is not clearly differentiated from the calyx. Stamens numerous or few, with long filaments, free or in bundles of various shapes, in some cases modified as staminodes. Gynœcium of 2—10 united carpels, with numerous ovules on parietal placentas. Integuments 2. Fruit a berry, drupe, or capsule. Seeds with endosperm, and often arillate. Species about 890, in 70 or more genera. Tropical.

Flacourtia, Azara, and *Lovyalis* yield horticultural forms. *Flacourtia indica* (Governor's Plum or Batoko-Plum), of southern Asia and of Madagascar, is a shrub or small tree, sometimes beset with axillary spines; it yields a plum-sized fruit with juicy pulp. *Azara microphylla,* a Chilean shrub, bears shiny leaves and orange-colored berries; petals are wanting in the flowers, and stamens number only 5. *Lovyalis caffra* is the Kei-Apple, a tall shrub with long, rigid thorns and greenish, clustered flowers from which spring yellow, plum-like, juicy fruits.

STACHYURACEÆ. Based on the solitary genus *Stachyurus* of the Himalayas and Japan. The plants are shrubs or small trees with alternate, stipulate leaves and regular, bisexual or unisexual, polygamo-diœcious, 4-merous, hypogynous (?) flowers. Stamens 8. Ovary 4-celled, or 1-celled by the incomplete intrusion of the placentas. Seeds numerous, arillate, with fleshy endosperm. Hypocotyl of the embryo elongate.

TURNERACEÆ. Comprises a hundred or more species of herbs and shrubs of tropical America. Leaves simple, alternate, stipulate or exstipulate, often lobed. Flowers regular, bisexual, perigynous, 5-merous except for a 3-merous gynœcium. Styles distinct, sometimes unequal (heterostyly). Fruit a 1-celled capsule. Seeds numerous, arillate, with endosperm. Integuments 2. *Turnera,* 70 species.

MALESHERBIACEÆ. Thirty or more species of hairy South American herbs and shrubs, mainly of the genus *Malesherbia.* Leaves simple, alternate, exstipulate. Flowers regular, bisexual, perigynous, 5-merous. Stamens 5. Gynœcium 3-carpellate, with distinct styles, and numerous ovules on parietal placentas. Fruit a capsule, invested by the hypanthium—which here is an elongate vasiform tube. The gynœcium and the andrœcium are raised on an elongate stalk (androgynophore) upon which the stamens are inserted immediately below the ovary.

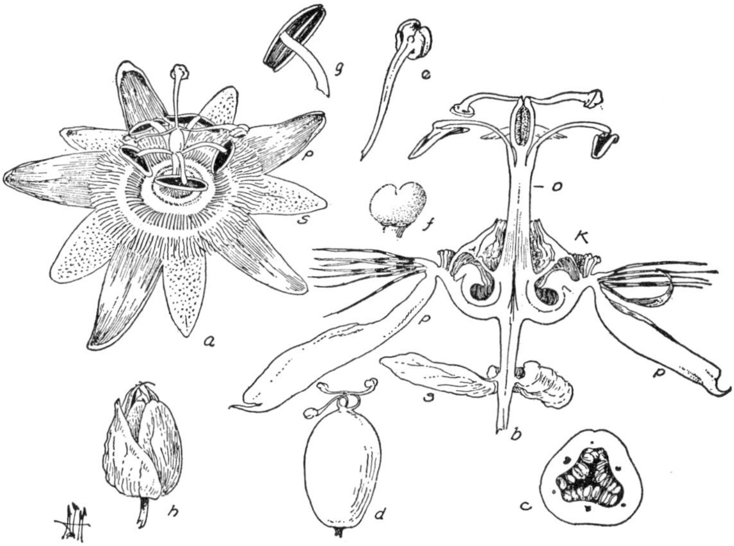

FIG. 296.—*Passiflora cærulea: a,* flower; *b,* median section of flower; *o,* androgynophore, with stamens and gynœcium above; *k,* corona; *p,* petals; *s,* sepals; *c,* transverse section of ovary; *d,* immature fruit (gynœcium); *e,* style and stigma; *f,* surface of stigma; *g,* stamen; *h,* flower-bud.

PASSIFLORACEÆ (*Passion-Flower Family*). Flowers regular, bisexual or unisexual, diœcious or monœcious, mostly perigynous, and mostly 5-merous. Petals are sometimes wanting. Stamens mostly 5, sometimes numerous, free or monadelphous. Gynœcium 3—5-carpellate, with numerous ovules on parietal placentas. Integuments 2. Fruit a berry, or an indehiscent or loculicidally dehiscent capsule. Seeds arillate, with endosperm and a large, straight embryo. A rudiment of the opposite sex is often present in unisexual flowers.

A distinctive feature of the flowers is the peculiar fringed or filamentous or otherwise configurated outgrowths of the hypanthium into the center of the flower, in superposed planes. These structures form the so-called *corona*—or the corona may result from a peculiar modification of the corolla, which is then sympetalous. The gynœcium and the stamens are elevated on a prominent androgynophore from which spring annular flanges at intervals. In some cases the stamens are borne near the bottom of the hypanthium. The styles are more or less united, or distinct.

The plants are mostly herbaceous or woody tendril-bearing climbers, and some twiners, many of them with varied and curiously shaped leaves. Some have spiny, cactus-like stems (*Adenis*), others have thick, cormose stems (*Echinothamnus*). *Passiflora* (Passion-Flower) yields a number of horticulturals, the fruits of which are pulpy berries (edible). The great majority of the species (300 or more) of the family belong in this genus, most of them American, but a number in Asia, Australia, and Madagascar. *P. lutea* and *P. incarnata* range from Virginia to Missouri and southward. (Fig. 296.)

ACHARIACEÆ. Restricted to three or four species of South African herbs and subshrubs. Leaves simple, lobed. Flowers racemose, regular, perigynous, unisexual and monœcious, 5-merous or 3-merous. Gynœcium 3—5-carpellate, syncarpous. Fruit a capsule. Seeds few, with copious endosperm. *Acharia, Ceratosycios.*

CARICACEÆ (*Papaw Family*). Comprises 28 or 30 species of tropical American and one or two African trees with milky sap and mostly simple, straight, soft-wooded, palm-like trunks bearing large, long-petioled, deeply palmately lobed leaves in a crown at the summit. The flowers are regular, perigynous, unisexual, diœcious, and 5-merous. In the staminate flowers the petals are connate in an elongate tube. The calyx is much reduced. Stamens 10, in 2 whorls, epipetalous in the throat of the corolla. Gynœcium 3—5-carpellate, syncarpous, with as many style branches as carpels. Ovules numerous on parietal placentas. Fruit a berry.

Carica, with 20—25 species, is the principal genus. *Carica-Papaya* (Papaw, Pawpaw, Melon-Tree) is grown for its edible fruit, which sometimes attains a diameter of twenty inches. The sap is used as milk.

The Epigynous Families. Ovary Inferior

LOASACEÆ (*Loasa Family*). These are erect, twining or climbing herbs, or sometimes shrubs or trees, clothed with hairs of diverse shapes (often barbed and stinging), and bearing variously shaped, alternate or opposite leaves, and often showy flowers in cymose, capitate, or racemose clusters. Sepals and petals varying in number but commonly 5. Stamens numerous, in a number of cases accompanied by staminodes or nectariferous scales. Gynœcium of 3 or more carpels, syncarpous, with a single style which may be cleft above. Ovary 1-celled, with numerous ovules on parietal placentas. Integuments 1. Fruit (spurious) a capsule, which is sometimes twisted and which dehisces at the top or lengthwise. The arrangement and peculiar shapes of the stamens, staminodes, and nectariferous scales furnish important taxonomic characters. The receptacle is in several cases developed beyond the ovary as an hypanthium, but the top of the ovary is in all cases free.

Genera 13; species 200—250. Our genera are *Mentzelia, Eucnide,* and *Petalonix. Loasa vulcanica* and *Caiophora laterita* are plants of flower-gardens. *Mentzelia* includes annual or perennial herbs, or shrubs and trees, more or less densely clothed with barbed hairs and bearing alternate or sometimes opposite, lobed or pinnatifid leaves. The flowers are yellow, often showy, solitary or in clusters, and subtended by leafy bracts. The stamens are numerous. The fruit is an elongate capsule opening at the top. Many of our species are rough and rigid plants, in the arid West forming a characteristic element of the vegetation of gravelly or sandy flats and washes. *Mentzelia oligosperma* and *M. decapetala* range eastward into North Dakota, Iowa, and Illinois. *M. albicaulis* reaches British Columbia. A dozen or more species occur in California.

Eucnide is mainly Mexican, but six or seven species range into California and Utah. The plants are herbs clothed with stinging hairs. The numerous stamens are inserted on the base of the petals. The style terminates in five twisted branches. *Petalonix Thurberi* (Sandpaper Plant) and *P. nitidus* occur in the Colorado Desert region. These are rough-pubescent shrubs with 5-merous flowers. There are but 5 stamens, with long, slender filaments, and the petals are long-connivent. The ovary is linear, with a long, slender style, forming when ripe a small, irregularly dehiscent capsule. The few remaining species are Mexican. In *Cevalia* and *Gronovia* there is but one ovule produced, and this is suspended from the top of the ovary. The solitary *Kissenia spathulata* of Arabia and South Africa is the only member of the family occurring outside of America—a significant disconnection.

DATISCACEÆ (*Datisca Family*). A small family of shrubs or tall trees with large, simple, entire, digitately nerved or divided, or digitately compound leaves, and mostly unisexual, polygamous or diœcious, epigynous flowers. Sepals and petals varying in number, the petals sometimes wanting in the staminate flowers. Stamens few to many. Gynœcium 3—8-carpellate. Style various. Ovary 1-celled, with numerous ovules on parietal placentas. Fruit a capsule opening at the top, or down the sides. The family is limited to the following: *Tetrameles nudiflora*—the only species—a tall tree of India, Ceylon, and Java; *Octomeles,* with two species in the East Indies and the Philippines; *Datisca glomerata* (Durango-Root) of southern California and Mexico; and *D. canabina* of northern India and western Asia. The California plant is a tall stout herb with pinnately divided or compound leaves, and small, axillary, polygamous or diœcious flowers. The gynœcium is 3—5-carpellate, with as many distinct spreading, and forked styles. Petals none. Fruit (spurious) a small capsule. In habit the plants resemble the hemp (*Cannabis*).

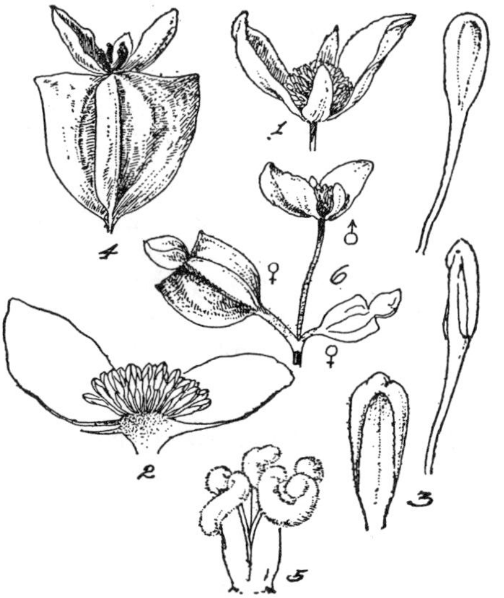

FIG. 297.—*Begonia* sp.: *1,* staminate flower; *2,* median section of staminate flower; *3,* stamens; *4,* carpellate flower; *5,* stigmas; *6,* inflorescence, with staminate flower above and carpellate below.

BEGONIACEÆ (*Begonia Family*). Flowers unisexual, monœcious, epigynous, more or less unsymmetrical. Sepals petaloid, petals sometimes wanting, both varying in number from 2 to 5 or 6. Stamens several to numerous, free or somewhat united; anthers of various shapes, opening by slits or pores, the connective frequently prolonged or broadened. Gynœcium commonly 3-carpellate, with as many styles—these free or united below and terminating in elongate, sometimes forked stigmatic branches which are densely papillate and coiled or twisted; in other cases the stigmas are capitate. The inferior ovary bears numerous ovules on parietal or axile placentas. Integuments 2. Fruit (spurious) an indehiscent berry, or a broadly winged, horned or ribbed, loculicidal capsule. The wings or horns, etc., are outgrowths of the receptacle. Seeds without endosperm at maturity. Staminodes are sometimes present in the carpellate flowers. Floral malformations are of frequent occurrence.

The plants are mostly perennial succulent herbs or subshrubs, with stems swollen at the nodes. The leaves are simple, alternate, oblique, more or less succulent and often large, and of diverse shapes. Of the 420 or more species in the family, 400 or thereabouts belong in *Begonia*. These are of bewilderingly diverse habits, including erect or creeping forms, some that climb by means of roots, others acaulescent and growing from rhizomes or thick stem tubers. A host of begonias are grown as potherbs; these are easily propagated by cuttings or even by parts of leaves, which freely form adventitious (secondary) roots and buds. The few remaining species of the family belong in *Hillebrandia* (Hawaii), *Symbegonia* (New Guinea), and *Begoniella* (Colombia). (Fig. 297.)

ANCISTROCLADACEÆ. One genus, *Ancistrocladus*, of about 8 species of tropical Asia and West Africa. The plants are lianas with hooked branches and simple, alternate leaves. Flowers small, regular, bisexual, epigynous, in panicles. Calyx and corolla 5-merous, the petals somewhat united at the base. Stamens 5 or 10, with short filaments which are expanded at the base. Gynœcium 3-carpellate, with a short, thick style, and 3-lobed stigma. Ovary 1-celled; ovule solitary, basal. Fruit a nut. Seeds with fleshy convoluted endosperm and a straight embryo with a thick hypocotyl and thick divergent cotyledons.

ORDER OPUNTIALES (CACTALES)

Flowers regular or sometimes irregular, bisexual, spirocyclic, epigynous, the sepals, petals, and stamens of indefinite number and inserted upon a more or less deeply tubular hypanthium. Gynœcium 4—8-carpellate, with 1 style. Ovary completely inferior, 1-celled, with numerous ovules on parietal placentas. Integuments 2. Fruit mostly a berry. Seeds with scanty endosperm or without endosperm. Embryo straight or curved, with foliaceous cotyledons or with organs poorly differentiated, the hypocotyl in some cases tuberous. Perennial succulent plants with simple or branched stems which are more or less solid or pulpy, spiny and glochidiate, cylindrical, globular, or flattened, and jointed or constricted at intervals. Leaves reduced to scales or wanting, or when present then more or less spindle-shaped and succulent and early deciduous; or when permanent then flat and succulent and of various shapes.

One family, *CACTACEÆ* (Cactus Family), with characters of the order. A distinctive group of problematical affinities, chiefly American, with some African. In America confined mainly to the arid West and Mexico. Species about 1,000, in 100 or more genera.

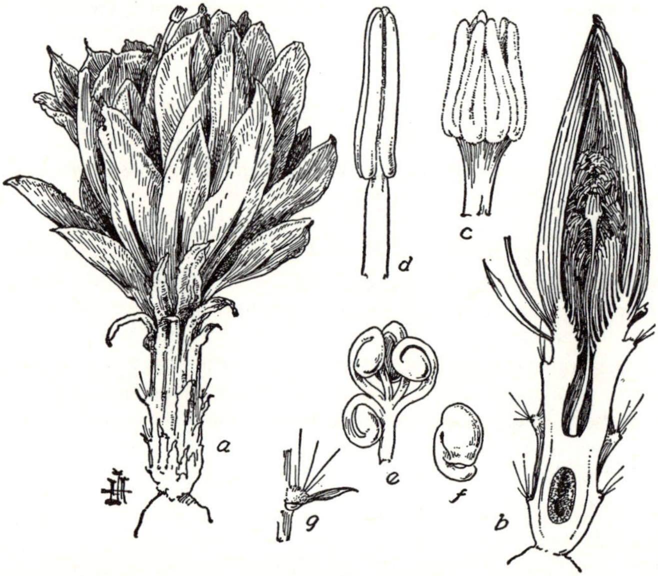

FIG. 298.—*Phyllocactus* sp.: *a,* flower; *b,* flower in section; *c,* stigma; *d,* stamen; *e,* cluster of ovules; *f,* ovule; *g,* bracteoid leaf and axillary bristle-bearing areoles.

The following genera are characteristic, but many others are represented in gardens and hot-houses: *Opuntia, Cereus, Echinocactus, Mamillaria, Carnegiea.*

The Opuntias are legion, numbering in the neighborhood of 250 species. The plants are of diverse habit, with cylindrical or flattened, jointed stems; many are very spiny. The genus is not only one of the largest of the family but also the most wide-spread, ranging across the continent from the Atlantic to the Mackenzie basin of Canada, and south to the Straits of Magellan. A half-dozen or more species occur east of the Rockies, in dry, sandy or rocky situations. The largest of our cacti is the giant Suwarro or *Carnegiea gigantea* (*Cereus giganteus*) of the Southwest, which produces a columnar trunk 50—60 feet in height.

In *Opuntia* cushion-like outgrowths (*areoles*) in the axils of the (early deciduous) leaves bear numerous barbed bristles (*glochids*) and woolly hairs, besides large, solitary or clustered spines (below the glochids). The fruit is an ovoid or ellipsoid, spiny or naked berry, which in many cases is edible. Flowers and barbed spines are produced by the areoles. In *Cereus, Echinocactus,* and *Mamillaria*

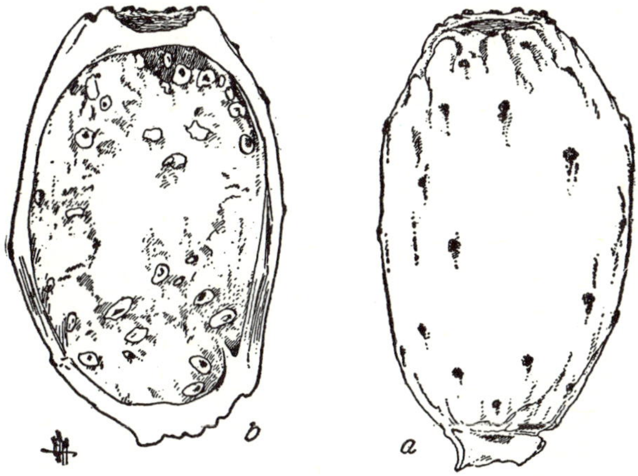

FIG. 299.—Fruit of prickly-pear (*Opuntia, a*); in section (*b*).

the stems are leafless and without joints, glochids, and barbed spines, and the floriferous and spiniferous areoles are distinct. *Mamillaria* bears its flowers between spiniferous tubercles, while its stems are globular to oblong. A common species of this genus is *Mamillaria tetrancistra,* the Fish-hook Cactus. The species of *Cereus* are of diverse habits—oblong or cylindrical or columnar, and fluted, with the ridges spiniferous. It differs technically from *Echinocactus* in having the floriferous areoles above the spiniferous ones; the flowers are funnel-form. *Echinocactus* has ovoid or oblong to cylindrical stems with spiniferous ridges; the flowers are short-tubular. (Figs. 298, 299.)

The cacti are usually readily propagated vegetatively by cuttings or even fragments of the stems, and these quickly form adventitious (secondary) roots. The seeds are said to germinate in many cases, if not in all, very soon after the fruit is ripe, in some cases even germinating inside the fruit (viviparous).

ORDER MYRTIFLORÆ (MYRTALES)

Flowers regular or sometimes irregular, cyclic, bisexual or sometimes unisexual, polygamous or diœcious, perigynous or epigynous, with distinct calyx and corolla. Stamens many or few. Gynœcium mostly 2—8-carpellate, occasionally more, rarely 1-carpellate; ovary superior or inferior, with numerous to few (sometimes 1) ovules on axils or parietal placentas. Integuments 1 or 2. Fruit various. Seeds with or without endosperm. Trees, shrubs, and herbs.

Perigynous Families. Ovary superior: *Geissolomataceæ, Penæaceæ, Thymelæaceæ, Elæagnaceæ, Lythraceæ, Heteropyxidaceæ, Melastomataceæ.*

Epigynous Families. Ovary inferior: *Oliniaceæ, Sonneratiaceæ, Crypteroniaceæ, Punicaceæ, Lecythidaceæ, Rhizophoraceæ, Nyssaceæ, Alangiaceæ, Combretaceæ, Myrtaceæ, Melastomataceæ, Hydrocaryaceæ, Œnotheraceæ, Haloragidaceæ, Hippuridaceæ, Cynomoriaceæ.*

Key to the Principal Families

A. Flowers perigynous. Ovary superior.
  Petals none.
    Ovule pendulous. Receptacle (hypanthium) petaloid. Trees and shrubs. *Thymelæaceæ.*
    Ovule erect. Trees and shrubs clothed with silvery peltate or stellate scales. *Elæagnaceæ.*
  Petals present.
    Anthers opening by a lateral slit. Petals sometimes none. Style 1. *Lythraceæ.*
    Anthers opening by a terminal pore, the connective often greatly elongated and modified. *Melastomataceæ.*
B. Flowers epigynous. Ovary inferior.
  Petals reduced or none.
    Aquatic or marsh herbs.
      Leaves whorled or pinnately dissected. Carpels 4. *Haloragidaceæ.*
      Leaves whorled. Carpels 1. *Hippuridaceæ.*
    Trees or shrubs. Petals minute, and fleshy when present. *Nyssaceæ.*
  Petals mostly present.
    Floating and submersed herbs with broad, ovate, floating leaves and finely dissected submersed leaves. Disk present. *Hydrocaryaceæ.*

Mostly land plants.
  Cells of ovary superposed. *Punicaceæ.*
  Cells of ovary not superposed.
    Stamens numerous.
      Anthers versatile or dorsifixed.
      Stamens 4-16. Fruit often winged. *Combretaceæ.*
      Stamens free or united into bundles or united at the base.
        Trees and shrubs, often aromatic. *Myrtaceæ.*
    Stamens as many or twice as many as the sepals or petals.
      Filaments geniculate or inflexed, the connective peculiarly modified, and a corona usually present between petals and stamens. *Melastomataceæ.*
      Filaments not geniculate; corona wanting. *Œnotheraceæ.*

The Perigynous Families. Ovary Superior

GEISSOLOMATACEÆ. A monotypic family based on *Geissoloma marginatum,* a shrub of the Cape region of South Africa. Leaves opposite, leathery. Flowers bisexual, 5-merous, in umbellate inflorescences. Petals none. Fruit a loculicidal capsule. Seeds with endosperm.

PENÆACEÆ. Five genera and 22 species of shrubs or subshrubs of the Cape region. Leaves opposite, leathery, evergreen. Flowers small, solitary or in pairs, axillary, bisexual, 4-merous. Fruit a loculicidal capsule, with 1 seed in each cell. Ovules with 1 integument. Endosperm none. Embryo with a thick, fleshy hypocotyl and small cotyledons. Hypanthium deep, tubular.

THYMELÆACEÆ (*Mezereum Family*). Flowers regular or irregular, bisexual or unisexual, polygamous or diœcious, heteromorphic or trimorphic, 5-merous or 4-merous, or sometimes 6-merous, commonly with a deep, vasiform, petaloid hypanthium. Sepals petaloid. Petals none. Stamens twice as many as the sepals or fewer (2), in two whorls, and inserted at different levels in the tube of the hypanthium. Below the ovary or at various levels in the hypanthium there are small processes growing out from the hypanthial wall. Gynœcium 2—many(12)-carpellate, or mostly consisting of a single carpel raised on a gynophore—but in all cases 1-celled and containing 1 ovule (anatropous) which is suspended from above. Style short or elongate terminating in a capitate, club-shaped, or cylindrical stigma. Fruit a nutlet or drupe, enclosed in the persistent hypanthium, or a loculicidal capsule. Seeds with scanty or no endosperm and a straight embryo with fleshy cotyledons.

The plants are trees or shrubs, or in some cases herbs, with alternate or opposite, exstipulate, crowded or scattered leaves. Flowers small, in terminal or axillary inflorescences of various sorts, or sometimes solitary. About 39 genera and 460 species of wide distribution through warm and temperate parts of the world but most abundant in Australia and South Africa. But two species of this

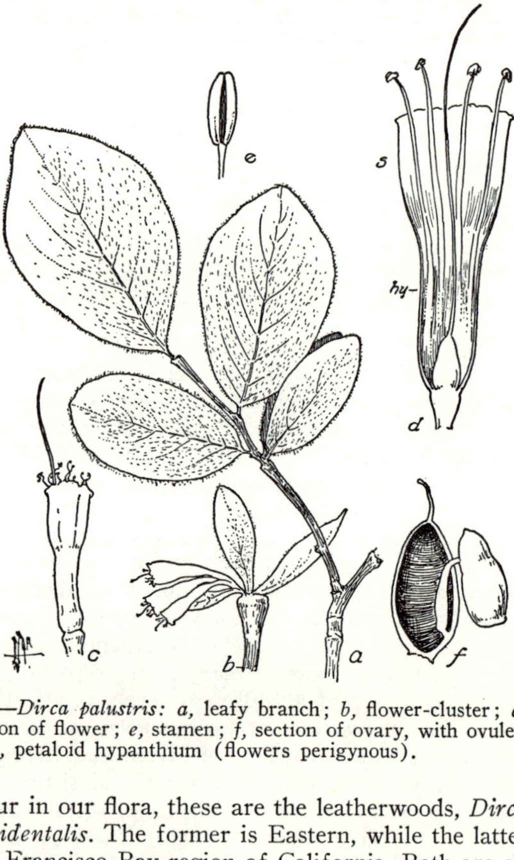

FIG. 300.—*Dirca palustris: a,* leafy branch; *b,* flower-cluster; *c,* flower; *d,* median section of flower; *e,* stamen; *f,* section of ovary, with ovule drawn out; *s,* calyx; *hy,* petaloid hypanthium (flowers perigynous).

family occur in our flora, these are the leatherwoods, *Dirca palustris* and *D. occidentalis*. The former is Eastern, while the latter is found in the San Francisco Bay region of California. Both are shrubs with very tough bark and soft, alternate leaves. The small, yellowish, apetalous flowers are borne in lateral clusters. Eight stamens arise from the throat of the deep, obliquely irregular hypanthium. The fruit is a drupe. The bark of the Eastern species was formerly used by the Indians for making thongs. The genus is limited to these two species. (Fig. 300.)

Several species of *Daphne* are grown as ornamentals. The Eurasian *Daphne Mezereum* occurs as an escape in northeastern parts of the country and in adjacent Canada. The plants are shrubs with evergreen or deciduous leaves and fragrant, clustered flowers. The hypanthium is tubular and regular. The fruit is a drupe. Species 40 or more, all native to Europe and Asia. A West-Australian shrub, *Pimelea ferruginea,* with opposite leathery leaves and rose-colored flowers, is grown as an ornamental in California.

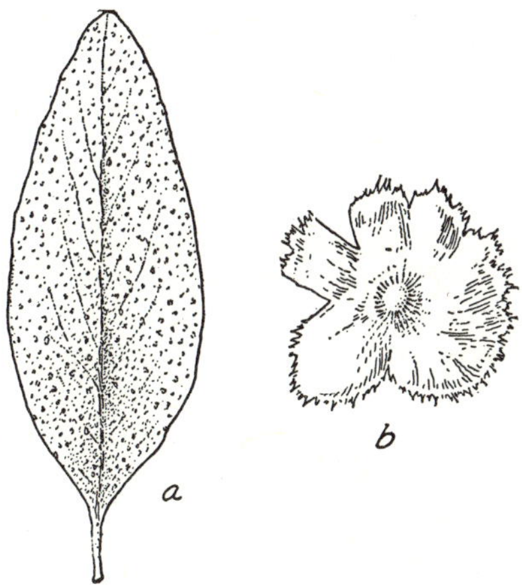

FIG. 301.—Leaf (*a*) and scale of leaf (*b*) of *Shepherdia.*

ELÆAGNACEÆ (*Oleaster Family*). Flowers regular, bisexual or unisexual, polygamous, monœcious or diœcious, with a cup-shaped or funnel-shaped hypanthium (shallow in staminate flowers). Calyx 4-merous or 2-merous; corolla wanting. Stamens as many or twice as many as the sepals and inserted in the throat of the hypanthium. Gynœcium 1-carpellate, with an elongate style terminating in a capitate or oblong stigma. Ovary containing but 1 ovule which arises from a basal placenta. A disk with as many lobes as stamens lines the throat of the hypanthium. Integuments 2. Fruit a nutlet or achene, invested by the basal portion of the hypanthium, which becomes fleshy. Seed with scanty or no endosperm. Embryo straight, with thick cotyledons.

Three genera and probably less than 20 species of shrubs or trees, chiefly of the north temperate zone but a few reaching into the tropics in southern Asia. Leaves simple, alternate or opposite. The plants are clothed with peculiar shield-shaped or stellate scales which give them a silvery appearance. (Fig. 301.)

Genera: *Shepherdia, Elæagnus, Hippophaë.*

Shepherdia is diœcious, with 8 stamens in the staminate flowers. The achene (or nutlet) is invested by the fleshy hypanthium, which is closed at the top by the 8-lobed disk. *Shepherdia canadensis* (*Lepargyrea canadensis*), a thornless shrub occurring mostly on stream-banks, ranges from Maine to British Columbia and Alaska and southward to New Mexico. It bears opposite leaves and an insipid red or yellow fruit. The thorny and more silvery *S. argentea* is found on the Western prairies and plains. It produces scarlet and edible but sour berries. In some localities it is common along the highways. Another species is said to occur in the Rocky Mountain region and westward.

Elæagnus is very similar to *Shepherdia* in appearance, but differs in having alternate leaves and bisexual flowers with only 4 stamens. Of the 20 species only one, *Elæagnus argentea* (Buffalo Berry), is found in North America. This is a beautiful silvery shrub found on the prairies and plains from Minnesota to Utah and far northward and northwestward. The small yellow flowers have a heavy, sweetish odor. The berry is silvery white and mealy, with a greenish pulp. The plant is an indicator of very poor soil. Eighteen or 20 other species occur in Europe, Asia, and Australia. A number of these may be seen in cultivation,—*e.g., E. multiflora* (Gumi), a shrub with acid fruit, and *E. angustifolia* (Oleaster).

Hippophaë comprises two species of willow-like deciduous shrubs or small trees of Europe and Asia. They have alternate leaves and thorny branches. The flowers have but 2 sepals and 4 stamens, but the fruit is similar to those of the preceding genera. *H. rhamnoides,* a small tree, is planted as an ornamental.

LYTHRACEÆ (*Loosestrife Family*). Flowers regular or irregular, bisexual, with a tubular, funnel-shaped, or campanulate, persistent hypanthium. Calyx segments 3—16, often with intermediate teeth. Petals as many or fewer (or none). Stamens many or 1, inserted at various levels in the hypanthium. Gynœcium 2—6-carpellate, with 1 style, and sometimes supported on a gynophore. Ovary 2—6-celled (rarely 1-celled), with numerous ovules on axile or thick basal placentas. Fruit a septicidal or irregularly dehiscent capsule which is completely invested by the persistent hypanthium. Seeds without endosperm. Embryo sometimes with auricled cotyledons. When the ovary is 1-celled, the ovules arise from parietal placentas.

The plants are herbs, shrubs, or trees (in tropics), with opposite or whorled, rarely alternate, entire, stipulate or exstipulate leaves, and axillary, solitary or variously clustered flowers, usually with 2-bracteolate pedicels. Heterostyly and dimorphic flowers are characteristic of some genera. Species about 400, in 20 or more genera, widely distributed over temperate regions, but most abundant in the American tropics.

Our genera are *Lythrum, Didiplis, Rotala, Decodon, Ammania, Cuphea.* Species of *Lawsonia* and *Lagerstroemia* are in cultivation.

In all of these but *Cuphea* the flowers are regular or nearly so. In *Lythrum* (Loosestrife) the hypanthium is funnel-shaped or campanulate and at different levels within it are 6—12 stamens in two sets—long and short. Petals 6. Some species are heterostylous. The two dozen or more species are annuals or perennials with quadrangular stems and opposite or whorled or alternate leaves. Some are cultivated. In ours the flowers are either small and solitary in the axils of the upper leaves or large and in whorls in an interrupted spike. The plants are widely distributed over the continent.

Didiplis is monotypic, the lone species *D. diandra* ranging from Minnesota and Wisconsin to Florida, Texas, and New Mexico. It is a small aquatic herb of ponds and lakes, but sometimes roots in the mud. It is characterized by quadrangular stems, opposite, linear, dimorphic leaves and very small greenish, apetalous flowers borne singly in the axils of the upper leaves. The submersed leaves are longer than the aërial ones and sessile by a broad base. Appendages are lacking in the sinuses of the calyx. Stamens 2 or 4. Gynœcium 3-carpellate, with a 2-celled ovary. The globose, indehiscent, capsular fruit projects slightly from the investing hypanthium and contains many small seeds. Somewhat similar is the low annual, *Rotala ramosior* (Toothed-Cup), which is widely distributed in swampy habitats over the continent. It has spread northward from South America and the West Indies. The genus contains about 30 species, mainly of the tropics. It differs from *Didiplis* in having 4 petals, 4 stamens, and a 4-carpellate gynæcium which forms a 4-celled fruit. *Decodon,* like *Didiplis,* is a monotypic genus of eastern North America. The species *Decodon verticillatus* (Swamp Loosestrife) is a tall, willow-like perennial aquatic with angular stem and long, lanceolate, commonly whorled leaves. Trimorphic flowers are borne in clusters in the axils of the upper leaves. The hypanthium is cup-shaped and between the 5—7 sepals are some linear appendages. Petals commonly 5, and stamens 8—10, in two sets—long and short. Gynœcium 3—5-carpellate, with a slender style and a 3—5-celled ovary. Fruit a globose, loculicidal, many-seeded capsule invested by the hypanthium.

Ammania comprises low annual swamp herbs with narrow, opposite, sessile, auriculate leaves, and the flowers solitary or clustered in the axils. Petals 4 or none. Stamens 4—8. Gynœcium 2—4-carpellate, with short or long style. Fruit similar to that of *Didiplis* and *Rotala*. Twenty or more species, widely distributed but most abundant in warm regions.

Cuphea stands apart from all the preceding genera by its distinctly irregular flowers; and in addition the hypanthium is spurred or saccate at the base and oblique in the throat, and the 6 calyx-segments are reduced to small teeth which alternate with as many appendages in the sinuses. The hypanthium is ribbed on the outside. The 6 petals are unequal. Stamens 11—12 or fewer, in two whorls. The 2-carpellate gynœcium is supported on a gynophore in some cases. There is 1 style and an unequally 2-celled ovary with few to many seeds in each cell. The globose, loculicidal capsule is invested by the hypanthium. In some cases there is developed a peculiar ring or collar on the gynophore at various levels below the ovary. The plants are herbs or shrubs with opposite or whorled leaves. A number are horticultural. About 200 species have been described. Three of these are found in the South, but one of them, *Cuphea petiolata,* has pressed north to New Hampshire, Illinois, and Kansas.

HETEROPYXIDACEÆ. Based on *Heteropyxis,* with 2 species of shrubs (or trees?) of southeastern Africa. Leaves alternate, lanceolate. Flowers regular, bisexual, 5-merous, in panicles. Hypanthium campanulate. Stamens twice as many as petals. Gynœcium 2-carpellate, with 1 style. Fruit a capsule. Lysigenous glands present.

The Epigynous Families. Ovary Inferior

OLINIACEÆ. One genus: *Olinia,* of two species; African. Shrubs or small trees with opposite entire leaves and small, white or red, regular, bisexual, 4-merous or 5-merous flowers in terminal cymose clusters. Gynœcium 3—5-carpellate, with 1 style, and a 3—5-celled ovary containing 2 or 3 ovules on axile placentas. Integuments 2. Endosperm none. Embryo with folded cotyledons. Fruit drupaceous.

SONNERATIACEÆ. Flowers unisexual or bisexual, regular, with differentiated perianth and campanulate receptacle, and perigynous or epigynous. Sepals 4—8, petals 4—8 or wanting. Stamens of indefinite number. Gynœcium of 4—15 carpels, with 1 style, and the cells of the ovary containing numerous ovules. Fruit capsular or berry-like, spurious in the epigynous members. Placentation basal or parietal.

Species about 12, in 3 genera, comprising trees and shrubs with leathery, mostly opposite, leaves with caducous stipules. In general the flowers are conspicuous and showy. The family is confined to tropical Asia and Africa. *Sonneratia,* with its 6 species, constitutes an important element of the mangrove vegetation of tropical Asia and East Africa. A striking feature of these members is the long aërial roots that grow up out of the mud of the shore.

As in other families of the order, the advance from perigyny to epigyny is to be noted, the epigyny being, however, incomplete. An inconstant feature is seen in the placentation, which is parietal, with a tendency toward a distinct basal type.

CRYPTERONIACEÆ. One genus, *Crypteronia,* of 4 species of trees of southern Asia to the Philippines. Leaves opposite, entire, exstipulate. Flowers regular, 4-merous or 5-merous, in racemes. Gynœcium 2—3-carpellate. Ovary half-inferior, in a cup-shaped receptacle. Fruit capsular. Seeds numerous.

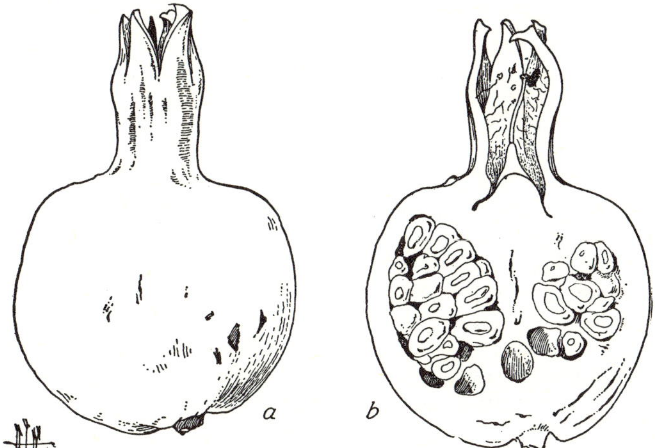

FIG. 302.—Fruit (*a*) of pomegranate (*Punica granatum*), with fleshy, prominent, persistent calyx at summit; *b,* sectional view,—style visible in calyx-tube (hypanthium).

PUNICACEÆ (*Pomegranate Family*). Flowers regular, bisexual, with a well-developed, thick-walled hypanthium. Sepals and petals 5—7. Stamens numerous, with slender filaments and versatile introrse anthers. Gynœcium of a varying number ($\pm$ 9) of carpels, with a long, basally thickened style terminating in a capitate stigma.

Ovary several-celled, the cells at unequal levels, and with numerous ovules on fleshy placentas. The spurious fruit is a many-seeded berry of the size of an apple, containing numerous large, angular seeds imbedded in a juicy pulp. Endosperm none.

The family comprises but one genus, *Punica*, of 2 species which are native from the Mediterranean countries to the Himalayas. The hypanthium and the calyx-segments persist in the fruit as a thick, fleshy, star-shaped crown. The bark is rich in tannin. *Punica granatum* (Pomegranate) is widely cultivated in the tropics for its edible fruit (known as a *balausta*) (Fig. 302).

LECYTHIDACEÆ (*Lecythis or Brazil-Nut Family*). Flowers regular, or irregular with respect to the andrœcium and corolla. Sepals and petals mostly 4—6. Stamens numerous, a part of them sometimes without anthers, or transformed into staminodes—in several series but in all cases more or less united at the base; anthers basifixed or versatile. Gynœcium 2—6-carpellate; ovary inferior or half-inferior, with as many cells as carpels, and with 1 to several (or many) ovules in each cell, variously suspended from axile placentas. Style simple, forked above; stigmas various. The summit of the ovary is commonly crowned by an annular disk. The fruits show a remarkable diversity in size, shape, and texture—they are woody capsules, or drupes or berries, are often very large, rough-walled or winged or pot-like, and often open by a lid. Seeds many to numerous, often large and hard-shelled, without endosperm. Embryo fleshy. The stamens are often curiously modified and clustered in a manner suggesting the passion-flowers. In many cases the flowers are very similar to those of the myrtles (see below).

The plants are tropical trees with large, striking, alternate leaves, and solitary or racemose, axillary or terminal flowers. Species about 130, in 18 genera. Largest genera, and number of species: *Barringtonia,* 60; *Japarandiba,* 20; *Lecythis,* 40; *Bertholletia excelsa* yields the Brazil-nuts of commerce, which are the seeds of a large, woody, capsular fruit. *Lecythis urnigera* produces a large woody urn-shaped fruit ("monkey-pot") which opens by a thick lid. Curiously shaped stamens occur in *Lecythis, Couropita, Cercophora,* and others.

RHIZOPHORACEÆ. Flowers regular, bisexual, or unisexual and polygamous, perigynous or epigynous. Sepals 3—16, commonly 4—8, fleshy or leathery. Petals when present as many as the sepals, The limb often lobed or fringed. Stamens 8—numerous, unequal, inserted on a perigynous or epigynous, lobed disk; anthers introrse.

Gynœcium commonly 2—5-carpellate. In epigynous forms the thick-walled hypanthium is produced beyond the ovary in a manner similar to the condition in the myrtles (below), but in perigynous forms the hypanthium is shallow and the gynœcium is supported on a gynophore. A disk is present in all cases and is infrastaminal, with the stamens usually borne on its outer margin. Ovary with as many cells as carpels, rarely 1-celled, and with 2, rarely 1 or 4, ovules in each cell on axile placentas. Fruit a berry, rarely a drupe or capsule. Seeds sometimes winged, with or without endosperm. Embryo small or large, straight or curved, often green, with 2 or sometimes 3 or 4 cotyledons which are more or less united or rudimentary. The hypocotyl is often enormously enlarged (tuberous).

The plants are trees or shrubs with opposite, evergreen, mostly entire, leathery, stipulate or exstipulate leaves. Flowers usually small, solitary or in cymose clusters in the axils of the leaves. The plants inhabit seashores and estuaries of the tropics, forming characteristic elements of the mangrove vegetation. Species about 60, fairly equally distributed among 17 genera. The following genera are characteristic: *Rhizophora, Bruguiera, Macarisia, Anisophyllum, Poga,* seeds of which yield a vegetable oil.

In *Rhizophora* the base of the stem soon dies away but is replaced by a mass of stout prop-roots which support the trees firmly in the mud of the shore. These prop-roots are abundantly provided with aërating organs which enable the trees to survive high tides. Germination of the seed often takes place while the fruit is still on the tree, in which case the hypocotyl develops to the enormous length of about a meter and finally becomes detached from the cotyledons and drops into the mud; as the radicle end is the heavier it drops in an upright position, thus burying the lower end in the mud, where development continues. In *Bruguiera* the seed falls with the fruit. *Rhizophora Mangle* is a characteristic element of the mangrove vegetation of the Floridan and West Indian coasts. The family is unique and interesting in every respect.

NYSSACEÆ (*Nyssa Family*). Flowers regular, bisexual or unisexual, polygamo-diœcious or diœcious, 5-merous, the stamens as many or twice as many as sepals or petals, or fewer, rarely more. Gynœcium 1-carpellate, with a 1-celled ovary containing 1 ovule, or 6—10-carpellate, with as many cells, each with 1 ovule; ovules pendulous from the top of the cell. Integuments 2. Style simple and stigmatic on the inner side, or with as many (radiating) branches as carpels, these stigmatic on the inner (upper) side. A thick, annular disk sometimes crowns the ovary. Endosperm copious, fleshy. Fruit drupaceous.

Trees and shrubs with alternate exstipulate leaves and small flowers in inflorescences of various sorts. Calyx sometimes greatly reduced, and petals sometimes wanting. Only the following genera: *Nyssa, Camptotheca, Davidia.*

This family has been variously interpreted, but the treatment here given is that of Hutchinson and of Engler and Gilg. By some writers *Nyssa* is placed with the dogwoods (*Cornaceæ;* see below), an interpretation which appears logical. As it stands the family appears to be a mixture of unrelated things. Compare the floral structures of the three genera.

In *Nyssa* the flowers are small, greenish, and polygamo-diœcious —the staminate numerous, the carpellate solitary or few in axillary clusters and appearing with the leaves. In the staminate flowers the sepals and petals are small, or the petals sometimes wanting, and inside the stamens is a prominent disk; a rudimentary gynœcium is sometimes present. The bisexual flowers have sepals and petals like those of the staminate flowers, besides 5—10 stamens. The inferior, 1-celled ovary is crowned by the disk (compare with *Cornus!*). There is no evident hypanthial tube. The simple style is stigmatic for some distance on one side, and the anatropous ovule is suspended from the top of the ovary (compare with *Cornus*). The spurious fruit is drupaceous, and contains a hard stone. The half-dozen or more species are native to eastern North America and eastern to central Asia. *Nyssa sylvatica* (Black Gum, Tupelo) is found in swampy habitats from Maine to Ontario and Michigan and southward; *N. aquatica,* in similar situations from Virginia to Illinois and Missouri and southward; *N. biflora,* in swamps and marshes from New Jersey southward and west to Louisiana, and *N. Ogeche* in swamps and marshes from South Carolina to Georgia and Florida. The pedicels in *Nyssa* bear two rather prominent bracteoles immediately below the flower. The embryo possesses a short hypocotyl and relatively large cotyledons and is surrounded by endosperm. *N. sylvatica* makes a brilliant display of foliage in autumn. The two other genera are Asiatic and monotypic. *Camptotheca* resembles *Nyssa* with respect to structure of the flower, but *Davidia* is entirely different—the staminate flowers consisting of but one stamen and being closely crowded on the enlarged areolate end of the peduncle which forms a common receptacle. On one side of this common receptacle there is usually a bisexual flower. The gynœcium of *Davidia* is 6—10-carpellate, with an inferior ovary supporting a thick style which terminates in a stigma of as many radiating branches as there are carpels. In each cell of the ovary there is one anatropous ovule. The perianth-segments are small, and there are several stamens at the base of the style. The fruit is a drupaceous berry.

ALANGIACEÆ. One genus, *Alangium*, of about 21 species of Asiatic, Australian, and African trees and shrubs with simple, alternate, exstipulate leaves, and sometimes bearing thorns. Flowers small, bisexual, regular, in cymes. Sepals, petals, and stamens 4—10 or more. Gynœcium 1—2-carpellate, with an inferior, 1—2-celled ovary crowned by a prominent fleshy disk. Style simple. Fruit drupaceous, 1-seeded. Endosperm fleshy.

COMBRETACEÆ (*Combretum Family*). Flowers regular, bisexual or unisexual, with 4—5 or more sepals and petals (sometimes wanting), and as many or numerous stamens. Gynœcium 1-carpellate (?), with inferior, 1-celled ovary. Ovules 2—6, pendulous from the top of the seed-chamber. Style 1, with dilated stigmatic lobes. The spurious fruit drupaceous or capsular. Seeds with endosperm investing a straight embryo. Fifteen or more genera and 450 species of tropical trees and shrubs with simple, alternate or opposite, entire, exstipulate leaves.

Principal genera: *Combretum*, 260 species of tropical Asia and Africa; *Terminalia*, 150 species of tropical East Africa and East Indies; *Quisqualis*, 4 species of climbing shrubs of tropical Africa and Asia. Species of each are found in cultivation. Some species yield tannins and edible seeds.

MYRTACEÆ (*Myrtle Family*). Flowers regular, bisexual, unisexual, or polygamous. Sepals and petals mostly 4 or 5, the sepals often greatly reduced, the petals rarely more than 5, often connivent, and sometimes wanting. Stamens commonly numerous, rarely few, with long, slender filaments, often in bundles, and generally forming the conspicuous part of the flower. Anthers small, opening by slits or in some cases by apical pores. Gynœcium of 2 to several carpels. Style 1. A deep or shallow hypanthium, the inner surface of which is often discoid, usually extends beyond the inferior ovary. Ovules numerous to 1 in each cell, mostly on axile, rarely parietal, placentas. The spurious fruit is baccate, drupaceous, capsular, or nutlike.

The plants are aromatic trees or shrubs with simple, opposite or alternate, often glandular-punctate leaves, with or without stipules.

Flowers often showy. One of the largest and most distinctive, but also one of the most technical, families of flowering plants, comprising over 70 genera and in the neighborhood of 3,000 species, mainly of the tropics of Australia and America. Many of the members yield valuable woods, gums, resins, oils, and edible fruits and choice ornamentals. None are native to the United States.

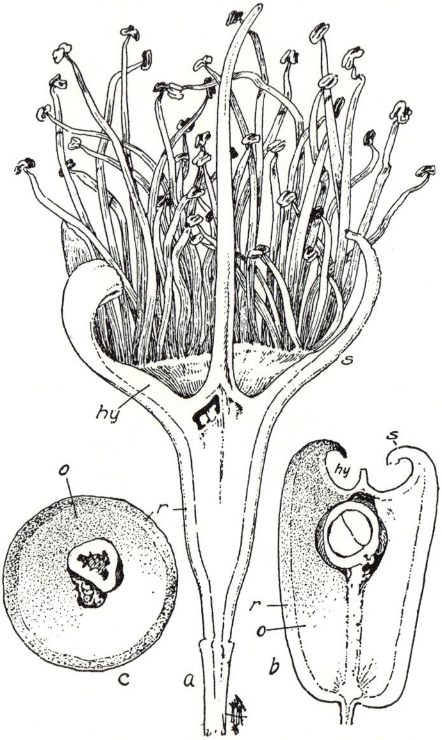

FIG. 303.—*Eugenia* sp.: *a*, median section of flower; *hy*, hypanthium; *o*, ovary; *r*, receptacle; *s*, sepal; *b*, section of fruit—hypanthium and sepal (*s*) above; *c*, transverse section of fruit.

The following genera are represented in cultivation, mainly in the South and Southwest: *Eucalyptus, Leptospermum, Myrtus, Eugenia, Pimenta, Feijoa, Psidium, Callistemon, Melaleuca.* These are distributed in two subfamilies, which are distinguished technically by their fruits, namely:

Myrtoideæ, in which the fruit is a berry or drupe. Including *Myrtus, Eugenia, Pimenta, Feijoa, Psidium.*

Leptospermoideæ, in which the fruit is a capsule. Including *Eucalyptus, Leptospermum, Callistemon, Melaleuca.*

Myrtus (Myrtle) comprises 100 or more species of evergreen shrubs and trees of warm regions. The leaves are opposite and entire, and the flowers solitary or in clusters in the axils of the leaves. Sepals and petals 4—5, but stamens numerous, with slender filaments and versatile anthers. The gynœcium is 2—3-carpellate, with 1 slender, elongate style. The numerous ovules arise on parietal placentas which are intruded into the cavity of the ovary until they meet and are then bent outward as false partitions, thus making the ovary appear two- or three-celled and the placentation axile. The spurious fruit is a berry. Embryo spiral or curved. *Myrtus communis,* the myrtle of literature, is a common ornamental, running into many forms.

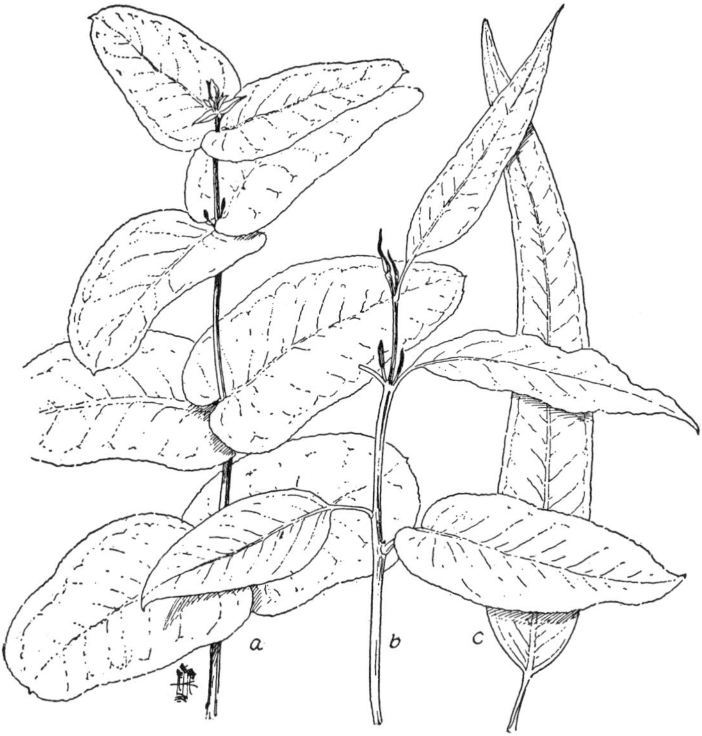

FIG. 304.—*Eucalyptus globulus:* *a,* juvenile leaves; *b,* transition form, and *c,* adult leaves.

Eugenia is very similar to *Myrtus* but is distinguished technically by its straight embryo. Authorities are at variance as to the limitations of this genus, but in the broadest conception it comprises perhaps 1,000 species (Bailey). The plants are shrubs or small trees, many of which are favorite ornamentals. The most important species is probably *Eugenia aromatica,* the dried flower-buds of which are the cloves of commerce. (Fig. 303.)

FIG. 305.—*Eucalyptus* sp.: *a,* branch with fruit-clusters; *b,* flower; *c,* transverse section of ovary below top; *d,* median section of flower; *s,* sepal; *p,* petal; *hy,* hypanthium; *r,* receptacle; *e,* flower after removing stamens; *f,* capsule; *g,* stamen; *h,* base of filament; *i, j, k,* different views of anthers; *m,* stigma; *n, o,* ovules; *p,* seeds.

Pimenta comprises about 5 species of tropical aromatic trees very similar to the myrtles and the eugenias, but differing from those in having fewer carpels and a spiral embryo. There are two important species, namely, *Pimenta officinalis* (Allspice or Pimento) and *P. acris* (Bay-Rum Tree). Allspice is derived from the dried unripe berry and bay-oil, the essential constituent of bay-rum, is distilled from the leaves.

Feijoa Sellowiana is a small tree which is grown for its luscious fruit, which is a pear-shaped berry crowned by the persistent calyx-

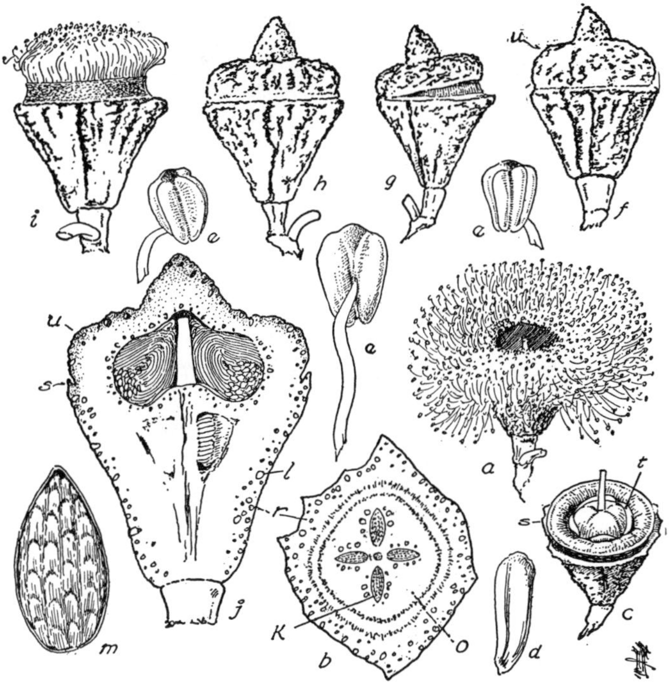

FIG. 306.—*Eucalyptus globulus: a,* flower; *b,* transverse section of "ovary,"—in center four locules, one shown enlarged at (*m*); *c,* young capsule with lid removed; *t,* gland; *s,* calyx border; *d,* seed; *e,* stamens; *f-i,* capsule viewed from different sides,—*i* with lid removed; *u,* lid; *j,* median section of flower-bud; *u,* corolla (?); *s,* sepal (?); *l,* oil-glands; *r,* receptacle; *k,* locule; *o,* ovary.

lobes. This and one other species are native to South America. *Psidium* is a large genus of about 150 species of shrubs and trees of tropical and subtropical America. *Psidium Guajava* (Guava) and *P. Cattleianum* (Strawberry Guava) are ornamentals. These and other species are grown in the tropics for their edible fruits, which are of the size of a plum. The flowers are 5-merous, except the numerous stamens, which are inserted on a disk inside the petals.

Eucalyptus, with 300 species, is one of the largest genera of the family; it ranges from the Malayan region through the East Indies to Australia and Tasmania. All are aromatic trees with alternate or opposite, broad or oblong to long, narrow, and sickle-shaped,

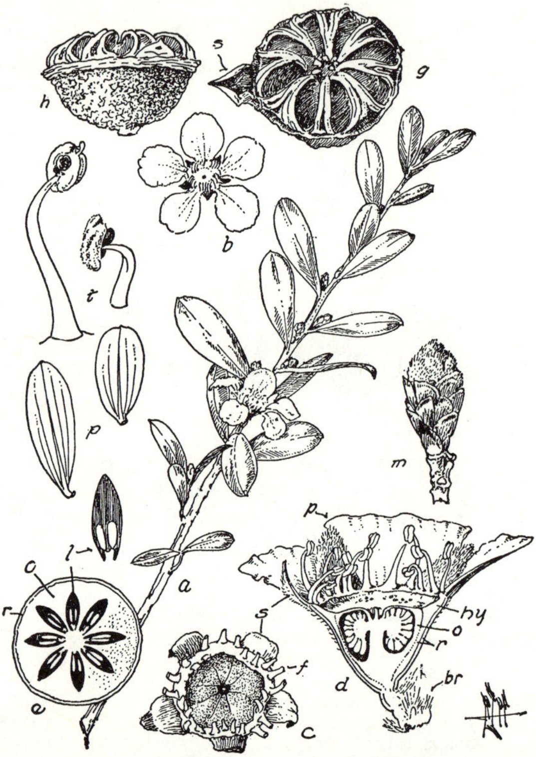

FIG. 307.—*Leptospermum lævigatum: a,* flowering branch; *b,* flower; *c,* flower with parts removed, showing arrangement of stamens (*f*) and discoid surface of ovary; *s,* sepals; *d,* median section of flower; *p,* petal; *hy,* hypanthium; *o,* ovary; *r,* receptacle; *br,* bracts; *e,* transverse section of ovary; *l,* locule; *m,* flower-bud; *g, h,* capsule after dehiscence.

leathery leaves. In many species the (juvenile) leaves of sprouts and young shoots differ greatly in shape from the ordinary leaves and are usually opposite (Fig. 304). A distinctive feature of the flower-buds is the peculiar lid or cap under which the numerous stamens

FIG. 308.—*Melaleuca hypericifolia: a,* stem and flowering branch; *b,* leafy branch; *c,* leaves; *d,* flower (bright red); *e,* bundle of stamens; *f,* anther; *g,* dorsal aspect of anther; *h,* median section of flower; *s,* sepal; *p,* petal; *y,* style; *i,* median section of ovary; *r,* receptacle (hypanthium none); *j,* stigma; *k,* inflorescence (spike) in early stage, with continuation of axis as a leafy shoot; *l,* mature fruits; *m,* immature fruit; *n,* mature individual fruit (spurious), with persistent thick calyx.

are crowded. When the buds expand, the lid dehisces around the base and releases the stamens, which then straighten and spread out radially. The lid has been interpreted as being evolved from the fusion of calyx and corolla. In all species the showy part of the flower is thus the stamens, which in some cases are brilliantly colored. The gynœcium is 3—6(or more)-carpellate, with 1 elongate style. In each cell of the ovary are numerous ovules on an axile placenta. The fruit is a smooth, rough, or angular capsule dehiscing loculicidally by slits or by pores at the top. The seeds are numerous and very small. The flowers are solitary or in heads or umbels in the axils of the leaves. The genus is a very technical one; fruit and seed are required for exact identification of species, except in cases where leaf-form or bark is especially distinctive. A very common species in California is *Eucalyptus globulus* (Blue Gum), of which many large specimens are to be seen. The bark comes off in long strips, leaving the trunk smooth. The ordinary leaves are sickle-shaped and drooping, the juvenile ones ovate-oblong and opposite. The stamens are white, and the fruit is a woody, warty, angular capsule. The tree is very aromatic. In its native home it is said to rival our sequoias in size. (Figs. 305, 306.)

About 25 species of *Leptospermum* occur from the Malay region to Australia and New Zealand. These are shrubs or small trees with small, flat, rigid, alternate leaves with a bluish tinge. The small 5-merous flowers are borne solitary or in twos or threes in the axils of the leaves or at the end of spur shoots. The hypanthium is shallow or bowl-shaped. The fruit is a small, broad capsule which opens by slits at the top. Several species are ornamental (Fig. 307). *Callistemon* and *Melaleuca* are Australian shrubs and trees. The leaves in both are lanceolate to linear, rigid, and often sharp-pointed. In *Melaleuca* the stamens are united in bundles, while in *Callistemon* they are free. The former comprises about 100 species, the latter 25. Many are ornamentals. The flowers are clustered along a portion of the stem, their long and numerous radiating stamens giving the stems the fancied appearance of bottle-brushes, as these plants are called. Fruit a capsule. (Figs. 308, 39).

MELASTOMATACEÆ (*Melastoma Family*). Flowers regular or irregular, bisexual, rarely unisexual, perigynous or epigynous. Sepals and petals commonly 4-merous or 5-merous. Stamens commonly twice as many as sepals or petals, the filaments bending inwards or downwards or otherwise variously curved, the anthers 2-celled and of various shapes, often with elongate and curved or curiously modified connectives and opening mostly by apical pores. Gynœcium commonly 4—5-carpellate, with straight or curved style. Hypanthium shallow to deep. Ovary inferior or in some cases superior, with as many cells as carpels and with several ovules in each cell

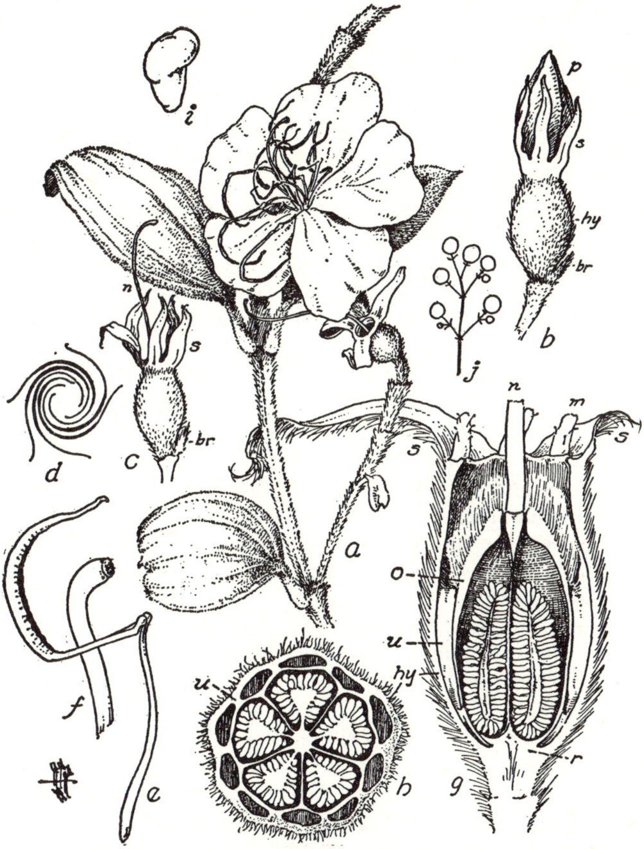

FIG. 309.—*Tibouchina semidecandra* (*Melastomataceæ*): *a,* flowering shoot; *b,* flower-bud; *c,* flower with petals removed; *d,* æstivation of corolla; *e,* stamen; *f,* summit of style with stigma; *g,* median section of flower,—petals, stamens, and part of style removed; *h,* transverse section of *g*; *i,* ovule; *j,* diagram of inflorescence; *br,* bracts; *m,* bases of filaments; *n,* style; *o,* ovary; *p,* petals; *r,* receptacle; *s,* sepals; *u,* septum connecting ovary (*o*) with hypanthium (*hy*).

on axile or sometimes basal placentas. Fruit a berry or a loculicidal or irregularly dehiscent capsule. Seeds often very small, without endosperm. Embryo straight, with large cotyledons. A large family of about 3,000 species of herbs, shrubs, or trees, distributed among about 175 genera, mainly of the tropics.

One genus, *Rhexia,* has pushed up out of the tropics into eastern United States, where it is represented by about four species. The plants are herbs, sometimes tuberiferous, with opposite, 3—5-nerved, ovate to lanceolate, mostly sessile leaves and showy 4-merous flowers borne solitary or in cymes. Stamens 8, the anthers straight or curved and sometimes spurred. The ovary is only partly inferior, and there is a well-developed tubular hypanthium. Fruit a capsule. Seeds spirally coiled. *Rhexia virginica* occurs in swamps from Maine to Ontario and Iowa, and southward.

According to Bailey the following genera are represented in cultivation: *Tibouchina, Centradenia, Heterocentron, Schizocentron, Sonerila, Bertolonia, Phyllagathis, Medinilla, Tococa, Miconia.* (Fig. 309.)

HYDROCARYACEÆ (*TRAPACEÆ*) (*Water-Nut Family*). Based on the genus *Trapa,* which includes about 3 species of aquatic herbs native to Europe, Asia, and Africa. The leaves are of two forms, namely, broad, floating ones with inflated petioles, and finely dissected (pinnatifid) submersed ones. The flowers are typically 4-merous except for a 2-merous gynœcium, the ovary of which is crowned by a disk. One of the carpels is suppressed, resulting in a 1-celled, 1-seeded nutlet. *Trapa natans* (Water-Nut or Water Caltrop), a Eurasian plant, has been naturalized in quiet streams and ponds in parts of New England. The seeds are said to be edible. By some writers the genus has been placed in the *Œnotheraceæ.*

ŒNOTHERACEÆ (*Evening Primrose Family*). Flowers regular or irregular, bisexual, epigynous, commonly with 4 sepals and 4 petals, but these organs often fewer and sometimes more, or the petals sometimes wanting. Stamens 4—8 or fewer or more, in some cases staminodial. Gynœcium generally 4-carpellate, rarely more or fewer (1-carpellate). Ovary completely inferior, with the hypanthium shallow or deep above it. Style 1, usually terminating in as many stigmatic branches as there are carpels. Ovules usually numerous in each cell, on axile placentas. Fruit a berry, nutlet, or capsule dehiscing loculicidally by valves so as to leave the central axis free with the seeds attached. The small seeds are naked, winged, or furnished with a tuft of hairs, and are without endosperm. The embryo is straight.

The members of this family are annual or perennial herbs or in some cases shrubs, with alternate or opposite, exstipulate leaves and often showy flowers borne singly or in spikes or racemes. Genera 38 or 40 and species about 500, chiefly American but widely distributed also in other parts of the world. The family is a difficult one taxonomically for the reason that characters intergrade to such an extent

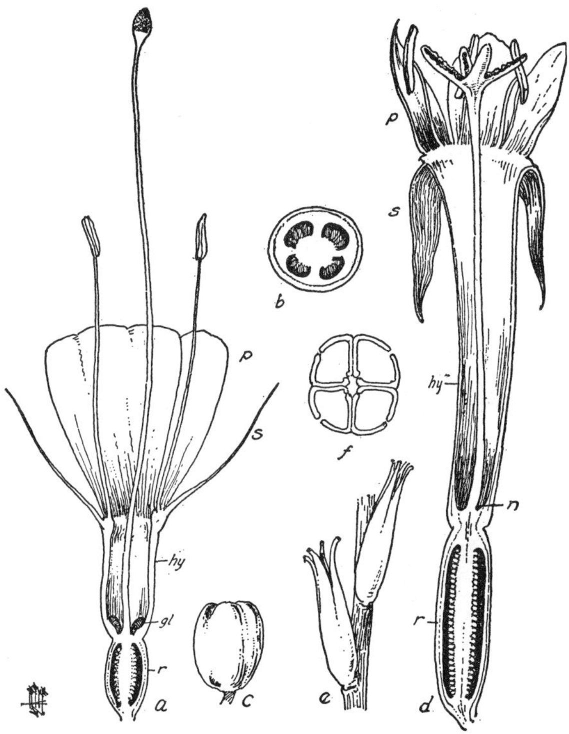

FIG. 310.—Median section of flower of a *Fuchsia* (*a*) and of an *Œnothera* (*d*); *b*, transverse section of ovary of *Fuchsia*, and *c*, the fruit (berry); *e*, capsules of *Œnothera*, and *f*, transverse section of a capsule; *hy*, hypanthium; *r*, receptacle; *gl*, gland; *p*, petals; *s*, sepals; *n*, nectary.

that generic and specific boundaries cannot be distinguished except on very technical characters. The family is abundantly represented in our flora, especially in the West, while a dozen or more genera are represented in gardens and hot-houses. The berries of some species are said to be edible. The following genera are characteristic:

Fruit a capsule; floral organs in 4's or more: *Œnothera, Clarkia, Epilobium.*

Fruit a nutlet; floral organs in 2's: *Circæa.*

FIG. 311.—*Epilobium angustifolium: a,* inflorescence; *b,* flower after opening; *c,* later stage of flower, with stigma exserted; *d,* flower-bud; *e,* median section of "ovary"; *f,* transverse section of ovary; *g,* stigma; *h,* stigma before expanding; *i,* ventral, and *j,* dorsal aspect of stamen; *br,* bract; *hy,* hypanthium; *o,* ovary; *p,* petal; *s,* sepal; *st,* stamen; *r,* receptacle; *gl,* gland; *u,* ovule.

Fruit a berry; floral organs in 4's: *Fuchsia.*

Œnothera (Evening Primrose). The evening primroses are annual or biennial, often tall and stout herbs with alternate leaves and showy flowers. The biennial ones produce a stout tap-root and a rosette of leaves close to the ground the first season. The flowers are typically 4-merous, with 8 stamens. In many species the hypanthium develops as an elongate tube from which the top of the style protrudes. The tube is nectariferous at the bottom. The style terminates in four spreading or recurving stigmatic branches. The fruit is an elongate capsule. Of the many native species the following are common: *Œnothera biennis, O. muricata, O. rhombipetala, O. laciniata,* and *O. pallida.* In its broadest interpretation the genus will probably aggregate 100 species. (Fig. 310.)

The clarkias (*Clarkia*) are rather fragile, low annuals, chiefly of the Far West. As in *Œnothera,* the parts of the flower are in 4's, with 8 stamens. But the plants differ from *Œnothera* in having conspicuously clawed and lobed petals, and the inner set of stamens is shorter than the outer, or is rudimentary. The capsule resembles that of *Œnothera.* The flowers are showy, in terminal racemes. There are probably less than a dozen species. One of the most widely distributed is *Clarkia pulchella* of the Rocky Mountain region.

Epilobium comprises 75 to 80 species of low to tall herbs (sometimes rather woody) with alternate or opposite leaves and small to large and showy flowers borne singly in the axils or in terminal axillary spikes or racemes. Structurally the flowers are like those of the preceding genera, but the capsule is longer and narrower, with the valves reflexing at dehiscence, and the seeds bear a tuft of hairs at one end. The numerous species in our flora are plants mostly of low, wet or moist grounds. *Epilobium angustifolium* (*Chamænerion angustifolium*), the Willow-Herb, is common across the continent as well as through Asia and Europe (Fig. 311). It is widely and easily disseminated by means of its light, tufted seeds, which are produced in abundance, and it is one of the first plants to take possession of burned-over forest areas (hence the name "fire-weed," by which it is frequently called). *Epilobium palustre, E. lineare,* and *E. adenocaulon* are common in bogs and swamps, particularly northward, the first-named reaching Alaska. In autumn they are conspicuous by their tangled masses of white, tufted seeds and dehisced capsules.

Circæa (Enchanter's Nightshade) is a genus of low, delicate herbs of deep, moist woods. The leaves are opposite, ovate, and soft, and the small white flowers appear in terminal or axillary racemes. The floral organs are in pairs—the stamens 2. The inferior ovary is

2-celled, and usually has 1 ovule in each cell, but 1-celled, 1-ovuled ovaries sometimes occur. A small nut-like fruit is produced, which is densely clothed with bristles.

Circæa lutetiana, the largest of our species, occurs in deep woods from Nova Scotia to Nebraska and Kansas, and also in the same latitudes in Asia and Europe. The smaller and more delicate *C. alpina,* which is found in deep, cold woods, has a much wider range—from Labrador south to Georgia and west to the Dakotas and thence to Alaska, and across Asia to Europe. It is a characteristic inhabitant of cold Northern cedar swamps. *C. intermedia* is found in the St. Lawrence and Great Lakes basins; and in the Sierras, *C. pacifica.*

Fuchsia (Fuchsia). The fuchsias are shrubs or trees or in some cases climbing plants, with simple, alternate, opposite, or whorled leaves and showy, mostly drooping, flowers in axillary panicles or racemes. The floral organs are 4-merous, the stamens 8. The hypanthium forms a tube of varying length which is variously colored. A prominent and variously lobed stigma terminates the long, slender style, which protrudes from the hypanthium. At the base of the style is an annular nectariferous disk of various shapes. The fruit of the fuchsias is a few- to many-seeded berry. The 70—80 species belong mainly in the American tropics, but a few are found in distant New Zealand—another illustration of a remarkably interrupted range. Several species are choice horticulturals. (Fig. 310.)

HALORAGIDACEÆ (*HALORRHAGACEÆ*) (*Water-Milfoil Family*). Flowers regular, bisexual or unisexual, monœcious or diœcious, epigynous, 4-merous, 3-merous, or 2-merous, or the gynœcium in certain cases reduced to 1 carpel. Stamens 2—8. Petals sometimes wanting. The inferior ovary 1—4-celled, usually with 2 ovules in each cell. Styles distinct, often with densely papillate or plumose stigmas. Fruit a drupe or an angular or winged nutlet. Aquatic or terrestrial herbs or somewhat woody plants of exceedingly diverse habits and remarkably wide geographic distribution, but mainly of the southern hemisphere. The leaves are alternate, opposite, or whorled, and of diverse shapes, sometimes pinnatifid. Species 100 or more, in about 8 genera.

Only the following genera are represented in our flora: *Myriophyllum, Proserpinaca.* Species of *Gunnera* are horticultural. These genera are distributed in two subfamilies which are distinguished as follows:

Halorrhagoideæ, in which there are 2—4 petals or sometimes none, and the gynœcium is 2—4-carpellate. *Myriophyllum, Proserpinaca.*

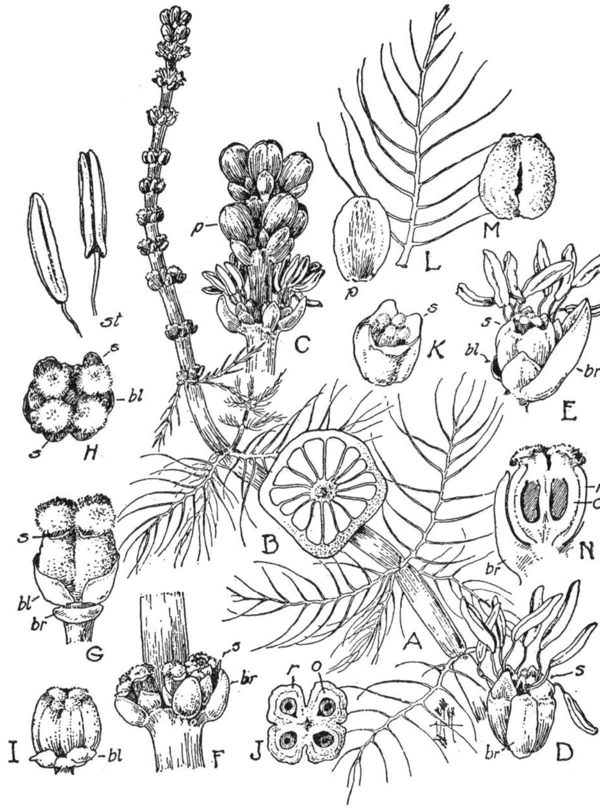

FIG. 312.—*Myriophyllum spicatum: A,* leafy shoot and terminal inflorescence; *B,* transverse section of stem; *C,* tip of inflorescence,—flower-buds above; *D,* bisexual flower; *E,* bisexual flower viewed from side; *F,* a whorl of flowers; *G,* a carpellate flower, and *H,* same viewed from above; *I,* immature fruit; *J,* transverse section of "ovary"; *K,* gynœcium and calyx of *E; L,* leaf; *M,* fruit (spurious), dorsal aspect; *N,* fruit in median section; *br,* bracts; *bl,* bracteoles; *p,* petals; *s,* sepals; *r,* receptacle; *o,* ovary.

Gunneroideæ, petals none, and the gynœcium 2-carpellate. *Gunnera.*

Myriophyllum (Water-Milfoil) (Fig. 312). The water-milfoils are aquatics with numerous alternate or whorled leaves of two sorts—aërial linear ones, and submersed ones which are finely dissected. The plants grow from creeping rhizomes. Polygamous or monœcious flowers are borne in the axils or in interrupted spikes, the staminate usually uppermost. The flowers are 4-merous, with 4 or 8 stamens; petals are sometimes wanting. At maturity the 4—1-seeded carpels split apart in the manner of a schizocarp. About 20 species have been described, of which probably less than a dozen occur in our flora, these being widely distributed in lakes, ponds, and streams; a few occur in Europe and Asia.

Proserpinaca (Mermaid-Weed). These are coarse, perennial aquatics from creeping rhizomes. The 3 or 4 species are confined to swamps of eastern North America, the West Indies, and Guatemala. The leaves are alternate, emersed or submersed, the former simple, elliptic or lanceolate and petiolate, the latter pinnately dissected into rigid segments. The flowers are bisexual, 3-merous or 4-merous, and apetalous. The fruit is a hard, 3—4-angled nutlet with as many cells, each 1-seeded. *Proserpinaca palustris* occurs in swamps from New Brunswick to Minnesota and southward into Central America, also in Bermuda.

Gunnera, with 25 species, is a genus of the southern hemisphere. Two species, *G. chilensis* (of Colombia, Ecuador, and Chile) and *G. manicata* (of Brazil), are grown as ornamentals. They are huge perennial herbs growing from rhizomes and bearing gigantic, palmately or digitately lobed leaves with petioles several feet (6!) long. Polygamous or monœcious flowers are produced in dense spikes or panicles, the staminate above, the carpellate mostly below, with bisexual ones in between. Sepals and petals are sometimes wanting, but when present number 2, accompanied by 1 or 2 stamens. The gynœcium is 1—2-carpellate; stigmas papillose. Fruit a drupe.

HIPPURIDACEÆ (*Mare's-Tail Family*). A rather anomalous family based on the genus *Hippuris* (Mare's-Tail), of which one or more species have been described, according to the point of view. The bisexual or polygamous flowers are distinctive in being 1-merous and apetalous, and the inferior ovary produces but 1 ovule, which is suspended from the top of the ovule-chamber. The spurious fruit is a small 1-seeded drupe or nutlet. The plants are perennial aquatics from creeping rhizomes, bearing mostly linear or lanceolate, verticillate leaves and small axillary flowers. *Hippuris vulgaris* is found in swamps all over the continent as well as in Asia and Europe, and in distant Patagonia.

CYNOMORIACEÆ. Based on the solitary *Cynomorium coccineum,* a curious succulent, root-parasitic herb of saline steppes of the

Mediterranean region and western Asia. Flowers unisexual or bisexual, epigynous, with 1—5 perianth segments. Staminate flowers with 1 stamen and a vestigial gynœcium. Carpellate flowers 1-carpellate and 1-ovulate. The thick aërial stems, a foot or more tall, spring from rhizomes, and bear numerous, spiral, scale-like leaves, and at the summit a thick, elliptical inflorescence of numerous closely crowded flowers. The sap is violet-colored and astringent.

ORDER UMBELLIFLORÆ (UMBELLALES)

Flowers regular, mostly bisexual, generally epigynous, 4-merous or 5-merous (rarely more). Sepals and petals small, the former commonly reduced to minute teeth. Stamens as many as sepals or petals. Gynœcium 5—1-carpellate (rarely more), with as many free styles. Ovary completely inferior. Ovules 1—2 in each cell. Endosperm copious. Integuments 1. Inflorescence mostly an umbel. Generally regarded as the most highly specialized order of the *Archichlamydeæ,* distinguished especially by the generally umbellate inflorescence, the small, completely epigynous flowers, and the generally constant number of floral organs.

Only three families: *Araliaceæ, Umbelliferæ, Cornaceæ.*

Key to the Families

Sepals, petals, and stamens 5.
  Carpels 2—5. Fruit a drupe or berry. *Araliaceæ.*
  Carpels 2. Fruit a schizocarp. *Umbelliferæ.*
Sepals, petals, and stamens 4. Fruit a 1—2-seeded drupe. *Cornaceæ.*

ARALIACEÆ (*Aralia or Ginseng Family*). Flowers mostly 5-merous, bisexual or polygamous. Calyx sometimes reduced to a mere ridge. Petals usually inflexed in the bud. Stamens alternating with the petals, rarely fewer or more than 5. Carpels commonly 5, but sometimes several or 1. Ovary generally inferior, with as many cells as carpels and crowned by a concave or convex disk, with 1 ovule in each cell. Styles distinct or partly united, with stigmatic lobes of various shapes. Fruit a berry or drupe. Seeds flattened or 3-angled, with a thin testa and containing a small embryo in copious endosperm.

Herbs, shrubs, or trees, with alternate, opposite, or whorled, simple or palmately, digitately, or ternately compound, often very large leaves. Stems in some cases beset with bristles or spines. Large members often palm-like in appearance. The plants are widely distributed in temperate and tropical regions. Authorities differ as to the number

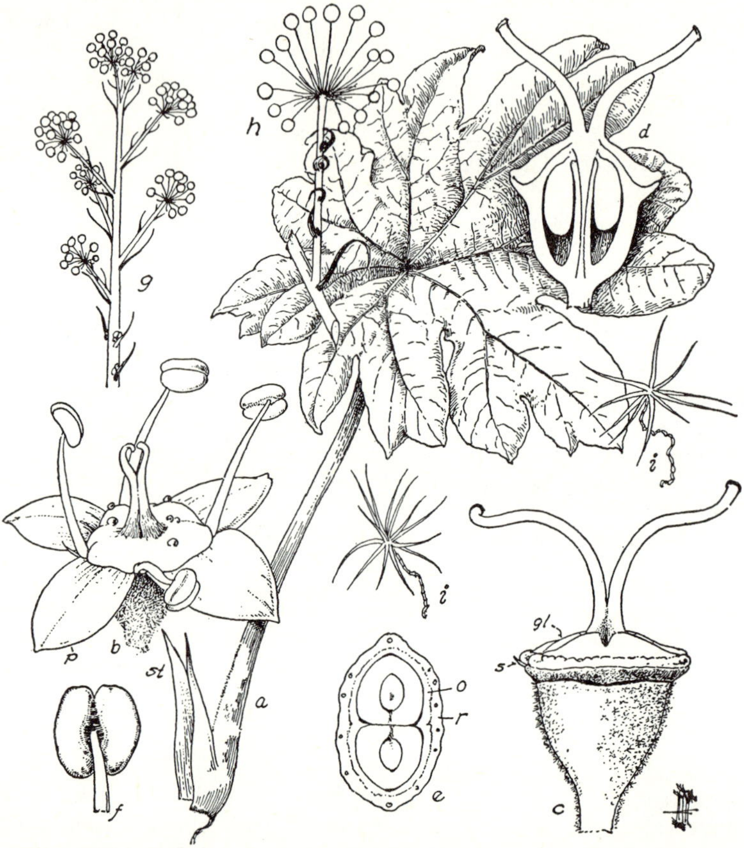

FIG. 313.—*Tetrapanax papyriferum: a,* leaf; *b,* flower; *c,* flower with petals and stamens removed; *d,* median section of same; *e,* transverse section of "ovary"; *f,* anther, dorsal aspect; *g,* inflorescence; *h,* umbel of inflorescence; *i,* stellate hairs of ovary; *p,* petals; *s,* sepal; *gl,* gland, *o,* ovary; *r,* receptacle.

of genera and species, these ranging from 50 to 63 and from 475 to 700, respectively.

Our native species belong to the genera *Aralia, Panax,* and *Echinopanax*. These and the following genera are represented by horticulturals: *Polyscias, Dizygotheca, Acanthopanax, Tetrapanax, Hedera.* (The suffix *-panax,* common to so many genera of this family, is derived from the Greek words πᾶς, "all" and ἄκος, "cure," meaning "all-healing"—the sap of the plants possessing certain healing properties.)

Aralia (Aralia). The aralias are aromatic herbs, shrubs, or small trees, with alternate, pinnately or ternately compound, often large leaves and often bristly or spiny stems. The plants produce small and numerous, whitish or greenish, 5-merous flowers in paniculate-umbellate clusters. The styles are distinct. The fruit, in ours, is a small, black, drupaceous berry crowned by the very small calyx-segments and the styles. *Aralia spinosa, A. racemosa,* and *A. nudicaulis* (Wild Sarsaparilla) are common in the East; in California and southwestern Oregon occurs the tall and handsome *A. californica. Aralia spinosa* (Hercules' Club or Angelica Tree) is a very spiny shrub or small tree. *A. racemosa* (Spikenard) is a large herbaceous form with widely spreading branches bearing ternately compound leaves. *A. nudicaulis* and *A. hispida* (Bristly Sarsaparilla) are small herbs with woody caudices. The former is abundant in the pine forests of the Great Lakes region; from the caudex arises a single, long-petioled, ternately compound leaf, beside which is a long, slender, flowering stem bearing the inflorescence at the summit. The "berries" are black when ripe.

In *Panax* (Ginseng) the flowers are polygamo-diœcious, with a 2—3-carpellate gynœcium, from which is developed a small, flattened, drupaceous fruit. The simple stems arise from one or two tuberous roots or from a fleshy tap-root and bear at the summit a whorl of digitately compound leaves and a terminal umbel. Two species, *Panax quinquefolium* (Ginseng) and *P. trifolium* (Dwarf Ginseng) occur in eastern North America, but both are now practically exterminated, if not entirely so, from many localities in which they were once common. They were gathered and sold, mostly in the Chinese markets, for the medicinal virtues of the roots. The true ginseng is the *Panax Ginseng* of China. *P. repens* occurs in Japan, and one or two other species are found in eastern Asia. From the standpoint of geographic distribution these plants are of interest as connecting the flora of eastern North America with that of eastern Asia.

Echinopanax horridum (Devil's Club) is a solitary species of the Pacific Coast and eastern Asia, with an isolated occurrence on Isle Royal in Lake Superior. It is a tall, densely prickly shrub bearing a crown of large leaves. Somewhat similar, but not spiny, is the handsome *Tetrapanax papyriferum,* the Japanese rice-paper plant, so common in yards in southern California (Fig. 313). The horticultural forms of *Dizygotheca* are unarmed woody plants with digitate leaves of 7—11 narrow segments. The gynœcium is 10-carpellate, with as many distinct styles. The cultivated forms of *Acanthopanax* are prickly or spiny shrubs or trees, with large, digitately lobed leaves. *Hedera Helix* is the common ivy, a creeping or climbing Asiatic species of polymorphic habit and leaf-form. It produces 5-merous flowers with a 3—5-carpellate gynœcium (Fig. 78).

UMBELLIFERÆ (*Carrot Family*). Flowers regular, bisexual or sometimes unisexual, the floral organs constantly 5-merous, except the gynœcium, which is constantly 2-carpellate. The sepals are reduced to small teeth, and the petals and stamens are inflexed in the bud. The two distinct styles are expanded below into a prominent *stylopodium* or disk, which thus crowns the ovary. The anthers are versatile. The inferior ovary is 2-celled, with 1 ovule in each cell. The fruit is a schizocarp crowned by the persistent stylopodium and persistent calyx-teeth. When fully ripe the two carpels separate as fruitlets or mericarps. The central axis of the fruit forms a carpophore from which the mericarps are suspended at the top. The mericarps are more or less winged or ridged on the sides, usually by five primary ridges or wings, and in the grooves between the ridges there are usually oil glands of various shapes and arrangements. These ridges, wings, and oil glands are of taxonomic value, but can be studied to best advantage only in ripe fruit; they serve a useful purpose as guides to genera (Figs. 314, 315, 316, 317).

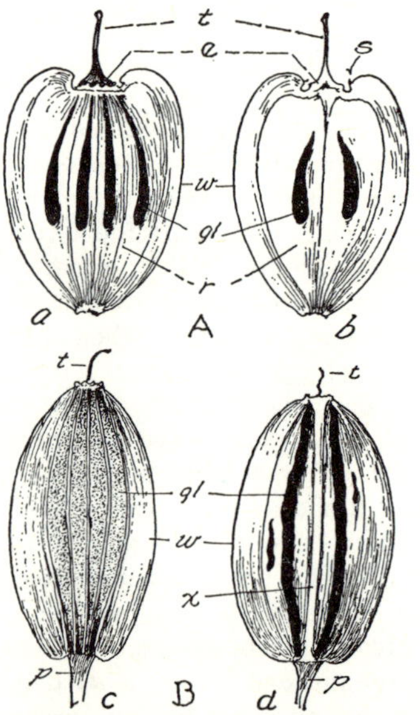

FIG. 314.—*A: Heracleum lanatum; a,* outer, *b,* inner face of fruit; *B: Lomatium Geyeri; c,* outer, *d,* inner face of fruit; *t,* style; *e,* stylopodium; *s,* sepal; *w,* wing; *gl,* gland; *r,* receptacle; *x,* axis; *p,* pedicel.

The plants are small to large herbs with simple or variously compound leaves, the petioles of which possess prominently sheathing bases. Stipules when present are minute. The stems are often hollow except at the prominent solid joints. A large, cosmopolitan family of 1,500—2,700 species in 250—270 genera, according to the point of view. The most obvious and characteristic feature is the umbellate inflorescence, the elements of which may or may not be subtended by bracts (involucre or involucel). The family yields a number of vegetables and medicinal products. Some are poisonous. (Fig. 317.)

Some of the characteristic genera are *Eryngium, Sanicula, Daucus, Osmorrhiza, Heracleum, Bupleurum, Lilæopsis, Hydrocotyle.*

In *Eryngium* (Eringo, Sea-Holly) the flowers are aggregated in

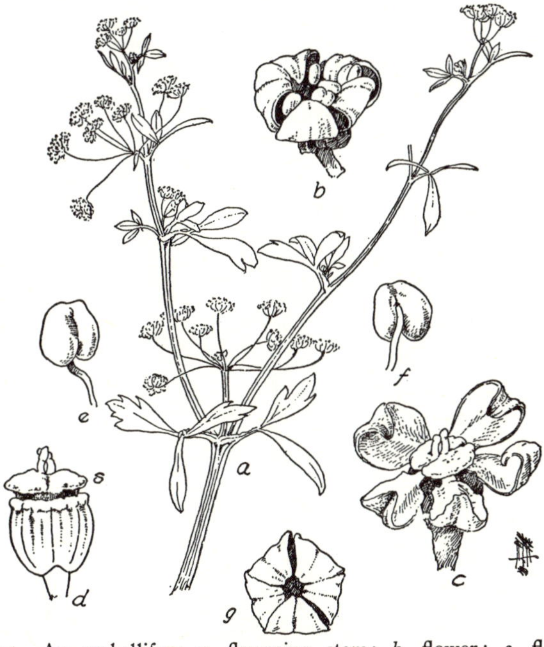

FIG. 315.—An umbellifer: *a,* flowering stem; *b,* flower; *c,* flower after pollination, with stamens removed; *d,* gynœcium; *e,* ventral, and *f,* dorsal, aspect of anther; *g,* corolla viewed from above; *s,* stylopodium.

involucrate heads or spikes, a marked digression from the ordinary umbellate type. The fruit is scaly or tuberculate. *Eryngium aquaticum* (Button Snakeroot) is a striking aberrant type as far as habit is concerned, as its leaves suggest a yucca. Other distinctive forms are *E. Leavenworthii,* which suggests a thistle, and *E. prostratum,* which might be mistaken for a buttercup. Between 175 and 220 species have been described for the genus, of which 30 or more are distributed over southern and western parts of the United States.

Sanicula (Snakeroot, Sanicle). These are biennial or perennial herbs with digitate leaves and small, yellowish, whitish, or purplish flowers in compound, few-rayed, umbels. The fruits are clothed with hooked prickles. The 25 or more species are mostly Northern, but a few are found in South America and in South Africa. Eighteen or 20 species are found in our woodlands. A typical species is *Sanicula*

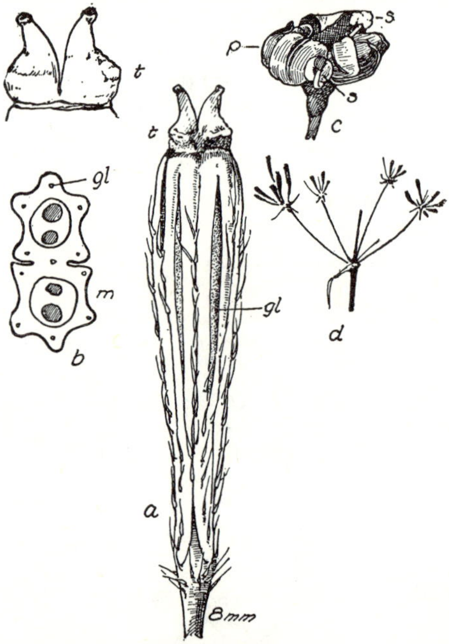

FIG. 316.—*Osmorrhiza longistylis: a,* immature fruit; *b,* transverse section of fruit; *gl,* oil gland; *m,* mericarp; *d,* inflorescence; *t,* stylopodium—enlarged at left; *c,* flower; *p,* petals; *s,* stamen.

marilandica, which occurs in rich woods from Newfoundland to Georgia and west to the Rockies. Several species are horticultural.

Daucus includes about 24 species scattered over various parts of the world. The most important species is *Daucus Carota,* the carrot, a tall, bristly biennial from a thick tap-root. In some localities it has spread as a weed. It is a native of Eurasia. The plant is distinguished by its pinnately dissected leaves and the usually purple central flower of the umbel, and the bristly or prickly fruit.

FIG. 317.—*Cicuta maculata* (Water Hemlock): *A,* leaf. *B,* inflorescence (compound umbel); *b,* involucral bracts. *C,* fruit (side view), with persistent stylopodium and styles at summit. *D,* dorsal, and *E,* lateral aspects of a mericarp, 3 millimeters in length, natural size; *g,* oil glands; *s,* stylopodium; *t,* style. *F,* tap-root.

Osmorrhiza (Sweet Cicely) is a group of soft, perennial, woodland herbs from fleshy aromatic roots. The decompound leaves have broad segments, but the most distinctive feature is the elongate, slender fruit, which is thickened above and bristly along the ribs. The 15 or more species are chiefly American and east-Asiatic. *Os,norrhiza longistylis* is common in rich woodlands from Nova Scotia to the Rocky Mountains, and within nearly the same range will be found *O. Claytoni,* and *O. divaricata.* (Fig. 316.)

One of the largest and stoutest of the native umbellifers is *Heracleum lanatum,* the cow-parsnip, which occurs in low grounds in rich woodlands across the continent into Alaska. The large, white umbels, which are sometimes a foot across, make the plant very conspicuous when in bloom; the leaves are very large and digitately compound, with large sheathing petioles. The stems are hollow except at the nodes. Of the 60 species this is the only one occurring in North America, the others being spread over other parts of the northern hemisphere.

In fields and waste grounds from New England to the Middle West and southward there is frequently found an introduced annual, up to two feet high, which differs markedly from our other umbellifers in having simple, clasping, perfoliate leaves and compound umbels of unusually large, yellowish or greenish-yellow flowers. This is *Bupleurum rotundifolium,* an immigrant from Europe. It produces an oval fruit on which oil tubes are lacking. A number of other species occur in the West. There are about 65 species in all, widely distributed. Very unusual also is the small *Lilæopsis lineata* of brackish and muddy shores of the Atlantic coast. This waif, scarcely more than 6 inches high, creeps along in the mud by means of rhizomes, from the nodes of which arise the flowering stems and one or two leaves which consist of no more than the petiole—which is hollow and jointed by cross partitions. The genus is monotypic.

Hydrocotyle is another distinctive genus, in which the leaves are orbicular, reniform, or peltate, or palmately lobed. Several species are scattered over the country, all creeping over low, wet, or moist grounds by means of rhizomes. From the nodes arise commonly solitary leaves and scapes bearing simple or proliferous umbels. A widely distributed species is *Hydrocotyle umbellata,* which occurs over most of the eastern United States to Texas and Mexico and the West Indies, and is also found in South America and in South Africa.

The following are common in cultivation: *Eryngium aquaticum* (Button Snakeroot), *Astrantia major* (Masterwort), *Daucus Carota* (Carrot), *Coriandrum sativum* (Coriander), *Ferula communis* (Giant Fennel), *Anethum graveolens* (Dill), *Fœniculum vulgare* (Fennel, common as a weed in waste grounds in California), *Levisticum officinale* (Lovage), *Petroselinum hortense* (Parsley), *Apium graveolens* (Celery), *Pimpinella sativa* (Anise), *Pastinaca sativa* (Parsnip), *Sium Sisarum* (Skirret), *Myrrhis odorata* (Myrrh, Sweet Cicely),

Anthriscus Cerefolium (Salad Chervil), *Chærophyllum bulbosum* (Turnip-rooted Chervil), *Carum Carui* (Caraway).

Poisonous are: *Cicuta maculata* (Fig. 317) and *C. virosa,* the Water-Hemlocks; *Æthusa cynapium,* and *Conium maculatum* (Poison Hemlock).

CORNACEÆ (*Dogwood Family*). Flowers regular, bisexual or unisexual, polygamous or diœcious, 4-merous or 5-merous, sometimes pleiomerous. Gynœcium 4—1-carpellate, with an epigynous disk. Style 1, slender or short and thick, or sometimes styles as many as carpels. Petals sometimes wanting. Ovary with as many cells as carpels, each cell containing 1 pendulous anatropous ovule. Fruit a drupe with a 1—2-seeded stone. Seeds with copious endosperm and a straight embryo with foliaceous cotyledons.

Shrubs or trees with opposite, alternate, or whorled, usually entire leaves and numerous small flowers in umbellate, cymose, or capitate inflorescences. About 85 species in 16 genera, chiefly of the northern hemisphere.

Principal genera: *Cornus, Aucuba, Helwingia, Griselina, Curtisia, Mastixia.*

Cornus (Dogwood). The dogwoods are handsome shrubs or trees with alternate, opposite, or whorled leaves and numerous small, white or purplish, uniformly 4-merous, flowers in cymose (umbel-like) or capitate clusters. The gynœcium, however, is 2-carpellate. The fruit is a red, white, blue, or black drupe with a 2-celled, 2-seeded stone. The sepals are reduced to minute teeth. Species with capitate inflorescence are distinguished by a showy involucre of four large bracts subtending the head (capitulum); the head itself is a compound inflorescence (Fig. 318). The 24 or more species are distributed over the northern hemisphere, but in the western hemisphere reach Peru. Many are grown as ornamentals. One of the handsomest of the native species is the large *Cornus florida* (Flowering Dogwood) of the Atlantic region, which has a showy involucre of four large white bracts. This has its counterpart in *C. Nuttallii* of the Pacific Coast, the five or six involucral bracts of which are often three inches long. Both yield scarlet drupes. In contrast to these is the dwarf *Cornus canadensis* of cold, damp woods from the north Atlantic coast to Alaska. Like the preceding, it possesses a showy involucre of four white bracts. The plants are commonly four to eight or ten inches high and grow in patches, from running rhizomes. The subherbaceous aërial stems bear a whorl of ovate or elliptic leaves. In autumn the patches of these plants make a colorful display in the pine region with their clusters of bright crimson fruits (Fig. 318). *Cornus suecica,* a similar but

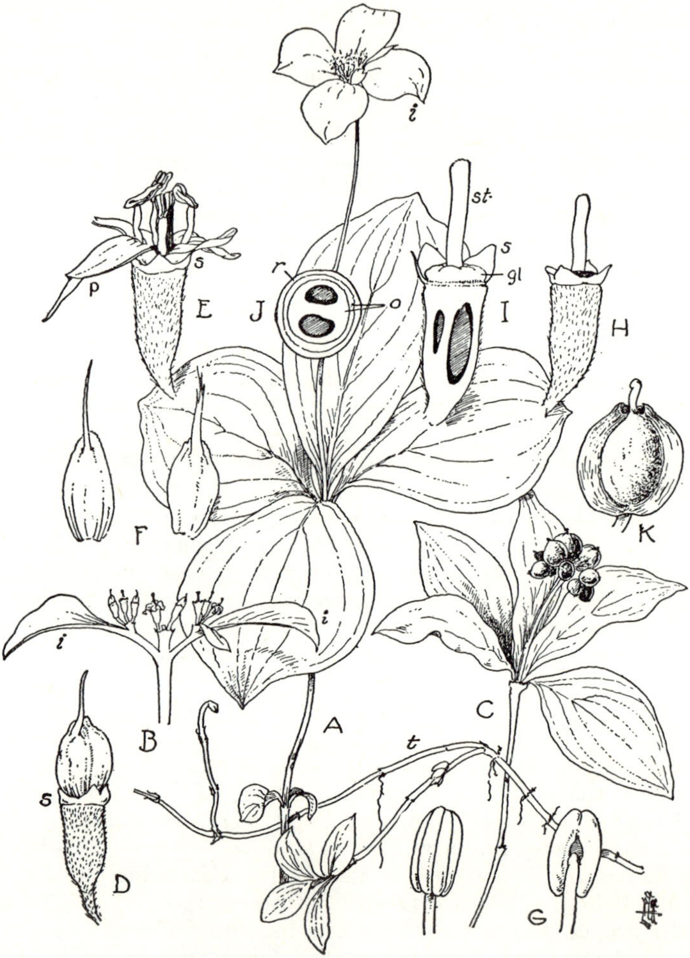

FIG. 318.—*Cornus canadensis: A,* habit sketch; *t,* stolon; *B,* inflorescence in sectional view; *i,* involucral bract; *C,* fruiting stage; *D,* flower-bud; *E,* flower; *F,* petals; *G,* dorsal, and at left, ventral, aspects of anther; *H,* flower with stamens and petals removed; *I,* same in sectional view; *J,* transverse section of "ovary"; *K,* median section of fruit; *i,* involucre; *p,* petals; *s,* sepals; *st,* style; *gl,* gland; *o,* ovary; *r,* receptacle.

smaller species of nearly the same range, bears three to six pairs of leaves and violet flowers; the fruits are red.

Among the dogwoods that do not possess an involucre, the com-

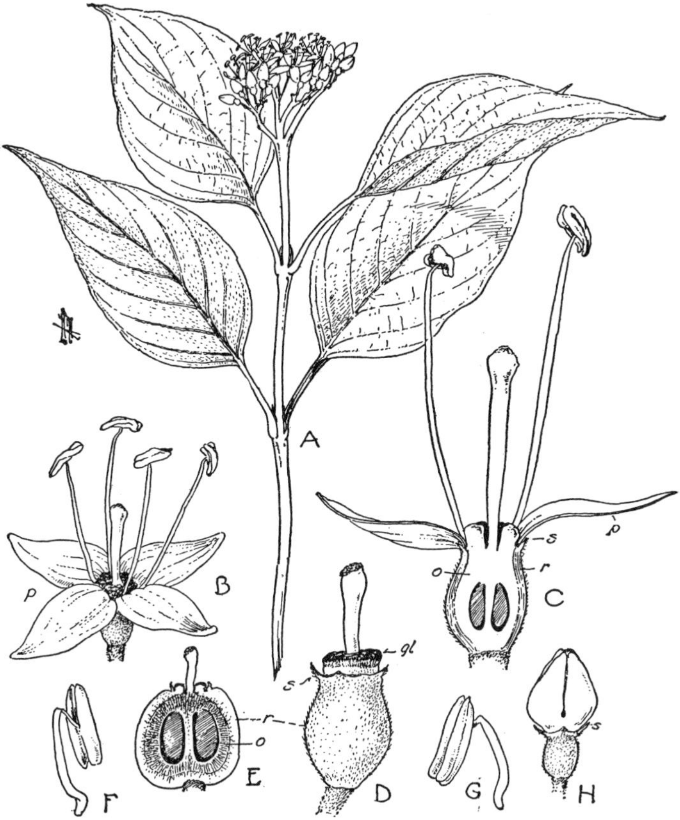

FIG. 319.—*Cornus Baileyi: A,* flowering stem; *B,* flower; *C,* median section of flower; *D,* flower with stamens and petals removed; *E,* median section of young fruit; *F,* dorsal, and *G,* lateral aspect of stamen; *H,* flower-bud; *gl,* gland; *p,* petals; *s,* sepals; *o,* ovary; *r,* receptacle.

monest is perhaps *Cornus stolonifera* (Red Osier), which is widespread in low grounds. It is distinguished by its reddish-purple bark and whitish or blue drupes. The handsome, round-leaved *C. rugosa* (*C. circinata*) is found in cold, shady woods northward. It is especially common in the pine woods of the upper Great Lakes region. The drupes are light blue. Most of these dogwoods are rather difficult to distinguish, but one of them stands out prominently from all the rest in that it bears alternate leaves. This is the tall, handsome *C. alternifolia* of Eastern woodlands. (Figs. 319, 99.)

One or two species of *Aucuba* are found in cultivation as ornamentals. These are evergreen shrubs with opposite, coarsely dentate leaves, thick branches, and diœcious flowers in compound cymes. The gynœcium is 1-carpellate, with a short, thick style which ends in a broad, one-sided, discoil stigma. The fruit is a drupaceous berry. A 1-carpellate gynœcium occurs in *Mastixia*. *Helwingia* and *Mastixia* are Asiatic, *Curtisia* is South African, and *Griselina* occurs in New Zealand.

CHAPTER XV

THE METACHLAMYDEÆ (SYMPETALÆ)

This is a heterogeneous assemblage of orders conveniently grouped together mainly because, with few exceptions, all possess a sympetalous corolla. Sympetaly is not restricted to these orders, however, for sympetalous forms are to be found also among the *Dialypetalæ*. Some members show affinities with certain orders and families of the *Dialypetalæ*. The sympetalous corolla, associated with zygomorphy and reduction in the number of stamens, has been interpreted as an "adaptation" to insect pollination, and in these particulars the group marks an advance over the *Dialypetalæ*. In general the herbaceous habit predominates and there is a notable absence of aquatics. The orders are relatively few and clearly defined.

To summarize the principal characters, the *Metachlamydeæ* are distinguished by (1) a sympetalous corolla, (2) a generally persistent calyx, (3) mostly epipetalous stamens, (4) cyclic arrangement of the floral organs, (5) one integument in the ovule (exceptions), (6) dominance of herbaceous forms, and (7) notable absence of aquatic forms.

The general evolutionary tendencies have advanced along the following lines: (1) attainment of sympetaly; (2) from actinomorphic to zygomorphic flowers; (3) to complete syncarpy; (4) reduction of the number of whorls—a whorl of stamens; (5) reduction in the number of members in a whorl; (6) epipetaly of the stamens; (7) reduction in the number of ovules; (8) from hypogyny to epigyny; (9) crowding of the flowers into compact inflorescences associated with increase in the number of flowers and corresponding reduction in their size.

On the basis of the number of whorls in the flower, the *Metachlamydeæ* are conveniently divided into two groups, (1) the *Pentacyclicæ* and (2) the *Tetracyclicæ,* the former with five, the latter with four whorls. To the *Pentacyclicæ* belong the first eight orders, the last six falling in the *Tetracyclicæ*. Hypogyny is the general rule in the *Pentacyclicæ,* but epigyny is attained in certain members of the *Ericales, Primulales,* and *Ebenales*. With respect to the *Tetracyclicæ,* the orders are equally divided, the lower three being hypogynous and the higher three epigynous (see syllabic chart).

Pentacyclicæ

Flowers generally 5-cyclic, conforming to the general formula $S5\ P5\ A5+5\ G5$, but often $S5\ P5\ A5+0\ G5-4-3$. Corolla regular, rarely irregular, in some cases with distinct petals. Stamens distinct or sometimes epipetalous, and opposite the petals (obdiplostemonous) when only one whorl is present. Ovary superior or inferior. Ovule in some cases with 2 integuments.

This is the less specialized group of the *Metachlamydeæ,* retaining in certain places characters of the *Dialypetalæ,* such as a choripetalous corolla, distinct and in some cases many stamens, and two integuments in the ovule. In the *Diapensiales* the carpels have been reduced to three, while an advance to epigyny is seen in the *Ericales, Primulales,* and *Ebenales.* Exceptions to the pentacyclic condition are frequent (*Ericales, Primulales, Ebenales*). In the *Theophrastaceæ* the outer whorl of stamens persists in the form of staminodes.

ORDER DIAPENSIALES

Flowers regular, or with an irregular calyx, bisexual, hypogynous, 5-merous except for the 3-carpellate gynœcium; style 1, ending in a 3-lobed or 3-branched stigma. Sepals free or somewhat united; the petals united towards the base; the 5 stamens alternate with the petals, and between them and the ovary sometimes a whorl of staminodes alternating with the stamens. Ovules numerous, on thick, projecting axile placentas. Fruit a loculicidal capsule. Seeds with copious endosperm surrounding a slightly curved embryo.

One family, *DIAPENSIACEÆ* (*Diapensia Family*) of 6 genera and about 12 species of the northern hemisphere, three of the genera in eastern North America, *viz., Diapensia, Pyxidanthera, Galax;* these belong in two tribes, distinguished as follows:

Diapensieæ: dwarf, woody, tufted evergreens with simple, alternate, small, and numerous leathery exstipulate leaves, and solitary flowers with sympetalous corolla and the 5 stamens alternating with the lobes of the corolla. Calyx with 2 or more bracteoles at the base. Staminodes none. *Diapensia, Pyxidanthera.*

Galacineæ: low, tufted, shrubby plants from a creeping rhizome, producing from a short caudex a few long-petioled, orbicular to broadly ovate leaves, and a scape of few to many small flowers in a racemose-spicate inflorescence. Stamens monadelphous, with 1-celled anthers, and inside the stamens a whorl of petaloid staminodes. Calyx with 2 minute bracteoles at the base. *Galax.*

Diapensia lapponica forms dense tufts on alpine summits of Labrador, Quebec, New England, the Adirondacks, Arctic America, and thence across Asia and Europe. Another species occurs in the Himalayas. *Pyxidanthera* and *Galax* are monotypic genera of the Atlantic States, where *Pyxidanthera barbulata* occurs in sandy pine-barrens from southern New Jersey to North Carolina, and *Galax aphylla* in mountain woods from Virginia to Georgia.

ORDER ERICALES

Flowers regular or sometimes irregular, bisexual, hypogynous or epigynous, 5-merous or 4-merous. Petals free or the corolla sympetalous. Stamens 5 or 10, free or epipetalous and, along with the corolla, inserted on the edge of an hypogynous disk; when 5 then opposite the petals (obdiplostemonous). Gynœcium of 2—5 carpels; these, when 5, opposite the petals. Ovary superior or inferior, with a simple style. Ovules with 1 integument. Seeds with a straight embryo surrounded by mealy endosperm. Herbs or shrubs.

Families: *Clethraceæ, Pyrolaceæ, Ericaceæ, Epacridaceæ.*

Key to the Families

Ovary superior.
  Trees, shrubs, or subshrubs.
    Flowers in elongate cylindrical racemes. Fruit a 3-valved capsule. Deciduous trees and shrubs. *Clethraceæ.*
    Flowers in inflorescences of various sorts. Fruit a capsule, berry, or drupe. Ours evergreen or deciduous shrubs, subshrubs, or trees. *Ericaceæ.*
  Low evergreen herbs, or low to tall and stout, parasitic or saprophytic plants without chlorophyll and with the leaves reduced to scales. Corolla more or less choripetalous. *Pyrolaceæ.*
Ovary inferior. Corolla more or less urn-shaped. Fruit (in ours) a berry or drupe. Mostly evergreen shrubs. *Ericaceæ.*

CLETHRACEÆ (*White-Alder Family*). Flowers regular, bisexual, hypogynous, 5-merous, except a 3-merous gynœcium. Stamens 10 (or 12?); anthers sagittate, opening by an apical pore. Ovary superior, 3-lobed, 3-celled, with numerous ovules in each cell on axile placentas. Disk obsolete. Style 1, terminating in 3 stigmatic branches. Fruit a 3-lobed, 3-angled, 3-celled, subglobose, loculicidal, 3-valved capsule. Embryo cylindrical.

One genus, *Clethra* (Pepper-Bush), of shrubs and trees with alternate deciduous leaves and small, fragrant flowers in cylindrical racemes, and more or less clothed with stellate-canescent pubescence. About 30 species, native to eastern North America, the West Indies, Mexico, South America, and Japan. Besides one Southern species the following extend into the North Atlantic States: *C. alnifolia, C. acuminata.* Both are grown as ornamentals (Fig. 320).

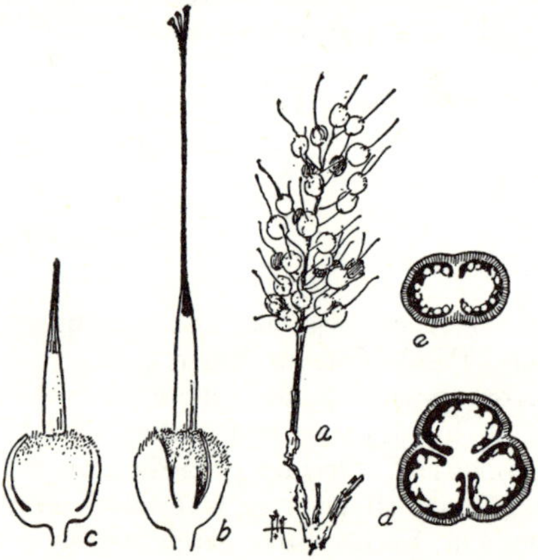

FIG. 320.—*Clethra alnifolia: a,* fruit-cluster; *b,* gynœcium and calyx; *c,* gynœcium (part of calyx removed); *d,* transverse section of a tricarpellate and *e,* of a bicarpellate ovary.

PYROLACEÆ (*Wintergreen Family*). Flowers regular or nearly so, bisexual hypogynous, 5-merous or 4-merous, the petals free or united. Stamens 10 (or 8) with introrse anthers which become inverted when the flowers open, and opening by pores or slits. Ovary superior, with as many cells as carpels, the cells with numerous anatropous ovules on axile placentas. Style short or elongate and slender, with as many stigmatic lobes as carpels. Disk obsolete. Fruit a globose capsule loculicidally dehiscent into 4—5 valves from the base. Seeds minute. Embryo undifferentiated. Low herbaceous or woody plants from rhizomes; leaves various. Flowers solitary or cymose (corymbose).

A small family of about 30 species in 10 genera, distributed over the colder regions of the northern hemisphere. They are plants of cold, deep to open woods. Two subfamilies, *viz.:*

Pyroloideæ, in which the petals are distinct and the anthers open by pores. Pollen-grains in tetrads (4's). *Pyrola, Moneses, Chimaphila.*

Monotropoideæ, in which the petals are distinct or united, the anthers open by curved or longitudinal slits, the pollen-grains are simple, and the plants are saprophytes or root-parasites. *Monotropa,*

FIG. 321.—*Pyrola americana: a,* habit sketch, with inflorescence at right; *b,* flower viewed from beneath; *c,* side view of flower with three petals removed; *d,* flower with several parts removed; *e,* tip of style, with stigma; *f,* transverse section of ovary; *g,* stamen; *h,* stamen after shedding of pollen; *r,* receptacle; *s,* sepals.

Allotropa, Hypopitys, Pterospora, Sarcodes, Newberrya, Pleuricospora.

Pyrola (Shin-Leaf, Wintergreen). The shin-leafs are low herbs with mostly basal leaves and scapose, nodding, white to purplish flowers; chiefly of northern woodlands. The species are easily confused because of their similarity. Our species may be grouped according to the orientation of the style, associated with certain characters of the sepals, petals, and stamens. In one group the style is straight (*Pyrola minor, P. secunda*); in the other group it is declined and then upcurved (*P. chlorantha, P. americana, P. elliptica, P. asarifolia*). *Pyrola secunda* has a one-sided (secund) inflorescence, while *P. asarifolia, e.g.*, is distinguished by its pink or purplish flowers. In all, the anthers open by terminal pores and are inverted at flowering time. (Fig. 321.)

Moneses is very similar to the shinleafs but is easily distinguished by its solitary flower. The floral organs are in 4's or 5's, the stamens 8 or 10. The anthers open by pores. The solitary species, *Moneses uniflora*, is a low perennial herb of cold northern coniferous woods. Also in Asia and Europe.

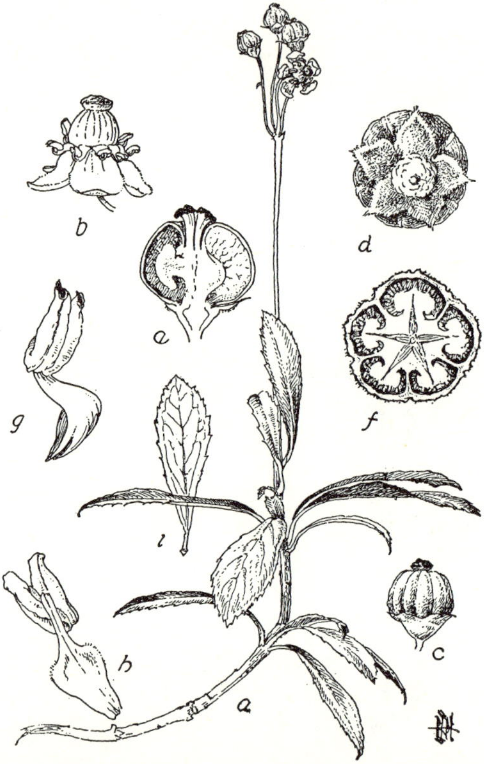

FIG. 322.—*Chimaphila umbellata: a,* habit sketch; *b,* flower; *c,* gynœcium; *d,* immature fruit, viewed from beneath, showing persistent calyx; *e,* median section of fruit, with open locule on left and septum on right; *f,* transverse section of ovary; *g,* stamen, after anthesis; *h,* dorsal aspect of stamen; *l,* leaf.

Chimaphila (Pipsissewa, Prince's Pine). The Pipsissewas are natives of North America (including Mexico and Santo Domingo) and Asia. Species about 6—one, *C. umbellata,* encircling the northern hemisphere; in North America down into Mexico and the mountains of Santo Domingo. It is common in sandy pine woods of our Northern States. The plant is somewhat woody and grows from a rhizome. The leaves are elongate-oblanceolate, serrate, rounded at the tip, shiny-leathery, and opposite, and more or less crowded towards the top of the low stem. A raceme of a few pinkish, nodding flowers terminates the stem. *C. maculata* occurs in the Eastern States and *C. Menziesii* on the Pacific Coast. The flowers are typically 5-merous throughout, with two whorls of stamens. Fruit a capsule, differing from that of *Pyrola* in splitting from above downwards (Fig. 322).

Monotropoideæ. Monotropa uniflora, the beautiful snow-white Indian pipe of our rich woods, is a typical example of this group. The plants usually grow in clusters from a thick coral-like mass of brittle mycorrhizal "roots." A large, solitary, nodding flower is borne at the summit of the scaly-leaved stem. The floral organs are somewhat inconstant in number, for instance, 2—4 sepals, 5—6 petals, 10—13 stamens, but the gynœcium conforms definitely to the 5-merous type. At the base of the ovary is a 10—12-toothed annular disk. The fruit is a 5-valved capsule. The species occurs throughout the continent, except in the farthest north, and recurs in Japan and the Himalayas. There is said to be one other species, absent, however, from North America.

The solitary *Allotropa virgata* (Sugar-Stick) is a low saprophytic herb of coniferous woods of the Pacific Coast, at high altitudes. The stems are red, streaked with white. The brownish-red flowers are racemose and apetalous. *Hypopitys* (Pine-Sap) includes a half-dozen species of whitish, yellowish, or brownish-red saprophytes with slender, simple stems from a coral-like cluster of mycorrhizal roots as in *Monotropa,* and similarly showing marked inconstancy in the number of floral organs: sepals 2—4, petals 3—4—5, stamens 6—10, carpels 4—5. An 8—10-toothed annular disk subtends the ovary. *Hypopitys lanuginosa* (*Monotropa Hypopitys*) occurs around the northern hemisphere.

Pterospora and *Sarcodes* possess sympetalous corollas, but lack the hypogynous disk. In *Pterospora* the anthers bear awn-like appendages at the base, but these are absent in *Sarcodes. Pterospora andromedea* (Pine-Drops), a solitary species, is a low to tall brownish saprophyte of rich northern woods in the East. In the West it reaches Arizona and New Mexico. The slender scaly-leaved stem, springing from a mass of mycorrhizal roots and bearing an elongate raceme at the summit, sometimes grows to a height of 3 to 5 feet.

Sarcodes sanguinea, the stout, brilliant-crimson, saprophytic Snow-

Plant of the California mountains, is perhaps the most striking member of this group. It, too, is a solitary species. *Newberrya* and *Pleuricospora* are Pacific Coast plants, and both are monotypic. In both the ovary is 1-celled.

ERICACEÆ (*Heath Family*). Flowers regular or irregular, hypogynous or epigynous, bisexual, 5-merous or 4-merous, generally with sympetalous corolla. Sepals often much reduced and mostly persistent. Stamens as many as the segments of the corolla or twice as many, free or epipetalous; anthers 2-celled, innate or versatile, often prolonged into tubes or with awn-like appendages (horns), opening by chinks, pores, or slits. Carpels 5—2; style 1, short or elongate, with capitate or peltate stigma. Ovary superior or inferior, with as many cells as carpels, producing usually numerous ovules on thick axile placentas. Integuments 1. Fruit a capsule, berry, or drupe. Seeds numerous (sometimes 1), usually minute, with a central cylindrical, often very short, embryo in copious fleshy endosperm; cotyledons short. About 1,350 species in 80 genera. Shrubs, trees, and perennial herbs of warm and cold-temperate regions and high mountains of the tropics. Twenty or more genera in our flora. The family is rich in choice ornamental plants, while some are cultivated for their edible fruits.

There are four subfamilies, which are conveniently grouped according to the position of the ovary, *viz.:*

The Hypogynous Heaths, in which the ovary is superior:

Rhododendroideæ, in which the fruit is a septicidal capsule, the petals free or united, and the anthers without appendages. *Ledum, Rhododendron, Menziesia, Kalmia, Phyllodoce.* Seventeen genera and 850 species.

Arbutoideæ, in which the fruit is a loculicidal capsule, the corolla sympetalous, the anthers provided with bristle-like appendages, or saccate below and tubular above. Twenty genera and 250 species. *Gaultheria, Arbutus, Arctostaphylos, Andromeda, Epigæa.*

Ericoideæ, in which the fruit is a septicidal or loculicidal capsule or a nut, the corolla sympetalous, and the anthers opening by pores and often provided with caudate appendages. Seventeen genera and 600 species. *Calluna, Erica.*

The Epigynous Heaths, in which the ovary is inferior:

Vaccinioideæ, in which the ovary is inferior, the corolla sympetalous and variously urn-shaped, the stamens distinct (in ours) or united. Twenty-three genera and 330 species. *Vaccinium, Gaylussacia, Chiogenes.*

Rhododendroideæ. From Greenland to the headwaters of the Mississippi we find *Ledum groenlandicum* or Labrador Tea forming dense and often extensive societies in cold bogs of the coniferous belt. It is a low shrub with oblong, spiral, evergreen leaves which are densely brownish-tomentose on the lower side, with strongly revolute margins and crowded towards the ends of the branches. The plants

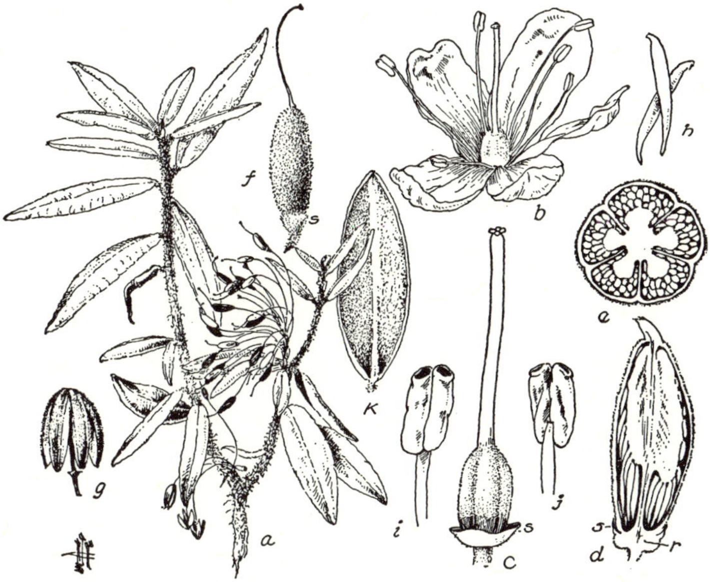

FIG. 323.—*Ledum grœnlandicum: a,* branch with fruit-clusters; *b,* flower; *c,* gynœcium, with calyx below; *d,* median section of ovary; *s,* calyx; *r,* receptacle; *e,* transverse section of ovary; *f,* immature fruit; *g,* dehisced fruit (capsule); *h,* seeds; *i, j,* ventral and dorsal aspects of stamen; *k,* leaf (lower surface).

are resinous and the leaves fragrant, at least when rubbed. The rather small, white flowers are crowded in dense terminal clusters, and are typically 5-merous. The anthers open by terminal pores. *Ledum palustre* ranges through boreal and arctic regions, and two other species are found on the Pacific Coast. (Fig. 323.)

Rhododendron, by far the largest genus (800 or more species—over half the family), is mainly Asiatic but is represented by several species in eastern and western North America, Europe, and Asia

Minor. The plants are handsome shrubs or small trees with mostly alternate, evergreen, leathery leaves and showy, somewhat irregular (bilaterally symmetrical in some), flowers in terminal umbellate clusters from large buds. Many are choice ornamentals. The flowers are typically 5-merous, the calyx greatly reduced, the stamens often twice as many as the petals. The fruit is a 5-valved capsule. The species with irregular funnel-shaped corollas and deciduous leaves are by some authorities treated as a separate genus—*Azalea*. Except for about 3 species of the Pacific Coast, our rhododendrons are restricted mainly to the Appalachian region; one, *R. viscosum,* reaches Arkansas; *R. maximum* (Great Laurel), Ontario, Quebec, and Ohio; a third, *R. canadense* (Rhodora), low shrub, Quebec and Newfoundland.

FIG. 324.—*Kalmia polifolia: a,* leafy twig; *b,* twig with fruit-clusters; *c, d,* leaves (lower surface); *e,* fruit (capsule); *s,* calyx.

The menziesia are North American and Japanese. Of the seven species, *Menziesia pilosa* is Appalachian; *M. glabella* of the Northwest has an isolated station on Minnesota Point on Lake Superior; *M. ferruginea* belongs to the Pacific Coast. The flowers are uniformly 4-merous, with the calyx reduced to small teeth. The kalmias or American laurels are mostly low evergreen shrubs of eastern to far northern parts of the continent. The leaves are leathery, opposite, alternate or whorled, and mainly of an oblong shape. The flowers are showy, in dense umbellate or corymbose clusters and typically 5-merous, with a crenate annular hypogynous disk. Stamens 10; anthers opening by pores, and without appendages. Fruit a somewhat globular capsule splitting septicidally from above. The corolla is rotate and is furnished with 10 pouches below the limb. Calyx persistent. The largest and handsomest species, which is often a small tree, is *Kalmia latifolia* (Mountain Laurel) of the Appalachians. *Kalmia polifolia,* a low shrub of peat-bogs, is a Northern species which pushes South in the mountains. The plant is somewhat similar to *Andromeda polifolia,* from which it is distinguished by its broader leaves, rotate corolla, and septicidal capsule. The leaves are dark green above and glaucous beneath. (Fig. 324.)

Phyllodoce is a circumboreal genus of about 8 species. The plants are low shrubs with small, crowded, linear, leathery, evergreen leaves and small, nodding, long-pediceled, 5-merous flowers in terminal umbels. The corolla is ovoid urn-shaped, and the fruit is a globose

FIG. 325.—*Gaultheria procumbens: a,* leafy branch; *b,* twig with fruit; *c,* flowering shoot (left) and young leafy shoot (right); *d,* flower-bud; *e,* flower; *f,* median section of flower; *g,* stamen; *h,* section of fruit; *r,* receptacle; *s,* calyx; *i,* summit of fruit; *l,* leaf (upper surface), *co,* corolla.

septicidal capsule. Stamens 10. *Phyllodoce cœrulea* of the Arctic appears on high summits of New England and Quebec, and in Greenland, Europe, and Asia as well.

Arbutoideæ. Gaultheria, one of the largest genera of the family, numbering about 100 species, only a few of which are North American and Asiatic, the great majority native in the Andean region of South America. They are small to large shrubs with alternate, leathery, evergreen leaves and small, axillary or clustered, 5-merous flowers with an urn-shaped or campanulate corolla. Stamens 10. The fruit is distinctive in that the calyx becomes very fleshy and juicy or pulpy and invests the capsule (true fruit). Our eastern species, *Gaultheria procumbens,* the common Wintergreen, is a low subshrubby plant creeping by rhizomes from which arise low aërial shoots bearing the shiny, leathery, evergreen leaves more or less clustered at the ends with drooping solitary flowers in the axils. The calyx is much reduced and the corolla urn-shaped. The "berries," which are bright crimson, ripen in the autumn and remain on the plant over the winter. Both berries and leaves possess a spicy and characteristic wintergreen flavor. About 3 other species occur on the Pacific Coast, one of which, *G. Shallon,* the Salal, reaches a height of 6 or 8 feet and forms impenetrable tangles in the moist coniferous forests. The berries are edible, having the flavor of black currants. (Figs. 325, 45A.)

Our only species of *Arbutus* is *A. Menziesii,* the Madroña of the Pacific Coast. The plant is a shrub or tall evergreen tree (100 ft.) with smooth dark-red bark which exfoliates yearly. The leaves are shiny, oval, and leathery, and the berries bright orange-red. This and *A. Unedo* (Strawberry Tree) of Europe are grown as ornamentals. All told there are about 20 species, native to the Mediterranean region, the Canary Islands, and western North America.

Arctostaphylos is a west-American genus of about 40 species, with the exception of the boreal *A. Uva-ursi,* the Bear-Berry or Kinnikinic, which encircles the northern hemisphere. This exception is a prostrate plant with small, leathery, spatulate leaves, spreading by rhizomes and usually forming dense mats in dry, sandy pine woods of the North. It bears insipid red berries (drupaceous). The two dozen or more Western species are characteristic of the arid mountain and chaparral belts of southern California into Lower California, where the plants are known by the Spanish name Manzanita (little apple). These are rigid shrubs usually growing in clusters, with bark very similar to that of the Madroña but with much thicker and smaller rigid leaves of a bluish tinge. The berries dry up at maturity. For the most part the species are distinguishable only on technical characters. By some writers *Arctostaphylos Uva-ursi* is separated from these forms as a distinct genus (*Uva-Ursi*).

Andromeda polifolia (Wild Rosemary) and *A. glaucophylla* (Bog Rosemary) are low shrubs of Northern peat-bogs. Both resemble *Kalmia polifolia* (*q.v.*). *Epigæa repens,* the Trailing Arbutus, is a hirsute trailing plant of the pine woods of the North. The flowers are sometimes diœcious, and heteromorphic as to style and stamens. Fruit a globose pubescent capsule. Its fragrant flowers have led to its extermination in some localities, and with the opening of roads into much of our formerly inaccessible Northern wilderness it is in need of rigid protection. *Chamædaphne calyculata* (Leather-Leaf) occurs in bogs from Labrador to the Mississippi and southward. (Fig. 326.)

Ericoideæ. Calluna is a monotypic genus of Europe and Asia. Its sole representative, *Calluna vulgaris,* the Heather, is a low shrub with opposite, crowded, minute, leathery leaves in 4 ranks, and small 4-merous flowers in elongate interrupted racemes. The species occurs from Newfoundland to New Jersey, where it was supposedly long ago introduced. *Erica,* the Heath, the second largest genus (500 species) is South African and Mediterranean, but many are grown as ornamentals in this country. As in *Calluna,* the leaves are minute and crowded and the flowers are 4-merous, with an urn-shaped corolla. The anthers are of diverse shapes, opening by pores, chinks, or slits. An annular hypogynous disk is present. Fruit a capsule.

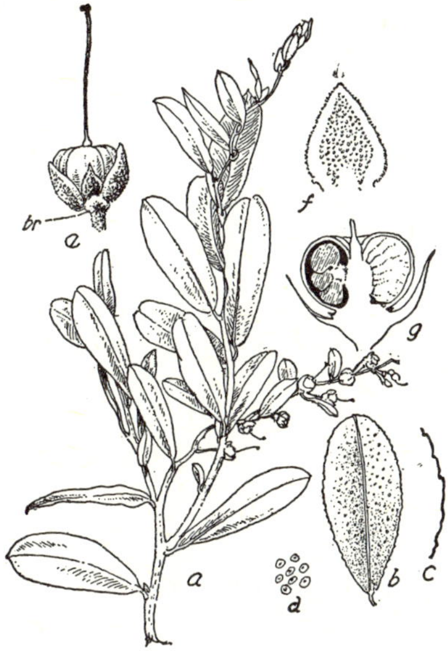

FIG. 326.—*Chamædaphne calyculata: a,* branch with fruits; *b,* leaf (lower surface); *c,* leaf-margin; *d,* scales from lower surface of leaf; *e,* gynœcium and calyx; *f,* sepal (dorsal view, with scales); *g,* section of ovary, *br,* bract.

Vaccinioideæ. Vaccinium (Blueberries, Bilberries, Cranberries); 130 or more species of erect or creeping shrubs of the northern hemisphere, with 25 or more in our flora. Leaves alternate, deciduous or evergreen, and the flowers 5-merous or 4-merous, solitary or in axillary or terminal racemes. Stamens 8—10. Corolla urn-shaped, campanulate, rotate, or cylindrical. Fruit a berry (spurious) crowned by the persistent calyx-lobes. The 4—5-celled ovary is sometimes 8—10-celled by the intrusion of false partitions from the wall to the axis. *Vaccinium arboreum* (Farkleberry) of the Central and Southern States is the largest of our species (25 ft.). Other tall species are *V. corymbosum* (12 ft. or more), Maine to Minnesota and southward; and *V. parvifolium* of the Pacific Coast. Our common blueberries of the eastern flora are *V. pennsylvanicum* (Low Sweet Blueberry) with shiny, leathery leaves and wiry stems; and *V. canadense* (Velvet-leaf Blueberry), with the leaves downy on both sides and a drier, mealier fruit. The Bilberries are more northern, the arctic *V. uliginosum* (Bog Bilberry) reaching alpine summits of New England and New York, and rocky barrens of northern Michigan, and accompanying it are *V. cæspitosum, V. membranaceum,* and the tall *V. ovalifolium.* In Eastern and Northern peat-bogs the two cranberries, *V. Oxycoccus* and *V. macrocarpon,* are often common. These are inconspicuous plants with wiry stems and very small spreading, leathery leaves, spreading over and through the peat moss. The flowers are 4-merous and nodding, on long, filiform pedicels. The anthers are without appendages and end in long tubes. In the former species the leaves are ovate with an acute tip, in the latter they are oblong and blunt, and the berries are larger. (Figs. 327, 328.)

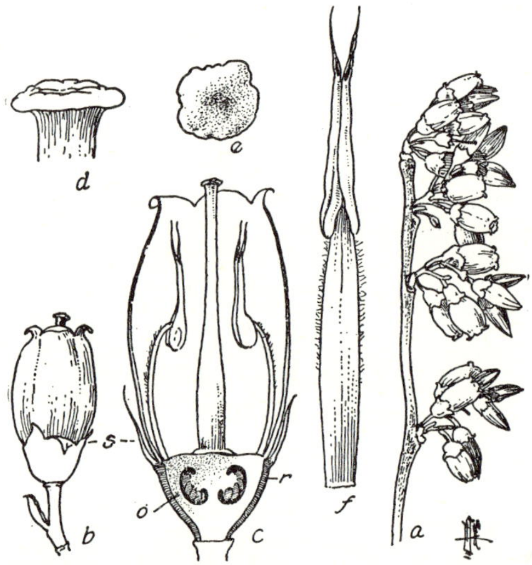

FIG. 327.—*Vaccinium corymbosum: a,* inflorescence; *b,* flower; *c,* median section of flower; *o,* ovary; *r,* receptacle; *s,* sepals; *d, e,* stigma; *f,* stamen.

The huckleberries (*Gaylussacia*) are chiefly South American. Of the 40 or more species, only about a half-dozen have become a part of our Eastern flora. The flowers are 5-merous, with an urn-shaped or campanulate corolla. The fruit is a berry-like drupe.

Chiogenes hispidula (Snowberry) is a solitary slender species, somewhat suggesting the cranberries, but with closely crowded, oblong, leathery leaves; the plants form dense mats on mossy deadfalls or in peat moss, a habit which together with their small but few snow-white berries make them quickly recognizable. The flowers are 4-merous.

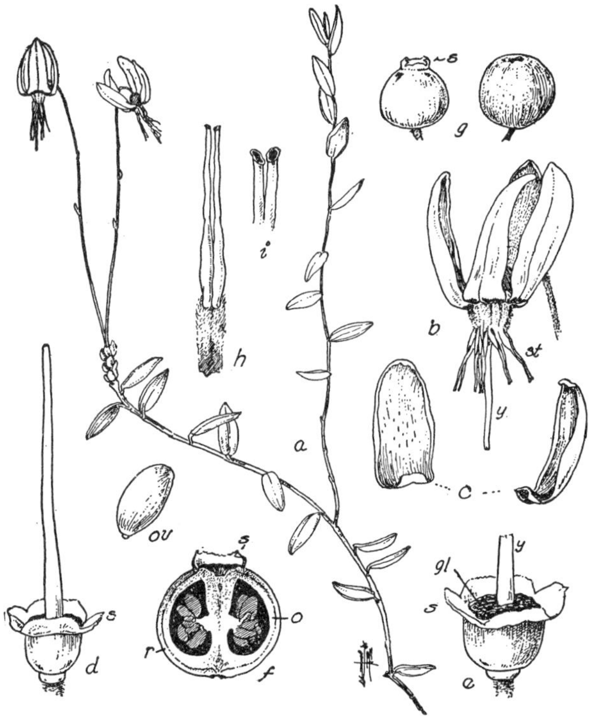

FIG. 328.—*Vaccinium Oxycoccus: a,* habit sketch; *b,* flower; *c,* petals; *d,* flower with stamens and petals removed; *e,* same, showing disk (*gl*); *f,* section of fruit; *g,* immature and mature fruits; *h,* stamen; *i,* apex of anther, showing pores; *ov,* ovule; *o,* ovary; *r,* receptacle; *s,* sepals; *st,* stamens; *y,* style.

EPACRIDACEÆ. Flowers regular, hypogynous, bisexual, 5-merous or sometimes 4-merous, differing from the heaths mainly in having but 1 whorl of stamens which are opposite the sepals, and 2-celled, versatile, unappendaged anthers opening by slits. The corolla is tubular to vasiform, with triangular or linear, spreading or reflexed segments. The stamens are epipetalous or borne on the edge of an annular hypogynous entire or lobed disk. The gynœcium is 5-carpellate, the ovary 5-ridged or 5-lobed; style 1, terminal or rising from a deeply depressed ovary; stigma capitate or lobed. Fruit a 5-valved capsule producing numerous stony seeds. Embryo cylindrical or slightly curved, in copious endosperm. Shrubs or subshrubs with spiral, rigid, short-petioled, or basally sheathing, leaves. Flowers solitary and axillary, or in terminal racemes or panicles, white or bright-red, subtended by 2 bracteoles, and bract, or by a series of bracteoles, the uppermost calyx-like. Twenty-one genera and about 340 species, chiefly Australian.

ORDER PRIMULALES

Flowers regular, rarely irregular, bisexual or unisexual, hypogynous or epigynous, 5-merous or sometimes 4-merous or 8-merous, with a sympetalous corolla, a single whorl of epipetalous stamens, and sometimes a whorl of epipetalous staminodes. Corolla sometimes of distinct petals. Gynœcium syncarpous, usually of as many carpels as petals, with 1 terminal style. Ovary superior or inferior, 1-celled, bearing numerous to few (1) ovules on a basal columnar placenta. Fruit drupaceous. Seeds with endosperm. Integuments 2.

Families: *Theophrastaceæ, Myrsinaceæ, Primulaceæ.*

Key to the Families

Fruit a capsule. Stamens 5, staminodes rarely present. Placenta free, columnar-globose, arising from the base of the ovary. Herbs with alternate, opposite, or verticillate leaves. *Primulaceæ.*
Fruit a drupe. Trees or shrubs.
  Ovary superior. Staminodes present. Drupe several-seeded. *Theophrastaceæ.*
  Ovary superior or inferior. Staminodes wanting. Drupe 1-seeded. *Myrsinaceæ.*

THEOPHRASTACEÆ. Flowers regular, hypogynous, bisexual or unisexual, 5-merous, with 5 stamens and 5 staminodes. Ovary superior. Fruit a drupe with 2 to numerous seeds. Shrubs or trees with opposite, alternate, or whorled, exstipulate leaves which are often crowded at the ends of the stems or branches. Forty-two to 70 (?) species and 12 genera of tropical America and the Sandwich Islands. Largest genera *Clavija* (33 species) and *Jacquinia* (6 species), both of tropical America. Some clavijas are grown in hothouses. The fruits are said to be edible.

MYRSINACEÆ (*Myrsine Family*). Flowers regular, bisexual or unisexual, hypogynous or epigynous, 4-merous or 5-merous. Stamens 5 and sometimes a whorl of 5 staminodes. Ovary superior or inferior, 1-celled, with numerous ovules on thick placentas of diverse shapes supported on a column rising from the base of the ovary. Style 1. Fruit a few- to many-seeded drupe. Embryo cylindrical, in fleshy or horny endosperm. Shrubs or trees with alternate, entire, exstipulate, soft or leathery, often evergreen leaves and small to large, usually bracteolate flowers in axillary or terminal, simple or compound, racemose clusters. About 1,000 species and 32 genera of the tropics and subtropics. Two subfamilies:

Myrsinoideæ, in which the ovary is superior, staminodes are wanting and the fruit is 1-seeded. *Ardisia* (240 species), *Rapanea* (140 species), *Embelia* (60 species).

Mæsoideæ, in which the ovary is inferior and the fruit is many-seeded.

Mæsa (102 species).

Ardisia crispa is an element of the mangrove vegetation of the Indo-Malayan region. Often cultivated. The embryo germinates while the seed is within the fruit.

PRIMULACEÆ (*Primrose Family*). Flowers regular or irregular, bisexual, hypogynous or sometimes epigynous, uniformly 5-merous, without bracteoles. Stamens 5 and opposite the lobes of the corolla (obdiplostemonous), epipetalous, and inserted on the base of the corolla-tube; filaments distinct or connate at the base; anthers introrse, 2-celled. Carpels completely fused. Ovary superior or partly inferior; ovules usually numerous, on a more or less thickened placenta borne on a short or elongate column arising from the base of the ovary; integuments 2. Fruit a 1-celled, variously dehiscent (sometimes circumscissile) capsule. Seeds few to many, with a small, straight embryo enclosed in copious horny endosperm. Herbs with alternate, opposite, or verticillate, exstipulate, often basal leaves, and small to large and showy flowers in terminal or axillary inflorescences of various types. Number of genera and species variously given as 22—30 and 400 to 700 respectively. A cosmopolitan family, best developed in the northern hemisphere, ranging into the Arctic and ascending to alpine summits southward, with one group considerably developed in the southern hemisphere.

For convenience our genera may be grouped as follows:

Flowers regular.

Ovary superior: *Primula, Androsace, Hottonia, Soldanella* (Europe), *Dodecatheon, Cyclamen* (Eur.), *Lysimachia, Steironema, Trientalis, Glaux, Anagallis, Centunculus.*

Ovary inferior. *Samolus.*

Flowers irregular; calyx bristly. *Coris* (Mediterranean).

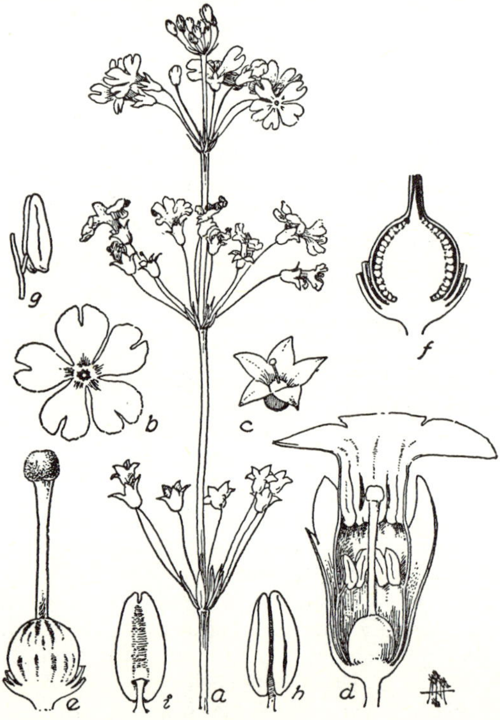

FIG. 329.—*Primula Forbesii: a,* inflorescence; *b,* corolla, seen from above; *c,* calyx,—gynœcium within; *d,* median section of flower; *e,* gynœcium, with base of corolla and calyx; *f,* median section of ovary; *g,* stamen; *h, i,* ventral and dorsal aspects of anther.

Primula (Primrose). The largest genus of the family, variously estimated at 150 to 300 species, chiefly of the northern hemisphere, a few in Java and southernmost South America. Numerous species and their varieties are horticultural. The plants are perennial, scapose, rhizomatous herbs with basal leaves of diverse shapes, and small to large flowers in umbels or heads, or in superimposed involucrate whorls. The calyx is campanulate, funnel-form or tubular, and shorter than the corolla tube; the corolla funnel-form or salver-form, with spreading lobes. The filaments are short and inserted at various levels on the corolla-tube or at its throat and included. Capsule globose, dehiscing by 5 or 10 valves. The flowers are typically 5-merous and the union of the carpels is complete, even to the stigma. *Primula farinosa* and *P. mistassinica,* both inhabiting wet situations on shores and cliffs, are native in the St. Lawrence and Great Lakes basins. In the crevices of rocks in the high Sierras is another species, *P. suffrutescens.* The primulas are noted for their heteromorphic flowers, in which the styles in some species are of different lengths and the stamens, correspondingly, are inserted at different levels —supposedly an "adaptation" to facilitate insect pollination. (Fig. 329.)

Androsace (Androsace, Rock Jasmine) differs technically from *Primula* in possessing a corolla-tube shorter than the calyx. The species, according to various accounts, range from 60 to 85. *Androsace occidentalis,* a low tufted annual, ranges from the Central States to the Southwest. From a basal rosette of leaves arise several stems, each bearing an umbel subtended by large, leafy bracts. Twelve or 15 species are reported from the West and Northwest. In shallow, stagnant ponds of the Eastern States is found the peculiar *Hottonia inflata,* the flowering-stems of which are remarkably inflated between the joints and are hollow. The flowers are borne in whorls at the joints, and the basal leaves are finely pinnately dissected. Another species is found in Europe and eastern Asia. The soldanellas are European plants (6 species) growing from thick rhizomes.

Cyclamen and *Dodecatheon* (Shooting-Star) differ from the other genera in possessing strongly reflexed and more or less twisted or contorted, elongate corolla-lobes. In both the corolla-tube is very short, with the stamens monadelphous, included in *Cyclamen* but long-exserted in *Dodecatheon,* and connivent by their anthers. The cyclamens grow from thick corms, the shooting-stars from rhizomes. The 20 or more species of the former are European; the latter (30 or more species) is American (chiefly Western) and east-Asiatic. *Lysimachia* and *Steironema,* the Loosestrifes, are very similar leafy-stemmed herbs with yellow flowers. The former is distinguished from the latter by its 4-merous or 5-merous calyx and corolla, the folding of the corolla-lobes, absence of staminodes, and the oblong or oval anthers (linear in *Steironema*) and punctate leaves. *Lysimachia* is much the larger genus (70 or more spp.), chiefly of the northern hemisphere, with a few in Africa and Australia; our half-dozen or more are all Eastern. *Steironema* comprises but 5 species, mostly all east of the Rockies. *Steironema ciliatum,* however, crosses the continent from Nova Scotia to British Columbia in the north and from Georgia to Arizona in the south. (Fig. 330.)

Anagallis and *Centunculus* have circumscissile capsules (compare with *Amaranthus* and *Plantago*). *Anagallis arvensis,* the common Pimpernel, of Eurasia, is a common weed (Fig. 331). It is a low, diffuse annual with sessile, ovate, opposite leaves, bearing a long-pedicelled flower in the axils. Species 15—25, Eurasian. The 3 species of *Centunculus* (Chaffweed) are widely distributed, *C. minimus,* a low annual with small, alternate leaves and sessile, axillary flowers, occurring native in eastern North America. *Glaux maritima,* a solitary species, inhabits salt-marshes and sea-beaches of the northern hemisphere. The axillary flowers have no corolla, but instead the calyx is petaloid. Four species of *Trientalis* are spread over the northern hemisphere; one species, *T. americana,* occurs in eastern North America and two others in the Northwest; one, *T. europæa,* in Europe. *Trientalis americana* occurs in damp woods from Labrador to Manitoba, Minnesota, and Illinois. The plants grow from a deep, slender, creeping rhizome, from which arises a slender stem bearing a whorl of long narrow leaves at the summit, and above the leaves one or more white, star-shaped, solitary flowers on slender peduncles. *Samolus floribundus* is the only one of the ten or more species native to our flora; it is an erect perennial with alternate oblong-ovate leaves and axillary racemes of small, long-pedicelled flowers. It is distinguished by its inferior ovary and 5 staminodes. It is widely distributed in swamps and brackish wet places except in the Far North; southward it

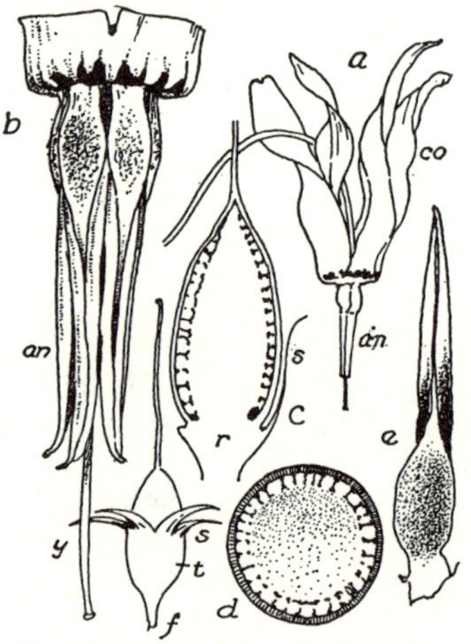

FIG. 330. — *Dodecatheon* sp.: *a,* flower; *co,* corolla; *an,* stamens; *b,* base of corolla with stamens (*an*), enlarged; *y,* style; *c,* median section of ovary; *s,* calyx; *r,* receptacle; *d,* transverse section of ovary; *e,* stamen, dorsal aspect; *f,* gynœcium and calyx (*s, t*). Compare flower with that of FIG. 328.

inhabits the West Indies and South America. Most of the species are South African and Australian.

In the small South American genus *Pelletiera* we find an anomalous situation with respect to the structure of the flower: this is 3-merous throughout, and the corolla is choripetalous.

FIG. 331.—*Anagallis arvensis: a,* habit sketch; *b,* flower, seen from above; *c,* sectional view of flower; *d,* calyx, seen from below; *e,* three stamens and base of corolla; *f,* gynœcium; *g,* median section of gynœcium; *h,* transverse section of ovary; *i,* immature fruit; *k,* gland-tipped hairs of stamens; *m, n,* dorsal and ventral aspects of stamens; *o,* ovules; *p,* floral diagram.

In general as regards the *Primulaceæ* we may note these important features: (1) absence of bracteoles, (2) isomerous whorls (except in *Pelletiera, Lysimachia, Centunculus,* and others), (3) absence of corolla (*Glaux*), (4) presence of a whorl of staminodes, (5) epigyny (*Samolus*), (6) heteromorphic flowers (*Primula*), (7) a 1-celled ovary with a central basal placenta, (8) bilateral symmetry (*Lysimachia, Dionysia, Coris*). With respect to the placentation compare with *Caryophyllaceæ*.

FIG. 332.—*Limonium vulgare: a,* leaf; *b,* inflorescence; *c,* cymules; *d,* flower; *br,* bract; *p,* corolla; *s,* calyx; *e,* corolla enlarged; *f,* gynœcium, with four of the styles cut off; *g,* stigmatic apical part of style; *h,* stamen; *i, j,* ventral and dorsal aspects of anthers; *k,* base of andrœcium and corolla; *l,* bract (*x*) in (*c*).

ORDER PLUMBAGINALES

Flowers regular, bisexual, hypogynous, 5-merous, and bracteolate, generally with a membranous calyx and a choripetalous or sympetalous corolla. Stamens as many as the petals and opposite them or the lobes (hence the inner whorl), free or connate at the base, or epipetalous. Gynœcium 5-carpellate, with as many styles, distinct or united at the base. Disk wanting or present. Ovary superior, 1-celled, with 1 basal anatropous ovule. Integuments 2. Fruit a utricle, achene, or capsule, invested by the calyx. Embryo straight, in mealy endosperm.

FIG. 333.—*Plumbago capensis: a,* inflorescence; *b,* corolla, seen from above; *c,* flower-bud; *d,* calyx, enlarged, with bracts below; *e,* median section of flower; *s,* calyx; *f,* base of flower, showing gynœcium, united bases of stamens, and annular disk (*gl*); *g,* median section of (*f*); *s,* calyx; *r,* receptacle; *h,* stigma; *i, j,* dorsal and lateral aspects of anther; *k,* gland-tipped hair of calyx.

One family: *PLUMBAGINACEÆ* (*Leadwort Family*). Characters of the order. Ten genera and 260—300 species of herbs or shrubs (some climbing), world-wide in their distribution, but best developed in saline steppes and on seashores.

Our genera are: *Limonium* and *Statice,* each represented by a few native species, but by several garden ones. Species of *Plumbago*

(Leadwort), a tropical genus, are grown as ornamentals and hothouse plants.

Limonium carolinianum, the Sea-Lavender, is found in saltmarshes from Labrador to Florida and Texas. It is an herb with basal oblanceolate, downwardly long-tapering leaves, and small, purplish flowers in spikelets on one side of the branches of a cymose-paniculate inflorescence. The genus is variously estimated to contain 120 to 180 species. A few other species occur in the South and one on the Pacific Coast. (Fig. 332.)

Statice Armeria (Thrift) is a low, tufted plant of the seacoast and mountains from Labrador to Alaska and thence southward to California. It differs from the preceding in having small, narrow, linear, and tufted leaves, and larger flowers in a terminal bracteate umbel. It is also said to occur at the Straits of Magellan—a distant and isolated occurrence. The other species (19) are found in Europe, north Africa, and western Asia.

Plumbago capensis is planted in California and other parts of the Southwest and is common in greenhouses. It is a straggling, sometimes half-climbing shrub with oblong-spatulate leaves and spicate clusters of pale-blue flowers. The flowers are characterized by a narrow, cylindrical, glandular calyx and a long-tubular corolla with broad, spreading lobes. The long, slender filaments of the stamens are connate by broad bases. The small, somewhat stipitate ovary supports a long, slender style which terminates in five stigmatic branches. A narrow, undulate disk is found at the base of the filaments. (Fig. 333.) The purplish-flowered Asiatic *P. indica* may be found as an ornamental.

ORDER EBENALES

Flowers regular, bisexual or unisexual, polygamous or diœcious, hypogynous, perigynous, or epigynous, mostly 5-merous, but 2-merous and 7-merous whorls common. Stamens as many as the petals, twice as many, or more numerous, epipetalous. Staminodes present in some cases. Gynœcium 2—8-carpellate, with 1 style or with as many styles as carpels. Ovary superior or inferior, 2- to several-celled, with few to several ovules in each cell on axile placentas. Integuments 1 or 2. Fruit fleshy or dry.

Families: *Sapotaceæ, Hoplestigmataceæ, Ebenaceæ, Diclidantheraceæ, Symplocaceæ, Styracaceæ, Lissocarpaceæ.*

Key to the Principal Families

I. Ovary superior, completely chambered.
  Stamens as many as the lobes of the corolla. Style 1. *Sapotaceæ.*
  Stamens twice as many as the lobes of the corolla. Flowers bisexual or unisexual.
    Styles 2—8. Flowers mostly unisexual, monœcious or diœcious. *Ebenaceæ.*
    Style 1. Flowers bisexual. *Styracaceæ.*
II. Ovary inferior.
  Ovary completely chambered. Stamens in 1-3 whorls. *Symplocaceæ.*
  Ovary incompletely chambered. Stamens in 1 whorl. *Styracaceæ.*

This is an anomalous order of problematical affinities. The outstanding characters are the commonly indefinite number of floral organs, as for instance the numerous stamens, which are arranged in several whorls, and the retention of two integuments in the ovule in all but two of the families (*Sapotaceæ, Hoplestigmataceæ*). An advance to monœcism and diœcism appears in the *Ebenaceæ,* while in the *Symplocaceæ* and the *Styracaceæ* epigyny is attained.

SAPOTACEÆ (*Sapodilla Family*). Flowers regular, hypogynous, bisexual or unisexual, polygamous. Sepals 4—6—8, in two isomerous whorls of 2, 3, or 4—free or united at the base, sometimes only 5. Petals of the same arrangement but commonly in a single whorl, and more or less united. Stamens in one or more whorls, the outer sometimes transformed into petaloid staminodes, or in some cases wanting; anthers nearly basifixed or sometimes versatile. Gynœcium of as many or twice as many carpels as there are stamens in one whorl. Style 1, short or elongate, stout or slender, with an insignificant stigma. Ovary superior, with as many cells as carpels and with 1 anatropous or amphitropous ovule in each cell on a basal or axile placenta. Fruit a berry. Seeds large, with a bony or crustaceous testa and a straight embryo in fleshy endosperm—or endosperm wanting.

The members of this family are shrubs or trees with milky sap and simple, alternate, pinnately veined, mostly leathery, exstipulate leaves and small flowers in axillary clusters. Genera 35—40, species 425—600, according to different writers. Chiefly of the tropics around the world. The genera coming into our flora are *Sideroxylon, Dipholis, Bumelia, Chrysophyllum,* and *Mimusops.* Of these *Bumelia* alone reaches farther north than southern Florida.

Sideroxylon Mastichodendron (Mastic) is found in southern Florida. It is a tall tree with long-petioled, evergreen leaves clustered at the ends of the branches and bearing small flowers in axillary clusters. The flowers appear in autumn, early spring, and summer; they have 5 sepals and a 5—6-lobed corolla with as many staminodes in the sinuses. The fruit is an oblong, dry berry with a thin, coriaceous pulp. The trunk of the tree is massive, sometimes attaining a diameter of three or four feet. Of the 60 or more species this is the most northern, the others being tropical.

Dipholis, with 3 species, is distributed over the West Indies to southern Florida. The Floridan species, *D. salicifolia,* is a tall tree with oblong-lanceolate, petiolate leaves. The flowers appear in March and April, in axillary clusters. At the base of the campanulate, 5-lobed corolla are a number of linear or awl-shaped appendages and, besides, there are present 5 petaloid staminodes. The fruit is a globular black berry. *Bumelia,* with 35 species, ranges from central United States southward to the West Indies and Brazil, and on the west side through Mexico to Central America. Two species, *B. lycioides* and *B. lanuginosa,* reach the latitude of Illinois and Virginia. These are trees with alternate, lanceolate to oblong, leathery leaves, and bearing typically 5-merous flowers in axillary clusters. Besides the 5 stamens there are 5 petaloid staminodes. The fruits are small, oblong, black berries.

Chrysophyllum, with 70 species, is tropical, chiefly American, but with a few in Africa, Asia, Australia, and the Hawaiian Islands. Pentamerous flowers are produced, but these lack both staminodes and the characteristic appendages of the corolla. *C. oliviforme* occurs in Florida; it is a tree with alternate, oblong-lanecolate, coriaceous leaves covered with a brilliant copper-colored pubescence. *Mimusops,* with about 160 species, spans the tropics of both hemispheres, one species, *M. Sieberi,* reaching southern Florida. This is a small tree with elliptic-oblong, leathery leaves clustered at the ends of the branches. Among the leaves are the axillary flowers, which are 6-merous to 8-merous. In each sinus of the corolla are two appendages, and there are 5 staminodes. The fruit is a berry with spongy pulp and milky sap, covered on the outside with rough brown scales.

According to Bailey the following members of the family are cultivated: *Sapota Achras* (Sapodilla)—a tall, evergreen tree; *Achras Zapota* (Sapote or Marmalade-Plum)—a tree; *Lucuma nervosa* (Canistel or Egg-Fruit)—a small tree, yielding an orange-yellow, juicy, and egg-shaped fruit, 2—4 inches long; *Palaquium Gutta*—a tree; and *Chrysophyllum Cainito* (Star-Apple)—a tree yielding an apple-sized fruit with a whitish translucent pulp.

HOPLESTIGMATACEÆ. One genus, *Hoplestigma,* of two species of tropical Africa. The plants are trees with large leaves, producing numerous flowers in helicoid panicles. Stamens numerous. Gynœcium 2-carpellate.

EBENACEÆ (*Ebony Family*). Flowers regular, unisexual and diœcious, rarely bisexual and polygamous. Calyx and corolla usually 3—7-lobed, the calyx persistent, the lobes of the corolla convolute or contorted in the bud. Stamens as many or twice as many as the lobes of the corolla or more, epipetalous on the base of the corolla-tube, the filaments free or connate at the base. Gynœcium of 2—8 carpels, the cells often divided by false partitions. Ovary superior, 2—16-celled, with 1—2 (or 3) suspended ovules in each cell. Styles 2—8, free or united at the base. Stigmas simple or forked or papillate-plumose. Fruit a berry, or drupaceous, or capsular, 1- to several-seeded. Embryo small, straight, with large, foliaceous cotyledons, in copious cartilaginous endosperm. Shrubs or trees with very hard wood and alternate, entire, exstipulate leaves. Flowers solitary or clustered in the axils. Five or 6 genera and about 320 species, chiefly of the tropics and subtropics.

Genera: *Diospyros, Royena, Euclea, Maba, Tetracyclis.*

The great majority of the species belong in *Diospyros. D. virginiana,* the common persimmon, has spread north to Connecticut and west to Iowa and Kansas. It sometimes grows to a height of 100 feet. A much smaller species, *D. texana,* occurs in Texas. In this genus the flowers are 4-merous to 6-merous, with 60—80 stamens—mostly aborted or wanting in the carpellate flowers. The staminate flowers are clustered and possess a rudimentary gynœcium. The carpellate flowers, which are usually solitary, have a 2—6-carpellate gynœcium with as many styles. The fruit of the Eastern species is a large, pulpy berry, which is smaller in the Western species. The calyx is persistent. The two species differ strikingly in foliage, the leaves of the Eastern species being large and ovate-lanceolate, those of the Western species spatulate and much smaller. *Diospyros virginiana* is commonly cultivated, as is also *D. Kaki* of China and Japan. The carpellate flowers of both bear 8 staminodes.

DICLIDANTHERACEÆ. One genus, *Diclidanthera,* of two species of Brazil. The plants are shrubs and trees with alternate, entire leaves and showy 5-merous flowers in axillary racemes. The fruit is a berry.

SYMPLOCACEÆ (*Sweet-Leaf Family*). Flowers regular, bisexual or rarely unisexual, epigynous. Calyx and corolla 5-merous, the latter often deeply lobed, the lobes twice as many as the calyx-segments. Stamens in 1—3 whorls or more, epipetalous, connate in

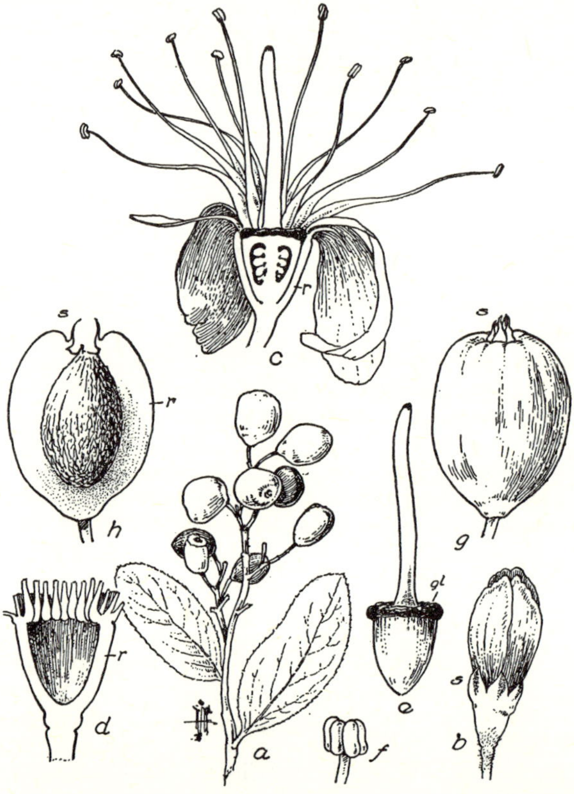

FIG. 334.—*Symplocos paniculatus: a,* twig with fruits; *b,* flower-bud; *c,* flower with ovary in sectional view; *r,* receptacle; *d,* sectional view of receptacle; (*r*), loosely adnate to ovary; *e,* gynœcium; *gl,* gland (disk); *f,* anther; *g,* fruit, with persistent calyx (*s*) at top; *h,* median section of fruit.

clusters, and the filaments usually broadened below; anthers globular or elongate. Gynœcium 2—5-carpellate, with a filamentous style; stigmas capitate or 2—5-lobed. Ovary more or less inferior, 2—5-celled, with 3—4 pendulous ovules in each cell. Disk none. Fruit drupaceous, usually 1-seeded. Seeds with straight embryo and fleshy endosperm. Trees or shrubs with alternate, simple, exstipulate, mostly petiolate leaves. Flowers mostly yellow, in axillary clusters.

One genus: *Symplocos,* of about 200 species of the tropics and subtropics but none in Africa and western Asia. Only the following in North America outside of Mexico.

Symplocos tinctoria (Sweet-Leaf). A shrub or small tree with alternate, broadly lanceolate leaves and bright-yellow flowers in scaly-bracted lateral clusters, inhabiting woodlands from Delaware to Arkansas and southward. In the North the flowers appear before the leaves. Fruit a nut-like drupe crowned by the calyx-segments. (Fig. 334.)

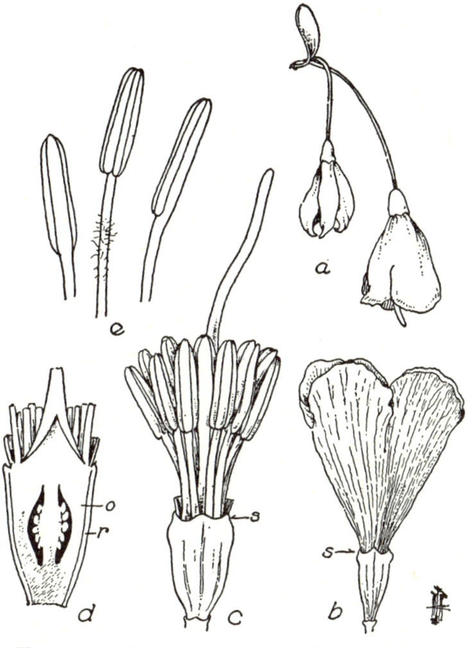

FIG. 335.—*Halesia carolina: a,* flower-cluster; *b,* flower; *s,* sepals; *c,* flower with corolla removed; *d,* median section of "ovary"; *o,* ovary; *r,* receptacle; *e,* dorsal, ventral, and lateral aspects of stamen.

STYRACACEÆ (*Storax Family*). Flowers regular, bisexual, or unisexual and polygamo-diœcious, in some cases appearing before the leaves. Calyx-segments 5 or 4, often reduced to small teeth. Corolla-lobes 5 or 4, sometimes as distinct petals. Disk none. Stamens twice as many as the lobes of the corolla or more, rarely as many, inserted at the base or on the tube of the corolla, often in clusters; filaments filiform; anthers elongate. Gynœcium 2—5-carpellate, with elongate, stout or slender style terminating in a capitate or slightly lobed stigma. Ovary more or less inferior, 2—5-celled, with 1 to few anatropous ovules in each cell on a basal, columnar, stipitate, or axile placenta. In some genera the ovary is incompletely septate—1-celled above but several-celled below. Trees or shrubs with simple, alternate, exstipulate leaves, bearing the flowers in axillary clusters.

Genera 6 or 8, including *Halesia, Styrax, Pterostyrax, Alniphyllum, Hua, Foveolaria, Pamphilia.* Species, according to various writers, 75—120, mainly of the tropics, with an isolated species in the Mediterranean region, and a few representatives of *Halesia* and *Styrax* in Atlantic North America.

Halesia çarolina (Snow-Drop or Silver-Bell Tree). This is a shrub or small tree of woodlands from Virginia to Illinois and southward. One or two other species are found farther south. The flowers are 4-merous or 5-merous, with a deeply lobed corolla. Stamens 8—16, more or less monadelphous, with filaments flattened downwards and adnate to the corolla-tube below. Gynœcium 2—4—5-carpellate, with 1 slender style. The 2—4—5-celled ovary is 1-celled by incomplete partitions above. Ovules 2 or more in each cell. Fruit an elongate-oblong capsule broadly winged on the angles. (Fig. 335.)

Styrax, with 75 or more species of America, Asia and the Mediterranean region (1), has about 5 species in eastern United States from Virginia to Illinois and southward, and 1 in California. Calyx and corolla are 5-merous. Stamens 10—16. The calyx-segments are much larger than in *Halesia* and the corolla is deeply lobed into long, narrow segments (almost choripetalous). Gynœcium 3-carpellate, with a slightly inferior ovary which is completely 3-celled below but 1-celled above. Ovules several in each cell. Style stout or slender. Fruit a globose capsule dehiscing above and partly invested by the persistent calyx.

LISSOCARPACEÆ. A monotypic family of northern Brazil, based upon *Lissocarpa Benthami.* Flowers regular, 4-merous, bisexual. Corolla salver-form, with broad, spreading limbs. Stamens 8, epipetalous, with anthers connivent as a tube. Gynœcium 4-carpellate, the ovary superior, 4-celled, with 2 pendulous ovules in each cell. Fruit dry, indehiscent. Seeds with cylindrical embryo and copious endosperm. Small trees with simple, alternate, entire, leathery leaves.

CHAPTER XVI

TETRACYCLICÆ

General characters: (1) Flowers 4-whorled, outer whorl of stamens suppressed, often with reduction in the inner whorl, conforming to the general formula $S5\ P5\ A5 — 0\ G5 — 2$, and (2) hypogynous or epigynous; (3) stamens generally epipetalous, sometimes reduced to 4, 2, 1; (4) gynœcium in more specialized forms reduced to 2 carpels; (5) ovule generally with 1 integument.

This is the more specialized group of the *Metachlamydeæ,* reaching its climax in the *Campanulatæ* (*Compositæ*). The notable evolutionary advances lie in the direction of irregularity of the corolla, inferior ovary (epigyny), suppression in the andrœcium and gynœcium and in the number of ovules and of one of the integuments, associated with aggregation of numerous small flowers into compact inflorescences (capitula). The climax of the evolutionary stages is reached in the ligulate corollas, the scale-like or bristly calyx, the one-celled and one-seeded inferior ovary, and the capitulum of the *Compositæ.* For incipient stages or parallel types of these climactic forms we may look back to the *Rubiales* and the *Tubifloræ,* or even to the *Umbellifloræ.* Compare for example, the inflorescences of composites and labiates (*Monarda*), *Caprifoliaceæ* and *Cornaceæ,* as well as modifications of the floral organs. The composites are peculiar in producing both regular and irregular flowers on the same individual and even in the same inflorescence, except in the most specialized forms (*Cichorieæ*) in which all the corollas are irregular (ligulate).

In the lower orders (SUPERÆ: *Contortæ, Tubifloræ, Plantaginales*) of the *Tetracyclicæ* the ovary is superior; in the higher orders (INFERÆ: *Cucurbitales, Rubiales, Campanulatæ*) the ovary is inferior.

Tetracyclicæ with Superior Ovary. Superæ

ORDER CONTORTÆ

Flowers regular, bisexual or unisexual, hypogynous, mostly sympetalous, 5-merous or varyingly 2-merous to 6-merous. Stamens as many as sepals or fewer (2—1), mostly epipetalous. Gynœcium

2-carpellate, rarely more, the carpels sometimes only slightly united. Fruit various, dry or fleshy. Trees, shrubs, or herbs, with simple or pinnately compound, exstipulate leaves. The characteristic feature of the order lies in the twisted or contorted æstivation of the corolla in the bud.

Families: *Oleaceæ, Desfontaineaceæ, Loganiaceæ, Gentianaceæ, Apocynaceæ, Asclepiadaceæ.*

Key to the Principal Families

Plants without milky sap.
- Trees or shrubs.
  - Stamens fewer than the lobes of the corolla, mostly 2. Leaves opposite, simple or compound. *Oleaceæ.*
  - Stamens as many as the lobes of the corolla. Leaves usually opposite, simple. *Loganiaceæ.*
- Herbs.
  - Style simple, with entire or 2-lobed stigma. *Gentianaceæ.*
  - Style 2-cleft (rarely 4-cleft). *Loganiaceæ.*

Plants with milky sap.
- Filaments of stamens distinct. Style 1. *Apocynaceæ.*
- Filaments united. Styles united by the stigmas. *Asclepiadaceæ.*

OLEACEÆ (*Olive Family*). Flowers regular, bisexual or unisexual, polygamous or diœcious, mostly sympetalous, 2-merous to 6-merous. Calyx small or none. Corolla sometimes none. Stamens commonly 2, sometimes 4, distinct, and commonly with short filaments; anthers stout, oblong to ovoid, 2-celled. Gynœcium 2-carpellate; style 1, short or elongate, or obsolete. Ovary superior, 2-celled, with 1—2—4—8 ovules in each cell on an axile placenta. Fruit a capsule, samara, berry, or drupe. Seeds with straight embryo, and with or without endosperm. Trees, shrubs, or subshrubs, with simple or pinnately compound, opposite, rarely alternate, leaves. Flowers small to large, often fragrant, in axillary or terminal cymes, compound panicles, or fascicles. Genera 20—22, species 390—525, according to different writers. Plants of temperate, tropical, or subtropical regions. Family important for its useful and ornamental plants.

The following characteristic genera are represented in our flora or in cultivation: *Fraxinus, Syringa, Adelia, Menodora, Chionanthus, Ligustrum, Olea, Forsythia, Jasminum.* These belong in two subfamilies which are distinguished as follows:

Oleoideæ. Ovules pendulous from the top of the placental axis. Fruits not constricted vertically. *Fraxinus, Forsythia, Syringa, Chionanthus, Olea, Adelia, Ligustrum.*

Jasminoideæ. Ovules erect from a basal placenta. Ovaries usually 2-lobed by a vertical constriction. *Jasminum, Menodora.*

Fraxinus (Ash). A widely distributed genus of 50 or more species, 20 or more occurring in our flora. The plants are trees with opposite, odd-pinnate leaves and usually numerous, greenish or brownish, polygamous or diœcious flowers which appear before or with the leaves at the nodes of the previous season's shoots. The floral organs are in 2's or 4's, although calyx and corolla may be absent. The superior ovary is 2-celled, with 2 pendulous ovules in each cell. Fruit a samara, the wing of which is developed from the upper edge of the ovary and carrying the style up with it as a minute peg. A very technical genus, the species of which are best distinguished on characters of the samaras. There is often considerable variation in the leaves of the same species, a feature especially noticeable in young shoots or sprouts. Perhaps the most readily distinguishable is the Northern *Fraxinus nigra* (Black Ash), which is common in low, rich, deciduous woods, the habitat alone being usually sufficient to identify it; the samara is rather broad and notched at the apex. The Southern *F. caroliniana* is easily recognized by its very broadly winged samaras, and the square twigs of the Eastern *F. quadrangulata* will set off this species. Four species

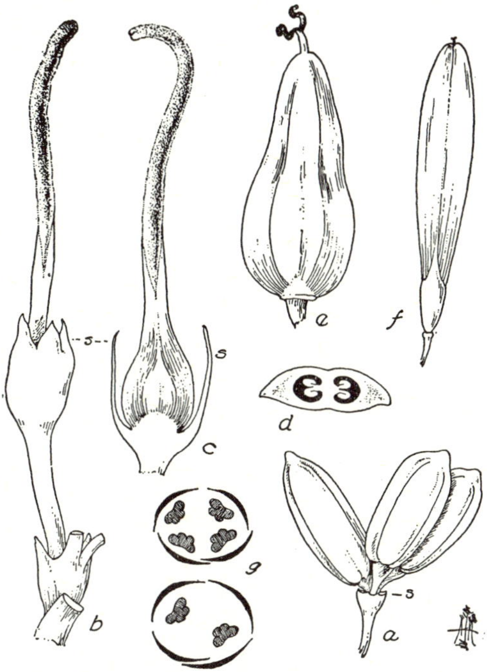

FIG. 336.—*Fraxinus* sp.: *a,* staminate flower; *b,* carpellate flower; *c,* gynœcium and calyx (*s*) in section; *d,* transverse section of ovary; *e,* young fruit; *f,* samara (fruit), with style at summit; *g,* floral diagram of a staminate flower with four stamens, and one with two.

occur in the Pacific States, of which *F. oregana* ranges north to British Columbia. Several species yield valuable lumber (*F. americana,* White Ash). (Fig. 336.)

Syringa vulgaris (Lilac). This favorite ornamental is native to the Balkans and the Orient. With us it is ordinarily a shrub, but in its native home it may attain tree size. The leaves are simple, opposite, ovate and entire, and the numerous small and fragrant violet, purplish, or white flowers, which are borne in showy thyrsoid inflorescences, are 4-merous, with a salver-form corolla-tube bearing 2 short stamens below the throat. The elongate-oblong capsular fruit dehisces loculicidally by 2 valves. Besides this species, four or five others and their varieties are in cultivation, mainly in northern parts of the country. The total number of species is in the neighborhood of 30, all Eurasian. (Fig. 337.)

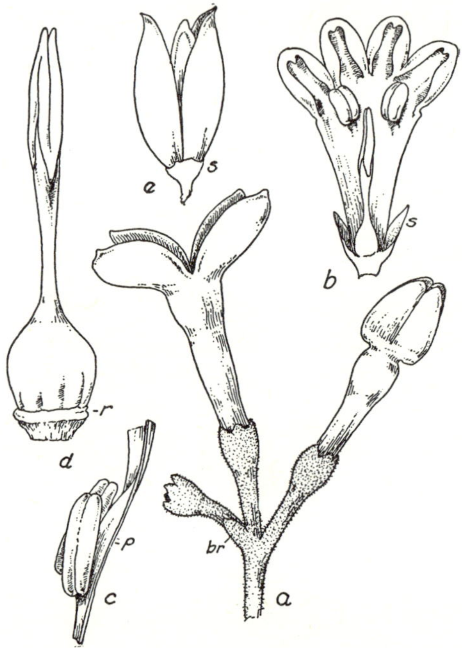

FIG. 337.—*Syringa vulgaris: a,* cymule; *b,* flower, in sectional view; *c,* stamens, epipetalous; corolla (*p*); *d,* gynœcium; *r,* receptacle; *e,* dehisced capsule, with calyx (*s*) below; *br,* bracts.

A solitary species of *Chionanthus* occurs in Atlantic North America. This is *C. virginica* (Fringe Tree), a shrub or small tree of rich woodlands. It is distinguished by rather large, elliptic-lanceolate leaves and bisexual flowers, the corolla of which is deeply segmented. The 2 stamens are inserted at the base of the corolla. The 2-carpellate gynœcium ripens as a small, oblong, 1-seeded drupe. Two other species occur in China.—Note the connection here between the floras of eastern North America and eastern Asia.

Ligustrum, with 35 or more species, is mainly Asiatic. Among a dozen ornamentals is the European *L. vulgare* (Privet). This is a shrub with simple, alternate, elliptic-lanceolate leaves, bearing small, white, bisexual flowers in terminal thyrsoid panicles. Calyx reduced to four small segments. Corolla short-tubular, with four spreading lobes. Stamens 2, in the throat of the corolla. Fruit a 1—3-seeded, globose berry. (Fig. 338.)

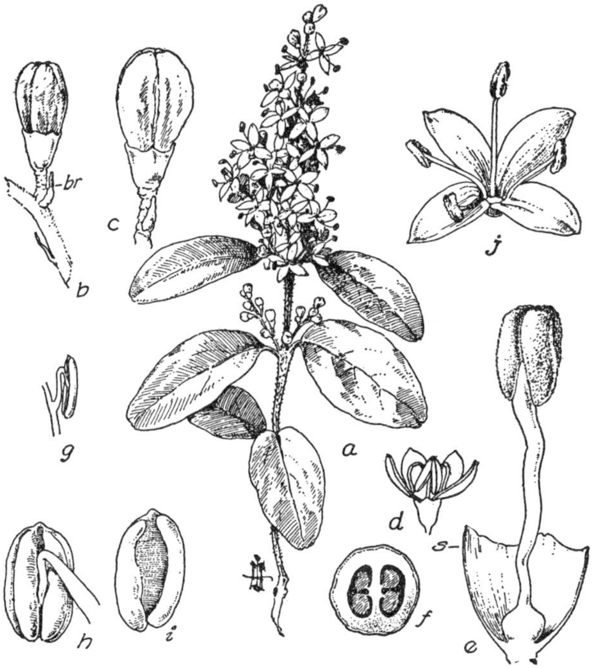

FIG. 338.—*Ligustrum* sp.: *a,* flowering branch; *b,* flower-bud; *br,* bract; *c,* anterior view of flower-bud; *d,* a flower opening, with two stamens; *e,* gynœcium and calyx (*s*); *f,* transverse section of ovary; *g,* stamen, epipetalous; *h, i,* dorsal and ventral aspects of anthers; *j,* a corolla bearing four stamens.

Forsythia (Forsythia). Three species of east-Asiatic shrubs commonly planted as ornamentals under the name golden-bells. Flowers yellow, with deeply segmented, 4-merous calyx and corolla. Stamens 2, inserted at the base of the corolla-tube. The heterostylous gynœcia are a special feature of the flowers. Fruit a woody septicidal capsule. The flowers appear before the leaves. *Forsythia suspensa* is distinguished by the hollow internodes of the stems. *F. viridissima* and *F.*

intermedia have lamellate pith. Ternate leaves occur sometimes in *F. suspensa* and *F. intermedia.* All are deciduous.

Olea (Olive) comprises about 40 species of evergreen shrubs and trees of the tropics and subtropics. Leaves opposite: Flowers white, bisexual, polygamous or diœcious, with 4-merous calyx and corolla. Stamens 2. Fruit a drupe. *Olea europæa* is the most common cultivated species.

Adelia (*Forestiera*), with 15 species, is mainly American, reaching our limits from the West Indies and Brazil. Several species occur in the South and Southwest, among which are *A. acuminata,* ranging from Indiana to Texas, and *A. neo-mexicana* of southern California. The flowers, which are polygamous or diœcious, sometimes lack both calyx and corolla, but when present these organs are very small. There are 4 sepals, but the corolla is reduced to 1 or 2 *distinct* petals. Stamens 2, 3, or 4. Gynœcium 2-carpellate, with a 2-celled ovary which becomes an oblong or globose drupe.

Jasminum (Jasmine, Jessamine), with about 200 species, is the largest genus of the family. It is distributed over the tropics and subtropics. Many are popular ornamentals. The plants are erect or climbing shrubs with odd-pinnate leaves, sometimes reduced to one leaflet. The generally showy flowers are aggregated in terminal cymes or occasionally solitary. Calyx campanulate or funnel-form, with 4—9 bristly teeth, and the corolla with as many lobes and a deep, narrow tube. Stamens 2. Gynœcium 2-carpellate; ovary constricted between the carpels. Fruit a 2-lobed berry.

DESFONTAINEACEÆ. One genus, *Desfontainea,* with 2 or 3 species in the Andes. Low shrubs with opposite, leathery, spinescent, holly-like leaves. Flowers large, axillary, 5-merous, at the ends of the branches. Fruit a globular berry.

LOGANIACEÆ (*Logania Family*). Flowers regular, bisexual or unisexual, 4-merous or 5-merous, hypogynous. Calyx campanulate; corolla funnel-form, campanulate, or rotate. Stamens as many as the lobes of the corolla and inserted in the corolla-tube or at its throat. Gynœcium commonly 2-carpellate, sometimes 3—5-carpellate; style commonly 1, sometimes 2, short and thick or elongate, terminating in as many stigmatic lobes or branches as there are carpels. Ovary superior, with as many cells as carpels, and producing numerous to few (or 1) ovules on axile placentas. Fruit a septicidal capsule, or a berry or drupe. Herbs, shrubs (sometimes vines), or trees, with simple opposite or whorled leaves which are stipulate or provided with interpetiolar stipular connections. Flowers small to large, in cymose paniculate or superposed whorled inflorescences. Thirty or more genera and 400—550 species of the tropics, subtropics, and temperate regions.

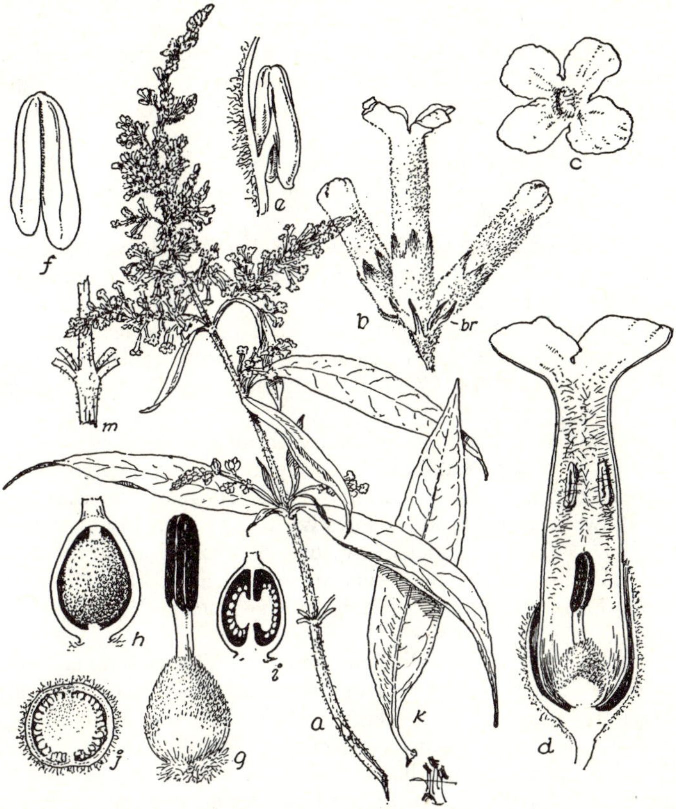

FIG. 339.—*Buddleia officinalis: a,* flowering branch; *b,* cymule; *c,* corolla; *d,* median section of flower; *e,* stamen, epipetalous; *f,* ventral aspect of anther; *g,* gynœcium; *h,* section of ovary; *i,* median section of ovary; *j,* transverse section of ovary; *k,* leaf, upper surface.

Our genera are: *Gelsemium, Spigelia, Cynoctonum, Polypremum.* Several species of *Buddleia* are ornamentals. *Strychnos* is an important tropical genus. *Strychnos nux-vomica* yields the powerful drug strychnine. From *S. toxifera* some South American Indians obtain an arrow poison. *S. colubrina* of the East Indies yields an antidote for snake-bite.

Gelsemium (False Jasmine) has one species, *G. sempervirens,* in southeastern United States, and another in eastern Asia (!). The American plant is a climbing or trailing vine bearing simple lanceolate leaves and large bright-yellow dimorphic flowers, solitary or in cymose inflorescences. Calyx very small; corolla funnel-form with broad, spreading lobes. The floral organs are 5-merous throughout. Beneath the ovary is a short, thick, stipitate disk. The oblong superior ovary supports a slender to filiform style which is 4-cleft at the summit. The fruit is an elliptical, flattened, septicidal capsule which splits into boat-shaped valves. Several flattened, winged seeds are produced in each cell.

Buddleia, with about 100 species, is distributed over the tropics of America, Asia, and Africa. Several species are popular ornamentals. The plants are shrubs or small trees, or in some cases herbs, with opposite, entire or dentate, ovate to lanceolate leaves. Flowers small, lilac-like, 4-merous, in paniculate, cymose, capitate, or superposed whorled clusters. Calyx campanulate; corolla campanulate or funnel-form, sometimes constricted at the throat, bearing the stamens in the tube. Gynœcium 2-carpellate, with an ovoid ovary supporting a thick style which terminates in a fleshy elongate stigma. Ovules numerous, on thick axile placentas. The transverse section of the ovary shows a placentation suggestive of the next following order (*Tubifloræ*). Fruit a septicidal capsule. Because of apparent similarities in the flowers the buddleias are often called lilacs. Their fragrance, too, is strongly suggestive of the lilacs. (Fig. 339.)

GENTIANACEÆ (*Gentian Family*). Flowers regular or in some cases irregular, bisexual or sometimes unisexual, polygamous or diœcious, 4-merous or 5-merous or even 6- to 12-merous, hypogynous or epigynous. Stamens as many as the lobes of the funnel-form or campanulate corolla and inserted at the base of its tube; filaments filiform or dilated at the base; anthers innate or versatile, 2-celled, opening lengthwise or by apical pores and commonly twisting on opening (a good recognition-mark). Gynœcium 2-carpellate; style mostly elongate, stout to slender, sometimes strongly dilated above or below, and divided at the summit into 2 flat or terete stigmatic limbs which are often conspicuously papillate or plumose, or the stigma is capitate or discoid or otherwise curiously formed. Ovary superior or inferior, conspicuously stipitate (borne on a gynophore) in some cases; 1-celled or nearly 2-celled with numerous anatropous or amphitropous ovules on simple, branched or lobed, parietal or central placentas. Disk when present inconspicuous. Fruit capsular, splitting by 2 valves. Seeds with a small cylindrical or conical embryo in copious endosperm. Bitter-tonic, annual or perennial herbs, shrubs, or subshrubs or small trees, with opposite or alternate, mostly entire, exstipulate leaves, and small to large, often showy flowers with or without bracteoles and bracts, in cymose or racemose inflorescences, or solitary. About 75 genera and 800 species, widely distributed over the world, especially in the northern hemisphere, where certain ones reach boreal and even Arctic latitudes. The family is especially notable for the remarkable diversity of shape of the corolla-tube, anthers, style, and stigmas, and the many curious sculpturings of the seed-coats, besides its tonic properties. Two subfamilies, *viz.*:

Gentianoideæ, in which the ovary is superior, the leaves opposite, and the lobes of the corolla convolute (twisted) or imbricate in the bud.

Sabatia, Centaurium, Eustoma, Gentiana, Pleurogyne, Frasera (Swertia), Halenia, Bartonia, Obolaria, Microcala.

Menyanthoideæ, in which the ovary is inferior or superior, the leaves alternate, and the lobes of the corolla induplicate-valvate in the bud. *Nephrophyllidium, Menyanthes, Nymphoides.*

Gentianoideæ. In all but *Bartonia* and *Obolaria* the corolla is convolute in the bud, while the style is filiform and the anthers are twisted or curving in age in *Sabatia, Centaurium, Eustoma.* A rotate corolla and recurved anthers are characteristic of *Sabatia,* while the parts of the flower are in 5's or 12's. *Eustoma* resembles some Sabatias, but the corolla is campanulate-funnel-form, and the flowers are 5-merous or 6-merous. In *Centaurium* the Corolla is salver-form, the flowers are 5-merous or 4-merous, and the anthers twist on dehiscence. These genera are further distinguished by their stigmas, which in *Sabatia* run to long, linear, spreading branches, in *Centaurium* appearing as 2 capitate to short, broad, and flattened lobes, and in *Eustoma* as 2 lamellate spreading lobes. In all three the gynœcium is 2-carpellate.

Sabatia, with about 18 species, ranges in eastern North America from Canada to Mexico and the West Indies. *Sabatia angularis* reaches Ontario in the North. *Centaurium* with about 25—30 species is chiefly Western, pushing down into South America and eastward to the West Indies, with a further development in the eastern hemisphere. *Eustoma* with 4 species is native to the southern and southwestern United States and the West Indies. One species, *Eustoma Russellianum,* spreads north to Nebraska and Colorado.

In *Gentiana, Pleurogyne, Frasera,* and *Halenia,* the style is stout and persistent, or wanting in *Pleurogyne* and some others. Except for the gynœcium, which is definitely 2-carpellate in all cases, the flowers are 4-merous in *Frasera,* 4-merous or 5-merous in *Pleurogyne* and *Halenia,* but 4-merous to 7-merous in *Gentiana.* Within the corolla-

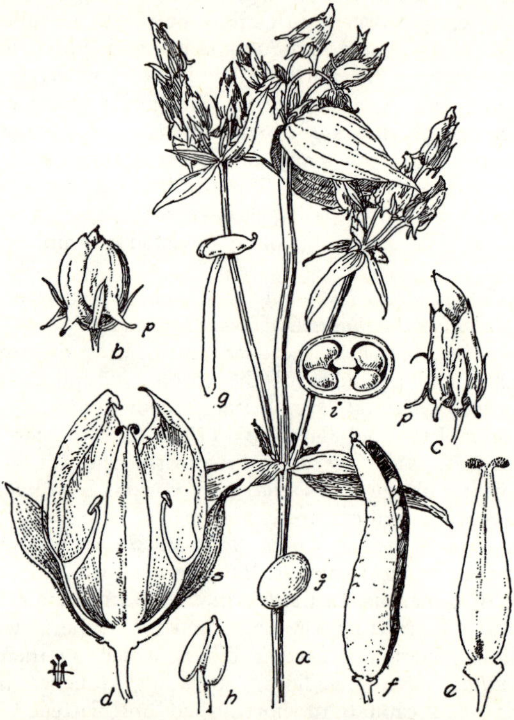

FIG. 340.—*Halenia deflexa: a,* inflorescence; *b,* flower; *p,* spur of corolla; *c,* flower at later stage; *d,* flower in sectional view; *e,* gynœcium; *f,* capsule dehiscing; *g,* stamen, lateral view; *h,* anther, dorsal view; *i,* transverse section of ovary; *j,* ovule.

tube in *Gentiana* and *Pleurogyne* are some small scale-like appendages, which in *Frasera* and *Halenia* are replaced by pits or glands at the base of the lobes. The deep and often large corolla-tube, and the frequently fringed corolla-lobes of *Gentiana* distinguish it from *Pleurogyne* with its short-tubed rotate corolla. *Frasera* is furnished with peculiar fringed glandular bodies on the inner side of the corolla-lobes. In *Halenia* (at least in ours) the corolla-lobes are prolonged downwards between the calyx-segments as prominent hollow, gland-tipped spurs, very much as in *Aquilegia* (*Ranunculaceæ*); this single character is a prominent earmark of our lone species, *Halenia deflexa,* aside from its rather small greenish to purplish flowers, which are clustered and pedicelled in the axils of the upper leaves. The plant is an inhabitant of deep, cold, moist woods in the North. *Frasera,* with 15 or more species, is best developed in the western United States, the only Eastern species being *F. carolinensis*. Eight or more occur in California. The 30 or more species of *Halenia* are natives of the Americas and Asia. (Fig. 340.)

Bartonia (4 spp.) and *Obolaria* (monotypic) are native to eastern North America. They differ from the preceding genera in having the corolla-lobes imbricated in the bud and in these lacking both appendages and glands. Vegetatively, in response to a saprophytic habit, their leaves are reduced to mere scales, giving their flowering stems a rather naked appearance. The flowers are 4-merous in both, except that *Obolaria* has but 2 sepals. The solitary *Obolaria virginica* (Pennywort) occurs as a low perennial in woodlands from New Jersey and southward and west to Illinois and Texas. The small, purplish flowers are borne near the ends of the branches in the axils of the upper leaves, which are enlarged and obcordate, and taper downwards. Of the four species of *Bartonia, B. virginica* reaches Nova Scotia and Newfoundland. *Microcala quadrangularis* is an insignificant annual (1—2 inches high) of coastal California. It is of particular interest, however, as a connecting link between the flora of southern California and that of the Mediterranean (north to Germany) where the only other species, *M. filiformis,* is found.

Menyanthoideæ. A notable representative of this group is *Nephrophyllidium crista-galli,* a solitary perennial from creeping rootstocks, inhabiting marshes from western Washington to Alaska, with a recurrence in Japan. The basal leaves with reniform, broadly notched, palmately-veined blades and long petioles arise from a scaly caudex, along with a scapose inflorescence subtended by conspicuous bracts. An advance is noted in the epigynous flowers in which the ovary is half-inferior. The campanulate corolla with its broad triangular lobes bears 5 stamens with short filaments and oblong anthers. The 5 lanceolate sepals are shorter than the corolla-tube, while the gynœcium is 2-carpellate, with a short, thick style forking into 2 thick, spreading stigmatic lobes. The 1-celled ovary contains numerous ovules on parietal placentas, and at maturity becomes a capsule which opens irregularly at the top. At the base of the corolla are five glandular structures. This odd species is one among a number of other plants that connect the flora of the north Pacific Coast with that of Japan.

Menyanthes is another monotypic genus, with *M. trifoliata* (Buck Bean) inhabiting cold bogs around the northern hemisphere. Like the preceding it is a perennial herb from a creeping scaly rootstock, but the leaves are trifoliolate, and the inflorescence is a bracteate raceme or panicle of white to purplish, 5-merous, hypogynous flowers. It is usually found growing in water at flowering time. The corolla-tube is short and funnel-form with the lobes fimbriate or bearded within. The 5 stamens are filiform with sagittate anthers. The ovary is 1-celled, with a slender style terminating in 2 lamellate stigmatic lobes. The numerous ovules arise from parietal placentas which are sometimes intruded into the ovary-cavity. The fruit is an irregularly dehiscent capsule.

Nymphoides, with 20 or more species, encircles the tropics and subtropics, with four species pressing into eastern North America; these are perennial aquatics from rootstocks, with leaves resembling those of the water-lilies. The flowers are 5-merous, and the ovary is superior.

APOCYNACEÆ (*Dogbane Family*). Flowers regular, bisexual, 5-merous, or in some cases 4-merous, hypogynous or sometimes epigynous. Calyx-segments free almost to the base, with or without glands within, persistent. Corolla sympetalous, mostly salverform or funnel-form, the lobes convolute or twisted in the bud, rarely valvate, the tube often provided with downwardly directed hairs at the base or on the throat within, or with scales, gibbosities, or other outgrowths. Stamens as many as the lobes of the corolla and alternate with them, the filaments short, rarely united, inserted at different levels in the corolla-tube; anthers introrse, generally elongate, linear-oblong, often sagittate; each anther 2-celled, often connivent around and sometimes adherent to the stigma, the connective often prolonged at the apex; pollen generally in simple grains and granular but sometimes glutinous. Gynœcium generally 2-carpellate but sometimes more, apocarpous or syncarpous, the carpels usually united above the styles but separating at maturity, the styles thus distinct or united into 1, often with curious annular thickenings below the stigma, which is of exceedingly diverse and curious shapes. Ovary superior to half inferior, 2-celled, each cell with few to numerous, anatropous or amphitropous, generally pendulous ovules on 2 parietal placentas —or the placentas may arise from the partition-walls; or when the carpels are distinct, then from the ventral suture. The nectariferous disk, when present, is cup-shaped or variously lobed. The fruit in apocarpous forms may be fleshy and indehiscent or dry and follicular, or a schizocarp with winged mericarps; in syncarpous forms it is commonly a berry or drupe, or sometimes capsular. Seeds from dry fruits are generally furnished with wings or caudate appendages or with a tuft of hairs at one or both ends. Endosperm, when present, is scanty or copious. The embryo has a short hypocotyl and thin, flat or variously folded cotyledons. The plants are herbs, shrubs, vines, or trees, with milky sap and opposite, alternate, or verticillate leaves without stipules. The notable features are the diverse growth-forms, the often incomplete or slight union of the carpels, the strikingly peculiar configurations of the stigma and its tendency to connect with the anthers, and the diverse types of fruits. The milky sap, however, is characteristic also of the next following family. The sap is often very poisonous, but of commercial importance in some species. Many species of different genera are ornamentals or hot-house plants and a few are grown for their edible fruits. Only a few genera are represented in our flora.

Genera 130, species about 1,100, widely distributed but mainly tropical, although some press far into north-temperate regions and others into temperate South America and into New Zealand. The genera with us are *Amsonia,* with alternate leaves; *Vinca, Trachelospermum, Apocynum,* with opposite leaves; *Cycladenia,* with whorled leaves.

Amsonia with 8 species has two in eastern Asia, the others distributed over the eastern, southern, and southwestern United States. *Amsonia Tabernæmontana,* a low to tall Eastern perennial resembling a milkweed, bears lanceolate leaves and conspicuous bluish flowers in a terminal cyme. The salver-form corolla with spreading pointed lobes bears the stamens in the throat. The 2 carpels are connected above by the styles, but separate at maturity as two slender follicles. From the capitate stigma there is suspended a cup-like annular membrane. There is no disk.

Vinca minor, the common Periwinkle, one of about a dozen Old World species, sometimes escapes from cultivation. Its trailing stems root and send up leafy shoots at the nodes and only one or two large blue flowers on long axillary pedicels are produced on a shoot. The corolla-tube is salver-form with large, spreading lobes, below which, where the stamens are inserted, it is dilated. The stamens, which are included within the tube, are furnished with elongate anthers on short geniculate filaments more or less thickened above. The 2 carpels are distinct to within a short distance above the ovaries, where they unite into a single terete or upwardly thickened style. The stigma varies considerably with the species, but usually consists of deflexed penicillate lobes supported on a thick column which is subtended by a thick annular disk; or in some cases it is spool-shaped, with the upper margin penicillate and with a bell-shaped appendage suspended from the lower ledge. From the base of the ovary there arise a pair of oblong or narrowly triangular glandular structures. The fruits are follicles. The 12 species are natives of the Old World. (Fig. 341.)

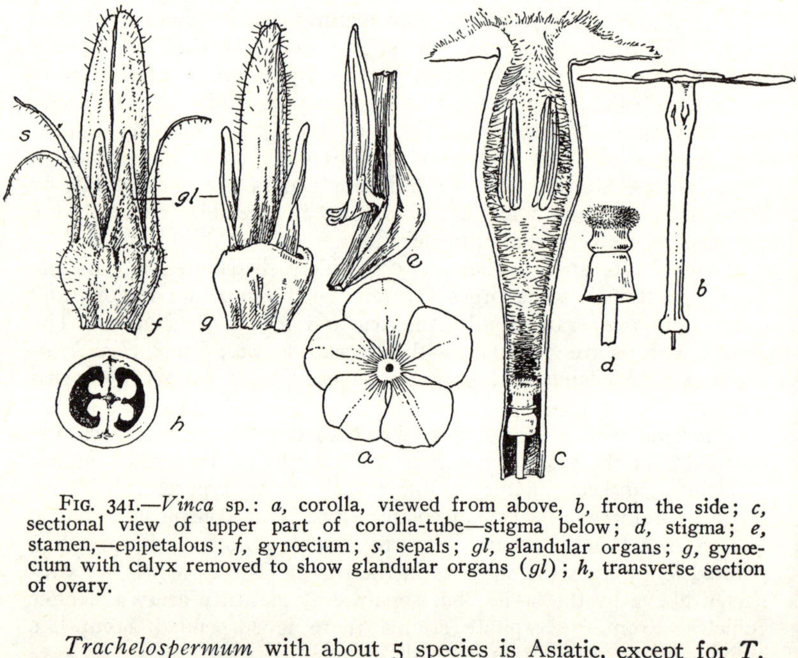

FIG. 341.—*Vinca* sp.: *a,* corolla, viewed from above, *b,* from the side; *c,* sectional view of upper part of corolla-tube—stigma below; *d,* stigma; *e,* stamen,—epipetalous; *f,* gynœcium; *s,* sepals; *gl,* glandular organs; *g,* gynœcium with calyx removed to show glandular organs (*gl*); *h,* transverse section of ovary.

Trachelospermum with about 5 species is Asiatic, except for *T. difforme* of moist woods from Delaware to Missouri and southward. This is a climbing vine with small yellow flowers in axillary cymose clusters. Inside of the whorl of the awl-shaped calyx-segments are found peculiar glandular structures, while the funnel-form corolla is without appendages. There is an annular disk of 5 glandular lobes. The follicles are very much longer and more slender than in the other genera and the seeds bear a tuft of hairs at one end (comose).

Apocynum (Dogbane), with about 11 species, encircles the north temperate zone, but the majority of the species are found in North America. Our species are rather technical and are frequently confused outside of herbaria, *e.g.*, *Apocynum androsæmifolium* and *A. canna-*

FIG. 342.—*Apocynum androsæmifolium*: *a*, branch with inflorescence; *b*, flower-bud; *c*, flower; *d*, median section of flower; *e*, lobe of corolla and a stamen; *f*, gynœcium; *g*, anther; *h*, a pair of follicles; *s*, sepal; *gl*, glands; *i*, seed.

binum, the former characterized by a pinkish corolla with spreading lobes, the latter by a greenish corolla with ascending lobes. The inflorescences are axillary or terminal cymes, but the flowers are quite small, although the clusters are quite showy. The corolla is urn-shaped to campanulate. An hypogynous disk is present, but there is no clearly defined style. The fruits are elongate follicles producing seeds with a tuft of hairs at one end. The plants are rather ornamental and have a very tough fibrous bark (whence the name Indian Hemp commonly applied to them). (Fig. 342.)

Cycladenia (monotypic) is represented by *C. humilis,* a low, dif-

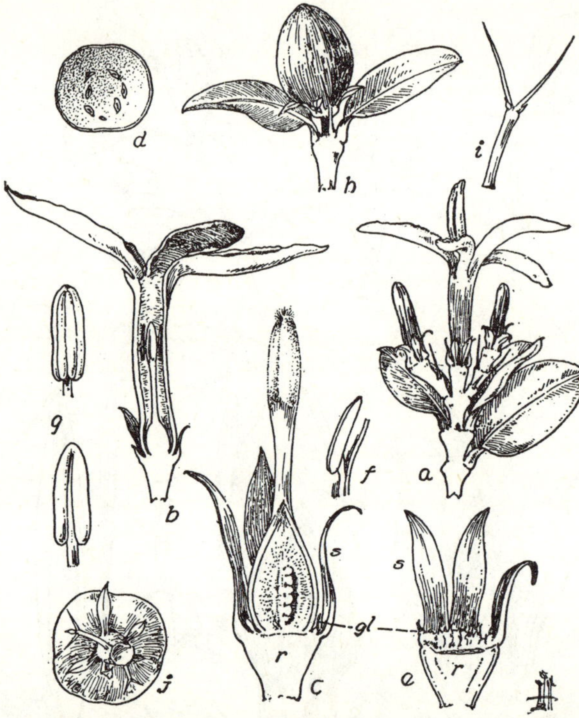

Fig. 343.—*Carissa grandiflora: a,* twig and cymule; *b,* median section of flower; *c,* section of flower with corolla removed; *e,* same with gynœcium removed; calyx (*s*), receptacle (*r*), and glandular lobed disk (*gl*); *f,* stamen—epipetalous; *g,* ventral and dorsal views of anther; *h,* fruit; *i,* bifurcate spine; *j,* fruit viewed from beneath, with spines and persistent calyx; *d,* transverse section of ovary (several-seeded).

fuse herb of the mountains of California. As in *Vinca* the stigma is furnished with a thin collar suspended from below it, and an hypogynous disk is present. The follicles are unusually long for the size of this very small plant.

Among the horticultural forms the choicest is perhaps the Oleander (*Nerium Oleander*), a native of the Mediterranean countries and of the Orient to Japan. The anthers, here, are appendaged above and below and united into a tube around the stigma, to which they adhere. There is no disk. The fruits are elongated follicles. Only two other species are known.

About three species of *Carissa* are in cultivation; these are spiny shrubs with opposite leathery leaves and bifurcate spines. The fruit in this case is a berry. (Fig. 343.)

Here the features to be particularly noted are the uniformly 5-merous flowers, the persistence of an apocarpous gynœcium associated with remarkable configurations of the stigma, the diverse forms of fruits, and the comose seeds.

ASCLEPIADACEÆ (*Milkweed Family*). Flowers uniformly regular, bisexual, 5-merous, and hypogynous. Calyx usually deeply lobed, sometimes with glandular processes within. Corolla rotate, campanulate, urn-shaped, funnel-form, or vasiform, the lobes twisted at the tip in the bud, often with a simple or double *corona* of free or united segments, which is sometimes affixed to the stamens or to both stamens and corolla. Stamens with short filaments or none, sometimes monadelphous, the anthers free or sometimes connate into a tube which may unite with the style to form a *gynostegium*. Pollen-grains in tetrads, granular, or in waxy masses (*pollinia*). The pollen is transferred in one group by a special organ, the *translator,* situated between the anthers, which is dilated into a pouch at one end while at the other end is a sticky disk; the pollen when discharged is poured into the pouch. In another group the translator consists of two arms joined at one end to a small body, the *corpusculum,* which is adnate to the complex stigma, while the opposite ends of the arms go to adjacent anthers and there become affixed to pollinia (Fig. 344). The gynœcium is apocarpous, consisting of 2 carpels with the styles united at the stigma, which is of discoid shape. The superior ovary is 1-celled, with numerous pendulous, anatropous ovules superposed and overlapping in series on a placenta which follows the ventral suture. Fruit a follicle, one from each carpel. Seeds ovoid or oblong, compressed, usually furnished with a tuft of hairs at one end. Embryo with short hypocotyl and flat cotyledons. Endosperm scanty. Perennial herbs, vines, shrubs, or sometimes trees, mostly with milky sap, and simple opposite or sometimes alternate or whorled leaves. The herbaceous forms are of diverse habits, some being succulent, others cactus-like. There is frequent reduction of the leaves among those of arid climates in correlation with succulence, although in other cases succulent leaves are retained. Flowers small to large, often showy and fragrant, in cymose or racemose clusters.

General 220—280; species 1,700—2,200; widely distributed from tropical to cold-temperate regions, those of xerophytic habit being characteristic of arid parts of Central and South Africa. Many are ornamentals. Our genera, *Asclepias, Asclepiodora, Acerates,* are erect or decumbent herbs, while *Gonolobus, Cynanchum,* and *Vincetoxicum* are twining vines. A few twiners or climbers of *Hoya, Ceropegia, Stephanotis, Cryptostegia,* and *Periploca,* of the tropics, are horticultural. *Stapelia,* with about 60 species of cactus-like African plants, is represented in hothouses or cactus gardens. The plant produces large mottled flowers which give off a fœtid odor.

Asclepias (Milkweed) is represented in our flora by 50 or more species, the remaining 50 or 100 being South American and African. The plants are upright, often tall and attractive herbs with opposite alternate or whorled leaves and small flowers of diverse colors (often brilliant) in axillary or terminal umbels. The lobes of the corolla are strongly reflexed and the corona consists of 5 hooded bodies which usually bear an incurved horn within them. The filaments, which arise from the base of the corolla, are united into a tube (monadelphous) around the gynœcium, with the anthers adnate to the stigma; the anthers are furnished with appendages, and each of their cells contains one or two pollinia. The pollinia are connected to the stigma by the translator and the corpusculum. The two carpels (one sometimes abortive) ripen as large (often very stout), soft follicles containing numerous superposed seeds which are furnished with a large tuft of hairs at one end (Figs. 345, 42). *Asclepias syriaca,* the Common Milkweed, which grows to a height of 5 feet, is common from New Brunswick to Saskatchewan, mostly in waste grounds. It bears large, velvety leaves and is pubescent above. The corolla is pale purplish to greenish-white. The follicles are very large (up to 5 or 6 inches) and thick and covered with soft, spine-like processes (Fig. 345). *Asclepias speciosa* is very similar as to fruit but differs in the usually lower habit and the shorter broadly ovate leaves and longer, narrower corona-segments. *A. tuberosa* is readily distinguished by its orange-colored flowers and *A. quadrifolia* by its lanceolate leaves which are in whorls over the middle portion of the stem, and by its long, slender follicles. The leaves of some species are very narrow or linear, as in *A. galioides, A. pumila,* and *A. verticillata* (with whorled leaves). The sap of all milkweeds is very bitter.

Asclepiodora is very similar to *Asclepias,* particularly as to fruit, but its corolla is characterized by conspicuously spreading (rotate)

FIG. 344.—*Asclepias syriaca: a,* upper part of stem with flower-clusters; *b,* flower-bud; *c,* flower viewed from above; *d,* median section of flower; *e,* flower after pollination; *f,* horned hood of corolla; *g,* andrœcium; *h,* portion of andrœcium with pollinia at right; *i,* section of andrœcium with pollinia at right and left; *j,* pollinia; *k,* gynœcium (ovary in section); *s,* sepals.

lobes, and the lobes of the corona are crested within instead of horn-bearing. When in fruit the plants may be quickly distinguished from most of the asclepiases by the alternate leaves. The half-dozen known species are natives of the southern United States and Mexico.

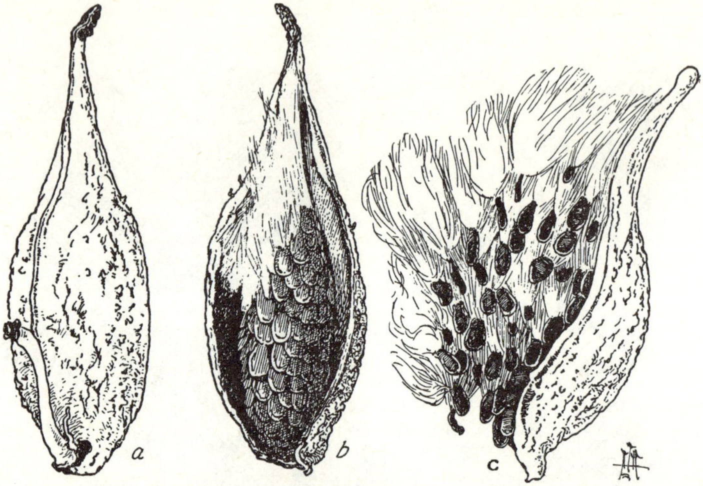

FIG. 345.—Follicles of *Asclepias syriaca*—*a*, with pedicel attached; *b* and *c* dehiscing.

FIG. 346.—Follicles of *Araujia sericifera*.

Acerates resembles the preceding genera in habit and has the reflexed corolla of *Asclepias,* but unlike these genera the corona has neither crest nor horn. The leaves are alternate or opposite and in some cases long and linear (as in some asclepiases), and tubercles are wanting on the pods (follicles). *A. auriculata* and *A. lanuginosa* occur on the Western prairies.

Of the three vines, *Cynanchum* and *Vincetoxicum* possess rotate corollas, whereas in *Gonolobus* the corolla-lobes are erect and the corona consists of 5 flat membranous lobes tipped with 1 or 2 awns. In the other two the corona is of the form of a ring or annular disk, of distinct pattern in each. In *Cynanchum* the anthers are tipped with a scarious membrane and the pollinia are pendulous, in contrast to the unappendaged anthers and horizontal pollinia of *Vincetoxicum. Vincetoxicum,* with about 75 species, is American; *Cynanchum,* with 100 species, is native both to the Old and the New World; and *Gonolobus,* with 3 species, is American. *Philibertia* is represented by 2 and *Solanoa* by 1 species in California, the former climbing, the latter prostrate plants. *Araujia sericifera* and *Stephanotis floribunda* are among the horticulturals (Fig. 346).

ORDER TUBIFLORAE

Flowers regular or irregular (bilaterally symmetrical in the median plane), bisexual, rarely unisexual, hypogynous. Calyx and corolla constantly 5-merous, the stamens generally so but sometimes fewer. Gynœcium commonly 2-carpellate, syncarpous, but in the lower families 3—5-carpellate or in rare cases 6—14-carpellate. Stamens epipetalous. Ovary superior, with as many cells as carpels; ovules few to many on axile, parietal (marginal), or basal, often thickened and intruded placentas. Integuments 1. Mostly herbs. A tremendous order of close to 16,000 species. The numerous families may be grouped according to the form of the corolla as follows:

Corolla regular. *Convolvulaceæ,* most *Polemoniaceæ, Fouquieraceæ, Lennoaceæ, Hydrophyllaceæ,* most *Boraginaceæ,* some *Verbenaceæ,* most *Solanaceæ, Nolanaceæ,* most *Columelliaceæ,* some *Acanthaceæ,* some *Myoporaceæ.*

Corolla irregular. Some *Polemoniaceæ,* some *Boraginaceæ,* most *Verbenaceæ, Labiatæ,* some *Solanaceæ, Scrophulariaceæ, Bignoniaceæ, Pedaliaceæ, Martyniaceæ, Orobanchaceæ, Gesneriaceæ,* some *Columelliaceæ, Lentibulariaceæ, Globulariaceæ,* some *Acanthaceæ,* some *Myoporaceæ, Phrymaceæ.*

Key to the Principal Families

I. Corolla regular.

A. Plants with tall, slender, wand-like stems, heavily armed with stout, spreading spines; soon leafless. Style terminal. Plants of the Southwestern deserts. *Fouquieraceæ.*

B. Brown root-parasites with scale-like leaves. Plants of the Southwestern deserts. *Lennoaceæ.*

C. Plants not as above.

a. Style terminal.

1. Leaves generally alternate.

(°). Glandular, woolly, or glabrous trees and shrubs. Carpels 2. Stamens 4, the fifth a staminode. Fruit a drupe. *Myoporaceæ.*

(°°). Herbs, vines, or shrubs (trees in tropics). Ours trailing, twining, or ascending vines, in some forms leafless; yellowish, orange, or whitish parasites. Carpels 2, rarely 3 or 5. Fruit a capsule. *Convolvulaceæ.*

(°°°). Erect herbs or shrubs, rarely vines.

(*). Leaves alternate or opposite. Gynœcium 3-carpellate, sometimes 2- or 5-carpellate, with as many stigmatic branches on style. Fruit a loculicidal capsule. *Polemoniaceæ.*

(**). Leaves alternate, rarely opposite or whorled.

(×). Ovary separating at maturity into four 1-seeded nutlets.

Stamens 5. Gynœcium 2-carpellate. Inflorescence often scorpioid. *Heliotropium* in *Boraginaceæ.*

Stamens 4, didynamous, rarely 2. Flowers in spikes arranged in corymbs or panicles. *Verbenaceæ.*

(××). Ovary not separating into nutlets at maturity. An annular disk often present. Gynœcium 2-carpellate.

Style mostly deeply 2-cleft at the summit, rarely merely lobed; or the styles distinct. Fruit a capsule. *Hydrophyllaceæ.*

Style mostly with fleshy stigmatic lobes. Fruit a capsule or berry. *Solanaceæ.*

2. Leaves opposite. Anther-bearing stamens 4 or 2. Disk annular or cup-shaped. Capsule 2-valved. *Acanthaceæ.*

b. Style gynobasic. Ovary deeply lobed.

Fruit separating into four 1-seeded nutlets. *Boraginaceæ.*

Fruit of 2 separate capsules, or indehiscent. *Convolvulaceæ.*

II. Corolla irregular.
  A. Ovary separating at maturity into 1-seeded nutlets.
    Style terminal. Stamens 4. *Verbenaceæ.*
      Stamens 5. *Boraginaceæ.*
    Style gynobasic.
      Stamens 5. Inflorescence often scorpioid. *Boraginaceæ.*
      Stamens 4, sometimes 2. Leaves opposite. Stems often square. *Labiatæ.*
  B. Ovary not separating at maturity into 1-seeded nutlets. Style terminal.
    a. Ovary superior.
      1. Yellowish-brown or -purple root-parasites with scale-like leaves. Flowers solitary or in spikes. Corolla 2-lipped. *Orobanchaceæ.*
      2. Aquatics with finely dissected leaves, those of the flowering stems scale-like; or plants of wet places, with broad leaves in a basal rosette; flowers solitary on scapes. Leaves of the aquatics often bearing small bladders. *Lentibulariaceæ.*
      3. Plants not as above.
        (°). Carpels 3—5, with as many style-branches. *Polemoniaceæ.*
        (°°). Carpels 2, except in *Pedaliaceæ.*
          (*). Fruit a drupe. Glandular, woolly, or glabrous trees or shrubs. Stamens 4, the fifth often staminodial. *Myoporaceæ.*
          (**). Fruit capsular, commonly furnished with stout, curved or hooked spines or prongs.
            Placentas axile. *Pedaliaceæ.*
            Placentas parietal, but intruded. *Martyniaceæ.*
          (***). Fruit a nutlet enclosed in the persistent calyx. *Globulariaceæ.*
          (****). Fruit a capsule or berry, rarely prickly or spiny.
            (+). Stamens 5, the fifth sometimes a staminode. Fruit a capsule or juicy berry. *Solanaceæ.*
            (++). Stamens 4 or 2, rarely 5.
              (×). Placentas parietal or on the septum of the ovary. Ours vines or trees with large, elongate pods or prominently winged fruits. *Bignoniaceæ.*
              (××). Placentas axile.
                Anthers generally free. Disk annular or lateral. Fruit a capsule or sometimes a berry. *Scrophulariaceæ.*
                Anthers connivent or connate in pairs or all connate at the apex.

Seeds attached to curved outgrowths of the placentas. Fruit a capsule, elastically dehiscing from the top. *Acanthaceæ.*

Seeds not attached as above. Fruit a loculicidal or septicidal capsule, or a pulpy berry. *Gesneriaceæ.*

(°°°). Carpels 1. Corolla 2-lipped. Stamens 4. Fruit an achene enclosed in the spinescent-toothed calyx. *Phrymaceæ.*

b. Ovary inferior. Leaves opposite, alternate, or whorled. Trees, shrubs, vines, or herbs from rhizomes or tubers. Fruit fleshy. *Gesneriaceæ.*

CONVOLVULACEÆ (*Morning-Glory Family*). Flowers regular, bisexual, rarely unisexual (diœcious), 5-merous (except gynœcium), hypogynous. Calyx persistent. Corolla mostly funnel-form, campanulate, or urn-shaped, plaited, flaring above, 5-lobed—the lobes convolute or twisted in the bud, rarely imbricate. Stamens 5, usually inserted deep down in the corolla-tube and alternating with the corolla-lobes; anthers ovoid to elongate, introrse. In some cases a ring of fimbriate lobes or flaps on the corolla-tube below the stamens (*Cuscuta*). Gynœcium of 2 to 4 carpels, united or distinct; styles distinct. Ovary superior, with as many cells as carpels, or twice as many by false partitions; ovules two to several in each cell, anatropous or nearly so, on axile placentas, and more or less basal at inner angle of locules. Stigmas various—globose, linear, filiform, or club-shaped. Disk when present annular to cup-shaped. Integuments 1. Seeds with endosperm. Embryo straight or curved, with foliaceous cotyledons. Fruit capsular, dehiscent or indehiscent, or a berry.

The plants of this family are herbs (sometimes parasitic), shrubs, subshrubs, or trees, with alternate, entire or variously lobed or divided, exstipulate leaves. Flowers often showy, solitary or in cymose inflorescences. The herbaceous forms are often twining or prostrate. Some members possess milky sap. Several species of different genera are cultivated. About 1,000 species in 40—50 genera, mainly tropical but also widely distributed over temperate parts of the world. The characters of the recognized subfamilies are added below. The genera given are those represented in our flora.

Dichondroideæ. Green plants with normal leaves but lacking infrastaminal scales or lobes. Embryo straight or curved, with folded cotyledons. Including *Dichondra, Convolvulus, Ipomœa, Evolvulus, Breweria, Cressa, Quamoclit.*

Cuscutoideæ. Yellowish or orange-colored parasitic plants without green leaves—these greatly reduced, and the flowers with infrastaminal scales. Embryo filiform, coiled, without cotyledons or these rudimentary. *Cuscuta.*

In *Dichondra* the carpels are distinct, a feature in which this genus differs outstandingly from the other genera here included, and a feature which is of rare occurrence among the *Metachlamydeæ.* The ovaries are expanded so that the styles occupy a position between them. The stigmas are capitate. There is also present a conspicuous

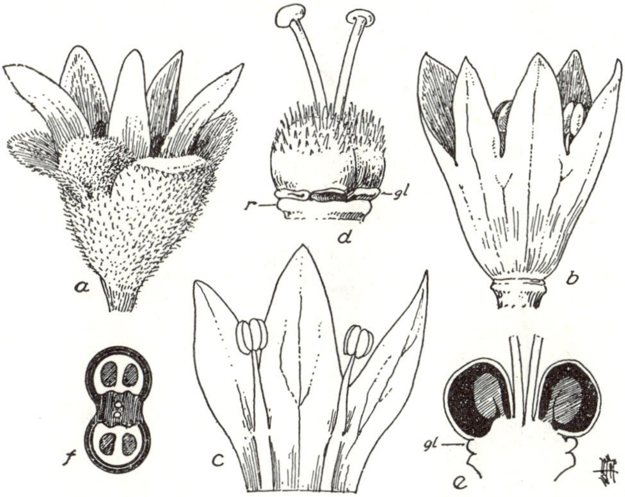

FIG. 347.—*Dichondra argentea: a,* flower; *b,* corolla; *c,* section of corolla with stamens; *d,* gynœcium, with gland (*gl*) below; *r,* receptacle; *e,* median section of ovary; *f,* transverse section of ovary.

annular disk. Fruit a capsule. There are 5 species, divided between the Americas, Asia, Japan, and Australia. They are all prostrate, creeping plants, rooting at the nodes. The white-flowered *Dichondra repens* occurs from Virginia to Texas, California, and Mexico; it bears kidney-shaped leaves on long, slender petioles (Fig. 347).

Ipomœa (Morning-Glory) is distinguished by a funnel-form, plaited, broadly flaring corolla which is only slightly lobed. The ovary is 2—4-celled, with an elongate, slender style which terminates in a globose or capitate stigma. At the base of the ovary is a prominent annular disk. The ovules arise basally in the inner angle of the cells.

Fruit a globose dehiscent capsule. Our species are all twining or trailing, annual or perennial herbs with cordate to palmately lobed leaves. Tropical forms are shrubs and even trees. The genus, as broadly interpreted, is rich in species—400 or thereabouts. A dozen

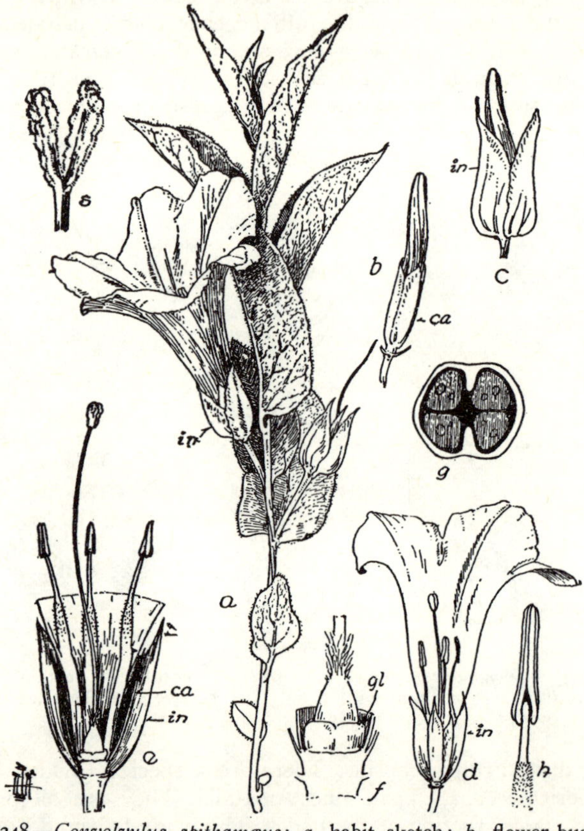

FIG. 348.—*Convolvulus spithamæus: a,* habit sketch; *b,* flower-bud; *c,* involucel (*in*) investing flower; *d,* flower with half of corolla removed; *e,* sectional view of flower; *f,* ovary and gland (disk, *gl*); *g,* transverse section of ovary; *h,* stamen; *ca,* calyx; *s,* stigma.

or more are horticultural. *Ipomœa pandurata* and *I. lacunosa* are Eastern; the former develops an enormous root weighing several pounds. *I. hirsutula* of tropical America is naturalized in southern California (Jepson).

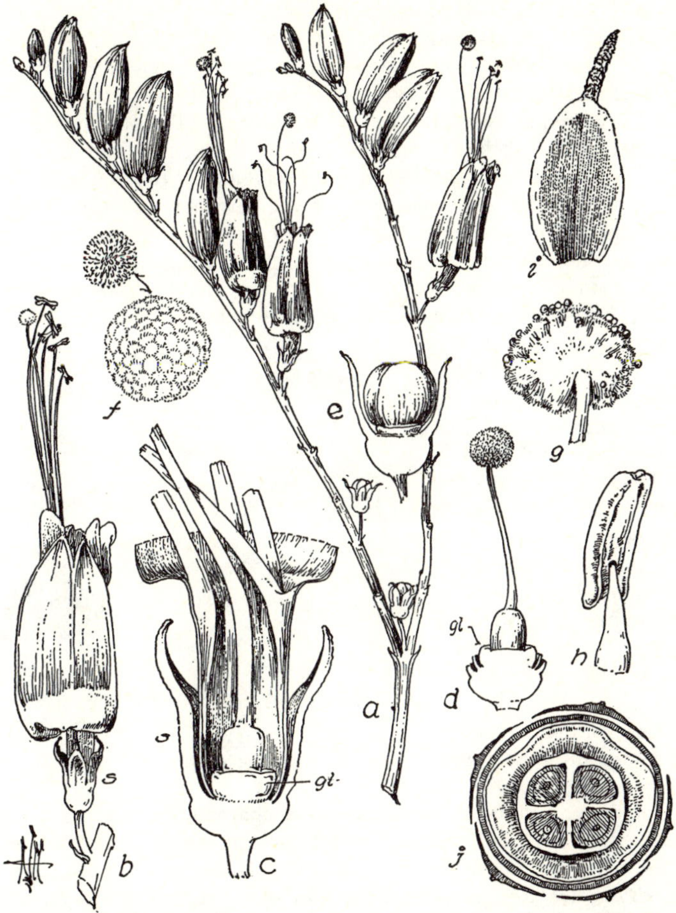

FIG. 349.—*Quamoclit lobata: a,* inflorescence; *b,* flower; *s,* calyx; *c,* section of flower with upper part of corolla and stamens removed; *gl,* disk; *d,* gynœcium and disk (*gl*); *e,* ovary and calyx, at later stage; *f,* stigma,—above, a globose projection of surface of stigma; *g,* stigma in sectional view; *h,* dorsal aspect of stamen; *i,* sepal, inner face; *j,* floral diagram, the thick disk investing ovary.

Convolvulus (Bindweed) resembles *Ipomœa* as regards corolla, but differs, at least with respect to our species, in having a 1—2-celled ovary and filiform to oblong stigmas, and variously sagittate to oblong leaves. Here also there is an annular cup-shaped disk present.

The sepals are generally unequal, the outer being the larger. A common species in woods and thickets and fields in the East is *Convolvulus sepium* (Hedge Bindweed), a trailing or twining form with hastate leaves—the auricles of which are somewhat truncate. The calyx is invested by two large bracts (involucre). The species encircles the northern hemisphere. The European *C. arvensis,* a common weed here, is somewhat similar but differs technically in having filiform stigmas and in the lower position of the involucre. In the northern coniferous region and southward to Florida, in sandy or dry, rocky situations, one will find the low, ascending *C. spithamæus* which often forms considerable patches and makes a conspicuous showing when in bloom. The genus contains at the most about 200 species, of which about 36 come within our limits. On the prairies and plains occurs the low, villous *Evolvulus pilosus,* which is further characterized by small leaves and flowers; the two stigmas are long-filiform. This genus contains about 85 species, most abundant in warm-temperate and tropical regions, about 8 coming within our range. (Fig. 348.)

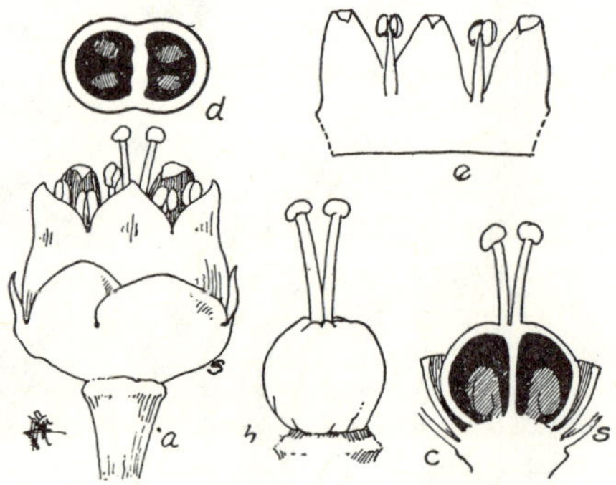

FIG. 350.—*Cuscuta arvensis: a,* flower; *b,* gynœcium; *c,* section of gynœcium; *d,* transverse section of ovary; *e,* section of corolla, with stamens (epipetalous).

Breweria includes prostrate spreading plants with oblong to linear leaves (in ours) and small flowers. The flowers are characterized by a 2—3-carpellate gynœcium with distinct styles, or the styles united below or to near the top. Three species—*B. humistrata, B. aquatica,* and *B. Pickeringii,* are found in the East and the South; the others, about 4, are Mexican. (See *Stylisma* in Britton and Brown's *Illustr. Flora.*) *Cressa,* a tropical genus of about 3 species, is represented in California by *C. cretica* (Alkali Weed). *Quamoclit,* with about 10 species, also of the tropics, yields a number of horticulturals. These are twining vines which produce numerous small, brightly colored, *irregular* flowers. Both filaments and style are very long and slender and long-exserted, and the ovary is seated in a cup-shaped disk. The large and globose stigma is very distinctive, being a mass of small, spherical, densely papillate, hair-like processes (Fig. 349).

Cuscuta (Dodder) comprises about 100 species widely distributed over the world. Probably over two dozen species occur in our flora. The plants are destructive parasites on other plants, being especially injurious in clover and alfalfa fields. Some are abundant in the chaparral hills of California. Their whitish to orange-colored stems twine in dense masses around the stems of the host plants, to the tissues of which they attach themselves by means of specialized sucking-organs called *haustoria*. The small flowers are borne in clusters. At the base of the bell-shaped corolla-tube is a ring of fimbriate flaps. The small, capsular fruit dehisces irregularly or circumscissilly by a lid. The disk is absent, and the embryo either lacks cotyledons or these are vestigial (Figs. 350, 85).

The most important member of the family is *Ipomœa Batatas,* the sweet-potato plant, a trailing tropical species which under cultivation yields large tuberous roots. Enormous tuberous roots are produced also by the native Eastern *I. pandurata,* and by several of the tropical species.

POLEMONIACEÆ (*Phlox Family*). Flowers regular or sometimes irregular, hypogynous, constantly 5-merous (except gynœcium). Calyx more or less deeply 5-cleft, commonly into narrow, often rigid, teeth. Corolla funnel-form, campanulate, rotate, or saucer-shaped, the 5 lobes convolute in the bud. Stamens 5, inserted in the corolla-tube and alternating with its lobes; filaments commonly long and slender; anthers short to elongate, versatile, 2-celled. Gynœcium 3-carpellate, rarely 2- or 5-carpellate, with a simple, elongate, slender style terminating in 3 stigmatic branches. Ovary superior, with as many cells as carpels and with 2—many ovules in each cell. Placentas at inner angle of each locule. Fruit a loculicidally 5-valved capsule. Seeds with or without wings, sometimes mucilaginous, with spiral tubes pushing out from the mucilage when wetted. Mostly herbs, sometimes woody below. Leaves alternate or opposite, entire or variously lobed or pinnately compound. Flowers few to numerous, often very showy, in cymose, corymbose, paniculate, or capitate clusters. Genera 12, species about 270, native to all the continents except Africa and Australia; most abundant in western North America. Taxonomically a very difficult family, in which genera and species are not readily differentiated as a rule except on rather technical characters.

The genera are conveniently grouped in two subfamilies on the basis of growth-form, thus:

Cobœoideœ. Trees and shrubs. *Cantua, Cobœa.*

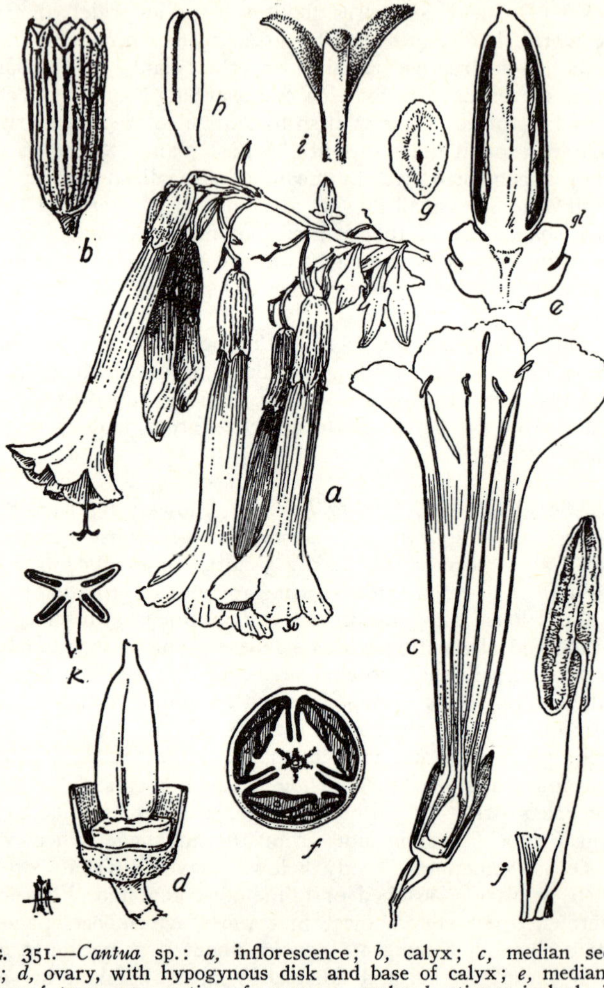

FIG. 351.—*Cantua* sp.: *a,* inflorescence; *b,* calyx; *c,* median section of flower; *d,* ovary, with hypogynous disk and base of calyx; *e,* median section of ovary; *f,* transverse section of ovary; *g,* ovule; *h,* stigma, in bud; *i,* stigma at anthesis; *j,* stamen; *k,* transverse section of anther; *gl,* disk.

Polemonioideæ. Herbs, of which the following have proper leaves: *Phlox, Microsteris, Gilia, Linanthus, Langloisia, Polemonium, Leptodactylon, Collomia, Navarretia, Hugelia, Bongplandia. Gymnosteris* is leafless.

Cantua buxifolia, one of 6 or more South American shrubs or trees, is an ornamental shrub with small, crowded, opposite leaves and variegated pendulous flowers. *Cobœa,* a tropical American genus of 10 species of tendril-climbing vines, has pinnately compound leaves ending in a branched tendril. The flowers are large and are characterized by a foliaceous calyx and campanulate corolla. *Cobœa scandens* is a hothouse plant but thrives out of doors in warm countries. (Fig. 351.)

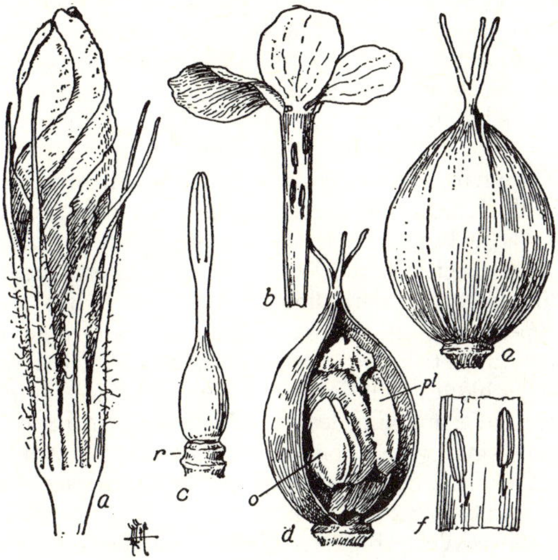

FIG. 352.—*Phlox pilosa: a,* flower-bud; *b,* corolla; *c,* gynœcium; *r,* receptacle; *d,* section of ovary; *o,* ovule; *pl,* placenta; *e,* capsule; *f,* section of corolla-tube with stamens.

The student will probably recognize most of our polemoniaceous plants as "phloxes," if by no other sign than that the flowers resemble the common garden phlox. But to distinguish such genera as *Phlox, Gilia, Linanthus,* and *Leptodactylon,* for example, requires detailed diagnosis. The gilias are particularly difficult because of the diverse shapes of the corolla and the leaves, the latter being commonly pinnately dissected and alternate or opposite, although the upper leaves of some phloxes are also alternate. Some of our gilias are rather outstanding, however, as the scarlet-flowered *Gilia aggregata,* or the tri-colored *G. tricolor,* with flowers suggesting those of some nemophilas or phacelias among the borages; or again, *G. capitata,* which bears its flowers in a dense head. In *Phlox* the corolla is rather uniformly salver-form, with a more or less elongate tube and spreading lobes. The leaves vary from large to small and linear and are closely crowded. Compare with *Linanthus* and *Leptodactylon*. *Microsteris*, with a half-dozen small Western annuals, resembles *Phlox*, except for its small flowers and mucilaginous seeds—which, however, do not release spiral tubes as do similarly mucilaginous seeds of other genera (*Gilia, Collomia, Polemonium, Navarretia*). *Collomia*, a Western genus of about 15 species, and rather technical, resembles *Microsteris* somewhat in habit, but the calyx is shorter in proportion to the length of the corolla-tube. *Polemonium*, with 15 or more species distributed over north-temperate regions, is distinguished by pinnately compound leaves as well as by an herbaceous or soft calyx. (Figs. 352, 77B.)

Linanthus, with 26 species in California (Jepson), is a genus of low, slender annuals with opposite palmately filiform leaves which appear as if whorled. Large or small flowers are produced at the ends of slender cymose branches or in capitate clusters. *Leptodactylon*, with probably 6 species in the West and Southwest, includes perennial herbs or subshrubs with flowers very much like those of the phloxes in appearance, but the leaves are small, alternate or opposite, and palmately dissected and rigid (pungent). Leaves of this sort occur in some phloxes, but in that genus they are simple and not linear segments of compound leaves. *Hugelia*, with 6 species in California, includes small woolly plants with simple, filiform or filiform-segmented leaves; the flowers, which are in heads, are distinguished by a densely woolly calyx and bracts.

Navarretia is a Western genus of about 24 species. The plants are low, rigid, viscid or glabrous annuals with very spiny pinnatifid leaves. Long spiny bracts subtend the heads of numerous small flowers. A further distinctive feature is the spiny calyx-segments. Spinescent leaves and calyx-segments are also peculiar to *Langloisia*, but the more distinctive character here is the bilaterally symmetrical (irregular) flowers. The plants are diminutive annuals of the Western desert regions. Species probably less than a half-dozen.

Navarretia, Collomia, and *Polemonium* have one feature in common which sets them off from the other genera, and that is that the calyx is not ruptured by the growing capsule. In some genera (*Phlox*) the sinuses of the calyx are scarious, but they are not so in others (*Leptodactylon*). A few species of *Gymnosteris* occur in Western mountain districts; these are diminutive leafless annuals. The persistent, connate-perfoliate cotyledons are a special feature of the embryo. *Bongplanda* is a Mexican subshrub with *irregular* flowers.

FOUQUIERACEÆ (*Candlewood Family*). Flowers regular, bisexual, hypogynous, 5-merous with respect to calyx and corolla and generally as to stamens (10—17), but gynœcium 3-carpellate. Sepals unequal. Corolla tubular, with short recurved lobes. Filaments appendaged at base. Styles united to middle. Ovary superior, incompletely 3-celled by the intrusion of the parietal lamellate placentas; ovules 4—6 in each cell. Fruit a capsule. Endosperm oily.

One genus, *Fouquiera,* of the Colorado Desert region to Texas and Mexico. Species probably less than a half-dozen. *Fouquiera splendens* (Ocotillo) is a striking plant of the Colorado Desert. Its tall (5—25 ft.), wand-like stems, which usually grow in clusters, are heavily armed with stout spines. At flowering time it produces a brilliant paniculate mass of scarlet flowers at the summit of the stems. The spines are probably the petiolar bases of leaves which shed their blades; another set of leaves forms later on spur shoots in the axils of the spines. To the uninitiated the plants are easily mistaken for a cactus (Figs. 353, 98).

LENNOACEÆ. Flowers regular, bisexual, 5-merous or pleiomerous. Corolla tubular. Gynœcium of 6—14 carpels, each 2-seeded. Cells of the ovary divided by false partitions. Fruit drupaceous. Three genera of chlorophylless parasitic plants of the far Southwest and Mexico, *viz., Pholisma* (monotypic), *Ammobroma* (monotypic), *Lennoa* (2 species of central Mexico). Says Jepson of *Ammobroma sonoræ:* "The stems and tubers are boiled or roasted by the Gulf Indians and taste like sweet-potatoes."

HYDROPHYLLACEÆ (*Water-Leaf Family*). Flowers regular, bisexual, hypogynous, generally 5-merous. Calyx deeply cleft or divided, persistent, the segments equal or unequal, sometimes with stipular appendages in the sinuses. Corolla tubular, funnel-form, salver-form, campanulate, or rotate, the lobes imbricate or twisted in the bud. Stamens rarely fewer nor more than 5, alternating with the corolla-lobes and arising at different levels in the corolla-tube; filaments often hairy below and expanded at the base; anthers ovoid, oblong, or linear-sagittate, versatile, 2-celled. Paired infrastaminal appendages at base of corolla-tube in some genera. Disk prominent only in a few cases. Gynœcium 2-carpellate, rarely more, with free or united styles, these when united forking above into stigmatic lobes or branches. Ovary superior, generally 1—2-celled, often septate by false partitions, each cell with 2 to numerous, anatropous or amphitropous ovules on thin, lamellate, fleshy or spongy, parietal and

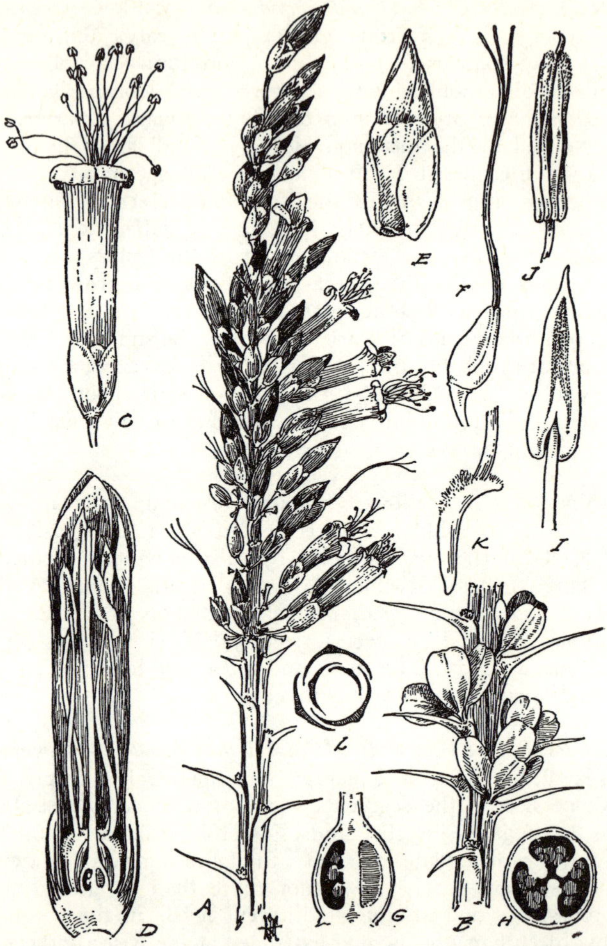

FIG. 353.—*Fouquiera splendens: A,* inflorescence; *B,* section of stem; *C,* flower; *D,* median section of flower; *E,* flower-bud; *F,* gynœcium; *G,* median section of ovary; *H,* transverse section of ovary; *I, J,* dorsal and ventral aspects of stamen; *K.* base of filament.

intruded or axile placentas. Fruit a loculicidal or septicidal, or irregularly dehiscent capsule. Embryo small, in fleshy or cartilaginous endosperm.

The members of this family are herbs or subshrubs with watery sap, rough, bristly pubescence, and simple or often pinnately divided or dissected exstipulate leaves, often in basal rosettes. Flowers small, few to numerous, in racemes, spikes, or scorpioid cymes, or sometimes solitary, with or without bracteoles. The family is almost exclusively restricted to the western hemisphere, where it is richly developed in the drier western parts of North and of South America. A few occur in Asia and Africa. The members are very diverse in habit as a result of the great range of ecological conditions under which they grow—from sea-level to high altitudes. There is a marked prevalence of endemism and of monotypic genera. Intergrading forms are common, so that generic and specific lines are difficult to draw.

Peter (*Pflanzenfamilien,* IV, 3a) has divided the family into the following tribes. His key is followed with slight modifications.

I. Style 1, simple or 2-cleft above.
  a. Capsule 1-celled. Placentas broad and fleshy, eventually becoming thin and investing the ovary-cavity. Corolla mostly convolute in the bud. Tribe I. *Hydrophylleæ.*
  b. Capsule 1—2-celled. Placentas narrow. Lobes of the corolla imbricate in the bud. Tribe II. *Phacelieæ.*

II. Styles 2, distinct. Capsule 1—2-celled. Corolla-lobes imbricate in the bud.
  a. Placentas divided, more or less intruded. Tribe III. *Nameæ.*
  b. Placentas large, thick, spongy, at length consolidated into one. Ovules numerous. Tribe IV. *Hydroleæ.*

The genera group themselves according to the tribes thus:

Hydrophylleæ:
  Hydrophyllum, Nemophila, Ellisia (*Nyctelea*).
Phacelieæ:
  Phacelia, Emmenanthe, Hesperochiron, Romanzoffia, Draperia, Tricardia, Conanthus.
Nameæ:
  Nama, Eriodictyon, Lemmonia, Wigandia.
Hydroleæ:
  Hydrolea.

Hydrophylleæ: Hydrophyllum (Water-Leaf). Species 12. These are perennial or biennial, mostly hairy herbs with pinnately compound leaves. Flowers whitish to purplish, in cymose or scorpioid clusters. The exserted filaments, which are hairy, distinguish the genus from its tribal allies, and a further distinction is seen in the linear, versatile anthers. In the sinuses of the calyx are small appendages. The rather conspicuously hairy ovary develops into a 2-valved capsule. A common species in moist Eastern woods is *Hydrophyllum virginicum* (Fig. 354). *Nemophila* (10 or more species) is best developed in the West. It differs technically from the preceding in having stamens

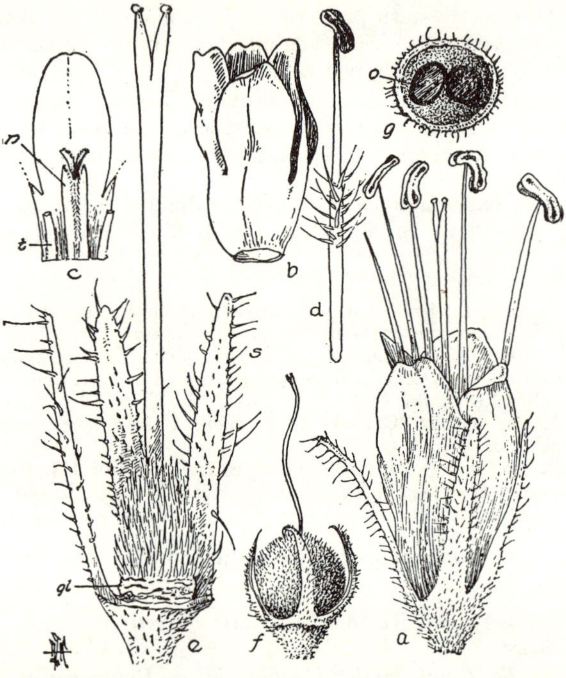

FIG. 354.—*Hydrophyllum virginicum:* *a,* flower; *b,* corolla; *c,* lobe of corolla with appendage (*n*); *t,* filament; *d,* stamen; *e,* gynœcium and calyx; *f,* immature fruit, with persistent calyx; *g,* transverse section of ovary; *o,* ovule; *s,* sepal; *gl,* gland.

shorter than the corolla-tube and in bearing ovoid to cordate anthers. The calyx is enlarged in the fruit and is furnished with reflexed appendages in the sinuses. The flowers are solitary or in racemes. The plants are low, diffuse, rather fragile, often retrorse-hairy, and prickly annuals. In *Ellisia* the stamens are included as in *Nemophila,* but appendages are absent from the sinuses of the calyx—which is much enlarged in the fruit. Species 3.

Phacelieæ: Phacelia. This is not only the largest genus by far but also a very polymorphic and technical one. The numerous species (55 in California alone—Jepson), while showing many diverse growth-

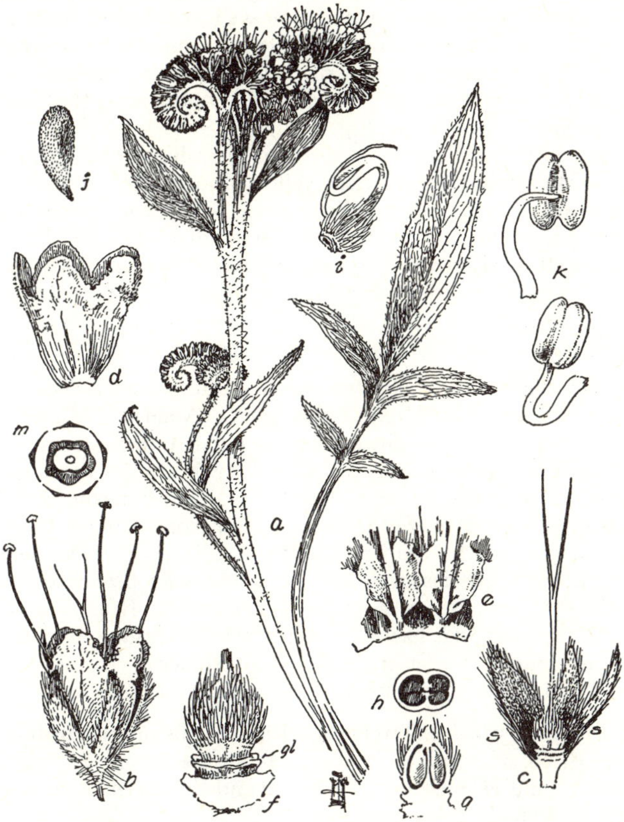

FIG. 355.—*Phacelia californica: a,* habit sketch; *b,* flower; *c,* gynœcium and part of the calyx (*s*); *d,* corolla; *e,* appendages at base of corolla; *f,* ovary and hypogynous disk (*gl*); *g,* median section and (*h*) transverse section of ovary; *i,* gynœcium, as in the bud; *j,* seed; *k,* dorsal and ventral aspects of stamen; *m,* floral diagram, corolla and stamens omitted.

forms, at the same time show most perplexing intergradations. The plants are annuals or perennials, mostly hispid or hairy, with alternate (or the lower sometimes opposite), simple, entire or pinnately divided or dissected leaves. Flowers in loose to scorpioid cymes, colored as in *Hydrophyllum.* In many species there are present small lamellate, infrastaminal appendages or scales on the corolla-tube. But the most reliable signs of the genus are the placentas, which are intruded into the ovary-cavity as thin partition-walls thickened on the margin (the ovule-bearing portion). Ovules 2 to numerous on each placenta. Fruit a 1-celled capsule, or 2-celled by the intrusion of the placentas. The genus ranges from British Columbia to the Straits of Magellan. Species 80—114. (Fig. 355.)

Emmenanthe, while similar to some phacelias in habit, is distinguished particularly by pendulous, yellowish or cream-colored flowers with persistent corolla but deciduous style. *Hesperochiron* is a genus of dwarf perennials of Western mountain districts. The leaves are in basal rosettes, from which spring the naked flower-stalk bearing a solitary flower. Two small arctic-alpine species of *Romanzoffia* are found in the Pacific region, one crossing over into Asia. *Draperia, Tricardia,* and *Conanthus* are monotypic genera of the Far West.

Nameæ: Nama. Probably two dozen species, mostly Western, extending down through Mexico into South America; a few in the Sandwich Islands. Low herbs of diverse habit. Capsule membranous, containing few or numerous seeds. *Eriodictyon* is a genus of aromatic shrubs with lanceolate leaves. Flowers small, in a terminal cymose-paniculate inflorescence. Seeds few or numerous from a coriaceous capsule. Mostly of the chaparral belt of California. *Lemmonia californica,* the only species, is a low, dichotomously branched plant of the Mojave Desert. Seeds 2—3 on each placenta. *Wigandia* belongs in tropical America, but has one species, *W. caracasana,* in cultivation as an ornamental. It is a striking tall shrub with very large leaves and bicolored flowers—the limb of the corolla violet, the tube white.

Hydroleæ: Hydrolea. Twelve species of herbs or subshrubs of wet grounds, often with spines in the axils of the leaves. *H. affinis, H. quadrivalvis,* and *H. ovata* in the Central and Southern States. (Plants referred to *Nama* by some writers.)

BORAGINACEÆ (*Borage Family*). Flowers regular, rarely irregular, bisexual, hypogynous, 5-merous, exceptionally 4-merous. Calyx more or less deeply lobed or cleft, and persistent. Corolla generally blue, often plaited, campanulate, funnel-form, salver-form, or rotate, often prominently lobed, the lobes variously folded in the bud, the corolla-tube often with appendages or crests in the throat. Stamens as many as the lobes of the corolla and alternating with them; filaments short to long and slender, inserted at various levels in the corolla-tube; anthers ovoid, oblong, elongate, sagittate, versatile, 2-celled, the connective often prolonged. Gynœcium 2-carpellate, with 2 distinct styles, or with 1 style which forks above or below the middle into 2 slender elongate stigmatic branches which are often themselves forked; stigma mostly globose-capitate and small; or the style is short and stout, terminating in a large, thick, conical, more or less 2-lobed stigma, and with a more or less prominent, fleshy, basal ring (or fleshy annular base). Ovary superior, often bristly-hairy, 2-celled or at length 4-celled by false partitions; each of the four cells containing 1 anatropous or amphitropous ovule. In one division the ovary is either lobed or entire on the sides, with a terminal style; in the other division it is deeply depressed in the center so as nearly to separate the two carpels or the four sections when the ovary is divided by false partitions. Between these sections thus separated the style arises. Placentas axile and mostly basal at inner angle of locule. In some cases ovules suspended. Disk annular and more or less prominent. Fruit drupaceous or mostly separating (in ours) into 4 small, 1-seeded nutlets. Genera 85—88; species 1,500—1,660.

These plants are herbs, shrubs, or trees (in tropics), with alternate or sometimes opposite or verticillate, simple, exstipulate leaves. Flowers generally small, mostly in scorpioid cymes, racemes, or spikes, which uncoil as flowering proceeds, with bracts on the lower side. The plants are generally clothed with rough or coarse hairs—including calyx and corolla. Except in white-flowered forms the corolla is reddish in the bud but turns blue on expanding. Distribution is world-wide in tropical and temperate regions, but the best development is in the Mediterranean region and in western North America.

Four subfamilies have been established (Gürke, *Pflfam.*), distinguished briefly as follows:

I. Style terminal.
- a. Style twice forked. Ovary not lobed. I. *Cordioideæ.*
- b. Style simple and once-forked, or styles 2 and distinct.
  - 1. Ovary not lobed. II. *Ehretioideæ.*
  - 2. Ovary mostly 4-lobed. Stigmas subtended mostly by a hairy ring. III. *Heliotropioideæ.*

II. Style arising between the nearly separated ovaries. Fruit separating into 1-seeded nutlets. IV. *Boraginoideæ.*

Cordioideæ. One cultivated shrub, *Cordia Sebestena* (Geiger Tree).

Ehretioideæ. Four species of *Coldenia* in the Far West and Southwest.

Heliotropioideæ. Including *Heliotropium,* and *Euploca albiflora* of the Mojave Desert region to Nebraska and south to Mexico—a diminutive herb.

Boraginoideæ. Including (among ours) *Cynoglossum, Lappula, Allocarya, Cryptanthe, Oreocarya, Mertensia, Amsinckia, Asperugo, Myosotis, Lithospermum, Onosmodium, Symphytum, Borago, Lycopsis, Echium, Plagiobotrys, Harpagonella* (horticultural).

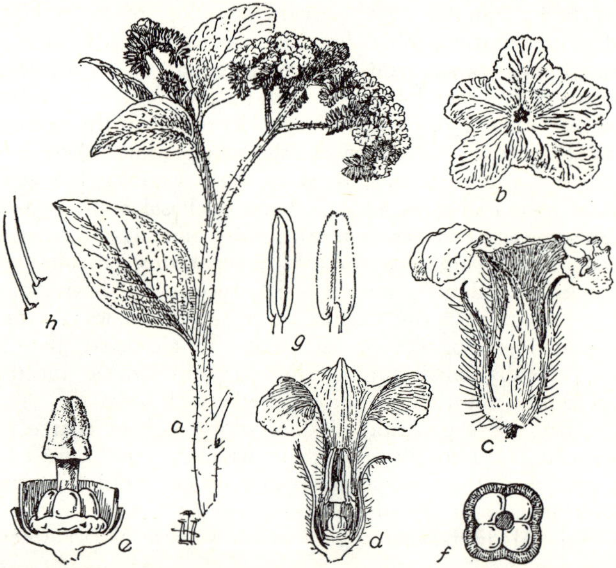

FIG. 356.—*Heliotropium peruvianum: a,* inflorescence; *b,* corolla, viewed from above; *c,* flower; *d,* median section of flower; *e,* gynœcium, with hypogynous disk, and bases of corolla and calyx; *f,* ovary and disk, viewed from above; *g,* lateral and dorsal aspects of stamen; *h,* trichomes of pubescence.

For a satisfactory study of this rather technical family it is necessary to have at hand not only as much of the plant as possible but especially the ripe fruit, for in the fruit lie the characters that set off the genera most precisely. Species are often difficult to identify. In the coating of rough or bristly hairs as well as in the scorpioid inflorescences the borages resemble the water-leafs, while in the splitting-up of the fruit into four nutlets they are strikingly like the mints.

Among the genera listed above, two, *viz., Lycopsis* and *Echium,* stand out rather clearly by reason of their irregular curved corollas.

In *Lycopsis* the stamens are exserted and the throat of the corolla is nearly closed by a crown of lobe-like appendages—characters which do not appear in *Echium*. The significance of the fruit will be appreciated in trying to distinguish between *Cynoglossum* and *Lappula,* for example. In either case the fruit is covered with barbed prickles, but in the former the four nutlets of the fruit are flattened, while in the latter they are of an ovoid shape and upright. In *Harpagonella* of the Far Southwest we have again a very distinctive fruit: here the hardened calyx persists as an investment of the one or two nutlets, the whole becoming a sort of bur. The color of the corolla serves as a guide to some genera, *e.g., Oreocarya* and *Plagiobotrys,* in which it is white, or it will set off certain species of a genus from others. Absence of bristly hairs is notable in *Mertensia.* A rotate corolla is present in *Borago,* a Mediterranean genus. Besides, in this genus, the anthers are connivent in a conspicuously projecting cone, and the leaves are broad. *Borago officinalis* frequently occurs as an escape from gardens. *Heliotropium,* in addition to the characters given in the key on a preceding page, has a characteristic conical stigma, the base of which is modified into an annular, collar-like structure (Fig. 356). This is by far the largest genus, variously estimated at 125—250 species.

One or more species of the following genera are represented in gardens: *Borago* (Borage), *Anchusa* (Alkanet), *Myosotis* (Forget-me-not), *Symphytum* (Comfrey), *Pulmonaria* (Lungwort), *Mertensia* (Bluebells), *Echium* (Bugloss), *Heliotropium* (Heliotrope).

VERBENACEÆ (*Verbena Family*). Flowers irregular (bilaterally symmetrical), rarely regular, bisexual or unisexual, hypogynous. Calyx persistent, generally deeply cleft into 4 or 5, or sometimes 6 or 8, segments. Corolla 2-lipped, with a cylindrical to beaker-shaped curved tube and usually 4 or 5 lobes which are imbricated in the bud. Stamens generally 4, rarely 5, sometimes 2, or modified as staminodes, or as many as the lobes of the corolla and alternate with them, frequently arising at different levels in the corolla-tube; filiments filiform or very short, sometimes thickened at the base; anthers introrse, 2-celled, dehiscing lengthwise, the connective sometimes furnished with a glandular appendage. Gynœcium of 2 or sometimes 4 or even 5 carpels. Ovary superior, often lobed according to the number of carpels, 2—4-celled, or cells twice as many by false partitions. Style mostly terminal, short or elongate, divided at the summit into 2 fleshy stigmatic lobes, or short and slender or filiform, stigmatic branches, one of which is sometimes reduced. Ovules generally 2 for each carpel, variously suspended, at first from parietal placentas with reflexed lobes, which later become central and placed at different levels. Fruit separating septicidally into 2 or 4 1-seeded nutlets, or drupaceous with succulent exocarp and dry, hard endocarp, enclosing as many seeds. Endosperm generally absent but when present fleshy. Embryo straight, mostly with flat cotyledons. Genera variously placed at 67—89 and species 750—1,300.

The plants are of diverse habits, including herbs, shrubs or subshrubs, lianas, and trees. Leaves opposite or whorled, sometimes alternate, exstipulate, simple or palmately or pinnately compound. Flowers generally small, in cymose, racemose, spicate, or paniculate inflorescences.

The subdivision into subfamilies is based on rather technical characters but for the few represented in our flora, and one or two horticulturals, the following scheme will suffice.

Fruit dry, separating into nutlets.
 Flowers in spikes. *Verbena.*
 Flowers in heads. *Lippia, Callicarpa, Duranta.*
Fruit a drupe. Flowers in axillary, cymose clusters. *Vitex, Clerodendron, Lantana.*

Verbena, 80—100 species. The verbenas are mostly natives of the American tropics, with a few extensions into temperate regions. The plants are mostly hairy herbs with opposite, simple, and coarsely serrate or pinnately lobed leaves. The small flowers are aggregated in elongate, often interrupted spikes. The tube of the salver-form or funnel-form corolla is straight or curved, with the 4 (didynamous), or rarely 2, stamens included in the tube. The ovary is 4-celled. The fruit, which is usually invested by the persistent calyx, separates into four 1-seeded nutlets, the surfaces of which are variously rough-rugose—a character useful in distinguishing species. The flowers are seated in the axil of a small bract; they are generally blue, purple, lilac, or white in our wild species. Several species are horticultural. Two of our common wild species, namely *Verbena urticifolia* and *V. hastata,* often grow to five feet or over in height. They are common in fields and waste grounds. *V. hybrida* is a common garden plant. (Fig. 357.)

Lippia, with over 100 species, is chiefly at home in the American tropics, but a few species occur in Africa. Less than a dozen species are scattered over our range. The plants are pubescent perennial herbs or shrubs with opposite, alternate, or whorled, downwardly tapering, sessile leaves. Flowers small, in terminal long-peduncled heads or spikes. The floral structure is of the same plan as in *Verbena.* Our three species, *Lippia cuneifolia, L. lanceolata,* and *L. nodiflora* are decumbent or creeping plants. The ovary is 2-celled.

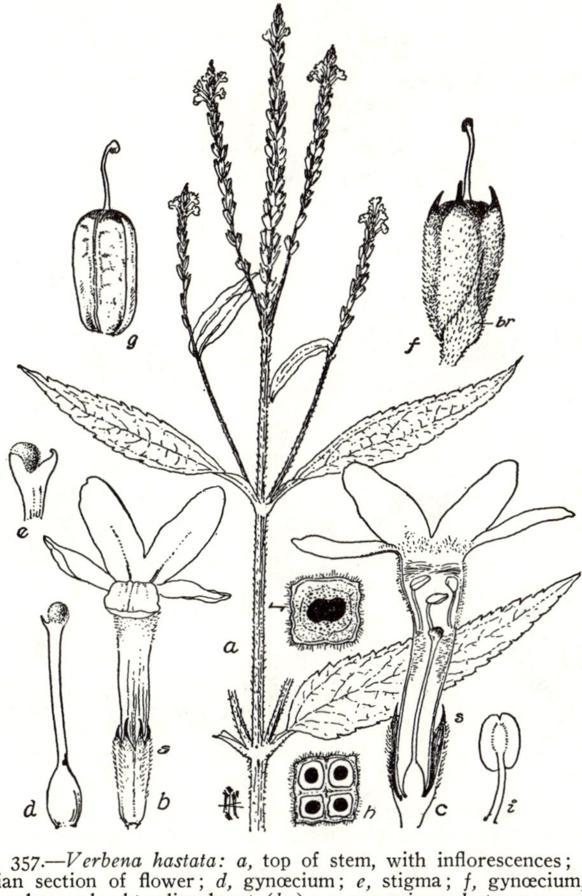

FIG. 357.—*Verbena hastata: a,* top of stem, with inflorescences; *b,* flower; *c,* median section of flower; *d,* gynœcium; *e,* stigma; *f,* gynœcium, with investing calyx and subtending bract (*br*); *g,* gynœcium; *h,* transverse section of ovary. At right of stem a transverse section of the internode.

Lantana. The lantanas are rough-hairy or tomentose shrubs, some of them half-climbing. Leaves opposite or whorled, rugose. Flowers small, bracteate, often brightly colored, in dense heads or spikes.

Calyx very small, cup-shaped and 2-lipped, with obscure lobes or teeth. Corolla-tube obliquely curved, the lobes in more or less unequal pairs. Style in some cases stout, supporting a 2-lobed fleshy

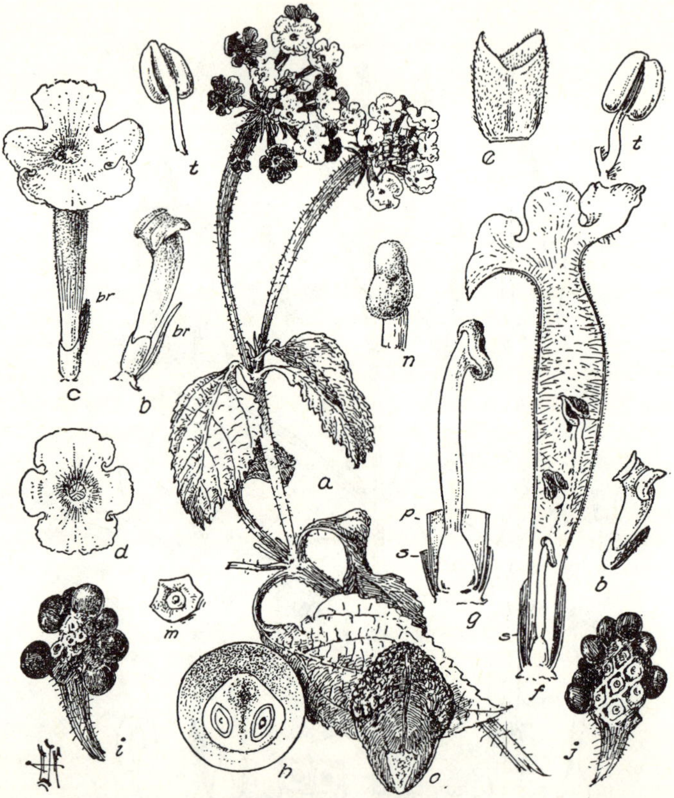

FIG. 358.—*Lantana Camara: a,* stem and inflorescences; *b,* flower-bud; *br,* bract; *c,* flower; *d,* corolla, seen from above; *e,* calyx; *f,* median section of flower; *g,* gynœcium; *p,* corolla-tube; *s,* calyx-tube; *h,* transverse section of ovary; *i,* fruit-cluster; *j,* fruit-cluster, with several drupes removed to show polygonal disks in which drupes are seated (shown enlarged at *m*); *t,* dorsal and ventral aspects of stamens; *o,* seed.

stigma. Stamens arising at different levels in the corolla-tube. The small drupes form a cluster on the conical apex of the peduncle, and each contains two bony nutlets (Fig. 358). There are 50 or more species in tropical and subtropical America, of which several are popular ornamentals in warm parts of the country.

Callicarpa is a genus of tropical America, Asia, and Africa. There are about 45 species, of which *C. americana* occurs in Eastern and Southern parts of this country This is a shrub with opposite ovate leaves and bearing small, bluish flowers in axillary cymes. The fruit is a bluish drupe. *Duranta, Vitex,* and *Cleradendron* are represented by ornamental shrubs or small trees.

LABIATÆ (*Mint Family*). Flowers irregular, in some cases regular, bisexual or unisexual, sometimes gynodiœcious, hypogynous. Calyx persistent, 3- or rarely 4-toothed or -lobed, sometimes 2-lipped, generally with longitudinal nerves or ribs. Corolla with a short or elongate tube which is often dilated (gibbous) below, 5- or 4-lobed, generally strongly 2-lipped, or in a few cases regular. Stamens 4, didynamous, or 2, with others sometimes in the form of staminodes, the filaments free, inserted on the corolla-tube and alternating with its lobes, rarely united; anthers introrse, 2-celled or confluently 1-celled above, in some cases with only one anther. Gynœcium 2-carpellate, subtended by a 2—4-lobed nectariferous disk. Ovary superior, 4-lobed or 4-parted by a false septum, each part with 1 commonly anatropous ovule on an axile or parietal (intruded) placenta; style often elongate-filiform and curved, bifid at summit, commonly gynobasic, *i.e.,* arising from the depression between the lobes of the ovary. In some cases the ovary is supported on a gynophore, which includes the disk. Fruit separating mostly into four 1-seeded nutlets, rarely drupaceous (with succulent exocarp and hard endocarp), mostly smooth, hairy, or wrinkled, rarely keeled or winged. Endosperm, when present, scanty. Embryo straight or curved, with flat or folded cotyledons.

The mints are aromatic herbs, shrubs, subshrubs, or (in the tropics) trees, generally with square stems and mostly opposite, simple, entire, serrate or palmately lobed, pinnately divided or dissected leaves, and are often coarsely and copiously pubescent. Flowers small to large and showy, sometimes solitary, but commonly in cymose clusters which are often arranged in whorls at the nodes and forming dense heads or interrupted spikes. Bracts and bracteoles usually present. Genera 160—170 or more. Species 3,000—3,200, widely distributed over temperate and tropical regions, and abundantly represented in our flora both as to genera and species. Horticulturals numerous.

The subdivision into subfamilies, of which there are about eight,

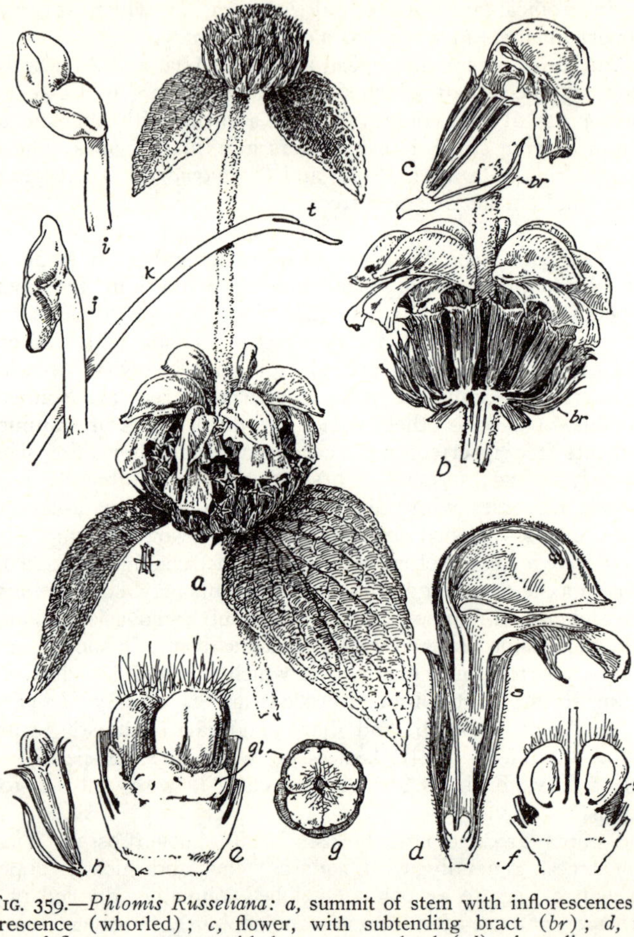

FIG. 359.—*Phlomis Russeliana: a,* summit of stem with inflorescences; *b,* an inflorescence (whorled); *c,* flower, with subtending bract (*br*); *d,* median section of flower; *e,* ovary with hypogynous gland (*gl*); *f,* median section of ovary; *g,* summit of ovary; *h,* flower-bud; *i, j,* stamens; *k,* upper part of style; *t,* stigma.

is based on rather technical characters. The following are representative.

Ajugoideæ, in which the ovary is merely lobed and the style not basal. Nutlets attached laterally. *Ajuga, Teucrium, Rosmarinus.*

In the following subfamilies the style arises basally between the lobes of the deeply 4-parted ovary—that is, the style is gynobasic. Nutlets attached basally.

Scutellarioideæ, in which the calyx is crested on the upper side. Stamens 4. *Scutellaria,* a divergent form.

Stachyoideæ, in which the calyx is not crested on the upper side and the stamens extend straight forward, ascending or spreading.

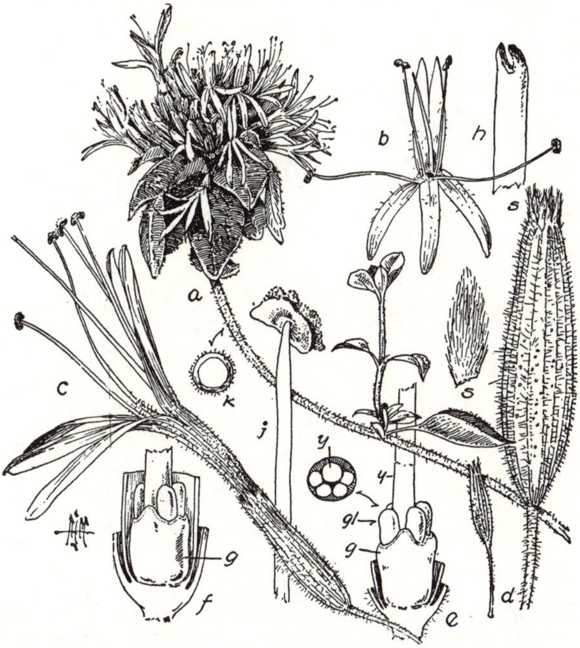

FIG. 360.—*Monardella villosa: a,* inflorescence; *b,* face view of flower; *c,* lateral view of flower; *d,* calyx, after flowering,—enlarged at right; *e,* ovary (*g*) and glands (*gl*),—at left a sectional view of glands (*y,*—style); *f,* ovary (*g*) and glands at somewhat later stage; *h,* stigma; *j,* stamen; *k,* transverse section of stem (*a*); *s,* apical segment of calyx.

Genera numerous, including *Marrubium, Nepeta, Glechoma, Lamium, Stachys, Phlomis* (Fig. 359), *Salvia, Monarda, Monardella* (Fig. 360), *Hyssopus, Thymus, Lycopus, Mentha, Satureia, Agastache.*

Ocimoideæ, in which the calyx-teeth are nearly equal and the stamens number 4, the upper pair the shorter, all declined and included within the saccate lobe of the lower lip of the corolla. *Hyptis.*

Teucrium (Germander) is represented by six or nine species. It is a tropical and temperate genus of 100 species or more. The plants are annual or perennial herbs or shrubs with simple, serrate or pinnately lobed or dissected leaves, and white to purplish flowers in heads or whorled in interrupted spikes. Corolla very irregular, the lower lip much the larger, and the 4 didynamous stamens exserted between the lobes of the lower lip, the anterior pair much the longer. Anther-

FIG. 361.—*Teucrium fruticosum: a,* flowering stem; *b,* flower; *c,* corolla; *d,* upper half of corolla showing insertion of filaments; *e,* lower lip of corolla in surface view; *f,* calyx; *g,* calyx viewed from above, with four lobes of ovary visible in center; *h, i,* stamens; *j,* ovary; *s,* calyx; *k,* transverse section of ovary; *m,* leaves.

cells confluent. Ovary merely 4-lobed, ripening into 4 rugose-reticulate nutlets. Style slender, bifid at summit. *Teucrium canadense* (American Germander or Wood Sage) is typical; it is widely distributed through the Mississippi Basin, east to the Atlantic, and southward and westward. (Fig. 361.)

Scutellaria (Skull-Cap) is readily distinguished by the crested, 2-lipped calyx, a character unique for the family. In addition the ovary is raised more or less on a gynophore (gynobase). The 4 stamens, of which the posterior pair is the shorter, arise under the upper lip of the corolla, and all bear ciliate anthers, but the anthers of the posterior stamens are 2-celled, while those of the anterior pair are 1-celled. The nutlets are rounded or depressed, papillate or tuberculate. Species about 100, of which about 30 with us. (Fig. 60b.)

Salvia (Sage) is the largest genus, with 500 or more species of temperate and tropical regions. Thirty or more species occur in our flora, the majority of which are Western. Most of ours are herbs, but some of the Western ones are shrubs or subshrubs. The campanulate calyx is 2-lipped, with a 3-toothed upper lip and 2-toothed lower lip. The upper lip of the corolla is usually concave and arched, the lower lip larger, drooping and spreading; both lips are lobed. The most striking feature is the unusually modified connective of the anthers; this organ is broadened or sometimes attenuated, with a functional anther at one end and a rudimentary one at the other end; or the end bearing the rudimentary anther is otherwise modified and devoid of anther. As in *Scutellaria,* the ovary is deeply 4-parted, ripening as four nutlets, which usually become mucilaginous and emit spiral tubes when wetted. The shape of the connective is a helpful guide to the identity of species. *Salvia Pitcheri, S. lanceæfolia,* and *S. urticifolia* are widely distributed through the East. One of the most striking of our sages is *S. carduacea* of southern California, a woolly species so remarkably like a thistle in habit as aptly to justify its specific as well as its common name—thistle sage. Numerous species are horticultural, among which are this last-named, and *S. splendens* (Scarlet Sage), *S. fulgens* (Cardinal Salvia), and *S. sclarea.*

Lycopus (Bugle-Weed) is represented by 8 or 10 species. These are perennial herbs from thick or slender rhizomes. Flowers small, white to purplish, densely clustered in the axils of the leaves. The campanulate corolla is not conspicuously irregular, and only 2 of the stamens are functional, the other pair being rudimentary or wholly suppressed. The nutlets are truncate above and contracted downwards. Species about 15, of northern regions. *Lycopus virginicus* is common Eastward; it also occurs in northern Asia.

Ocimoideæ. Hyptis, one of the largest of the genera—with 300 or more species, is native to tropical America, chiefly Brazil. However, one species, *H. Emoryi* (Bee Sage) is found in the desert region of the Far Southwest. This is a fragrant, scurfy-tomentose shrub up to 12 feet high, bearing whitish to purplish flowers. The lobes of the corolla are nearly similar, except the middle lobe of the lower lip, which is slender-clawed and strongly saccate at the tip and invests the stamens. Stamens 4, didynamous, the upper the shorter.

Coleus. A genus of 90 or more species, distributed over the tropics. Varieties of *Coleus Blumei* are common pot and garden plants with variegated foliage.

NOLANACEÆ (*Nolana Family*). Flowers regular, bisexual, typically 5-merous, hypogynous, the corolla plicate in the bud. Stamens equal, inserted at the base of the corolla-tube and alternating with its lobes. Gynœcium 5-carpellate. Style 1. Ovary many-celled by invaginations of the ovary-wall; ovules 1 or more in each cell. The base of the ovary is surrounded by a more or less prominent lobed disk. Fruit a schizocarp which breaks up into 1- or several-seeded nutlets or follicles. Genera 3, species about 50, of western South America from Peru to Chile. Herbs or shrubs with alternate or opposite simple leaves and solitary flowers. Genera: *Nolana, Alona, Dolia. Nolana atriplicifolia,* a prostrate annual with succulent spatulate leaves and large, blue, morning-glory-like flowers, white within, is grown in gardens.

In the plan of the flower, especially with respect to the 5-carpellate gynœcium, the family suggests the *Convolvulaceæ,* while as to fruit it is very similar to the *Boraginaceæ*. It is usually regarded, however, as being closest to the *Solanaceæ,* which family it resembles in the æstivation of the corolla, in the branching, in the arrangement of the leaves, and in the structure of the seed.

SOLANACEÆ (*Potato Family*). Flowers regular or sometimes irregular; bisexual or sometimes unisexual, 5-merous. Calyx gamosepalous, the lobes various. Corolla rotate, campanulate, funnel-form or salver-form, rarely 2-lipped, plicate or induplicate-valvate in the bud. Stamens as many as the lobes of the corolla and alternate with them, inserted in the tube—in irregular flowers often of different lengths or sometimes rudimentary; anthers 2-celled, rarely confluently 1-celled, opening by longitudinal slits or by apical pores. Gynœcium 2-carpellate, obliquely placed, with a simple, slender style supporting a 2-lobed, more or less fleshy stigma of diverse shapes. Ovary superior, 2-celled, or 3—5-celled or more, by the intrusion of false septa or rarely 1-celled, each cell with numerous anatropous or amphitropous ovules on very thick axile placentas. A more or less prominent, fleshy, annular disk surrounds the base of the ovary. Fruit a berry or capsule. Seeds with rough to reticulate testa and a terete straight or curved embryo in fleshy endosperm. Cotyledons somewhat plano-convex.

The members of this interesting family are annual or perennial herbs, shrubs, or vines, or small trees. They are mainly tropical but are well represented in temperate regions. Leaves alternate or sometimes opposite—especially on the upper parts of the stem (in herbs), exstipulate, simple and often variously lobed, or pinnately compound or dissected. Herbaceous forms sometimes producing tuberous roots or tuberous rhizomes (stem-tubers). Flowers small to large (often unusually so) and showy, solitary or in cymose clusters; in either case sometimes supra-axillary, that is, the pedicel or peduncle arises from the stem at some distance above the axil of its proper leaf. Genera 75—85; species 1,700—2,000.

This family yields a considerable number of useful plants, namely: the potato, tomato, egg-plant, ground-cherries, red peppers, and drug plants—such as belladonna, tobacco plant, henbane, Jimson weed. Some members are popular horticulturals. Our native species are not numerous, and these belong to but a few genera.

Wettstein long ago divided the family into tribes, essentially as follows (*Pflfm.*, IV, 3b):

A. Embryo conspicuously curved. Stamens all functional or of nearly equal lengths.
  a. Ovary 3—5-celled. Tribe I. *Nicandreæ.*
  b. Ovary 2-celled. Tribe II. *Solaneæ.*
  c. Ovary 4-celled. Tribe III. *Datureæ.*

B. Embryo straight or nearly so.
  a. Stamens all functional and of the same length or 1—3 shorter. Tribe IV. *Cestreæ.*
  b. Only 2—4 stamens functional, and always of different lengths. Tribe V. *Salpiglossideæ.*

Some of the genera of these tribes are:

Nicandreæ:
  Nicandra.
Solaneæ:
  Lycium, Atropa, Hyoscyamus, Physalis, Chamæsaracha, Capsicum, Solanum, Lycopersicon, Mandragora.
Datureæ:
  Datura.
Cestreæ:
  Cestrum, Nicotiana, Fabiana.
Salpiglossideæ:
  Salpiglossis, Petunia, Schizanthus.

By using the fruit instead of the embryo as the basis of classification into major categories, the tribes will be rearranged thus (the *Solaneæ,* it will be observed, now become split up):

A. Fruit a berry. Tribe II. *Solaneæ,* including all the genera listed except *Hyoscyamus.*

B. Fruit a capsule. Including all the remaining tribes above listed, as well as *Hyoscyamus* of the *Solaneæ.*

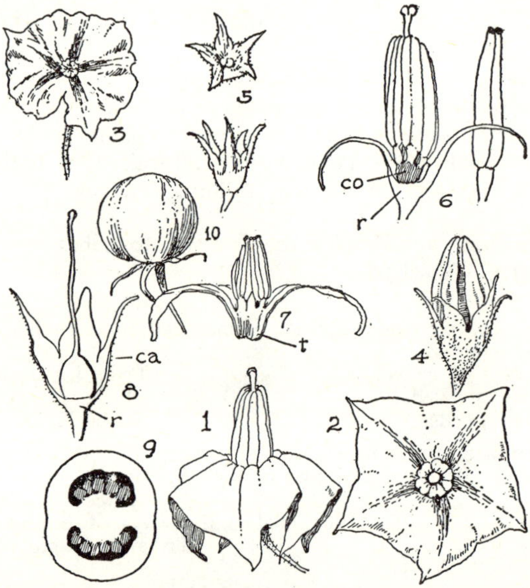

FIG. 362.—*Solanum tuberosum* (Potato): *1,* flower; *2,* corolla seen from above; *3,* late stage of flower; *4,* flower-bud; *5,* calyx; *6,* andrœcium and section (*co*) of corolla; *r,* receptacle; at right an anther, showing apical pores; *7,* half of andrœcium and corolla; *8,* gynœcium and calyx (*ca*); *9,* transverse section of ovary; *10,* fruit (berry).

By using the shape of the corolla as the basis, still another scheme of grouping the tribes can be devised in which all members possessing a rotate corolla will fall into one group and those possessing a funnel-form of salver-form corolla will fall into a second group. Compare with *Liliaceæ, Rosaceæ, Myrtaceæ, Umbelliferæ, Ericaceæ, Boraginaceæ, Labiatæ, Scrophulariaceæ, Cucurbitaceæ, Compositæ,* and other families.

Tribe I. *Nicandreæ: Nicandra* (Apple of Peru). A monotypic genus based on *Nicandra physaloides* of Peru, a coarse, glabrous herb with ovate, large-toothed leaves and bearing rather large flowers with campanulate corolla. The berry, which becomes dry with age, is invested by the large, deeply lobed calyx, the broad lobes of which are sagittate at the base. The species is the only member of its tribe; it sometimes occurs as an escape from gardens.

Tribe II. *Solaneæ: Solanum* (Nightshade). A tremendous genus of about 1,200 species, widely distributed, but most abundant in the

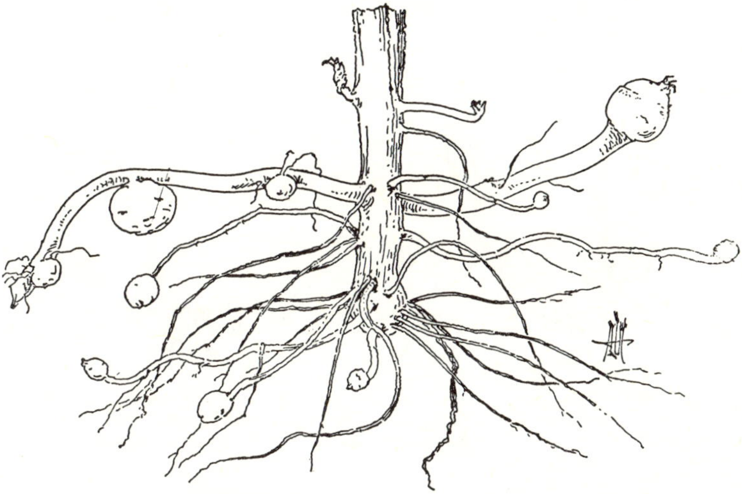

FIG. 363.—Underground parts of potato showing developing tubers.

American tropics. About 30 species are found on the prairies and plains and in the mountains of the West. The flowers are characterized by a 5-toothed or 5-cleft calyx and a rotate corolla with a short tube and plaited 5-lobed limb. The stamens are inserted on the tube of the corolla; they are distinguished by short filaments and elongate anthers which are connivent into a tube around the style; the anthers open by lateral slits or by apical pores. The berry resembles a small tomato. The calyx persists and in some cases becomes a thin bladdery pouch which completely encloses the berry. Most of the species are erect herbs or shrubs but some are climbing, and many are very

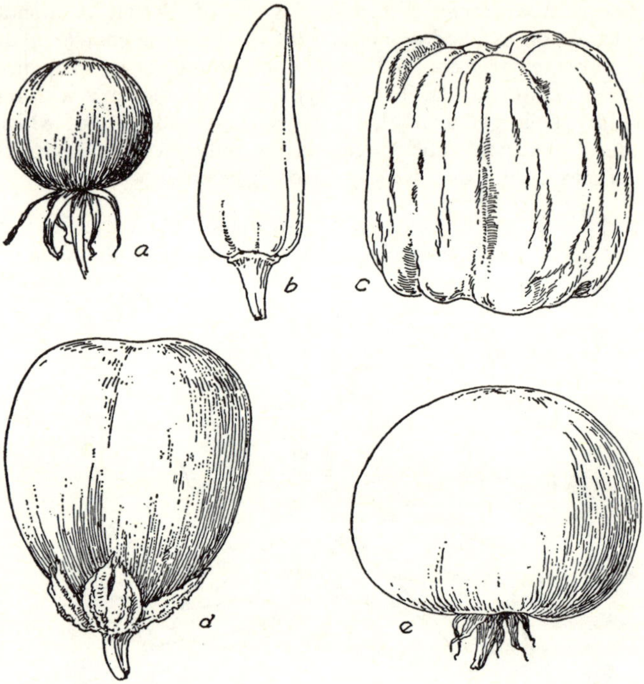

FIG. 364.—Fleshy fruits of various members of the potato family: *a,* potato; *b, c,* two varieties of red pepper (*Capsicum fruiticosum*); *d,* egg-plant (*Solanum Melongena* var. *esculentum*); *e,* tomato (*Lycopersicon esculentum*).

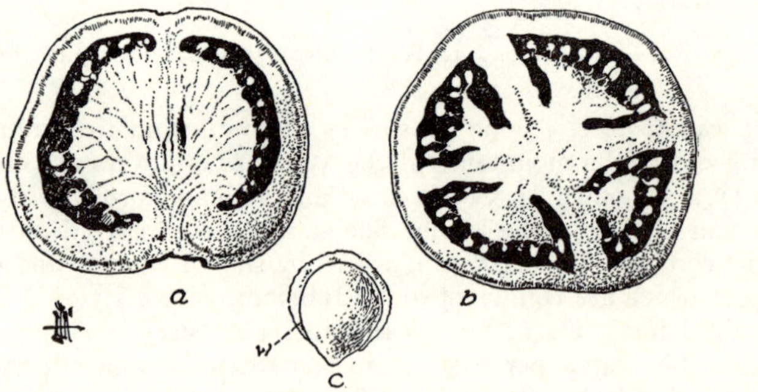

FIG. 365.—Fruit of tomato in median (*a*) and transverse (*b*) sections; *c,* seed; *w,* wing of seed.

spiny or prickly. The leaves are simple and often pinnately lobed or incised, or are pinnately compound. The flowers are white or various shades of blue, purple, or yellow, in cymose, paniculate or

FIG. 366.—*Solandra guttata: a,* corolla in sectional view; *b,* calyx, with protruding style; *e,* gynœcium; *d,* summit of ovary, with style removed; *gl,* gland; *e, f,* stamens.

umbellate, often supra-axillary clusters. (See also *Sparganium.*) A stellate pubescence is characteristic of some species. To this family belong a number of important cultivated plants, such as *Solanum tuberosum* (Potato), *S. Melongena* (Egg-Plant), *S. muricatum* (Pepino), *S. Pseudo-Capsicum* (Jerusalem Cherry), *S. Dulcamara* (Climbing Bittersweet), *S. nigrum* (Black Nightshade). (Figs. 362, 363, 364.)

Among other cultivated members of this tribe are *Atropa Belladonna* (Belladonna), the berries of which are very poisonous; *Hyoscyamus niger* (Henbane), also poisonous, the many-seeded capsular fruit (pyxidium) of which dehisces by a lid; *Lycopersicon esculentum* (tomato); *Physalis Alkekengi* (Chinese Lantern), the berry of which is enclosed in the large, brightly colored, saccate calyx; *Capsicum frutescens* and its varieties (red peppers, cherry-

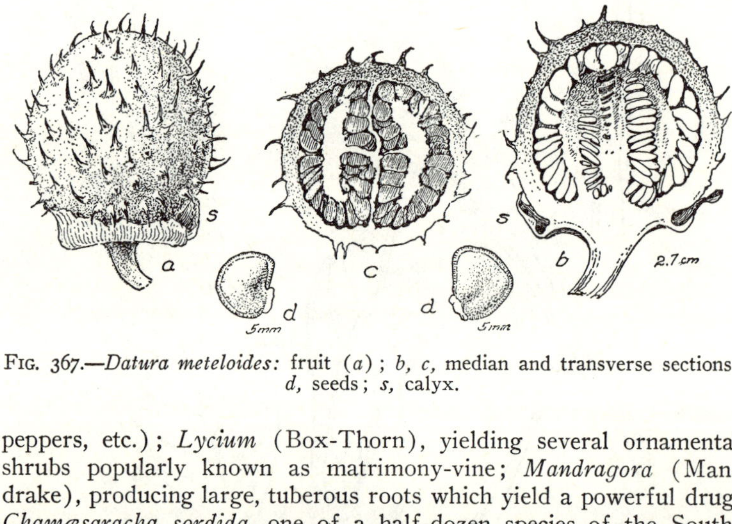

FIG. 367.—*Datura meteloides:* fruit (*a*); *b, c,* median and transverse sections; *d,* seeds; *s,* calyx.

peppers, etc.); *Lycium* (Box-Thorn), yielding several ornamental shrubs popularly known as matrimony-vine; *Mandragora* (Mandrake), producing large, tuberous roots which yield a powerful drug. *Chamæsaracha sordida,* one of a half-dozen species of the Southwest and of Mexico, resembles *Physalis* except for the absence of a bladder-like calyx. (Fig. 365.)

Tribe III. *Datureæ. Datura* comprises about a dozen widely distributed species of coarse narcotic herbs, shrubs, or trees, with alternate leaves and large, showy flowers. The funnel-form plicate corolla is often 6—10 inches long and several inches broad, with broad lobes. The plants produce a large, often prickly or spiny capsule from a 2-celled or falsely 4-celled ovary, which is subtended by the persistent base of the calyx. Two robust herbs, *Datura Stramonium* and *D. Metel,* are frequent in waste grounds. Several are cultivated as ornamentals. (Figs. 366, 367.)

Tribe IV. *Cestreæ. Cestrum* is represented by several ornamentals. Species about 150, including shrubs and trees of the American tropics. Leaves simple and generally narrow. Flowers of diverse

FIG. 368.—*Nicotiana glauca: a,* flowering stem; *b,* flower-bud; *c,* section of flower; *d,* stigma; *e, s,* calyx; *f,* ovary; *g,* median section of flower with upper part of corolla removed; *gl,* disk; *h,* transverse section of ovary showing large placentas; *i,* stamens; *j,* leaf.

colors. Ovary 2-celled, containing few ovules, developing into a fleshy berry.

Nicotiana yields *N. tabacum,* the tobacco plant, and several stately ornamental herbs or subshrubs, often bearing showy flowers of diverse colors. Calyx tubular-campanulate. Corolla usually with a long, narrow tube which is often oblique. Capsule 2—4-valved. The tall, handsome *Nicotiana glauca* of South America has run wild in California. Left undisturbed, the plant attains the proportions of a small tree, blossoming practically all the year round. The yellow flowers are characterized by a long-tubular corolla with small, spreading lobes. *Fabiana imbricata,* of Peru, is a shrub of a strikingly heath-like

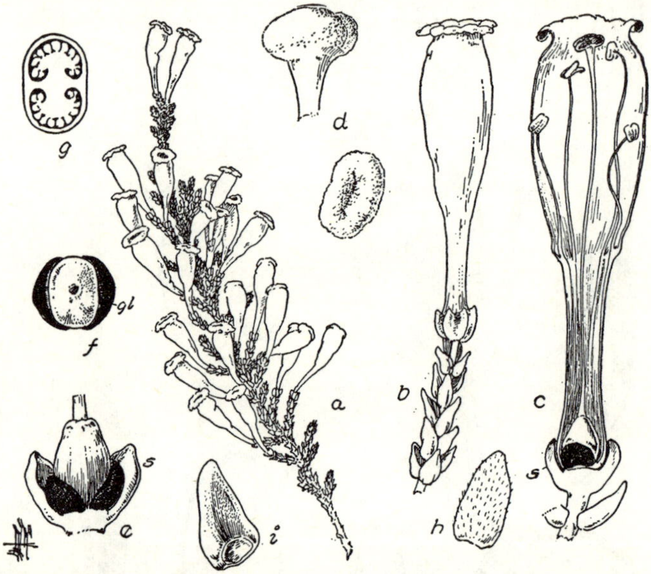

FIG. 369.—*Fabiana imbricata: a,* flowering branch; *b,* flower,—*c,* in median section; *d,* stigma,—in surface view below; *e,* ovary, glands (black) and sepal (*s*); *f,* summit of ovary, with glands (*gl*); *g,* transverse section of ovary; *h,* sepal, *i,* sepal.

habit. The corolla-tube in this species is strongly contracted below the middle, and the lobes are very short and reflexed. (Figs. 368, 369.)

Tribe V. *Salpiglossideæ: Petunia.* Species about 14. The petunias are mostly annual South American herbs, several of which are common garden and pot plants. The flowers are usually distinguished by a large funnel-form or salver-form corolla with flaring plicate limb (sometimes slightly irregular), the tube contracted below the insertion of the stamens. The calyx-segments are small and narrow.

The ovoid ovary supports a long, slender style which terminates in a fleshy 2-lobed stigma. An annular disk is present. All the stamens are functional. *Salpiglossis* has a flower of somewhat similar structure except that the limb of the corolla is more irregular and, besides, the disk is absent, and only four of the stamens are functional. The calyx-tube is also deeper. In *Schizanthus* (Butterfly-Flower), of which there are several horticulturals, there are only 2 functional stamens. (Figs. 370, 371.)

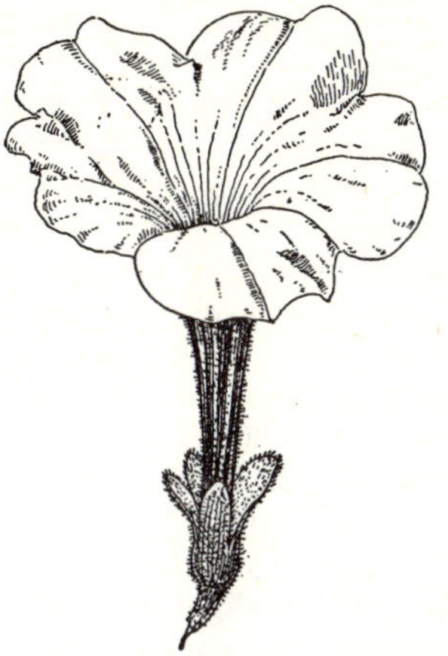

FIG. 370.—Flower of *Petunia.*

SCROPHULARIACEÆ (*Figwort Family*). Flowers irregular (bilaterally symmetrical), bisexual, hypogynous, usually with bracts and bracteoles. Calyx 4—5-toothed, -cleft, or -divided, sometimes split on the lower side or on both sides, persistent.

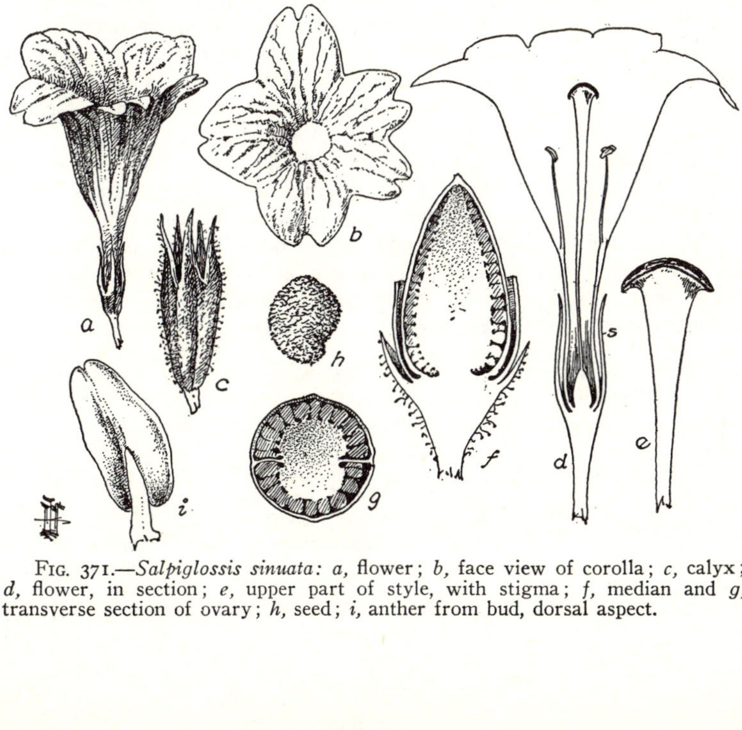

FIG. 371.—*Salpiglossis sinuata: a,* flower; *b,* face view of corolla; *c,* calyx; *d,* flower, in section; *e,* upper part of style, with stigma; *f,* median and *g,* transverse section of ovary; *h,* seed; *i,* anther from bud, dorsal aspect.

Corolla bilaterally symmetrical, 2-lipped, upper lip 2-lobed, lower 3-lobed, the tube often curved. Stamens, 2, 4, or 5, didynamous or nearly equal, when 4 then the posterior one suppressed, when 2 then the two anterior ones suppressed, inserted on the corolla-tube and alternating with its lobes; anthers of various shapes, equal or unequal, 2-celled, or 1-celled by confluence above, sometimes appendaged, dehiscing by lateral slits. Gynœcium uniformly 2-carpellate, the carpels median, generally with a simple, slender style terminating in a 2-lobed stigma of various shapes. Ovary superior, subtended by an annular or sometimes one-sided disk, 2-celled, rarely 1-celled; ovules numerous, small or sometimes large, anatropous or amphitropous, on thick, intruded, axile placentas (compare with *Solanaceæ*). Fruit a septicidal or loculicidal capsule or in some cases a berry. Seeds with various reticulate or striate sculpturings. Embryo small, straight or slightly curved, in fleshy endosperm. Herbs, shrubs, or trees, mainly of temperate climates but ranging into the tropics and the colder regions. Genera 165—205; species 2,600—2,700.

The figworts are of diverse habits, some being prostrate or creeping, some climbing by leaf-stalks, many aquatic, others parasitic or hemiparasitic; many are glabrous, others pubescent. Leaves of various shapes, commonly alternate, although the lower may be opposite, in some cases whorled. A wide diversity of leaf-form may be found in the same genus, and marked heterophylly may obtain in aquatics. In parasitic members the leaves are reduced to scales. Stipules are wanting. Flowers small to large, often showy, corolla often ringent, sometimes spurred or saccate. The inflorescences are chiefly racemose or cymose, but sometimes the flowers are solitary and axillary. Abnormal flowers are to be found in some cases.

The family is of little economic importance except as it yields a host of garden plants. A few forms are medicinal, among which are some poisonous ones. Over 3,000 species, distributed among 50 or more genera; many in our flora.

Wettstein has distinguished the following subfamilies (*Pflnfam.*). The technical characters on which the major divisions are based may be readily seen in tranverse sections of young flower-buds.

Group A, in which the posterior lobes of the corolla are covered by the lateral lobes in the bud. Capsule generally septicidal.
- a. Leaves all alternate. Stamens usually 5. Subfamily I. *Pseudosolanoideæ.*
- b. At least the lower leaves opposite. Posterior stamen wanting or modified. Subfamily II. *Antirrhinoideæ.*

Group B, in which the posterior lobes of the corolla are covered in the bud by one or more of the lateral lobes. Capsule generally loculicidal. Subfamily III. *Rhinanthoideæ.*

Of the numerous genera only a few can be mentioned; these are distributed among the subfamilies as follows:

Pseudosolanoideæ:
Verbascum.
Antirrhinoideæ:
Antirrhinum, Linaria, Collinsia, Scrophularia, Chelone, Pentstemon, Mimulus, Gratiola, Ilysanthes, Calceolaria.
Rhinanthoideæ:
Veronica, Syntheris, Digitalis, Gerardia, Castilleja, Orthocarpus, Melampyrum, Euphrasia, Pedicularis, Odontites, Rhinanthus.

Using the number of stamens as the basis of major grouping, an artificial but convenient rearrangement may be devised in which the separate genera may then be keyed out on other characters. For example: stamens 2, *Veronica;* stamens 4, *Antirrhinum;* stamens 5, *Verbascum;* etc., etc.

Pseudosolanoideæ. The name of this subfamily suggests an affinity with the *Solanaceæ.* The principal genus is *Verbascum* (Mullein), with 125—200 species of the Mediterranean region. A few have become established in our flora, such as *Verbascum Thapsus* (Great Mullein), a tall, vigorous, densely woolly plant with oblong leaves, which during the flowering period puts forth a long, wand-like inflorescence. It is frequent in fields and waste grounds and in dry, sandy stream-beds in certain parts of the Northwest. The flowers are distinguished by a rotate corolla (compare with *Solanum*) which is but slightly irregular. All the stamens are functional and some or all of them have bearded filaments. The calyx is deeply cleft. The fruit is a septicidal, 2-valved, many-seeded capsule. (Fig. 372.)

Antirrhinoideæ. Antirrhinum (Snapdragon). In the snapdragon the corolla-tube is gibbous in front and so developed as nearly to close the throat; the upper lip is erect. The 4 stamens arise under the upper lip and are included in the tube. The lower leaves are opposite, the upper alternate. Species 90—100, of the northern hemisphere. *Antirrhinum majus* is a popular garden plant. In *Linaria* the corolla-tube is prominently spurred at the base. *Linaria vulgaris* (Butter-and-Eggs) is common along roadsides and in fields throughout the country. *Kickxia Elatine,* a plant sometimes referred to *Linaria,* is distinguished by its hastate, convolvulus-like leaves. In *Scrophularia*

(Figwort) the corolla is small, with an inflated globose tube and broad, rigidly erect upper lobes. The 4 functional stamens are didynamous, while the fifth (posterior) is reduced to a mere scale on the corolla-tube. A small ovoid capsule is developed. Our species (6 or fewer) are very similar in appearance but can be distinguished by the color of the stamens and the serration of the leaves. *Scrophularia marilandica* and *S. leporella* are common in Eastern woodlands. (Figs. 373, 374.)

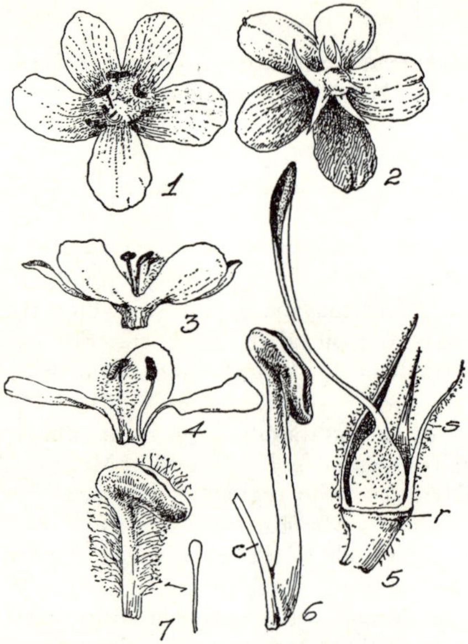

FIG. 372.—*Verbascum longifolium:* *1*, flower seen from above; *2*, flower, viewed from beneath; *3*, side view; *4*, section of corolla, showing insertion of stamens; *5*, gynœcium; *s*, sepals; *r*, receptacle; *6*, naked stamen; *c*, corolla-tube; *7*, bearded stamen; at right a glandular trichome of filament.

Pentstemon (or *Penstemon*) (80—100 species) and *Mimulus* (50—60) species are abundantly represented in the West. In the former the leaves are opposite or verticillate, in the latter they are all opposite and the fifth stamen is completely suppressed. The outstanding feature in *Calceolaria* is the strongly inflated ladies'-slipper-like lower lip of the corolla; in horticultural varieties this organ is greatly enlarged and showy. The genus is one of the largest in the family, numbering about 200 species. The plants range from Mexico to Chile. (Fig. 375.)

Rhinanthoideæ. Veronica (Speedwell), another very large genus (about 200 species), ranges over the temperate and cold-temperate parts of the world. The New Zealand species, which are numerous and by some students of them regarded as distinct from our American members, are of special interest in that they are shrubs or trees, the latter sometimes of great size. A number of the shrubs are common ornamentals in warmer parts of the country; these bear 4-ranked evergreen leaves. Our 12—15 native species are herbs with opposite or verticillate leaves. The small and numerous flowers are borne in terminal or axillary spikes or racemes, or are sometimes solitary in the axils; they are distinguished by a somewhat rotate corolla which is 4-lobed by the union of the two upper lobes; the lower lobe is usually narrow, and the tube short. One of the handsomest of our species is the tall *Veronica virginica* (Culver's Root)—the *Leptandra virginica* of some books—of Eastern parts of the country. The plant bears whorled lanceolate leaves and numerous small, white flowers in showy, racemose-spicate inflorescences. (Fig. 376.)

The largest genus of the subfamily is *Pedicularis*; its 250 or more species are spread widely over the northern hemisphere, 35 or 40 of them within the United States. They are low plants with pinnately lobed or pinnately dissected leaves, bearing their red to yellow flowers

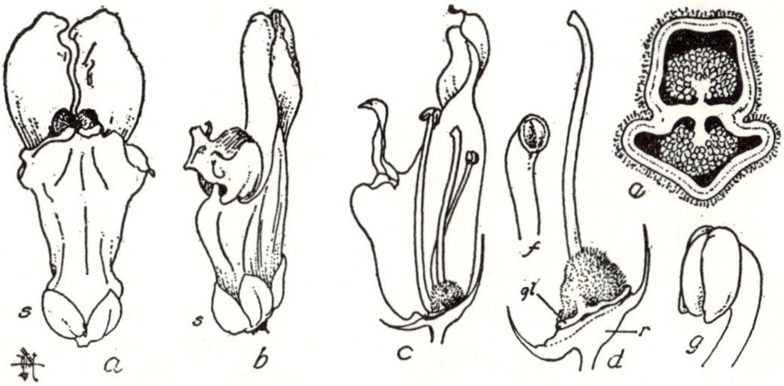

FIG. 373.—*Antirrhinum majus: a, b, c,* different aspects of the flower; *d,* gynœcium; *gl,* annular disk; *r,* receptacle; *e,* transverse section of ovary; *f,* stigma; *g,* anther.

in terminal spikes or racemes. They are found in a variety of habitats. The upper lip of the strongly 2-lipped corolla is usually conspicuously arched and compressed over the 4 didynamous and included stamens. *Castilleja* (50 species) and *Orthocarpus* (30 species) are somewhat similar in plan of the flower, but their most marked feature is the showy, leafy-bracted spikes, the bracts of which are brilliantly colored. The species of *Castilleja* (Indian Paint-Brush) are hemiparasitic on roots of other plants. Several species of *Digitalis* are horticultural, amongst them *Digitalis purpurea* (Foxglove), from which a powerful drug is obtained. The plant is poisonous to livestock. In the Puget Sound region the plant has run wild in pastures and waste grounds. The genus is Eurasian, with about 25 species. *Euphrasia,* with over 100 species, is distributed over temperate and cold regions of the northern and southern hemispheres. It is distinguished by a 4-cleft calyx, or this when 5-cleft is irregular—one lobe smaller than the others.

The family contains a considerable number of large genera (with 100 or more species), a number which will be materially increased if we include genera with 50 or more species. It is of interest, too, that the plants are found in a great variety of habitats. Thus species of *Chelone, Mimulus, Conobea, Gratiola, Ilysanthes,* and *Limosella* will be found in swamps, on muddy shores or other wet grounds, while such as *Micranthemum micranthemoides* and *Gerardia maritima* occupy salt-marshes along the Atlantic Coast. A few, such as *Veronica Wormskjoldi,* are boreal or ascend to alpine summits farther south, while others are arctic, as *Bartsia alpina* of Labrador, Greenland, and Europe. Certain species of *Pedicularis* occur in dry situations, others in wet. *Antirrhinum cirrosum* climbs by its leaf-stalks, while an opposite extreme is found in certain South African forms (*Harveya, Hyobanche*) which are parasitic on other plants, and as a consequence have their leaves reduced to scales. Species of *Melampyrum,* of which we have two—*M. lineare* and *M. latifolium*—are said to be hemiparasitic. *Paulownia tomentosa,* a solitary species of Japan, is a tall tree. It is said to have escaped from cultivation in the Eastern States.

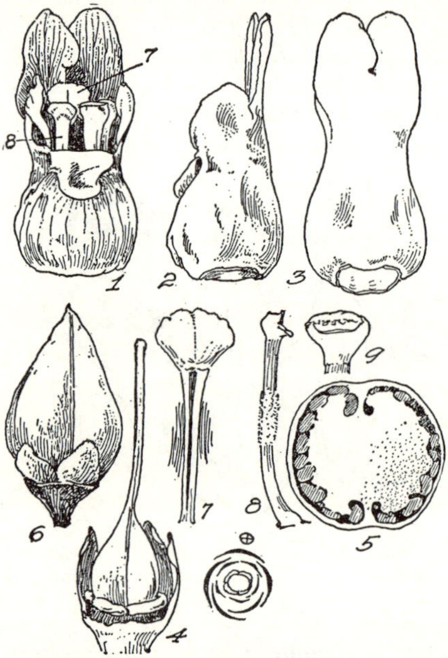

FIG. 374.—*Scrophularia leporella: 1, 2, 3,* front, side, and dorsal views of corolla; *4,* gynœcium, with hypogynous disk, and calyx; *5,* transverse section of ovary; *6,* capsule, with persistent calyx; *7,* staminode; *8, 9,* stigma.

Genera and species of the *Scrophulariaceæ* are sometimes difficult to distinguish. Thus on general appearance *Pentstemon, Mimulus, Collinsia, Gerardia,* and *Gratiola* are likely to be confused; on the other hand certain genera stand out quite clearly, as do *Verbascum,*

Scrophularia, Kickxia, Limosella, Syntheris, and *Pedicularis.* In *Pentstemon* and *Collinsia* the student may find difficulty in identifying some species. A helpful guide to the family affinities will be found in the placentation, which can be readily ascertained from transverse sections of the ovary. The family connects with the *Solanaceæ* through the tribe *Pseudosolanoideæ,* in which the corolla

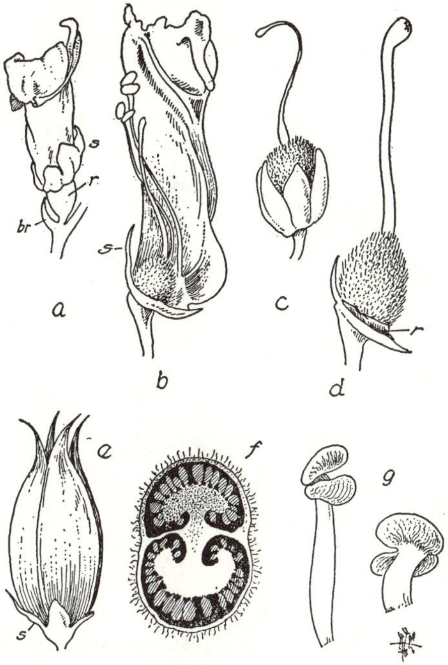

FIG. 375.—*Pentstemon* sp.: *a,* flower,—*b,* in section; *c,* gynœcium and calyx; *d,* gynœcium; *r,* receptacle; *br,* bract; *s,* calyx; *e,* capsule; *f,* transverse section of ovary; *g,* stamens.

is rotate and almost regular. In other directions it connects with the tribe *Salpiglossideæ* of the *Solanaceæ.* Furthermore the ovary is uniformly 2-celled and median, while it will be recalled that in the *Solanaceæ* the ovary is oblique and frequently several-celled. In the next following families the student will find certain characters, too, that seem to tie these up with the *Scrophulariaceæ,* and very likely he will on general principles classify the representatives he may meet with as scrophulariaceous forms.

BIGNONIACEÆ (*Bignonia Family*). Flowers irregular (bilaterally symmetrical), bisexual, hypogynous. Calyx gamosepalous, 5-cleft, 2-lipped, or truncate. Corolla campanulate, funnel-form or cylindrical, 5-lobed, 2-lipped, the lower lip the larger, the tube straight or curved, often much expanded above and gibbous in front, and strongly contracted below, the lobes imbricate or valvate in the bud. Anther-bearing stamens 2—4, didynamous, with elongate slender filaments arising from the lower part of the corolla-tube and often thickened and hairy or papillate below; the posterior stamen stamino-

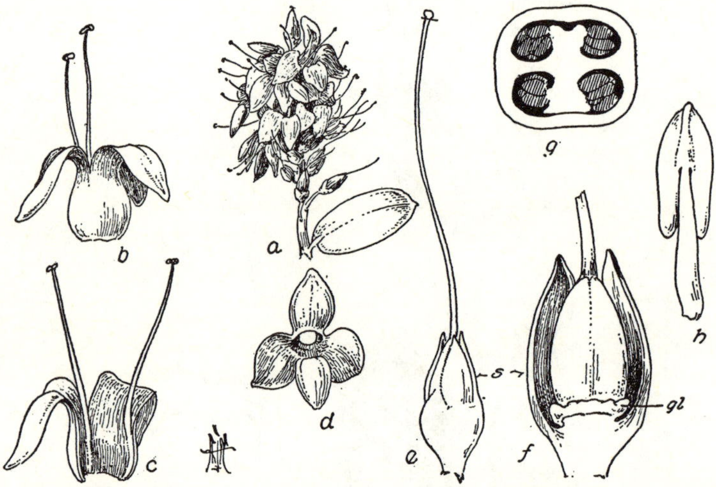

FIG. 376.—*Veronica buxifolia: a,* inflorescence; *b,* corolla and stamens in superficial view; *c,* in section; *d,* face view of corolla; *e,* calyx and style; *f,* ovary, calyx (*s*), and disk (*gl*); *g,* transverse section of ovary; *h,* stamen, dorsal aspect.

dial, and sometimes only the two anterior stamens functional. Gynœcium uniformly 2-carpellate, sessile or stipitate (on a gynophore), with a slender style terminating in 2, usually thick or broad, stigmatic lobes. Ovary superior, 2-celled or 1-celled, or 2-celled below and 1-celled above, producing numerous, erect or horizontal, anatropous ovules on single or double axile placentas, or on parietal placentas which are intruded until they meet. (Compare with immediately preceding families.) At the base of the ovary there is usually an annular disk. Fruit a loculicidal, septicidal, or septifragal capsule, or sometimes fleshy and with a fibrous or juicy pulp. Seeds often large, flat, transverse, more or less compressed and winged. Endosperm none. Embryo with a short, straight hypocotyl and broad and flat, 2-lobed or emarginate cotyledons.

Bignoniaceous plants are trees, shrubs, woody vines, or in some cases herbs, with opposite, simple or compound—once- or twice-pinnate—exstipulate leaves and generally large, showy flowers in paniculate or racemose clusters. Chiefly of the tropics around the world but spreading into temperate regions. A few are alpine. The shrubs are especially well developed in the South American *campos,* while the woody vines form a characteristic element of the liana vegetation of the American tropics. Three genera have pushed northward in eastern North America, while another has reached the deserts of the far Southwest. Genera 69—110, species 500—600.

The shrub forms of the *campos* show special adaptations to a xerophytic habitat in their reduced or modified stems. The vines climb by means of the leaf-stalks, or by leaf-tendrils, which are often furnished with adhesive disks at the ends. Large trees are common to several genera. A marked feature of the family is the great diversity of fruit forms; these serve as useful guides to the identification of genera.

Our genera are: *Bignonia, Catalpa, Tecoma, Chilopsis.* These are all represented in cultivation. Among a considerable number of other genera represented in cultivation the following are characteristic: *Eccremocarpus, Pandorea, Jacaranda, Crescentia.*

The technical subdivision of the family into tribes, of which there are five, is based mainly on the number of cells in the ovary. For our immediate purpose, however, our few genera may be keyed out on characters of the leaves and the fruits. Thus:

I. Leaves simple. Pod long, slender, and terete.	
1. Leaves large, broadly ovate.	*Catalpa.*
2. Leaves much smaller, lanceolate-linear.	*Chilopsis.*
II. Leaves compound.	
1. Leaves pinnate.	*Tecoma.*
2. Leaves 2-foliolate, with terminal tendril.	*Bignonia.*

Catalpa. The catalpas, of which there are 5 or 6 species, two in eastern North America, two in eastern Asia (!), and possibly two in Cuba, are trees with large ovate-cordate leaves and large, showy, white or mottled flowers in terminal corymbs or panicles. The corolla-tube is rather inflated and the three lower lobes are much the larger. The anthers diverge strongly, a feature that is facilitated by their being attached to the filaments by their upper ends. An annular disk surrounds the base of the rather conical ovary. Style elongate, terminating in a stigma of two broad, flat, oblong lobes which are fleshy-stigmatic on the edges. The style and the stamens are arched under the upper side of the corolla-tube. The fifth (upper) stamen is

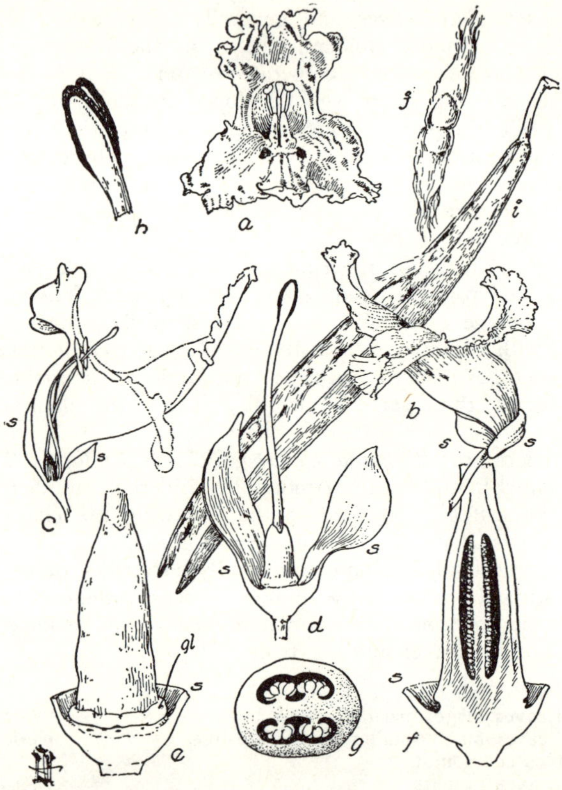

FIG. 377.—*Catalpa speciosa: a,* face of corolla; *b,* flower,—in section at *c; s,* calyx; *d,* gynœcium and segments of calyx (*s*); *e,* ovary and basal gland (*gl*),—in median section at *f,* and transverse section at *g; h,* stigma; *i,* pod; *j,* seed.

transformed into a staminode, while the other four are functional and didynamous. Calyx small, and 2-lipped after the opening of the flower. Fruit a long, slender, bean-pod-like, loculicidal capsule, enclosing numerous flat, light seeds furnished with elongate fringed wings.

Our two species, *Catalpa speciosa* and *C. Catalpa*, are mainly Southern but are commonly planted in the North. (Fig. 377.)

The solitary *Chilopsis linearis* (Desert Willow) is a characteristic tall, willow-like shrub of sandy washes of the Mojave and Colorado deserts and adjacent Mexico and east to Texas. It is easily recog-

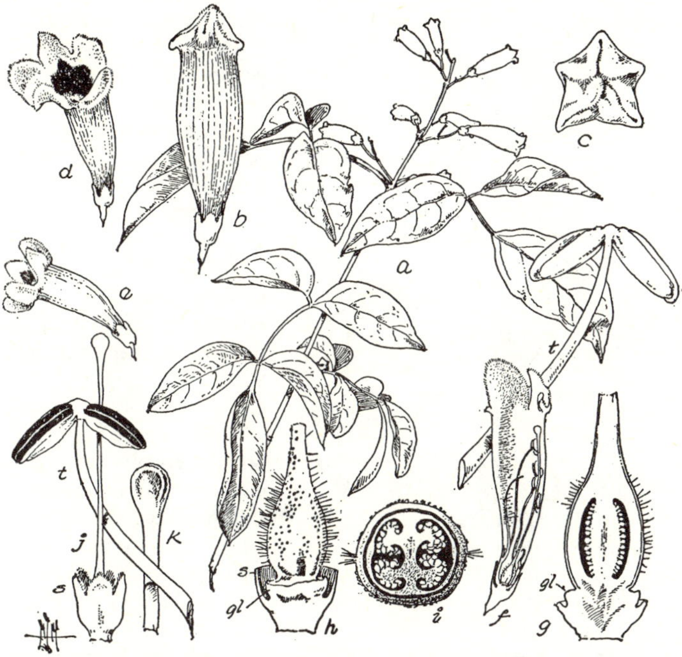

FIG. 378.—*Pandorea australis:* *a*, flowering branch; *b*, flower-bud; *d*, *e*, views of flower; *f*, section of flower,—*g*, of ovary; *gl*, disk; *h*, ovary, disk (*gl*), and calyx-tube (*s*); *i*, transverse section of ovary; *j*, calyx and projecting style; *k*, stigma; *t*, ventral and dorsal aspects of stamen; *c*, bud viewed from above.

nized by its linear-lanceolate, willow-like leaves and catalpa-like pods and seeds. The showy pink flowers are very fragrant. *Tecoma* is represented in eastern United States by *T. radicans* (Trumpet-Creeper), a woody vine which climbs by means of aërial roots or grows prostrate; it is common along roadsides and fence-rows and is easily recognized by its showy scarlet flowers. It produces an elongate capsule which is compressed contrary to the partition. The seeds are borne in several rows and are furnished with broad, translucent wings. The plant is a popular ornamental. The generic affinities of the species have been variously interpreted. *Bignonia capriolata* (Cross-Vine) is a woody vine of moist Eastern and Southern woodlands. The species produces orange-colored flowers and a long, broad, capsular fruit. Two species of *Jacaranda* and a few of *Pandorea* are common ornamentals in southern California. These are handsome trees with twice-pinnate leaves and showy blue flowers produced in profusion, especially in early spring. It produces a very distinctive flat, discoid fruit. (Figs. 378, 379, 114.)

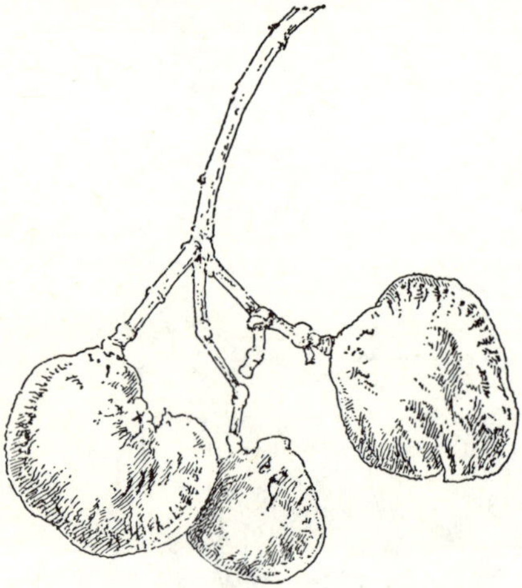

FIG. 379.—Fruit of *Jacaranda.*

PEDALIACEÆ (*Pedalium Family*). Flowers irregular (bilaterally symmetrical), bisexual, hypogynous or epigynous, with calyx and corolla mainly as in the preceding family, but the corolla-tube in some cases is spurred. Stamens 4, didynamous, the fifth (upper) staminodial, or only two stamens functional. Gynœcium 2-carpellate, or sometimes 3- or 4-carpellate. Style elongate, with as many stigmatic lobes as carpels. Ovary superior or sometimes inferior, 2—4-celled or falsely 1-celled, with few to many ovules on axile placentas. Fruit a capsule, or dry and indehiscent, winged or monstrously shaped, with numerous hooked spines or with elongate, spreading, stout, spinous branches. Embryo straight, in scanty endosperm. About 50—60 species of tropical and subtropical herbs or shrubs, distributed among 14 genera. One species, *Sesamum orientale* (Sesame) is a garden plant.

MARTYNIACEÆ (*Martynia Family*). Flowers irregular (bilaterally symmetrical), bisexual, hypogynous, with a 4—5-lobed calyx. Corolla not strongly 2-lipped. Stamens 4, didynamous, but only two functional, the others staminodial; anthers divergent. Gynœcium 2-carpellate, subtended by an annular disk. Ovary superior, 1-celled, with several ovules in each cell on parietal placentas. Fruit a monstrous 2-horned or 2-beaked woody capsule. Embryo straight, in scanty endosperm. Three genera and about 10 species of stout, glandular-hairy herbs of tropical and subtropical America. One species, *Martynia louisiana* (or *Proboscidea louisiana*), the Unicorn Plant, is native as far north as Indiana and west to New Mexico. The

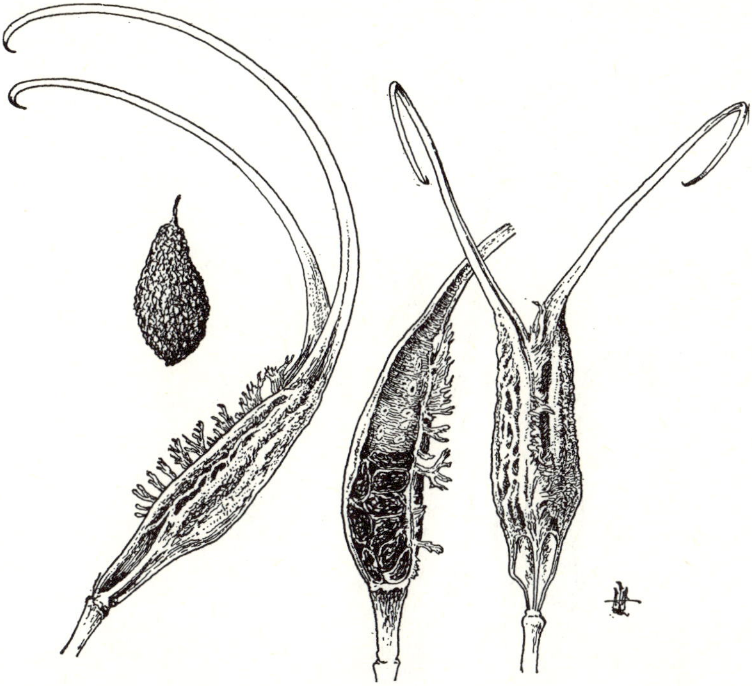

FIG. 380.—Capsules and seed of *Martynia* (*Proboscidea*) *louisiana.*

plant is grown mostly as a curiosity. Six or seven other species occur in tropical America. (Fig. 380.)

OROBANCHACEÆ (*Broom-Rape Family*). Flowers irregular (bilaterally symmetrical), bisexual, sometimes cleistogamous, hypogynous, with a 4—5-toothed calyx (often split on the sides), and a tubular, more or less oblique and curved, conspicuously 2-lipped corolla. Stamens 4, didynamous, with sometimes a fifth, borne on the corolla-tube and alternating with its lobes; anthers 2-celled, parallel, one sometimes not functional. Gynœcium 2- or sometimes 3-carpellate, with simple style terminating in a 2—4-lobed discoid or variously shaped stigma. Ovary superior, 1-celled, with numerous anatropous ovules on double (four) parietal placentas. Fruit a 1-celled, 2-valved capsule. Embryo straight, in fatty endosperm. Simple or branched root-parasitic herbs with leaves reduced to scales and bearing pale flowers in terminal bracteate spikes, or solitary in the

FIG. 381.—*Thalesia uniflora: a,* habit sketch; *b,* flower; *c,* face of corolla; *d,* gynœcium and calyx.

axils of the scale-leaves. About 11 genera and 200 species, widely distributed over the northern hemisphere.

Our species belong to the following genera: *Epifagus, Conophila, Orobanche, Thalesia, Boschniakia.*

The special features to be noted here are the reduction of the leaves in association with the parasitic habit, the presence of cleistogamous flowers, and the 3-carpellate gynœcium and the 1-celled ovary.

In *Epifagus* the flowers are borne in racemes or spikes. The upper flowers, which are sterile, are characterized by long filaments and style, while the lower ones, which are cleistogamous, have short stamens, style, and corolla. *Epifagus virginiana* (*Leptamium virginianum*), the only species, is Eastern. *Conopholis americana* (Squaw-Root), also an Eastern species, is stouter; its flowers have a spathe-like calyx which is 3-toothed above and split on the lower side. The corolla is strongly 2-lipped. The filaments are long-exserted and bear bristly-hairy anthers. Flowers all fertile.

In *Orobanche* all the flowers are fertile; they are borne in spikes or racemes and are subtended by bracts and bracteoles. The calyx is split nearly to the base above and below, while the strongly 2-lipped corolla is more or less oblique. Species about 90; ours most abundant in the West. *Thalesia* (Broom-Rape) has generally larger flowers, solitary or in long-pedunculate clusters. Calyx-segments sharp and corolla oblique. The stamens are included in the corolla-tube. *Thalesia uniflora* and *T. fasciculata* are generally found over most of the continent. *Boschniakia tuberosa,* a rather stout tuberiferous species, ranges from the mountains of southern California northward to Alaska, and across into Asia. (Fig. 381.)

GESNERIACEÆ (*Gesneria Family*). Flowers more or less irregular, bisexual, hypogynous or epigynous. Calyx more or less deeply lobed or cleft. Corolla with short or elongate tube, more or less 2-lipped, and limb unequally lobed. Stamens 4, rarely 5, didynamous, or sometimes only 2 fertile, the others staminodial; anthers mostly 2-celled, or 1-celled by confluence above, sometimes with prolonged connective. Gynœcium 2-carpellate, with a simple elongate style supporting a 2-lobed stigma. Annular disk present. Ovary superior or slightly to wholly inferior, 1-celled, or 2—4-celled by intrusion of the placentas, or sometimes only completely 2-celled below. Ovules numerous, on parietal placentas when the ovary is 1-celled, otherwise on axile placentas. Fruit capsular or fleshy. Endosperm present or absent. Embryo straight, terete, with small cotyledons. Rhizomatous, tuberiferous, or prostrate herbs, or shrubs, with alternate, opposite or whorled, simple leaves. Flowers often large and showy, solitary or in clusters. About 84 genera and 1,100 species, mostly of the tropics. According to Bailey several genera are represented in cultivation, among which are those mentioned in the following key.

Ovary superior:

Trichospermum, with 4 fertile stamens and a cup-shaped disk. *Streptocarpus,* with 2 fertile stamens and low disk.

Ovary inferior:

Naegelia, with annular disk. *Isoloma,* with a 5-lobed disk. *Sinningia* (*Gloxinia*), with 5-lobed disk and anthers confluent above. *Sinningia* (*Gloxinia*) *speciosa* is a common potherb.

COLUMELLIACEÆ. A very small family based on *Columellia,* a genus of two species of Colombia and Peru. Flowers nearly regular, bisexual, 5-merous to 8-merous, with 2 stamens and a 2-carpellate gynœcium. Fruit a 4-valved capsule. Seeds with endosperm and a small embryo. Shrubs or small trees.

LENTIBULARIACEÆ (*Bladderwort Family*). Flowers irregular, bisexual, hypogynous. Calyx 2—5-lobed, nearly regular or 2-lipped, persistent. Corolla prominently 2-lipped, the lower lip the larger, 3-lobed, with a raised, commonly bearded palate which nearly closes the throat, and spurred or gibbous at the base in front. Stamens 2, rarely 5, anterior, inserted at the base of the corolla-tube; anthers 1-celled. Gynœcium 2-carpellate, median. Style short or absent; stigma 2-lobed, prominent, the posterior lobe more or less obsolete. Ovary superior, 1-celled, bearing numerous anatropous ovules on a central globose placenta which arises from the base of the ovary and occupies most of the cavity. (Compare with *Primulaceæ, Caryophyllaceæ.*) In some cases only 2 ovules and these connate. Fruit a globose, ovoid, or elongate capsule, opening by 2—4 valves or by a slit. Seeds small, with endosperm. Mostly perennial, aquatic (or in wet soil), often epiphytic and rootless herbs with simple or finely dissected leaves. Flowers solitary, racemose, cymose, or spicate. Genera, in the broadest sense, about 5; species about 250, widely distributed over temperate and tropical regions. Only *Utricularia* and *Pinguicula* are represented.

Utricularia (Bladderwort). The bladderworts are submersed aquatics with finely dissected leaves, these bearing small bladders which entrap minute organisms. The bladders open by a trap-like lid actuated by tactile hairs which line the orifice; after the organism has entered the bladder, the lid closes. Some species (*Utricularia inflata, U. radiata*) bear inflated leaves which serve as floats, and some bear cleistogamous flowers (*U. clandestina*). The flowers are borne solitary or several together; they are distinguished by the erect upper lip of the corolla and the strongly curved stamens. The great

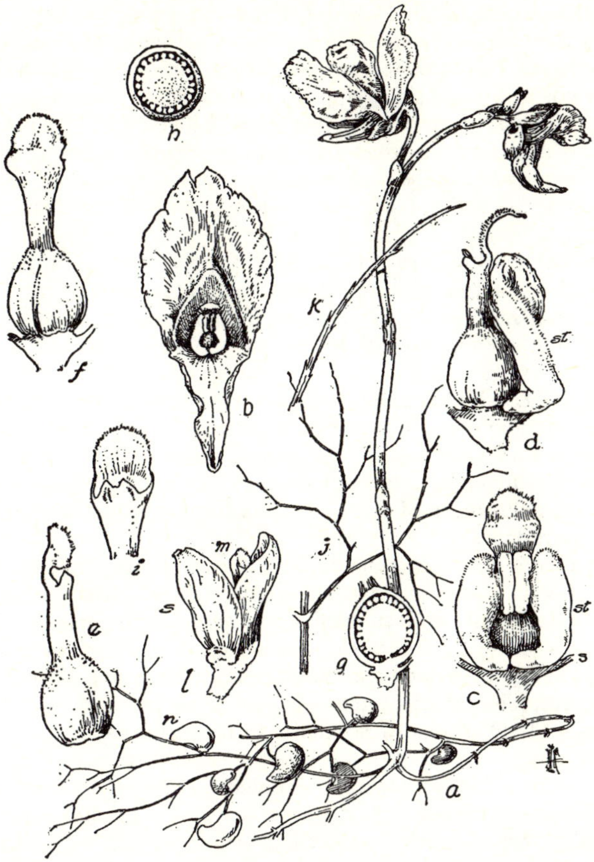

FIG. 382.—*Utricularia cornuta: a,* habit sketch; *b,* face of flower; *c,* stamens (*st*), gynœcium in center, (*s*) calyx; *d,* gynœcium and stamen (*st*); *e,* gynœcium, lateral view; *f,* dorsal view of gynœcium; *g,* median, and *h,* transverse section of ovary; *i,* stigma; *j,* branched leaf; *k,* tip of leaf segment, enlarged; *l,* calyx, style (*m*) protruding; *n,* bladders.

majority (200 or more) of the species of the family belong in this genus. (Fig. 382.)

Pinguicula (Butterwort). These are low, acaulescent, fibrous-rooted herbs with leaves of an oblong to lanceolate shape and somewhat fleshy, in a basal rosette. They are found in bogs or in wet situations among rocks. Viscid-glandular hairs on the upper surface of the leaves serve to catch insects (compare with *Drosera*). Besides the terrestrial habit and 1-flowered scapes, the plants differ from the bladderworts in having a 2-lipped calyx with a 3-cleft upper lip and a 2-cleft lower lip and spreading corolla-lobes. Our plants are referred to probably less than half a dozen species; among these the arctic *Pinguicula villosa* is circumpolar in its distribution. The genus as a whole is widespread over the northern hemisphere. Species about 25.

The outstanding features of the family are the insectivorous habit and the large, globose, central placenta. In the symmetry of the flowers, the number of stamens, and the median position of the gynœcium, they suggest affinity with the *Scrophulariaceæ.*

GLOBULARIACEÆ (*Globularia Family*). Flowers irregular, bisexual, hypogynous. Calyx campanulate, deeply lobed, persistent, with narrow to awl-shaped segments. Corolla more or less strongly 2-lipped, the upper lip 2-lobed and shorter than the 3-lobed lower lip. Stamens 4 or 2, inserted on the corolla-tube and alternating with its lobes. Staminodes none. Anthers introrse, 2-celled, opening transversely. Gynœcium 2-carpellate. Style slender, with capitate or lobed stigma. Ovary superior, 2-celled (or 1-celled), with 1 suspended ovule in each cell, and subtended by an annular disk or this reduced as a gland in front. Fruit a nutlet. Three genera and 20 or more species of herbs and shrubs of Europe, the Cape Verde Islands, and Socotra. The principal genus is *Globularia,* with about 17 species of herbs and subshrubs of southern Europe and southwestern Asia. *Globularia trichosantha* (Globe Daisy) is horticultural. Certain tendencies suggest affinities with the *Labiatæ.* Special features are the persistent deeply lobed calyx, the absence of stiminodes, and the sometimes 1-celled ovary, and nut-like fruit.

ACANTHACEÆ (*Acanthus Family*). Flowers irregular (bilaterally symmetrical) or nearly regular, bisexual, hypogynous. Calyx more or less deeply cleft, the segments broad to linear, or two of them sometimes united. Corolla 5-lobed or 2-lipped, oblique, sometimes nearly regular, the tube straight or curved and often strongly inflated above the insertion of the stamens. Stamens 4—2, rarely 5, didynamous, the filaments free or connate in pairs, often strongly curved or geniculate, and inserted at different levels in the corolla-tube; anthers 2-celled or 1-celled, dehiscing lengthwise or by pores, often hairy, spurred or horned, one anther-sac often rudimentary; the connective various, often attenuated in the form of arms by which the anthers are parted and held at different levels. Staminodes commonly 1 or 2, rarely 3. Gynœcium 2-carpellate. Style slender, often strongly bent above; stigmas very diverse—capitate, lobed, spatulate, or otherwise curiously modified, and large. Ovary superior, 2-celled, with 2 to several (rarely 1) ovules in each cell on axile placentas and commonly beset with hairs, at least at the summit. Cleistogamous flowers are produced in some cases. Fruit a loculicidal capsule, in some cases drupaceous. Seeds smooth, hairy, scaly, or prickly, borne on outgrowths from the placentas, often mucilaginous when wetted. Embryo mostly with flat, orbicular, or heart-shaped cotyledons and short hypocotyl. Genera 170—180; species about 2,000.

The members of this large and distinctive family are herbs, shrubs, or rarely trees, with opposite, rarely verticillate, exstipulate, often very large leaves, producing often showy flowers in cymose or racemose inflorescences. Rosette-leaves are present in some cases and the stems are often swollen at the nodes. Taxonomically the family is a very technical one, but is of special interest because it offers an illustration of a group in which important distinguishing characters reside in the pollen-grains—which show exceedingly diverse sculpturings. There are about four subfamilies, distinguished on very technical characters. The family is mainly tropical but with outposts in temperate regions. The plants occupy a great variety of habitats, from wet to dry. Horticultural forms are legion, from many genera. (Figs. 383, 384.)

The genera represented in our flora may be keyed out as below:

Stamens 2.
  Herbs.
    (a). Lower lip of corolla 3-cleft or 3-lobed. *Dianthera.*
    (b). Lower lip of the corolla entire or merely 3-toothed. *Diapedium.*
  Shrub. Corolla red, the lower lip 3-lobed. Stamens exserted. *Beloperone.*
Stamens 4. (a). Calyx-lobes long-filiform. Ovules 2 in each cell. Herbs. *Dyschoriste.*
  (b). Calyx-lobes linear-lanceolate. Ovules 6-20 in each cell. Herbs or shrubs. *Ruellia.*

Among the many genera represented by horticulturals are *Acanthus, Thunbergia, Strobilanthes, Barleria, Fittonia, Justicia, Beloperone, Jacobinia, Pachystachys.* Some of these are especially rich in species, as *Strobilanthes,* with 200; *Ruellia,* 200; *Barleria,* 150; *Justicia,* 250—300; *Dianthera,* 100.

The special features to be noted in the family are the sculpturings of the pollen-grains, the union of two of the calyx-segments, the oblique 5-lobed and sometimes nearly regular corolla, the presence of 5 stamens and the curved or geniculate and sometimes connate fila-

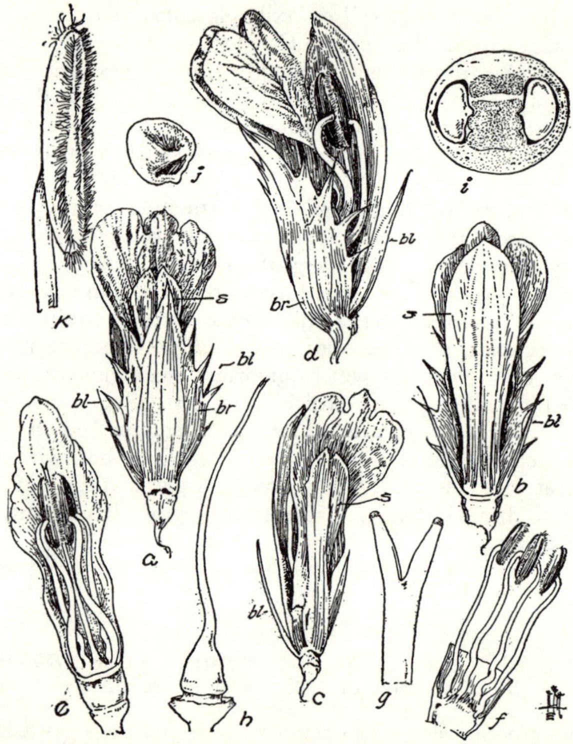

FIG. 383.—*Acanthus mollis:* dorsal (*a*), ventral (*b*), lateral (*d*) views of flower; *s*, sepal; *br*, bract; *bl*, bracteole; *c*, flower with bract (*br*) removed; *e*, calyx and corolla removed, showing stamens and style; *f*, stamens and base of corolla-tube; *g*, stigmas; *h*, gynœcium; *i*, transverse section of ovary; *j*, seed, *k*, anther.

ments, the peculiarly modified connectives and the 1-celled anthers, hairy ovules, drupaceous fruits and mucilaginous seeds, and the occurrence of cleistogamous flowers.

Dianthera is represented by probably a half-dozen species, of which *D. americana* is widely distributed Eastward and Southwestward.

About ten species of *Ruellia* occur through Eastern and Southern parts of the country, with *R. strepens* reaching its northern limits in Wisconsin. In the Colorado Desert there is one species of *Beloperone,* namely *B. californica,* a shrub which becomes leafless and broom-like with age. Thirty or more other species of this genus are native to Mexico and the American tropics to Brazil and the West Indies. *Diapedium* (*Dicliptera*), a tropical genus of about 60 species, has less than a half-dozen species within our limits; one of these, *D. brachiatum,* comes north to the latitude of North Carolina and Kansas. *Dyschoriste,* with about 30 species in the tropics and subtropics, is represented by about five species, mainly in the South; of these

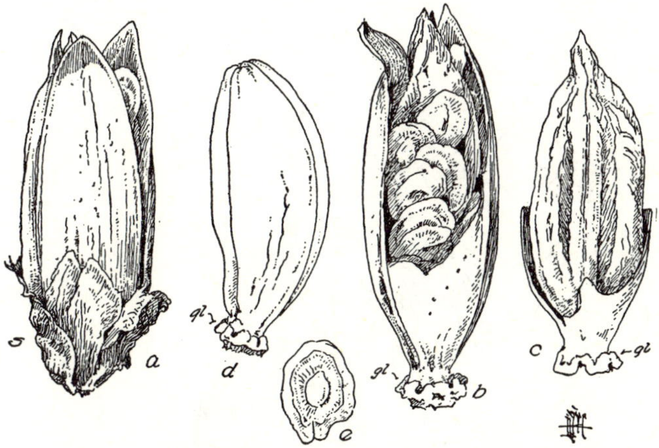

FIG. 384.—*Thunbergia: a,* capsule dehiscing, with calyx below; *b,* capsule with a valve removed, exposing seeds; *c,* placenta; *d,* gynœcium, without style; at base the lobed disk, *gl; e,* seed.

D. oblongifolia occupies sandy barrens from Florida to Virginia. In the warm-temperate and tropical genus *Blepharis* there is an interesting approach to the composites in that the corolla-tube is attenuated and sometimes coarsely toothed on the margin, and the calyx-segments are often exceedingly narrow, while the stamens show a tendency to become connivent; furthermore, the plants are very thistle-like in appearance, with the flowers aggregated in spikes.

MYOPORACEÆ. Flowers irregular or nearly regular, bisexual, hypogynous, with 5-lobed or 5-cleft calyx and 2-lipped corolla. Stamens 4, didynamous, inserted on the corolla-tube, the fifth stamen staminodial; anthers 2-celled or 1-celled. Gynœcium 2-carpellate, with simple style ending in a 2-lobed stigma. Ovary superior, 2-celled and then with 1—8 ovules in each cell, or 3—10-celled by false partitions and then with 1 ovule in each cell; ovules suspended, anatropous, with a strongly developed funicle. Fruit drupaceous, with a 2—10-celled stone, rarely separating into 1-seeded parts. Endosperm copious. Trees, shrubs, or subshrubs, clothed with glandular or woolly pubescence. Leaves simple, alternate or opposite, exstipulate. Flowers ebracteolate, solitary or clustered. Five genera and about 90 species

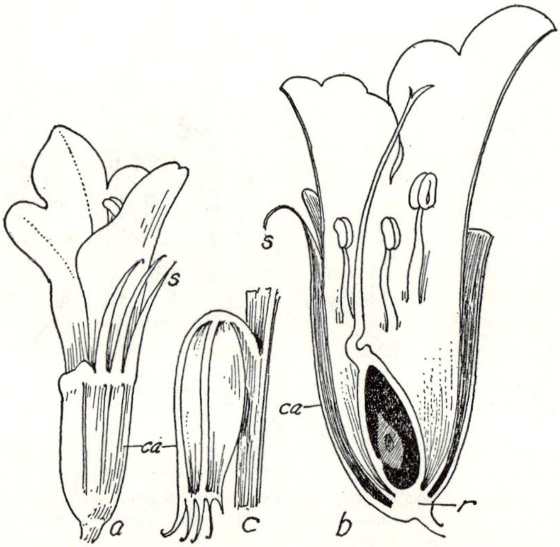

FIG. 385.—*Phryma leptostachya:* flower (*a*), in section (*b*); *ca,* calyx-tube; *s,* segments of calyx; *c,* fruit.

of tropical Africa, Australia, Asia, and the West Indies. Largest genera: *Pholidia* with 60 species, and *Myopora* with about 25.

PHRYMACEÆ (*Lopseed Family*). A monotypic family based on *Phryma leptostachya* (Lopseed), a perennial herb of Eastern woodlands and of eastern Asia and Japan (!). Flowers irregular, bisexual, hypogynous, 2-bracteolate. Calyx persistent, 2-lipped, with awl-shaped or bristly segments (compare with *Valeriana* and *Diervilla*). Corolla 2-lipped, with cylindrical tube, a short 2-lobed upper lip and 3-lobed lower lip. Stamens 4, didynamous, inserted in the upper part of the corolla-tube and included within it; anthers kidney-shaped, divergent, introrse, 2-celled. Gynœcium 1-carpellate and somewhat obliquely placed, with long, slender style ending in a 2-lobed stigma. Ovary superior, 1-celled, with 1 erect orthotropous ovule on a basally parietal placenta. Fruit dry and indehiscent (achene) or somewhat capsular, invested by the enlarged calyx. Referred by some writers to the *Verbenaceæ,* to which it may be profitably compared. Note the persistent calyx, the solitary carpel, achenaceous fruit, and solitary seed (Fig. 385).

ORDER PLANTAGINALES

Flowers regular, bisexual or unisexual, hypogynous, polygamous or diœcious, uniformly 4-merous, without bracteoles but each flower subtended by a prominent bract. Calyx gamosepalous, 4-lobed or 4-parted. Corolla membranaceous, 4-lobed. Stamens 4, rarely 1 or 2, with filiform filaments inserted on the corolla-tube and alternating with its lobes; anthers versatile, opening lengthwise. Gynœcium 2-carpellate, with simple style. Ovary superior, 1—2-celled or falsely 3—4-celled, with 1—several amphitropous, shield-shaped ovules in each cell. Fruit a circumscissile capsule (pyxidium) or an indehiscent nutlet. Seeds mostly with straight embryo in fleshy endosperm. Annual or perennial herbs with alternate or opposite, broad to narrow, basal or cauline leaves tapering to a sheathing petiolar base. Flowers small and numerous, in scapose heads or spikes, rarely solitary. One family, *Plantaginaceæ* (with characters of the order), comprising about 200 species, and 3 genera—*Plantago, Litorella,* and *Bougueria* (in the Andes).

The great majority of the species belong in *Plantago* (Plantain), with centers of distribution in the Mediterranean region, the Asiatic steppes, and abundant in the Andean region of South America. Probably 18 or 20 species occur within our limits; these are herbs with broadly ovate, cordate, or lanceolate to linear leaves. The very small dull-colored flowers are arranged spirally in short or elongate terminal heads or spikes. The flowering proceeds from below upwards at such rate that the fruits of the lowermost flowers are often mature before the uppermost flowers open. *Plantago major,* the door-yard plaintain, is common almost everywhere throughout the country. It is distinguished by broad, ovate leaves and long, slender spikes. Says Asa Gray about it: "According to tradition the wayside plantain was called by the American Indians 'white man's foot,' from its springing up everywhere that foot had been planted." *P. Rugelii, P.*

FIG. 386.—*Plantago lanceolata: a,* habit sketch; *b,* flower, anterior view; *c,* flower with bract removed; *s,* sepal; *p,* petal; *br,* bract; *d,* gynœcium and stamens; *e,* ovary; *f,* corolla, with style protruding; *g,* face of corolla; *h,* anther.

media, and *P. eriopoda* are somewhat similar. *P. Purshii,* a species mainly of the prairies and plains, is easily distinguished by its woolly habit and long, narrow leaves. Easily recognizable also is the other characteristic species of the plains, *P. aristata,* which is distinguished by long aristate bracts subtending the flowers. *P. lanceolata* is a common weed. In some of the plantains the lobes of the corolla are spreading or reflexed, in others erect and closing more or less over the fruit. This character, associated with the shape of the leaves, the shape and mode of dehiscence of the capsule, and the number of stamens and seeds, is useful in classification or in keying of the species. The plants are found in a variety of habitats, from wet to dry. (Fig. 386.)

Litorella uniflora, the only representative of its genus in our flora, resembles some of the narrow-leaved plantains but differs in producing unisexual and monœcious flowers with spreading corolla-lobes, and a small nut-like fruit. The plant occurs along the shores of lakes and ponds in the Great Lakes and St. Lawrence basins. Only one other species is known and that is found in the southern hemisphere—in distant antarctic South America! *Bougueria* has a solitary species in Colombia and Peru—an herb with polygamous flowers in spikes.

There exists considerable doubt as to the proper affinities of the *Plantaginaceæ,* and it may be suggested, in passing, that the floral characters be compared with those of *Veronica* in the *Scrophulariaceæ.*

CHAPTER XVII

THE EPIGYNOUS TETRACYCLICÆ

Tetracyclicæ with Inferior Ovary. Inferæ

ORDER RUBIALES

Flowers epigynous, regular or irregular, bisexual or sometimes unisexual, 5-merous or 4-merous. Stamens free. Carpels 2 or in some cases 1 or several (7—8). Ovary inferior, with as many cells as carpels, and 1 to several ovules in each cell. Endosperm present.

The few families may be grouped according to the following characters:

I. Stamens as many as the lobes of the corolla.
  a. Leaves opposite and stipulate, or whorled and destitute of stipules. *Rubiaceæ.*
  b. Leaves opposite or perfoliate, never stipulate. *Caprifoliaceæ.*
II. Stamens twice as many as the lobes of the corolla. Low herb. *Adoxaceæ.*
III. Stamens fewer than the lobes of the corolla (1—3).
  a. Ovary 3-celled, with two cells empty. *Valerianaceæ.*
  b. Ovary 1-celled. Flowers in dense involucrate heads. *Dipsacaceæ.*

RUBIACEÆ (*Madder Family*). Flowers regular or nearly so, epigynous, generally bisexual, or unisexual by abortion, often with rudiments of the other sex, often dimorphic or trimorphic, the staminate in clusters, the carpellate solitary and terminal, 4-merous or 5-merous, rarely pleiomerous but with reduction in the gynœcium. Calyx regular or irregular, persistent and much reduced. Corolla funnel-form, tubular, campanulate, or rotate, 4—5-lobed, glabrous or hairy in the throat. Stamens as many as the lobes of the corolla and alternate with them, rarely fewer, sometimes united at the base and inserted at different levels in the corolla-tube; anthers elongate, splitting lengthwise, 2-celled. Gynœcium 2-carpellate, with short to elongate slender style, sometimes heterostylous, with capitate to branched stigma. Ovary inferior. Commonly 2-celled, sometimes 1- to several-celled, with 1 to several ovules in each cell on axile or intruded placentas of different shapes. Fruit a loculicidal or septicidal capsule, berry, or drupe, generally exceedingly diverse. Seeds with membranous or crustaceous seed-coat or sometimes winged. Embryo straight or curved, at the base of fleshy, horny, or cartilaginous endosperm.

Trees, shrubs, and some herbs, with entire, opposite or whorled leaves and interpetiolar, sometimes leaf-like, stipules, or with interpetiolar lines. Genera 340—380, species 4,500—6,000, according to different writers. A considerable number of genera are monotypic and relatively many are rich in species, these running into the hundreds. The family is important as yielding dyes, medicinal and aromatic compounds, edible fruits, and ornamentals. Among the most useful plants of the family are varieties of *Coffea,* which yield the coffee of commerce, and species of *Cinchona,* which yield quinine or Peruvian bark. *Rubia tinctorium* yields the dye alazarine, a brilliant red pigment used by artists. Horticulturals are found among *Galium, Asperula, Coprosma, Rubia, Serissa, Cinchona, Genipa,* and others. *Coprosma Baueri,* a handsome shrub with shiny leaves, is a common ornamental in California. The leaves are furnished with peculiar open pits in the angles of the veins along the midrib on the lower side (Figs. 387, 388). An interesting symbiotic relationship with ants (myrmecophily) is found among some genera. There is a notable diversity as to habit, expressed in the many herbaceous and woody climbers, aside from trees and shrubs; ordinary herbs are relatively few.

Subfamilies two: *Coffeoideæ* and *Cinchonoideæ,* distinguished by the number of ovules. The grouping of our genera according to these subfamilies is shown in the accompanying key.

I. *Coffeoideæ*. Ovules solitary in each cell.
- Herbs. Leaves in whorls. *Galium, Sherardia, Asperula.*
  - Leaves opposite, never in whorls.
    - Flowers separate in the axils and fruit dry. *Spermacoce, Diodia.*
    - Flowers twin, the ovaries united and forming a twin-berry. *Mitchella.*
- Shrubs or small trees. Flowers in dense globular heads. *Cephalanthus.*

II. *Cinchonoideæ*. Ovules numerous in each cell. Leaves opposite.
- Top of the ovary free from the receptacle and projecting. Low herbs. *Houstonia.*
- Ovary wholly immersed in the receptacle. Herbs. *Oldenlandia.*

Galium (Bedstraw, Cleavers), with 200—250 species of annual or perennial herbs, is distributed over all the continents but is especially well represented in North America and Europe. There are about 55—60 species with us. These are erect to straggling herbs with quadrangular stems (in some cases prickly on the angles) growing in different kinds of habitats. The "leaves" appear whorled owing to the size of the stipules, but the true leaves are only those with branches in their axils. The flowers are very small and of various colors (white in many), with calyx segments reduced to minute teeth

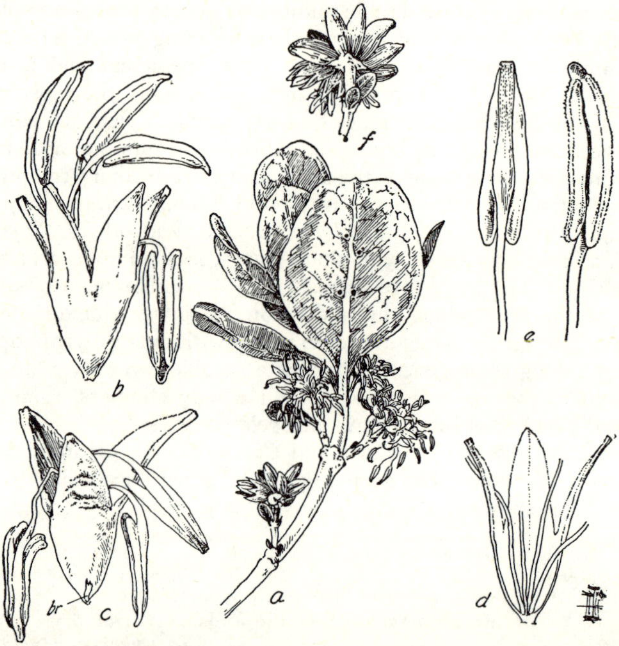

FIG. 387.—*Coprosma Baueri: a,* branch with staminate flowers; *b,* a tetramerous, *c,* a trimerous staminate flower; *d,* section of *b,*—anthers removed; *e,* stamens.

or suppressed, and the corolla with 4 spreading lobes. Stamens 4 or sometimes 3, with short filaments and small, introrse anthers. The 2-carpellate gynœcium is furnished with a 2-branched style and capitate stigmas. Around the base of the style is a fleshy, annular, continuous or lobed disk, the shape of which is of some taxonomic value. (Compare with *Umbelliferæ.*) The (spurious) fruit is very small and often clothed with hooked prickles; at maturity the twin (didymous) carpels separate in the manner of a schizocarp as indehiscent mericarps (compare with *Umbelliferæ*). In some cases only one of the carpels matures. On the whole the species are rather technical. One of the commonest bedstraws is the erect, profusely white-flowered *Galium boreale* (Northern Bedstraw), with prickly fruits. As a garden plant it is quite showy. *Galium circæzans* (Wild Licorice) is readily recognized by its unusually large and broad "leaves," which are in 4's; its fruit is hispid rather than prickly. *G. trifidum* and *G. Claytoni* are

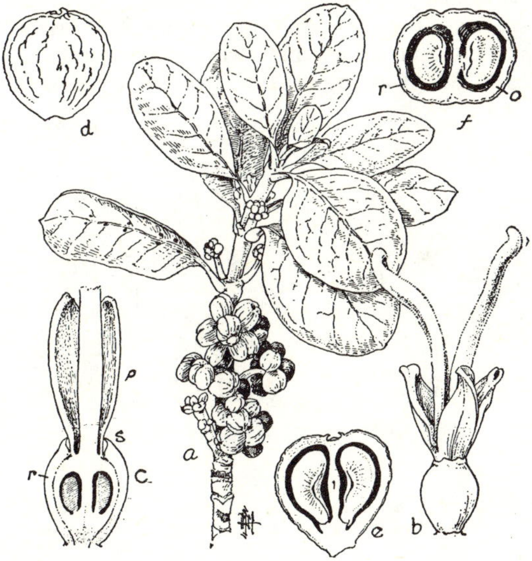

FIG. 388.—*Coprosma Baueri: a,* branch with carpellate flower-buds (above) and fruits (below); *b,* flower, in section at *c; p,* corolla; *s,* calyx; *r,* receptacle; *d,* fruit, calyx at summit; *e,* median, and *f,* transverse, section of ovary; endocarp stony; *o,* ovary.

weak, straggling, profusely branched, prickly stemmed species leaning on other plants and forming tangled masses in swampy thickets and woods. (Fig. 389.)

Asperula is very similar to *Galium,* differing chiefly in the shape of the corolla. Our 2 or 3 species are adventives from Europe and have not so far advanced beyond the Mississippi. *Sherardia arvensis* (Field Madder), a solitary European species, occurs here and there in the North Atlantic States. *Spermacoce glabra* (Buttonweed), a stout perennial, is native from Ohio and Illinois southward; its stems are quadrangular and the small flowers (with funnel-form corollas) are densely crowded in the axils. The species is said to range south to Uruguay. Another species is found in Central America. *Diodia,* with 35 or more species of diverse habit, is mainly tropical American, but two species, *D. teres* and *D. virginiana,* have penetrated far northward in the United States. The latter possesses a long, slender corolla-tube. The fruits are indehiscent. In *Spermacoce* one carpel is dehiscent and the other indehiscent.

Mitchella repens (Partridge-Berry, Twin-Berry), the only North American species, is a creeping or trailing herb with evergreen, broadly ovate, opposite leaves and twin, white, terminal or axillary flowers, usually found at the base of trees in woodlands of eastern

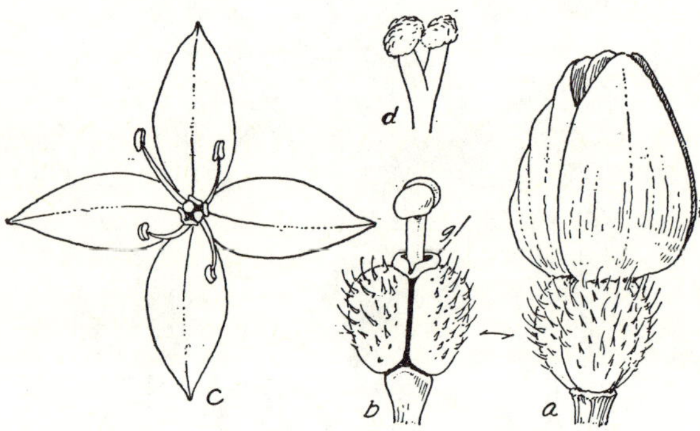

FIG. 389.—*Galium* sp.: *a,* flower-bud; *b,* flower with corolla and stamens removed; *gl,* disk; *c,* face of flower; *d,* stigmas.

North America. The calyx is very small, but the corolla-tube much elongated and funnel-form, with 4 spreading lobes. The style is exserted and terminates in 4 short, spreading, stigmatic branches. The flowers are dimorphic, with united ovaries which ripen into 2 united drupelets (spurious fruits). The only other species occurs in Japan. (Compare with other similar interrupted ranges.) *Houstonia,* with about 25 species ranging from Canada to Mexico, is represented by about 10 species east of the Rocky Mountains. These are mainly low, diffuse, or tufted herbs with generally small, basal, and cauline leaves and dimorphic flowers in few-flowered cymes, the corollas generally long-tubular. *Oldenlandia,* a tropical Asiatic genus, mainly, of about 175 species, is represented eastward, here, and in the West Indies, by a single species, *Oldenlandia uniflora,* a weak, tufted annual of low grounds. The solitary *Kelloggia galioides,* a weak Galium-like diffuse herb, is found on the Pacific Coast.

Cephalanthus, with about 6 species, all but the following Asiatic, comprises shrubs and small trees with simple opposite petiolate leaves and small flowers densely crowded in globular heads—whence the name "Button-Bush" for our species (*Cephalanthus occidentalis*). The flowers have tubular corollas and a long-exserted slender style. The fruit is an obpyramidal nut crowned by the persistent calyx, the whole mass of nuts of the globular head forming an aggregate or multiple fruit. The plant is found in swampy or low grounds from the Atlantic Coast to Arizona and California. It is commonly grown as an ornamental. In *Sarcocephalus* of the tropics the individual fruits are consolidated into one mass. (Compare with the Osage Orange (*Maclura*) and (*Anona*).)

CAPRIFOLIACEÆ (*Honeysuckle Family*). Flowers regular or irregular, bisexual, epigynous, the calyx much reduced—to 3—5 small teeth or bristly, the corolla regular or nearly so or very irregular, campanulate, funnel-form or tubular, often gibbous, 5-lobed or 2-lipped. Stamens 5 (or 4 and didynamous), inserted on the corolla-tube and alternate with its lobes; anthers introrse or sometimes extrorse, elongate, versatile. Gynœcium 2—5-carpellate (rarely more), the style short, elongate, and simple, or styles as many as carpels, with capitate stigmas. Ovary inferior, 1—5-celled, rarely several-celled, with 1—many anatropous ovules in each cell, suspended on axile placentas. Fruit a 1- to several-celled berry, drupe, or capsule (spurious fruits), usually crowned by the persistent calyx-teeth. Seeds of various shapes and sculpturings, with a straight embryo in endosperm of diverse textures.

Trees, shrubs, vines, or perennial herbs, with opposite, compound, or perfoliate, exstipulate leaves, and often small and numerous flowers in cymose-umbellate showy clusters. The plants range over the temperate regions of the northern hemisphere but occur also at higher altitudes in the tropics, with a few in New Zealand. A few yield useful woods, others edible fruits, and many are grown as ornamentals for their flowers and foliage. The family is fairly well represented in our eastern woodlands. Genera 10—11, species 300—350.

The characters of the leaves and the gynœcium and the number of ovules furnish bases for classification into tribes and genera.

A. Leaves compound; anthers extrorse; fruit a drupaceous berry.
Tribe I. *Sambuceæ: Sambucus.*

B. Leaves simple but sometimes lobed; anthers introrse; fruit a berry or drupaceous.
  a. Carpels 1—5, each with 1 ovule.
    Tribe II. *Viburneæ. Viburnum* (shrubs) and *Triosteum* (herbs).
  b. Carpels 2—5—8, each mostly with several ovules.
    1. Carpels 3—4. Tribe III. *Linnæeæ. Symphoricarpos* (4 carpels), *Linnæa* (3 carpels).
    2. Carpels 2—5—8, each with several ovules.
      Tribe IV. *Lonicereæ,* including
      (°). Carpels 2. *Diervilla.*
      (°°). Carpels 2—3. *Lonicera.*
      (°°°). Carpels 5—8. *Leycesteria* (Himalayas).

The shape of the corolla may also be used as a basis for grouping into major categories. Thus *Sambucus* and *Viburnum* are distinguished from all our other genera by their rotate, nearly regular corollas. Differences in habit, leaves, and the fruits, may similarly be used as group distinctions. In the last case *Linnæa* and *Diervilla* will form a dry-fruited group, while the remaining genera will constitute the fleshy-fruited group.

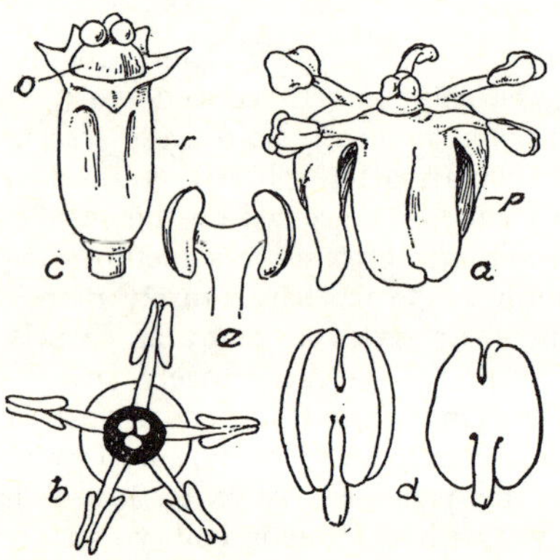

FIG. 390.—*Sambucus racemosus: a,* flower; *p,* corolla; *b,* flower with corolla removed, viewed from above; *c,* developing fruit; *o,* upper part of ovary projecting above receptacle (*r*) and capped by three globose stigmas; *d,* ventral and dorsal aspects of stamens; *e,* stamen with forked connective.

Sambucus (Elder). The distinctive features of *Sambucus* are the compound odd-pinnate leaves and large, flattened to thyrsoid inflorescences of numerous small, white flowers. Our two Eastern species are distinguished by the shapes of the flower-clusters and the color of the fruit. In *S. canadensis,* the inflorescence is broad and flattened and the fruit black to deep purple; in *S. pubens,* the inflorescence is thyrsoid and the fruit red. On the Pacific Coast are found the two tall shrubs or small trees, *S. glauca* and *S. velutina.* The ovary of the commonly 3-carpellate gynœcium rises somewhat above the receptacular portion and is crowned by 3 capitate sessile stigmas. The stamens spread laterally, and the anthers are separated more or less by a thick connective. A spreading to reflexed corolla and the very small calyx-teeth complete the principal distinctions. Species about two dozen, widely distributed; about 10 with us. The stems are weak, with ample pith. (Fig. 390.)

Three species of *Triosteum* range over the eastern parts of the continent. These are stout herbs with narrow to broad, sessile, perfoliate leaves and axillary yellowish-green flowers. The calyx-segments are exceptionally elongate and more or less foliaceous. In *T. perfoliatum* some or all of the leaves are perfoliate, but in *T. aurantiacum*

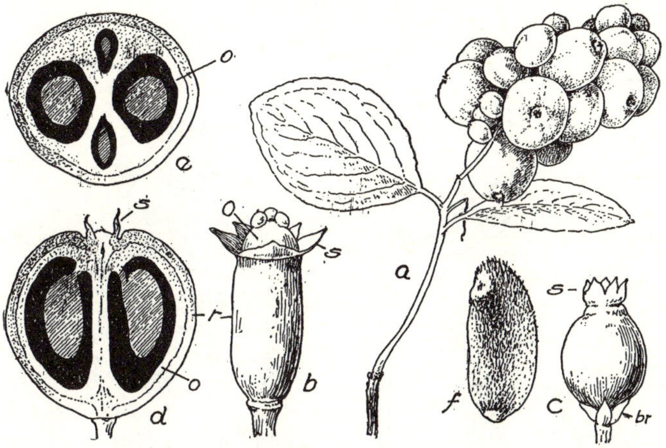

FIG. 391.—*Symphoricarpos racemosus: a,* branch with fruits; *b,* flower after removing corolla and stamens; *c,* young fruit; *d,* median, *e,* transverse section of young fruit; *br,* bracts; *o,* ovary; *r,* receptacle; *s,* calyx-lobes.

and *T. angustifolium* they are merely sessile (narrower in the latter).

Frail and diminutive in contrast with its allies is the slender, creeping *Linnæa borealis* var. *americana,* the Twin-Flower of our cold Northern coniferous woods. It often forms extensive patches, sending up numerous slender, erect, twin-flowered scapes of pinkish, fragrant flowers from lateral branches of the creeping stems. Only 1 of the 3 carpels of the small, globular, viscid fruit produces a seed. The small leaves are somewhat truncated and toothed above and round-tapering below. The few other species are Eurasian.

Symphoricarpos (Partridge-Berry, Snow-Berry), with about 10 species, is exclusively North American, ranging from far northward down into Mexico. The plants are generally low shrubs with broad. oblong-ovate, entire leaves and numerous white to pinkish flowers in the axils of the upper leaves. The berries are white to reddish or purplish, sometimes connate. The commonest are *Symphoricarpos racemosus* and *S. occidentalis*. Three or four others are found on the Pacific Coast. Several are ornamentals. (Fig. 391.)

FIG. 392.—*Diervilla Lonicera: a,* flowering stem; *b,* flower; *s,* calyx-segments; *bl,* bracteoles; *br,* bracts; *p,* pedicel; *c,* stamen and part of corolla; *d,* ventral and dorsal aspects of stamens; *f,* young fruit; *g,* section of young fruit; *h,* transverse section of ovary; *i,* flower-bud; *o,* ovule.

Diervilla Lonicera, the Bush-Honeysuckle, is a ubiquitous undershrub of Northern pine forests. The yellowish, funnel-form corolla is gibbous at the base and the fruit is a distinctive elongate capsule narrowed to a beak upwards and crowned by the persistent bristle-like calyx-segments, rather strikingly foreshadowing the "pappus" of the composites. (Fig. 392).

The 100 or more species of *Viburnum* are trees and shrubs, mainly of North America and Asia, with two or three others occurring in Europe and northern Africa. There are about 20 in our flora. The leaves are simple but sometimes palmately lobed and coarsely serrate, and the numerous small white to pinkish flowers are clustered in rounded to flattened compound cymes. Structurally the flowers

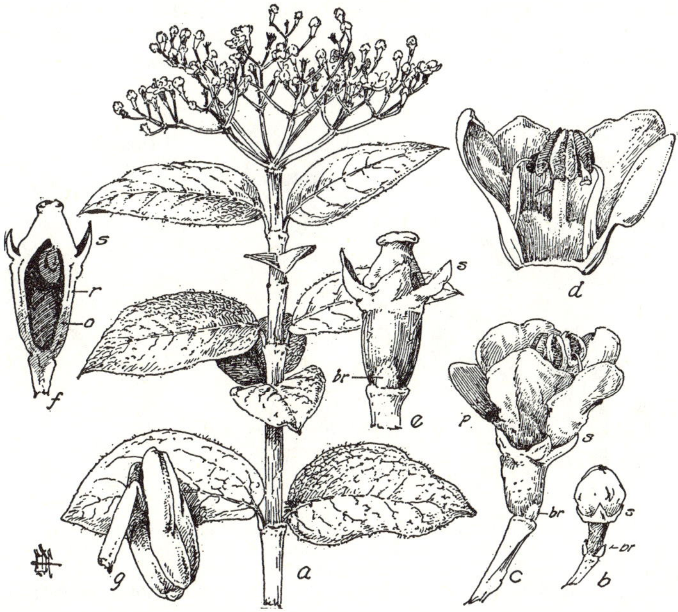

FIG. 393.—*Viburnum Tinus: a,* branch with terminal inflorescence; *b,* flower-bud; *c,* flower; *br,* bract; *s,* calyx; *p,* corolla; *d,* corolla laid open; *e,* flower with corolla removed, and in sectional view at *f; r,* receptacle (Compare with *Cornus*); *g,* stamen.

are very similar to those of *Sambucus* and *Symphoricarpos,* except that the gynœcium is sometimes 1-carpellate (Compare with the dogwoods, *Cornaceæ*). The drupes are red to black, and pulpy or juicy. In *V. alnifolium* (Hobble-Bush, Wayfaring Tree) and *V. trilobum* (*V. Opulus* var *americanum*), the High-Bush Cranberry, the outer flowers of the cluster are neutral and much the larger and function as "ray-flowers," again simulating a condition that we find so perfected in the composites. In common with *V. pauciflorum* and *V. acerifolium,* this species has palmately lobed leaves. Finely serrate leaves occur in *V. Lentago* (Black Haw, Nanny-Berry); coarsely toothed ones in *V. pubescens, V. dentatum,* and others; and entire ones in *V. nudum.* The fruits of the High-Bush Cranberry and the Black Haw are edible and particularly juicy after a heavy frost; they often are found on the bushes in winter. Two dozen or more species of *Viburnum* are ornamentals. (Fig. 393.)

Lonicera (Honeysuckle), with about 160 species, is mainly north-temperate, only a few occurring in the tropics. The plants are erect shrubs, or vines, often with connate-perfoliate leaves and often fragrant flowers.

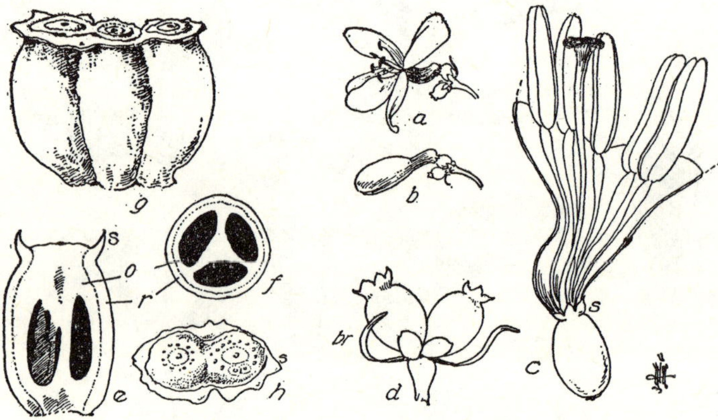

FIG. 394.—*Lonicera tatarica: a,* flower; *b,* flower-bud; *c,* flower with corolla-tube in section; *s,* calyx; *d,* a pair of young fruits, with two sets of bracts (*br*) below; *e,* median, *f,* transverse section of young fruit; *o,* ovary; *r,* receptacle; *g,* a triplet of coalesced fruits; *h,* discoid summit of a twin fruit.

The corolla-tube is often elongate, gibbous, and furnished with a 2-lipped limb. The stamens are borne on the corolla-tube and are exserted. The berries are frequently connate or in pairs. The upper leaves are connate-perfoliate in *Lonicera Caprifolium, L. dioica, L. glaucescens, L. Sullivantii, L. hirsuta, L. flava,* and others, and the flowers in these forms are often in large, showy, terminal clusters. The plants are climbing, trailing, or leaning vines. *L. canadensis, L. tatarica, L. involucrata,* and others, are erect shrubs without connate-perfoliate leaves, and with the flowers in pairs on bracted peduncles in the axils of the uppermost leaves. In the last-named the bracts are greatly enlarged and foliaceous, persisting around the purple berries. This species ranges across the continent northward as far as Alaska and thence south into California and Arizona. Twenty-eight or 30 species are found in our flora, and over 30 are found in cultivation. The Himalayan *Leicesteria* is of special interest because of several-carpellate gynœcium. (Figs. 394, 395.)

ADOXACEÆ. This monotypic family is based on the strange *Adoxa moschatellina,* a low, glabrous, musky-odorous, perennial herb of rich woodlands around the northern hemisphere. It bears opposite, basal, ternately compound leaves and a scape with a pair of similar but smaller leaves, from a scaly, rhizomatous base. The flowers are regular, bisexual, epigynous, with a 2—3-toothed calyx and a 4—6-lobed corolla bearing in its tube, in pairs, 8—12 spreading stamens with short filaments and 1-celled spreading anthers. The gynœcium is 3—5-carpellate, with as many styles, and with 1 ovule in each carpel. The fruit is a small drupaceous berry. Seeds with cartilaginous endosperm. An isolated plant, the exact affinities of which have never been definitely determined, but which in its foliage suggests some crowfoots and in the structure of the flower approaches the saxifrages.

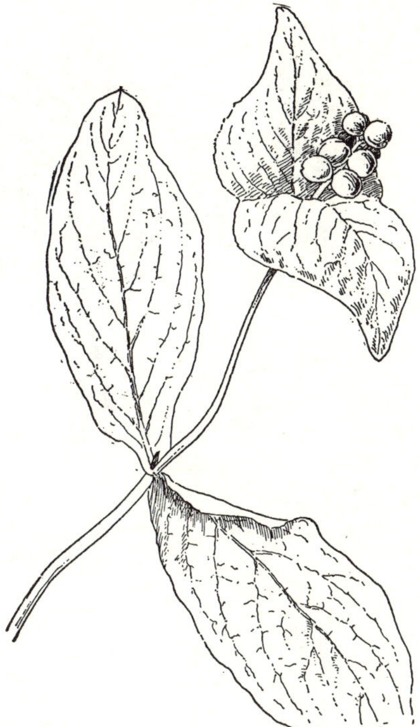

FIG. 395.—Foliage and fruit of *Lonicera dioica.*

VALERIANCEÆ (*Valerian Family*). Flowers irregular, asymmetrical, bisexual or unisexual, polygamous or diœcious, epigynous. Calyx inconspicuous, becoming prominent only in the fruit. Corolla 5- or sometimes 3- or 4-lobed, irregular, spurred or gibbous at the base, with spreading limb. Stamens 1—4, borne on the corolla-tube and alternating with its lobes; anthers introrse. Gynœcium 3-carpellate or with only 1 carpel fertile, with 1 suspended anatropous

FIG. 396.—*Centranthus ruber: a,* inflorescence; *b,* flower-buds and flower; at left (*y*) a transverse section of corolla-tube showing pulpy lining (*p*) of tube investing style; *c,* face of corolla; *d,* throat of corolla showing pulpy investment (*p*) of style, and insertion of stamen; *e,* sides of "ovary" with bract (*br*) and spur of corolla-tube at left; *f,* stigmas; *g,* anthers; *h,* section of ovary; *i,* calyx.

ovule. Style 1; stigma with 1—3 segments. Ovary inferior, 1—3-celled, 2 cells usually empty. Fruit dry (spurious), 1-seeded. Endosperm scanty or none. Embryo straight, with oblong cotyledons.

Herbs with opposite exstipulate leaves, and mostly small flowers in cymes or panicles. About 9 genera and 300 species, mainly of the northern hemisphere. Only two of the genera are represented in our flora, *viz.*, *Valeriana* and *Valerianella*. Species of *Centranthus* and *Fedia* are horticultural. *Plectritis* is represented by 4 species in California. (Fig. 396.)

Over half the species (about 175) are embraced in *Valeriana*. These are widely distributed over the northern hemisphere, with a dozen or more within our limits. The plants are stout, perennial, strong-smelling herbs with simple or pinnately compound leaves and paniculate inflorescences. The calyx is especially distinctive, being modified as 5 to many plumose-bristly segments, very suggestive of the pappus of composites. The corolla-tube is gibbous at the base, and borne within it are 3 slender stamens. The fruit is a cypsela. *Valeriana edulis* and *V. uliginosa* are widespread Eastward. The roots of the former are edible (tobacco-root). *Valerianella*, with about 50 species of the northern hemisphere and a dozen or more with us, differs in having a capitate inflorescence, a calyx reduced to small teeth, a 3-celled ovary, and an annual habit. (Fig. 397.)

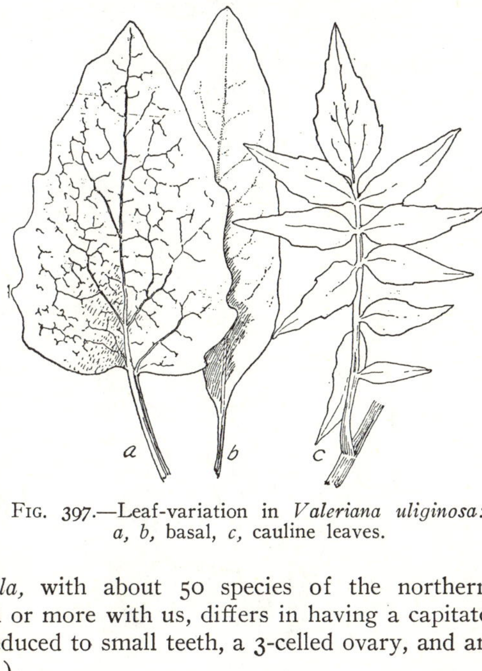

FIG. 397.—Leaf-variation in *Valeriana uliginosa*: *a*, *b*, basal, *c*, cauline leaves.

DIPSACACEÆ (*Teasel Family*). Flowers more or less irregular, bisexual, epigynous. Calyx various, often bristly, surrrounded by a cup-shaped or bristly epicalyx. Corolla irregular, 5-lobed or somewhat 2-lipped, dilated at the throat. Stamens 2—4, inserted on the corolla-tube and alternating with its lobes; anthers introrse, versatile, splitting lengthwise. Gynœcium 2-carpellate, with a filiform style; stigma of 1—2 lobes or segments. Ovary inferior, 1-celled, with 1 pendulous anatropous ovule. Fruit a cypsela, often crowned by the persistent calyx (Compare with composites). Endosperm scanty to copious. Embryo straight, with prominent cotyledons.

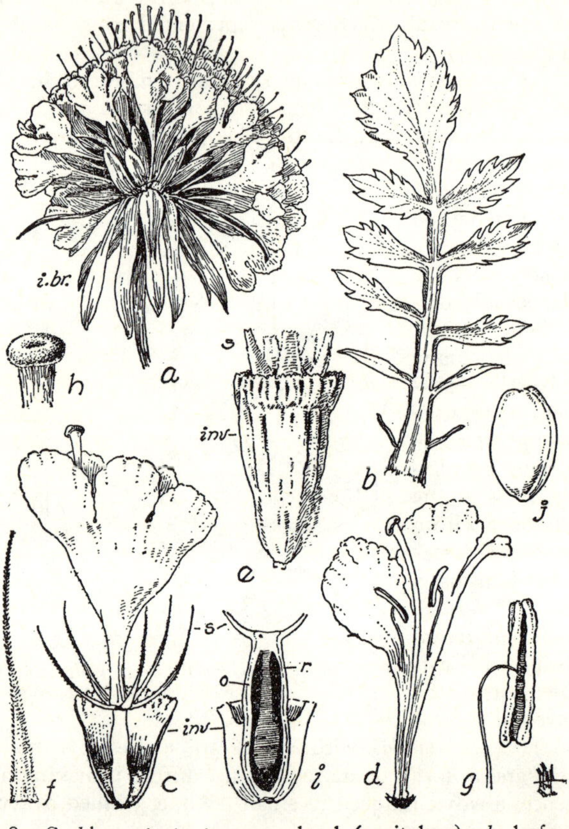

FIG. 398.—*Scabiosa atropurpurea: a,* head (capitulum); *b,* leaf; *c,* flower, seated in involucel (*inv*), with which it is adnate below; *d,* section of corolla; *e,* involucel; *s,* sepals; *f,* sepal enlarged; *g,* stamen; *h,* stigma; *i,* median section of ovary and involucel; *o,* ovary; *r,* receptacle; *j,* ovule; *i, br,* involucral bracts of capitulum.

Annual, biennial, or perennial herbs (but the stems sometimes woody) with opposite to whorled exstipulate leaves, and numerous small flowers densely crowded in large or small, oblong or globose heads subtended by an involucre of conspicuous bracts. Genera 7, species 140—155, mostly Eurasian; only a few in North America and these not native, belonging in *Dipsacus, Succisa, Scabiosa,* and *Knautia,* all represented in flower-gardens along with species of *Cephalaria* and *Morina.* (Fig. 398.)

Dipsacus (Teasel). *D. sylvestris, D. fullonum,* and *D. laciniatus* occur in scattered localities; these are tall, stout, thistle-like plants. They are distinguished by a cup-shaped, 4—5-lobed calyx, a 2-lipped, 4-lobed corolla, and 4 epipetalous stamens. In *Scabiosa* the heads are flattened or globose, the calyx is awn-like or bristly, and the receptacle of the inflorescence lacks scales (floral bracts). Species 70 or more, of the Old World. (Figs. 398, 399.)

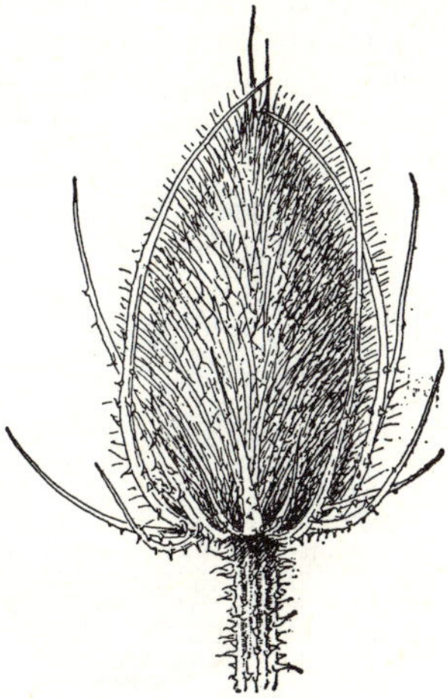

FIG. 399.—Capitulum of *Dipsacus palustris,* in fruiting stage, — with spinescent involucral bracts.

ORDER CUCURBITALES

Flowers regular or sometimes irregular (bilaterally symmetrical), bisexual or unisexual and monœcious or diœcious—with rudiments of the other sex, epigynous, 5-merous (except gynœcium in general). Corolla sometimes choripetalous, plicate-valvate in the bud. Stamens epipetalous, united into a synandrium or in pairs (one odd), or free. Anthers 1-celled in staminate flowers, extrorse, and often curved or S-shaped. Stamens free in *Fevillea.* Gynœcium 3-carpellate, in some cases 1—4—5-carpellate. Ovary completely inferior, with as many cells as carpels. Hypanthium more or less prominent. Ovules numerous on each side of commonly thick, fleshy, bifid placentas which often extend out to the wall of the ovary so as to be bent back, and fill most of the space of the ovary-chambers. Style short and thick, with as many branches or lobes as carpels and these bearing a U-shaped stigma. Fruit a many-seeded (rarely 1-seeded) berry (gourd, pepo). Seeds without endosperm. Embryo straight, with large, flat, foliaceous cotyledons containing considerable oil; hypocotyl very short.

One family, *CUCURBITACEÆ* (Gourd Family), with characters of the order. Genera about 90; species 700—760, mainly tropical but ranging far into temperate regions. The Old World has most of the genera, but most of the species are in the New World; seven of the genera are common to both. The plants are herbaceous, coarse, and mostly rough-pubescent vines with weak stems and watery sap, climbing by means of simple or branched tendrils or trailing over the ground. A few are shrubs, as in *Gurania* of tropical America, and the tree-habit is attained in *Dendrosicyos* of Socotra—in which the branches are clustered at the top of the stem. The tendrils are in certain cases supposed to be modified bracteoles; they are absent in *Ecballium* of the Mediterranean region. A curious leafless, thorny habit is attained by *Acantosicyos,* a plant inhabiting the sand-dunes of southwestern Africa. This genus also develops a huge, fleshy taproot which at times grows to a length of several meters. An enormous water-storing root is developed similarly in certain Californian species of *Echinocystis.*

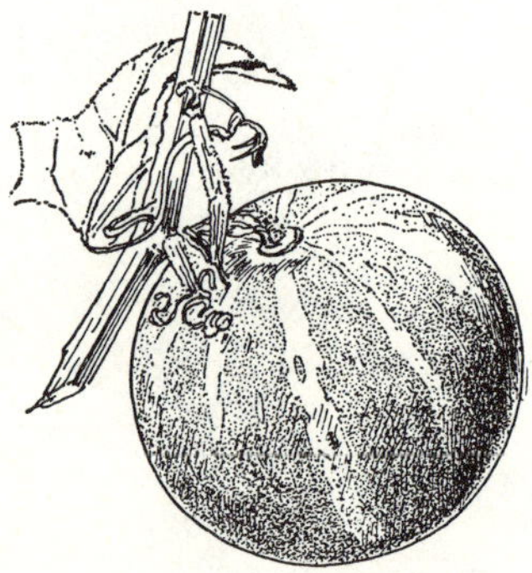

FIG. 400.—Fruit (pepo) of *Cucurbita fœtidissima.*

The leaves of cucurbits are alternate, simple, petiolate, palmately lobed or divided, or deltoid or cordate. In our cultivated forms they are generally very large and coarse. The flowers are small to large, commonly yellow, and solitary or racemose, the staminate generally more numerous than the carpellate (often solitary). Calyx usually small, with narrow segments. Corolla rotate to campanulate or tubular, with prominent lobes in the sympetalous forms. Staminodes are present in the carpellate flowers, and in the staminate flowers there is usually present a rudimentary gynœcium. The fruit in cultivated forms is often developed to enormous proportions, as in pumpkins, squashes, and watermelons—the largest of all known fruits. In some genera the fruit is prickly or spiny (*Echinocystis*); it is rarely dry (*Luffra*). A 1-celled fruit, developed from a 1-carpellate gynœcium, occurs in *Cyclanthera, Sicyos,* and *Sechium* (Fig. 403), and in the two last named this fruit is 1-seeded. In shape the fruits vary from globular to oblong to elongate, or dorsiventrally flattened (some squashes). The seeds are usually flat and ovoid in shape, white to black. The cotyledons are usually epigean and function for a length of time as green leaves. In some members (*Cucurbita*) a peculiar protuberance of the hypocotyl below the cotyledons, called the *peg,* serves to dislodge the testa of the seed as the seedling rises from the ground (Fig. 66). Vivipary occurs in squashes and pumpkins and in *Sechium* (Fig. 403)—that is, the seeds germinate inside the fruit before this disintegrates or before it is detached from the plant (*Sechium*).

The herbaceous cucurbits are mostly annuals. Bilateral symmetry of the flowers occurs in the Asiatic *Thladiantha,* and in this genus as well as in *Fevillea* the stamens are free or united at the base. In *Cyclanthera,* an American genus, the anthers are united into a transverse ring at the summit of the style, the anthers and the style together forming a dumb-bell-shaped or spool-shaped column. In general the configuration of the andrœcium is of taxonomic value in distinguishing the various tribes. A number of species of different genera are cultivated for their edible fruits, *e.g.,* pumpkins, squashes, watermelons, musk-melons, cantaloupes; others are grown as ornamentals or merely as curiosities. A few yield bitter pharmaceutical compounds.

Our genera may be keyed out as follows:

A. Flowers large, yellow. Fruit globular, smooth, rather large. Plants trailing. *Cucurbita.*
B. Flowers small to medium-sized.
  1. Plants climbing or leaning vines.
    a. Fruit prickly.
      Fruit indehiscent. *Sicyos.*
      Fruit bladdery, bursting at top. *Echinocystis.*
      Fruit bursting irregularly. *Cyclanthera.*
    b. Fruit smooth, small. *Melothria.*
  2. Plants trailing. Fruit smooth or somewhat prickly, 1-seeded. *Brandegea.*

Cucurbita fœtidissima (Calabazilla or Fetid Wild Pumpkin), a coarse, trailing plant with triangular-cordate leaves, occurs from Missouri to Texas, southern California and Mexico. "Also called Chili Coyote and used by Spanish Californians, the root as a cleanser, the leaves medicinally."—Jepson. Another species, *C. palmata,* of similar habit but with palmately cleft leaves, is also found in California, south to the Colorado Desert. The corolla in these is campanulate-rotate. The small, orange-sized fruits resemble a watermelon. Among the cultivated forms of *Cucurbita* are *C. pepo* (pumpkin), *C. moschata* (winter crookneck squash), *C. maxima* (winter squash), and *C. ficifolia* (Malabour gourd). (Fig. 400.)

Sicyos angulata (One-Seeded Bur-Cucumber), a viscid, angular-stemmed vine, inhabits river banks in Eastern parts of the continent.

FIG. 401.—*Echinocystis macrocarpa: a,* habit sketch; *b,* staminate flower-bud; *c,* staminate flower; *d,* stamens (*synandrium*); *hy,* hypanthium; *p,* corolla; *e,* stamens after anthesis; *f,* carpellate flower-bud; *g,* carpellate flower, enlarged at *h,*—with large discoid stigma in center; *p,* corolla; *i,* transverse section of fruit (*j*).

The corolla is rotate or campanulate. The 35 species of the genus are American and Australasian. *Echinocystis,* an American genus of about 25 species, is represented in eastern North America by *E. lobata* (Squirting Cucumber, Wild Balsam-Apple) and by about 5 other species on the Pacific Coast, mostly in California, among which are *E. macrocarpa* (Chilicothe), *E. oreganus* (Hill Man-Root), and *E. fabacea* (Common Man-Root). In these the segments of the more or less rotate corolla are parted nearly to the base, or to below the middle, with an evident hypanthium in the staminate flowers. *Cyclanthera dissecta*, one among about 40 American species of this genus, ranges from Kansas to Louisiana and Mexico; the corolla is rotate. A few species of *Melothria* occur in the Eastern and Southern States, and about 65 others in the tropics and subtropics; these are distinguished by a campanulate corolla with deeply parted segments. *Echinocystis lobata* and *Sicyos angulatus* cross the Canadian boundary, thus pushing farther north than any other American members of this family. (Fig. 401.)

Cucumis is represented in cultivation by *C. Melo* (Common Cucumber) and its varieties *C. Melo inodorus* (Cassaba Melon), and *C. Melo Conomon* (Oriental Pickling Melon). The flowers of these have a rotate or campanulate corolla. *Ecballium elaterium* (Squirting Cucumber), a solitary Mediterranean species with rotate-campanulate corolla, is a common horticultural. *Bryonia dioica* is the Bryony of Europe. (Fig. 402.)

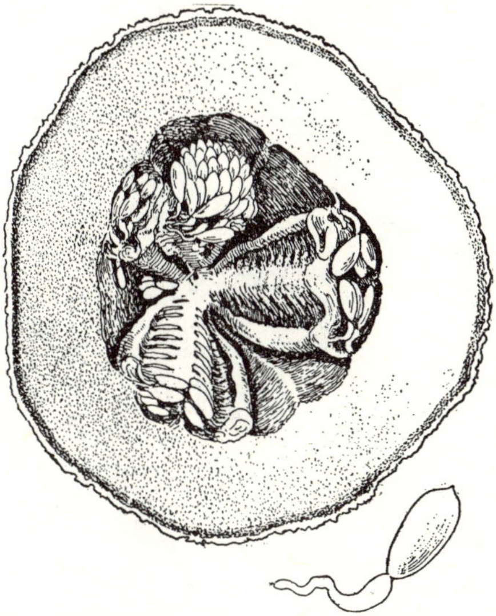

FIG. 402.—Transverse section of fruit of muskmelon (*Cucumis Melo*); at right a seed.

The *Cucurbitaceæ* is by some writers referred to the *Archichlamydeæ*, but in order not to confuse the student the family is here retained in the *Metachlamydeæ*.

ORDER CAMPANULATÆ (CAMPANULALES)

Flowers regular or irregular (bilaterally symmetrical), epigynous, bisexual or sometimes unisexual, mostly 5-merous but sometimes pleiomerous or oligomerous. Calyx in higher groups much reduced, modified, or obsolete. Corolla rarely choripetalous. Filaments free or epipetalous; anthers free or connate into a tube around the style.

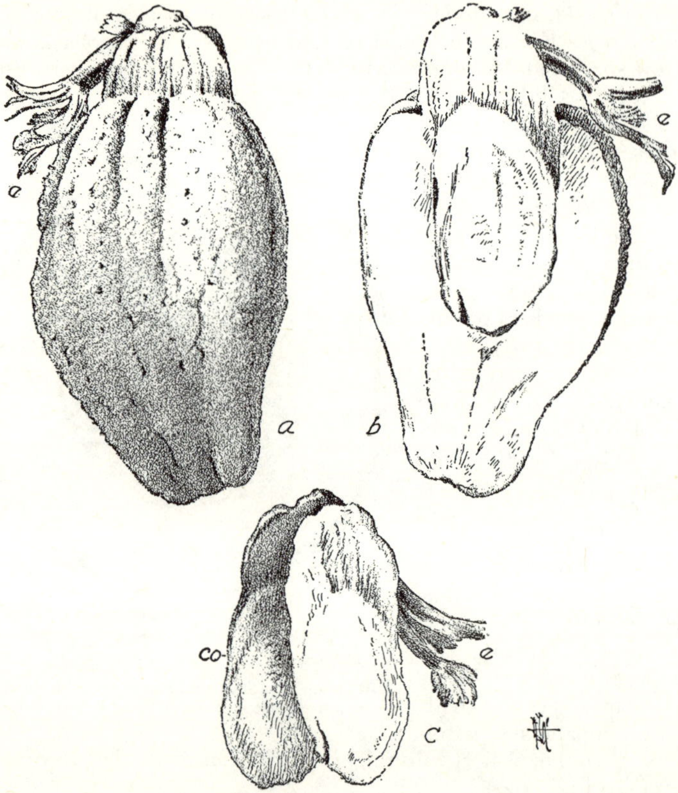

FIG. 403.—Fruit of *Sechium edule* (*a*); in section (*b*); with large germinating embryo above; *e,* embryo; *co,* cotyledon; *e,* plumule.

Gynœcium 5—2-carpellate, rarely pleiomerous. Fruit a capsule or a cypsela, rarely baccate. A tremendous order of about 15,000 species.

Families: *Campanulaceæ, Lobeliaceæ, Goodeniaceæ, Brunoniaceæ, Stylidiaceæ, Calyceraceæ, Compositæ.*

KEY TO THE FAMILIFS

A. Flowers in heads surrounded by a common involucre.
  a. Ray-flowers mostly present. Ovary 1-celled, with 1 erect ovule. Fruit a cypsela. Calyx mostly of scaly segments, or a pappus. *Compositæ.*
  b. No ray-flowers.
    Fruit a nutlet, invested by the persistent hardened calyx, with 1 basally attached seed. *Brunoniaceæ.*
    Fruit a cypsela (achene), with 1 pendulous seed. *Calyceraceæ.*
B. Flowers not in heads nor surrounded by a common involucre.
  a. Stamens 5.
    1. Flowers regular. Anthers mostly free. *Campanulaceæ.*
    2. Flowers irregular.
      Ovules numerous in each cell. Anthers united into a tube around the style. *Lobeliaceæ.*
      Ovules 1—4 in each cell. Anthers free, or united around the style. *Goodeniaceæ.*
  b. Stamens 2—3, free or united to style. Carpels 2. *Stylidiaceæ.*

CAMPANULACEÆ (*Bellflower Family*). Flowers regular, bisexual, epigynous, mostly 5-merous, rarely 3-, 4-, 6- or 10-merous. Calyx persistent, much reduced, the segments mostly narrow to awl-shaped. Corolla campanulate to tubular or rotate, the lobes valvate in the bud. Stamens as many as the lobes of the corolla, inserted on its tube and alternating with its lobes; filaments free on the disk or epipetalous; anthers free or connate, 2-celled, introrse. Gynœcium 5—2-carpellate, the carpels when 5 alternating with the stamens; when 2, then median; when 3, the odd one posterior. Style simple, mostly elongate and slender, with as many stigmatic branches above as there are carpels. Surrounding the base of the style there is usually an epigynous, saucer-shaped to tubular or flattened, convex or concave disk. Ovary more or less inferior, the top free, with as many cells as carpels. Ovules numerous, anatropous, on thick, axile, projecting or stalked placentas. Fruit a capsule, rarely baccate, crowned by the persistent calyx-segments, opening at the top, on the sides or at the base by valves, chinks, or pores. Seeds with minute embryo in fleshy endosperm.

Annual or perennial herbs, subshrubs, or in the tropics shrubs or trees, generally with milky sap and with simple, usually alternate, exstipulate leaves. Flowers small to large and showy, in cymose, paniculate, or racemose clusters. Genera 40—61; species 1,000—1,500, mostly of temperate regions but ranging scantily into the subtropics and tropics. They are plants of diverse habitats but preferring mostly higher elevations. Some press far northward—to Greenland, Aleutian Islands, and Lapland. They are of slight value except as ornamentals or horticulturals. We have the following genera:

A. Capsule dehiscing by lateral pores or valves.
  a. Corolla campanulate to rotate.
    Flowers all of one kind. Calyx-segments small and narrow. *Campanula.*
    Flowers of two kinds, ordinary and cleistogamous, calyx-segments foliaceous. *Heterocodon.*
  b. Corolla rotate; flowers of two kinds. *Specularia.*
B. Capsule opening by valves or pores at the top.
    Flowers in dense involucrate heads. *Jasione.*
    Flowers terminal on the branches. *Githopsis.*

Campanula (Harebell, Bellflower). This large genus—250 species—is spread over the northern hemisphere but is best developed in the Mediterranean region. A dozen or more species are native to our flora; two dozen or more are horticultural. The plants are annuals or perennials with alternate, basal, or cauline leaves, the latter mostly of the lanceolate or linear type. The plants bear small to medium-sized, blue, violet, or white flowers, solitary or generally in loose to compact clusters. In our species the gynœcium is generally 3-carpellate, with a correspondingly 3-branched stigma on a long or short style. The more or less obpyramidal capsule opens at the top, on the sides, or at the base. The commonest and most familiar species is *Campanula rotundifolia* (Harebell or Bluebell) which encircles the boreal regions of the world and is particularly common throughout the St. Lawrence and Great Lakes basins. The species produces dimorphic leaves, an early set in a basal rosette and strongly suggestive of the leaves of some violets, while the flowering stems which arise from the rosette bear linear leaves. Two weak stragglers, *C. aparinoides* and *C. uliginosa,* which have smaller flowers and narrow leaves, are common in Northern swamps and meadows. Our largest and stoutest species, *C. americana,* which often grows to six feet in height, occurs in Eastern parts of the country. (Fig. 404.)

Specularia (Venus's Looking-Glass), with 10 species scattered over the northern hemisphere, and of which there are less than a half-dozen in our flora, differs from *Campanula* in producing two kinds of flowers—an early set of cleistogamous flowers, followed later by a set of the typical form; in addition, at least in ours, the capsule is more elongate. In certain cases, too, there are but three or four segments to the calyx, or the corolla is 4-lobed. Besides, a considerable range of variation is seen in the 2-, 3-, or 4-carpellate gynœcium. Our most distinctive species is *Specularia perfoliata,* which is easily recognized by its somewhat obliquely funnel-shaped, clasping leaves. It is widely distributed over the continent and is said to occur also in the mountains of Jamaica and of Santo Domingo. *Jasione montana* (Sheep's-Bit) occurs as an adventive from Europe

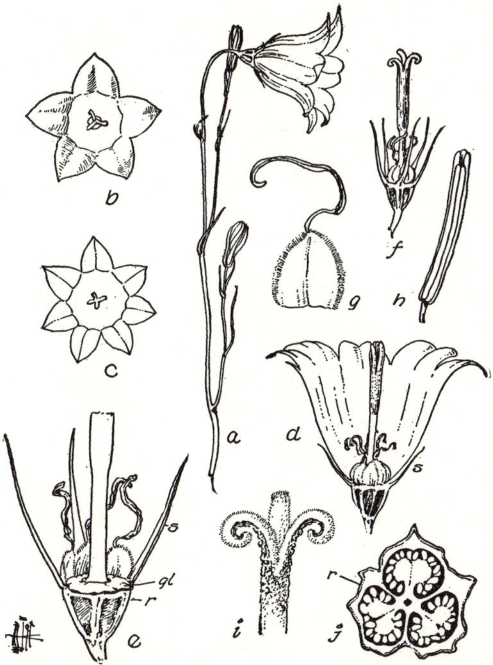

FIG. 404.—*Campanula rotundifolia: a,* stem with flower and flower-buds; *b,* face of a corolla with five corolla-lobes, and (*c*) with seven corolla-lobes; *d,* median section of flower; *e,* flower with several parts removed,—*s,* sepal; *gl,* disk; *r,* receptacle; *f,* flower with corolla removed; *g,* stamen; *h,* anther; *i,* stigma; *j,* transverse section of ovary; *r,* receptacle.

in waste grounds in the Atlantic States. *Heterocodon* and *Githopsis* are represented on the Pacific Coast. *Platycodon* (Balloon-Flower), *Symphyandra,* and *Adenophora* are represented in flower-gardens. (Fig. 405.)

Some members of the family produce tuberous roots or a fleshy tap-root (*Campanula*), others rhizomes and rosette-leaves. In some species the sepals are swollen or bear lateral appendages (*Campanula, Platycodon,* Figs. 403, 404). Irregular flowers and a vine-like habit are characteristic of the African *Cyphia,* and in *Pentaphragma* of the East Indies the inflorescence is scorpioid. In the Eurasian *Phyteuma* the flowers tend to become aggregated in involucrate heads.

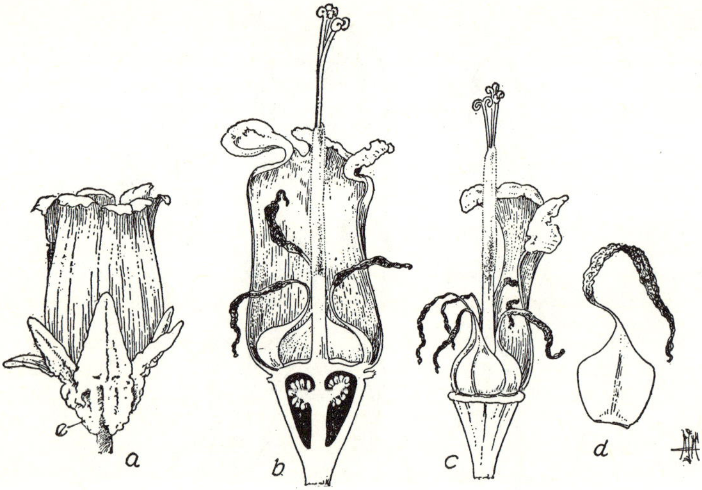

FIG. 405.—*Platycodon grandiflorum: a,* flower; *e,* auricles of calyx; *b,* median section of flower; *c,* flower showing andrœcium, style, and a section of the corolla; *d,* stamen.

LOBELIACEÆ (*Lobelia Family*). Flowers irregular, epigynous, bisexual or unisexual, mostly 5-merous, resupinate. Calyx regular or irregular, persistent, the lobes unequal, with a tendency to become bilabiate. Corolla irregular, bilabiate or unilabiate, tubular, often with considerably elongated tube, split on one side often nearly to the base, more or less deeply lobed, and in some cases the lobes nearly distinct. Stamens epipetalous or free, the filaments sometimes united into a tube; anthers mostly united into a tube around the style, often furnished with an apical brush of hairs; three of the anthers sometimes larger than the other two. Gynœcium mostly 2-carpellate, in some cases 3- or 5-carpellate. Style simple, elongate, terminating in a fringed stigma of as many lobes as carpels. Ovary inferior, with as many cells as carpels; ovules numerous in each cell, anatropous, on axile placentas. The spurious fruit a capsule opening at the top or sometimes on the sides, in some cases a berry. Seeds small, smooth or furrowed, with a small straight embryo in copious endosperm.

Herbs, shrubs, or trees, often with milky sap, containing a narcotic poison, and bearing alternate, exstipulate, simple or pinnately divided leaves and solitary, paniculate, or racemose flowers. About 20 genera and 600 species, mostly of the tropics and subtropics but well represented in north temperate regions. Tree forms occur in the Sandwich Islands. The family has a limited representation in flower-gardens. A few species yield narcotic compounds. The accompanying key, taken from Jepson, will serve as a guide to our few genera.

Anthers distinct. Flowers minute.
  Flowers solitary on filiform pedicels borne on zigzag, erect branches. *Nemacladus.*
  Flowers subsessile in capitate clusters. Stems prostrate. *Parishella.*
Anthers united and forming a tube. Flowers larger.
  Corolla-tube cleft to the base on one side. *Lobelia.*
  Corolla-tube not at all cleft.
    Filaments distinct below. Corolla-tube several times longer than the limb. *Palmerella.*
    Filaments united.
      Tube of the corolla equaling or nearly as long as the limb; ovary thick (oblong or conical). *Laurentia.*
      Tube of the corolla equalling or nearly as long as the limb; ovary long and slender (stalk-like).
        Corolla conspicuously 2-lipped. *Downingia.*
        Corolla segments only slightly unequal. *Howellia.*

Lobelia (Lobelia). Species about 250, widely distributed, with 30 or more in our flora. The lobelias are generally tall and stout herbs with alternate, broadly to narrowly lanceolate, cauline, and basal leaves, and bearing small to large and showy flowers in paniculate, racemose, spicate, and commonly leafy-bracted inflorescences. From the 2-carpellate gynœcium is developed a loculicidally 2-valved capsule. Some species, such as *Lobelia Dortmanna* and *L. paludosa,* inhabit the borders of small lakes or ponds. The former is also found in Europe; in this country it ranges from Newfoundland to Washington and British Columbia. The tall and stout perennials, *L. syphilitica* and *L. cardinalis* (with scarlet flowers), are found from the Atlantic Coast to the Rocky Mountains. The handsome but smaller *L. splendens,* with brilliant red corolla, is found in southern California. The other genera are rather sparingly represented in southern California and adjacent regions.

GOODENIACEÆ (*Goodenia Family*). Flowers mostly irregular, hypogynous or epigynous, bisexual, 5-merous. Calyx small, in some cases absent. Corolla bilabiate or unilabiate, with valvate or induplicate lobes. Stamens free or epipetalous; anthers introrse, free or connivent around the style. Gynœcium 2—1-carpellate. Style simple, commonly with as many stigmatic lobes as carpels and furnished with a peculiar 2-lobed, cup-shaped pollen-receptacle ("indusium") beneath the stigma. Ovary superior or more or less inferior, 2-celled or rarely 1-celled, with 1—2 or numerous, ascending, or pendulous ovules in each cell. Fruit a capsule, drupe, or nut. Seeds small and flat, with a small, straight embryo, mostly in fleshy endosperm.

Annual or perennial herbs, subshrubs or shrubs, without milky sap. Leaves alternate, simple, sometimes lobed, exstipulate. Genera about 11; species about 300, mainly tropical and Australian, with some in Asia, Africa, and tropical America. Useful mainly for their bitter-tonic properties. Largest genera: *Goodenia* (70 or more species), *Scævola* (60), *Dampiera* (34 or more), *Deschnaultia* (30 or more). In *Villeia* the ovary is superior or nearly so.

BRUNONIACEÆ. Based on the solitary *Brunonia australis,* a perennial, silky-hairy herb of Australia and Tasmania. Leaves in a basal rosette, and the small, blue flowers are crowded into a scapose involucrate head. Flowers almost regular, bisexual, 5-merous, hypogynous (or epigynous?). Calyx-segments small and awl-shaped. Corolla nearly regular, with long funnel-form tube and narrow, spreading, involute lobes. Stamens free or united above; anthers connate into a tube around the style. Gynœcium 1-carpellate (!) with a simple style. Stigma as in preceding family. Ovary superior, 1-celled (!), with a solitary basal ovule. Fruit a small nut invested by the persistent calyx, the segments of which become spinous and efficacious in securing dissemination.

The outstanding features of this family are the 1-carpellate gynœcium with its 1-celled, 1-ovuled ovary, and the nut-like fruit.

STYLIDIACEÆ (*Stylidium Family*). Flowers mostly irregular, rarely regular, bisexual or unisexual, epigynous. Calyx-lobes free or connate or bilabiate. Corolla-lobes imbricate, mostly unequal, the lower forming a lip. Stamens 2 or 3; filaments united, forming a tube around the style; anthers extrorse, dehiscing transversely. An epigynous disk sometimes present. Gynœcium 2-carpellate, with simple style. Ovary inferior, 2-celled, or 1-celled at the base, with numerous ovules in each cell on axile or basal columnar placentas.

Fruit capsular, spurious, septicidal, or indehiscent. Seeds small, with small embryo in fleshy endosperm.

Herbs or shrubs with numerous basal or cauline and fascicled, or closely crowded, spirally arranged, exstipulate or scaly-stipulate leaves. Flowers small, solitary or in racemes or corymbs. Four or 5 genera and about 160 species. Largest genus: *Stylidium,* with 100 or more species. Other genera are *Phyllachne, Levenhookia.* The family is Australian.

CALYCERACEÆ. Flowers regular or irregular, bisexual or unisexual, epigynous, 4-merous to 6-merous. Calyx reduced to very small segments, persistent. Corolla mostly with an elongate slender tube and spreading lobes. Stamens as many as the lobes of the corolla; filiments more or less united; anthers connate, forming a tube around the style, 2-celled, dehiscing lengthwise. Gynœcium 1-carpellate (!), with long, slender style. Ovary inferior, 1-celled, with a solitary pendulous ovule. The fruits are cypselas which sometimes become connate, often crowned by the persistent spreading and hardened calyx-segments. Embryo small, straight, in scanty or copious endosperm.

Annual or perennial herbs with alternate, entire or pinnately lobed, basal or cauline leaves. Flowers small and numerous, crowded in composite-like heads subtended by involucral bracts. Three genera and about 24 species, all South American. *Boöpis, Calycera, Acicarpha.* Here as in the *Brunoniaceæ* the outstanding features are the 1-carpellate gynœcium with 1-celled, 1-ovulate ovary, connate anthers, and the composite-like inflorescence.

COMPOSITÆ (*Composite Family*). Flowers regular or irregular, epigynous, bisexual or unisexual, monœcious or diœcious or neutral, typically 5-merous. Calyx very diverse, greatly reduced or modified as scales or bristles (pappus), persistent or deciduous, or in some cases wanting. Corolla regular and then tubular-campanulate, funnel-form, or vasiform, often much constricted below, the base of the tube forming a nectary, or irregular and split on one side and strap-shaped or ligulate, commonly with a short tube, or rarely bilabiate. Stamens as many as the lobes of the corolla and alternating with them, rarely fewer, epipetalous; filaments free; anthers 2-celled, linear, introrse, connate into a tube around the style, the connective more or less prolonged. Gynœcium 2-carpellate, the carpels median—one in front and one in the rear. Style simple, elongate, cleft above into 2 (rarely more) spreading-recurving stigmatic branches which are papillate or brush-like on the inner side, or the entire stigma sometimes capitate or brush-like. Ovary completely inferior, 1-celled, with 1 anatropous ovule arising from the base. Receptacle more or less prolonged above the ovary as a solid neck or stipe fused with the style and supporting the calyx, corolla, and stamens. Fruit a cypsela (achene) which is generally small and often crowned by the calyx-teeth or the pappus. Testa of the solitary seed consolidated with the ovary-wall. Embryo varying in size with the fruit, straight, with prominent, flat, oblong cotyledons and short hypocotyl.

The composites are annual, biennial, or perennial herbs of diverse habit, or subshrubs, shrubs, or trees, with milky sap in one tribe. Leaves alternate or opposite, simple or compound, sometimes connate-perfoliate, of various shapes and cuttings, often varying on the same plant. Vegetative propagation by tuberous roots or tuberiferous or non-tuberiferous rhizomes is general in biennial or perennial forms. A limited number are aquatics or occupants of wet grounds (*Bidens, Cotula*). The majority occupy moist to arid habitats. Basal rosette-leaves and stout tap-roots are characteristic on biennials. Many are aromatic.

The generally small and numerous flowers are arranged in heads (*capitula*) subtended by an *involucre* of few to numerous bracts of diverse shapes and textures, often foliaceous. Morphologically the capitulum is an indeterminate inflorescence which has become compacted by the extreme shortening of the axis. In an ecological sense the capitulum is analogous to a single flower. The shortened axis is commonly called the "common receptacle" or *disk;* it varies in shape according to the genus or even species, but is commonly more or less flattened or concave or convex to conical. In most composites the heads or capitula bear two kinds of flowers, namely *ray-flowers* and *disk-flowers.* The former when present occupy the margin of the capitulum, next to the involucral bracts; the latter occupy the main body of the common receptacle or disk. The disk-flowers are often very numerous, as in the common garden sunflower, where the number runs into the hundreds. The disk-flowers, too, are regular, while the ray-flowers are irregular, with ligulate corollas. In some cases the ray-flowers are absent, but in the *Cichorieæ* all the flowers are ligulate. Compound heads, composed of a number of one-flowered smaller heads, occur in *Echinops* (one of the thistles). The capitula of composites are described as *radiate* or *discoid* according to the presence or absence of ray-flowers, respectively. In all composites the flowers are arranged spirally in the heads, and often each flower is subtended by a more or less conspicuous (floral) bract which invests the flower. These bracts constitute the *chaff* of the disk, and when this chaff is absent the disk is said to be *naked*. Thus the presence or absence of the chaff is of taxonomic value, as are also any special features it may possess. (See *Helianthus, Coreopsis, Aster.*) Flowering proceeds from the margin of the capitulum to the center, *i.e.,* in centripetal or acropetal succession; and thus it often happens that the outermost flowers are far along towards mature fruit while the innermost flowers are still in the bud stage (see garden sunflower). Ray-flowers are often neutral, that is they do not set fruit, since style and stigma are wanting (and stamens as well). The individual flowers of a capitulum are usually called *florets*. When the florets of a capitulum are alike in form and function, the capitulum is described as *homogamous;* when the florets are unlike in these respects, the capitulum is *heterogamous.*

The individual fruits (cypselas) usually drop from the capitula at maturity, but in certain cases they are retained in the capitula, which are then so modified as to act as special organs of dissemination (*Xanthium, Arctium*). In other cases the pappus is specially modified as a parachute or grappling organ, effective in dissemination by wind, water, or animals. In many cases where the fruits fall individually, the bare disk shows a pitted or reticulate surface. In all cases the seed is retained within the fruit. It may be emphasized in passing that in studying the composites it is important to distinguish between the "common receptacle" or *disk* of the capitulum and the receptacle (*floral axis*) of each individual flower—the one is a part of an inflorescence, the other a part of a flower.

The *Compositæ* is one of the largest families of the flowering plants; it embraces in the neighborhood of 13,000 species, distributed among 850 or more genera, some of which are rich in species (Rendle puts the genera at 1,000 and the species at 23,000!). As a whole the family is believed to be of recent evolutionary origin, and probably is still in a state of flux. Most composites are easily recognized as such in the field, but certain genera and especially certain species of particular genera are often too technical for anyone but a specialist. (Note particularly *Aster, Erigeron, Gnaphalium, Anaphalis, Solidago, Antennaria, Senecio.*) Outside of horticultural forms, which are legion, the family is, out of all proportion to its size, of slight economic value. Numerous species occur as noxious weeds. The plants are widely distributed over the world; in our flora they surpass in number all other plants, both as to genera and species, with the possible exception of the *Gramineæ*. They are especially abundant in western

North America. Asters, goldenrods, and sunflowers dominate the autumn colorings of our landscape, while over wide stretches of the Western plains the sagebrushes form the characteristic feature of the vegetation. A single species, *Balsamorrhiza sagittata,* tinges with yellow the hills of eastern Washington and adjacent Idaho and Oregon in early summer, and farther south the arid washes of New Mexico, Arizona, and California are ablaze in autumn with golden yellow composites.

From the evolutionary standpoint the composites constitute a climax group of the flowering plants. The outstanding features are the aggregation of the numerous small flowers into involucrate capitula, the differentiation of the flowers into ray-flowers and disk-flowers, the connate anthers, the 1-celled ovary of the 2-carpellate gynœcium, the brush-like bifid stigmas, the reduction of the ovules to one, the 1-seeded achenaceous fruit, and the peculiarly modified calyx. These characters are so constant throughout the entire family that one can hardly fail to recognize a composite at sight. Ecologically the floral structures as well as the capitula are regarded as a high type of specialization to insure pollination by insects, while the modifications of the calyx and in many cases of the capitula and the involucral bracts, associated with the peculiar type of fruit, are regarded as "adaptations" for insuring effective dissemination. The effectiveness of the various devices for insuring dissemination are well illustrated in such cases as the burdocks, the beggar-ticks, and the chicories (dandelion, prickly-lettuce).

The following key * will give a prospectus of the family as it applies to our flora.

A. Plants without milky sap; heads with disk flowers only (discoid) or with disk flowers and ray-flowers (radiate).
Series I. *TUBULIFLORÆ.*

1. Heads discoid; branches of the style (*i.e.,* stigmas) long and slender, minutely bristly hairy all over; flowers never yellow. Tribe I. *Vernonieæ.*
2. Heads discoid; branches of the style thickened upward or club-shaped, obtuse, minutely and uniformly pubescent; flowers never yellow. Tribe II. *Eupatorieæ.*
3. Heads discoid or radiate; anthers not caudate at base; branches of style in disk-flowers flat, smooth up to where the conspicuous marginal stigmatic lines abruptly terminate, and prolonged above this into a flattened lance-shaped or triangular appendage which is evenly hairy all around or pubescent outside. Receptacle naked. Leaves alternate. Tribe III. *Astereæ.*

* Adapted in part from Hoffman, in *Pflanzenfamilien,* IV, 5, in part from Gray's *Manual.*

4. Heads discoid (radiate only in *Inula*); carpellate flowers mostly filiform and truncate; style-branches with unappendaged obtuse or truncate naked tips. Anthers sagittate, the basal lobes attenuate into tails. Pappus capillary or none.
Tribe IV. *Inuleæ.*
5. Heads radiate or discoid; style-branches truncate or hairy-appendaged. Pappus never capillary, sometimes absent. Receptacle (disk) chaffy. Involucre not scarious. Tribe V. *Heliantheæ.*
6. Heads radiate or discoid. Pappus a row of several chaffy or bristly-dissected scales. Receptacle mostly not chaffy. Involucre hardly at all imbricated, in some cases scarious.
Tribe VI. *Helenieæ.*
7. Heads radiate or discoid. Involucre mostly of dry and scarious imbricated bracts. Pappus a short crown or none. Cypselas (achenes) small. Flowers white. Aromatic plants.
Tribe VII. *Anthemideæ.*
8. Heads radiate or discoid. Involucre little or not at all imbricated, not scarious. Receptacle naked. Anthers tailless. Pappus capillary. Tribe VIII. *Senecioneæ.*
9. Heads discoid, without conspicuous ray-flowers. Involucre strongly imbricated. Anthers caudate or long-appendaged at tip. Style-branches short or united, obtuse, unappendaged, smooth, often with a pubescent ring below. Pappus mostly bristly. Tribe IX. *Cynareæ.*
10. Heads discoid. Corolla bilabiate, upper lip 2-lobed, lower lip 3-lobed and elongate in marginal flowers. Anthers long-caudate at base. Style-branches long, unappendaged in bisexual flowers. Pappus of capillary bristles. Tribe X. *Mutisiæ.*

B. Plants with milky sap; all the flowers of the heads ray-flowers.
Series II. *LIGULIFLORÆ.* Tribe XI. *Cichorieæ.*

Some of the commoner genera of these tribes are:

Vernonieæ:
Vernonia, Elephantopus.

Eupatorieæ:
Eupatorium, Mikania, Brickellia, Kuhnia, Liatris.

Astereæ:
Grindelia, Chrysopsis, Solidago, Bellis, Aster, Erigeron, Baccharis.

Inuleæ:
Pluchea, Antennaria, Gnaphalium, Anaphalis, Inula, Adenocaulon.

Heliantheæ:
Silphium, Parthenium, Iva, Ambrosia, Xanthium, Heliopsis, Rudbeckia, Brauneria, Lepachys, Helianthus, Encelia, Coreopsis, Bidens, Galinsoga, Dahlia.

Helenieæ:
Hymenopappus, Helenium, Gaillardia.

Anthemideæ:

Achillea, Anthemis, Matricaria, Chrysanthemum, Tanacetum, Cotula, Artemisia.

Senecioneæ:

Tussilago, Petasites, Arnica, Senecio.

Cynareæ:

Arctium, Echinops, Carduus, Cirsium, Onopordum, Silybum, Centaurea, Cnicus.

Cichorieæ:

Lapsana, Cichorium, Krigia, Hypochœris, Leontodon, Tragopogon, Taraxacum, Sonchus, Lactuca, Lygodesmia, Agoseris, Crepis, Prenanthes, Hieracium.

Of these numerous genera only a few representative types of each tribe need be discussed.

Tribe I. *Vernonieæ: Vernonia* (Ironweed). These are tall (5—12 ft.), robust, perennial herbs, generally with alternate, lanceolate leaves, bearing small to medium-sized discoid heads in cymose-paniculate inflorescences. The pappus is double, consisting of an inner series of rough capillary bristles and an outer series of scales or short bristles. The receptacle (disk) is flat and naked. The fruits are 8—10-ribbed and truncate. A common species in the Central States is *Vernonia fasciculata.* The genus is one of the largest in the family, numbering around 500 species—most abundant in South America, but otherwise widely distributed. In *Elephantopus* the small, few-flowered heads are grouped in glomerules, each glomerule subtended by large leaf-like bracts.

Tribe II. *Eupatorieæ: Eupatorium* (Thoroughwort). The thoroughworts are also tall (1—8 ft.) perennials, but mostly with opposite or whorled leaves and much smaller and few-flowered heads. The flowers are white, blue, or purple, and they are provided with a pappus of capillary, usually rough bristles arranged in a single series. The cypselas are 5-angled and truncate. The disk is flattened, convex, or conical, and naked. Also a large genus—500 or more species, mostly of the tropics and subtropics. *Mikania* is exceptional in that the plants are twining herbaceous vines. The 150 or more species are native to the American tropics, but *Mikania scandens* pushes north across the Canadian boundary in Ontario. *Kuhnia* is distinguished by an andrœcium of distinct or nearly distinct anthers.

Liatris (*Lacinaria*) (Blazing-Star). This is a genus of handsome perennial herbs of eastern and central North America, growing mostly in dry soils. Species about 35. They are plants with erect, simple stems bearing lanceolate to linear leaves. The small to large, obconical, purple and showy heads are arranged in spicate or racemose, leafy-bracted inflorescences. The involucral bracts are generally large and broad, becoming shorter downwards. The plants are easily recognized in the field. They produce globose or oblong corms or "root-crowns." This is a very technical genus, in which the species are difficult to separate.

Tribe III. *Astereæ: Grindelia* (Grindelia). The grindelias are rigid biennials or perennials, with alternate, spinulose-serrate, glutinous leaves—characters which help to distinguish the plants fairly readily, and, besides, the glutinous feature extends also to the involucral bracts. The heads are yellow-flowered, the ray-flowers being fertile, while the disk-flowers are bisexual or staminate ("sterile"). There is a pappus of 2—8 awns or bristles, which are deciduous. The cypselas are short, thick, and 4—5-ribbed. *Grindelia squarrosa* is common throughout the country in waste grounds and pastures and along roadsides. There are about 30 species in the genus, mainly of western North America and western South America.

Solidago (Goldenrod). Mainly a North American genus, with only a scanty representation in Mexico, South America, and Europe. Species about 125. The genus is quite clearly defined, and one experiences little difficulty in recognizing a goldenrod in the field, but the several species are hard to differentiate. The goldenrods are an outstanding element in our autumn-flowering vegetation. The plants are low to tall, leafy-stemmed perennials from rhizomes, with spirally arranged leaves mostly of the narrow, lanceolate type, although broad leaves occur in certain species. The numerous small and commonly yellow heads are relatively few-flowered and are aggregated in showy inflorescences of various forms—paniculate, thyrsoid, wand-like, racemose, capitate, or truncate. The pappus consists of numerous roughish bristles arranged in one or two series. *Solidago serotina,* a widespread species, sometimes grows to eight feet in height. *S. missouriensis* is characteristic of the Western prairies and plains and mountain districts, but also extends far Eastward. *S. ulmifolia,* as the specific name indicates, is distinguished by broad, elm-like leaves. A few species are grown as ornamentals. The genus *Euthamia* of Nuttall is by some writers incorporated with *Solidago. Bellis* is represented in gardens by the common *Bellis perennis* (True or English Daisy).

Aster. Next to the goldenrods the asters are the most abundant composites in our Northern flora. About 200—250 or more species have been described, the great majority being North American. As in the case of *Solidago* the species are very confusing (compare with *Erigeron*). The plants are mainly perennials from rhizomes, with leaves of diverse shapes and sizes. The flowers are blue, violet, purple, or white, and aggregated in loose to compact, racemose, corymbose, or paniculate inflorescences; rarely are the heads solitary. There is a pappus of numerous roughish bristles arranged in one, rarely two, series. *Aster macrophyllus,* a tall, robust, pubescent species, is found in Eastern parts of the country and is especially abundant in the pine districts of the Great Lakes region. One of the tallest and handsomest of our asters is the bright-purple flowered Eastern *A. novæ-angliæ.*

Erigeron (Fleabane) is very similar to *Aster,* but differs technically in the absence of bracts on the peduncles and in the possession of looser and but slightly imbricated involucral bracts. Among the species that are widespread over the country are *Erigeron ramosus, E. acris, E. pulchellus,* and the ubiquitous weed, *E. canadense* (*Leptilon canadense* of some books). *E. uniflorus,* among others, is found on alpine summits and far into arctic regions. *E. compositus* is common in the Northwest.

Tribe IV. *Inuleæ: Antennaria* (Everlasting, Ladies' Tobacco). These are low, perennial, woolly, diœcious or polygamo-diœcious herbs with alternate, simple, more or less spatulate leaves, the basal ones much the largest, usually, and generally in rosettes. The plants spread vegetatively by means of leafy stolons. The flowers are distinguished by a copious pappus of capillary bristles borne in a single series; in the bisexual or carpellate flowers the bristles are united into a ring at the base, but in the staminate flowers they are thickened upward so as to become club-shaped. The staminate plants are usually smaller than the others and sometimes abundant. Some species are parthenogenetic. About 50 species have been described, belonging mostly in north Temperate regions, but a few are found in Chile and Argentina, and in Australia.

Anaphalis (Everlasting) and *Gnaphalium* (Cudweed) are very similar in appearance and difficult to distinguish. The principal distinction is that *Anaphalis* is generally diœcious, while *Gnaphalium* is bisexual.

Tribe V. *Heliantheæ: Silphium* (Rosin-Weed). These are tall (2—12 ft.), robust perennials of eastern North America. The special features are the two or three series of carpellate and fertile ray-flowers and the bisexual but sterile disk-flowers. Quite outstanding, too, are the broad and flat involucral bracts. In addition the fruit is flattened dorsally and winged laterally, with a notch at the top. When the pappus is present its two awn-like segments are fused with the wings of the cypsela. *Silphium perfoliatum,* one of the larger species, is distinguishable by its connate-perfoliate, cup-shaped upper leaves. In the tall (12 ft.) *S. laciniatum,* the leaves are pinnatifid or bipinnatifid. There are about 12 species in all.

Iva, Ambrosia, Xanthium, and a few other genera are by some writers separated as a distinct family—*Ambrosiaceæ.* Technically these genera are distinguished by the distinct or only slightly connivent anthers, the bur-like spurious fruits—composed of entire capitula, and wind-pollination. *Iva* (Marsh-Elder) and *Ambrosia* (Ragweed), both popularly called ragweeds, are decidedly "hay - fever" plants. The pollen is produced in abundance. *Ambrosia trifida* (up to 15 ft. in height), *A. artemisiifolia,* and *A. psilostachya* are especially common as weeds in the prairie and plains States. Associated with these one will also find the tall (6 ft.) *Iva xanthiifolia,* a robust annual with large ovate leaves.

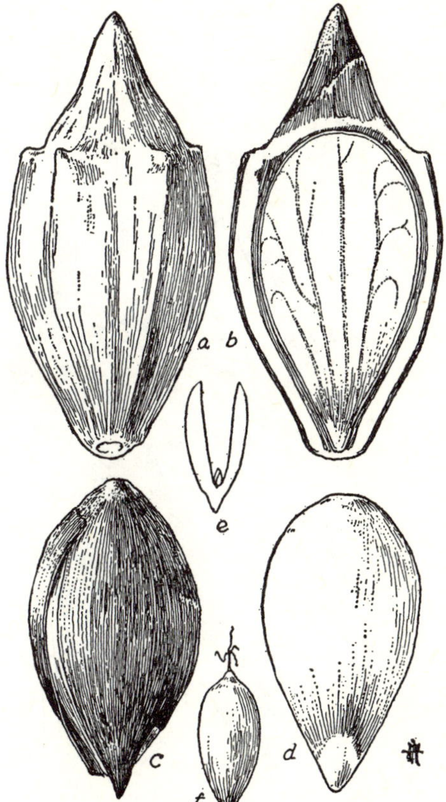

FIG. 406.—*Ambrosia* sp.: *a,* fruit; *b,* section of fruit, embryo in center;—the wall of the fruit consisting of the involucre, receptacle, and ovary; *c,* seed; *d,* embryo, side view (enlarged); *e,* embryo; *f,* gynœcium.

Species of *Xanthium* will be met with frequently on sandbars or sandy shores of rivers and lakes. These are low, coarse, and rough annuals with large, broad leaves; the flowers are unisexual and monœ-

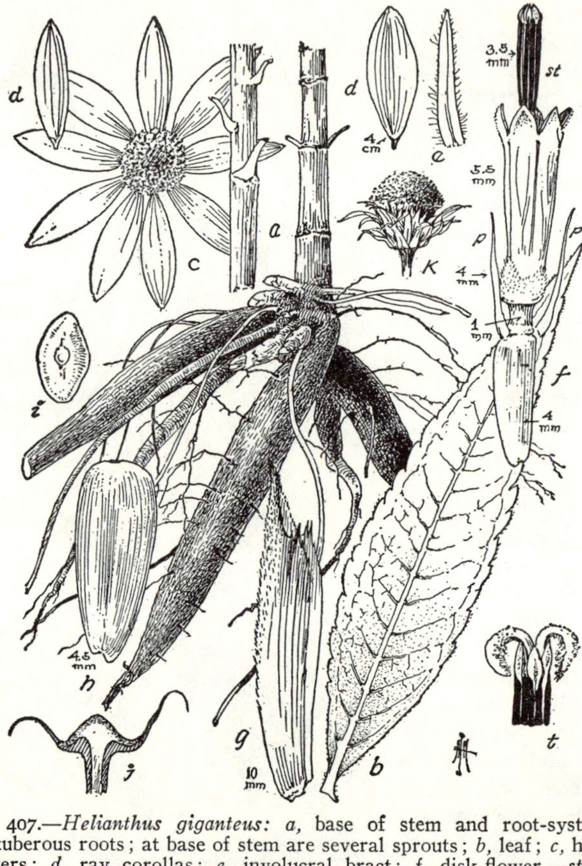

FIG. 407.—*Helianthus giganteus: a,* base of stem and root-system, with several tuberous roots; at base of stem are several sprouts; *b,* leaf; *c,* head, with ray-flowers; *d,* ray corollas; *e,* involucral bract; *f,* disk-flower,—*st,* anther-tube; *p,* pappus (sepals); *g,* floral bract; *h,* fruit (cypsela, achene); *i,* summit of fruit; *j,* disk of capitulum (head); *t,* stigma and summit of anther-tube.

cious. Aside from their growth-form their most distinctive feature is the bur-like fruit, thickly beset with stout hooked spines (modified involucral bracts). The burs are modified capitula. In the ambrosias the burs are much smaller and achene-like; they are but 1-flowered and

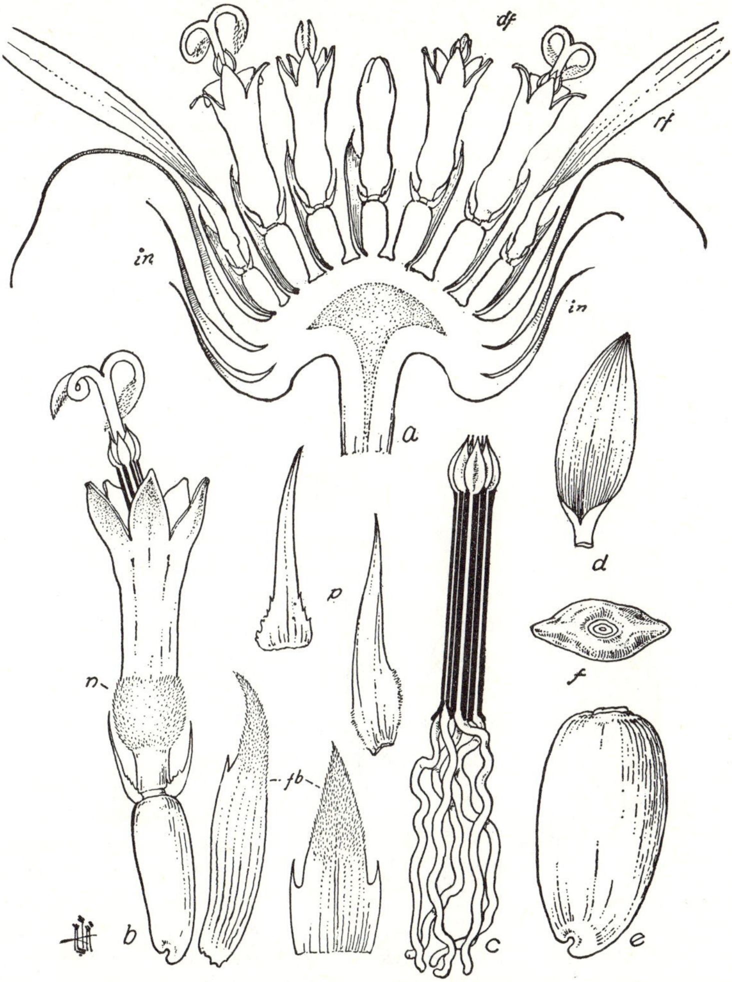

FIG. 408.—*Helianthus: a,* median section of capitulum; *b,* disk-flower; at right floral bract (*fb*),—side view, and dorsal view of apex; *c,* stamens; *d,* ligule; *e,* fruit; *f,* summit of fruit; *p,* pappus (sepals); *in,* involucre; *df,* disk-flowers; *rf,* ray-flower.

possess but a few involucral bracts, these forming a ring of stout tubercles at the top of the spurious fruit. No bur is formed in *Iva,* the fruit being a typical composite cypsela. Besides, staminate and carpellate flowers occur together in the same capitula. *Iva* and *Ambrosia* are strictly American, while *Xanthium* is common to both the Old and the New World. It may be noted in passing that in the tropical American genus *Acanthospermum* there are to be seen the beginnings of a spurious bur-fruit such as that of *Xanthium,* except that here only the inner bracts of the involucre form the bur around the cypsela.

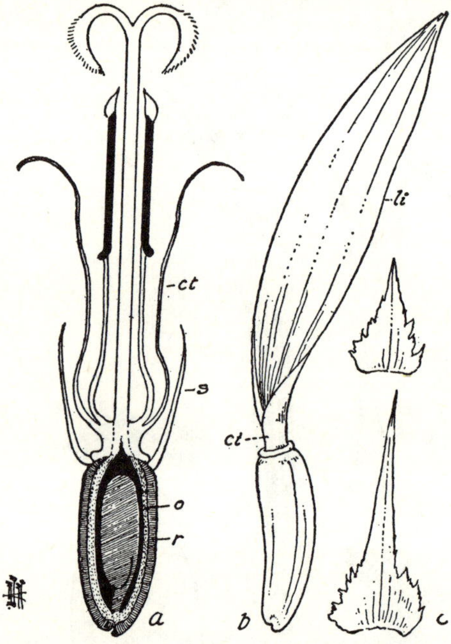

FIG. 409.—Median section of disk-flower of a sunflower, *a; ct,* corolla-tube; *s,* sepal; *o,* ovary; *r,* receptacle; ovule within; *b,* ray-flower; *li,* ligule; *ct,* corolla-tube; *c,* sepals.

Acanthospermum australe is of interest as ranging as far north as Virginia. Only the ray-flowers form fruit (compare with *Silphium*). (Figs. 406; 49A, B.)

Rudbeckia (Cone-Flower), *Brauneria* (Purple Cone-Flower) and *Lepachys* (*Ratibida*) (Prairie Cone-Flower) are characteristic composites of the region east of the Rockies to the Atlantic. There are several horticulturals, such as *Rudbeckia trilobata* and *R. hirta,* both known as Black-Eyed Susan, and *R. laciniata* (Golden-Glow).

Helianthus (Sunflower). This characteristic composite and outstanding American genus is especially well represented in the prairie and plains States. Most of the species are perennials from tuberiferous or non-tuberiferous rhizomes or from tuberous roots; a few are annuals. The leaves range in shape from large, broad, ovate-cordate forms to narrow and lanceolate; they are alternate or opposite, but sometimes both arrangements appear on the same plant. The heads are mostly aggregated in loose, branching inflorescences, but in some species they terminate long, slender branches; they are small to rather large, but in the common cultivated sunflower reach enormous proportions. The ray-flowers are neutral, *i.e.*, neither staminate nor carpellate (no style nor stigma). Each disk-flower is seated in the axile of a prominent floral bract (the *chaff*) which partly invests the "ovary." The pappus consists of 2—5 thin, translucent, scale-like segments which are usually unequal in size and deciduous. The cypselas are thick, oblong or obconical, with oblique base, and, except in the common garden sunflower, are so similar in appearance that specific differences are obscure. Some species are exceedingly variable, as *Helianthus tuberosus,* while on the other hand others are quite stable (*H. Maximiliani, H. orgyalis, H. mollis*). *H. Maximiliani* and *H. orgyalis* can be readily distinguished by their leaves. In the former the leaves tend to fold up with age, while in the latter they are conspicuously linear. *H. mollis* is a distinctive woolly species. Any one of the species makes a showy ornamental when in flower. *H. annuus* is abundant Westward; literally acres of it may be seen across the plains States. (Figs. 407, 408, 409, 410, 411.)

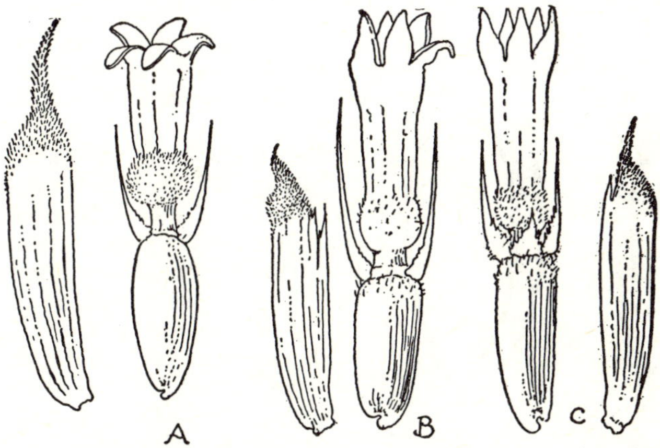

FIG. 410.—Similarities and differences in disk-flowers and subtending floral bracts of three species of sunflowers; *A, Helianthus doronicoides; B, H. lævigatus; C, H. lætiflorus.*

Bidens (Bur-Marigold) is readily distinguished by its barbulate-horned cypselas—on which the pappus persists in the form of two rigid, retrorsely barbed teeth or prongs, to which the genus owes its name. The bur-marigolds are among the very few composites that inhabit wet grounds, being often common in low, marshy hollows or on wet shores of lakes and streams. *Coreopsis* is somewhat similar as to appearance of the heads and rays. Both genera yield horticulturals. *Galinsoga parviflora,* a small, innocent-looking tropical species, has spread almost over the entire country as a pernicious weed in yards and waste grounds. (Figs. 412, 413.)

Tribe VI. *Heleniece. Hymenopappus* is characterized by small discoid heads with a naked receptacle. In *Helenium* the heads are also rather small and the receptacle is naked, but the rays are reflexed with age. From this genus *Gaillardia* differs in the possession of larger heads, spreading rays, and a bristly receptacle; in some species rays are wanting. Both genera yield horticulturals.

Tribe VII. *Anthemideæ.* These are strongly aromatic plants, mostly (at least in ours) with pinnately divided or finely dissected leaves. In *Achillea* (Yarrow, Milfoil) and *Anthemis* (Chamomile) the receptacle is chaffy, in the other genera it is naked or hairy. Except in *Achillea ptarmica* of the St. Lawrence Basin (and Asia), the leaves are finely dissected and the heads very small and in corymbose clusters at the summit of the stems. In *Anthemis* the heads are rather large and the cypselas terete instead of flattened as in the yarrows. *Matricaria* (Wild Chamomile) has conspicuously radiate or discoid heads of yellow flowers subtended by a few series of involucral bracts. Pappus none or merely a thin ring. The cypselas are ribbed. *Matricaria inodora,* with radiate heads, and *M. matricarioides,* with discoid heads, are common weeds in yards and waste grounds. Species: *Achillea,* about 75, of the northern hemisphere; *Anthemis,* about 60, of Europe, Asia, and Africa; *Matricaria,* about 20, mainly of the northern hemisphere, a few in South Africa.

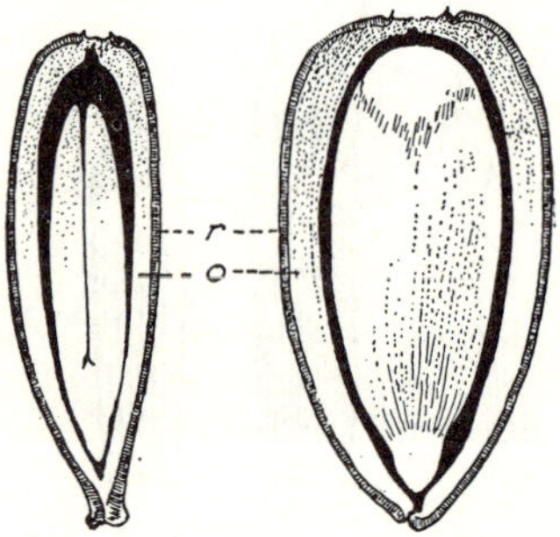

FIG. 411.—Sections of fruit (cypsela) of a sunflower; *o,* ovary; *r,* receptacle; embryo within.

Tanacetum (Tansy). The tansies are coarser plants, with dissected leaves and generally small discoid heads. The 30 or more species are distributed over the northern hemisphere; probably a half-dozen occur within our limits. *Tanacetum vulgare* (Common Tansy, introduced) and the native *T. huronense* are common eastward and northward. Two or three others occur in California. The

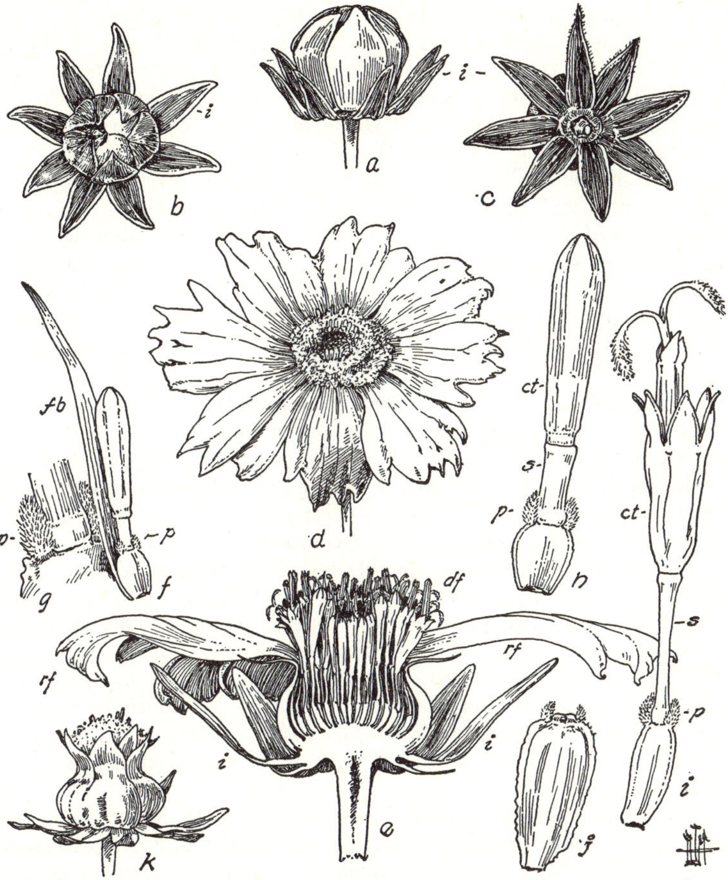

FIG. 412.—*Coreopsis* sp.: *a,* capitulum in bud stage; *b,* capitulum viewed from above, and *c,* from below; *d,* capitulum at flowering time; *e,* median section of capitulum; *i,* involucral bracts; *df,* disk-flowers; *rf,* ray-flowers; *f,* bud stage of disk-flower; *fb,* floral bract; *g,* pappus (*p*) of *f* (*p*) enlarged; *h,* bud of disk-flower, enlarged; *i,* disk-flower in bloom; *j,* immature fruit; *k,* capitulum in fruiting stage; *rf,* ray-flower; *ct,* corolla-tube; *s,* stipe; *p,* pappus (sepals).

large genus *Chrysanthemum* (100 or more species) is widely distributed over the northern hemisphere. The most familiar species is probably *C. Leucanthemum* (Field Daisy) which has become established in meadows and pastures. The heads are distinguished by a somewhat flattened receptacle, and the flowers lack a pappus or this is reduced to a shallow cup. Numerous species are horticultural.

Artemisia (Wormwood). A large genus of about 200 or more species of the northern hemisphere and South Africa. The plants are strongly aromatic shrubs or subshrubs. The shrubs are especially abundant over wide areas of the West. The leaves are small and lobed or dissected. The heads are small, discoid, often drooping, in inflorescences of various sorts. The flowers are without a pappus. A common Western species is *Artemisia tridentata* (Sagebrush). In the North *A. canadensis* is rather common, while over the plains and prairies occurs the tall, slender, narrow-leaved *A. dracunculoides.* In the Mississippi Basin and westward the white-woolly-leaved *A. ludoviciana* is quite common in dry soil along streams and in thickets.

Tribe VIII. *Senecioneæ: Senecio* (Groundsel). This is an enormous genus of 1,200—1,300 species, and one of the largest of all genera of flowering plants. The members are spread all over the world, and naturally, in so large a group, are of very diverse habits. Between 50 and 75 species occur in North America north of Mexico. The characteristic features are the comparatively few bracts of the cylindrical or campanulate involucres, all arranged in one series. Rays are sometimes absent, but there is a pappus of numerous smooth or rough capillary bristles. The heads are usually relatively small and aggregated in corymbose or paniculate inflorescences. The cypselas are terete or compressed (in marginal flowers), and more or less pubescent or papillose. Characteristic Northern species are such as *Senecio Balsamitæ* (*S. pauperculus*), *S. canus,* and *S. aureus.* There are several horticulturals, *e.g., S. cruentus* (*Cineraria cruenta,* Cineraria).

Petasites sagittaria (Arrow-Leaved Colt's-Foot), a stout perennial with large hastate leaves, white-tomentose beneath, is characteristic of sedgy or willowy sloughs northward. In *Arnica* the leaves are opposite and the heads showy and suggestive of a sunflower. The pappus is capillary and rough-barbellate, the cypselas linear to fusiform and pubescent.

Tribe IX. *Cynareæ.* This tribe comprises the thistles and their allies. The plants are mostly stout, vigorous biennials or perennials with simple or pinnately lobed or incised leaves and small to rather large flowers in showy heads.

Arctium (Burdock). The burdocks are coarse, large-leaved herbs of Eurasian origin. Species probably 6, of which three occur here as weeds. The first leaves appear in huge rosettes. The characteristic feature is the bur-like "fruit," which consists of an entire head, the involucral bracts of which form an armature of hooked prickles which are very efficient devices for effecting dissemination, the burs

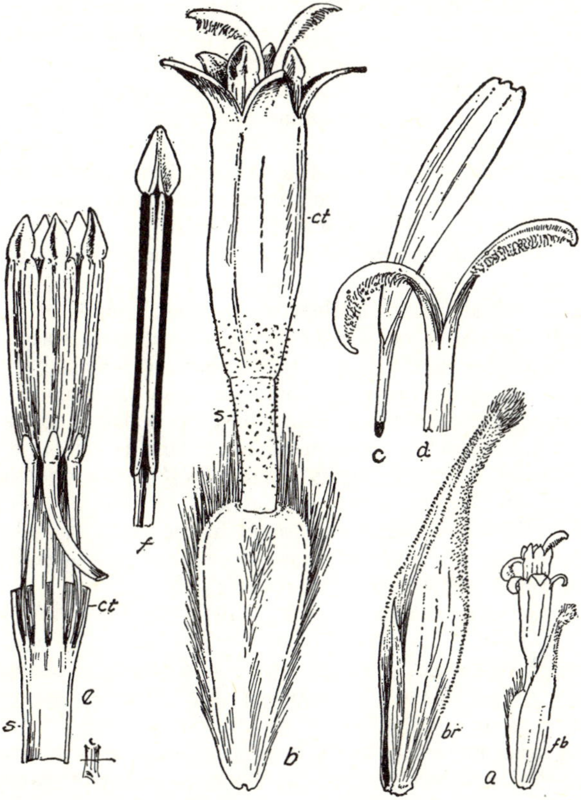

FIG. 413.—*Encelia californica: a,* disk-flower with floral bract (*fb*); *b,* disk-flower, much enlarged; *s,* stipe; *c,* ray-flower; *d,* stigma; *e,* stamens; *ct,* corolla-tube; *s,* stipe; *f,* anther-tube; *br,* floral bract, enlarged.

thereby becoming securely lodged in the fur of animals and in clothing. Each bur contains a number of distinct cypselas. The pappus consists of several scales. The two species commonly seen here are *Arctium Lappa* (Great Burdock) and *A. minus* (Common Burdock);

both occur as weeds. The former is medicinal and has been grown also for its edible roots.

Cirsium (Common Thistle). The thistles are stout, erect, prickly, glabrous or woolly herbs with incised to pinnatifid spiny-edged leaves. The first leaves usually form a large rosette. The heads are small to large, globose, discoid, with spreading to closely appressed and imbricate, usually very spiny-tipped involucral bracts. Flowers white, yellow, purple or violet, scarlet or orange. Many of the thistles are handsome plants and in various parts of the country form a characteristic element of the late summer- and autumn-flowering vegetation. *Cirsium arvense* (Canada Thistle) is a noxious weed. Some species are diœcious. The corollas are much narrowed below and the filaments usually pilose. The pappus consists of numerous, slender, plumose bristles. The receptacle is bristly and convex or flattened. Species about 200, all of the northern hemisphere; 60—70 in our flora.

Carduus (Plumeless Thistle), a Eurasian genus of about 80 species, only two or three of which have become established here. In contrast with *Cirsium,* the pappus is not plumose. The solitary *Cnicus Benedictus* (Blessed Thistle) is found in waste grounds in Atlantic North America and on the Pacific Coast. It is native to the Mediterranean region and Asia Minor. The plant is an annual, with yellow flowers in large solitary heads. *Silybum marianum* (Milk Thistle) of the Mediterranean region and *Onopordum Acanthium* (Cotton or Scotch Thistle) are found in the Atlantic and Eastern Lakes States. In the latter the broad involucral bracts end in stout spreading tips.

Echinops sphærocephalus (Globe Thistle), a European plant, occurs in the East as an escape from gardens. It is distinguished by 1-flowered heads which are aggregated into dense globose clusters. *Centaurea* (Star-Thistle) a genus of 330 or more species, comprises annuals or perennials without prickly leaves but with bristly to prickly or spinous involucral bracts. *C. americana* occurs on the plains of the Southwest. Over a dozen species have come to us from Europe, occurring as weeds or escapes from gardens. *C. Cyanus* is the common bachelor's button. *Cynara Cardunculus* (Cardoon) is grown for its edible root and *C. Scolymus* for its large, edible heads (involucral bracts).

Tribe XI. *Cichoreæ* (Chicories). The chicories differ from all other composites in having milky sap and completely radiate heads. The group is sometimes set apart as a distinct family. This group yields the bulk of the economically important composites, such as *Cichorium Intybus* (Chicory), *C. Endivia* (Endive). *Lactuca sativa* (Lettuce), *Tragopogon porrifolius* (Salsify, Fig. 414), *Taraxacum officinale*

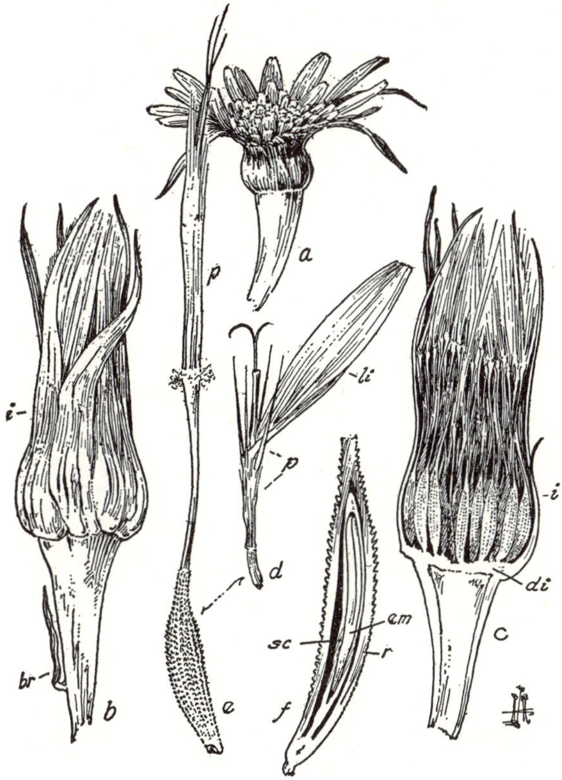

FIG. 414.—*Tragopogon porrifolius: a,* capitulum; *b,* capitulum closed; *br,* bract; *i,* involucre; *c,* median section of capitulum; *d,* flower; *li,* ligule; *p.* pappus,—of coarse and fine bristles; *e,* fruit, with pappus above; *f,* median section of fruit; *r,* receptacle; *sc,* seed-coat (testa); *em,* embryo.

(Dandelion). Besides the dandelion, a number of other members are noxious weeds, *e.g., Lactuca scariola* (Prickly Lettuce), *Hieracium florentinum, Crepis runcinata, Hypochæris radicata,* and the sow thistles *Sonchus asper* and *S. oleraceus.* (Fig. 415.)

In *Taraxacum* (Dandelion) the copious bristly pappus is raised on a slender stipe as a "parachute" for disseminating the fruit. The heads close at night or with a fall in temperature and continue to open and close until all the flowers have bloomed. When the fruits

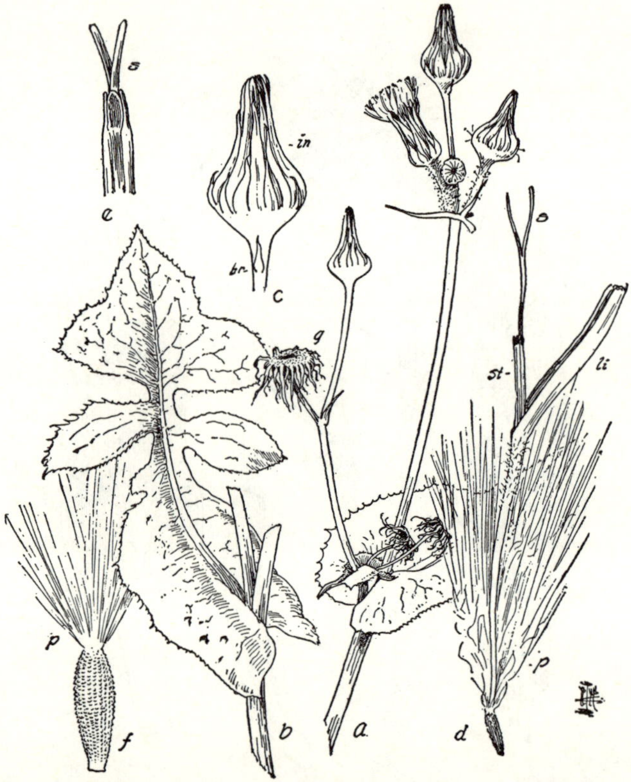

FIG. 415.—*Sonchus arvensis: a,* inflorescence (compound); *b,* leaf; *c,* capitulum, closed; *br,* bract; *d,* flower; *li,* ligule; *st,* anther-tube; *s,* stigma; *e,* stigma and summit of anther-tube at an early stage; *f,* fruit; *g,* disk of capitulum after shedding fruits; *p.* pappus; *in,* involucre.

are mature the heads become reflexed, thus allowing the fruits to be blown away. The young leaves of the dandelion are often used as greens.

CHAPTER XVIII

MONOCOTYLEDONES

Embryo with one lateral cotyledon (*scutellum*), generally small and associated with copious endosperm. Cotyledon remaining within the seed (that is, it is *hypogeal*) and serving as an organ for absorbing the nutritive substance of the endosperm. Primary root of relatively short duration, being replaced by secondary (adventitious) roots developed from the epicotylary portion of the seedling. Plants mostly of an herbaceous perennial habit, growing from bulbs, corms, or rhizomes, but woody tree-like forms with definite trunk occur in some groups. Vascular bundles of the closed type and scattered through the parenchymatous ground tissue of the stem. Cambium absent. Secondary growth in thickness in the tree-like forms, where it takes place by means of meristematic cells near the periphery (pericycle) of the stem; otherwise growth in diameter is effected simply by the enlargement of the cells. Branching habit poorly developed in general owing to the failure of the axillary buds to develop. Leaves of varied shapes, in some members growing to enormous size, with narrow to broad blades, parallel venation and slight development of transverse veins, and with a sheathing (*amplexicaule*) base. Petiole well-developed in some members. Parts of the flowers generally in threes, but in some cases numerous, in others stamens and carpels solitary. Seeds numerous to few or one.

The generally accepted view at the present time is that the Monocotyledons have sprung from the Ranalian complex of the Dicotyledons. This view is emphasized particularly by Bessey and by Wettstein (pp. 11, 10). It rests primarily on the evidence of the anatomical structure of the stem, supported by evidence from the structure of the flower of certain members of the *Ranales* (cf. Engler, p. 9) as compared with the Monocotyledons. The Ranalian members that are especially significant in this connection are the water-lilies (*Nymphæaceæ*), the buttercups (*Ranunculaceæ*), and certain barberries (*Berberidaceæ*). As to whether the Monocotyledons are of monophyletic or polyphyletic origin there is no conclusive evidence. Outside of the Ranalian complex there appears to remain no other Dicotyledonous group from which they might possibly have been derived.

As a rule the Monocotyledons inhabit low, moist or wet grounds, and numerous forms are aquatic, while others occupy arid areas where they often form a distinctive feature of the vegetation. In general they are rapid growers.

Key to the Monocotyledons

I. Plant body reduced to a minute, flattened, free-floating structure (thallus). Flowers unisexual, with one stamen and one carpel, respectively. *Lemnaceæ* (*Spathifloræ*).

II. Plant body of the ordinary type.

A. Flowers numerous on a fleshy axis or spadix which is seated in a conspicuous bract or spathe, or flowers sometimes solitary.

Perianth undifferentiated, often wanting. Ovary superior.

Leaves long and sword-shaped. *Pandanales.*

Leaves mostly very broad and often very large.

Leaves mostly of the "palm" type, fan-shaped or feather-shaped.

Plants mostly large and tree-like. *Principes* (Palms).

Plants stemless, but with subterranean caudex, often climbing or decumbent. *Synanthæ.*

Leaves not of the palm type but mostly large and broad with definite blade and petiole. *Spathifloræ.*

B. Flowers bisexual or unisexual, without perianth or with a perianth of scales or bristles, seated in the axil of a scale-like bract and aggregated in spikelets of heads. Ovary superior. *Glumifloræ.*

C. Flowers with an undifferentiated or a differentiated perianth.

a. Gynœcium apocarpous. Ovary superior.

Stamens and carpels numerous to one. *Helobiæ* (except *Hydrocharitaceæ*).

Stamens 3—6. Carpels indefinite. Perianth of 3—8 segments. *Triuridales.*

b. Gynœcium syncarpous.

1. Ovary mostly superior.

Stamens varying from one to six. Carpels 1, 2, 3, or several. *Farinosæ.*

Stamens 6 or 3. Carpels 3, rarely fewer or more. *Liliifloræ.*

2. Ovary inferior.

(*). Flowers mostly regular. *Liliifloræ.*

(**). Flowers mostly irregular.

Seeds arillate. *Scitamineæ.*

Seeds not arillate. *Microspermæ.*

ORDER PANDANALES

Plants with unisexual hypogynous flowers, naked or with a perianth of similar bract-like segments in one or two whorls. Staminate flowers with numerous stamens or sometimes only one. Carpellate flowers with numerous distinct carpels or only one. Both kinds of flowers borne separately in club-shaped (spadiciform) or globose inflorescences. Seeds with endosperm.

Families: *Typhaceæ, Sparganiaceæ, Pandanaceæ.*

KEY TO THE FAMILIES

Aquatic or marsh herbs, perennial from rootstocks.
  Flowers and fruits very small and numerous, in large terminal cylindrical spikes, the staminate above the carpellate. Tall plants. *Typhaceæ.*
  Flowers and fruits in globose heads at the upper nodes of a zigzag simple or branched flowering-stem. Smaller sedge-like plants. *Sparganiaceæ.*
Tree-like or climbing plants of the tropics, with sword-shaped, spiny-toothed leaves crowded at the ends of the stems or branches. *Pandanaceæ.*

TYPHACEÆ (*Cat-Tail Family*). The cat-tails are common stout marsh plants from stout, scaly-leaved rhizomes, with long, sword-shaped leaves sheathing the tall flowering stem which bears at the summit the characteristic club-shaped inflorescence. The plants form dense stands covering acres of marshland. The staminate and carpellate flowers are borne in separate parts of the cylindrical inflorescence, the staminate above the carpellate, each part subtended by a leaf-like bract (spathe). The staminate flowers consist of 2—5 stamens, the filaments sometimes united and the connectives of the anthers projecting. The carpellate flowers consist of a single carpel with an elongate style and a more or less spindle-shaped stigma; they are elevated on an elongate stipe from which spring a number of long hair-like processes (perianth?). The fruit is a nutlet or caryopsis. Seeds with very thin perisperm and fleshy endosperm, enclosing an axial embryo.

One genus, *Typha,* of about 10 species, widely distributed in temperate and tropical regions. *Typha latifolia* and *T. angustifolia* occur with us, the former the more common and the larger. Ordinarily the two species are distinguishable by their inflorescences: in *T. latifolia* the staminate and carpellate spikes are commonly con-

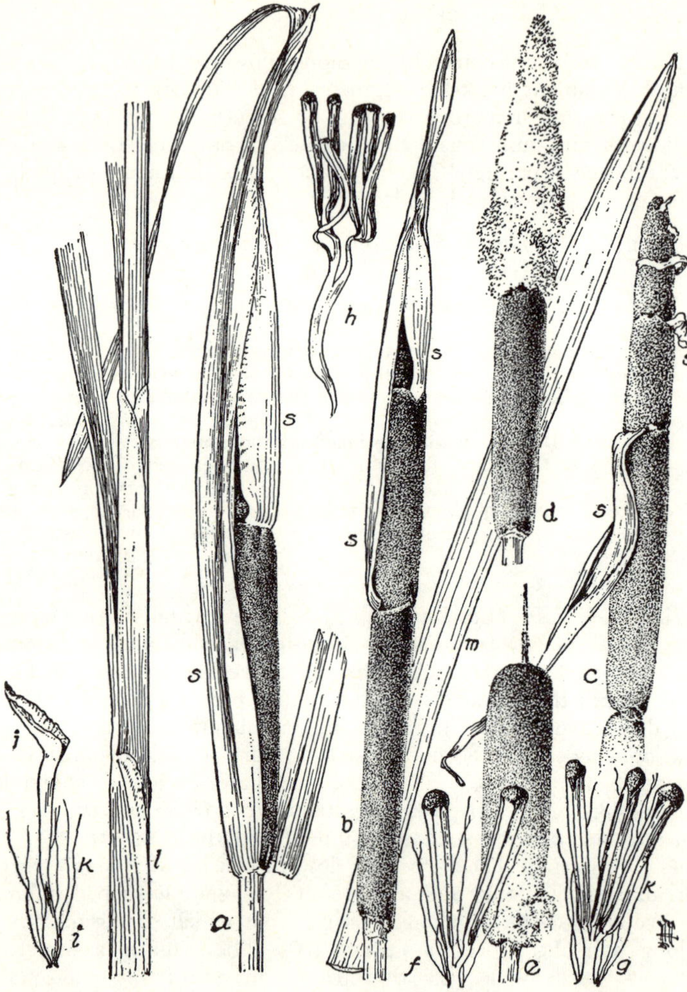

FIG. 416.—*Typha latifolia: a, b, c,*—different stages in the unfolding of the inflorescence; *s,* spathes; *d,* spike with staminate flowers above (past flowering) and carpellate below; *e,* mature stage of carpellate part of spike, the axis of the staminate part above; *f, g, h,*—staminate flowers with unequal numbers of stamens; *k,* bristles (perianth ?); *i,* carpellate flower; *k,* perianth (?); *j,* stigma; *l,* section of culm with leaf-sheaths; *m.* leaf. Spike (*b*) with three, and spike (*c*) with seven carpellate sections (separate spikes).

FIG. 417.—*Sparganium eurycarpum: a,* base of plant, with sheathing leaf-bases, secondary roots, rhizome (*r*), and bud (*s*); *b,* inflorescence; *c,* carpellate heads; *d,* staminate heads; *e,* transverse section of staminate head; *f,* stamens; *g,* one of scales interposed between stamens; *h,* carpellate flower, with perianth; *i,* gynœcium; *j,* section of gynœcium; *k,* transverse section of ovary.

tiguous, while in *T. angustifolia* they are separated by an interval of the axis. But this distinction does not always hold, as distantly separated spikes frequently occur also in *T. latifolia* (Fig. 416).

SPARGANIACEÆ (*Bur-Reed Family*). These are smaller plants, of shallow waters and marshes, growing from stout rootstocks and bearing alternate, 2-ranked, sword-shaped leaves sheathing at the base. The flowers are small, unisexual, monœcious, in sessile or pedunculate globose heads, the staminate above the carpellate. Both kinds of flowers are subtended by slender scale-like segments (perianth) somewhat irregularly interposed. The staminate flowers are not always well defined as the stamens often appear irregularly placed. The gynœcium is 1-carpellate or 2-carpellate (syncarpous), tapering into a stout style bearing one or two elongate stigmas. The carpellate heads when ripe appear bur-like from the numerous projecting beaks (styles).

One genus, *Sparganium,* with about 15 species of wide distribution in cold climates. Our plants are common along lake-shores, growing among bulrushes, spike-rushes, and arrowheads. *S. eurycarpum,* the plant illustrated, is the only American species with a 2-carpellate gynœcium (Fig. 417).

PANDANACEÆ (*Screw-Pine Family*). A small family of 3 genera and 200—400 species of tropical plants, the largest and most important of which are *Pandanus* (Screw-Pine), with 200—250 species, and *Freycinetia* with 60 or more species of climbing plants. The plants are mainly shrubs or trees with rigid leaves and mostly naked unisexual flowers. Staminodes are sometimes present. Fruit aggregate. Species of *Pandanus* are planted.

ORDER HELOBIÆ

Flowers cyclic or partly so, hypogynous or epigynous, with or without a perianth—this when present simple and bract-like or differentiated into calyx and corolla. Stamens numerous to 1. Carpels numerous, free or united. Seeds mostly without endosperm or this only slightly developed.

Hypogynous families: *Potamogetonaceæ, Najadaceæ, Aponogetonaceæ, Scheuchzeriaceæ, Alismataceæ, Butomaceæ.*

Epigynous family: *Hydrocharitaceæ.*

Key to the Families

A. Flowers hypogynous.
  a. Perianth none or rudimentary. Aquatics.
    1. Flowers bisexual or unisexual, monœcious or diœcious.
      Carpels and stamens 4—1. *Potamogetonaceæ.*
      Carpels and stamens 1. Leaves toothed. *Najadaceæ.*
    2. Flowers bisexual. Stamens 6 or more. Ovules numerous, parietal. *Butomaceæ.*
  b. Perianth present, in two whorls.
    Perianth not clearly differentiated into sepals and petals. Carpels few (3—6). Rush-like marsh herbs. *Scheuchzeriaceæ.*
    Perianth differentiated into sepals and petals. Carpels numerous, in globose heads. Aquatics and marsh herbs with broad, hastate to linear, sheathing, basal leaves. *Alismataceæ.*
B. Flowers epigynous, bisexual or unisexual. Aquatics with numerous very small, crowded, opposite or whorled, leaves, or the leaves basal, grass-like or petiolate, with broad blades. *Hydrocharitaceæ.*

The Hypogynous Helobiæ

POTAMOGETONACEÆ (*Pond-Weed Family*). The pondweeds are aquatic, marsh, and bog herbs of the coasts and the interior. Flowers regular, bisexual or unisexual and monœcious, with 1-merous or 4-merous whorls. Perianth none, or if present then in the form of a beaker-shaped sheath (gamophyllous). Stamens 1—4; anthers extrorse, 1—2-celled, sometimes furnished with a perianth-like connective. Carpels 1—4, distinct, 1-ovulate. Fruit a nutlet or drupelet. Endosperm none.

Genera in our flora: *Potamogeton, Ruppia, Zannichellia, Zostera.*

In *Potamogeton* and *Ruppia* the flowers are bisexual, in *Zannichellia* and *Zostera* unisexual and monœcious.

Potamogeton (Pond-Weed, Pickerel-Weed). The pond-weeds are submersed aquatics with jointed leafy stems which commonly root freely at the nodes. The flowers are borne in spikes which in the bud are enclosed in the sheathing stipules of the leaves; they are 4-merous—an anomalous condition among the Monocotyledons. There is a perianth of 4 segments, and 4 stamens—one opposite each segment of the perianth. Carpels 4, distinct, rarely 1, each with a short thick style or sessile stigma. Fruit drupaceous (Fig. 418). Species about 65, north-temperate; about 40 in North America. The plants are common in fresh waters.

Potamogeton is a technical genus. There are two groups of

species according to the foliage. In one group the leaves are of two sorts—one thick, leathery, and *floating,* the other narrow or filiform and *submersed.* In the second group the leaves are narrow or filiform to bristly and *all submersed.* Some species have developed

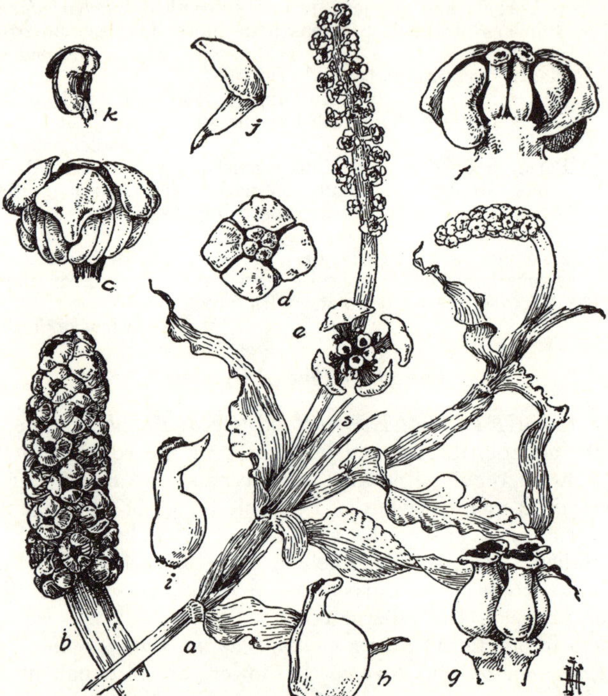

FIG. 418.—*Potamogeton Richardsonii: a,* end of stem with inflorescences; *b,* inflorescence, enlarged; *c,* lateral aspect of a flower; *d,* flower viewed from above; *e,* flower open; *f,* section of flower, exposing two carpels and two large anthers; *g,* gynœcium; *h,* carpel; *i,* younger carpel; *j,* perianth-segment; *s,* stipules; *k,* stamen, and behind it a perianth-segment.

special means of vegetative propagation by winter buds or by fragmentation of the branches (*P. Robbinsii*).

Ruppia (Ditch-Grass). A genus comprising, in the broadest interpretation, the solitary species *Ruppia maritima.* The plants grow submersed in saline or brackish waters along the coast or in the interior. They develop long, slender stems bearing filiform, basally sheathing leaves. The flowers are borne at the surface of the water, but after pollination they are drawn beneath the surface by the coiling of the stems. The fruit is a drupe (Fig. 419).

Zannichellia (Horned Pondweed). A genus of 2 or 3 species of slender, submersed plants of fresh or brackish waters. Leaves linear

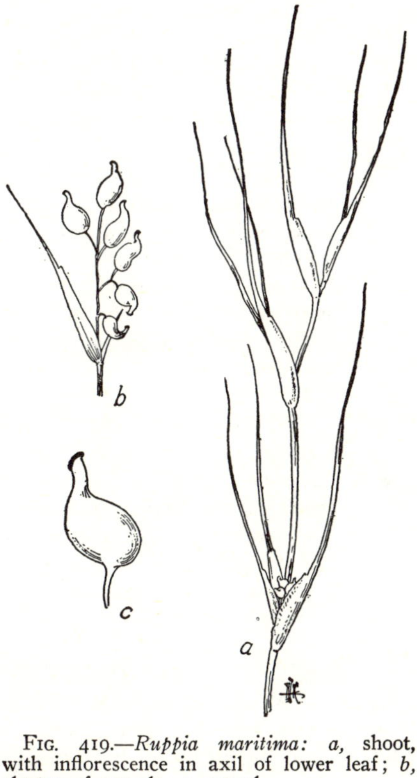

FIG. 419.—*Ruppia maritima: a,* shoot, with inflorescence in axil of lower leaf; *b,* cluster of carpels; *c,* carpel.

FIG. 420.—*Zannichellia palustris: a,* shoot with flowers in axils of lower leaves; *b,* young carpel; *c,* immature fruit (carpel).

or filiform, with membranaceous sheathing stipules enclosing the flower-buds and leaf-buds. Carpels flask-shaped, with thick, broad stigmas. Fruit a small, elongate, pod-like nutlet. *Zannichellia palustris* is widely distributed in North America and the Old World, in fresh or brackish waters (Fig. 420).

Zostera (Eel-Grass, Grass-Wrack). About 6 species of grass-like marine plants with flattened, branching stems from slender rootstocks. Staminate and carpellate flowers alternating in two rows on a linear spadix contained in a spathe. Perianth none. Carpels 1, with an elongate forking style. *Zostera marina* occurs along the Atlantic Coast from Greenland to Florida and on the Pacific Coast from Alaska to California.

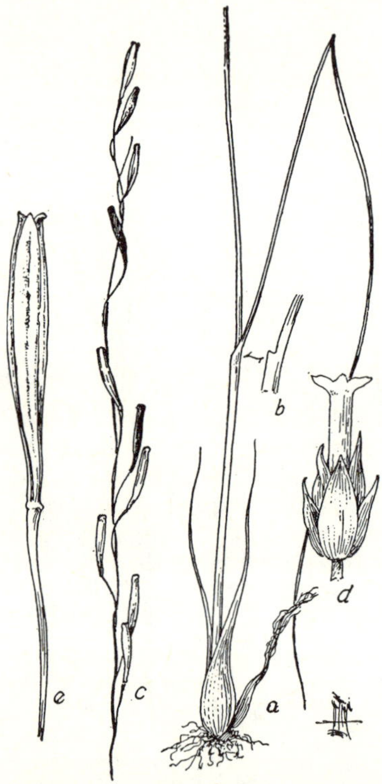

FIG. 421.—*Triglochin palustris: a,* habit sketch; *b,* summit of leaf-sheath; *c,* inflorescence,—fruiting stage; *c,* fruit; *d,* flower, with two whorls of perianth-segments, and projecting style.

NAJADACEÆ. One genus, *Najas* (Naiad) of about 10 species of world-wide distribution in fresh waters. About 5 species in North America. The slender stems bear filiform or spiny-margined leaves, and grow submersed. Staminate flowers with 1 stamen and carpellate flowers with 1 ellipsoid carpel. *Najas marina* occurs in lakes across the continent.

APONOGETONACEÆ. One genus, *Aponogeton,* of about 22 species of the tropics and subtropics from Africa to Australia. The plants are aquatics with tuberous stems. Perianth of 3—2—1 petaloid segments. Gynœcium 3—6-carpellate.

SCHEUCHZERIACEAE (JUNCAGINACEÆ) (Arrow-Grass Family). Marsh herbs of wide distribution, all rush-like plants with basally sheathing leaves. Flowers bisexual or unisexual in spicate or racemose inflorescences. Perianth of 3—6 segments in two whorls. Carpels 3—6, sometimes abortive. About 10 species in 4 genera—*Triglochin, Scheuchzeria, Tetronicum, Lilæa.* Our species are *Triglochin palustris, T. maritima, T. striata,* and *Scheuchzeria palustris* (the only species of its genus). (Figs. 421, 422.)

ALISMATACEÆ (*Water-Plantain Family*). The water-plantains are marsh plants with arrow-shaped or lanceolate sheathing leaves. Flowers bisexual or unisexual, and monœcious or diœcious. Perianth of 3 sepals and 3 petals. Stamens 6 or more; anthers 2-celled, extrorse. Gynœcium of numerous to few distinct carpels, with 1-ovulate ovaries. Fruit (in our species) achenes, often prominently beaked by the style. Endosperm none. Ovule campylotropous. Latex tubes present. Species about 65 in 10 (13?) genera—of which the following are represented in our flora: *Alisma, Echinodorus, Lophotocarpus, Sagittaria,* and *Damasonium* (California).

In *Alisma, Echinodorus,* and *Damasonium* the flowers are bisexual; in *Lophotocarpus* they are polygamous—the lower bisexual, the upper staminate; in *Sagittaria* unisexual—monœcious or diœcious.

FIG. 422.—*Scheuchzeria palustris: a,* leaves, rhizome, and inflorescence (in fruiting stage); *b,* flower; *c,* carpel (mature); *d,* dehisced carpel.

Alisma (Water-Plantain). Less than 10 species of wide distribution in temperate and tropical regions; represented in our flora by the species mentioned below, all common on wet shores and in ditches. The flowers are distinguished by a definite trimerous calyx and corolla, generally 6 stamens, and numerous carpels arranged in a characteristic ring which is more or less open in the center according to the species. Thus in *Alisma Geyeri* the carpels touch so that the ring is closed, and this feature alone will readily distinguish this species from *A. Plantago-aquatica* in which the carpels do not touch in the center of the ring. At maturity the carpels become flattened discoid achenes without any conspicuous beak at the summit. The receptacle is broad and flattened as a support for the broad ring of carpels. The flowers are borne in distinctive and usually elongate, paniculate or umbellate-paniculate inflorescences in which

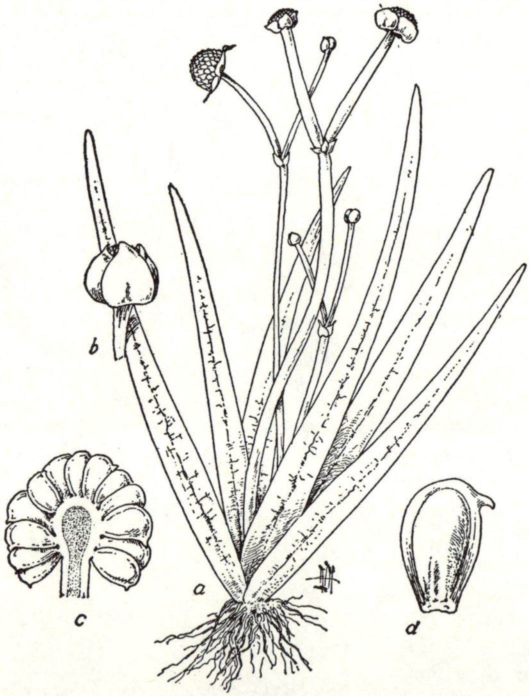

FIG. 423.—*Lophotocarpus spongiosus: a,* habit sketch; *b,* flower-bud; *c,* section of gynœcium (polycarpellate) ; *c,* carpel.

the branches are whorled. The inflorescences and the characteristic fruits at once distinguish the plants. The leaves are broad to narrow, with petioles prominently sheathing at the base.

Echinodorus. Species about 12, chiefly of the western hemisphere, of which probably two or three come within our limits. Our plants are soft herbs inhabiting wet shores and ditches. Leaves mostly

FIG. 424.—*Sagittaria arifolia: a,* habit sketch; *b,* form of plant growing submersed; *c,* staminate flower with rudimentary carpels; *d,* section of same enlarged; *e,* stamen; *f,* gynœcium and calyx of carpellate flower; *g,* median section of gynœcium of carpellate flower; *h,* carpel of flower (*d*); *i,* carpel of flower (*g*); *s,* sepal; *r,* rhizome.

broad, with elongate petioles sheathing at the base. Inflorescence racemose or paniculate, with whorled branches (flower-stalks). Flowers similar to those of *Alisma* but the carpels arranged in globular heads. Sepals persistent; stamens few to many. *Echinodorus tenellus* ranges from Massachusetts to Minnesota and southward into South America.

Lophotocarpus ranges from tropical America to north central parts of the United States. Species about 7. All are soft plants inhabiting swamps, bogs, and ponds, in some cases growing in brackish waters. The leaf-blades are hastate to cordate, on long petioles which are sheathing at the base. Inflorescences similar to those of the preceding genera but smaller and fewer flowered. Flowers like those of *Echinodorus,* but stamens 9—15, with flattened filaments. The fruits are winged or beaked achenes. Our most widespread species is *Lophotocarpus calycinus.* (Fig. 423.)

Sagittaria (Arrow-Head). This is the largest genus, with about 40 species, mostly temperate and tropical; about 35 in the United States and a few in Canada. They are plants of shallow waters and muddy shores, growing from tuberous or jointed rootstocks. Sap milky. Leaf-blades hastate to linear, on long petioles with sheathing base. Flowers monœcious or diœcious, borne on elongate petioles in whorls of three. Sepals persistent. Stamens mostly numerous. Carpels numerous, distinct, forming conspicuous, globular heads, becoming at maturity flattened, winged, and prominently beaked achenes. The shape and orientation of the beaks are of taxonomic value in distinguishing the various species. (See excellent figures in Gray's *Manual,* 7th ed.) Useful characters are found also in the tuber-bearing habit. The tubers constitute an important article of food of water-fowl and no doubt of beavers and muskrats. In common with other members of the family the arrowheads show a wide diversity of foliage according to the habitat, features which are sometimes confusing in distinguishing species. Such variations can only be fully appreciated when ample material of the species is assembled from all parts of its range. The members of this family are of special interest from the fact that their flowers bear a striking resemblance to those of the buttercups among the Dicotyledons. (Fig. 424.)

BUTOMACEÆ. A family of but a few species of tropical America and temperate Europe and Asia. In its affinities it appears close to the preceding family. Flowers bisexual, with differentiated perianth. Carpels 6 to numerous, more or less united at the base, and especially distinctive in bearing numerous ovules on the sides of the ovary-chambers and in the ovaries becoming follicular fruits. Genera: *Butomus* (Eur., Asia, N. Amer.), *Hydrocleis* (trop. Amer.), *Limnocharis* (trop. Amer.). *Butomus umbellatus* is said to occur on the shores of the St. Lawrence River near Montreal.

FIG. 425.—*Vallisneria americana: a,* carpellate plant; *b,* staminate plant; *c,* median section of staminate head of (*b*); *d,* carpellate flower; *s,* spathe; *e,* stolon with terminal bud.

The Epigynous Helobiæ

HYDROCHARITACEÆ (*Frog's Bit Family*). These are submersed or floating plants with whorled or spiral leaves. Flowers regular, bisexual or unisexual, monœcious or polygamous. Perianth simple or differentiated, *gamophyllous* (*!*) in the carpellate flowers. Stamens

3—12, distinct or monadelphous, some of them in certain cases staminodial. Carpels 2—15, united. Ovary inferior. Ovules numerous, on parietal placentas. Over 60 species, in 6 genera; represented in our flora by species of the genera *Elodea* (*Philotria*), *Vallisneria,* and *Limnobium*. *Elodea canadensis* (Water-Weed) is common in ponds and slow streams from Quebec to Virginia and west to the Mississippi. It is a long, slender, submersed plant with numerous small, oblong leaves arranged in whorls. The flowers are raised to the surface on long, slender stalks. In shallow running waters in North Central and Eastern parts of the country one will find *Vallisneria americana* (Tape-Grass), a small, soft plant with basal rosettes of grass-like leaves. The carpellate flowers are raised to the surface on long slender stalks (scapes) which coil up after pollination and draw the flowers beneath the surface. Pollination is effected in both *Vallisneria* and *Elodea* by the breaking of the staminate flowers from their stalks and the shedding of their pollen on the stigmas as they float on the surface. *Limnobium Spongia* (Frog's Bit), a plant with broad petiolate leaves, inhabits shallow stagnant waters of Eastern and Southern parts of the country. It bears unisexual, monœcious flowers (Figs. 425, 426).

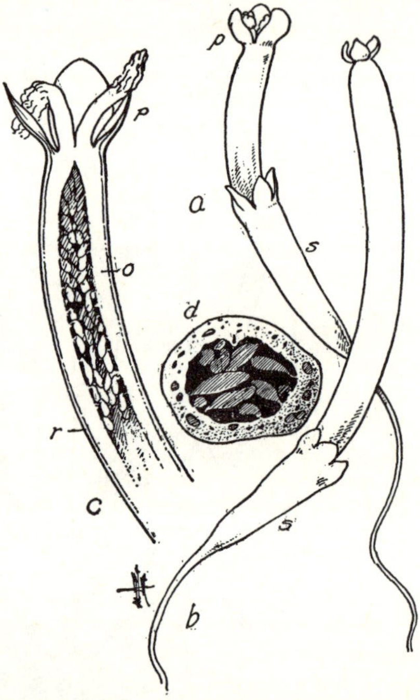

FIG. 426.—*Vallisneria americana: a, b,* carpellate flowers; *s,* spathe; *c,* median and (*d*) transverse section of ovary; *p,* perianth,—styles within; *o,* ovary; *r,* receptacle.

ORDER TRIURIDALES

One family—*TRIURIDACEÆ,* comprising a small number of saprophytic plants of the tropical forests of Brazil and of the Indo-

Malayan region and West Africa. Flowers bisexual or unisexual, with a petaloid perianth of 3—8 segments more or less united at the base (gamophyllous). Stamens 2—6. Carpels numerous, vestigial in staminate flowers. About 40 species in 2 genera: *Triuris, Sciaphila.* The order is of doubtful systematic status.

ORDER GLUMIFLORÆ (GRAMINALES)

Flowers hypogynous, in spikelets, without a perianth or with a perianth of bristles. Spikelets aggregated in racemose, paniculate, or spicate inflorescences (compound). The unit of the inflorescence is the spikelet. Stamens mostly 3, rarely fewer or several. Gynœcium 2—3-carpellate, syncarpous—only one carpel functional, the others aborted. Ovary superior. Stigmas 2 or sometimes 3 or 1. Fruit a 1-celled caryopsis or grain (grasses) or an achene (sedges). Endosperm copious. Embryo lateral in the grasses, axile in the sedges.

This order comprises two of the largest families of flowering plants, the *Gramineæ* and the *Cyperaceæ.* In the number of individuals these plants exceed all other flowering plants; they form the dominant vegetation of prairies, plains, lowlands, and open hills, and are not excluded from even heavy forests. The sedges are essentially aquatics or inhabitants of low, moist or wet grounds. In economic importance the grasses stand unrivaled among plants, but no such importance attaches to the sedges.

GRAMINEÆ (*Grass Family*). Species about 4,500, in about 400 genera, of which 100 or more are represented in the United States. Flowers bisexual, rarely unisexual, arranged in spikelets. No definite perianth. Stamens commonly 3, rarely 1 or 6 or more; filaments filiform; anthers 2-celled, versatile. Carpels 2, rarely 3 or 1, with united ovaries, short styles, and elongate plumose stigmas of the same number as the carpels. Fruit a caryopsis enclosing a single seed, the testa of which is completely fused with the ovary-wall (pericarp). Endosperm copious. Embryo lateral, with large, fleshy cotyledon (*scutellum*). (Figs. 427, 57.)

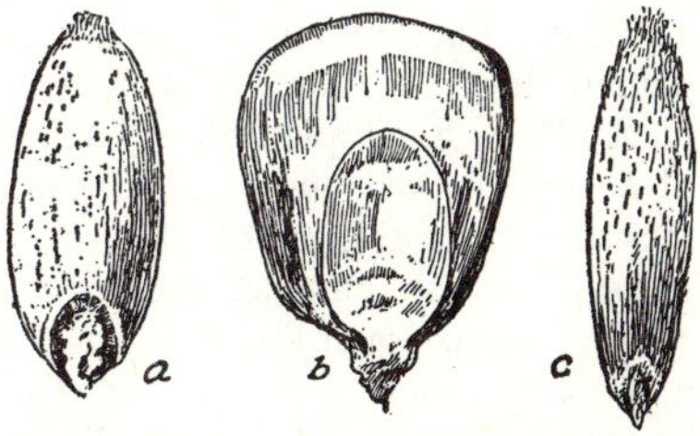

FIG. 427.—Caryopses of (*a*) wheat, (*b*) maize, (*c*) oat,—embryo side.

The grasses are annual or perennial herbs (in some cases treelike) with jointed stems (culms) and spiral sheathing leaves. The internodes are usually more or less elongate and hollow with age, but the joints (nodes) are solid. The leaves are especially distinctive; these consist of a prominent elongate *sheath* which is split on one side; from the opposite side springs a usually long linear blade which is sometimes prominently auriculate at the base; and at the base of the blade there is present a more or less prominent annular, membranaceous or fimbriate organ called the *ligule*. Perennial grasses are usually propagated rapidly from stolons, rhizomes or "suckers"; the plants may be simple or branched (especially from the base). (Fig. 428.)

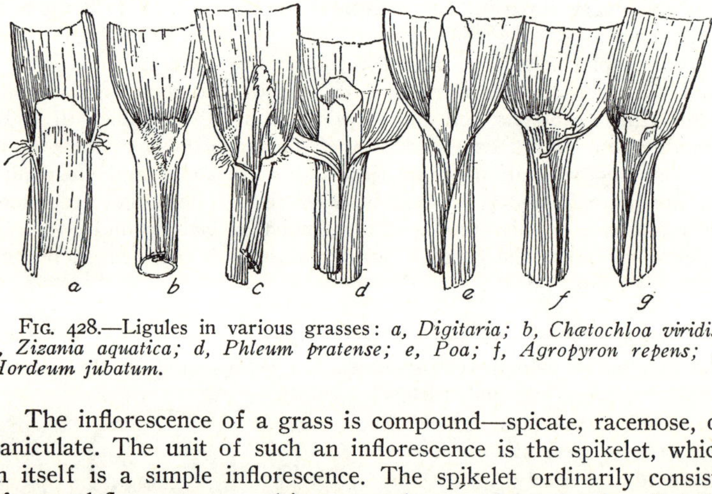

FIG. 428.—Ligules in various grasses: *a, Digitaria; b, Chætochloa viridis; c, Zizania aquatica; d, Phleum pratense; e, Poa; f, Agropyron repens; g, Hordeum jubatum.*

The inflorescence of a grass is compound—spicate, racemose, or paniculate. The unit of such an inflorescence is the spikelet, which in itself is a simple inflorescence. The spikelet ordinarily consists of several flowers arranged in two ranks on a jointed axis called the *rachilla.* At the base of the rachilla is a pair of bracts called the *glumes*—which envelop more or less the whole cluster of flowers. At the base of each flower is inserted another pair of bracts—one, the *lemma,* in front of the flower, the other, the *palea,* at the rear (between the flower and the rachilla). Actually the flower is seated in the axil of the lemma, which is therefore a floral bract. The palea usually invests the flower more or less tightly; morphologically it belongs to the perianth. This is the situation in general, but it is often the case that the upper lemmas of a spikelet are empty or rudimentary, or that both lemma and palea are absent. (Figs. 429, 430, 431.)

The flowers (or florets) are bisexual or sometimes unisexual and the species monœcious or diœcious. Stamens commonly 3 but occasionally 1 or 6, with elongate, 2-celled, versatile anthers on filiform filaments. Stigmas elongate and plumose, as many as the carpels. In front of the flower there are to be seen from one to three minute scale-like structures which are called *lodicules;* these are believed to

FIG. 429.—*A,* flower of a grass (schematic); *lo,* lodicules; *r,* rachilla; *le,* lemma; *pa,* palea; *B,* longitudinal diagram of flower; *C,* spikelet; *D,* floral diagram.

belong morphologically to the perianth. Their shape and prominence varies with the kind of grass (Fig. 429).

The grass family is divided into two subfamilies, namely, the *Poacoideæ* and the *Panicoideæ.* This systematic grouping is based on certain important taxonomic features which should be carefully noted.

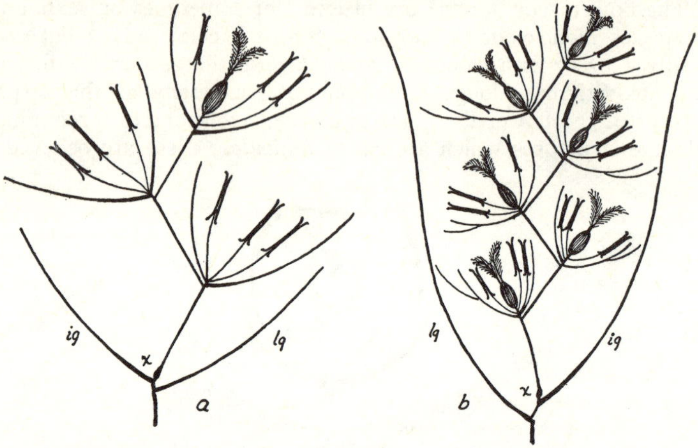

FIG. 430.—Diagrams of spikelets in *Poacoideæ: a, Hierochloa; b, Danthonia; lg,* outer glume; *ig,* inner glume; *x,* point of articulation of spikelet.

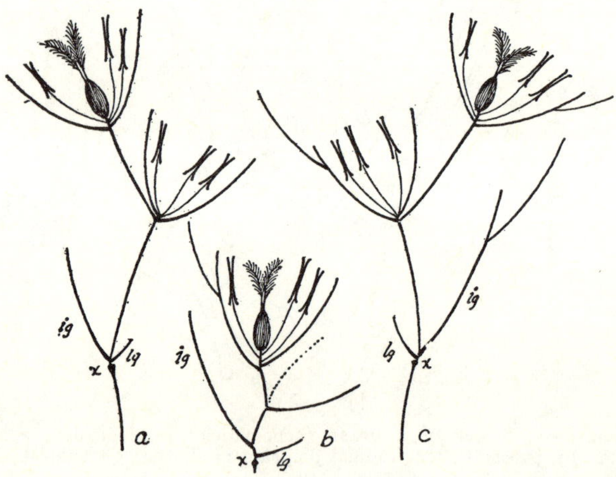

FIG. 431.—Diagrams of spikelets in *Panicoideæ: a, Panicum; b, Andropogon; c, Echinochloa; lg,* outer (lower) glume; *ig,* inner glume; *x,* point of articulation of spikelets.

These points are as follows: There are *two* ways in which the spikelets break away (disarticulate) from the pedicels at maturity. In one case the spikelets disarticulate *above* the glumes, in the other case *below* the glumes. The first condition is characteristic of the subfamily *Poacoideæ,* the second condition of the *Panicoideæ.* Furthermore, the spikelets in the *Poacoideæ* are *laterally* compressed, while in the *Panicoideæ* the spikelets are *dorsally* compressed. However, there are a number of exceptions to the rule concerning the disarticulation of the spikelets: Thus in the *Poacoideæ* the spikelets disarticulate *below* the glumes in the tribe *Oryzeæ* and in the genera *Alopecurus, Polypogon, Cinna, Holcus, Sphenopholis, Spartina,* and *Beckmannia,* but the laterally compressed spikelets of these will distinguish them from the *Panicoideæ* (Hitchcock). (Figs. 430, 431, 432, 433.)

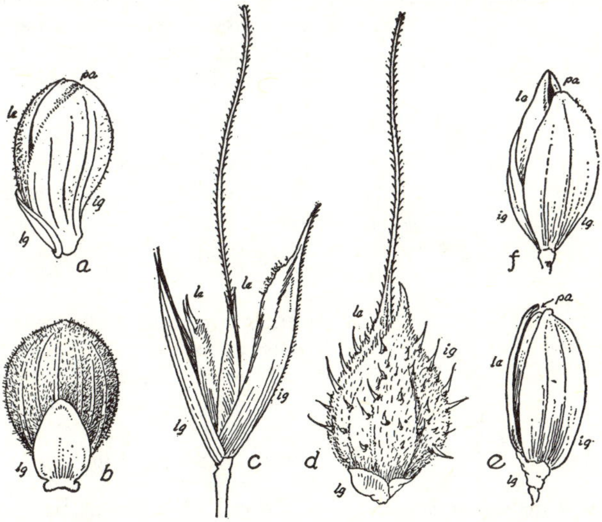

FIG. 432.—Spikelets in grasses (*Panicoideæ*): *a, b, Panicum xanthophysum; c, Andropogon furcatus; d, Echinochloa crus-galli; e, Chætochloa viridis; lg,* outer glume; *ig,* inner glume; *le,* lemma; *pa,* palea. See figures 434, 435, 436 for several-flowered spikelets (*Poacoideæ*).

The subfamilies are subdivided into tribes on characters of the inflorescence and the spikelets, as shown below.*

Subfamily *Poacoideæ.*

Tribe I. *Oryzeæ.* Spikelets 1-flowered, unisexual or bisexual, laterally compressed, in panicles, and articulated below the glumes. Stamens often 6.

* After Hitchcock.

Including: *Zizania, Zizaniopsis, Leersia* (*Homalocenchrus*). (Fig. 80.)

Tribe II. *Phalarideæ.* Spikelets 1- or sometimes 3-flowered, in panicles or racemes. Fertile floret subtended by two sterile lemmas which fall with the floret. Palea 1-nerved. *Phalaris, Anthoxanthum, Hierochloë.*

Tribe III. *Agrostideæ.* Spikelets 1-flowered, in open panicles or racemes, Rachilla sometimes evident as a bristle. Palea sometimes absent; when present, 2-nerved or sometimes 1-nerved. *Phleum* (*P. pratense,* Timothy), *Agrostis* (Bent-Grass), *Sporobolus* (Dropseed), *Calamagrostis* (Reed Bent-Grass), *Aristida* (Triple-awned Grass), *Cinna* (Wood Reed-Grass, has one stamen), *Stipa.*

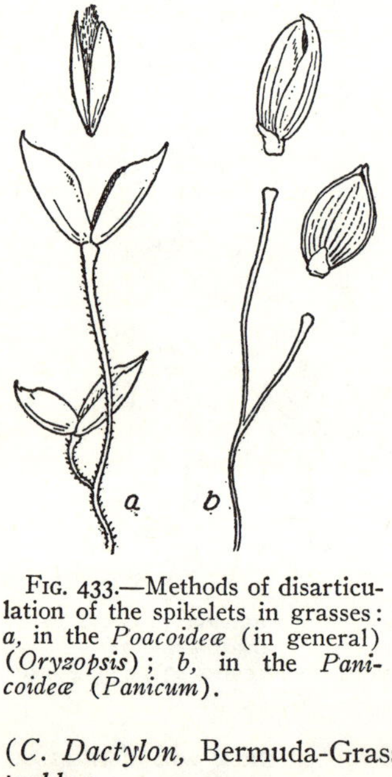

FIG. 433.—Methods of disarticulation of the spikelets in grasses: *a,* in the *Poacoideæ* (in general) (*Oryzopsis*); *b,* in the *Panicoideæ* (*Panicum*).

Tribe IV. *Aveneæ* (Oat-Grasses). Spikelets 2—several-flowered, in racemes or open panicles; florets mostly awned on the back; glumes usually elongate, and rachilla prolonged behind the uppermost floret. *Avena* (Oat), *Holcus* (*H. halepensis,* Johnson-Grass). (Fig. 434.)

Tribe V. *Chlorideæ.* Spikelets 1—several-flowered, in two rows in one-sided spikes mostly forming racemes or panicles, or solitary or digitate. *Spartina* (Marsh-Grass), *Cynodon* (*C. Dactylon,* Bermuda-Grass), *Bouteloua* (Mesquite-Grass), *Leptochloa.*

Tribe VI. *Nazieæ.* Spikelets 1-flowered, naked, spiny, solitary or in spikes. *Nazia.*

Tribe VII. *Festuceæ* (Fescue-Grasses). Spikelets 2—many-flowered, in racemes or panicles; glumes very short; lemmas awned or awnless.

Phragmites (Reed), *Eragrostis, Dactylis* (*D. glomerata,* Orchard-Grass), *Poa* (Meadow-Grass), *Glyceria* (Manna-Grass), *Festuca*

FIG. 434.—*Avena fatua* (*Aveneæ*): *a*, panicle; *b*, several-flowered spikelet; *c*, diagram of spikelet; *d*, lowest floret; *e*, second floret; *f*, upper reduced floret; *le*, lemma; *pa*, palea; *p*, awn of lemma; *g*, transverse section of palea; *h*, anterior view of flower; *lo*, lodicules; *i*, *j*, lateral and posterior views of flower, stamens removed; *n*, node of panicle, with bristles.

(Fescue-Grass), *Bromus* (Brome-Grass), *Melica* (Melic-Grass), *Cortaderia* (*C. Selloana*, Pampas-Grass), *Arundo*. (Fig. 435.)

Tribe VIII. *Hordeæ* (Wheat-Grasses). Spikelets 1—several-flowered, sessile, on opposite sides of a zigzag axis, forming a spike, and often prominently awned.

Lolium (Darnel), *Agropyron* (Wheat-Grass), *Hordeum* (Barley-Grasses), *Elymus* (Rye-Grasses), *Triticum* (Wheat), *Secale* (Rye). (Figs. 336, 337, 63, 68.)

Tribe IX. *Bambuseæ* (Bamboos). Spikelets few- to many-flowered, in panicles or racemes. Tall grasses with broad-bladed, petiolate leaves.

FIG. 435.—*Eragrostis Purshii* (*Festuceæ*): *a*, panicle; *b*, many-flowered spikelet; *lg*, outer glume; *ig*, inner glume; *r*, rachilla, with glumes at base; *c*, section of rachilla enlarged; *le*, lemma; *pa*, palea; *le*, below,—inner face of lemma.

Bambusa (Bamboo), *Arundinaria* (Cane). (Fig. 126.)

Subfamily *Panicoideæ*.

Tribe X. *Maydeæ*. Flowers unisexual, monœcious or diœcious. Glumes hard and awnless. *Zea* (*Z. Mays,* Indian Corn), *Euchlæna* (*E. mexicana,* Teosinte), *Coix* (*C. Lacryma-Jobi,* Job's Tears).

Tribe XI. *Andropogoneæ*. Spikelets in pairs or threes, in spike-like racemes. One spikelet sessile and bisexual, the other bisexual, staminate, or neutral. Glumes usually hardened. Awned lemmas sometimes present.

Andropogon (Beard-Grass), including *A. Sorghum* (Sorghum), *Saccharum* (*S. officinarum*, Sugar-Cane).

Tribe XII. *Paniceæ*. Spikelets 1-flowered, in racemes or panicles. Glumes often very unequal, one large, the other very small. Lemma and palea hardened and tightly enclosing the grain.

Panicum (Panic-Grass), including *P. miliaceum* (Broom-Corn Millet), *Pennisetum* (*P. glaucum*, Pearl Millet), *Chætochloa* (*C. italica*, Foxtail Millet, and *C. glauca*, Foxtail-Grass), *Echinochloa* (*E. crus-galli*, Barnyard-Grass), *Cenchrus* (Bur-Grass).

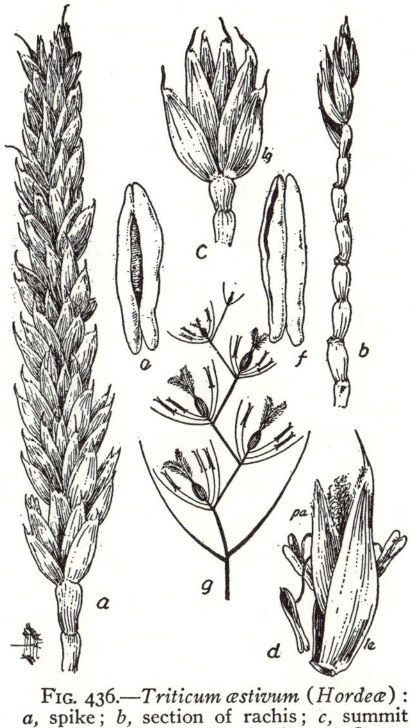

FIG. 436.—*Triticum æstivum* (*Hordeæ*): *a*, spike; *b*, section of rachis; *c*, summit and terminal spikelets of spike; *d*, floret; *le*, lemma; *pa*, palea; *e*, dorsal and *f*, ventral view of anther; *g*, diagram of spikelet (several-flowered); *lg*, first (outer) glume.

For satisfactory study of grasses, fruiting material as well as flowering material is essential, together with as much of the plant as possible, including the underground parts. Few other plants offer such a fascinating field for study, unless one prefers the sedges, an equally fascinating group. Herbarium specimens are easily prepared and kept, and the larger the collection the quicker does one begin to discern generic differences.

CYPERACEÆ (*Sedge Family*). This is a large family of grass-like or rush-like, annual or perennial, fibrous-rooted herbs, mostly with solid (pithy), often 3-angled stems and 3-ranked leaves with closed sheaths. Perennials are rhizomatous. Flowers bisexual or unisexual, solitary in the axile of a glume-like bract of the spikelet. Spikelets 1—several-flowered, arranged in capitate, spicate, or umbellate compound inflorescences. Perianth when present consisting of a number of bristles or scales. Stamens generally 3, rarely more or 1. Gynœcium 3—2-carpellate, syncarpous, with a correspondingly cleft style. Ovary superior. Fruit a nutlet or achene which is triangular in 3-carpellate forms and lenticular in 2-carpellate forms, often beaked by the persistent style. Embryo minute, axile, in the basal part of the mealy endosperm. Species about 3,200, in 70—75 genera, widely distributed over the world. (Figs. 438, 439, 440.)

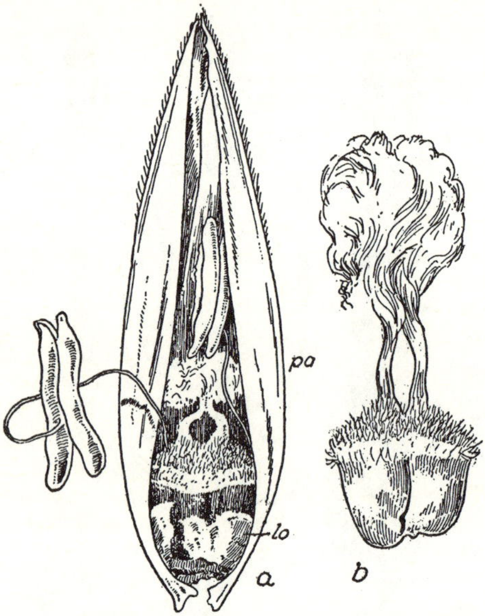

FIG. 437.—Flower of wheat (*Triticum æstivum*): *a,* flower, with one stamen removed; *lo,* lodicules; *pa,* palea; *b,* gynœcium.

There are three subfamilies, characterized as follows:

Subfamily I. *Scirpoideæ.* Flowers bisexual, mostly in many-flowered spikelets. Bristles present or absent. Spikelets flattened or terete.

Including: *Kyllingia, Cyperus, Dulichium, Eleocharis, Dichromena, Psilocarya, Stenophyllus, Scirpus, Fimbristylis, Eriophorum, Fuirena, Hemicarpha, Lipocarpha.*

Subfamily II. *Rhynchosporoideæ.* Flowers bisexual or unisexual, mostly in 1—2-flowered spikelets, with two or more of the lower bracts empty. Bristles present or absent.

Including: *Rhynchospora, Schoenus, Mariscus* (*Cladium*).

Subfamily III. *Caricoideæ.* Flowers unisexual, monœcious, in the same or in separate spikes, or sometimes diœcious.

Including: *Scleria, Kobresia, Carex.*

The following described types will illustrate the general characters of each subfamily.

Subfamily I. *Scirpoideæ: Cyperus* (Gallingale). In this genus few to many flowers are arranged in usually elongate, slender, flattened spikelets aggregated in loose or compact, umbellate or racemose-umbellate inflorescences. Bristles are absent and the achene is not beaked. The common Eastern *Cyperus filicinus,* of brackish marshes of the Atlantic Coast, is a typical form. Besides the many flattened, elongate spikelets arranged in a characteristic umbellate inflorescence, the umbels are subtended by conspicuous bracts (involucre), a characteristic feature of the genus. The bracts or scales of the spikelet, each of which encloses a single flower (floret), are in two ranks, on opposite sides of the somewhat zigzag rachis of the spikelet. The achene is lenticular, since it is developed from the ovary of a 2-carpellate gynœcium. In this genus the long, slender bifid style does not persist. In some species the gynœcium is 3-carpellate, with a corresponding 3-cleft style, and forms a triangular achene. A few species are horticultural, as *Cyperus Papyrus* (Papyrus), *C. alternifolius* (Umbrella Plant), and *C. esculentus* (Chufa). (Fig. 441.)

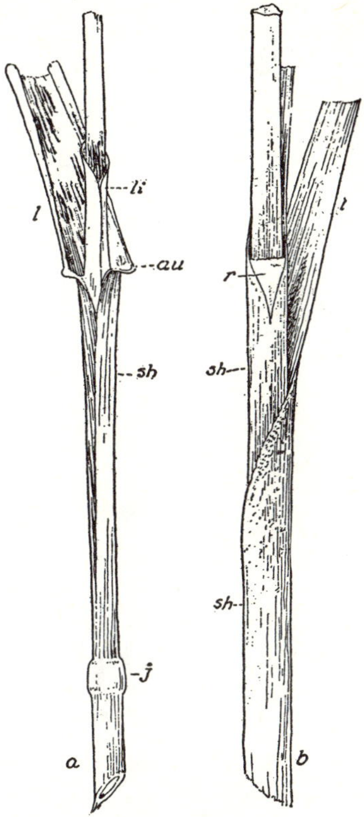

FIG. 438.—Section of culm and leaf of a grass (*a*) and of a sedge (*b*), showing important differences: *l,* blade; *li,* ligule; *au,* auricle; *sh,* sheath,—open in grass, closed in sedge; *j,* joint; *r,* thin membrane closing throat of sheath.

Dulichium arundinaceum (Dulichium), the only species of this genus, is an anomalous sedge possessing certain characters of a grass. For this reason it is very distinctive and easily recognized in the field. It possesses the terete hollow culms and solid nodes of a grass but the closed-sheathed, 3-ranked leaves and the spikelets of a sedge. It occurs in marshy grounds across the continent from Newfoundland to the Pacific Coast. The somewhat flattened elongate and

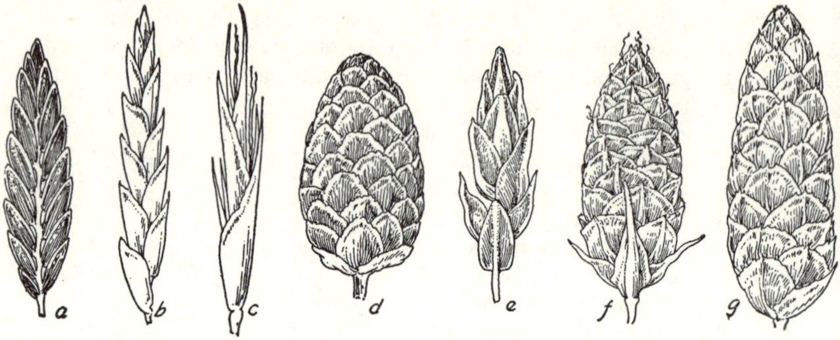

FIG. 439.—Spikelets of various sedges: *a, Cyperus rivularis; b, Cyperus esculentus; c, Dulichium arundinaceum; d, Eleocharis obtusa; e, Stenophyllus capillaris; f, Fimbristylis castanea; g, Scirpus acutus.*

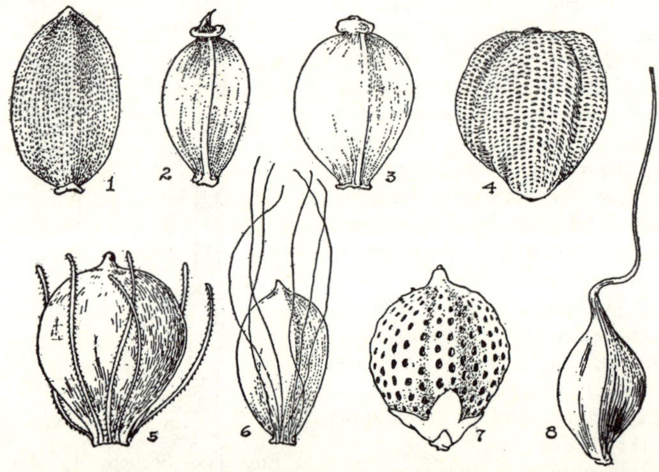

FIG. 440.—Achenes in various sedges: *1, Cyperus Schweinitzii; 2, Eleocharis Wolfii; 3, Stenophyllus capillaris; 4, Fimbristylis castanea; 5, Scirpus validus; 6, Scirpus polyphyllus; 7, Scleria reticulata; 8, Carex lurida.*

sharply pointed spikelets are few-flowered and clustered in elongate flattened inflorescences which spring from the axils of the leaves. The scales of the spikelets are slender and rather closely appressed. Unlike *Cyperus, Dulichium* has a perianth of 6—9 bristles. There are 3 stamens, and a 2-carpellate gynœcium with a long, slender style which forks into two short, stigmatic limbs. The style persists as a beak on the triangular achene. Because of its peculiar characters it has been suggested that *Dulichium* possibly represents a link between the sedges and the grasses (Fig. 442).

Scirpus (Bulrushes). These are common plants of marshes and shallow waters. Some are tall, with terete or triangular culms, and spikelets in 2's or 3's or more in loose or dense umbel-like inflorescences terminating the culms and subtended by an involucre of one or more leaves. Other species are low and sometimes inconspicuous, with simpler inflorescences and without involucre. The spikelets are short and stout, ovoid or elongate; in some species solitary. The scales of the spikelets are spirally arranged and usually all but one or two of the lowermost are flower-bearing. Bristles present or absent. Stamens 2—3. Gynœcium 2—3-carpellate, with as many styles—not swollen at the base. The achene is triangular or lenticulate, and sometimes beaked by the persistent base of the style. For purposes of identification and classification of the species the achenes furnish useful diagnostic characters. Thus the achene of *Scirpus Hallii,* for example, is plainly distinguished by the corrugated surface, and that of *S. rufus* is clearly set off by its unusually long beak. The diminutive *S. nanus* of brackish marshes of the Atlantic Coast, with its bristle-like culms and small, solitary, terminal spikes, bears a close resemblance to *Eleocharis,* and possibly connects *Scirpus* with that genus. (Fig. 443.)

FIG. 441.—*Cyperus filicinus* (*Scirpoideæ*): *a,* habit sketch; *b,* spikelet; *c,* gynœcium.

One of the commonest of the bulrushes is the tall, slender-culmed *Scirpus validus,* which occurs generally throughout the country, except in the extreme South, in shallow waters. The dark-green, smooth, terete culms, sometimes nine feet high, grow from stout rhizomes which can be seen trailing along on the bottom, with rows of culms rising from them. At the summit of the culm, and borne laterally but appearing terminal, is the compound inflorescence of numerous oblong-ovoid spikelets in clusters of two or three or more, or solitary. The few involucral bracts are erect and shorter

FIG. 442.—*Dulichium arundinaceum* (*Scirpoideæ*) : *a*, upper part of stem, with inflorescence; *b*, spikelet; *c*, gynœcium, with bristles (perianth).

FIG. 443.—*Scirpus Hallii* (*Scirpoideæ*) : *a*, habit sketch; *b*, spikelet; *c*, achene, with base of style at summit. The only bulrush with wrinkled achene.

than the inflorescence. There are many bulrushes similar to *Scirpus validus*, more or less, but in some cases with very much larger spikelets (an inch or more long). The achenes are relatively large and subtended by conspicuous barbellate bristles. In the so-called woolrushes of this genus, which owe their name to the long, filiform bristles, the spikelets and achenes are very much smaller, relatively. There are about 150 species in all. Many species possess a pleasing ornamental habit and should be worth cultivating in water-gardens. (Fig. 444.)

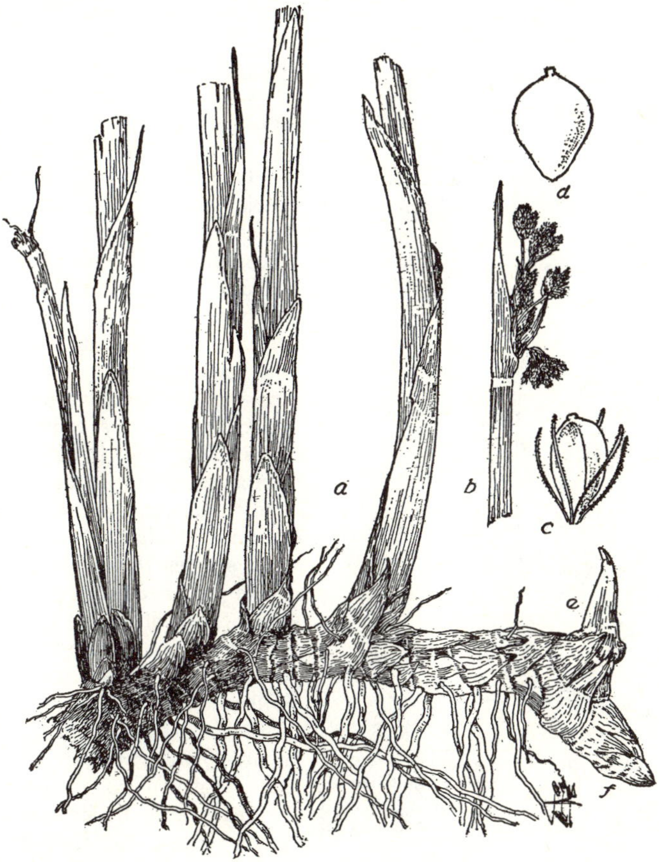

FIG. 444.—*Scirpus validus* (*Scirpoideæ*): *a*, basal part of plant, with scaly rhizome; *e*, lateral bud; *f*, terminal bud; *b*, inflorescence; *c*, achene and bristles, lateral aspect; *d*, achene, dorsal aspect.

Eleocharis (Spike-Rushes). The spike-rushes differ conspicuously from the sedges just described, except some of the smaller bulrushes. A pronounced feature is the appearance of the generally leafless and tufted culms, which are invested at the base by the sheaths of the bladeless leaves. The plants inhabit marshes and wet shores, growing from creeping or densely matted rhizomes. Another distinctive feature is the terminal cylindrical, ovoid or capitate spikelets (or spikes here) of numerous spirally arranged, flower-bearing scales. There are 2—3 stamens and a 2-carpellate or 3-carpellate gynœcium, with correspondingly cleft style, and there is a perianth of generally 6 but sometimes 3—2 bristles which are minutely downwardly barbellate. But a very distinctive feature of the genus is the peculiar tuberculate base of the style which persists at the top of the lenticular or triangular achene as a conspicuous beak. This beak is of peculiar shape according to the species and thus very valuable as a diagnostic character. In certain species the tubercle is as large as or even larger than the achene itself (Fig. 445). There are also cases where certain vegetative characters will identify a species with some certainty; thus *Eleocharis quadrangulata,* a rather rare Eastern spike-rush, can be recognized by its quadrangular culms—as the specific name suggests. The spike-rushes number about 140 species, widely distributed, probably over 40 in the United States. (Fig. 446.)

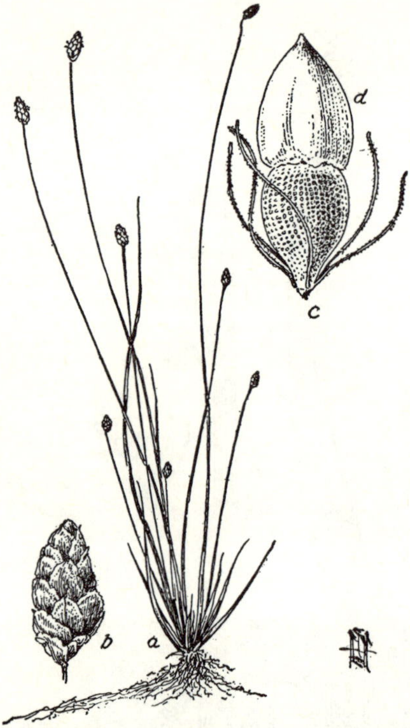

FIG. 445.—*Eleocharis tuberculosa* (*Scirpoideæ*): *a,* habit sketch; *b,* spikelet; *c,* achene, with bristles (perianth), and at summit the large persistent base (*d*) of style (beak).

Eriophorum (Cotton-Grass). The cotton-grasses are plants of Northern bogs but are often conspicuous in road-side ditches. They are relatively few in number, probably 15 species all told. They are mostly tufted perennials growing from matted rhizomes. Their culms are slender and low, triangular or somewhat terete, with very narrow leaves, the uppermost leaves reduced to sheaths. The spikelets are terminal, solitary, or in heads or umbels which are naked or more or less involucrate. Flowers bisexual, with a perianth of 6 to numerous, white or brownish, greatly elongate bristles which at maturity give the heads the appearance of a ball of cotton and make the plants so conspicuous even at a distance. Stamens 1—3. Gynœcium 3-carpel-

FIG. 446.—*Eleocharis acicularis* (*Scirpoideæ*): *a,* habit sketch; *b,* spikelet; *c,* section of spikelet with scale and axillary flower (at right); *d,* gynœcium, early stage,—this specimen possessed four stigmas; *e,* gynœcium after anthesis; *f,* anther; *g,* scale (bract) of spikelet.

late. Achene triangular (trigonous). A widely distributed species is *Eriophorum gracile,* which occurs in cold bogs and swamps from the North Atlantic Coast to the Pacific Coast. (Fig. 447.)

Fimbristylis, a large genus of about 125 species of wide distribution, is represented in North America by about a dozen species. These are annuals or perennials, with culms leafy below. In some respects they remind one of the bulrushes, as the heads are similar in appearance. There is no perianth, however, and the base of the style is bulbous, but does not persist. The inflorescence is an involucrate, simple or compound umbel. *Stenophyllus* comprises about 90 species, mostly of warm latitudes, but the slender, tufted annual, *S. capillaris,* invades the North as far as Canada. The plant resembles some of the small species of *Fimbristylis*. A few species are found in the South. *Dichromena,* with about 20 species, range from tropical to temperate America. *Kyllingia, Psilocarya, Lipocarpha, Hemicarpha,* and *Fuirena* are small genera of low, mostly tufted sedges poorly represented in our flora.

Fig. 447.—*Eriophorum tenellum* (*Scirpoideæ*) : *a,* habit sketch; *b,* inflorescence; *c,* involucral bracts.

Fig. 448.—*Rhynchospora alba* (*Rhynchosporoideæ*) : *a,* inflorescence and leaf; *b,* spikelet; *c,* gynœcium and bristles (perianth).

Subfamily II. *Rhynchosporoideæ: Rhynchospora* (Beak-Rushes).

Spikelets elongate and slender or short, in loose or dense clusters. The lower flowers of the spikelets are bisexual, the upper unisexual. A perianth of bristles is present. The smooth or variously wrinkled achene is capped by the persistent base of the style, or in some of the coarser species by the whole style, which forms a very prominent and characteristic beak. The shape of the beak and the surface markings of the achenes furnish reliable diagnostic characters. Species about 200, widely distributed and well represented in our flora. (Fig. 448.) Stamens 3 and gynœcium 2-carpellate. In *Mariscus,* which somewhat resembles *Rhynchospora,* the perianth is absent, stamens 2 or sometimes 3, and the gynœcium 2—3-carpellate, with 2—3-cleft style—which is deciduous. The achene is thus without a tubercle.

FIG. 449.—*Scleria reticularis* (*Caricoideæ*): *a,* habit sketch; *b,* inflorescence; *c, d, e,*—different views of achene, with persistent perianth; *f,* gynœcium.

Subfamily III. *Caricoideæ: Scleria* (Nut-Rushes). The nut-rushes are leafy sedges with 3-angular culms, and mostly perennial from rhizomes. The spikelets are borne in terminal or axillary clusters. The distinctive features of these sedges are the globose, crustaceous or stony, smooth or wrinkled or warty achenes with a peculiar disk or cushion at the base—a structure which possibly represents a modified perianth. There are no bristles. Stamens 1—3. Gynœcium 2—3-carpellate, with correspondingly cleft style, which is persistent or deciduous. Flowers unisexual and monœcious, the carpellate spikelets 1-flowered and usually intermixed with the staminate spikelets. Species about 200, temperate and tropical; about 15 in our flora. (Fig. 449).

Kobresia comprises about 30 species of small sedges of arctic and alpine regions. Flowers unisexual, in 1-flowered spikelets, or sometimes in 2-flowered spikelets—in which one flower is staminate and the other carpellate. A notable feature is that the carpellate flower is borne in the axil of a spathe-like bract which is subtended by a scale of the spikelet. (Compare with *Carex,* below!)

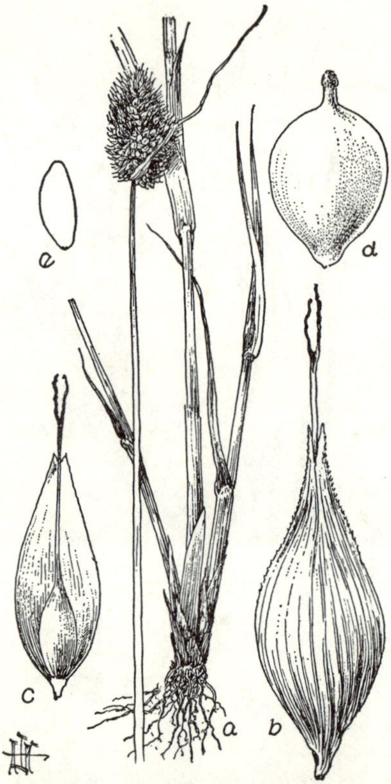

FIG. 450.—*Carex adusta* (*Caricoideæ*): *a,* habit sketch; *b,* perigynium with protruding style; *c,* section of perigynium, with enclosed gynœcium; *d,* achene, with persistent base of style (beak); *e,* transverse section of achene.

Carex (Sedge). This is a tremendous genus of about 1,000 species, and by far the largest genus of the family, as well as one of the largest genera of all flowering plants. In habit the plants range from low and slender to tall and robust forms, perennial from rhizomes. They are all grass-like and generally pass for grasses, but they possess the typical 3-angular culms and 3-ranked leaves of the sedges. The small unisexual flowers are rarely diœcious, but are borne together in the same spike or in separate spikes on the same culm. Stamens 3, in the axil of a scale. Gynœcium 2—3-carpellate, with 2—3-cleft style. The outstanding feature of *Carex* is the *perigynium*—the sac-like closed *spathe* which encloses the achene and is subtended by a bract of the spikelet. To this perigynium one looks for the important diagnostic characters. (Fig. 450.) Our species fall conveniently into two subgenera—one characterized by a 2-carpellate gynœcium and a correspondingly lenticular achene, the other by a 3-carpellate gynœcium and triangular (trigonous) achene. (Figs. 450, 451.)

Certain species of *Carex* exhibit a remarkable behavior when subjected to injury. The effect is to cause them to produce a second crop of spikelets with flowers which are often abortive and of fantastic structure. Sometimes several flowers are produced on a single stalk (rachilla!) within the perigynium. This several-flowered stalk with its investing perigynium may be interpreted as a spikelet—a reversion to an ancestral condition. From this we may infer that the present condition in *Carex* was brought about by suppression of all but one of the flowers (the terminal?) of such a primitive spikelet. And it is to be noted further that it is but a step from the *open* spathe of *Kobresia* to the *closed* perigynium of *Carex*. With respect to the genus as a whole, those species that possess a simple spicate inflorescence may be regarded as the more primitive.

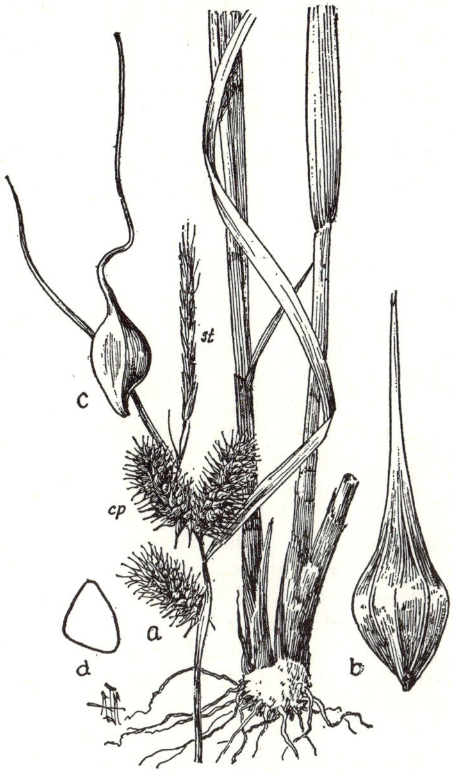

FIG. 451.—*Carex lurida* (*Caricoideæ*) : *a,* inflorescence and base of plant; *st,* staminate and *cp,* carpellate, spike; *b,* perigynium; *c,* gynœcium; *d,* transverse section of achene.

The species are too numerous to mention any one in particular, but *Carex stricta,* a species with wiry leaves and forming extensive pure stands in Northern marshes or bogs, is utilized in the making of mats ("Crex"). This is the only sedge that possesses any utilitarian value to speak of.

ORDER PRINCIPES

Flowers cyclic, trimerous, hypogynous, regular or in some cases slightly irregular, unisexual, monœcious or diœcious, rarely bisexual, in simple or branched, often gigantic inflorescences usually subtended by a large spathe. Perianth of two whorls, sometimes differentiated into calyx and corolla. Stamens mostly 6, but sometimes 3, 9, or many. Gynœcium 3-carpellate, apocarpous or syncarpous, with a 1—3-celled ovary; ovules mostly 1 in each cell. Abortion of one or two carpels occurs occasionally. Fruit a berry, drupe, or nut. Endosperm copious, oily or bony. Embryo small, lateral, the cotyledon enlarging greatly at germination. Only the following family.

PALMACEÆ (*PALMÆ*), *the Palms.* The palms are only scantily represented outside the tropics and subtropics. They are exceedingly numerous and a technically difficult group: species about 1,200, in about 140 genera. Many of them are tree-like, with graceful trunks ending in a crown of large leaves, among which are borne the huge flower-clusters. Some species grow from subterranean stems and a few are climbing (*Calamus*). Branching is not uncommon, but sprouts are frequently put forth from the base. The most striking feature is the usually enormous leaves which are of two distinct sorts—in the one case fan-shaped (palmate), in the other feather-shaped (pinnate); in the former the petiole is very prominent. In the bud the leaves are folded in two ways, thus: *induplicate* when folded in V-shaped fashion and *reduplicate* when folded back like an inverted *V* — Λ. The juvenile leaves may be very different from the adult form. The flowers are very small, but the fruit in certain cases grows to enormous size (*Lodoicea*). Among the palms are said to be found some of the most useful of all plants (*e.g.,* Date Palm). Only in southern Florida and in southern California do the palms reach our flora. (Figs. 452, 453, 454.)

For the sake of convenience the palms may be grouped on the basis of leaf-form as "fan-palms" and "feather-palms," whereas a technical grouping would take into consideration the structure of the flower and the fruit.

Under the technical classification five subfamilies of palms are recognized. These are:

Subfamily I. *Coryphoideæ.* Carpels 3, distinct or slightly united. Fruit a berry. Diœcious forms occasional.

Including: *Phœnix dactylifera* (Date Palm), *P. canariensis* (Canary Island Palm), *Chamærops, Rhapis, Corypha, Copernicia, Livistona, Washingtonia.*

Subfamily II. *Borassoideæ.* Carpels 3. Ovary 3-celled. Fruit drupaceous.

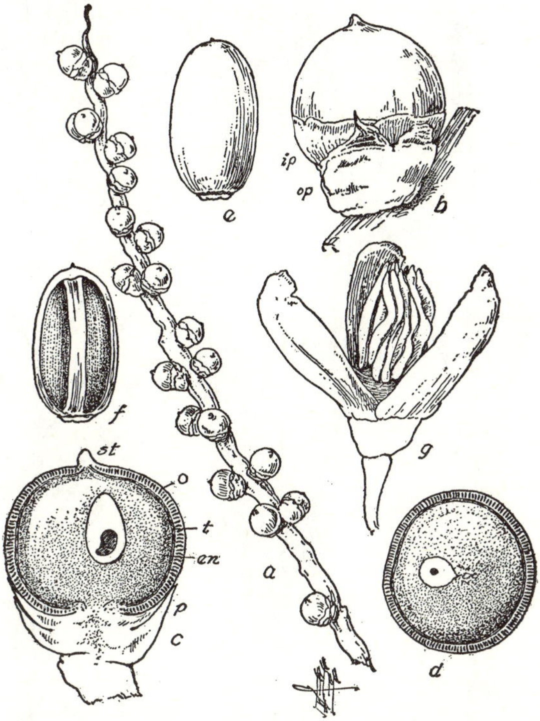

FIG. 452.—*Phœnix canariensis* (*Palmaceæ*) : *a,* branch of carpellate inflorescence; *b,* young fruit (ovary) seated in inner (cup-like) and outer perianth (*ip, op*) ; *c,* median, *d,* transverse section of young fruit; *st,* base of style; *o,* ovary-wall (pericarp) ; *t,* testa; *en,* endosperm, with embryo within; *e,* fruit; *f,* section of fruit, showing stone (seed).

Including: *Hyphæne, Borassus flabelliformis* (Palmyra Palm), *Lodoicea* (fruit the largest known in the wild state).

Subfamily III. *Lepidocaryoideæ.* Flowers bisexual or unisexual. Carpels 3, united. Ovary 3-celled. Fruit covered with scales. Dimorphic flowers and diœcious forms occurring.

Including: *Mauritia, Raphia* (leaves gigantic—up to 20 meters),

FIG. 453.—*Phœnix canariensis:* a, seedling; b, juvenile leaf; c, later leaf of seedling.

Metroxylon, Cœlococcus (yields "vegetable ivory"), *Calamus rotang* and *C. araco* (yields "dragon's blood").

Subfamily IV. *Ceroxyloideæ.* Carpels 3. Ovary 3—1-celled. Fruit a berry or drupe. Monœcious and diœcious forms occurring.

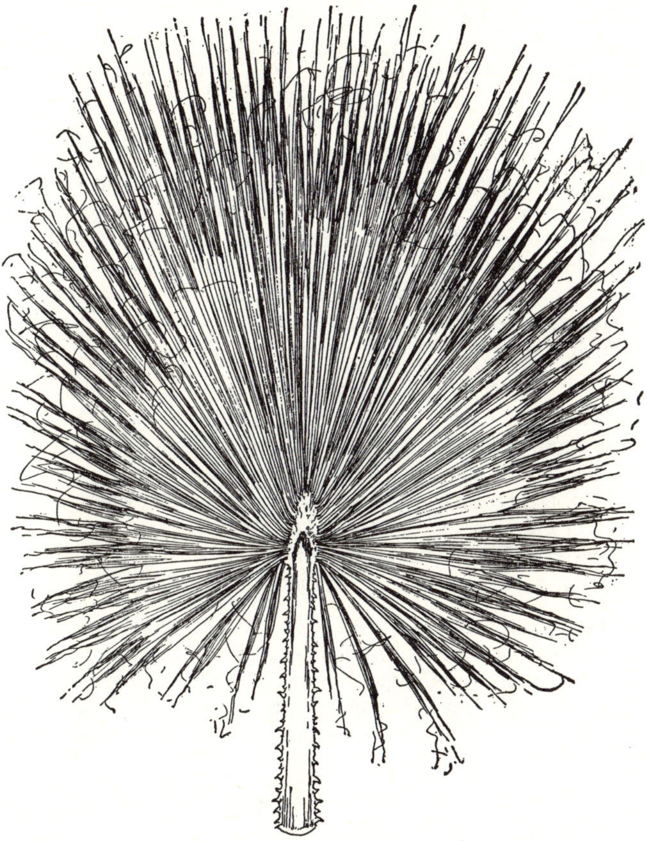

FIG. 454.—Leaf of *Washingtonia filifera;* shaded conical area at base of blade is the ligule.

Including: *Arenga, Caryota, Ceroxylon, Kentia, Oreodoxa regia* (Royal Palm), *Areca catechu* (Betel-Nut Palm), *Elæis, Cocos nucifera* (Cocoanut Palm), *Jubæa, Bactris.*

Subfamily V. *Phytelephantoideæ.* Carpels several (?). Ovary

4—9-celled. Flowers unisexual, the carpellate naked, the staminate with numerous stamens.

Including: *Phytelephas* (yields "vegetable ivory").

Subfamily VI. *Nipoideæ*. Carpels 3 (?). Ovary 1-celled. Flowers unisexual, the carpellate naked, the staminate with united stamens forming a synandrium. Including *Nipa.*

Washingtonia filifera (Washington Palm). This tall, stately fan-palm is native to southern California. At its best it is said to reach a height of 70 feet, with a trunk-diameter of 25—30 inches. At the summit of the trunk is borne a huge crown of pale-green leaves which become more or less frayed by wind. A striking feature, besides the crown of green fan-leaves, is the dense shroud of dead leaves of previous years that envelop the trunk below the crown. In southern California this palm is one of the stateliest of avenue trees. The small flowers are produced in enormous inflorescences (up to ten feet long). The fruit is a small, black, elliptical berry containing an elongate brownish stone. At the base of the leaf-blade is a prominent ligule (Fig. 454).

ORDER SYNANTHÆ

Flowers unisexual, monœcious, with or without a perianth. Staminate flowers with 6 to numerous stamens. Carpellate flowers with a staminode opposite each segment of the perianth. Carpels 2 or 4, united. Ovary imbedded in the spadix. Fruit (spurious) a berry; berries aggregated in a multiple fruit along with the spadix. Seeds numerous, on 2—4 parietal placentas.

One family: *CYCLANTHACEÆ* (Cyclanthus Family). Palm-like, acaulescent or caulescent plants of the American tropics. Inflorescence a spadix subtended by several bract-like spathes. About 44 species in 6 genera. Greenhouse plants in the north. Most important genus: *Carludovica,* with about 34 species. *Cyclanthus,* 4 species.

ORDER SPATHIFLORÆ (ARALES) (AROIDS)

Flowers cyclic, bisexual or unisexual, 3-merous or 2-merous, hypogynous. Perianth undifferentiated, sometimes wanting. Stamens 3—1. Gynœcium 3—1-carpellate, syncarpous. Inflorescence a spadix enclosed in a leaf-like bract called the *spathe*. Only the two following families.

Plants large, with ordinary leaves. Flowers numerous on a prominent spadix, more or less enclosed in a prominent spathe. *Araceæ.*

Plants aquatic, the plant-body reduced to a minute thalloid structure. Flowers reduced to a single stamen or carpel. *Lemnaceæ.*

The aroids are among the most unique of flowering plants. Many of them are decidedly weird in appearance. Some grow to large size,with thick stems and large leaves, while others are the smallest flowering plants known. The most distinctive feature of the plants is the inflorescence.

ARACEÆ (*Arum Family*). A large tropical family of over 1,500 species in about 100 genera. A characteristic of the plants is the pungent sap and the presence of biting crystals (rhaphides) in the tissues. Some members grow to very large size, bearing huge leaves. Among primitive peoples certain aroids were used for food. The family is very well defined, and the characteristic inflorescence enables a person to recognize the members with comparative ease.

The following genera are represented in our flora: *Arisæma, Peltandra, Calla, Symplocarpus* (*Spathyema*), *Lysichiton, Orontium, Acorus.* These will be sufficiently characterized by a description of a few representative species.

One of the commonest of our aroids is the quaint *Arisæma triphyllum* (Jack-in-the-pulpit) which is found in springy places in cool, deciduous woods. The plants are distinguished by one or two trifoliolate leaves and the single scape arising from a stout corm. The juice of the corm is exceedingly biting, which may explain why the plant is sometimes called Indian turnip. At the summit of the scape is the characteristic spathe, the broad end of which forms a sort of canopy over the enclosed spadix. The upper part of the spadix is prominently club-shaped, with the flowers borne below. As the species is often diœcious, it is necessary, when collecting, to look for both individuals. There is no perianth. The filaments are very short and bear 4-celled anthers which open at the top by means of pores. The gynœcium is 1-carpellate (!). Five or six seeds are formed in the 1-celled ovary. At maturity the spadix bears a mass of bright-scarlet berries. The genus contains about 50 species, mostly native to Asia. In our flora there are probably 6. The spathe usually furnishes the significant diagnostic characters; however, *Arisæma Dracontium,* for example, is readily set off by the many elongate and narrow segments of its solitary leaf.

Symplocarpus fœtidus (Skunk Cabbage) is, with the exception of the similar appearing but larger *Lysichiton kamtschatcense* of the Pacific Coast, the largest aroid in our flora. Like *Arisæma* it inhabits wet or springy places in cool, deciduous woods. Its thick, stout,

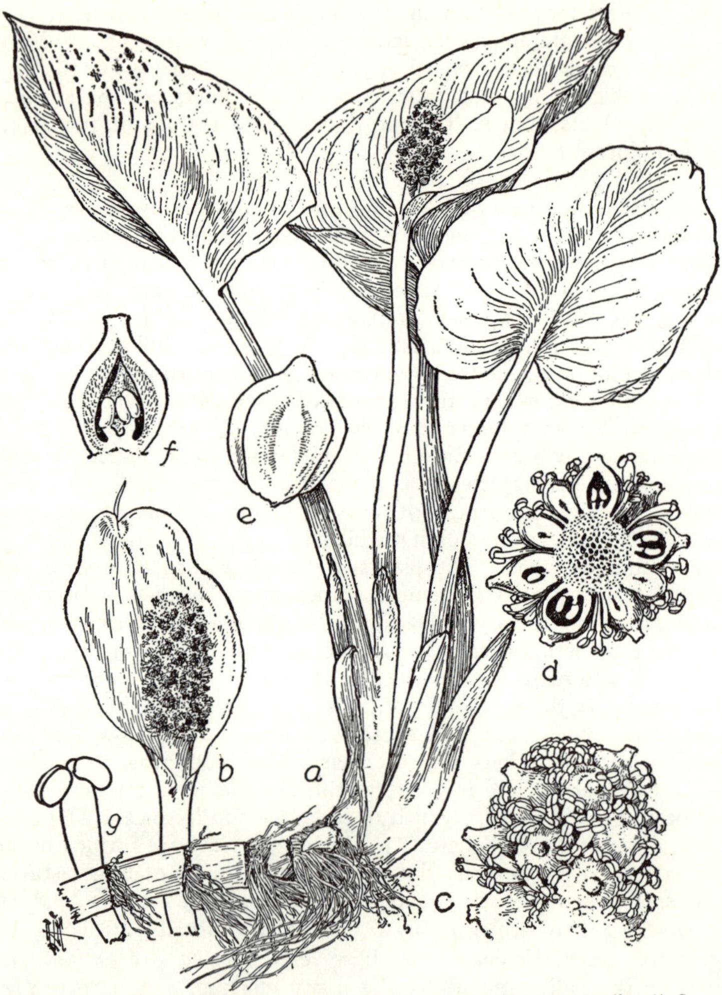

FIG. 455.—*Calla palustris: a,* habit sketch; *b,* spathe and spadix (inflorescence); *c,* surface view of summit of spadix, showing gynœcia and stamens; *d,* transverse section of spadix; *e,* gynœcium; *f,* median section of gynœcium; *g,* stamen.

purplish-mottled spathes can be seen coming up out of the ground early in the spring, these being followed some time later by the large green leaves. The plants possess a characteristic and unmistakable mephitic odor. The bisexual flowers are 2-merous with respect to the perianth and the andrœcium; the 4 perianth-segments are similar. And here as in *Arisæma* the gynœcium is 1-carpellate, but quadrangular-pyramidal, with a thick sessile stigma, and the fruits are achenes instead of berries.

The solitary *Calla palustris* (Water-Arum or Wild Calla Lily) encircles the globe in temperate and subarctic latitudes. It is an inhabitant of low marshy or boggy places, often occurring among cattails. From a thick soft rootstock it sends up a few long-petiolate leaves with broad ovate to cordate blades and a single scape which terminates in a conspicuous snow-white spathe. Bisexual flowers are produced which cover most of the thick spathe, and from these spring bright-red berries. There is no perianth, but the gynœcium is 1-carpellate as in the two preceding genera. (Fig. 455.)

In striking contrast to these forms is the common sweet-flag, *Acorus Calamus,* a stout sedge-like plant growing in marshy grounds throughout the northern parts of the world. Its long, rather rigid, sword-shaped, ascending leaves set it off from all the other aroids. Very distinctive, too, is its rather large cylindrical spadix borne at the summit of the tall scape and to one side, with the spathe extending erect beyond it as a continuation of the scape. The whole plant, including the thick, fibrous-coated and fibrous-rooted but spongy rootstock, is pleasantly aromatic—a feature which enables one to detect the presence of the plant when walking through a marsh. In this plant the flowers are bisexual, with 6 stamens and a 2—3-carpellate gynœcium, the ovary of which is 2—3-celled, with 2—8 ovules in each cell. The fruits are gelatinous and berry-like. The plant has been used for a long time in medicine, and the rootstock is said to have been used to some extent as a confection by the Puritans. A smaller species, *Acorus gramineus,* is found in central and eastern Asia and in Japan.

A considerable number of aroids are popular hothouse and potherb plants, and many are grown in gardens in southern parts of the country. Among these may be mentioned *Zanthedeschia æthiopica* (the "calla lily"), and species of *Arum, Alocasia, Colocasia, Monstera, Spathiphyllum, Scindapsus.* These are prized chiefly as curiosities on account of their foliage, odd inflorescences ("flower"), and often grotesque habit. *Monstera deliciosa,* a very large plant, is prized for its huge leaves. The fruits are edible but contain bitter rhaphides.

LEMNACEÆ (*Duck-Weed Family*). The duck-weeds are the smallest of the flowering plants. They are minute, perennial, floating aquatics, with a plant-body reduced to a flattened or discoid, more or less elliptical, oblong, or irregular, thallus-like structure; in some forms one or more rootlets are suspended from the lower surface of the thallus. The minute, naked flowers, which are unisexual and monœcious, are borne on the edge or the upper surface of the thallus. The staminate flowers are reduced to a single stamen, and the carpellate to a single carpel which yields one or more seeds. There are 4 genera, namely: *Spirodela, Lemna, Wolffia,* and *Wolffiella,* all represented in our fresh waters. The minutest forms, not much larger than a pinhead, are found in species of *Wolffia.* The plants multiply very rapidly by vegetative propagation. Species of *Lemna* and *Spirodela* cover ponds and ditches in profusion.

ORDER FARINOSÆ (XYRIDALES)

Flowers cyclic, 3-merous or 2-merous, regular or irregular, hypogynous or epigynous. Perianth simple or differentiated as calyx and corolla, sometimes bracteoid, and in some cases gamophyllous. Stamens 6—1, sometimes modified into staminodes. Gynœcium 3-carpellate, rarely 1-carpellate. Seeds mostly with starchy endosperm. Embryo straight. (Order owes its name to the mealy endosperm).

The plants of this order are herbs of wet or dry habitats, or aquatics; a few are epiphytic (*Tillandsia*).

Families: (1) Flowers hypogynous: *Flagellariaceæ, Restionaceæ, Centrolepidaceæ, Mayacaceæ, Xyridaceæ, Eriocaulaceæ, Thurniaceæ, Rapateaceæ, Bromeliaceæ* (in part), *Commelinaceæ, Pontederiaceæ, Philydraceæ.*

(2) Flowers epigynous: *Bromeliaceæ* (in part), *Cyanastraceæ* (African, *Cyanastrum,* 4 species).

The following key covers the families represented in our flora.

A. Flowers hypogynous (ovary superior).
  a. Carpels 3.
    Plants moss-like, with solitary terminal flowers. *Mayacaceæ.*
    Plants not moss-like.
      Plants with tufted, basal, grass-like leaves and flowers in ovoid or oblong rush-like heads. *Xyridaceæ.*
      Plants epiphytic, with linear leaves, hanging in loose tufts from the branches of trees. Atlantic Coast from Virginia to Florida. *Bromeliaceæ.*

Bog or marsh plants with hastate, kidney-shaped, or linear leaf-blades, and solitary or clustered, white or blue flowers. *Pontederiaceæ.*

b. Carpels 2—3.

Plants with grass-like leaves in basal rosettes. Flowers in globose heads on scapose stems. *Eriocaulaceæ.*

Plants with broad to narrow, lanceolate or linear, grass-like, sheathing leaves not in rosettes. Flowers often showy, in cymose-umbellate clusters subtended by (often long) leafy bracts. Calyx green. Stamens 3—6. *Commelinaceæ.*

B. Flowers epigynous (ovary inferior). Carpels 3. Inflorescence a scape, not subtended by a leafy involucre, or an involucre of reduced leaves is present. Foliage-leaves in rosettes, generally sword-shaped and rigid, with spiny or spinulose margins. *Bromeliaceæ.*

The Hypogynous Farinosæ

FLAGELLARIACEÆ. Three genera and about 7 species of tropical Asia, Australia, and Africa. *RESTIONACEÆ* includes about 250 species and 14 genera of temperate and subtropical climates. Flowers mostly unisexual and diœcious, with a differentiated perianth resembling that of *Juncus*. Gynœcium 3—2—1-carpellate, forming a 3—2—1-celled capsule. The largest genus, *Restio,* is represented by about 100 species in Australia and South Africa. The *CENTROLEPIDACEÆ* are small, rush-like bog plants of Australia, Tasmania, and New Zealand. Species about 30, in 6 genera. The *MAYACACEÆ* are small, moss-like plants of wet habitats, chiefly of the Americas. One genus—*Mayaca,* of about 8 species. Gynœcium 3-carpellate but with 1-celled ovary containing several ovules on parietal placentas. Perianth differentiated as calyx and corolla. Fruit a capsule. *Mayaca Aubleti* occurs in fresh-water pools and streams from Virginia to Florida and Texas.

XYRIDACEÆ (*Yellow-eyed-Grass Family*). Small annual or perennial, tufted, grass-like plants of the tropics. Absent from Europe. The perianth is differentiated into an *irregular* calyx and a regular, gamophyllous corolla. The three outer stamens are wanting or transformed into staminodes, the three inner are epipetalous. Gynœcium 3-carpellate, with a 1-celled or incompletely 3-celled ovary containing numerous or few orthotropous ovules on parietal placentas. The style is three-branched above. Anthers extrorse. Fruit a capsule dehiscing by three valves and invested by the withered corolla. Species 60 or more; genera two—*Xyris* and *Albolboa.* About 17 species of *Xyris* (Yellow-eyed-Grass) occur within our limits. These inhabit mostly low, moist grounds, bogs, or swamps. The flowers are mostly yellow, in the axils of coriaceous bracts forming oblong-cylindrical or ovoid heads borne solitary at the summit of a slender scape. The plants are all Eastern and Southern. The most widely distributed are *Xyris flexuosa, X. montana,* and *X. caroliniana. X. montana* crosses into Canada from Nova Scotia to Ontario. Compare with the next following family, especially with respect to the type of inflorescence.

ERIOCAULACEÆ (*Pipewort Family*). A family of about 560 species in 9 genera, widely distributed in the tropics, with a few in Temperate climates, but chiefly South American. The plants are mostly small perennial herbs of bogs and wet shores, forming tufted, grass-like basal leaves, and bearing small unisexual, monœcious or diœcious flowers in solitary heads at the summit of a slender scape. The heads are subtended by a bracteate involucre, and each flower is borne in the axile of a scale, the whole inflorescence (head) very suggestive of that of the composites among the Dicotyledons. In monœcious forms the staminate and the carpellate flowers are often mixed together in the same head. The perianth is regular or irregular and somewhat differentiated, mostly in 2 whorls. The outer whorl of stamens is frequently absent. Gynœcium 2—3-carpellate, with corresponding style-branches. The 2—3-celled ovary contains 1 suspended ovule in each cell on axile placentas. Fruit a loculicidal capsule. Endosperm present. Affinities probably with the *Centrolepidaceæ* and the *Restionaceæ* with which compare. Our genera are: *Eriocaulon, Syngonanthus, Lachnocaulon.*

The family has been divided into two subfamilies or tribes on the basis of the number of cells in the anthers, thus:

I. *Diplantheræ* (meaning "2-anthered") in which the anthers are 2-celled. Stamens distinct and perianth in two whorls. Including *Eriocaulon* and *Syngonanthus.*

II. *Haplantheræ* (meaning "half-anthered") in which the anthers are 1-celled. Stamens united at the base and perianth in one whorl. Including *Lachnocaulon.*

About 200 species of the family belong in *Eriocaulon,* of which 6 or 7 occur in the United States. *Eriocaulon septangulare* reaches Canada, from Newfoundland to Ontario; this species is of interest in connection with the study of plant distribution, as it occurs also in Ireland and Scotland, thus tying up the flora of Europe with that of eastern North America. A specific feature of the plant is the presence of seven parallel ridges of the scape ("septangular"). The genus is mainly tropical. *Syngonanthus,* with 80 species, is a genus of tropical America, but *S. flavidulus* has succeeded in pushing north along the Atlantic Coast from Florida to Virginia. *Lachnocaulon* is strictly American, 8 species occurring in southern United States, one of them, *L. anceps,* reaching Virginia.

The *THURNIACEÆ* are perennial herbs with narrow leaves and numerous flowers aggregated in globose-capitate involucrate heads, each flower seated in the axil of a bract (scale). One genus, *Thurnia,* with two species in British Guiana. The plants indicate affinities with the *Juncaceæ.*

RAPATEACEÆ. Perennial herbs with linear or lanceolate basal leaves, growing from thick rhizomes. The inflorescence consists of numerous 1-flowered spikelets aggregated in a head or in a 1-sided spike and subtended by 1 or 2 more or less connate bracts at the summit of a scape. Perianth differentiated into a calyx and a gamophyllous corolla. The anthers are peculiar in being 4-celled at the base, with the cells fusing into a tube at the summit. Gynœcium 3-carpellate, with a simple style ending in a very small stigma; ovary 3-celled with 2 or more ovules on axile placentas, or with 1 or 2 basal ovules. Fruit a 3-celled, loculicidal capsule, or 1-celled and 2-valved by abortion. Seeds sometimes with a spongy cap at one end. The 20 or more species are grouped in about 6 genera, all South American. *Rapatea* and *Saxo-Fridericia* the largest genera, each with 5 species.

COMMELINACEÆ (*Spiderwort Family*). The spiderworts are annual or perennial herbs with jointed stems and broad to narrow, grass-like, clasping or sheathing leaves, and generally more or less succulent. Flowers regular or irregular, bisexual, often showy, in cymes subtended by spathe-like or involucre-like, often conspicuous, leafy bracts. Perianth differentiated as a green calyx and colored corolla. Stamens 6, some of them in certain cases transformed into staminodes or wanting; filaments filiform or somewhat flattened, often clothed with moniliform hairs; anthers often with an expanded or variously modified connective. Gynœcium 3—2-carpellate, with 1 style. Ovary superior, 3—2-celled; ovules solitary or several in each cell, on axile placentas. Fruit a loculicidal or indehiscent capsule. Species about 300, in 25 or more genera, of the tropics and subtropics but with some members reaching well into northern latitudes. The only genera with us are *Commelina* and *Tradescantia,* the former represented by 9 or 10, the latter by 14 or 15 species, mostly Eastern and Southern. Species of *Zebrina, Dichorisandra,* and *Rhœo* are horticultural.

Zebrina pendula (Wandering Jew) and its varieties are common hot-house plants and potherbs. (Fig. 125c). In *Commelina* (Day-flower) the flowers are distinctly irregular both as to corolla and calyx, and there are usually 3 staminodial stamens. The filaments are all glabrous. The seed-coats are rough, smooth, or reticulate.

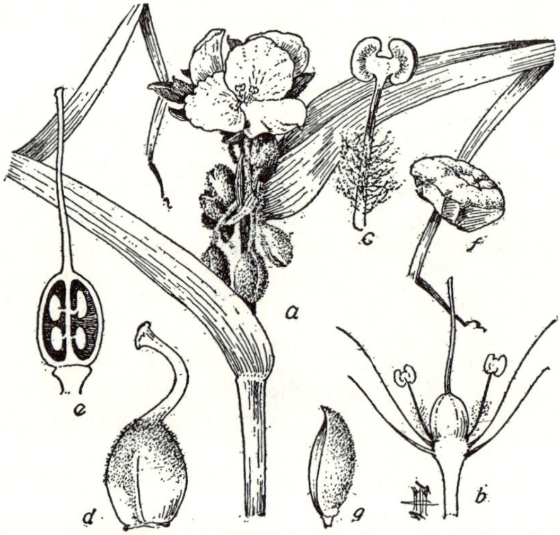

FIG. 456.—*Tradescantia bracteata: a,* inflorescence and leaf-like bracts; *b,* section of flower, semi-diagrammatic; *c,* stamen; *d,* gynœcium; *e,* gynœcium in sectional view; *f,* seed.

In *Tradescantia* (Spiderwort) there is a clearly differentiated green calyx and a showy blue or violet corolla, and the flowers are regular. Besides these well-marked features, the anthers possess greatly expanded connectives, while the filaments are clothed with long moniliform hairs. The numerous flowers are borne in umbellate clusters subtended by conspicuous elongate involucral leaves of a grass-like type. A further prominent feature is seen in the thick-jointed stems. Species about 35. (Fig. 456.)

PONTEDERIACEÆ (*Pickerel-Weed Family*). Five genera and 20—25 species of fresh-water herbs, distributed from tropical America to Canada. The plants float on the surface or grow fixed in the mud of bogs or shores. Leaf-blades large and hastate, arrow-shaped or kidney-shaped, or in some cases grass-like. Flowers irregular, bisexual, solitary or in spikes, with undifferentiated, gamophyllous perianth forming a prominent tube. Gynœcium 3-carpellate, with simple style. Stamens 3—6, inserted on or at the base of the perianth-tube. Fruit a capsule or utricle. Seeds with copious endosperm. Genera with us: *Eichhornia, Pontederia, Heteranthera.*

The tropical *Eichhornia crassipes* (Water-Hyacinth) is a common plant of hot-houses and water-gardens. It has become established in parts of the Gulf States, where, owing to its rapid vegetative propagation, it becomes so dense as to obstruct navigation. The plants float by means of strongly inflated petioles. The leaf-blades are reniform (Fig. 125a). *Pontederia cordata* (Pickerel-Weed) is found in shallow waters of ponds and streams from the Gulf States to Nova Scotia and Minnesota. It is a polymorphic species with sagittate to narrow leaf-blades on sheathing petioles, often with stipular appendages. Flowers numerous, blue, trimorphic, in a spathe. Stamens 6. Gynœcium 3-carpellate; ovary 3-celled, but two of the cells abortive and empty. Fruit a 1-seeded utricle. The outstanding feature of the flowers is the 2-lipped perianth, each lip 3-lobed. *Heteranthera* is represented by 4 species, one of which, *H. dubia,* ranges across the continent. About 10 species are known, two of them occurring in tropical Africa(!). The plants grow in mud or shallow water. Leaf-blades round-ovate to reniform or grass-like. One to several small flowers are borne in a spathe. Perianth only slightly differentiated. Stamens 3. Fruit a capsule containing numerous seeds.

PHILYDRACEÆ. Flowers irregular, bisexual, with petaloid perianth. Andrœcium consisting of but 1 stamen, the anterior. Gynœcium 3-carpellate, with 1 style. Ovary superior, 3-celled or 1-celled, in the former case with axile, in the latter with parietal placentation. Ovules numerous. Fruit a capsule. Embryo small, surrounded by the endosperm.

Herbs with two-ranked leaves, and flowers in spicate inflorescences. A small family of less than a half-dozen species. The solitary *Philydrum lanuginosum* occurs as a bog or marsh plant in the Indo-Malayan region.

BROMELIACEÆ (*Pineapple Family*). The hypogynous *Bromeliaceæ* may be illustrated by *Tillandsia,* a genus of about 250 species of tropical and subtropical America. These are mainly epiphytic, caulescent or acaulescent herbs forming prominent basal rosettes of very narrow elongate leaves which are usually much broadened or thickened at the base. The inflorescence is a scape bearing few to several flowers of various colors. The perianth is differentiated as calyx and corolla, the segments of both distinct. Stamens 6. A capsular fruit is developed from the 3-carpellate superior gynœcium. A superior ovary occurs also in *Guzmania* and *Puya,* species of which are horticultural. In *Vriesia* a transition to the epigynous group is seen, the flowers of some species being hypogynous, of others epigynous—the ovary being more or less inferior. A similar situation obtains in *Pitcairnia.*

The Epigynous Farinosæ

BROMELIACEÆ (*Pineapple Family*). Genera 35—40; species around 900, of tropical and subtropical America. Mostly scurfy epiphytes with stiff, entire or spinulose-serrate, sheathing leaves in basal rosettes or equitant. Some members inhabit steep slopes or clefts of rocks. After fruiting, which occurs but once, the plants generally die, but some forms perennate by suckers from the base. The flowers, which are borne solitary, in spikes, or in panicles, are regular and bisexual, with the perianth differentiated into calyx and corolla, the segments of which may in each case be free or united. Stamens 6, in two whorls, distinct or united at the base, often epipetalous. Gynœcium 3-carpellate, with 1 style. Ovary (in these members) partly or wholly inferior. Fruit (spurious when the ovary is inferior) a capsule or berry. Ovules and seeds on axile placentas which are frequently forked or variously lobed. Embryo small, in copious endosperm.

The following epigynous genera are represented in cultivation (Bailey): *Ananas, Æchmea, Billbergia, Cryptanthus, Nidularium, Canistrum.*

A familiar representative of this group is the robust *Ananas sativus* (Pineapple), a native of the West Indies and Central America but extensively cultivated in Hawaii and somewhat established in the wild state in tropical Asia and Africa. The outstanding feature of the pineapple is the "fruit"; this is in reality a multiple fruit which consists of the thickened axis of the inflorescence over which are crowded numerous flowers, each subtended by a broad, somewhat coriaceous floral bract. At maturity the individual fruits developed from the inferior ovaries, together with the axis of the inflorescence and the floral bracts, form the succulent mass called the "pineapple." The inflorescence develops a leafy shoot at the summit, which is used in propagating the plants. The genus comprises probably a half-dozen species.

ORDER LILIIFLORÆ (LILIALES)

Flowers cyclic, 3-merous, rarely 2-merous or 4-merous, regular or irregular, hypogynous, perigynous, or epigynous. Perianth of two whorls of distinct or sometimes united (gamophyllous), mostly similar, segments, these in some cases differentiated as calyx and corolla. Stamens 3—6. Gynœcium 3-carpellate, rarely 2-carpellate, syncarpous. Endosperm present.

This is one of the most distinctive orders of the flowering plants as well as a very representative Monocotyledonous order.

Hypogynous families: *Juncaceæ, Stemonaceæ, Liliaceæ, Hæmodoraceæ* (in part).

Epigynous families: *Hæmodoraceæ* (in part), *Liliaceæ* (in part), *Amaryllidaceæ, Velloziaceæ, Taccaceæ, Dioscoreaceæ, Iridaceæ.*

Key to Our Families

A. Flowers *hypogynous;* ovary superior—except in *Aletris* in *Liliaceæ.*
- Plants grass-like, mostly tufted, wiry or soft, with small, inconspicuous flowers in cymose-umbellate clusters at top of culms. Perianth of two very similar whorls of brownish, oblong or lanceolate segments. Carpels 3. Stamens generally 3 or 6, rarely 4 or 5. *Juncaceæ.*
- Plants not as above, growing from bulbs, corms, or rootstocks. Flowers generally showy, with perianth-segments nearly similar (except in *Trillium*); sometimes united (gamophyllous). Stamens 6. Carpels generally 3. *Liliaceæ.*

B. Flowers *epigynous;* ovary inferior (superior in some exotic *Hæmodoraceæ*). Stamens 3—6. Carpels 3.
- a. Twining, climbing, or decumbent vines. *Dioscoreaceæ.*
- b. Plants not as above.
  - 1. Stamens 6, anthers mostly introrse. Flowers regular or nearly so; showy. Plants growing from bulbs or rootstocks. *Amaryllidaceæ.*
  - 2. Stamens 3, in some cases 6.
    - Stamens opposite the three inner perianth-segments, or when 6 then alternating with the perianth-segments. Perianth persistent. Plants stoloniferous. *Hæmodoraceæ.*
    - Stamens alternating with the perianth-segments; anthers extrorse. Flowers large and showy or small, regular or irregular. Plants growing from rootstocks. *Iridaceæ.*

The Hypogynous Liliifloræ

JUNCACEÆ (*Rush Family*). The rushes are mostly tufted wiry or grass-like herbs of wet or dry habitats. The small brownish flowers, mostly bisexual, are borne loosely or in compact clusters (*glomerules*) aggregated in cymose-umbellate inflorescences at the summit of the culms—in this respect resembling the bulrushes or some of the spike-rushes (*Cyperaceæ*). Most of the rushes are perennials from rootstocks or stolons. They are cosmopolitan in their distribution. Species about 300, distributed among about 8 genera. The perianth consists of two only slightly differentiated, 3-merous whorls of

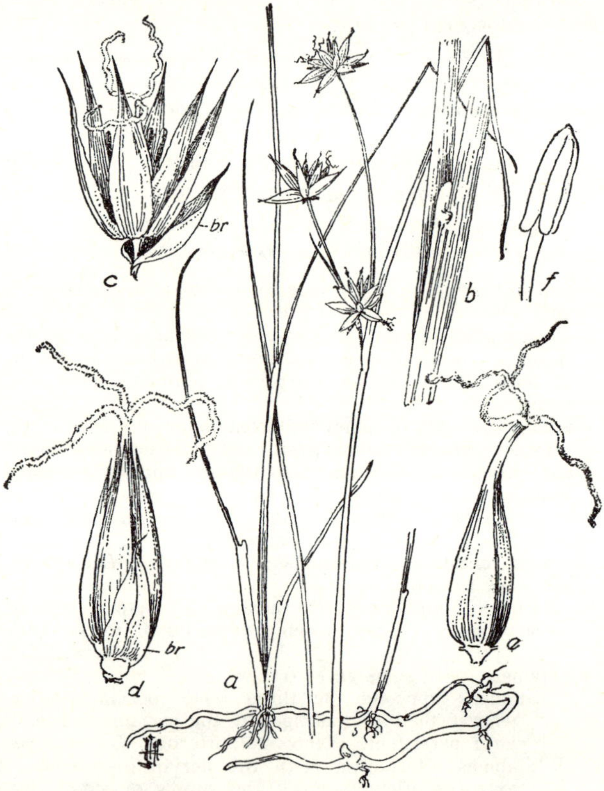

FIG. 457.—*Juncus nodosus: a,* habit sketch, with rhizome; *b,* section of culm and leaf-sheath with auricle; *c,* flower; *br,* bract; *d,* later stage of flower, anterior view, with bract (*br*); *e,* gynœcium; *f,* stamen.

segments of a glumaceous texture. Stamens 3 or 6, rarely 4 or 5, with introrse anthers dehiscing lengthwise by a slit. Gynœcium 3-carpellate, syncarpous, with an elongate or short slender style terminating in 3 stigmatic branches. The superior ovary is 1—3-celled,

FIG. 458.—*Juncus Torreyi: a,* habit sketch, with tuberiferous rhizome; *b,* inflorescence; *c,* part of leaf-blade and sheath; *d,* flower; *e,* seed.

with 3—many ovules in each cell. Fruit an ovoid loculicidal capsule. Seeds small, cylindrical, spindle-shaped, or nearly spherical, the seed-coat often forming characteristic tail-like appendages. Ours belong to the two genera, *Juncus* and *Luzula* (*Juncoides*).

The principal genus, *Juncus* (Rush), comprises both low and tall wiry plants, in aspect like the smaller bulrushes. Taxonomically *Juncus* is a rather difficult genus because of certain obscure characters which are detectable only on close inspection. Two well-defined subgenera have been recognized, both readily distinguishable. In one subgenus, the *Prophyllatæ* (meaning species with *prophylls, i.e.,* bractlets or bracteoles), there are two small bracts, the *prophylls,* below the flower, besides the bract (floral) at the base of the pedicel; in the other subgenus, the *Eprophyllatæ* (meaning without prophylls), these bracts are absent at the base of the flower or below it. These are helpful distinguishing characters and should be looked for when trying to identify a specimen. Besides these features, other characters residing in the inflorescence, the leaves, the capsule, and particularly in the seeds—such as the tail-like appendages—must be taken into consideration. The leaves are flat, channeled or terete, grass-like or filiform, septate or unseptate (by visible transverse partitions). The capsule is very distinctive generically but varies in shape and size according to the species. The seeds are strikingly beautiful and fascinating objects for study; they are very small but conspicuously bright-cinnamon colored, oblong, spindle-shaped, or rounded, with more or less prominent tail-like appendages at the ends. *Juncus bufonius* is commonly found along roadsides or in open, hard, moist ground. (Figs. 457, 458, 90.)

Luzula (Wood-Rush) differs from *Juncus* in being mostly soft, hairy, and more grass-like plants. Inflorescence and flowers similar to those of *Juncus,* except that the ovary is 1-celled and 3-seeded instead of 3-celled and many-seeded as in that genus. Both genera are large, *Juncus* especially, with well over 200 species as compared with about 65 for *Luzula.* In the plan of the flower these plants are very similar to the lilies, and would undoubtedly be so called were the perianth colored.

STEMONACEÆ. A small Asiatic and tropical American family of climbing or twining herbs with regular, bisexual flowers. Gynœcium 2-carpellate (!) but with 1-celled ovary. *Stemona* (*Roxburghia*) and *Stichoneuron* are Asiatic, *Croomia* American.

LILIACEÆ (*Lily Family*). Flowers regular or irregular, hypogynous, perigynous, or epigynous, bisexual or in some cases unisexual, polygamous or diœcious, 3-merous (with rare exceptions), small to large and showy, rarely with the perianth differentiated clearly as calyx and corolla, in some cases the segments united (gamophyllous) to form a lobed tube. Stamens 6, in two whorls; anthers introrse or extrorse; filaments sometimes borne on the perianth-tube (perigynous). Gynœcium 3-carpellate, syncarpous, with distinct styles or with 1 style. Fruit a loculicidal capsule or a berry, with many seeds on axile placentas. Species around 2,600, in over 200 genera, widely distributed over the world.

This is perhaps the most representative family of Monocotyledons. The plants grow from bulbs, corms, rootstocks, or a woody caudex. In habit they are very diverse. While the herbaceous habit is quite general and attains considerable proportions as to size, a woody tree-like habit is attained in certain cases (*Dracæna Draco,* the Dragon-Tree of the Canary Islands, and *Yucca brevifolia,* the Joshua-Tree of the Mojave Desert). The family yields a host of popular horticulturals, besides a few food plants (onions, garlics, asparagus). Irregular flowers are rare, as are also staminodes. The inflorescence is of various types; in most bulbiferous forms it is relatively few-flowered.

Among the numerous horticulturals the following may be mentioned: *Allium* (onion, garlic), *Asparagus* (asparagus), *Lilium* (lily), *Hyacinthus* (hyacinth), *Tulipa* (tulip), *Yucca* (yucca), *Aloe* (aloe).

Among the genera represented in our flora are: *Allium, Brodiæa, Zygadenus, Erythronium, Lilium, Fritillaria, Calochortus, Yucca, Trillium, Maianthemum, Medeola, Vagnera, Disporum, Uvularia, Streptopus, Polygonatum, Smilax.*

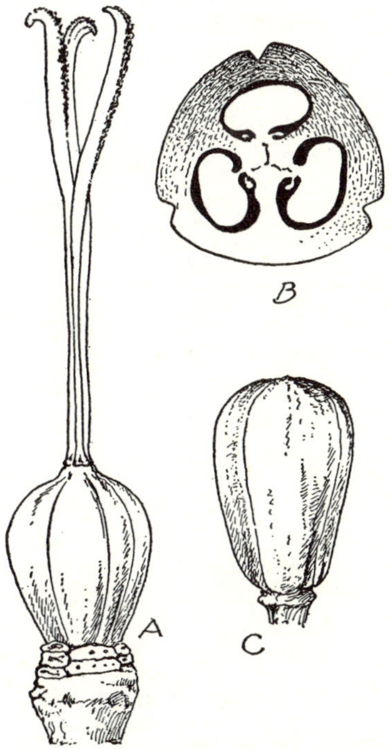

FIG. 459.—*Uvularia* sp.: *A,* gynœcium; *B,* transverse section of ovary; *C,* immature fruit.

There are several subfamilies, of which the following may be mentioned, together with a number of their characteristic genera, including some of those represented in our flora.

The Hypogynous Liliifloræ

I. *Melanthoideæ.* Mostly leafy-stemmed herbs from bulbs or rhizomes. Inflorescence scapose. Style simple or with thick or slender stigmatic branches. Anthers extrorse. Fruit a loculicidal or septicidal capsule, never a berry. *Tofieldia, Narthecium, Xerophyllum, Zygadenus, Melanthium, Veratrum, Uvularia, Colchicum, Chamælirium.*

II. *Asphodeloideæ.* Rhizomatous or bulbous herbs mostly with basal leaves. Inflorescence scapose, racemose, paniculate, or spicate. Perianth of distinct or united segments (gamophyllous). Anthers introrse. Fruit a capsule or in some cases a berry. *Asphodelus, Hosta, Hemerocallis, Phormium, Hesperocallis, Kniphofia, Aloe, Schoenolirion, Chlorogalum, Odontostomum, Leucocrinum.*

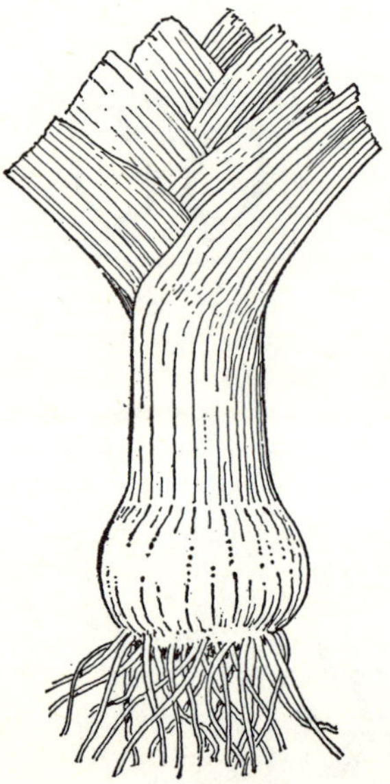

FIG. 460.—Bulb of *Allium Porrum* (Leek), with sheathing equitant leaf-bases.

III. *Allioideæ.* Bulbous or rhizomatous herbs with basal leaves or sometimes leafy stems. Inflorescence mostly umbellate and sometimes involucrate. Perianth-segments sometimes gamophyllous. Anthers introrse. Fruit a loculicidal capsule. *Allium, Agapanthus, Muilla, Brodiæa, Nothoscordum.*

IV. *Lilioideæ.* Bulbous, leafy-stemmed, basal-leaved herbs. Inflorescence scapose, with flowers solitary or racemose. Anthers introrse. Fruit a loculicidal or in some cases a septicidal (*Calochortus*) capsule. *Lilium, Fritillaria, Tulipa, Erythronium, Calochortus, Ornithogalum, Hyacinthus, Muscari, Scilla.*

V. *Dracænoideæ.* Rhizomatous herbs or tree-like, never bulbous. Leaves numerous, long and sword-shaped or linear and grass-like, more or less rigid, in a terminal crown or in dense basal tufts. Perianth-segments distinct or united at the base. Anthers introrse. Fruit a capsule or berry. *Yucca, Nolina, Dasylirion, Cordyline, Dracæna.*

VI. *Asparagoideæ.* Rhizomatous or cormose leafy-stemmed herbs, or the leaves radical or reduced to small scales bearing phylloclades in their axils, in other cases the leaves in a whorl near the middle of the stem. Flowers solitary, racemose or in umbellate clusters, or axillary. Perianth undifferentiated, or differentiated as calyx (green) and corolla. Fruit a berry. *Asparagus, Semele, Ruscus, Polygonatum, Clintonia, Smilacina, Maianthemum, Disporum, Streptopus, Convallaria, Aspidistra, Medeola, Scoliopus, Trillium, Paris.*

VII. *Smilacoideæ.* Ours climbing or half-climbing herbs or subshrubs, with leafy, sometimes tendril-bearing and prickly stems. Flowers diœcious, in umbellate clusters. Fruit a berry. *Smilax.*

The Epigynous Liliaceæ

VIII. *Ophiopogonoideæ.* Rhizomatous or stoloniferous herbs with rigid, sword-shaped or linear, grass-like or basally tufted leaves. Flowers in racemose or spicate clusters. Perianth-segments distinct or united (gamophyllous). Ovary generally more or less inferior. *Sansevieria, Liriope.*

IX. *Aletroideæ.* Rhizomatous herbs with basally tufted, grass-like leaves. Perianth gamophyllous. Fruit capsular. Inflorescence racemose. *Aletris.*

There is space for a description of but a few representative forms of the *Liliaceæ.*

Melanthoideæ. Tofieldia (False Asphodel). About 12 species in northern parts of the world, one or two in the Andes, and about four in North America. *T. palustris,* a widespread species, ranges from Greenland to Alaska and south to the Great Lakes region, and is also found in Europe and Asia. Another species is found in the southeastern United States and two in the Northwest. The plants are distinguished by linear two-ranked basal leaves and small bracted flowers in a terminal racemose cluster. Below each flower is a gamophyllous scarious involucre. In some species the flowers are solitary in the axils of the bracts, in others in threes. Certain taxonomic differences also are to be noted in the seeds. The style is rather thick, and 3-lobed at the summit. The capsule is septicidal. *Narthecium* is similar in habit, but differs particularly in having woolly filaments and a conical loculicidal capsule. The small, scarious involucre is absent. One species, *N. americanum,* occurs in swampy localities in the New Jersey and Delaware pine-barrens, and another in the Northwest. A few others are Asiatic. *Xerophyllum* comprises about 3 species of perennials from stout rootstocks, producing a dense tuft of long, grass-like, basal leaves and a tall, leafy, flowering stem terminating in a showy cluster of numerous small, white flowers. Perhaps the showiest of the species is the tall *X. tenax* (Bear-Grass) which forms dense snowy patches on open mountain sides in the Far West. When in full bloom these snowy patches are visible from a long distance. Another tall species occurs in the South Atlantic region. The mostly loculicidal capsules contain but very few seeds. In *Zygadenus* the flowers are bisexual or polygamous. The style-branches are rather prominent and spreading and the perianth-segments bear peculiar small glands. The plants grow from rootstocks or a bulbous base. They bear long grass-like leaves and small to medium-sized flowers in racemes or panicles. Some species have been reported as poisonous to sheep. By some writers the genus has been broken up into several genera.

Melanthium resembles *Zygadenus* somewhat in habit, but the leaves are sometimes broader and the flowers are monœcious or polygamous. About 3 species occur in Atlantic United States. Six or 7 species of *Veratrum* are divided between Eastern and Western parts of the country, with about 5 others in Asia and Europe. They are all tall, stout, broad-leaved perennials from stout, poisonous rootstocks, and produce numerous small, greenish flowers in a more or less dense and large terminal panicle. In this genus, also, the flowers are monœcious or polygamous. Thus *Zygadenus, Melanthium,* and *Veratrum* illustrate a tendency towards unisexuality of the flowers as expressed in the polygamous or monœcious conditions. When we come to *Chamælirium* we note that not only has complete unisexuality been attained but diœcism as well.

In *Uvularia* we see a group of plants that outwardly resemble *Disporum* or *Streptopus;* however, they differ from these in having a three-angled or three-winged capsular fruit—which dehisces loculicidally. They are perennial from rootstocks. All are restricted to the Eastern half of the country. About a half-dozen species have been described, at least three of them ranging into Canada. The flowers are quite large, yellow, drooping, and bell-shaped (hence the common name Bellwort), and are usually borne solitary or two together at the ends of branches. At the base of the lanceolate perianth-segments there is sometimes a small nectary. The style is similar to that in *Disporum* and *Streptopus*—elongate and slender with three spreading stigmatic branches at the summit. In at least two of the species, *Uvularia perfoliata* and *U. grandiflora,* which are found in rich woods as far north as Canada, the leaves are perfoliate (amplexicaule) at the base—a good recognition-mark of these species. It should be noted that in some books those species of *Uvularia* which do not possess perfoliate leaves are referred to the genus *Oakesia* (or *Oakesiella*). (Figs. 459, 116.)

Colchicum. Thirty or more species of cormose herbs with crocus-like flowers inhabiting Europe, western Asia, and North Africa. The large, showy, mostly purple or white flowers are borne on a very short subterranean scape so that they appear to spring right out of the ground. An outstanding feature of the flowers is the gamophyllous perianth which forms a long, slender tube below, and is thus crocus-like. The three styles are also very long and slender. The 6 stamens are inserted on the perianth. The fruit is a 3-lobed septicidal capsule. Strap-shaped to linear leaves appear along with the flower or later. In the familiar *Colchicum autumnale* (Autumn Crocus), the flower alone appears above the ground in autumn, and the following spring the capsule is lifted above the ground along with the leaves. A tunicate corm is produced, bearing a few scaly and sheathing leaves which invest the base of the scape. The plant is poisonous and medicinal. It is noteworthy that among the genera of this group mentioned, *Zygadenus, Veratrum,* and *Colchicum* contain poisonous species. Compare *Colchicum* with *Crocus* (see below) and note that while the flowers are similar in outward appearance, the distinction lies in the fact that the flowers of the former are hypogynous, while those of the latter are epigynous. A tendency in the direction of irregular (zygomorphic) flowers is to be seen in some genera of this group (*e.g., Chionographis, Schoenocaulon*).

Asphodeloideæ. Hemerocallis (Day-Lily). The day-lilies are natives of Europe and Asia, but *H. fulva* occurs as an escape from cultivation in the Atlantic States. The plants are large perennials with stout, fibrous roots and elongate, narrow, keeled leaves from a rhizomatous or bulbous base. Large and showy red, yellow, or orange flowers are produced several together at the summit of tall, leafless, flowering stems. Aside from their size, the marked feature of the flowers is the funnel-form or campanulate perianth with broad, spreading, or recurved segments which are gamophyllous in a narrow tube below. The long, slender stamens arise from the throat of the tube. Anthers versatile. The oblong ovary supports a very long and slender, declining and ascending style terminating in a small capitate stigma. The fruit is an elongate loculicidal capsule. Species 5—7. The two whorls of perianth-segments differ somewhat in shape and, besides, the flowers show a marked tendency toward bilateral symmetry (zygomorphy). Practically all the species are horticultural.

Aloe (Aloe). The aloes are robust perennials with large basal rosettes of rigid, succulent, mostly sharp-pointed leaves arising from a stout, simple or branched caudex, although in certain species there is a prominent aërial stem developed bearing the leaves in a crown at the summit (or at the ends of branches). The leaves are mostly beset with stout and often spiny teeth along the margins. The most showy red or yellow flowers are produced in dense racemes or umbels which are often raised on elongate, bracteate or leafy flowering stems, Each flower is subtended by a scarious bract. The perianth is tubular, straight, or curved, with the segments free or connate (gamophyllous) below. Anthers versatile, on slender hypogynous filaments. Fruit a 3-angled loculicidal capsule. Species about 100, tropical but mostly South African. Many are prized ornamentals, among which are *Aloe vera,* and *A. ciliaris.*

Phormium. Large perennials with long, sword-shaped, equitant leaves from a short or branched, fleshy, rhizomatous caudex. Flowers large, red or yellow, in a paniculate inflorescence. Perianth tubular and curved, bilaterally symmetrical, the inner and outer whorls notably different. Fruit a 3-angled loculicidal capsule. Only 2 species, in New Zealand. *Phormium tenax* (New Zealand Flax) is a popular ornamental and hothouse plant. The leaves, which sometimes grow to a length of nine feet, are very tough and yield a strong fiber. Species of *Asphodelus* (Asphodel), *Kniphofia, and Hosta* are represented in cultivation or in hothouses. *Hesperocallis, Schoenolirion, Chlorogalum, Odontostomum,* and *Leucocrinum* are represented in the Far West and Southwest. In *Leucocrinum* the ovary is "seated below the ground at the base of the perianth-tube."—Jepson.

The members of this subfamily may be profitably compared with the amaryllises, remembering that the *Melanthoideæ* are hypogynous, while the amaryllises are epigynous. Not infrequently students and others will confuse the amaryllises with the melanthoideous lilies because of the apparent outward similarities of the flowers.

Alliodieæ: Allium (Onion, Garlic). Cultivated onions are known the world over, though the garlic and the wild onions are not so familiar. Characteristic features are the odor and the bulb—which in cultivated forms attains considerable size in some cases. The genus is tremendous—around 300 species having been described, from wide parts of the world. And not only is the genus large, but it is very technical as well. Thus in classifying our numerous American species, of which there are between 60 and 70, mostly Western, it has been found that useful group, or even specific, characters reside in the textures or sculpturings of the scaly coats of the bulbs. Hence in working with these plants with a view to classification or keying, bulbs of the various species should be at hand. Specific floral distinctions are often difficult to detect without close study. The flowers, which are of various colors, are aggregated in involucrate umbels at the summit of the scapose stems. The involucre is sometimes conspicuous, consisting often of prominent papery bracts (spathes). The 6 perianth-segments are distinct or united at the base, and at various levels upon them the stamens are inserted. Filaments sometimes wing-margined. The 3-celled ovary becomes a loculicidal capsule. In the next following genus to be described it will be seen that perianth-segments are to a much greater degree gamophyllous. (Figs. 460, 131.)

Brodiæa (Brodiæa). The brodiæas are scapose herbs with grass-like leaves arising from a subterranean corm. The inflorescence is umbellate as in the onions and is similarly subtended by a bracteate,

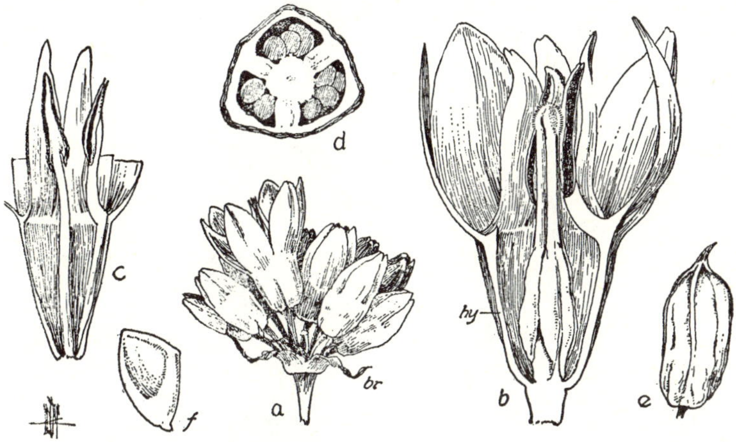

FIG. 461.—*Brodiæa* sp.: *a,* umbel; *br,* bracts; *b,* longitudinal section of flower; *hy,* hypanthium; *c,* section of flower showing stamens adnate to hypanthium; *d,* transverse section of ovary; *e,* immature fruit; *f,* seed.

but less conspicuous, involucre. The outstanding feature of the flowers is the gamophyllous perianth, with respect to this condition much advanced from the perianth in the onions. Furthermore, the 3 or 6 stamens are borne at or near the throat of the perianth, their filaments being of diverse shapes and sometimes winged as in those of the onion, with the wings prolonged beyond the anthers in some cases. In some species staminodes are present, while as a further distinction the ovary is sometimes stipitate (*i.e.,* raised on a gynophore). The style persists on the capsular fruit. In some species the inflorescence is condensed into a capitate form as in the onions. Species between 25 and 30, mainly in the Pacific States. (Fig. 461.) A few species of *Muilla*

FIG. 462.—*Lilium umbellatum* (Prairie Tiger Lily): *a*, habit sketch; *b*, longitudinal section of flower; *c*, gynœcium; *d*, ovary, in section above; *e*, summit of style, and stigma; *f*, transverse section of ovary; *g*, ovule; *h*, stamen.

are found in California. *Nothoscordum bivalve* occurs in Eastern parts of the country. Species of *Agapanthus* are horticultural.

Lilioideæ: Lilium (Lily). This genus of popular plants may be said fairly to typify the Monocotyledons. They are mostly graceful, leafy-stemmed herbs from bulbs, with the narrowly or broadly lanceolate leaves arranged spirally or in whorls. At the summit of the stems there is usually a few- to several-flowered racemose cluster of showy, red or scarlet or whitish, mottled, erect or drooping flowers. The two whorls of the perianth are but slightly dissimilar, while the long, slender filaments, with their elongate versatile anthers, add another distinctive feature. Supporting a prominent elongate style capped by a thick prominent stigma is the elongate 3-lobed ovary. The fruit is a rather large, oblong, loculicidal capsule which is often strongly contracted at the base; it contains numerous flat seeds. The

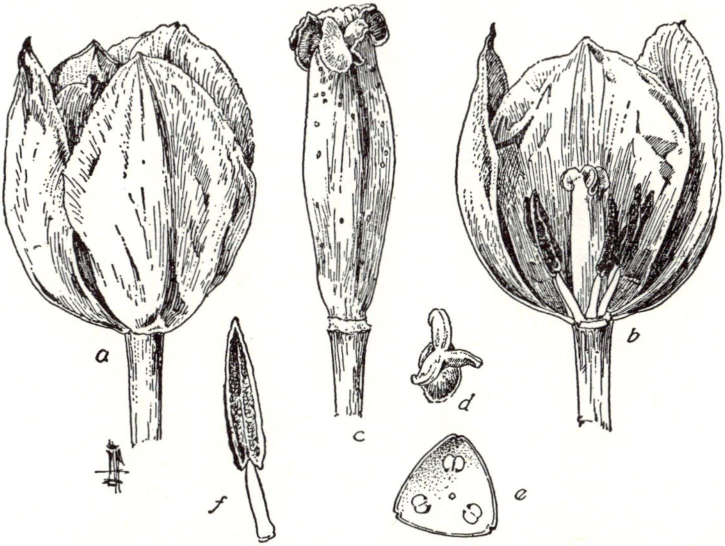

FIG. 463.—*Tulipa: a,* flower, (*b*) in sectional view; *c,* gynœcium; *d,* stigma, viewed from above; *e,* transverse section of ovary; *f,* stamen.

40—45 species are spread over north temperate parts of the world, 15 or 16 occurring with us—about equally divided between the Eastern and Western parts of the country. Several species are favorite horticulturals. *Lilium umbellatum,* the Prairie Tiger-lily, which sometimes bears solitary flowers, is common in the prairie States and Westward. In the early eighties and the nineties it was abundant on the virgin prairies along the Canadian border. (Fig. 462.)

Tulipa (Tulip) differs from *Lilium* in the broader perianth-segments and in possessing a more strongly 3-angled ovary and sessile stigma. As to fruit it is similar to *Lilium*. Species 50—60, native to the Mediterranean region, Asia, and Japan. Varieties and crosses of various species are favorite plants of the flower-gardens. *Tulipa sylvestris* is said to have become established in Pennsylvania (Fig. 463). *Fritillaria* is a genus of beautiful lilies, of which we have

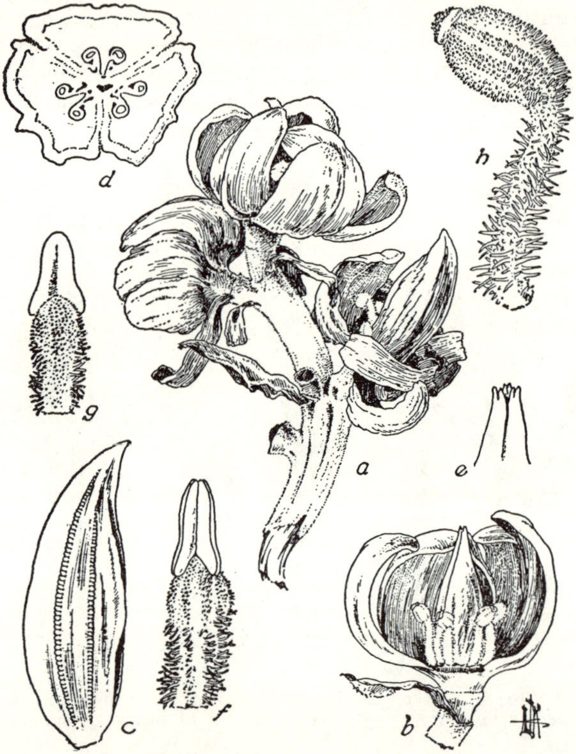

FIG. 464.—*Yucca brevifolia: a,* branch of inflorescence; *b,* section of flower; *c,* median section of gynœcium; *d,* transverse section of ovary; *e,* stigma; *f, g,* ventral and dorsal views of stamen; *h,* filament, enlarged.

about a dozen or more species, mostly in the West, but *F. atropurpurea* reaches North Dakota and Nebraska. All are represented in California. The dull-colored perianth-segments are often beautifully mottled with purple. *Fritillaria pudica* (Doe-Lily) is a characteristic and widespread species, reaching north into British Columbia. *Calo-*

chortus (about 35 species) is a Western genus, with showy tulip-like flowers of various colors, but the perianth is well differentiated into calyx and corolla, the calyx being much the smaller and consisting of narrow segments. A distinctive feature of the petals is seen in the peculiar glandular structures situated at the base of the limb. These glands are fringed with hair-like processess which are distinctive of the species. *Erythronium* (Dogtooth Violet) comprises Clintonia-like lilies of Eastern and Northern woods and of Western mountains. Species about a dozen, all American except the Eurasian *E. dens-canis*. *Ornithogalum* comprises about 75 species of Europe, Asia, and Africa; in habit they suggest *Zygadenus* or *Aletris*, having narrow tufted basal leaves and racemose inflorescence. *O. nutans* and *O. umbellatum* have become established in the North Atlantic States.

Hyacinthus (Hyacinth) is distinguished by its terminal racemes of erect or pendulous flowers with a campanulate or funnel-form, gamophyllous perianth. The 30 or more species are native to Mediterranean countries and to South Africa. *Hyacinthus orientalis* and its varieties are popular flower-garden forms. In *Muscari* the perianth is also gamophyllous but smaller, and urn-shaped to globose. Species 40—50, from Mediterranean countries into Asia. In all cases the flowers are in a dense terminal raceme. In *Scilla* the perianth-segments are free or united at the base. Species about 80, of Europe, Asia, and Africa; several horticulturals.

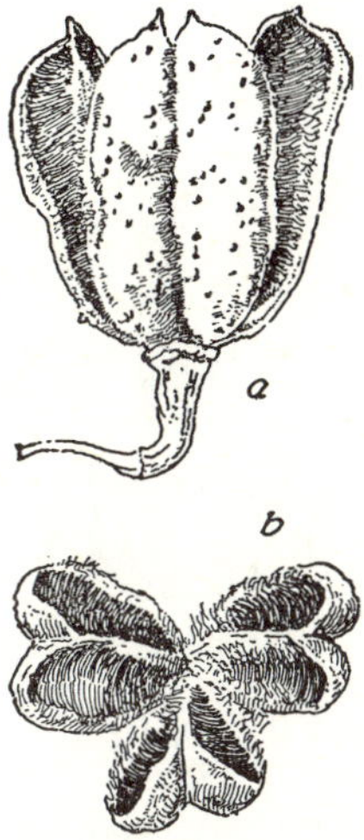

FIG. 465. — Capsule of *Yucca brevifolia: a,* after dehiscence; *b,* viewed from above.

Dracænoideæ: Yucca (Yucca). The outstanding features of the yuccas are their large size and long, rigid, sword-shaped or linear, needle-pointed leaves which form a dense, basal tuft, or in the tree-like species are crowded into tufts at the ends of the branches. The plants are strictly North American, ranging from the Atlantic to the Pacific and south into Central America. They form a characteristic element of the vegetation of the plains and arid mountains of the West. When in full bloom the huge clusters of whitish flowers are striking objects in the landscape. The most striking of our species is the tall, tree-like, branching *Yucca brevifolia* or Joshua-tree of the Mojave Desert. In the yuccas the perianth-segments are usually thick and free or partly united. The filaments are also thickened, especially so at the top, and bear small versatile anthers. The pods (capsules) are usually large and many-seeded. Species about 15 or 16. *Nolina* comprises about 10 species of large, very small-flowered, yucca-like plants ranging from Texas to California and into Mexico. The fruit (capsule), though similar to that of the yuccas, differs in being mostly 1—2 celled and containing but one or two seeds. About 10 species of *Dasylirion* are found from western Texas to Mexico. These are tall, tree-like plants with a crown of long linear leaves, the stem terminating in a leafy, densely flowered, racemose or spicate inflorescence of considerable length. The fruit is distinctive in that it yields but very few seeds, sometimes only one. *Cordyline* includes about 10 species of Asia, Australia, and New Zealand. They are tall tree-like plants resembling the dracænas; one species is said to occur in America. Several are horticultural. The capsular or berry-like fruit contains from 6 to 15 seeds. *Dracæna,* an Old World genus of about 50 species, consists of large tree-like or shrubby plants with sword-shaped leaves crowded at the summit of the stem or at the ends of the branches. The fruit in this case is a berry or sometimes a capsule, 3—1-celled, and containing 1 or 3 seeds. *Dracæna Draco* (Dragon-Tree) of the Canary Islands is one of the most ponderous of Monocotyledons. (Figs. 464, 465.)

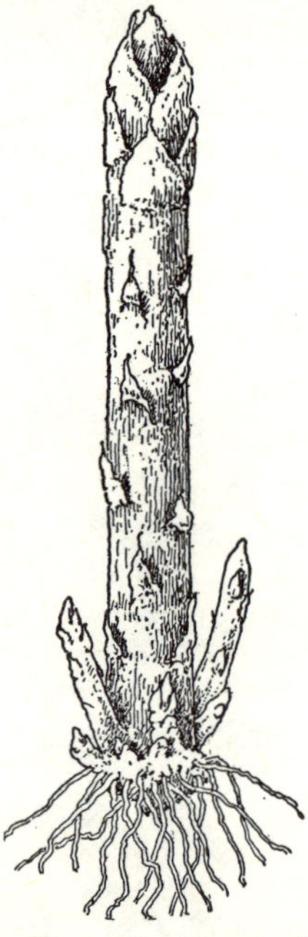

FIG. 466.—Sprout of asparagus *(Asparagus officinalis)*, with smaller sprouts at base.

Asparagoideæ: Asparagus (Asparagus). The distinguishing marks of the asparaguses are their minute scale-like leaves and the broad and leaf-like or narrow and needle-like phylloclades. With age the stems usually become hardened and woody in texture. The small flowers are borne solitary or in racemose or umbellate clusters. The stamens are inserted at the base of the distinct or united perianth-segments. The plants bear red scarlet, purplish, or blackish berries. Species about 150, ranging from Siberia to South Africa. Several are in cultivation—among which are *Asparagus officinalis,* the common garden asparagus, and the common *A. plumosus* of the florists. (Figs. 466, 95a, b.)

In *Ruscus* the phylloclades are thick and very rigid and spiny-tipped, thus forming a formidable armature for the plants. In *Semele* the phylloclades are very large and leaf-like and bear the flowers on their margins. Subtending each phylloclade is a small scale-like leaf. Both are limited Old World genera. (Fig. 95.)

Twenty-five or more species of *Smilacina* (False Solomon's Seal), or *Vagnera* as it is written in some books, are divided between North America and Asia. These are leafy-stemmed herbs from running

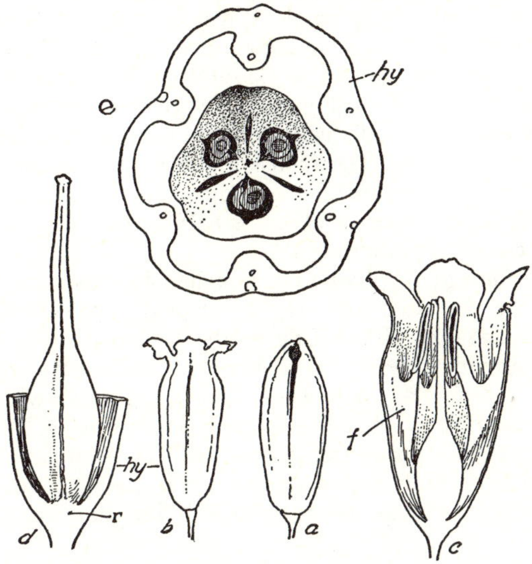

FIG. 467.—*Polygonatum commutatum: a,* flower-bud; *b,* flower; *c,* longitudinal section of flower; *f,* base of filament; *d,* gynœcium and base of hypanthium (*hy*); *e,* transverse section of ovary, including hypanthium (*hy*).

rootstocks and bearing usually numerous small flowers in racemes or panicles. In certain respects these plants remind one of *Polygonatum* (Solomon's Seal) as one sees them in the field, except that in the latter the flowers are borne two or more together in axillary clusters and the perianth is gamophyllous. (Compare with *Hyacinthus, Brodiæa, Allium, Muscari.*) The 20 or more species of *Polygonatum* (Fig. 467) belong mainly in Europe and Asia; *P. commutatum* (Solomon's Seal) and *P. biflorum,* a pubescent species, occur over most of the Eastern half of the United States. *Disporum* and *Streptopus* are rhizomatous branching herbs with foliage of broad leaves similar to those of the two preceding genera. In the former the flowers are clustered in terminal umbels and are quite large; in the latter they are borne solitary or in pairs in the axils of the leaves, and are smaller and nodding. Six or seven species of *Disporum* are divided between Eastern and Western parts of the continent, while as many or more are found in Asia. *D. lanuginosum* is a characteristic Eastern woodland species, while *D. trachycarpum* is common Westward. *Streptopus* contains probably a half-dozen species, three being North American, the others Asiatic. *S. amplexifolius,* distinguished by its amplexiform or cordate clasping leaves, ranges across the continent Northward. *Convallaria* is a monotypic genus of Europe. The species, *C. majalis* (Lily of the Valley), is a popular horticultural. A closely similar species occurs in the Middle Atlantic States. The small, campanulate, gamophyllous flowers are borne in an open raceme. The fruit is a pulpy, globose berry. *Aspidistra* (Aspidistra) is a genus of stout, rhizomatous, large-leaved herbs of Asia and Japan. Species about a half-dozen, one of which, *A. elatior,* is a common potherb.

Maianthemum (*Unifolium*) is anomalous in possessing "4-parted" flowers (or the parts in 2-merous (?) whorls). The common Eastern species, *M. canadense* (False Lily of the Valley), is a low rhizomatous herb bearing one or two broad ovate-cordate leaves and a terminal raceme of numerous, small, greenish-white flowers from which spring a cluster of greenish purple-mottled berries. Both glabrous and pubescent forms will be found. The species is common in the pine region of the Great Lakes States. Another very similar species has been described for the Far West and Asia.

Medeola, Scoliopus, Paris, and *Trillium* all possess a perianth differentiated into calyx and corolla, and all but *Paris* are represented in our flora. *Scoliopus* resembles *Erythronium* or *Clintonia* in producing a pair of broadly lanceolate radical leaves from a rhizomatous base, but the rather small flowers are borne in an umbellate cluster. In the other three genera the leaves are whorled—one whorl in *Trillium,* two in *Medeola* and *Scoliopus,* the second whorl appearing as a prominent involucre below an umbellate inflorescence. In *Trillium* a solitary, pedicellate or sessile and usually showy flower terminates the stem, and the subterranean stem is cormose. Only one species of *Medeola* is known, namely, *M. virginiana* (Indian Cucumber-Root), which is confined to Eastern woods. The three long, slender style-branches form a conspicuous feature of the flowers. About two dozen species of *Trillium* have been described, 16 or 17 of which are North American, the others Asiatic. In this genus the styles are shorter and stouter, and the ovary 3-lobed or 3-angled. One of the showiest species is *T. grandiflorum* of Eastern parts of the country. *T. petiolatum*, a species of the Northwest, bears a sessile purple flower. (Fig. 468.) *Scoliopus Bigelowii* and *S. Hallii* occur in the Pacific states. *Paris*, with about 6 species, is Eurasian.

FIG. 468.—*Trillium* sp.: *a*, leaves and flower; *b*, stem and corm; *c*, corm.

Smilacoideæ: Smilax (Smilax). A large genus of about 225 species, mostly American and Asiatic; around two dozen in our flora, with one exception all Eastern and Southern. The small and numerous flowers are diœcious and borne in prominent globose umbels, usually long-stalked, from the axils of the leaves. Only a few are herbaceous (*e.g.*, *Smilax herbacea*), the others being woody in texture and often tendril-bearing and prickly. Well-marked characters are also to be noted in the generally broad, ovate or deltoid and petiolate leaves. The lower leaves are reduced to scales. Abortive stamens are present in the carpellate flowers of some species. *Smilax hispida, S. glauca,* and *S. rotundifolia* are prickly-stemmed species. The woody species are all perennials.

Aletroideæ: Aletris (Colic-Root, Star-Grass) is represented by two species in Atlantic and Central parts of the country. These are the only epigynous lilies in our flora and, furthermore, the perianth is gamophyllous. The flowers are aggregated into a narrow spike-like raceme. *Aletris farinosa* ranges from Maine to Minnesota and south to Florida and Arkansas. *A. aurea* is Southern. Several other species in Asiatic countries.

The Epigynous Liliifloræ

HÆMODORACEÆ (*Bloodwort Family*). A small family, mostly of the tropics of the Southern Hemisphere. Genera 9, species about 35. Flowers regular or irregular, epigynous in most of the genera. Stamens 3, inserted at the base of the inner perianth-segments (hence obdiplostemonous). Ovary 3-celled, with relatively few ovules in each cell, on axile placentas (in rare cases ovules numerous). Fruit a loculicidal capsule. Represented in the Atlantic States by the solitary *Lachnanthes tinctoria* (or *Gyrotheca tinctoria* of some books), a plant up to two or more feet tall, with grass-like leaves and small flowers in a cymose cluster at the summit of a leafy flowering-stem, occurring in swamps. The genus *Lophiola* is by some authorities included in this family, but the numerous ovules and the type of placentation, especially, seem to connect it more closely to the next following family, as other authorities prefer.

AMARYLLIDACEÆ (*Amaryllis Family*). Flowers bisexual, regular or irregular. Stamens 6, inserted at the base of the perianth-segments or in the throat of the hypanthial tube. Anthers introrse. Ovary completely inferior (except in *Lophiola*), the receptacle continued above it as an elongate tube (hypanthium) or consolidated with the basal part of the style as a short or elongate neck. Ovules numerous (rarely reduced to 1 or 2), anatropous, generally in two rows in each cell, on axile placentas. Fruit mostly a capsule, rarely a berry. Genera about 70; species about 800, mostly of the tropics and subtropics with a few in temperate parts of the world. The plants grow from bulbs, corms, or rootstocks. Extrose anthers occur only in a couple of Tasmanian species belonging to a distinct subfamily. Aside from this group three other subfamilies are recognized, which are characterized below.

Subfamily I. *Amaryllidoideæ.* Plants growing from bulbs. Inflorescence an umbel subtended by an involucre, or only 1-flowered.

Including among many others, the following genera: *Clivia, Galanthus, Leucojum, Amaryllis, Zephyranthes, Cooperia, Crinum, Narcissus, Pancratium, Hippeastrum, Lycoris.*

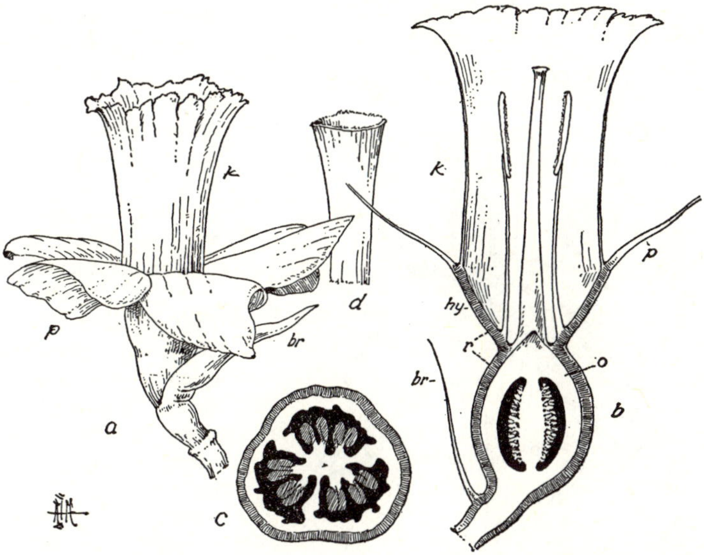

FIG. 469.—*Narcissus Pseudo-Narcissus: a,* flower; *p,* perianth; *k,* corona; *br,* bract (spathe); *b,* median section of flower; *hy,* hypanthium; *r,* receptacle; *o,* ovary; *c,* transverse section of "ovary"; *d,* stigma and summit of style.

Subfamily II. *Agavoideæ.* Plants growing from rhizomes, and bearing thick, hard, rigid leaves or thinner sword-shaped leaves in dense rosettes, either basally or in a crown at the summit of the stem. Inflorescence racemose or paniculate, sometimes enormous in size.

Including among others: *Agave, Fourcroya.*

Subfamily III. *Hypoxidoideæ.* Plants growing from rhizomes. Leaves soft, broad to narrow and grass-like. Inflorescence racemose, cymose, or cymose-paniculate. Including: *Alstroemeria, Hypoxis, Curculigo, Lophiola, Anigosanthus.*

Only the following genera are represented in our flora: *Hymenocallis, Zephyranthes, Cooperia, Agave.*

Amaryllidoideæ: Narcissus (Narcissus, Jonquil). About 20 Old World species with long, linear leaves arising from a globose bulb. The flowers, which are subtended by a spathe, are borne solitary or several together at the summit of a leafless scape. The most outstanding feature of the flowers is the prominent crown (paracorolla) of the perianth, which in some species is developed as a conspicuous tube, often mistaken for a corolla. This crown is very prominent in the common *Narcissus Pseudo-Narcissus* of the gardens (Fig. 469). The plant frequently occurs as an escape in some localities. In the equally common *N. poeticus* the crown is very short. In *Hymenocallis* the crown is vasiform, flaring above into a broad, irregularly fringed or lobed funnel. The lower part of the filaments is connected to the tube of the crown. The perianth-segments are long, narrow, and spreading from the greatly elongated and very slender hypanthial tube (or neck). The 30 or more species, mostly of tropical America, with *H. occidentalis* occurring from Georgia and Alabama to Illinois and Missouri. (Figs. 469, 470, 132.)

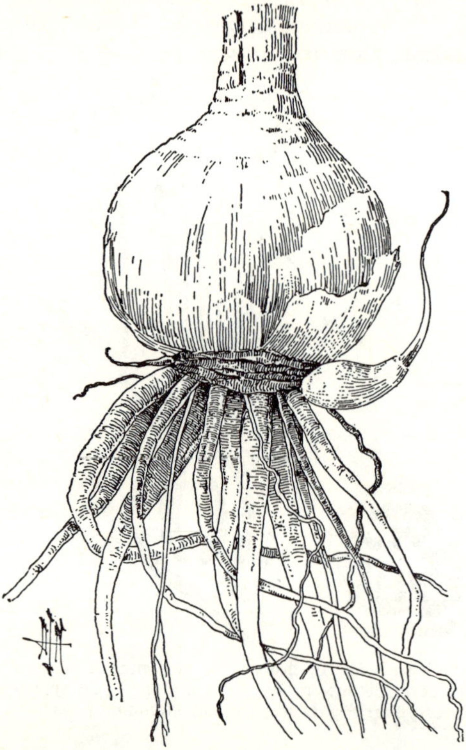

FIG. 470.—Bulb of a *Hymenocallis,* with bulbil at right below.

In *Zephyranthes Atamasco* (Atamasco Lily), which is found from Florida to Pennsylvania, the crown is absent, the hypanthium is shorter, and the perianth-segments are spreading and more lily-like. The genus, with about 30 species, is native to tropical and subtropical America. In *Cooperia* the perianth is much smaller, but the hypanthial tube is very long and slender and is subtended by a membranous sheathing spathe which is inflated more or less at the base. Two species occur in the Southwest and Mexico, *C. Drummondii* spreading Eastward into Kansas. The other genera listed are represented more or less in hot-houses and gardens.

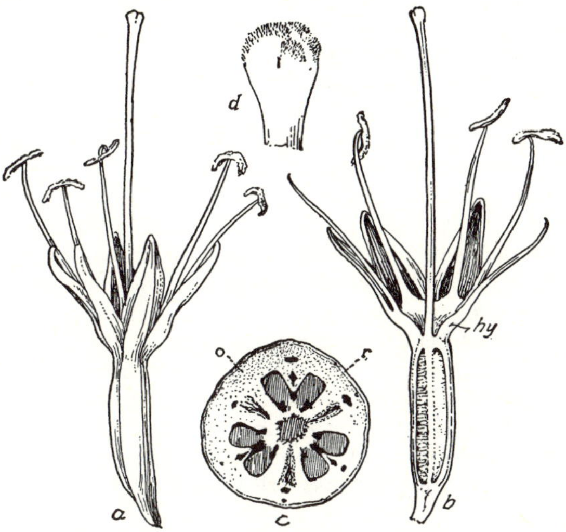

FIG. 471.—*Agave* sp.: *a,* flower; *b,* median section of flower; *hy,* hypanthium; *c,* transverse section of ovary; *o,* ovary; *r,* receptacle; *d,* stigma.

Agavoideæ: Agave. An enormous genus of 300 or more species of southwestern United States, Mexico, and South America, characteristically inhabiting arid districts. These plants rank among the largest of the Monocotyledons. The thick, hard leaves, which taper to a point and are usually armed with stout, sharp, marginal teeth, radiate from a huge basal rosette, although some species bear them in a crown at the summit of a definite trunk-like stem. The numerous flowers are borne in enormous towering panicles or spikes. Some species bloom annually, others after intervals of several years, others again bloom but once and then die. The perianth is generally funnelform, from a usually short, solid neck of the receptacle above the ovary. The style is stout and terminates in a three-lobed, somewhat capitate stigma. The fruit is a 3-valved, loculicidal capsule, containing numerous thin, black seeds. Several species occur in the arid Southwest, and a few are grown as ornamentals, *e.g., Agave americana* (Century Plant). Certain Mexican species of *Agave* are the source of sisal hemp, pulque, and mescal. (Fig. 471.) *Fourcroya* (Furcræa) is somewhat similar both as to size and habit, but has a sharply 3-angled capsule. Species about 15, in subtropical America. Cultivated specimens are seen in southern California.

Hypoxidoideæ: Hypoxis (Star-Grass). A genus variously estimated as containing 50 to 100 species, widely distributed in the tropics and subtropics, but three or four species coming into our range, one of which, *H. hirsuta,* crosses the Canadian border. The plants are low herbs with grass-like leaves arising from a cormose or rhizomatous base and bearing small, mostly yellow, star-shaped flowers in a few-flowered cyme at the summit of a short scape. In most cases the plants are villous-pubescent. The perianth-segments are in all cases distinct, there being no hypanthial tube or neck developed above the ovary. *H. hirsuta* usually occurs in grassy meadows or in woods, and is said to occur at high altitudes in Virginia.

Lophiola aurea (or *L. americana,* as it is sometimes called), the only member of its genus, occurring in bogs on the Atlantic Coast, is a tall, woolly herb with grass-like leaves and small, yellow flowers in a terminal cyme. The principal feature of the flowers is the half-inferior ovary, wherein it differs from the other genera above listed. *Alstroemeria, Curculigo,* and *Anigosanthus* are represented in gardens. *Lophiola* and *Anigosanthus* are by some writers placed in the *Hæmodoraceæ.*

VELLOZIACEÆ. A small South American group of 70 or more species in two genera—*Vellozia* and *Barbacenia.* The family is exceptional among Monocotyledons in that the stamens, while normally 6, are several times 6 or very numerous in the genus *Vellozia,* and, moreover, are united in bundles by their bases. A further distinctive character appears in the placentas—which are lamellate. Some forms are suggestive of the aloes.

TACCACEÆ. A small family of the tropics of the Old and the New World. Two genera, *Tacca,* with about 9 species, and *Schizocapsa* with 1. In habit the plants resemble some aroids, but the flowers are aggregated into umbellate clusters. The ovary is in some cases one-celled. The fruit is a capsule or berry. The plants spring from rhizomes or corms.

DIOSCOREACEÆ (*Yam family*). Flowers bisexual, or unisexual and monœcious or diœcious, regular, 3-merous. Perianth not clearly differentiated as calyx and corolla (persistent in the carpellate flowers). Staminate flowers with 3 or 6 stamens, and sometimes with a rudimentary gynœcium. Carpellate flowers sometimes with 3 or 6 staminodes. Ovary inferior, 3-celled, with 3 styles. Ovules 2 or sometimes 1 in each cell. Fruit mostly a 3-angled capsule or a berry. In some cases only one cell of the ovary is developed. Genera 9—10, species about 200. Twining, mostly herbaceous vines of the tropics, growing from fleshy or woody rootstocks. Leaves broad, alternate or opposite, entire or digitate, petiolate, in form and venation suggesting the dicotyledonous type. Embryo small, in copious endosperm.

The great majority of the species belong in the genus *Dioscorea,* one species of which, *Dioscorea villosa* (Yam), ranges from the Gulf States to Canada. The plants produce a thick or slender rootstock. Several species are cultivated for their tuberous, subterranean or aerial rootstocks, called yams, a name also applied to the sweet potato. *Dioscorea Batatas* is the Chinese Yam. Large tubers are produced by species of the West Indian genus *Rajania.* In this genus the flowers are unisexual and monœcious; in *Dioscorea* they are unisexual and monœcious or diœcious. Significant evolutionary features in this family are the advances to monœcism and diœcism and the presence of a rudimentary gynœcium in the staminate flowers of some forms and of staminodes in the carpellate flowers.

IRIDACEÆ (*Iris Family*). Flowers regular or irregular, bisexual, often large and showy, usually clustered, and subtended by more or less prominent bracts (spathes). Perianth undifferentiated or more or less differentiated, contracted into the receptacular tube or neck below. Stamens 3, placed opposite the outer segments of the perianth (the inner whorl of stamens thus being absent) or opposite the inner segments of the perianth—obdiplostemonous; filaments usually long and slender, free or united (more or less); anthers 2-celled, extrorse. Ovary wholly inferior, the receptacle forming a short or elongate tube or neck above it. Style elongate, 3-cleft—each limb sometimes forked into two slender branches. Fruit a globose or oblong, 4-angled or 3-lobed capsule, containing numerous seeds in one or two rows in each cell. Genera around 60, species about 1,000, all perennials from rhizomes or corms, with narrow, equitant leaves. Tropical and temperate parts of the world. The family yields many choice horticulturals. The following subfamilies are recognized:

Subfamily I. *Crocoideæ.* Flowers solitary or clustered and axillary, on a very short (or somewhat elongate) leafy flowering-stem (scape). *Crocus.*

Subfamily II. *Iridoideæ.* Flowers generally in clusters subtended by more or less prominent spathes. Plants rhizomatous. *Iris, Tigridia, Nemastylis, Belamcanda, Sisyrinchium.*

Subfamily III. *Ixioideæ.* Flowers solitary in the spathes, and often irregular. Plants rhizomatous or cormose. *Gladiolus, Tritonia, Babiana, Freesia, Ixia.*

Crocoideæ: Crocus (Crocus). The crocuses bear leaves with linear grass-like blades and elongate sheathing bases arising from a characteristic corm, the internodes of which are usually so shortened that the leaves appear to spring from the same point. The flower-bud is pushed up through the investing leaf-sheaths by the elongation of the flowering stem, which is sometimes quite short, so that the flower appears to spring directly from the ground. The flowers appear before, along with, or after the leaves, according to the species. The outstanding features of the flowers are the sometimes very much elongated hypanthium ("perianth-tube")—which in some species is more than twice the length of the perianth and very slender—and the peculiar funnel-shaped or ramiform or fringed-filiform stigmas which terminate the three style-branches. There is also to be noted a considerable differentiation of the two whorls of the perianth. The seeds are numerous and on axile placentas. The 3 short stamens arise from the throat of the hypanthium. Fruit a small, oblong, loculicidal capsule. Species about 75, of the Mediterranean region and southwestern Asia. Several are popular horticulturals, *e.g., Crocus sativus* (Saffron Crocus) and *C. susianus* (Cloth-of-Gold Crocus).

Iridoideæ: Iris (Iris, Fleur-de-Lis). The irises are stout perennials with sword-shaped or linear, equitant leaves, closely sheathing at the very base, and arising from stout, often tuberous, jointed rhizomes which are covered more or less by the fibrous remnants of scale-leaves. The large, showy, regular flowers arise from the axils of a pair of bracts (spathes), usually in a panicle but in some cases solitary at the top of the stout flowering-stem (sometimes leaf-bearing). The receptacle extends beyond the oblong ovary as a solid stipe or neck which is consolidated with the lower portion of the long, slender style and supports the perianth and elongate stamens. The perianth-whorls are considerably differentiated but petaloid, and are contracted downward to form a "claw." The segments of the outer whorl are usually reflexed and often conspicuously bearded over the

FIG. 472.—*Iris versicolor: a,* leaves and rhizome; *b,* flower; *s,* spathes; *c,* flower with perianth removed, exposing stamens and three broad petaloid styles,—with stigma (*st*) beneath, near apex; *d,* median section of "ovary"; *e,* transverse section of ovary; *o,* ovary; *r,* receptacle; *f,* developing fruits; *g,* fruit (capsule) after dehiscence.

middle portion of the upper surface, but the segments of the inner whorl are erect or ascending and arched outward. A very distinctive character is seen in the peculiar style-branches: These are petaloid and broadly expanded upward, 2-lobed at the upturned tip, and arched outward over (and thus *opposite*) the 3 stamens so as to conceal them from view. On the lower surface, near the tip of each style-branch, will be found a thin projecting flap—the stigma. The fruit (spurious) is an oblong loculicidal capsule containing numerous seeds on axile placentas. The pods of the irises may be profitably compared with those of *Lilium* and other liliaceous forms, which they closely resemble, at the same time remembering that in the irises the ovary is inferior and the resulting fruit consequently spurious, while in the lilies (except the epigynous forms) the ovary is superior and the fruits therefore are true fruits. Thus in the lilies the receptacle—at least the part of it that bore the perianth and stamens—will appear as a slight expansion of the fruit-stalk just below the pod (Fig. 465); in the irises no such structure will be found.

The number of species of *Iris* that have been described is variously quoted as 100—150, widely distributed over northern parts of the world. Numerous species and varieties are grown in flower-gardens. Probably over 20 species occur in the wild state throughout the country; among these, *Iris versicolor* is widespread over the Eastern half of the continent from Canada to Florida and Arkansas. The very similar *I. missouriensis* occurs over a large part of the West. (Fig. 472.)

Sisyrinchium (Blue-Eyed Grass) is a large North American genus of about 150 species. The plants are low, tufted herbs from rhizomes, with narrow, grass-like leaves, and flattened flowering-stems bearing an umbel of small blue, yellow, or white flowers in the axils of one or two bracts (spathes) at the summit. In this genus the filaments are more or less united around the style and alternate with the style-branches (compare with *Iris*). In the large-flowered, Far Western *Sisyrinchium grandiflorum*, the filaments are united only at the base, a character which has led to its separation in some botanical works under the name *Olsynium grandiflorum*. In all cases the hypanthium is very short, so that the perianth appears sessile on top of the ovary. The perianth-segments are generally of an oblong shape, with a bristle-bearing notch at the tip; there is but slight differentiation, if any, of the two whorls. All produce a small, globose, loculicidal capsule containing relatively few seeds. Our species are found generally in low, moist, grassy ground. One of the most widespread species is *Sisyrinchium angustifolium*, which occurs from Newfoundland to the Rocky Mountain region. (Fig. 473.)

The Asiatic *Belamcanda chinensis* has been naturalized in Eastern woodlands (the name is sometimes written *Gemmingia chinensis*).

This species bears rather large flowers in an umbellate-paniculate cluster. *Nemastylis,* a genus of about 10 species of tropical and temperate America, is represented in Eastern and Southern parts of the United States by about four species. Of these, *N. acuta* reaches the

FIG. 473.—*Sisyrinchium bellum: a,* branch with inflorescence; *b,* petal; *c,* flower, with part of perianth removed, showing column of stamens investing style; *d,* immature capsule; *e,* transverse section of immature capsule.

latitude of Kansas and Tennessee. In these genera, as in *Sisyrinchium,* the style-branches alternate with the stamens.

Tigridia Pavonia (Tiger-Flower), a Mexican and Central American plant, is grown in gardens. Here, as in *Iris,* the style-branches are opposite the stamens. Several other species are found from Mexico to Chile.

Ixioideæ: Gladiolus (Gladiolus). This is one of the largest genera of the family, if not the largest, numbering around 200 or more species. The plants are native to the Mediterranean countries and South Africa—where they are especially abundant. In the sword-shaped or linear leaves, tall flowering stems (which are sometimes leafy), and spicate or racemose inflorescences, the plants resemble somewhat the irises, but the flowers are *irregular*. Each flower springs from the axil of a spathaceous bract. In horticultural forms the flowers are unusually large and showy. A tubular hypanthium rises from the summit of the ovary, merging into the spreading lobes of the perianth, the upper three of which are the larger, and bringing about a rather conspicuous median bilateral symmetry of the flower. The 3 stamens arise from the throat of the hypanthium. The style is slender and 3-branched at the summit. The inferior ovary becomes a capsule (a spurious fruit) at maturity, containing numerous winged seeds and dehiscing loculicidally.

Tritonia (*Montbretia*), *Babiana, Freesia,* and *Ixia* are represented in gardens.

The notable evolutionary advances to be observed in the *Iridaceæ* are: (1) epigyny, (2) zygomorphy, (3) the suppression of one whorl of the andrœcium, either the inner or the outer, as the case may be, and (4) union of the filaments in some forms. As regards affinities, the family stands probably closest to the *Amaryllidaceæ*.

ORDER SCITAMINEÆ (SCITAMINALES)

Flowers regular or irregular (zygomorphic or asymmetrical), bisexual or unisexual by abortion, epigynous. Perianth undifferentiated or differentiated as calyx and corolla, but petaloid. Andrœcium normally of two 3-merous whorls, but commonly with either 1 or 5 stamens transformed into staminodes. The inferior ovary is 3-celled or in some cases 1-celled, and commonly with numerous ovules (or 1) on axile placentas. Fruit a capsule or berry. Seeds with endosperm and perisperm.

Families: *Musaceæ, Zingiberaceæ, Cannaceæ, Marantaceæ.*

Key to the Families

Anther-bearing stamens 5, the sixth transformed into a staminode. Tall robust herbs. *Musaceæ.*

Anther-bearing stamens 1, the other 5 staminodial.

Anther-bearing stamen with a 2-celled anther. *Zingiberaceæ.*

Anther-bearing stamen with a 1-celled anther. Flowers very irregular.
 Ovules many in each cell of the ovary. *Cannaceæ.*
 Ovules 1 in each cell of the ovary. *Marantaceæ.*

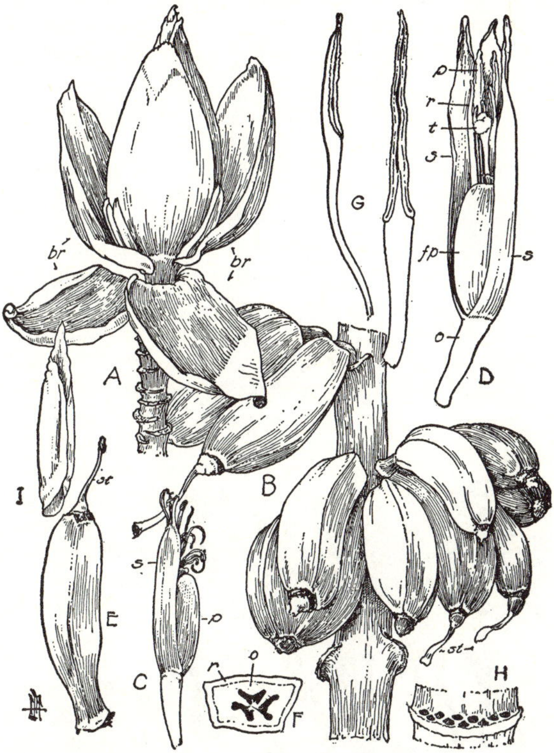

FIG. 474.—*Musa Cavendishii: A,* summit of inflorescence; *br,* bracts (dark purple), the two spreading upper ones with flowers in the axils; *B,* clusters of young fruits, three with persistent style; *C,* a flower, side view; *s,* sepal; *p,* free petal, stamens and style protruding above; *D,* flower with a sepal removed; *p,* petals; *s,* sepal; *r,* staminode; *t,* stigma; *fp,* free petal; *o,* ovary; *E,* young fruit; *F,* transverse section of "ovary"; *o,* ovary proper; *r,* receptacle; *G,* lateral and ventral aspects of stamen; *H,* node of inflorescence with leaf-scar, and above it the scars of the flower-stalks; *I,* free petal.

MUSACEÆ (*Banana Family*). Flowers irregular, more or less bilaterally symmetrical, bisexual, or unisexual and monœcious, sessile or pedicellate in the axils of spathaceous bracts and aggregated into spicate or paniculate, sometimes very large, inflorescences subtended by spathes. Perianth petaloid, differentiated or undifferentiated. One of the 6 stamens usually a staminode. The inferior ovary usually with numerous ovules in each cell, and supporting a single style which is 3-branched at the summit, the branches sometimes forked. Fruit a capsule or berry. Genera about 6; species around 60, all tropical.

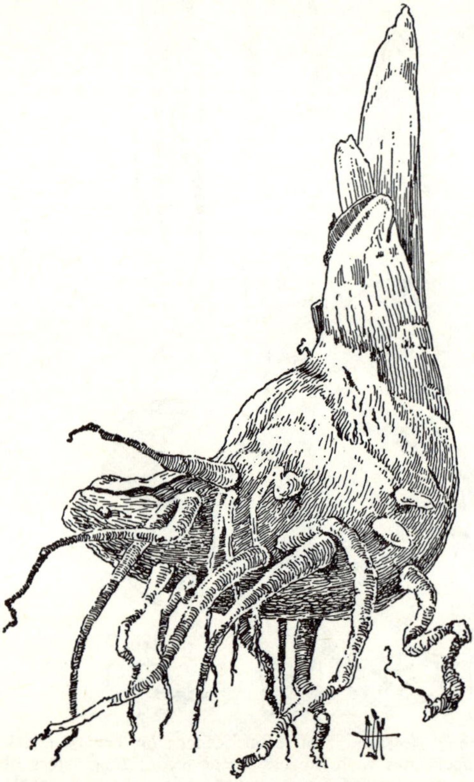

FIG. 475.—Bulb of a banana plant.

Some members of this family, such as the bananas and the ravenalas, are among the largest of the Monocotyledons. In general all are robust plants, sometimes of woody texture, growing from rhizomes or a bulbous base. The leaves are commonly enormous in size, with long, broad blades with a pinnate type of venation and long, stout petioles which are closely sheathing at the base.

The important genera are: *Musa, Ravenala, Strelitzia, Heliconia.*

Musa (Banana). The bananas are among the largest of all herbs. They are distinguishable at once by their large and characteristic leaves and huge fruit-clusters. The plants grow from a rhizomatous or bulbous base, from which bulbous sprouts are usually developed, so that several plants often appear in clumps. Owing to the type of venation the leaves are usually considerably frayed by wind. The bracts of the inflorescence are usually brightly colored and showy.

The flowers are unisexual by abortion, the staminate being usually found in the upper part of the inflorescence and the carpellate below. The perianth-segments are more or less connate and one of the six stamens is a staminode. The fruit is an elongate, rather angular berry —as typified in the familiar banana, the fruit of *Musa paradisiaca* var. *sapientum*. The largest of all the banana species is the Abyssinian *M. ensete,* the leaves of which grow at best to a length of twenty feet. Its chief use is as a towering ornamental. The fruit is dry and tough and inedible. *Musa textilis* yields a stout fiber which is the source of "Manila hemp." All the species, of which 50 or more are known, are native to the tropics of the Old World. (Figs. 474, 475.)

Ravenala. These are palm-like plants bearing a crown of huge leaves closely alternating in one flat plane and spreading like an enormous open fan. Only two species are known, these being native to Madagascar. The flowers are irregular, with distinct perianth-segments. Fruit an elongate capsule. A distinctive feature of the seeds is the brightly colored, fringed aril. *R. madagascariensis* may be seen in some hot-houses, or in gardens in Southern parts of the country.

Strelitzia Reginæ (Bird-of-Paradise Flower), one among a half-dozen South African species of this genus, is frequent in gardens in Southern parts of the country. It is a robust plant with large basal leaves and striking flowers resembling a bird about to alight. The most striking character about the flowers is the so-called "tongue"—an organ formed by the two lower "petals," which are folded over the anthers so as to conceal them completely. Pressure on the tip of the "tongue" causes the petals to spread and thus expose the anthers, but immediately on release of the pressure the petals snap back over the anthers again. The flower is pollinated by a bird. The other perianth-segments, which are showily colored, spread upward and laterally, thus enhancing the suggestion of a fluttering bird. Above the ovary is a well-marked stipe (receptacle) which is shaped so as to hold the "flower" in a horizontal position. All in all the flowers of these plants exhibit a most unique "adaptation" for bringing about pollination. *Heliconia* includes 40 or more robust banana-like plants of the American tropics. *H. Bihai* is a tall ornamental. (Fig. 476.)

ZINGIBERACEÆ (*Ginger Family*). Flowers bilaterally symmetrical or unsymmetrical, bisexual or rarely unisexual by abortion, generally curving downward and upward. Perianth mostly differentiated into calyx and corolla, the calyx more or less tubular or campanulate, the corolla funnel-form below, with spreading segments.

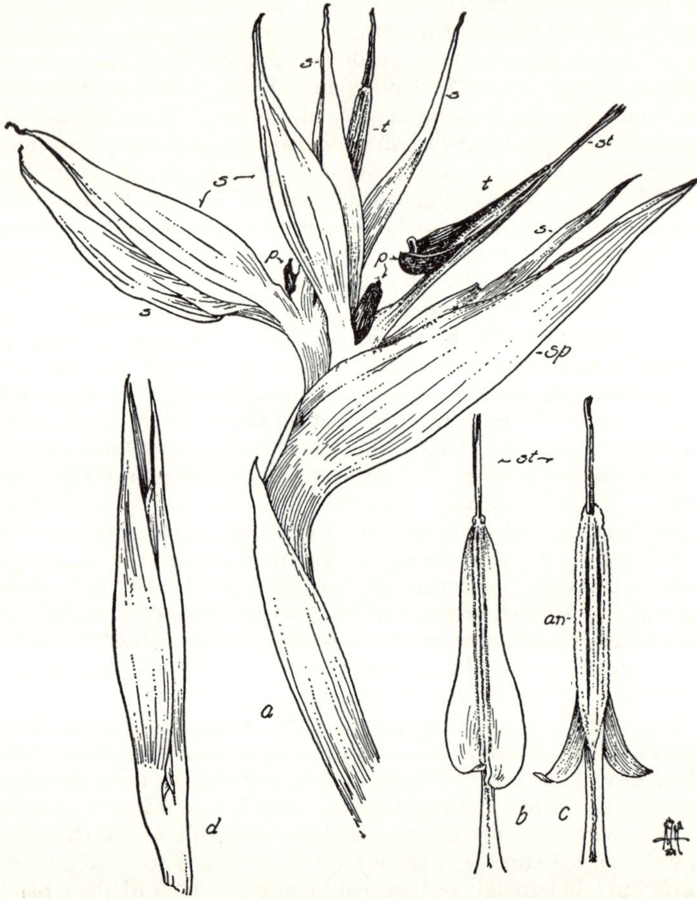

FIG. 476.—*Strelitzia Reginæ: a,* inflorescence; *sp,* spathe; *s,* sepals; *p,* petals; *st,* stigmas; *t,* "tongue," which encloses the stamens and style; *b,* tongue, viewed from above, when closed; *c,* tongue with petals folded back, exposing anthers (*an*); *d,* flower-bud.

Functional stamens only 1—the posterior of the inner whorl, the connective of which is variously modified, and the 4-celled anther folded in the form of a groove around the style. The two other stamens of the inner whorl are united and transformed into an anterior, often 2-lobed, staminode (*labellum*) which is folded around the functional stamen as well as the style and often forms the showiest organ of the flower. The anterior stamen of the outer whorl is absent, but the two others (posterior and lateral) are staminodial and invest more or less the functional stamen. The long slender style is funnel-shaped or otherwise modified above and terminates in a stigma of various peculiar shapes. Apparently there is no hypanthium developed, so that both calyx and corolla appear to arise from the top of the ovary. The stamens and staminodes arise from the throat of the corolla. Numerous ovules are produced in each cell, on axile placentas, or on parietal placentas if the ovary is 1-celled. The spurious fruit is a loculicidal, 3-valved capsule. Embryo straight, cylindrical, invested by a layer of endosperm surrounded by perisperm.

These plants are generally stout herbs growing from elongate or short tuberous rhizomes or tuberous roots from which arise leafy vegetative shoots and numerous-bracted flowering-stems bearing flowers in a spicate or racemose cluster at the top. Leaves generally with blades of the lanceolate type from a prominent sheathing base, with a more or less prominent ligule at the base of the blade—thus strongly suggestive of the leaves of the grasses (compare also with the palms, *e.g., Washingtonia*). The flowers are borne singly or several together in the axile of a conspicuous bract, either in an open racemose sort of inflorescence or in a compact spike. Genera about 40; species 400 or more, mostly of the tropics of the Old World. Several of the genera are rich in species and yield handsome garden plants.

For the purpose of classification of the genera the most useful characters are found in the number of cells in the ovary, the shape of the connective of the functional stamens, and the shape of the staminodes. According to Bailey the following genera are represented in gardens, namely, *Zingiber, Amomum, Alpinia*—in which the staminodes are not petaloid—and *Curcuma* and *Hedychium,* in which the staminodes are petaloid and showy.

Zingiber officinale, a native of Asia, is the source of ginger, for the production of which it is extensively cultivated in the tropics. The genus contains from 50 to 60 species. In some species the flowering-stem is very short so that the inflorescence appears to arise from the ground. *Amomum* has about 100 species in Asia. *Amomum Cardamon* is planted for its ornamental qualities and fragrant foliage. *Alpinia* is perhaps the largest genus—with about 150 species. *Alpinia speciosa* (Shell-Flower) is grown in gardens, *Curcuma* and *Hedychium* comprise about 40 or more species each in the tropics of the Old World.

CANNACEÆ (*Canna Family*). This family consists of the single genus *Canna* only, of which there are about 50 species known, mostly natives of tropical countries. The plants grow from thick, branching, and sometimes tuberiferous rhizomes, with the flowering-stem sheathed by the petioles of the large broad leaves. Flowers large and showy, of red or yellow hues, unsymmetrical, in spicate, racemose, or paniculate inflorescences. Each flower is short-pedicelled, in the axil of a con-

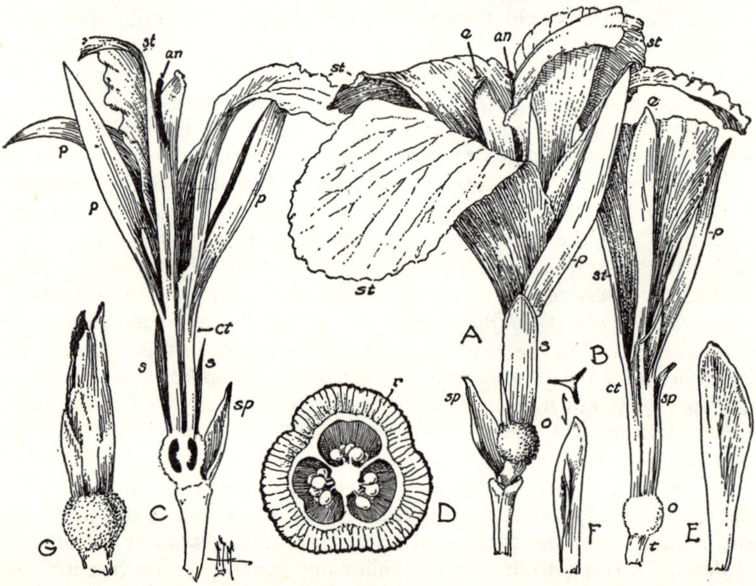

FIG. 477.—*Canna* (hybrid): *A,* flower; *B,* median section of perianth; *C,* median section of flower; *D,* transverse section of "ovary"; *E,* stigma and top of style; *F,* stigma viewed from edge, showing lateral wing; transverse section of stigma above; *G,* flower bud; *s,* sepals; *p,* petals; *st,* staminodes; *an.* anther; *e,* stigma; *ct,* corolla-tube; *o,* ovary; *r,* receptacle; *sp,* spathe; *t,* pedicel.

spicuous bract. The perianth is clearly differentiated into an outer whorl of lanceolate, erect, more or less greenish sepals and an inner whorl of much longer but similar, erect petals, both whorls arising from the summit of the obliquely globular ovary. But the most striking feature is the unusually modified andrœcium which, here, forms the showy part of the flower; that is, the stamens are all transformed into large petal-like segments (of varying number). One of these—the only functional one—bears a 2-celled anther on one edge near the top; the others are all staminodes, and the largest of these is the *labellum*. The style, which is also petaloid, is shaped like a scalpel-blade, with a wing on each side (reduced stigmatic lobes?) and stigmatic on one edge at the top. The ovary is tuberculate on the outer surface and produces numerous ovules in each cell on axile placentas; it matures into a warty, 3-valved capsule, with the persistent sepals at the top. The petals may be somewhat united at the base to form a short tube (sympetalous!) bearing the adnate segments of the andrœcium. "The garden cannas of the present day," says Bailey, "are the results of hybridization and breeding and are not referable to botanical species, although the marks of one or another species may be prominent in them." Dr. Bailey lists eight species as being represented in gardens. (Fig. 477.)

MARANTACEÆ (*Maranta* or *Arrow-Root Family*). Flowers unsymmetrical, bisexual, epigynous, in capitate or spicate inflorescences subtended by spathe-like bracts. Perianth differentiated into calyx and corolla, the sepals distinct but the petals united into a tube of various lengths—one petal larger than the others and hood-like. Andrœcium of 6 segments, only one of which is a functional stamen, the other 5 modified as petaloid staminodes. (Compare with preceding family.) In some cases one or two of the carpels are aborted. Style slender. Fruit a dehiscent or indehiscent capsule, or berry-like, with one seed in each cell. Seeds arillate, with a curved embryo surrounded by perisperm. Genera about 30; species about 260, tropical but mostly in the western hemisphere.

The arrow-roots are perennial herbs of diverse habits growing from elongate and branching or tuberous rhizomes. Leaf-blades pinnately veined, tapering more or less gradually or abruptly to a petiole or a sheathing base.

Species of *Maranta* and *Calathea* are cultivated. In the former the ovary is 1-celled, in the latter 3-celled. *Thalea dealbata* ranges North to the latitude of Missouri, with another species occurring farther South. In general aspect the plants suggest a dicotyledon.

ORDER MICROSPERMÆ (ORCHIDALES)

Flowers regular or irregular, cyclic, epigynous. Perianth more or less differentiated into calyx and corolla. Stamens 6, 3, 2, 1, one or more staminodial or variously modified and coalescent with the style. Gynœcium 3-carpellate; ovary inferior. Fruit a 3-celled or 1-celled capsule. Seeds very numerous and small, with or without endosperm.

One of the largest orders of flowering plants, chiefly of warm temperate and subtropical regions. Only the two following families.

Functional stamens 6 or 3 (inner whorl). *Burmanniaceæ.*
Functional stamens 1 or 2, rarely 3. *Orchidaceæ.*

BURMANNIACEÆ (*Burmannia Family*). A small family of about 10 genera and 60 or 70 species of the tropics, but poorly represented in our flora. The plants are small annuals or perennials from rhizomes or tubers and mostly of saprophytic habit. Leaves numerous, scale-like or linear, and sometimes in basal rosettes. Flowers relatively large, regular or irregular, bisexual, rarely unisexual by abortion, solitary or in spicate or racemose inflorescences. Perianth-segments commonly modified into many diverse and curious shapes, either seated on top of the inferior ovary or on the margin of a prominent campanulate hypanthium—which also bears the andrœcium. Connective of the anthers often expanded laterally or otherwise modified. Placentas axile or parietal. Style rather short, with 3 stigmatic lobes at the summit. Fruit a capsule. Integuments 2.

Burmannia and *Apteria* are the only genera represented in our flora, and these with one species each in southeastern United States, *Burmannia biflora,* a diminutive plant, being found in swampy areas as far north as Virginia. In this species the hypanthium is conspicuously winged.

ORCHIDACEÆ (*Orchid Family*). This is a vast assemblage of perennial herbs distributed over tropical and temperate parts of the world. The number of species is variously estimated at 15,000—17,000. In habit the plants are very diverse, in response to the different ecological conditions under which they grow. In general the plants grow from bulbs, corms, tubers, or tuberous roots, as erect, prostrate, or climbing, mostly green, geophytes or epiphytes, while on the other hand others grow as leafless saprophytes. The leaves are of diverse shapes but mostly of the liliaceous type. Flowers very irregular, bisexual, epigynous, solitary or in spikes or racemes, and often very showy. By a twisting of ovary the flower is inverted (*resupinate*). The most outstanding feature of the flowers is the *gynandrium* or *column,* an organ formed by the union in various ways of the stamens with the style. Only one stamen, the anterior of the outer whorl, is anther-bearing, the other five being transformed into staminodes; in some cases, however, two functional stamens are present. The pollen is formed mostly in masses called *pollinia*—which adhere to the bodies of insect visitors (compare with *Asclepiadaceæ*). The inferior and usually elongate ovary is twisted and 1-celled, with numerous minute ovules on three parietal placentas. The carpel, which faces the functional stamen, is sometimes transformed into a beak (*rostellum*) which functions in sticking the pollinia to the bodies of visiting insects. From the stylar portion of the other two carpels is formed a viscid stigma. Fruit a capsule dehiscing by 3 valves (or by 6). The seeds are minute, more or less spindle-shaped, with a loose hyaline coat. Endosperm none. The embryo is reduced to a fleshy oblong body lacking cotyledons. In some species the seeds are ejected by fine hygroscopic hairs produced inside the ovary. Surrounding the aërial roots of epiphytic orchids is a soft, water-holding tissue of various thicknesses called the *velamen*. Genera well over 400.

Subfamily I. *Pleonandreæ*. Functional anthers 2. Flowers irregular in our species, and the inner segment of the corolla modified into a saccate lip. *Cypripedium.*

Subfamily II. *Monandreæ*. Functional anthers 1. Style of the rudimentary carpel forming a beak (rostellum).

Green-leaved plants. *Habenaria, Calopogon, Pogonia, Spiranthes, Calypso.*

Leaves reduced to sheathing scales. Plants yellowish or purplish saprophytes. *Corallorrhiza.*

Cypripedium (Ladies' Slipper, Moccasin Flower). Species about two dozen, mostly of Eastern parts of the country. The characteristic feature is the yellow, white, pinkish, or purplish, strongly inflated (saccate) lip of the corolla, by which the plants may be recognized at once. The two functional stamens are situated one on each side of the thick column, while on the upper side of the column is a thick, petaloid staminode. In some species the sepals are distinct but in others two of them are united under the lip. *Cypripedium parviflorum* (Yellow Ladies' Slipper), a small-flowered form, is common in woods and thickets from Nova Scotia to Minnesota and southward. A larger-flowered form, the variety *pubescens,* occurs throughout the same region; it is quite common in the Northern prairie region. The beautiful *C. Reginæ* occurs in cold swamps. In many localities this species has been exterminated, and is becoming rare.

Habenaria has been broken up by some writers into a dozen or more genera. In general the plants are small to tall and leafy-stemmed, bearing few to many whitish to greenish flowers in a loose to compact terminal raceme or spike. The lip is entire, toothed, lobed or fringed, and produced into a more or less bent, short or elongate spur. The plants are mostly inhabitants of cold swamps or bogs, or

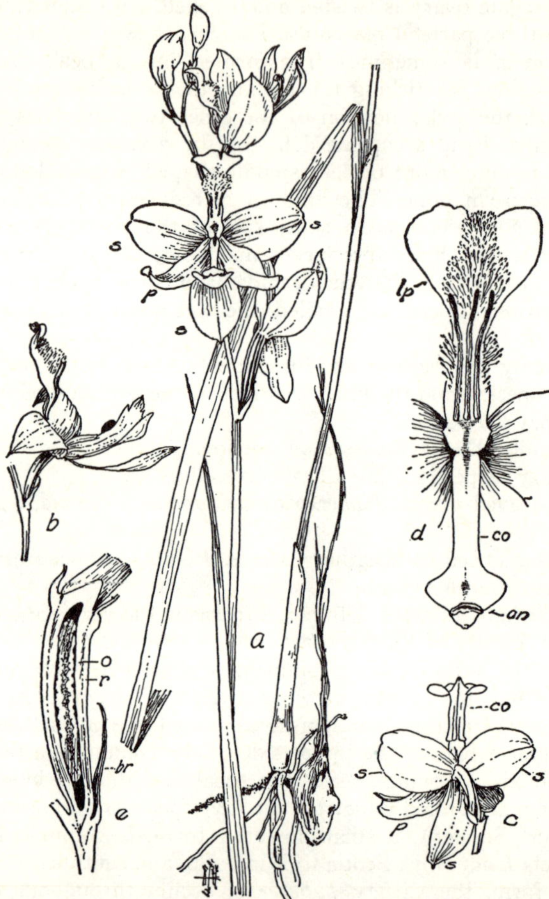

FIG. 478.—*Calopogon pulchellus: a,* habit sketch, with old bulb at base to right; *b,* side view of flower; *c,* flower viewed from below; *d,* inner part of flower viewed from above; *e,* median section of "ovary"; *s,* sepals; *p,* petals; *lp,* lip; *co,* column; *an,* anther; *br,* bract; *o,* ovary proper; *r,* receptacle.

wet situations. They are graceful and beautiful orchids, some of the larger species producing large clusters of very showy flowers, *e.g., H. lacera, H. psycodes, H. fimbriata. Habenaria* (*Lysias*) *orbiculata* is frequently met with in the Northern coniferous regions. It is readily recognized by its two large, glossy, orbicular leaves lying flat on the ground, and by the open spike of greenish flowers.

Calopogon (Calopogon). These are scapose orchids growing from bulbs. The bulb of the season is developed from the bulb of the previous year and produces a single leaf; the following season a leafy flowering-stem (scape) appears, bearing a racemose cluster of flowers at the top. Our few species are confined to the Central and Eastern parts of the country. The column is elongate and expanded at the summit where the transverse anther is attached. The lip is broadly expanded and lobed towards the apex, and densely bearded along the middle line on the upper surface. In the species figured three large, club-shaped hairs occur at the base (Fig. 478). In *Pogonia* the lip is either lobed or lacerate at the tip and the column is not winged. Probably less than a half-dozen species are found in Eastern parts of the country. In *Pogonia verticillata* and *P. affinis* the leaves are whorled at the top of the flowering-stem. These two are woodland species. In *Spiranthes* the upper sepal and the petals are more or less united. Both lip and column are short. The flowers are small, white or greenish, and twisted in an elongate, loose or compact spike terminating a bracteate scape with basal leaves. The genus comprises 50 or more species, ranging from the tropics into temperate latitudes, with about a dozen mostly in eastern United States. *Calypso* (*Cytherea*) *bulbosa,* the only species of the genus, is a beautiful low orchid found in low, cold, coniferous (cedar) woods from Maine to Alaska and southward in the mountains to Arizona, and also in Europe. From the solid corm arises a single broad leaf and a low bracted scape bearing a single purplish flower at the summit. The lip is saccate, somewhat as in *Cypripedium,* and the column petaloid.

Corallorrhiza (Coral-Root). The coral-roots are yellowish, brownish, or purplish orchids growing from coral-like masses of profusely branching rootstocks. From the rootstock arises a stout or slender flowering-stem bearing scale-like leaves and terminating in a racemose inflorescence. The lateral sepals are more or less united with the base of the lip and the column to form a spur. Species over a dozen, scattered over northern parts of the world, ten or more occurring in our flora. One of the most widespread is *Corallorrhiza trifida,* which ranges from Newfoundland to the Rocky Mountains and north to Alaska, and is also found in Asia and Europe.

GLOSSARY

Abaxial. On the side away from the axis.
Abortive. Non-functional, barren, or defective.
Acaulescent. Without an evident stem, or with stem very short and more or less subterranean.
Accrescent. Descriptive of organs that continue to grow after flowering.
Accumbent. Descriptive of the cotyledons of an embryo when their edges lie against the radicle.
Acerose. Sharp-pointed, as applied to leaves, bracts, or scales that have narrow or slender tips.
Achene (*achænium*). A small, dry, hard, one-chambered, one-seeded fruit.
Achlamydeous. Without a floral envelop (cloak)—that is, without a perianth or calyx and corolla.
Acicular. Needle-shaped.
Acotyledonous. Without cotyledons—as in orchids.
Aculeate. Set with prickles; prickly.
Acuminate. Drawn out at the apex into a long, slender point—usually applied to leaf-tips.
Acute. Terminating gradually in a sharp point.
Acropetal. Towards the apex or top.
Actinomorphic. Having radial symmetry; regular; star-shaped.
Acyclic. Not whorled.
Adaxial. Facing the primary axis.
Adherent. United—referring to the union of organs of different categories.
Adnate. United—synonymous with *adherent,* referring to the union of one organ with another, as ovary with receptacle, stamen with corolla.
Adventitious. Appearing in other than the usual place, as roots springing from cuttings.
Adventive. Not fully naturalized.
Æstivation. The arrangement or mode of folding of the perianth or of calyx or corolla in the bud.
Affinity. Relationship; phylogenetic connections.
Alate. Winged.
Albuminous. Provided with albumen or endosperm (and perisperm, or both).
Alpine. Inhabiting high mountains.
Alternate. Arranged in a zigzag manner, as when leaves are on opposite sides of the stem but at different levels.
Alveole. A small cavity or depression on the surface of an organ.
Alveolar. Marked with alveoles.
Ament. A catkin—a type of spicate inflorescence.
Amentiferous. Bearing aments or catkins.
Amphitropous. Curved upon itself—said of an ovule with the hilum in the middle.

Amplexicaule. Clasping or surrounding the stem—usually referring to the attachment of certain leaves.
Ampulla. A kind of hollow leaf, as in *Utricularia.*
Analogous. Similar in function but of different origin.
Anastomosing. Running together; becoming confluent, or united, as the veins of a leaf.
Anatropous. Bent over or inverted; referring to an ovule with the hilum and micropyle near together at one end, the chalaza at the other end.
Andro- and *-androus.* Of male or staminate character. Used in combinations.
Androdiœcious. With perfect (bisexual) flowers on one individual and staminate flowers only on another individual of the same species.
Andrœcium. Collective term for the stamens.
Androgynophore. Stalk bearing both andrœcium and gynœcium.
Androgynous. Bearing staminate and carpellate flowers in the same inflorescence.
Andromonœcious. Bearing staminate and bisexual flowers on the same plant.
Androphore. Stalk bearing the andrœcium.
Anemophilous. Wind-pollinated; literally "wind-loving."
Angiosperm. A plant bearing the ovules in a closed carpel (macrosporophyll).
Anisocarpous. Carpels unequal in number to the other floral organs.
Annual. Of one season's growth.
Annular. Ring-like.
Anterior. Referring to the front side of a flower—that is, the side away from the axis of the inflorescence, or the side on which the floral-bract stands.
Anther. The pollen-bearing organ of a flower; pollen-sac; microsporophyll.
Antheriferous. Anther-bearing.
Anthesis. The period of opening of the flower; the period of pollination.
Anthocarp. A fruit (achene) invested by the hardened calyx (*Nyctaginaceæ*).
Apetalous. Lacking petals.
Aphyllous. Leafless.
Apiculate. Ending in a sharp-pointed tip.
Apocarpous. Composed of distinct carpels (said of a gynœcium).
Appressed. Closely applied; lying close or flat against.
Arachnoid. Having the appearance of a cobweb; composed of entangled hairs.
Arcuate. Curved or bent like a bow.
Areolate. Divided into angular spaces; reticulate.
Areole. A small, angular space marked upon a surface.
Aril (arillus, arillode). A small appendage from or around the hilum of a seed.
Arillate. Furnished with an aril.
Aristate. Awned; bearing a stiff, bristly appendage.
Articulate. Jointed; easily separating at nodes or joints, as in spikelets of grasses.

Ascending. Growing obliquely upward, or curving upward during growth. Applied to branches, hairs of stems, and ovules.
Asperous. Rough; harsh.
Assurgent. Tending to grow upward; ascending.
Asymmetrical. Without symmetry.
Attenuate. Drawn out into a long, slender point.
Auricle. An ear-shaped appendage.
Auriculate. Furnished with auricles.
Autophytic. An independent—that is, a chlorophyllous plant.
Awn. A stiff bristle-like appendage.
Axial (*axile*). Belonging to the axis.
Axil. The angle formed by a leaf with the stem to which it is attached.
Axillary. Situated in the axil of a leaf.
Axis. The line running lengthwise through the center of an organ, as of a flower or a stem; also applied to the stem itself.

Bacca. A berry; a pulpy unilocular fruit with a soft outer coat or rind, the seeds imbedded in the pulp.
Baccate. Having the substance of a berry.
Balausta. The fruit of the pomegranate.
Barbed. Furnished with stiff, spine-like points or bristles bent like a fish-hook.
Barbellate. Furnished with small, fine barbs.
Barbulate. Finely bearded.
Basifixed. Affixed by the base, as an anther to the filament.
Basifugal. Proceeding away from the base—towards the top.
Basilar. Belonging to the base.
Basipetal. In the direction of the base; from the summit downwards.
Basiramous. Branching from the base.
Bast. The inner fibrous bark of dicotyledonous trees and shrubs.
Beaked. Furnished with a prominent tip, as the persistent base of the style in some spike-rushes.
Bearded. Furnished with long, stiff hairs or with a long awn, as in barley.
Berry. A many-seeded pulpy fruit.
Bi-, Bis-. A Latin prefix meaning "twice."
Bidentate. Two-toothed.
Biennial. A plant that lasts through two growing seasons, fruiting the second season.
Bifarious. In two rows, one on each side of an axis.
Bifid. Two-cleft; forked into two slender limbs to near the middle.
Bilabiate. Two-lipped, referring especially to the corolla.
Bilamellar. Having two, usually, flat lobes or divisions, as some stigmas.
Bilocular. Two-celled, as applied to an ovary or an anther.
Bipinnate. Twice-pinnate, referring to compound leaves.
Bipinnatifid. Twice-pinnatifid.
Bisexual. Having both stamens and carpels.
Biternate. Twice ternate.
Bole. Trunk of a tree.
Bract. A much reduced leaf, as of an inflorescence or rhizome.
Bracteate. Furnished with bracts.

Bracteolate. Bearing bractlets (bracteoles).
Bracteole. A small bract—usually referring to the small bracts on the pedicel of a flower.
Bud. An undeveloped shoot in a resting condition.
Bulb. A bud, most of the leaves of which are thickened with reserve food-materials.
Bulbiferous. Bearing bulbs.
Bulbil. Small bulbs borne singly or in clusters.
Bulbous. Of the character of a bulb.
Bullate. Blistered or puckered.
Byssoid. Slender and thread-like, as a cobweb.

Caducous. Falling off very early.
Cæspitose. Tufted; growing in tufts, like many grasses.
Caliciflorœ. A subdivision of the flowering plants, in which the stamens and petals are borne on the calyx—"calyx-flowers."
Calyculate. Bearing sepal-like segments on the calyx; furnished with an epicalyx.
Callosity. A hardened or leathery thickening of a part of an organ.
Callus. A hard protuberance or callosity.
Calyculus. An outer row of sepal-like segments or bracts borne on the calyx and giving the calyx the appearance of being double; an epicalyx.
Calyx. The outer set of perianth-segments of a flower.
Calyx-tube. The tube of a gamosepalous calyx; also the hypanthium of a perigynous or epigynous flower, or the receptacle of such flowers.
Campanulate. Bell-shaped.
Campylotropous. Bent so as to bring the hilum, micropyle, and chalaza near together—said of ovules.
Canaliculate. Longitudinally channeled or grooved.
Cancellate. Veined in the form of a lattice.
Canescent. Bearing a hoary, grayish pubescence.
Capillary. Hair-like, thread-like, filiform.
Capitate. Head-like; collected into a dense cluster. Like a pin-head, as in gland-tipped hairs.
Capitulum. The head of composites.
Capreolate. Bearing tendrils.
Capreole. A tendril.
Capsular. Belonging to or having the character of a capsule.
Capsule. A dry dehiscent fruit (pod).
Carina. A keel.
Carinal. Belonging to or having the character of a keel.
Carinate. Furnished with a carina or keel.
Carnose. Fleshy—referring to the endosperm of some seeds.
Carpel (*Carpidium*). The ovuliferous organ of the flower; a closed macrosporophyll (megasporophyll); a simple pistil.
Carpo-. *Carpos*. Greek, "fruit." Of the nature of or belonging to a fruit.
Carpophore. A stalk or prolongation of the floral axis which bears the gynœcium.
Caruncle (*Caruncula*). A thick, fleshy or spongy excrescence covering the hilum of some seeds (castor-bean).
Caruncculate. Furnished with a caruncle.

Caryopsis. The grain (fruit) of grasses.
Catkin. An ament; a spicate inflorescence.
Castaneous. Chestnut-colored; dark brown.
Caudate. Bearing a tail-like appendage.
Caudex. The persistent, usually subterranean base of a biennial or perennial herbaceous stem.
Caudicle (*caudicula*). Diminutive of *caudex*. The stalk-like process supporting the pollen-mass (pollinium) of orchids and milkweeds.
Caulescent. Having an evident stem, usually short and above ground.
Cauline. Belonging to the stem; borne upon the stem, above ground.
Cell. In taxonomy usually referring to the seed-cavity or locule of an ovary, or to the pollen-chambers of an anther, or to the air-spaces in the stem of some aquatics.
Centrifugal. Growing out from the center—referring to a cymose inflorescence.
Centripetal. Growing towards the center—referring to a racemose inflorescence.
Cernuous. Pendulous or nodding.
Cespitose. (See *Cæspitose.*)
Chalaza. The basal part of an ovule where the stalk or funiculus enters.
Chaff. A small, somewhat dry, membranaceous bract. Referring especially to the small bracts at the base of and usually investing the disk-flowers of some composites.
Character. In taxonomy, a special feature or features of an organ or plant.
Chlamys. Greek, a cloak or covering (mantle)—used in combination, *e.g., Archichlamydeæ;* heterochlamydeous.
Chori-. In combination meaning "separate," "distinct."
Choripetalæ. A subdivision of the flowering plants in which the petals are distinct (*Dialypetalæ,* in this volume).
Choripetalous. Having distinct petals (polypetalous).
Channeled. Having longitudinal grooves or furrows.
Chartaceous. Of the texture of writing or wrapping paper.
Cilia. The short, usually stiff, hairs on the margins of leaves.
Ciliate. Furnished with cilia.
Ciliolate. Minutely ciliate.
Circinnate. Rolled from the top downwards, as the leaves of ferns.
Circumscissile. Dehiscing or separating by a circular line, as the capsule of *Amaranthus, Plantago,* or *Hyoscyamus.*
Cladode. A leaf-like branch.
Cladophyll. A branch modified in the form of a leaf; a leaf-like stem.
Clathrate. Latticed.
Clavate. Club-shaped; gradually thickened upward, somewhat like an Indian club.
Claw. The narrow stalk of some petals (mustards), resembling a petiole.
Cleft. Divided or split about to the middle into two divisions.
Cleistogamous. Self-pollinated or self-fertilized in the bud-stage, the flowers not opening (violets).
Clinanthium (flower-bed). The common or capitular receptacle of the head of composites.
Coalescent. Fused or united.

Coccus. The sections of the capsule of spurges (*Euphorbiaceæ*).
Cochleate. Spirally coiled like a snail-shell—referring to a kind of æstivation.
Coherent. Fused or united, when the organs are of the same kind.
Coleorhiza. The sheath investing the hypocotyl of grasses.
Collateral. Placed side by side, as some carpels and ovules, or as the glumes of some grasses.
Collum. The internode or neck between the cotyledonary node and the node next above (first epicotylary node) of an embryo.
Columella. The carpophore of the fruit of umbellifers.
Column. That part of the flowers of orchids which is formed by the union of the style and the filaments and which supports the anthers and the stigma.
Coma. A tuft of hairs, as in some types of pubescence.
Commisure. The line of cohesion between two carpels, as in the *Umbelliferæ.* The two carpels (mericarps) separate along this line, leaving the carpophore between them free.
Comose. Having a tufted pubescence, or bearing a tuft of hairs, as the seeds of willows.
Complicate. Folded upon itself, as a leaf.
Compound. Composed of a number of similar united parts, as carpels in a syncarpous gynœcium (compound pistil), or divided into a number of similar parts or divisions, as the leaflets of "compound" leaves.
Compressed. Flattened laterally or lengthwise, as the carpels of the arrowheads or the spikelets of some grasses and sedges.
Conceptacle. A receptacle like that of the rose or of *Calycanthus.*
Conduplicate. Rolled or folded lengthwise upon itself, as a leaf.
Confluent. Running together or fusing or coalescing into one.
Congeneric. Belonging to the same genus.
Connate. United, referring to two or more similar organs.
Connective. The part of a stamen that lies between the anthers.
Connivent. Coming together; the converging or bending towards one another of two or more similar organs (anthers).
Conspecific. Belonging to the same species.
Contorted. Regularly twisted and folded in one direction, as the corolla of gentians.
Convolute. Rolled up lengthwise.
Coralline. Having the form and appearance of coral—said of the roots of some saprophytic plants (*Corallorrhiza*).
Cord. The elongate stalk of some seeds (cucumber).
Cordate. Heart-shaped—referring to the base of some leaf-blades.
Coriaceous. Of leathery texture.
Corm. A short, thick, underground stem (*Cyclamen, Brodiæa*).
Cormose. Of the nature of a corm.
Corneous. Of horny texture, *cornu,* L., a horn.
Corolla. The inner colored perianth-segments of a flower.
Corona. A collar-like or tubular appendage of the corolla of *Narcissus.*
Coroniform. Having the form of a corona.
Corrugate. Wrinkled in even or irregular folds.
Corymb. A flat-topped racemose inflorescence.
Corymbose. Of the nature or form of a corymb.

Costa. A rib; the longitudinal rib or ribs of some carpels, bracts, and glumes.
Costate. Ribbed; bearing prominent longitudinal ribs or nerves.
Cotyledon. A seed-leaf. The temporary, usually distinctive, first leaf or leaves of an embryo.
Cotype. A specimen of a plant associated with the type specimen from which a species is described.
Cowled. Hooded.
Crampons. Adventitious roots like those of the ivy—grappling-roots.
Crateriform. Crater-like; cup-shaped or saucer-shaped, as the hypanthium of perigynous and some epigynous flowers.
Creeping. Running along the surface of the ground and rooting from the nodes—said of stolons and runners.
Cremocarp. Composed of two achenes (mericarps) which separate at maturity—a term applied to the fruit of the *Umbelliferæ.*
Crenate. Furnished with rounded marginal teeth—said of leaves.
Crenatures. The separate teeth of a crenate leaf.
Crenulate. Finely crenate.
Crest. An apical appendage of some seeds. Also applied to projections or protuberances of floral organs.
Cristate. Bearing an elevated appendage of the shape of a crest.
Crown. The tubular or ring-like appendage of a corolla (*Narcissus*). The top or stem part of a fleshy tap-root (carrot, parsnip).
Cruciate (*cruciform*). Having the shape of a cross.
Crustaceous. Crusty; hard, thin, and brittle.
Crypto-. Greek, hidden, concealed. Used in combination.
Cucullate (from Latin *cucullus,* a hood). Hooded or hood-shaped.
Culm. The stem of grasses and sedges.
Cultigen. A plant known only in cultivation.
Cultivar. A variety or race that has originated and persisted in cultivation (Bailey).
Cultriform. Shaped like a knife-blade.
Cuneate (*cuneiform*). Wedge-shaped.
Cupula. The cup of the acorn or beach-nut.
Cuspidate. Prolonged into a gradually tapering sharp-pointed tip.
Cuticle. The outer layer of the outer free wall of epidermal cells.
Cyathium. The type of inflorescence in *Euphorbia.*
Cyathiform. Of the form of a cyathium, that is, like a funnel or hollow cone.
Cymbiform. Boat-shaped.
Cyme. A determinate inflorescence, in which the central flower opens first.
Cymose. Of the form or nature of a cyme.
Cypsela. The one-seeded, dry, indehiscent fruit of composites.

Dactyliferous. Bearing finger-like appendages.
Deca. Greek, meaning "ten"—used in combinations.
Decagynous. Having ten carpels.
Decandrous. Having ten stamens.
Deciduous. Falling off, not persistent.
Declinate. Directed downwards from the base.

Decompound. More than once compounded or divided. A compound leaf composed of numerous leaflets.
Decorticated. Deprived of bark (cortex).
Decumbent. Growing flat along the ground, but with the tip ascending.
Decurrent. Running down the sides of the stem from the point of insertion—said of leaves (leaf-bases).
Decussate. Alternating in two rows, in zigzag manner, on opposite sides of the stem.
Definite. Of constant number—referring to floral organs.
Deflexed. Bent or curving sharply downwards.
Dehiscent. Splitting open by valves or slits—said of capsules and anthers.
Deltoid. Having the shape of the Greek letter Δ.
Dentate. Toothed, with the teeth perpendicular to the margin of the leaf.
Dentation. The arrangement or direction of the marginal teeth of a leaf.
Denticulate. Finely dentate.
Depressed. The flattening of an organ downwards—in the direction of the axis.
Determinate. Arrested in development—referring to a cymose inflorescence.
Di- Dis-. Greek, meaning "two," "twice." Used as prefix.
Diadelphous. In two brotherhoods—said of stamens when united into two sets.
Diagnostic. Defining, determining, or distinguishing—referring to taxonomic characters.
Dialycarpous. Composed of distinct carpels.
Dialypetalous. Composed of distinct petals.
Diandrous. Having two stamens.
Dicarpellate (*dicarpellary*). Composed of two carpels—said of a gynœcium.
Dichasial. Belonging to or of the character or nature of a dichasium.
Dichasium. A two-branched inflorescence.
Dichlamydeous. Having distinct calyx and corolla.
Dichotomous. Forking, or forked branching, the limbs equal.
Diclinous. Bearing the andrœcium and the gynœcium in separate flowers.
Dicotyledonous. Having the character of a dicotyledon—that is, having an embryo with two seed-leaves or cotyledons.
Didymous. Twin; occurring in pairs.
Didynamous. Having the stamens in two unequal pairs.
Differentiate. To distinguish or separate on diagnostic characters.
Diffuse. Widely or loosely spreading.
Digitate. Spreading like the fingers of a hand.
Digynous. Composed of two carpels.
Dimerous. Having the parts in twos—said of flowers.
Dimorphic. Occurring in two forms.
Diœcious. Bearing the staminate flowers on one individual and the carpellate on another of the same species.
Diplo-. Greek, double. Used in combinations.
Diplostemonous. Having the stamens double the number of petals or sepals.
Dipterous. Two-winged, as seeds and fruits.
Discoid (*disciform*). Having the shape of a disk.

Disk. A glandular outgrowth of the receptacle or hypanthium, hypogynous or epigynous according to the position of the ovary. Also used in referring to the common receptacle of the head of composites.
Dissected. Cut into numerous narrow segments.
Dissepiment. The partition-wall or septum separating the seed-chambers or locules of an ovary.
Dissilient. Bursting in an elastic manner, as some dry dehiscent fruits.
Distichous. In two rows, on opposite sides of the stem—said of leaves—the two rows lying in the same vertical plane.
Distinct. Separate, not united.
Dithecal. Having two cavities (cells) or chambers—usually referring to anthers.
Divaricate. Widely spreading.
Divergent. Spreading apart.
Divided. Cut to the base into lobes or segments.
Dorsal. Belonging to or appertaining to the back of an organ.
Dorsiventral. In the median plane.
Double Flower. A flower in which the stamens and carpels are modified into petals.
Drupaceous. Of the nature or texture of a drupe.
Drupelet. Diminutive of drupe.
Dumose. Having a low or shrubby aspect.

E-, Ex-. Latin, a privative, meaning "without," "out of," "from." Used as prefix.
Ebracteate. Without bracts.
Echinate. Beset with bristles or spines.
Ecological. Referring to the habitat or surroundings of a plant.
Emarginate. Having a shallow notch at the end or tip, as leaves or petals.
Embryo. The undeveloped dormant plant within the seed.
Embryo-sac. The gametophytic part of the ovule in which the embryo develops.
Enate. Springing or growing out.
Endocarp. The inner layer of the pericarp of a fruit.
Endogenous. Growing from the inside—said of the growth of stems in thickness.
Endosperm. The reserve food-storage tissue or "albumen" of a seed.
Ensiform. Sword-shaped.
Entire. Without marginal serrations or teeth.
Entomophilous. Insect-loving—referring to insect pollination; insect pollinated.
Envelope (floral). The perianth, or calyx and corolla.
Ephemeral. Of short duration.
Epi-. Greek, "upon." Used in combinations.
Epicalyx. A set of bracts borne between the sepals or calyx-segments of some flowers, giving the calyx the appearance of being double. *Calyculus.*
Epicarp. The outer layer of the pericarp of fruits.
Epicotyl. That portion of an embryo above the cotyledonary node; plumule.

Epicotyllary. Belonging to the epicotyl.
Epidermis. The outermost layer of cells of an organ.
Epigeal (*epigean*). Above ground; borne above ground—referring to cotyledons.
Epigynous. Borne upon or on top of the "ovary" or gynœcium.
Epipetalous. Borne upon the petals or the corolla-tube—referring to stamens.
Epiphyllous. Borne upon a leaf, as some inflorescences.
Epiphyte. A plant growing upon another plant, but not necessarily a parasite.
Eprophyllate. Without prophylls or small bracts (bracteoles) below the flower, in addition to the floral bract at the base of the pedicel.
Equitant. Overlapping in two ranks (distichous), like the leaves of the iris.
Erose. Irregularly cut along the margin as if gnawed—said of leaves.
Exalbuminous. Without albumen or endosperm—said of seeds.
Excentric (*eccentric*). Not lying in the center—said of embryo in relation to endosperm.
Excurrent. Running beyond the margin or tip.
Exfoliated. Peeling off in thin layers (bark of birches).
Exocarp. The outer layer of a fruit.
Exserted. Projecting beyond the summit of, as stamens beyond the petals.
Exsiccati. Dried plants, as herbarium specimens.
Extrorse. Facing outward from the center of the flower, as anthers.
Eucyclic. Said of a flower in which the organs have the same number of whorls and the same number of members in each whorl.

Falcate. Scythe-shaped; *falciform.*
Farinaceous. Mealy or composed of much starch—said of endosperm.
Farinose. Covered with mealy powder.
Fasciated. Malformed; fused together—referring to stems or branches.
Fascicle. A dense flower-cluster or umbellate cyme.
Fasciculate. Formed of closely crowded clusters.
Fastigiate. Of pyramidal form—referring to branches growing upward, as in the Lombardy poplar.
Fenestrate. Window-like—referring to the translucent areas in some leaves.
Ferruginous. Rust-colored.
Fertile. Fertilized and capable of producing fruit.
Fibril. A slender, thread-like fiber.
Fibrillose. Composed of fine fibers.
Fid. Used in combination, meaning cut down to the middle (bifid).
Filament. The stalk supporting the anther of a stamen.
Filamentous. Thread-like.
Filiform. Thread-like.
Fimbriate. Fringed along the edge.
Fimbrillate. Minutely fringed.
Fissiparous. Divided into two parts by means of a septum.
Fistular (*fistulous*). Hollow, as culms of grasses.
Flabelliform. Fan-shaped; palmate.
Flaccid. Like a loose, papery sack; lax.

Flagellate. Having the shape of a flagellum or whip.
Flexuous (*Flexuose*). Bent in a zigzag manner; continuing in alternate directions like the rachilla of the spikelets of grasses.
Flocci. Woolly filaments.
Floccose. Having tufts or locks of soft, woolly hair.
Flosculous. Having tubular corollas like the florets of composites.
Floret. A small flower, as in the spikelets of grasses or the heads of composites.
Floriferous. Bearing flowers.
Foliation. The development of leaves.
Foliaceous. Having the character of a leaf.
Foliate. Having leaves. Used in combination—bifoliate.
Foliolate. Having leaflets. Used in combination—trifoliolate.
Foliole. A small leaf.
Foliose. Having numerous leaves.
Follicle. A fruit formed from a single carpel and usually dehiscing along the ventral suture.
Follicular. Of the nature of a follicle.
Forb. An herb other than a grass or sedge.
Forked. Divided into branches or limbs of equal rank.
Fovea. A pit or depression.
Foveate (*foveolate*). Furnished with pits or depressions on the surface.
Free. Distinct, not adnate or united—as floral organs.
Frondose. Having the character of a frond or leaf of a fern.
Fruit. The seed-bearing organ of a plant.
Frutex. A shrub.
Fruticose. Shrubby; shrub-like.
Fugacious. Falling off early—early deciduous; ephemeral; evanescent.
Fulvous. Tawny yellow.
Funiculus (*funicle*). The stalk of an ovule, by which it is attached to the placenta.
Furcate. Forked.
Furfuraceous. Scurfy or scaly.
Fuscous. Grayish brown.
Fusiform. Spindle-shaped.

Galea. A hooded or helmet-shaped sepal or petal, as in *Aconitum*.
Galeate. Shaped like a hood or helmet; furnished with a galea.
Gamo-. Greek, used in combinations—meaning "united" (gamosepalous).
Gamopetalous. Having united petals, referring to a sympetalous corolla.
Gamosepalous. Having united sepals—*synsepalous*.
Geminate. United in equal pairs.
Gemma. A leaf-bud; a bud by which a plant propagates itself vegetatively.
Gemmiferous. Bearing propagative buds (gemmæ).
Gemmiparous. Reproducing by buds (gemmæ).
Gemmule. Diminutive of *gemma*.
Generic. Belonging to a genus; characteristic of a genus
Geniculate. Bent like a knee.
Genus. An aggregation of species having certain characters in common.
Germination. The extraseminal development of an embryo; sprouting.

Gibbosity. A swelling or inflation, usually of the base of an organ, as in the disk-flowers of a sunflower, or of the outer petals of the bleeding-heart.

Gibbous. Swollen at the base.

Girt. Girdled; adnate at base, as ovary to receptacle, or disk to ovary.

Glabrate. Somewhat glabrous, or becoming glabrous.

Glabrous. Devoid of pubescence.

Gladiate. Sword-shaped.

Gland. An organ of secretion. A tissue or cells, or a single cell functioning as an organ of secretion (as in hairs). Hair glands may be unicellular or multicellular.

Glandular. Furnished with glands, or of the nature of a gland.

Glans. A nut, as the acorn, hazel-nut—usually seated in an involucre.

Glaucous. Covered with a bluish, whitish, or pale greenish bloom, as the leaves of cabbage.

Glochid. A small barbed spine or bristle.

Glocidiate. Bearing or furnished with barbed spines or bristles.

Glomerate. Crowded in a usually small, compact cluster (flowers).

Glomerulate. In very small, compact clusters (flowers).

Glomerule. A small, compact cluster of flowers.

Glumaceous. Bearing glumes; of the character of glumes.

Glume. A rigid, chaff-like or scale-like bract—referring especially to the two empty bracts at the base of the spikelet in grasses.

Glutinous. Sticky; mucilaginous; covered with a mucilaginous substance.

Grain. The fruit or caryopsis of grasses.

Granulate. Covered with granules or minute grain-like bodies.

Granulose. Granulate.

Grumose. Formed of clustered grains or granules.

Gynæcandrous. Having staminate and carpellate flowers in the same spikelet, as in some sedges.

Gynandrous. Having the stamens united with the gynœcium.

Gynobase. A stalk supporting the gynœcium or carpel; gynophore.

Gynodiœcious. Bearing carpellate flowers on one plant and bisexual flowers on another plant of the same species.

Gynœcium. A carpel or an aggregation of carpels, free or united.

Gynomonœcious. Bearing bisexual and carpellate flowers on the same plant.

Gynophore. A prominent stalk supporting the gynœcium.

Gynostemium. The column of an orchid, composed of the stamens and the style.

Gyrate. Twisted or coiled.

Habit. The aspect or general appearance of a plant; growth-form; life-form.

Habitat. The immediate environment of a plant.

Halberd-shaped. Shaped like an arrow-head, but with the two basal lobes directed outward at right angles to the blade of the leaf.

Halophyte. A "salt-loving" plant; a plant growing in a saline or alkaline habitat.

Halo-. Greek, a combining term referring to salt.

Haplochlamydeous. Descriptive of a perianth which is not differentiated into calyx and corolla.
Haplo-. Greek, combining term meaning "half."
Hastate. Halberd-shaped.
Haulm. A dead and dried herbaceous stem like that of grasses and sedges.
Haustorium. A sucker-like or disk-like absorbing organ of some parasitic plants (dodder).
Head. The inflorescence or capitulum of composites; a compact inflorescence.
Heart-shaped. Descriptive of the base of a leaf-blade which has a deep sinus and rounded basal lobes.
Heart-wood. The inner dead wood or duramen of a tree-trunk.
Haplostemonous (*haplostemony*). Having half the usual number of stamens.
Helicoid (*helicoidal*). Of the shape of a helix or snail-shell—descriptive of an inflorescence.
Helmet. The upper petaloid sepal of the aconite (*Aconitum*).
Hemi-. Greek, meaning "half"—a combining term.
Hemicarp. One of the halves (mericarps) of the fruit in the *Umbelliferæ.*
Hemicyclic. Having half the usual number of whorls of floral organs.
Hepta-. Greek, meaning "seven"—a combining term.
Heptamerous. Having seven stamens.
Herb. A plant which is not a tree or a shrub, and is usually of relatively soft texture, producing relatively little woody tissue.
Herbaceous. Herb-like; of the nature of an herb; succulent.
Hermaphrodite (*hermaphroditic*). Bisexual; having the andrœcium and the gynœcium in the same flower.
Hesperidium. The fruit of the orange and other citrus plants.
Hetero-. Greek, meaning "dissimilar" or "diverse"; a combining term.
Heterocarpous. Producing more than one kind of fruit.
Heterocephalous. Bearing staminate and carpellate heads on the same plant, as in some composites.
Heterochlamydeous. Having both calyx and corolla, or a perianth of two different whorls.
Heterocyclic. Differing in the number of whorls of floral organs.
Heterogamous. Bearing two different kinds of flowers, as the spikelets of some sedges; or the heads of some composites, in which the disk-flowers are bisexual and the ray-flowers are carpellate or sterile (neutral).
Heteromerous. Having whorls of floral organs which differ in the number of their component members.
Heteromery. A heteromerous condition or state.
Hexa-. Greek, meaning "six"—a combining term.
Hexandrous. Having six stamens.
Hilum. The scar on a seed, marking its point of attachment to the seed-stalk.
Hirsute. Clothed with long stiff hairs (pubescence).
Hirsutulous. Slightly hirsute.
Hirtellous. Minutely or finely hirsute.
Hispid. Clothed with stiff, bristly hairs or bristles.
Hispidulous. Minutely hispid.

Hoary. Clothed with a grayish-white, fine, close pubescence.
Holo-. Greek, meaning "whole"—a combining term.
Holosericeous. Clothed with minute silky hairs.
Homogamous. Bearing heads the flowers of which are bisexual, referring especially to the heads of composites.
Homogeneous. Having a uniform structure or composition. Uniform in character. Having constant or uniform characters, referring to the aggregate characters of a species, genus, family, or other category, as opposed to *heterogeneous*—of different characters.
Homoios (*homos*). Greek, meaning "similar" in composition—a combining term.
Homoiochlamydeous. Having a perianth the segments of which are similar —not differentiated into calyx and corolla.
Homologous. Of similar origin, although differing in function.
Homotypic. Of the same nature or character.
Horological. Opening and closing at certain hours, as the flowers of the night-blooming *Cereus.*
Humifuse. Spreading along the ground.
Hyaline. Transparent or translucent.
Hybrid. A cross between two species.
Hydrophyte. An aquatic plant; a plant growing in water. *Hydrophytic.* Of the nature of a hydrophyte.
Hygroscopic. Bending or turning under the influence of moisture conditions.
Hypanthium. The tube of the receptacle upon which the calyx, corolla, and stamens are borne; calyx-tube; perianth-tube (in monocotyledons).
Hypo-. Greek, meaning "under," "beneath," or "below"; a combining term.
Hypochil (*Hypochilium*). The lower part of the lip (*labellum*) of orchids.
Hypocotyl. The part of an embryo below the cotyledons and including the cotyledonary node; the stem of the embryo.
Hypogynous. Borne beneath or below the gynœcium or ovary—referring to stamens, petals, sepals, and disk.
Hypopodium. The base of the petiole of leaves.
Hypso-. Greek, meaning "high," "on high"—a combining term.
Hypsophyll. A leaf which is not a sporophyll; a bract or a foliage leaf.

Icosi- (*Icos-*). Greek, meaning "twenty"—a combining term.
Icosandrous. Having twenty stamens.
Imbricate. Overlapping like tiles of a roof—referring to sepals and petals in the bud.
Immersed. Growing under water.
Imparipinnate. Unequally pinnate; odd-pinnate, as a pinnate leaf having an odd terminal leaflet.
Inarticulate. Devoid of nodes or joints—referring to a stem or rachis.
Incised. Cut deeply, as a leaf, into regular or irregular lobes.
Included. Not protruding—as stamens from a corolla.
Incumbent. Bending inwards; in a seed referring to the orientation of the cotyledons—placed with their backs against the hypocotyl.

Indefinite. Inconstant in number; applied also to the continuous growth in length of a racemose inflorescence.
Indehiscent. Not splitting open; remaining closed, as an achene.
Indeterminate. Of indefinite growth—as a racemose inflorescence.
Indigenous. Native to a region.
Indumentum. Pubescence.
Induplicate. Inrolled, like the edges of some sepals and petals in the bud.
Indurate. Hardened.
Indusiate. Furnished with an indusium.
Indusium. A cover.
Inermous. Destitute of prickles or spines.
Inferior. Below, usually referring to the position of the ovary in an epigynous flower.
Inflated. Bladdery; baggy.
Inflorescence. A flower-cluster.
Infra-. Latin, "below," "beyond." Used as a prefix.
Infra-axillary. Below the axil (of a leaf).
Infra-petiolar. Beyond the base of the petiole—said of stipules.
Infundibulate (*infundibuliform*). Funnel-shaped—referring to the shape of some sympetalous corollas.
Innate. Affixed at the top of a support, as anther to filament.
Innovation. An offshoot from a stem; barren shoots of grasses; a mode of budding or of producing offshoots.
Inserted. Attached; growing upon.
Integument. The jacket of an ovule.
Inter-. Lat., "between"; used as a prefix.
Internode. Part of a stem lying between two successive nodes.
Interpetiolar. Standing between the petioles—as stipules.
Interrupted. Broken or irregular in arrangement—said of an inflorescence or leaves.
Intra-. Lat., "within," "inside of"; used as a prefix.
Intramarginal. Within or near the margin.
Intrastaminal. Inside of the whorl or whorls of stamens, or between the whorls of stamens.
Introduced. Brought in from another country or region.
Introrse. Facing inwards or facing the stigma—referring to anthers.
Involucel. A small involucre—as the bracts subtending the secondary umbels in the *Umbelliferæ.*
Involucellate. Bearing involucels.
Involucral. Belonging to the involucre.
Involucrate. Bearing an involucre.
Involucre. Bracts subtending an inflorescence—as in the heads of composites or in the umbels of umbellifers.
Involute. Having the edges rolled inwards.
Irregular. Differing in size and shape; unsymmetrical—referring to flowers.
Iso-. Greek, meaning "equal"; used as a prefix.
Isobilateral. The two sides equal, as in bilateral symmetry of a flower.
Isocyclic. Having an equal number of whorls, or having similar whorls.
Isomerous. Having whorls of an equal number of segments (flowers).

Jugate. Having a pair of leaflets, from, Latin, a yoke, *jugum*.

Keel. A prominent dorsal rib or ridge, as in some carpels or in glumes of grasses.

Labellum. A lip; one of the divisions of the corolla of orchids.
Labiate. Lipped, as the corolla of mints.
Lacerate. Irregularly torn along the edges—referring to leaves.
Laciniate. Cut or fringed into narrow segments.
Lacustrine. Inhabiting lakes.
Lagenate. Flask-shaped.
Lamella. A flattened or plate-like outgrowth or ridge.
Lamellate. Bearing lamellæ.
Lamina. A leaf-blade.
Laminate. Having a lamina; of the character of a lamina.
Lanate. Woolly.
Lanceolate. Lance-shaped—broadest over the middle and tapering to both ends.
Lanuginous. Clothed with long, entangled hairs.
Lanulose. Short-woolly.
Lateral. Borne on the side, as leaves and buds.
Latex. The milky sap of some plants.
Laticiferous. Containing latex.
Leaflet. A segment of a compound leaf.
Legume. The pod of members of the pea family—a pod dehiscent along both sutures.
Leguminous. Having the character of a legume or of a leguminous plant.
Lemma. The outer bract of the flower of grasses.
Lenticel. Aërating organ in the bark of twigs of trees and shrubs.
Lenticular. Lens-shaped.
Lentil. The fruit or seed of certain leguminous plants.
Lepidote. Beset with small, scurfy scales (from *lepis*, a scale, Lat.).
Lianas. Twining woody plants characteristic of tropical forests.
Liber. The inner fibrous bark.
Ligulate. Strap-shaped or tongue-shaped, like the corollas of a dandelion or the ray-flowers of a sunflower.
Ligule. The strap-shaped part of the corolla of composites; the annular collar-like process at the junction of the leaf-blade and sheath in grasses.
Liguliflorœ. Flowering plants having ligulate corollas (composites).
Limo. The blade of a petal or leaf.
Line. The twelfth part of an inch. An obsolete unit of measurement used in botany.
Linear. Long and narrow, as the leaf-blade of grasses.
Lip. One of the divisions of a labiate corolla; labellum.
Lobe. A division or segment of an organ, as leaf or ovary.
Locular. Furnished with locules.
Lodicule. A rudimentary organ at the base of the flower in grasses.
Loment. A leguminous fruit (pod) which is constricted between the seeds.
Lomentaceous. Of the nature of a loment.
Lunate. Crescent-shaped.

Lyrate. Lyre-shaped—descriptive of a compound leaf having a large terminal leaflet, the lateral leaflets becoming gradually smaller towards the base.

Macros- (*Macro-*). Greek, meaning "large"; used as a prefix.
Macrosporangium. An ovule. *Megasporangium.*
Macrosporophyll. A carpel.
Malpighiaceous. Applied to shield-shaped hairs of the sort characteristic of the *Malpighiaceæ.*
Marcescent. Withering but persisting, as leaves (grasses).
Mattulla. The fibers covering the petiole of palms.
Median. Descriptive of the middle plane as of a flower—passing through and bisecting the subtending floral bract.
Medulla. The pith of a stem. *Medullary.* Belonging to the medulla.
Membranaceous (*membranous*). Of the texture of a membrane or parchment.
Mericarp. One of the halves (*achenes*) of the fruit of an umbellifer.
Meso-. Greek, meaning "middle."
Mesocarp. The middle coat or layer of a fruit.
Mesophyte. A plant inhabiting moist or humid situations.
Meter. The standard unit of linear measurement of the metric system, equivalent to 39.37979 inches.
Micro-. Greek, meaning "small"; used as a prefix.
Micropyle. The opening leading into the nucellus of an ovule, formed by the incomplete closing of the integuments.
Microsporangium. An anther.
Microsporophyll. A stamen.
Millimeter. One one-hundredth of a meter or one-tenth of a centimeter, about 0.039 of an inch.
Mitriform. Mitre-shaped.
Monadelphous. Having the stamens united by their filaments into one body.
Moniliform. Like a string of beads.
Mono-. Greek, meaning "one"; used as a prefix.
Monocarpous. Having one carpel.
Monochlamydeous. Having one floral envelope or an undifferentiated perianth.
Monoclinous. Having the andrœcium and the gynœcium in the same flower.
Monocotyledon. A plant producing embryos with only one cotyledon.
Monocotyledonous. Having the characters of a monocotyledon.
Monopetalous. Having but one petal.
Monœcious (*monoicous*). Having the andrœcium and the gynœcium in separate flowers but on the same plant.
Monogynœcial. Descriptive of a fruit formed from the gynœcium of a single flower.
Monogynous. Having one carpel or gynœcium.
Monophyllous. Having but one leaf.
Monopodium. An inflorescence having one primary axis giving off branches to right and left. *Monopodial.* Belonging to or having the character of a monopodium.

Monstrosity. An abnormally developed organ, like a fasciated stem.
Mucro. A short, stiff, abrupt point at the end of an organ, as of a leaf.
Mucronate. Furnished with a mucro.
Mucronulate. Diminutive of mucronate.
Multi-. Lat., meaning "many"; used as a prefix.
Multicostate. Many-ribbed.
Multifid. Split or cleft into many "usually fine" divisions.
Multilocular. Having many locules—as an ovary.
Multiseriate. Forming many series.
Muricate. Beset with small, sharp points or excrescences; rough.
Muriculate. Finely muricate.
Muriform. Arranged like bricks in a wall—referring to the markings on the surface of small seeds and achenes.
Muticous. Awnless, not terminating in a pointed or awn-like process.

Naked. Not furnished with an envelope; destitute of a perianth.
Napiform. Turnip-shaped.
Nascent. Appearing to view, as a bud or root.
Naturalized. Running wild from cultivation and becoming established.
Nectariferous. Bearing or producing nectar.
Nervation. The system of venation of a leaf.
Nerve. The vein of a leaf or bract.
Netted. Net-like, reticulated—as applied to the venation of a leaf.
Neutral. Devoid of stamens or functioning gynœcium.
Nitidous. Having a smooth or glossy surface, as some seeds.
Node. The joint of a stem, at which the leaves are inserted.
Nodose. Having swollen nodes.
Nodulose. Having small knots or knobs.
Nut. A hard-shelled indehiscent fruit; diminutive, *nutlet.*

Ob-. Lat., meaning "the reverse," or "contrariwise"; used as a prefix.
Obcompressed. Flattened dorsiventrally.
Obconic. Inversely conical, with the apex directed downwards.
Obcordate. Inversely cordate.
Obdiplostemonous. Having the stamens opposite the petals.
Oblanceolate. Inversely lanceolate.
Oblique. Having the sides unequal at the base, as a leaf.
Oblong. Rectangular in general outline but with the ends rounded.
Obovate. Inversely ovate.
Obovoid. Like an inverted egg.
Obsolescent. Gradually disappearing or becoming rudimentary.
Obsolete. Rudimentary; verging on extinction or suppression.
Obtuse. Bluntly rounded at the end.
Ochroleucous. Yellowish-white.
Ocrea (*ochrea*). A sheathing stipule as in *Rumex* (*Polygonaceae*).
Ocreate. Furnished with sheathing stipules or ocreæ.
Octandrous. Having eight stamens.
Octo-. Lat. or Gr., meaning "eight"; used as a prefix.
Octogonous. Having eight carpels or gynœcia.
Officinal. Sold in apothecaries; medicinal.
Oligandrous. Having fewer than twenty stamens.

Oligo-. Greek, meaning "few" or "small in number"; used as a prefix.
Oligomerous. Having fewer than the usual number of parts. *Oligomery*.
Oligotaxy. Diminution of the floral organs or members of a whorl.
Opaque. Dull, not glossy.
Operculate. Furnished with a lid or cap (*Operculum*).
Ornithophilous. Pollinated by birds.
Ortho-. Gr., meaning "straight"; used as prefix.
Orthotropous. Straight, erect.
Ovary. Part of a carpel or gynœcium containing the ovules and seeds.
Ovate. Egg-shaped; *Ovoid*.
Ovulary. Ovary.
Ovule. The macrosporangium (megasporangium).
Ovuliferous. Bearing ovules.
Oxylophyte. A plant inhabiting acid soil.

Palate. The projecting part of the lower lip of a bilabiate corolla, which closes the throat—as in the snapdragon.
Palea (*palet, pale*). The upper or inner of the two bracts of the flower in grasses, often partly invested by the lemma.
Paleaceous. Chaffy; covered with small erect membranaceous scales.
Palmate. Shaped like a palm-leaf.
Palustrine. Inhabiting swamps.
Pandurate. Fiddle-shaped, as a leaf constricted in the middle.
Panicle. A compound racemose inflorescence, as in the oat.
Paniculate. Of the character of a panicle; bearing panicles.
Papilionaceous. Butterfly-shaped, referring to the corolla of some leguminous flowers.
Papilla. A nipple-like prominence on the surface of a leaf or petal.
Papillate. Bearing or furnished with papillæ.
Papillose. Of the nature of papillæ.
Pappus. The modified bristly or scale-like calyx of composites.
Paracorolla. The "crown" of *Narcissus*.
Parasitic. Growing upon and deriving nourishment from another living plant.
Parietal. Borne on the sides or wall of a locule—as ovules.
Paripinnate. Evenly pinnate; a compound leaf ending in two leaflets.
Parted (*partite*). Divided nearly to the base—as a leaf.
Patent. Widely spreading.
Patulent. Less spreading than patent.
Pectinate. Comb-shaped.
Pedate. Palmately divided into narrow parallel segments.
Pedatifid. More divided than pedate.
Pedicel. The flower-stalk.
Pedicellate. Furnished with a pedicel.
Peduncle. The stalk of an inflorescence.
Pedunculate. Furnished with a peduncle.
Pellucid. Clear, transparent.
Peloria. An abnormality—referring to floral organs.
Peltate. Shield-shaped.
Pendulous. Suspended, hanging more or less from the top of the locule—as an ovule.

Penta-. Greek, meaning "five"; used as a prefix.
Pentacyclic. Having five whorls of floral organs.
Pentagynous. Having five carpels or styles.
Pentamerous. Having five members in a whorl.
Pentandrous. Having five stamens.
Pepo. A gourd fruit.
Perennial. Living through more than one season.
Perfect. Having both andrœcium and gynœcium in the same flower; bisexual.
Perfoliate. Having the stem apparently passing through the leaf.
Peri-. Greek, meaning "around"; used as prefix.
Perianth. The floral envelop; perigone.
Pericarp. The wall of the ovary—referring to a fruit.
Pericladium. The large sheathing petiole of umbellifers.
Periclinium. The involucre of composites.
Perigone. An undifferentiated perianth.
Perigynium. The inflated sac (bract) enclosing the gynœcium or carpellate flower in *Carex*.
Perigynous. Standing around the gynœcium—borne on the margin of the hypanthium.
Perisperm. The nucellus of the ovule as an accessory food-storage organ.
Persistent. Lasting, not deciduous; remaining attached to the axis or part which bears it.
Personate. Two-lipped, bilabiate.
Pertuse. Having slits or holes.
Perulate. Bearing scales (*perulæ*)—as a bud.
Petal. One of the colored inner perianth segments.
Petaloid. Having the aspects of a petal.
Petiolate. Furnished with a petiole.
Petiole. A leaf-stalk.
Phænogam. A flowering plant. *Phanerogam*.
Phænogamous. Bearing true flowers. *Phanerogamous*.
Phænos. Greek, meaning "showy"; used as a prefix.
-phore. Greek, meaning "to bear," "to support"; used in combination.
Phylloclade. A stem modified as a leaf, and performing the functions of a leaf.
Phyllode (*phyllodium*). A leaf-stalk (petiole) modified as a leaf.
Phyllon, phyllum. Greek, meaning "leaf"; used in combination.
Phyllopodium. A phyllode; a leaf-base.
Phyllotaxy. Leaf-arrangement.
Phylogeny. Race-history of a plant or group; genealogy. *Phylogenetic*.
Phylum. A major plant group.
Physiognomy. The general aspect of a plant; growth-form; life-form.
Phyto, phyte. From Greek *phyton*, meaning "plant"; used in combination.
Pilose. Bearing soft hairs or pubescence.
Pinna. A leaflet of a compound leaf.
Pinnate. Having the form of a feather.
Pinnatifid. Pinnately cleft into narrow segments.
Pinnule. A pinna of the second order.
Pistil. The ovuliferous or seed-bearing organ or organs of a flower; carpel; gynœcium.

Pistillate. Bearing the gynœcium only; carpellate.
Placenta. That part of the locule of an ovary on which the ovules are borne.
Placentation. The arrangement or orientation of the placentas.
Platys. Greek, meaning "large and broad"; used as a prefix.
Platyphyllous. Broad-leaved.
Pleio-. Greek, meaning "several"; used in combination.
Pleiomerous. Having more than the usual number of parts.
Pleiomery. State of being pleiomerous. *Pleiotaxy.*
Plicate. Plaited; folded into plaits.
Plumose. Feathery, or plume-like.
Plumule. The epicotyl of an embryo.
Pluri-. Lat., meaning "several"; used as prefix.
Plurilocular. Composed of several locules—as an ovary.
Pod. A dehiscent fruit.
Podium. The base of an organ.
Pollination. Transferring of pollen from the anthers to the stigma.
Polliniferous. Bearing pollen.
Pollinium (*pollinia*). The mass of cohering pollen-grains of orchids and milkweeds.
Poly-. Greek, meaning "many"; used as prefix.
Polyadelphous. Stamens united into several bundles.
Polyandrous. Having numerous stamens.
Polygamous. Bearing unisexual and bisexual flowers on the same plant.
Polypetalous. Having separate petals.
Polysepalous. Having separate sepals.
Polystichous. Having the leaves in several ranks.
Pomaceous. Of the nature of an apple.
Pome. An apple-like fruit.
Pore. The orifice through which pollen escapes from the anthers of some flowers.
Porose. Marked with small openings or pores.
Porrect. Extended forwards.
Posterior. The side of the flower facing the axis of the inflorescence.
Pouch. A kind of saccate pod.
Pous, podos. Greek, meaning "foot" or "stalk"; used in combination. *Podium.*
Prefloration (*præfloration*). Æstivation; the arrangement of sepals and petals in the bud.
Prefoliation. Vernation; the arrangement of leaves in the bud.
Premorse (*præmorse*). Bitten off; truncate, referring to roots.
Procumbent. Trailing on the ground without producing roots.
Proliferate (*proliferation, proliferous*). Producing offshoots, sometimes abnormal, as when carpels or stamens give rise to leafy shoots.
Propagule (*propagulum*). A vegetative bud or shoot by which a plant propagates itself; any kind of propagating organ.
Prostrate. Lying flat on the ground.
Proterandrous. Maturing the stamens or shedding the pollen before the stigma is receptive.
Proterogynous. Having the stigma receptive before the anthers mature.
Pruinose. Covered with a coarse, granular, dust-like secretion.

Pseudo. Greek, meaning "false"; used as a prefix. *Pseudocarp.*
Puberulent. Finely pubescent.
Pubescent. Clothed with short, soft hairs. A general term for all sorts of hairy coatings.
Pulverulent. Covered with a fine, powdery substance.
Pulvinate. Cushion-shaped.
Pulvinus. The swollen base of the leaf-stalk in some leaves (*Mimosa*).
Punctate. Marked with very small dots or glands or translucent spots.
Puncticulate. Minutely punctate.
Pungent. Having a sharp terminal point.
Pyridium. A pome; a fruit like an apple or pear.
Pyriform. Shaped like a pear.
Pyxidium (*pyxis*). A capsule opening by a lid.

Quadrate. Four-sided—as some branches.
Quadri-. Lat., "four"; used as prefix.
Quadrifarious. In four rows.
Qaudrifid. Having four pairs of leaflets on the same axis.
Quadrilocular. Four-celled; having four locules.
Quadripartite. Deeply cut or divided into four parts or lobes.
Quinary. Five-parted.
Quincuncial. Having the petals arranged in a five-leaved whorl, in which two petals are interior, two are exterior, and the fifth has one edge covered by one of the exterior petals and the other edge covering one of the interior petals.

Race. A permanent variety.
Raceme. An indeterminate inflorescence composed of a primary axis bearing pedicelled flowers.
Racemose. Having the form of a raceme.
Radiate. Star-shaped; bearing ray-flowers.
Rachilla. The axis of a spikelet in grasses.
Rachis. The axis of an inflorescence or of a compound leaf.
Radicle. The undeveloped root of an embryo.
Rameal. Belonging to a branch.
Ramiform. Having the aspect of a branch or of a branching plant.
Ramose. Branched; having branches.
-Ramous. A suffix, referring to branches or arms or forkings.
Raphe. The ridge between the hilum and the chalaza of an anatropous ovule.
Raphides. Small, needle-like crystals found in the tissues of some plants.
Ray. The ligule of the corolla in flowers of composites.
Receptacle. The part of the floral axis supporting the floral organs, including the hypanthium.
Reclinate. Curved downwards from a horizontal position and then bending upwards, like branches or offshoots near the ground.
Recurved. Bent backwards.
Reduplicate. Having the edges rolled outward in æstivation—referring to leaves, sepals, and petals.
Reflexed. Bent abruptly downwards—referring to sepals.
Regma. The ovary of euphorbiaceous flowers (gynœcia), composed of elastic cocci.

Regular. Having radial symmetry; actinomorphic.
Reliquiæ. The remains of withered leaves attached to the stem.
Reniform. Kidney-shaped.
Repand. Having an undulating or sinuous margin.
Repent. Creeping along the ground and rooting at the nodes.
Replum. The partition (septum) between the two locules in the fruit of crucifers.
Resiniferous. Containing or producing resin.
Resupinate. Upside down. Inverted by the twisting of the pedicel, as the flowers of orchids.
Reticulate. Composed of a reticulum or network.
Reticulum. A network; the numerous small, anastomosing veins of leaves.
Retiform. Having the form of a network.
Retinaculum. The glandular viscid part and the extremity of the stalk of a pollinium (orchids and milkweeds).
Retrorse. Turned backwards.
Retuse. Having a shallow notch at the apex—as in some petals.
Revolute. Rolled backwards from the top or from the margins.
Rhachilla, Rhachis. See *Rachilla, Rachis*.
Rhiza, Gr., "root"; *rhizo-* in combination.
Rhizomatous. Producing rhizomes.
Rhizome. A subterranean stem or branch.
Rib. A prominent ridge or vein.
Ringent. Gaping—having an arched upper lip—as in bilabiate corollas.
Riparian. Inhabiting river banks.
Rootstock. A rhizome.
Rosette. A cluster of closely crowded radiating leaves appearing to rise from the ground.
Rostellate. Furnished with a small beak.
Rostellum. A little beak; a projection from the upper edge of the stigma in orchids.
Rostrate. Beaked.
Rosula. A rosette.
Rosulate. Bearing rosettes; of the nature of a rosette.
Rotate. Wheel-shaped—referring to a sympetalous corolla.
Rubellous. Reddish.
Rufescent. Reddish; tinged with red.
Rufous. Reddish-brown.
Rugose. Wrinkled.
Ruminate. Mottled.
Runcinate. Pinnatifid; cut into sharp triangular lobes, the points directed backwards.

Sac. A cavity or vesicle.
Saccate. Sac-shaped; bag-shaped.
Sagittate. Arrow-shaped, with the basal lobes directed downwards.
Salient. Projecting; protruding; springing up.
Samara. An indehiscent winged fruit.
Sarmentose. Twining and supporting itself on the stems of other plants.
Scabridulous. Slightly rough or scabrous.
Scabrous. Rough; rough-pubescent.

Scapose. Of the character of a scape.
Scalariform. Barred like a ladder.
Scandent. Climbing by means of special organs, as over a wall.
Scape. A stout or slender peduncle rising from the ground and bearing one or more flowers at the summit.
Scarious. Dry, thin, scale-like; membranaceous.
Scion. A cutting or slip; a graft.
Scobiform. Like sawdust.
Scobina. The flexuous rachis of grasses.
Scorpioid. Coiled like the tail of a scorpion—referring to an inflorescence.
Scrobiculate. Having small pits or depressions.
Scurfy. Clothed with small scales.
Scutellate. Shield-shaped.
Scutellum. Lat., "shield." The shield-shaped cotyledon of grasses.
Seculate. Sickle-shaped.
Secund. Turned to one side, as the flowers of an inflorescence.
Seed. An embryo and investing parts of the ovule.
Segment. A division or part of a leaf, or of calyx and corolla.
Segregate. To separate, distinguish, as one species from another.
Seminal. Appertaining to a seed.
Sepal. A segment of the calyx.
Sepaloid. Having the character of a sepal.
Septate. Divided by partitions or septa.
Septem. Lat., "seven"; Gr. *hepta;* used in combination.
Septicidal. Dehiscing in the plane of a septum.
Septifragal. A mode of dehiscence of a capsule by which the valves break away from the septa.
Septum. A partition—referring especially to an ovary.
Serrate. Having sharp teeth on the margin, and directed forward.
Serration. The cutting of a leaf-margin by which the teeth are directed forward.
Serratures. The marginal teeth of a leaf.
Serrulate. Finely serrate.
Seta. A bristle or stiff, sharp-pointed hair. A tail-like appendage.
Setaceous. Bristle-shaped; of the character of a bristle. Setiform.
Setose. Clothed with bristles.
Setulose. Clothed with minute bristles.
Sheath. The tubular basal portion of a leaf of a grass or sedge, which invests the culm.
Shrub. A woody plant which does not become tree-like and usually produces several stems from a common base.
Silicle (*silicula*), *silique* (*siligua*). The fruit (pod) of crucifers.
Siliquose. Of the nature of a silique; bearing siliques.
Silky. Clothed with a soft, straight-haired pubescence.
Simple. Not branched—referring to stems; not compounded—referring to gynœcia.
Sinistrorse. Directed towards the left; counter-clockwise—as the twining of a stem.
Sinuate. Having a wavy outline or margin, as a leaf.
Sinus. The cleft or indentation between the basal lobes of a leaf-blade.

Sobole. A creeping subterranean stem or rhizome.
Soboliferous. Bearing soboles.
Solitary. Single, like the flower of the tulip.
Sorosis. A succulent fruit like that of the mulberry or bread-fruit.
Spadix. A succulent axis supporting an inflorescence, as in the aroids.
Spathaceous. Of the nature of a spathe.
Spathe. The leaf-like colored bract investing the inflorescence (spadix) of aroids and palms.
Spatulate. Spoon-shaped.
Species. An aggregation of individuals possessing similar characters.
Spermatophyte. A seed-plant.
Sphagnous. Appertaining to peat-moss (*Sphagnum*) or peat-bogs.
Spicate. Having the form of a spike.
Spiciform. Spike-like.
Spike. An elongate floriferous axis bearing sessile flowers.
Spikelet. A small spike; the unit of the inflorescence of grasses and sedges.
Spine. A sharp-pointed thorn-like outgrowth from a stem.
Spinescent (*spinose*). Furnished with spines.
Spinule. A small spine.
Spiral. Arranged in a winding series, as leaves or floral organs.
Spontaneous. Appearing of itself without planting.
Sporophyll. A spore-bearing leaf.
Spur. A hollow nectariferous process, usually of a petal or a sepal.
Spurred. Furnished with a spur.
Squama. A scale, like the bracts of an inflorescence (willows).
Squamose. Covered with scales. *Squamate.*
Squarrose. Spreading at the tip, at right angles or more.
Squarrulose. Diminutively squarrose.
Stachys. Greek, "spike"; used in combination.
Stamen. The anther-bearing organ of a flower.
Staminate. Bearing stamens; consisting of stamens, as a flower.
Staminode (*staminodium*). A modified non-functional stamen.
Standard. The broad upper petal of a papilionaceous flower. Vexillum.
Staddle. Seed-trees left standing in a forest when logging.
Stellate. Star-shaped.
Sterile. Not productive. Not capable of producing seed, as a neutral flower.
Stichous. Greek, meaning "row"; used as a suffix (distichous).
Stigma. That part of the style which is modified for the reception and germination of the pollen.
Stigmatic. Belonging to or of the nature of the stigma.
Stipe. A stalk, as of a gynœcium or a flower; gynophore; androgynophore.
Stipel. A small stipule, like the small appendage at the base of the leaflets of some compound leaves.
Stipitate. Stalked.
Stipulary. Of the nature of or taking the place of stipules, as a tendril.
Stipulate. Furnished with stipules.
Stipule. A leaf-like appendage of the base of the petiole of leaves.

Stolon. A slender modified stem (rhizome) running along the ground or below the surface of the ground and rooting at the nodes.
Stoloniferous. Bearing stolons.
Stoma, stomata. The aërating organs on the surface of leaves or other green parts of plants.
Stomatiferous. Bearing stomata.
Stramineous. Straw-colored.
Stria, Striæ. Narrow lines or furrows. Striation.
Striate. Marked with fine lines or furrows, usually running lengthwise.
Strict. Rigid, straight, and erect.
Strigose. Clothed with sharp, straight, stiff, and appressed hairs.
Strobiloid. Having the characters of a strobilus.
Strobilus. A cone-like inflorescence; a flower.
Strophiolate. Appendaged from the proximity of the hilum.
Strophiole. An aril or swollen appendage at the hilum of some seeds.
Struma. A swelling at the base of the stalk of a leaflet of a compound leaf.
Stupose. Beset with tufts of hairs.
Style. The contracted upper part of a carpel or gynœcium which supports the stigma.
Stylopodium. The expanded base of the style in umbellifers.
Sub-. A Latin prefix, meaning "almost," "nearly," "somewhat," or "below," "underneath."
Subaërial. Below the surface of the ground or of a body of water.
Subalpine. Below the alpine zone of vegetation on high mountains.
Subarctic. Below the arctic zone of vegetation.
Suberous. Of corky texture.
Subinferior. Not wholly inferior—referring to an ovary.
Subfamily. A primary subdivision of a family.
Subgenus. A primary subdivision of a genus.
Suborder. A primary subdivision of an order.
Subsessile. Not quite sessile—stalkless.
Subspecies. A primary subdivision of a species.
Subterranean. Underground; growing beneath the surface of the ground.
Subtropical. Inhabiting regions bordering the tropics.
Subulate. Awl-shaped.
Succulent. Fleshy; juicy; herbaceous; having little woody tissue.
Sucker. An underground offshoot or branch. A vegetative propagative shoot.
Suffrutescent. Not evidently shrubby.
Suffruticose. Having the characters of a low shrub.
Sulcate. Grooved, furrowed, or channeled.
Superior. Wholly free from the receptacle—referring to the ovary.
Super, supra. A Latin prefix, meaning "above."
Supra-axillary. Situated above the axil of a leaf.
Surculate. Having the form of a sucker.
Surculus. A sucker.
Suspended. Hanging from the top of a locule—referring to an ovule.
Sutural. Dehiscing along a suture (carpel).
Suture. The line of dehiscence of dry fruits. The line of junction or of cleavage of two united organs.

Syconium. The spurious fruit of the fig.
Symmetrical. Regular in shape, size, and number of parts; actinomorphic. Capable of being divided into equal parts by planes of symmetry —referring to a flower.
Sympetalous. Having united petals, or a one-piece corolla; gamopetalous.
Sympodium. An axis formed of successive secondary axes of a dichotomous system of branching, one axis or arm of the fork being shorter than the other or suppressed. *Sympodial.* Belonging to or of the nature of a sympodium.
Syn-. Greek, meaning "together"; used as a combining term, meaning "united."
Synandrous (*synantherous*). Having the anthers united.
Synanthous. Having the flowers united.
Syncarpous. Having united carpels.
Syngenesious. Having the anthers united, as in composites.
Synocreate (*synochreate*). Having the stipules (*ocreæ*) united on the side of the stem opposite to the insertion of the petiole (*Polygonaceæ*).
Synsepalous. Having united sepals; gamosepalous.

Tap-root. A thickened primary root.
Tassel. The staminate inflorescence of maize.
Tassellate. Having the form of a tassel.
Tawny. Dull yellowish brown.
Taxonomic. Appertaining to taxonomy; classificatory.
Tegmen. The second coat of a seed.
Tegule. The sepal-like involucral bracts of composites.
Tepal. A segment of an undifferentiated perianth.
Tepaloid. Having the characters of a tepal.
Teratological. Appertaining to abnormalities or monstrosities.
Tercine. The third coat of the ovule or seed.
Terete. Cylindrical; circular in transverse section.
Ternate. Having the parts arranged in threes; *ternary.*
Testa. The outer, usually hard, coat of a seed.
Tetra-. Greek, meaning "four"; used as a prefix.
Tetracyclic. Having four whorls of floral organs.
Tetradynamous. Having four long and two short stamens.
Tetragynous. Having four carpels.
Tetramerous. Having the floral organs in fours or multiples of four.
Tetramery. A tetramerous condition or state.
Tetraploid. In fours.
Thalamifloral. Having the floral organs distinct from each other on the receptacle.
Thalamus. The receptacle of a flower.
Theca. A pollen-sac of an anther. *Thecal.* Belonging to a theca.
Throat. The orifice of a sympetalous corolla.
Thyrse, thyrsus. A cylindrical or conical paniculate inflorescence.
Thyrsoid. Having the character of a thyrsus.
Tomentose. Clothed with a cottony or woolly indumentum.
Tomentum. A cottony or woolly entangled pubescence.
Torose. Contracted or constricted at intervals along its length, as a root or rhizome.

Tortulose. Twisted.
Torulose. Diminutive of torose.
Torus. The receptacle of a flower; thalamus.
Tracheate. Of the nature or form of tracheæ.
Translator. A pollen-transferring organ of orchids.
Traumatic. Relating to injury of an organ.
Treis, tris. Greek, "three"; Latin, "tres." In combination *tri-*.
Triadelphous. Having the stamens united into three bundles.
Triandrous. Having three stamens.
Trichotomous. Dividing successively into three branches. *Trichotomy.*
Tricostate. Three-ribbed.
Tricuspidate. Three-toothed or three-pointed at the apex.
Tridentate. Three-toothed.
Trifarious. In three rows or series pointing in three directions.
Trifid. Three-parted, or deeply divided into three linear segments.
Trifoliate. Three-leaved. *Trifoliolate.* Having three leaflets.
Trigonous. Three-angled.
Trigynous. Having three gynœcia; tricarpellate.
Trijugate. Having three pairs of leaflets.
Trilocular. Having three locules.
Trimerous. Having the members of a whorl in threes—referring to flowers.
Trimery. A trimerous condition.
Trimorphic. Existing in three forms.
Tripartite. Deeply divided into three segments or lobes.
Tripinnatifid. Pinnatifid, with the segments twice divided in a pinnatifid manner.
Triplicate. Having three ribs or veins proceeding from the base—as a leaf.
Triploid. In threes.
Triquetrous. Having three angles, with the sides or faces concave.
Tristichous. In three rows or ranks, as leaves on a stem.
Triternate. Thrice divided in a ternate manner.
Truncate. Abruptly cut off at the end, crosswise.
Tuber. A thickened, usually subterranean stem (potato).
Tubercle. A small tuber-like body, or excrescence.
Tuberculate. Of the nature of tubercles; bearing tubercles.
Tuberiferous. Bearing tubers.
Tuberous. Of the nature of a tuber.
Tumid. Swollen.
Tunicate. Having concentric coats or scales—as the onion.
Turbinate. Top-shaped; inversely conical.
Turescent. Swollen.
Turio. A (scaly) shoot (sprout) springing from an underground stem, like the sprouts of asparagus.
Turion. A sprout from the base of a plant or from an underground stem.
Turgid. Swollen.
Type. "That specimen or individual plant in a collection of specimens of a supposed species which in the judgment of the taxonomist is the most ideal representative of the species as a whole; or the specimen or individual plant of a collection, which formed the basis of the description of the species. A type may be only a single specimen of

the plant, not previously described. In constructing a new genus the type is selected from the most representative species."
Typical. Most representative, as of a species, or genus, etc. Having the characters that embody the ideal concept of a species, or genus, etc.
Typify. To diagnose the ideal concept of a species, genus, etc.

Ubiquitous. Occurring everywhere.
Umbel. An umbrella-shaped inflorescence, in which the pedicels radiate from a common point at the summit of the peduncle.
Umbellate. Of the form of an umbel.
Umbellet. A secondary umbel; one of a set of smaller umbels of a compound umbel.
Umbellule. An umbellet.
Umbellulate. Having the form of an umbellule.
Umbelliferous. Bearing umbels.
Umbelliform. Umbel-shaped.
Umbo. A conical projection or protuberance from a surface.
Umbonate. Bearing an umbo or thick protuberance.
Uncinate. Bent inward at the apex, as the filament of a stamen.
Uncus. Latin, "hooked process." In combination *unci-*.
Undecem. Latin, "eleven."
Undulate. Having a sinuous or wavy margin, the "waves" being long; repand.
Unguis. Latin, "claw"; in combination *ungui-*.
Unguiculate. Clawed; contracted at the base into a claw—as some leaves and petals.
Unus. Latin, "one." In combination *uni-*.
Unicarpellate. Composed of one carpel—referring to a gynœcium.
Unilateral. Arranged on one side.
Unilocular. Composed of one locule.
Uniseriate. In one series or row.
Urceolate. Urn-shaped; constricted at the throat, like some tubular corollas.
Ustulate. Blackened.
Utricle. An inflated achene-like fruitlet.

Vagina. A sheath, as in leaves.
Vallecula, -æ. The groove or interval between the ridges or ribs in some dry fruits—like those of umbellifers.
Valvate. Dehiscing by valves or equal sections. In æstivation—when the segments of the perianth are so placed that their edges touch; not overlapping.
Valve. A section of a dry, dehiscent fruit (capsule) separated in dehiscence.
Valvular. Belonging to valves.
Variable. Inconstant as to taxonomic characters.
Variant. A plant form that digresses from the typical form.
Variation. A digression from the type.
Variety. A subspecific category. A group of similar individuals digressing insufficiently from the species to entitle them to specific rank.
Vasculum. A vessel or container. The "collecting can" of the botanist.

Vasiform. Having the form of a vase or elongated funnel-shaped vessel.
Veins. The nerves or fibrovascular bundles of a leaf.
Velutinous. Velvety in appearance. Having a velvety surface—referring to a sort of pubescence.
Venation. Nervation. The arrangement of the nerves or veins of a leaf.
Ventral. Belonging to the adaxial side (facing the primary axis) of an organ; the inward side of an organ, in relation to the primary axis.
Ventricose. Unequally or unevenly swollen; gibbous.
Vermiform (*vermicular*). Of the shape of a worm.
Vernal. Belonging to spring. Opening or blossoming in the spring.
Vernation. The arrangement or mode of folding of leaves in the bud.
Verrucose. Covered with wart-like protuberances or excrescences.
Versatile. Attached by the middle so as to swing freely—as an anther.
Vertical. Descriptive of the longitudinal section of an organ cut through the center.
Verticil. A whorl.
Verticillaster. A cluster or false whorl formed of two nearly sessile cymes arising from the axils of two opposite leaves.
Verticillastrate. Bearing verticillasters or false whorls; of the form of a verticillaster.
Verticillate. Arranged in whorls.
Vesicle. A small bladder-like sac or cavity.
Vesicular. Of the nature of a vesicle.
Vesiculose. Covered with vesicles.
Vexillary. Appertaining to the vexillum; having the vexillum folded over the other organs—referring to a sort of æstivation, as in the *Leguminosæ.*
Vexillum. The upper broad petal (standard) of a papilionaceous flower.
Villous. Clothed with long, soft hairs.
Vimineous. Having or producing long, slender twigs, shoots, or sprouts.
Virgate. Long and straight like a wand; wand-shaped.
Viscid, viscous. Clammy, sticky; glutinous.
Vitta, -æ. An oil-tube—referring to the longitudinal oil-tubes on the sides of the fruit of umbellifers.
Vittate. Having oil-tubes, as in the umbellifers; striped.
Volubile. Twining around a support—as a tendril or some kinds of stems.

Whorl. A circle or ring of organs inserted around an axis—as the organs of a flower or leaves on a stem; a cycle. *Whorled.*
Wing. An outgrowth from the side of an organ; a lateral petal in a papilionaceous flower.

Xantho-. Greek, meaning "yellow"; used in combinations.
Xanthophyllous. Yellow-leaved.
Xero-. Greek, meaning "dry," "arid"; used in combination.
Xerophyllous. Having leaves characteristic of plants of arid situations.
Xerophylly. The condition of the foliage characteristic of plants of arid regions.
Xerophyte. A plant characteristic of an arid environment. *Xerophytic.*

Zonal. Referring to zones or definite vegetational areas defined by climatic, or environmental, conditions.

BIBLIOGRAPHY

Floras, Manuals, and Special Treatises

ABRAMS, LEROY. Flora of Los Angeles and Vicinity. Stanford Bookstore, Palo Alto, 1917.

ABRAMS, LEROY. Illustrated Flora of the Pacific States. Vol. I. Stanford University Press, 1923.

ALMQUIST, ERNST. "The Origin of Constant Species." *Journ. Bot.*, 62:97-102, 1924.

AMES, OAKES. An Enumeration of the Orchids of the United States and Canada. The Amer. Orch. Soc., Boston, 1924.

ANDERSON, N. J. Conspectus Vegetationis Lapponicæ. Pars I. Upsala, 1846.

BAILEY, L. H. Manual of Cultivated Plants. Macmillan, New York, 1924.

BAILLON, H. L'Histoire des Plantes. 13 vols. Paris, 1872. English translation by M. M. Hartog: The Natural History of Plants. 8 vols. Reeve & Company, London, 1871-1878.

BARTLING, F. T. Ordines Naturales Plantarum. Göttingen, 1930.

BARTON, W. P. O. Flora Cestrica. Philadelphia, 1853.

BEAL, W. J. and WHEELER, B. S. Michigan Flora. Ann Arbor, 1892.

BECKWITH, FLORENCE and MACAULEY, MARY. Plants of Monroe County, New York, and Adjacent Territory. *Proc. Rochester Acad. Sci.*, 3: 1-150 (1894), 1896.

BENNETT, J. L. Plants of Rhode Island. Providence, 1888.

BENTHAM, G. Plantæ Hartwigianæ. London, 1839.

BENTHAM, G. and HOOKER, J. D. Genera Plantarum. 3 vols. London, 1862-1883.

BERGMAN, H. F. Flora of North Dakota. *Sixth Ann. Rep. of the Director, N. Dak. Soil and Geol. Surv.* Bismarck, 1912.

BIENERT, T. Baltische Flora. Dorpat, 1872.

BIGELOW, JACOB. Florula Bostoniensis. Boston, 1840.

BONNIER, GASTON and LAYENS, G. DE. Flore complète portative de la France et de la Suisse. Paris, undated.

BRANDEGEE, KATHERINE. Plants of San Francisco. *Zoë*, 2:335-386, 1891.

BREWER, W. H., WATSON, S. and GRAY, A. Geological Survey of California, Vol. 1 (Botany), 2nd. rev. ed.: i-xx, 1-628, 1880.

BRITTON, N. L. Flora of Bermuda. Scribner's, New York, 1918.

BRITTON, N. L. Manual of The Flora of The Northern States and Canada. Holt, New York, 1910.

BRITTON, N. L. and BROWN, A. Illustrated Flora of The Northern United States, Canada and British Possessions. 3 vols. Scribner's, New York, 1913.

BRITTON, N. L. and MILLSPAUGH, C. F. L. The Bahama Flora. New York, 1920.

BRITTON, N. L. and ROSE, J. N. *Cactaceæ.* Carnegie Inst., Washington, D. C., *Publ.* No. 248, 1922.

BROWN, S. Alpine Flora of The Canadian Rocky Mountains. Putnam, New York, 1907.

BUTTERS, F. K. The Vegetation of The Selkirks, in "Mountaineering and Exploration in The Selkirks" by Howard Palmer, Appendix A: 352-362. Putnam, New York, 1914.

CHAPMAN, A. W. Flora of The Southeastern United States. Cambridge, 1897.

CLEMENTS, F. E. and (MRS.) E. S. Flower Families and Ancestors. H. W. Wilson Co., New York, 1928.

CLUTE, W. N. Flora of The Upper Susquehanna. Binghamton, New York, 1898.

CLUTE, W. N. American Plant Names. Willard N. Clute Co., Joliet, Ill., 1923.

COULTER, J. M. Manual of The Phanerogams and Pteridophytes of Western Texas. *Contrib. U. S. Nat'l Herb.,* 2: 1-588, 1891.

COULTER, J. M. and NELSON, AVEN. New Manual of Rocky Mountain Botany. American Book Co., New York, 1909.

COULTER, STANLEY. A Catalog of The Flowering Plants, etc., Indigenous to Indiana. Lafayette, 1889.

COVILLE, F. V. Botany of The Death Valley Expedition. *Contrib. U. S. Nat. Herb.,* Vol. 4, 1893.

DARLINGTON, WM. Flora Cestrica. Philadelphia, 1853.

DAME, L. L. and COLLINS, F. S. Flora of Middlesex County, Massachusetts. Walden, Mass., 1881.

DAVIDSON, A. and MOXLEY, G. L. Flora of Southern California. *Times Mirror Press,* Los Angeles, 1923.

DAVIS, WM. M. Flora of Mt. Desert Island, Maine. Cambridge, Mass., 1894.

DEAN, WALTER. Flora of The Blue Hills, Middlesex Fells, etc. Boston, 1896.

DE CANDOLLE, A. Prodromus Systematis Naturalis Regni Vegetabilis. 17 vols. and 4 index vols. Paris, 1824-1873.

DURAND, TH. Index Generum Phanerogamorum. Brussels, 1888.

DYER, SIR WM. THISELTON. Flora of Tropical Africa. London, 1901.

EATON, AMOS. Manual of Botany of North America. Albany, New York, 1836.

ELLIOT, STEPHEN. A Sketch of The Botany of South Carolina and Georgia. 2 vols. Charleston, 1821.

ENDLICHER, S. L. Genera Plantarum. Vienna, 1836-1850.

ENGLER, A. Das Pflanzenreich. A compendious work in many volumes, dealing with the genera and species of the plants of the world. 1900 and continuing. Engelmann, Leipzig.

ENGLER, A. and GILG, E. Syllabus der Pflanzenfamilien. 9th and 10th editions in one volume. Borntraeger, Berlin, 1924. An indispensable treatise on the flowering plants.

ENGLER, A. and PRANTL, H. Die Natürlichen Pflanzenfamilien. 20 vols. Engelmann, Leipzig, 1897-1915. Second edition in progress.

FAWCETT, W. U. and RENDLE, A. B. Flora of Jamaica. London, 1910-1912.

FINK, BRUCE. Spermatophyta of The Flora of Fayette, Iowa. *Report Iowa Acad. Sci.*, Vol. 4. Des Moines, 1897.

FIORI, ADRIANO. Iconographia Floræ Italicæ ossia Flora Italiana. 2nd ed. Sancasciano Val di Pesa, 1921.

FITZPATRICK, T. J. Manual of The Flowering Plants of Iowa. Iowa City, 1899.

FRYE, T. C. and RIGG, G. B. Northwest Flora. University of Washington, Seattle, 1912.

GODET, C. H. Flore de Jura. Neuchatel, 1853.

GRAY, ASA. The Genera of The Plants of The United States. 2 vols. Putnam, New York, 1849.

GRAY, ASA. Manual of Botany of Northern United States. 5th ed. New York and Chicago, 1867.

GRAY, ASA. New Manual of Botany. 7th ed., revised by Benjamin Lincoln Robinson and Merritt Lyndon Fernald. American Book Co., New York, 1908.

GRAY, ASA and TORREY, JOHN. Flora of North America. 2 vols. Putnam, New York, 1838-1843.

GRAY, ASA and WATSON, SERENO. Synoptical Flora of North America. Continued and edited by B. L. Robinson. American Book Co., New York, 1878.

GRAY, S. F. Natural Arrangement of British Plants. London, 1921.

GREENE, E. L. Flora Franciscana. San Francisco, 1891.

GREENE, E. L. Manual of Botany of The Region of San Francisco. San Francisco, 1894.

HALES, B. J. Selected Western Flora: Manitoba, Saskatchewan, Alberta. Macmillan Company of Canada, Toronto, 1922.

HALL, H. M. and C. C. A Yosemite Flora. San Francisco, 1912.

HALL, H. M. and CHANDLER, H. P. Plants of The Sierra Nevada Mountains, 1900.

HAUSMANN, F. B. VON. Zur Flora Tirols. Innsbruck, 1855.

HEGI, G. Illustrierte Flora von Mitteleuropa. Lehmann, Munich, 1929. 13 vols.

HELLER, A. A. A Catalog of North American Plants North of Mexico. 3rd ed. (unfinished). Lancaster, Pa., 1909-1914.

HENDERSON, L. F. Flora of The Olympics. *Zoë*, 2:253-295, 1892.

HENRY, J. K. Flora of Southern British Columbia and Vancouver Island. Gage, New York, 1915.

HOLM, T. Catalog of Plants of The West Coast of Greenland. *Bull. Torr. Bot. Club*, 27: 65-68, 1900.

HOOKER, G. J. Icones Floræ Borealis Americanæ. Berlin, 1877.

HOOKER, SIR J. D. The Student's Flora of The British Islands. London, 1884.

HOOKER, SIR J. D. Flora of British India. London, 1897.

HOOKER, SIR WM. J. Icones Plantarum. Dulau, London, 1842-1897.

HOOKER, SIR WM. J. *et al.* British Flora. London, 1860.

HOWELL, THOMAS. Flora of Northwest America. Portland, Ore., 1903.

HUTCHINSON, J. The Families of Flowering Plants. Macmillan, London, 1926.

HUTCHINSON, J. and DALZIEL, J. M. Flora of West Tropical Africa. The Crown Agent for The Colonies. London, 1927.

JACKSON, JOSEPH. Flora of Worcester County, Massachusetts. Worcester, 1894.

JEPSON, W. L. A Flora of Western Middle California. Berkeley, 1901.
JEPSON, W. L. A School Flora for The Pacific Coast. Appleton, New York, 1910.
JEPSON, W. L. Manual of The Flowering Plants of California. Associated Students' Store, Berkeley, 1923-1925.
JUNCK, MAX. Flora Gleiwitz und Umgegend. Gleiwitz, 1889.
JUSSIEU, A. L. Genera Plantarum. Paris, 1791.

KEARNEY, T. H. Notes on The Flora of Southeastern Kentucky. *Bull. Torr. Bot. Club,* 20: 474-485, 1893.
KEARNEY, T. H. Pine Barren Flora of Eastern Tennessee. *Plant World,* 1:33-35, 1897.
KUNTZE, OTTO. Revisio Generum Plantarum. Leipzig, 1898.

LAMOTTE, M. Prodrome de la Flore du Plateau Central de la France. Paris (?), 1877.
LA PEYROUSE, P. P. DE. Plantes des Pyrenées. Toulouse, 1813.
LA PEYROUSE, P. P. DE. Figures de la Flore des Pyrenées. Paris, 1795-1801.
LAVIALLE, P. Les Plantes Médicinales. Mulhause, 1921.
LECOQ, H. H. and LAMOTTE, M. Plantes Vasculaires du Plateau Central de la France. Paris, 1847.
LEDEBOUR, C. F. Flora Altaica. Berlin, 1829.
LEDEBOUR, C. F. Flora Rossica. Stuttgart, 1844.
LINDLEY, J. An Introduction to The Natural System of Botany. London, 1830.
LINDLEY, J. The Vegetable Kingdom, etc. London, 1853.
LINNAEUS, L. Flora Lapponica. Amsterdam, 1737.
LINNAEUS, L. Genera Plantarum. Lund, 1737.
LINNAEUS, L. Species Plantarum. Stockholm, 1753.
LLOYD, J. D. Vegetable Drugs. The Caxton, Cincinnati, 1921.
LOWE, E. N. Plants of Mississippi. 1921.
LUERSSEN, C. Systematische Botanik. Leipzig, 1882.

MACKENZIE, K. K. and BUSH, B. F. Manual of The Flora of Jackson County, Missouri. 1902.
MACMILLAN, CONWAY. Metaspermæ of The Minnesota Valley. Harrison & Smith, Minneapolis, 1892.
MACOUN, JOHN. Catalog of Canadian Plants. Montreal, 1888.
MAIDEN, J. H. An Alphabetical List of Victorian Eucalypts. *Proc. Roy. Soc., Victoria (N. S.),* 34: 73-84, 1922.

Malmgren, A. J. Phanerogamic Flora of Spitzbergen. *Journ. Bot.*, 2 : 130-147; 162-176, 1864.

Martius, G. F. P. de. Flora Brasiliensis. Leipzig, 1840-1906.

Mc Clatchie, A. J. Flora of Pasadena and Vicinity. Los Angeles, 1895.

Meehan, Thos. The Native Flowers and Ferns of The United States, Boston, 1878.

Meisner, K. F. Plantarum Vascularum Genera. Leipzig, 1836-1840.

Merriam, C. H. Plants of The Pribiloff Islands. *Proc. Biol. Soc. Wash.*, 1892.

Merrill, E. D. An Enumeration of Philippine Flowering Plants. *Bur. Sci. Publ.*, 18 (Philippine Islands), 1 : vii + 463 pp., 1922-1925; 2 : 520 pp., 1923; 3 : 628 pp., 1923; 4 : 515 pp., 1926.

Micheaux, A. Flora Borealis Americana. Paris, 1803.

Millspaugh, C. F. and Nuttall, L. W. Flora of Santa Catalina Islands, California. *Field Mus. Nat. Hist.*, Publ. 212, 1923.

Milspaugh, F. S. Herbarium Organization. *Field Mus. Nat. Hist.*, Publ. 229, *Mus. Tech. Ser.*, 1 : 1-18, 1925.

Mohr, Chas. Plant Life of Alabama. *Contrib. U. S. Nat. Herb.*, Vol. 6, 1901.

Muenscher, W. C. Keys to Woody Plants. Cornell Publ. Print. Co., Ithaca, N. Y., 1922.

Muschler, Reno. A Manual Flora of Egypt. 2 vols. Berlin, 1921.

Nelson, A. Spring Flora of The Intermountain States. Boston, 1912.

Norman, J. M. Norges Arktiske Flora. Oslo, 1894.

Nuttall, Thos. The Genera of North American Plants. Philadelphia, 1818.

Nyman, C. F. Conspectus Floræ Europæ, 1854-1855; 1878-1882; 1883-1899. Örebro Seciæ.

Orcutt, C. N. Flora of Southern and Lower California. San Diego, 1885.

Patterson. Check-list of North American Plants North of Mexico. 1892.

Pepoon, H. S. Synopsis Plantarum. Paris and Tübingen, 1805.

Petersen, N. F. Flora of Nebraska. Publ. by author. Lincoln, 1923.

PFEIFFER, LUDOV. Nomenclatur Botanicus. Leipzig, 1871-1875.

PIPER, C. V. Flora of The State of Washington. *Contrib. U. S. Nat. Herb.*, Vol. 11, 1906.

PIPER, C. V. and BEATTIE, R. K. Flora of Southeastern Washington and Adjacent Idaho. Washington State College, Pullman, Wash., 1914.

PIPER, C. V. and BEATTIE, R. K. Flora of The Northwest Coast. Washington State College, Pullman, Wash., 1915.

PIROTTA, G. G. Flora del Modense e del Riggiano Modena. 1882.

PLUKENETII, LEONARD. Opera I-VI. London, 1691-1705.

POST, T. VON and KUNTZE, OTTO. Lexicon Generum Phanerogamarum. Stuttgart, 1904.

PRAIN, D. Index Kewensis Plantarum Phanerogamarum, 5th Supplement, 1909-1915. Clarendon Press, Oxford, 1921.

PRITZEL, G. A. Iconum Botanicorum Index. Berlin, 1866.

PRITZEL, G. A. Thesaurus Literaturæ Botanicæ. Leipzig, 1872.

PROSCHOWSKY, A. R. Palms of The Riviera. *Gard. Chron.*, 71 : 317 and 72 : 66, 1922.

PURSH, F. Flora Americana Septentrionalis. London, 1814.

RAFINESQUE, C. S. Medical Flora. Manual of Medical Botany of The United States of North America. Philadelphia, 1828-1830.

RATTAN, VOLNEY. West Coast Botany. San Francisco, 1898.

RENDLE, A. B. The Classification of The Flowering Plants. University Press, Cambridge. 2nd ed. Vol. 1, 1930; Vol. 2, 1925.

RHODE, ELEANOR S. The Old English Herbals. Longmans, London, 1922.

ROPER, I. M. The Earliest English Herbal. *Somersetshire Archæol. and Nat. Hist. Soc. Proc.*, 67 : 65-71, 1922.

RYDBERG, P. A. Flora of The Black Hills of South Dakota. *Contrib. U. S. Nat. Herb.*, Vol. III, No. 8, 1896.

RYDBERG, P. A. Flora of Montana. *Mem. N. Y. Bot. Gard.*, Vol. I, 1900.

RYDBERG, P. A. Flora of Colorado. *Bull.* 100, *Colorado Exp. Sta.*, 1905.

RYDBERG, P. A. Flora of The Rocky Mountains and Adjacent Plains. New York, 1917.

RYDBERG, P. A. *Ambrosiaceæ. North Amer. Flora,* 33 : 3-44, 1922. *Carduaceæ. North Amer. Flora,* 33 : 45-46, 1922.

SAVAGE, S. The Discovery of Some of Jacques Le Moyne's Botanical Drawings. *Gard. Chron.*, 71 : 44, 1922. Le Moyne accompanied a sixteenth century French expedition to America as artist.

SCHAFFNER, J. H. Field Manual of The Flora of Ohio and Adjacent Territory. R. G. Adams Co., Columbus, Ohio, 1928.

SIMMONS, H. G. The Vascular Plants in The Flora of Ellesmereland. *Report of the Second Norwegian Arctic Expedition in The Fram,* 1898-1902, No. 2. Oslo, 1906.

SMALL, J. K. The Summit Flora of King's Mountain, North Carolina. *Torreya,* 1 : 7, 1901.

SMALL, J. K. Flora of The Southeastern United States. New York Botanical Garden. 1903.

STANDLEY, PAUL. Flora of Glacier National Park. *Contrib. U. S. Nat. Herb.,* 22, 1921.

STANDLEY, PAUL. Trees and Shrubs of Mexico. *Contrib. U. S. Nat. Herb.,* 1 : 1-169, 1920; 2 : 170-515, 1922; 3 : 517-848, 1923.

STEUDEL, E. G. Nomenclator Botanicus. Stuttgart and Tübingen, 1841.

STEUDEL, E. G. Synopsis Plantarum Graminarum. Stuttgart, 1855.

STEUDEL, E. G. Synopsis Plantarum Cyperacearum. Stuttgart, 1855.

STEVENS, G. T. An Illustrated Guide to The Flowering Plants. Dodd, Mead, New York, 1910.

TAUSCH, I. F. Hortus Canalius. Prague, 1823.

TIDESTROM, I. Flora of Utah and Nevada. *Contrib. U. S. Nat. Herb.,* Vol. 25, 1925.

TORRE, C. G., DE DALLA, and HARMS, H. Genera Siphonogamarum. Leipzig, 1900-1907.

TORREY, J. and GRAY, A. A Compendium of The Flora of The Northern Middle States. New York, 1826.

TORREY, J. and GRAY, A. Flora of North America. New York, 1838-1840.

UPHAM, WARREN. Catalog of The Flora of Minnesota. Saint Paul, 1884.

WALPERS, W. G., and MUELLER, C. M. Repertorium Botanices Septendricæ. Leipzig, 1842-1848. Continued as Annales Botanices Systematices, 1848-1868.

WARMING, E. A Handbook of Systematic Botany. English translation by A. F. Potter. Macmillan, London, 1920.

WATSON, SERENO. Botany of California. *Geol. Surv.* Calif., 2: i-xv, 1-559, 1880.

WETTSTEIN, R. V. Handbuch der Systematischen Botanik. Leipzig, 1924.

WILLDENOW, C. L. Linnæus's Species Plantarum. Berlin, 1799.

WILLDENOW, C. L. Enumeratio Plantarum. Berlin, 1809.

MANUALS AND HANDBOOKS OF TREES AND SHRUBS

APGAR, A. C. Trees of The United States. American Book Co., New York, 1892.

BAILEY, L. H. Cyclopedia of Horticulture. 3 Vols. Macmillan, New York, 1924.

BAILEY, L. H. Manual of Cultivated Plants. Macmillan, New York, 1924.

BLAKESLEE, A. F. and JARVIS, C. D. Trees in Winter. Macmillan, New York, 1913.

BRITTON, N. L. North American Trees. Holt, New York, 1908.

BROWN, H. P. Trees of New York State. *Tech. Publ.* 15, New York State College of Forestry, 1921.

BURNS, G. P. and OTIS, C. H. The Trees of Vermont. *Bull.* 194, *Vt. Agric. Exp. Sta.,* 1916.

CURTIS, C. C. A Guide to The Trees. Greenburg, New York, 1925.

DAME, L. L. and BROOKS, H. Handbook of Trees of New England. Ginn, Boston, 1904.

DEAM, C. C. Trees of Indiana. State Board of Forestry, Indianapolis, 1919.

DYSON, Mrs. Stories of The Trees. Nelson & Sons, London, 1890.

EMERSON, A. I. and WEED, C. M.. Our Trees and How to Know Them. Lippincott, Philadelphia, 1908.

EMERSON, GEO. B. Trees and Shrubs of Massachusetts. Dutton, Boston, 1846.

FERNOW, B. H. The Care of Trees. Holt, New York, 1911.

GIBSON, H. H. American Forest Trees. Hardwood Record, Chicago, 1913.

GOING, MAUDE. With The Trees. Baker & Taylor, New York, 1903.

GREEN, SAMUEL B. Forestry in Minnesota. *Geol. and Nat. Hist. Survey of Minn.*, Saint Paul, 1902.

HOUGH, R. B. Handbook of The Trees of The Northern States and Canada East of The Rocky Mountains. Publ. by Author. 1907.

HUNTINGTON, A. O. Studies of Trees in Winter. Knight & Millet, London, 1905.

ILLICK, J. S. Pennsylvania Trees. *Bull.* 11, *Penn. Dept. of Forestry,* Harrisburg, 1914.

JEPSON, W. L. The Trees of California. Cunningham, Curtis & Welch, San Francisco, 1909.

JEPSON, W. L. Silva of California. *Memoirs,* University of California, The University Press, Berkeley, 1910.

KEELER, H. L. Our Native Trees and How to Identify Them. Scribner's, New York, 1915.

KELLOGG, ALBERT. Illustrations of West American Oaks. San Francisco, 1889.

LONGYEAR, BURTON O. Trees and Shrubs of The Rocky Mountain Region. Putnam, New York, 1927.

MACFARLAND, J. H. Getting Acquainted with The Trees. Outlook Co., New York, 1904.

MATHEWS, F. S. Fieldbook of American Trees and Shrubs. Putnam, New York, 1915.

MCCLATCHIE, A. J. Eucalypts Cultivated in The United States. *U. S. Dept. Agric., Div. For., Bull.* No. 35, 1902.

MOHLER, J. C., *et al.* Trees in Kansas. Kansas State Board of Agric., Topeka, 1928.

OTIS, G. H. Michigan Trees. University of Michigan, Ann Arbor, 1913.

POOL, R. J. Handbook of Nebraska Trees. *Botanical Surv. of Neb., N. S.,* No. 3, 2nd ed., 1929.

PRATT, M. B. Shade and Ornamental Trees of California. California State Board of Forestry, Sacramento, 1921.

RAMALEY, F. Wild Flowers and Trees of Colorado. Greenman, New York, 1909.

REHDER, A. Manual of Cultivated Trees and Shrubs. Macmillan, New York, 1926.

ROCK, J. F. The Indigenous Trees of the Hawaiian Islands. Pub. by Author. Honolulu. 1913.

ROCK, J. F. The Ornamental Trees of Hawaii. Pub. by Author. Honolulu, 1917.

ROGERS, J. E. The Tree Book. Doubleday, New York, 1907.

ROSENDAHL, C. O. and BUTTERS, F. K. Trees and Shrubs of Minnesota. University of Minnesota Press, Minneapolis, 1929.

SARGENT, C. S. Manual of Trees of North America Exclusive of Mexico, 2nd ed. Houghton Mifflin, Boston, 1922.

SARGENT, C. S. Silva of North America. 14 vols. Houghton Mifflin, Boston, 1891-1902.

SAUNDERS, C. F. Trees and Shrubs of California Gardens. McBride, New York, 1926.

SCHAFFNER, J. H. Field Manual of Trees. R. G. Adams Co., Columbus, Ohio, 1926.

SIMMONS, J. R. Historic Trees of Massachusetts. Marshall Jones, Boston, 1919.

SMALL, J. K. Florida Trees. Pub. by Author, 1913.

SPONSLER, O. L. A Bud and Twig Key. Geo. Wahr, Ann Arbor, 1916.

SPONSLER, O. L. Trees. Geo. Wahr, Ann Arbor, 1923.

STANDLEY, P. O. Trees and Shrubs of Mexico. *Contrib. U. S. Nat. Herb.*, 23:1-169, 1920; 23:170-515, 1922; 23:517-848, 1923.

SUDWORTH, G. B. Nomenclature of The Arborescent Flora of The United States. *U. S. Dept. Agric., Div. For., Bull.* No. 35, 1897.

SUDWORTH, G. B. Trees of the Pacific Coast. *U. S. Dept. Agric., For. Service,* 1908.

TRELEASE, WM. Winter Botany. Plant Materials for Decorative Gardening. Pub. by Author, 1918, Urbana, Illinois.

TRELEASE, WM. The American Oaks. *Mem. Nat. Acad. Sci.,* Vol. 22, 1924.

TRELEASE, WM. Trees and Shrubs of Illinois. Urbana, Illinois, 1929.

Ward, H. M. Trees. 5 vols. University Press, Cambridge, 1904-1910.

Wiegand, K. M. and Foxworthy, F. W. A Key to The Genera of Woody Plants in Winter. Ithaca, New York, 1908.

Wilson, E. H. The Romance of Our Trees. Doubleday, New York, 1920.

Wilson, E. H. America's Greatest Garden. Stratford, Boston, 1925.

GRASSES AND CEREALS

Barrett, O. W. The Tropical Crops. Macmillan, New York, 1928.

Beal, W. J. Grasses of North America. 2 vols. Holt, New York, 1896.

Burt-Davy, J. Maize. History, Cultivation, etc. Longmans, London, 1914.

Carlton, M. A. The Small Grains. Macmillan, New York, 1916.

Chase, Agnes. First Book of Grasses. Macmillan, New York, 1922.

Francis, M. E. The Book of Grasses. Doubleday, New York, 1912.

Gress, Ernest M. The Grasses of Pennsylvania. *Pa. Dept. Agric.,* 7: 1-245, 1924, and *Bur. Pl. Ind., Tech. Series* No. 2, Gen. Bull. 384.

Hackel, E. The True Grasses. Transl. by F. L. Scribner and E. A. Southworth. Holt, New York, 1890.

Hitchcock, A. S. A Textbook of Grasses. Macmillan, New York, 1914.

Hitchcock, A. S. The Genera of Grasses of The United States. *U. S. Dept. Agric., Bull.* 772, 1920.

Hunt, T. F. The Cereals of America. Orange Judd, Chicago, 1905.

Hunter, H. Oats: Their Varieties and Characteristics. E. Bern, New York, 1924.

Macself, A. J. Grass. A Practical Treatise on Lawn, Green, and Ornamental Grasses. C. Palmer, London, 1924.

Montgomery, E. G. The Corn Crops. Macmillan, New York, 1913.

ROBBINS, W. W. Botany of Crop Plants, 2nd ed. Blakiston, Philadelphia, 1924.

VASEY, G. See under "General Morphology," etc.

REFERENCES OF A POPULAR NATURE

ALLEN, GRANT. The Colours of Flowers. London, 1882.

ALLEN, GRANT. Flowers and Their Pedigrees. Appleton, New York, 1884.

ARBER, AGNES. Water Plants. Macmillan, London, 1920.

ARMSTRONG, MARGARET. Field-Book of Western Wild Flowers. Putnam, New York, 1915.

BAILEY, L. H. Lessons with Plants. Macmillan, New York, 1898.

BERRY, E. W. Tree Ancestors. Williams & Wilkins, Baltimore, 1923.

BLANCHAN, NELTJE. Nature's Garden. Doubleday, New York, 1905.

BLANCHAN, NELTJE. Are Wild Flowers Worth Knowing? Doubleday, New York, 1917.

BROTHERSTON, R. P. The Nettle as a Food. *Gard. Chron.*, 71 : 232. 1922.

CLUTE, WILLARD N. Agronomy. Ginn, Boston, 1913.

DOWNING, E. B. A Naturalist in The Great Lakes Region. Univ. Chicago Press, 1924.

DURAND, HERBERT. My Wild Flower Garden. Putnam, New York, 1930.

DUTHRIE, J. F. Field and Garden Crops of The Northwestern Provinces and Oudh, India. 3 parts. Roorkee, 1893.

DYER, T. F. THISELTON. Folklore of Plants. Appleton, New York, 1889.

FRANCIS, MARY E. Grasses. *Nature Library,* Doubleday, New York, 1912.

GAYE, SELINA. The Great World's Farm. Macmillan, New York, 1905.

HEDRICK, U. P. *et al.* Plums of New York. *N. Y. Agric. Exp. Sta.,* Albany, 1911.

JOHNSON, ARTHUR M. Wildings of The Land of Lakes. *Nature Magazine,* 15: 305 *et seq.,* May, 1930.

KNOWLTON, F. H. Plants of The Past. Princeton Univ. Press. 1927.

LUBBOCK, SIR JOHN. Flowers, Fruits, and Leaves. Macmillan, London, 1888.

LUBBOCK, SIR JOHN. A Contribution to Our Knowledge of Seedlings. Paul, Trench & Trübner, London, 1907.

NATIONAL GEOGRAPHIC MAGAZINE. The Book of Wild Flowers. Washington, D. C., 1924.

PARSONS, MARY E. Wild Flowers of California. California School Book Depository, San Francisco, 1925.

POWERS, JAS. W. Guesses at Purposes in Nature. Society for Promoting Christian Knowledge. London. Undated. An intriguing volume.

ROCKWELL, F. F. The Book of Bulbs. Macmillan, New York, 1927.

THOMSON, J. ARTHUR. See titles under "Ecology," etc.

WALLACE, ALFRED RUSSELL. The World of Life. Moffat, Yard & Co., New York, 1911.

GENERAL MORPHOLOGY, PHYLOGENY, EVOLUTION, TAXONOMY

AMES, OAKES. *Orchidaceæ.* Illustrations and Studies of The Family *Orchidaceæ,* 6: i-xiv, 1-335, 1920.

AMES, OAKES. A Discussion of *Pogonia* and Its Allies in The Northwestern States. *Orchidaceæ,* 7: 3-38, 1922.

ANONYMOUS. Index Kewensis. *Kew Bulletin,* 1923: 95-96, 1923. Contains directions for proper recording of proposed new species, etc.

ARBER, AGNES. Monocotyledons: A Morphological Study. University Press, Cambridge, 1925.

ARBER, AGNES. Studies in The *Gramineæ*. *Ann. Bot.,* 40: 447-469, 1926 et seq.

ARBER, E. A. N. Devonian Floras. A Study of The Origin of The *Cormophyta.* Cambridge University Press, London, 1922.

ARBER, E. A. N. and PARKIN, J. On The Origin of The Angiosperms. *Journ. Linn. Soc.,* 38: 29, 1907.

ARBER, E. A. N. and PARKIN, J. Studies in The Evolution of The Angiosperms. Relations to The *Gnetales. Ann. Bot.,* 22: 489-515, 1908.

AVEBURY, LORD. A Contribution to Our Knowledge of Seedlings. Paul, Trench & Trübner, London, 1907.

AVERY, G. S., JR. The Coleoptile of *Zea Mays* and Other Grasses. *Bot. Gaz.,* 86: 107-109, 1928.

BABCOCK, E. B. Studies in *Juglans,* I. *Univ. Calif. Publs. in Agric. Sci.,* 2: 1-46, 1913; II, ditto, 2: 47-70, 1914; III, ditto, 2: 71-80, 1916.

BAILEY, I. W. and SINNOTT, E. W. Investigations on The Phylogeny of Angiosperms. *Bot. Gaz.,* 58: 36-60, 1914.

BAILEY, L. H. Gentes Herbarum. Fascicles 8-IV. Ithaca. 1920 et seq.

BAILEY, L. H. The Cultivated Brassicas. Gentes Herbarum, 1: 53-108, Ithaca, 1922.

BAILLON, H. Monographie des *Amaryllidacées, Bromeliacées* et *Iridacées* (*L'Histoire des Plantes*). Paris, 1894. Reviewed in *Bot. Centralbl.,* 61: 166-170, 1895.

BALL, C. Notes on American Willows. *Bot. Gaz.,* 60: 45-54, 1915.

BATESON, W. Materials for The Study of Variation. Macmillan, New York, 1894.

BATESON, W. Mendel's Principles of Heredity. Putnam, New York, 1909.

BEAL, W. J. Grasses of North America. 2 vols. Holt, New York, 1896.

BECHTEL, A. R. The Floral Anatomy of The *Urticales. Amer. Journ. Bot.,* 8: 386-410, 1921.

BENSON, M. The Origin of The Flowering Plants. *New Phytologist,* 3: 49-51, 1904.

BERRY, E. W. Notes on The Phylogeny of *Liriodendron. Bot. Gaz.,* 34: 44-53, 1902.

Berry, E. W. The Middle and Upper Eocene Floras of Southeastern North America. *U. S. Geol. Surv., Prof. Paper,* 92: 1-206, 1924.

Bessey, C. E. The Phylogenetic Taxonomy of The Flowering Plants. *Ann. Mo. Bot. Gard.,* 2: 109-164, 1915.

Bonnier, G. and Du Sablon, L. Cours de Botanique. Paris, 1905.

Bower, F. O. Origin of A Land Flora. Macmillan, London, 1908.

Bower, F. O. Remarks on The Present Outlook on Descent. *Proc. Roy. Soc. Edin.,* 44: 1-7, 1924.

Britton, N. L. and Rose, J. N. The *Cactaceæ.* Descriptions and Illustrations of The Family *Cactaceæ,* Vol. 8. *Carneg. Inst., Wash., Publ.,* 248, 1922.

Buchenau, E. Monograph of The *Juncaceæ. Bot. Jahrb.,* 12, 1890.

Buckholz, J. T. Studies Concerning The Evolutionary Status of Polycotyledony. *Amer. Journ. Bot.,* 6: 106-119, 1919.

Campbell, D. H. Plant Life and Evolution. Holt, New York, 1911.

Campbell, D. H. The Phylogeny of The Angiosperms (A review of Engler's, Hutchinson's, and Wettstein's theories). *Bull. Torr. Bot. Club,* 55: 479-497, 1928.

Carpenter, Anna E. Further Cases of Inconstancy in Color-Forms. *Rhodora,* 27: 199, 1925.

Célakovsky, L. Linne's Anteil an der Lehre von der Metamorphose der Pflanze. *Bot. Jahrb.,* 6: 146-186, 1885.

Célakovsky, L. Gedanken zu einer Zeitgemässen Reform der Theorie der Blütenstände. *Bot. Jahrb.,* 16: 33-51, 1893.

Chamberlain, C. J. Growth-Rings in a Monocotyl. *Bot. Gaz.,* 72: 293-304, 1921.

Chase, Agnes. Notes on Genera of *Paniceæ,* I-III. *Proc. Biol. Soc., Wash.,* 19, 1906; 21, 1908.

Chase, Agnes. The Structure of The Spikelet of *Aphanelytrum. Bot. Gaz.,* 61: 340-343, 1916.

Chrysler, M. A. The Development of The Central Cylinder of *Araceæ* and *Liliaceæ. Bot. Gaz.,* 38: 161-184, 1904.

Chrysler, M. A. The Nodes of Grasses. *Bot. Gaz.,* 41: 1-16, 1906.

Church, A. H. Types of Floral Mechanism. Clarendon Press, Oxford, 1908, *et seq.* A work every student of taxonomy should see.

Church, A. H. Introduction to The Systematy of Indian Trees. *Bot. Mem.,* 12, 50 pp., Oxford, 1921.

CLEMENTS, F. E., HALL, H. M., and MASON, DAVID. Experimental Taxonomy. *Carneg. Inst., Wash., Yearbook,* 22: 308, 1924.

CLUTE, W. N. The Pentstemons. *Amer. Bot.,* 30: 47-49, 1924.

COLE, RUTH D. Imperfection of Pollen and Mutability in The Genus *Rosa. Got. Gaz.,* 63: 110-122, 1917.

COOK, O. F. Origin and Evolution of The Angiosperms. *Proc. Wash. Acad. Sci.,* 9, 1907.

COULTER, J. M. The Phylogeny of The Angiosperms. *Decennial Publication,* Univ. Chicago, First Series, Vol. 10, 1903.

COULTER, J. M. The Endosperm of Angiosperms. *Bot. Gaz.,* 52: 380-385, 1911.

COULTER, J. M. Reproduction in Plants. *Bot. Gaz.,* 58: 337-352, 1914.

COULTER, J. M., BARNES, C. R., and COWLES, H. C. A Textbook of Botany. American Book Co., New York, 1910-1911.

COULTER, J. M. and CHAMBERLAIN, C. J. Morphology of The Angiosperms. University Chicago Press, 1903.

COULTER, J. M. and CHAMBERLAIN, C. J. Seed Plants. Appleton, New York, 1908.

COULTER, J. M. and LAND, W. J. G. The Origin of Monocotyledony. *Bot. Gaz.,* 57: 509-519, 1914.

COULTER, J. M. and ROSE, M. Notes on The *Umbelliferæ. Bot. Gaz.,* 12: 12-16 *et seq.,* 1887.

COULTER, J. M. and ROSE, M. Monograph of The North American *Umbelliferæ. Contrib. U. S. Nat. Herb.,* 7: 1-258, i-xiii, 1900.

CROW, W. B. Symmetry in Organisms. *Amer. Naturalist,* 62: 207-227, 1928.

CROW, W. B. Contribution to The Principles of Morphology. Kegan, Paul & Company, Ltd., London, 1929.

DARWIN, CHARLES. Different Forms of Flowers on Plants of The Same Species. Appleton, New York, 1884.

DARWIN, CHARLES. The Power of Movement in Plants. Appleton, New York, 1885.

DARWIN, CHARLES. Origin of Species. Collier's, New York, 1909.

DAVENPORT, E. Domesticated Animals and Plants. Ginn, Boston, 1910.

DAVIS, B. M. The Relationships of Sexual Organs in Plants. *Bot. Gaz.,* 38: 241-264, 1904.

De Bary, A. Comparative Anatomy of The Vegetative Organs of Phanerogams and Ferns. Clarendon Press, Oxford, 1884. English Translation.

De Candolle, A. Géographie Botanique Raisonée. Paris and Geneva, 1855.

De Candolle, A. Origin of Cultivated Plants. Appleton, New York, 1908.

Demeter, Karl. Vergleichende Asclepiadeenstudien. (A Comparative Study of The *Asclepiadaceæ.*) *Flora,* 115: 130-176, 1922.

De Vries, H. Species and Varieties. Open Court Publ. Co., Chicago, 1905.

Diels, L. Käferblüten bei der *Ranales* und ihre Bedeutung für die Phylogenie der Angiospermen. *Berichte der Deutschen Bot. Gesellschaft,* 34: 758-774, 1921.

Dykes, W. R. The Genus *Iris.* Cambridge, 1913.

Eames, A. J. and MacDaniels, L. T. Introduction to Plant Anatomy. McGraw-Hill, New York, 1925.

Eames, A. J. and Wilson, C. L. Carpel Morphology in The *Cruciferæ. Amer. Journ. Bot.,* 15: 251-270, 1928.

East, E. M. Xenia and The Endosperm of Angiosperms. *Bot. Gaz.,* 50: 217-224, 1913.

Ehrenberg, R. Theoretische Biologie. Springer, Berlin, 1925.

Eichler, A. W. Blütendiagrame. Leipzig, 1875.

Engler, A. Beiträge zur Kenntniss der *Araceæ. Bot. Jahrb.,* 6: 273-275, 1875.

Engler, A. Versuch einer Entwichlungsgeschichte der Pflanzenwelt. Berlin. 1882.

Epling, Carl C. Monograph of The Genus *Monardella. Ann. Mo. Bot. Gard.,* 12: 1-106, 1925.

Epling, Carl C. Studies on South American *Labiatæ,* I. *Ann. Mo. Bot. Gard.,* 12: 107-132, 1925; II, 13: 35-70, 1926; III, 14: 47-86, 1927.

Farr, C. H. Origin of The Inflorescence of *Xanthium. Bot. Gaz.,* 59: 139-148, 1915.

Fassett, Norman C. A Study of The Genus *Zizania. Rhodora,* 26: 153-160, 1924.

Fedde, F. Monographie der Gattung *Mahonia* (*Berberidaceæ*). *Bot. Jahrb.,* 31: 30, 1902.

Fernald, M. L. Notes on Some Northwestern Castilejas of The *Parviflora* Group. *Erythea,* 6: 41-51, 1898.

FERNALD, M. L. The North American Species of *Eriophorum*. *Rhodora*, 7, 1905.

FERNALD, M. L. *Nymphozanthus*. The Correct Name for The Cow Lilies. *Rhodora*, 21: 183-188, 1919.

FERNALD, M. L. The Identity of The Sand Cherries of Eastern North America. *Rhodora*, 25: 69-74, 1923.

FERNALD, M. L. Isolation and Endemism in Northeastern America and Their Relation to The Age-and-Area Hypothesis. *Amer. Journ. Bot.*, 11: 558-572, 1924.

FISCHER, MARY J. Morphology and Anatomy of The Flowers of The *Salicaceæ*. *Amer. Journ. Bot.*, 15: 307-326, 372-394, 1928.

FITTING, H. *et al.* Lehrbuch der Botanik für Hochschulen. Fischer, Jena, 1929.

FRITSCH, K. Die Stellung der Monokotylen in Pflanzensystem. *Bot. Jahrb.*, 34: 22-40, 1905.

FRITSCH, K. and WIESNER, J. On The Classification of The *Cyperales* and The *Juncaceæ*. *Organographie und Systematic*, 3rd ed., 1909.

GAGER, C. S. Heredity and Evolution in Plants. Blakiston, Philadelphia, 1920.

GAISBERG, E. VON. Zur Deutung der Monocotylenblätter als Phyllodien. *Flora*, 115: 177-190, 1922.

GEDDES, P. and THOMSON, J. A. Evolution. Holt, New York, 1920.

GILG, E. Unsere Erfahrungen über die Brauchbarkeit der Serodiagnostik für die Botanische Verwandtschaftsforschung (gemeinsam mit P. Schürhoff). *Ber. Deutsch. Bot. Gesellsch.*, 45: 315-330, 602-606, 1927.

GOEBEL, K. Beiträge zur Morphologie des Blattes. III. Über die Anordnung der Staublätter in Einigen Blüten. *Bot. Zeit.*, 40: 353-364 *et seq.*, 1882. I and II in *Bot. Zeit.*, 38, Nos. 45-50, 1880.

GOEBEL, K. Organography of Plants. 2 vols. English translation. Univ. Press, Oxford, 1900-1905.

GOW, J. E. Observations on The Morphology of Aroids. *Bot. Gaz.*, 56: 127-142, 1913.

GRANT, ADELE L. A Monograph of The Genus *Mimulus*. *Ann. Mo. Bot., Gard.*, 11: 99-388, 1924.

GRAY, ASA. Scientific Papers. Houghton Mifflin, New York, 1899.

GRAY, ASA. Lessons in Botany. American Book Company, New York, 1887. Revised, 1899.

GREENE, E. L. The Genus *Ptelea* in The Western and Southwestern United States. *Contrib. U. S. Nat. Herb.*, 10, 1896.

HABER, JULIA M. The Anatomy and Morphology of The Flower of *Euphorbia*. *Ann. Bot.*, 39: 657-707, 1925.

HABERLANDT, G. Physiological Plant Anatomy. English translation. Macmillan, London, 1914.

HACKEL, E. Über Kleistogamie bei Gräsern. *Oest. Bot. Zeitschr.*, 1906.

HALLIER, H. Die Verwandtschaftsbeziehungen der *Ampelidaceen* und anderen Pflanzenfamilien. Reviewed in *Bot. Centralbl.*, 75: 140-141, 1898.

HALLIER, H. Über die Verwandtschaftsverhältnisse der *Tubifloren* und *Ebenalen*, den polyphyletischen Ursprung der *Sympetalen* und die Anordnung der *Angiospermen* überhaupt. *Abhandl. Wissenschaft. Vereins.* Hamburg. 1901.

HALLIER, H. Über den Umfang, etc., der *Hamamelidaceen*. *Beih. zur Bot. Centralbl.*, 14: 247, 1903.

HANSGIRG, A. Phyllobiologie nebst Übersicht der biologischen Blattypen von ein-und-sechzig Siphonogamenfamilien. Borntraeger, Berlin, 1903. A valuable work.

HEDRICK, U. P. Sturtevant's Notes on Edible Plants. *Rept. N. Y. Agric. Exp. Sta.*, pp. i-iv, 1-686, 1919. Geneva, N. Y.

HEMSLEY, W. BOTTING. On The *Julianiaceæ*, A New Natural Order of Plants. *Ann. Bot.*, 20: 467-471, 1906.

HENDERSON, I. F. and HENDERSON, W. D. Dictionary of Scientific Terms. 2nd ed. University Press, Cambridge, 1929.

HILDEBRAND, F. Die Gattung *Cyclamen* L; eine systematische und biologische Monographie. Jena. 1898. Reviewed in *Bot. Centralbl.*, 75: 131-132, 1898.

HILL, A. W. The Origin of The Monocotyledons. *Ann. Bot.*, 22: 713-714, 1908.

HIRMER, MAX. Zur Lösing des Problems der Blattstellung. Jena, 1922.

HOAR, C. S. Sterility as The Result of Hybridization and The Condition of The Pollen in *Rubus*. *Bot. Gaz.*, 62: 370-378, 1916.

HOBSON, E. W. The Domain of Natural Science. Macmillan, New York, 1923.

HOLDEN, RUTH. Some Features of The Anatomy of The *Sapindales*. *Bot. Gaz.*, 53: 50-58, 1912.

HOLM, T. Studies in The *Cyperaceæ*. *Amer. Journ. Sci.*, 166: 17-44, 1903 *et seq.*

HOLM, T. *Podophyllum peltatum.* A Morphological Study. *Bot. Gaz.,* 27: 419-433, 1899.

HOLM, T. Morphology of North American Species of *Polygala. Bot. Gaz.,* 88: 167-185, 1929.

HÖCK, F. Kurze Bemerkungen zur Systematik der *Cormophyten. Bot. Centralbl.,* 76: 171-175, 1898.

HUHN, H. Über die Verwertbarkeit der Serodiagnostik in der Botanik, Erläutert an den *Sympetalen.* Cohn's *Beiträge zur Biologie der Pflanze,* 15: 228-262, 1927.

HUMBERT, H. Les Composées de Madagascar. Caen, 1923.

IRISH, H. C. A Revision of The Genus *Capsicum,* with Especial Reference to Garden Varieties. *Ninth Annual Report, Mo. Bot. Gard.,* pp. 53-110, 1898.

JADIN, F. Récherches sur la Structure et Les Affinités des *Terebinthacées. Ann. Sc. Nat. Bot.,* Série 7, 19: 1: 1-51, 1894.

JEFFREY, E. C. The Anatomy of Woody Plants. Univ. Chicago Press, 1917.

JEPSON, W. L. Revision of The California Species of *Arctostaphylos. Madroña,* 1: 76-96, 1922.

JOHNSON, ARTHUR M. Revision of The North American Species of The Section *Boraphila* Engler, of The Genus *Saxifraga. Minn. Stud. in Pl. Sci.* 4: 1-110, 1923.

JOHNSTON, I. M. Taxonomic Records concerning American Spermatophytes. *Contrib. Gray Herb., N. S.,* 70: 61-92, 1924.

JOHNSTON, I. M. Studies in The *Boraginaceæ. Contrib. Gray Herb.,* I, 68: 43-80, 1923; II, 70: 3-61, 1924; III, 73: 42-78, 1924.

JUDSON, J. E. The Morphology and Vascular Anatomy of The Pistillate Flower of The Cucumber. *Amer. Journ. Bot.,* 16: 69-86, 1929.

KARSTEN, G. Die Vegetationsorgane der Palmen. *Abhandl. d. k. Akad. d. Wiss. Berlin* for 1847: 73-236, 1849.

KARSTEN, G. Ueber die Entwicklung der weiblichen Blüten bei einigen *Juglandaceen. Flora,* 1902.

KARSTEN, G. Zur Phylogenie der *Angiospermen. Zeitschr. f. Bot.,* 10, 1918.

KERNER, VON MARILAUN, ANTON and OLIVER, F. W. Natural History of Plants. Engl. Transl. Holt, New York. Undated.

KNOBLAUCH, E. Die Nomenclatur der Gattungen und Arten. *Bot. Centralbl.,* 61: 1-6, 1895.

KOEHNE, A. *Lythraceæ.* Bau der Blüten. *Bot. Jahrb.,* 6: 1-48, 1885.

KRAUSE, K. Die Botanische Literatur über die Türkei. In Fedde *Repert. Spec. Nov.,* 24: 113-126, 1927.

KUSNETZOW, N. J. Les Principes, etc., du Système Phylogenetique. *Bull. Jard. Republ. Russe.,* 21, 1922.

LE MAOUT, E. and DE CAISNE, J. Traite Général de Botanique. Paris. 1868. English translation by Mrs. Hooker, Longmans, London, 1916.

LEYEL, MRS. C. F. A Modern Herbal. Jonathan Cape, London, 1929.

LINDMAN, C. A. M. Über die Floral Syndimorphismus einiger *Festucaceen. Ark. für Bot.,* 8: 1-17, 1909.

LODGE, SIR OLIVER. Science and Hypothesis. *Nature,* 123: 645-648, 1929.

LOESENER, O. Übersicht über die Arten der Gattung *Veratrum.* Fedde, *Repert. Spec. Nov.,* 24: 61-72, 1927.

LONES, T. E. Aristotle's Researches in Natural Sciences. West, Newman & Co., London, 1912.

LOTSY, J. B. Phylogeny of Plants. *Bot. Gaz.,* 49: 460-461, 1910.

LOTSY, J. B. Vorträge über Botanische Stammesgeschichte. Jena, 1911.

LOWIG, E. Beiträge zu Sterilitätsfragen. *Flora,* 23: 62-103, 1928.

LUBBOCK, SIR JOHN. Flowers, Fruits, and Leaves. Macmillan, London, 1886.

MACKENZIE, K. K. Proper Use of The Name *Leontodon. Rhodora,* 27: 47, 1925.

MAIDEN, J. H. A Critical Revision of The Genus *Eucalyptus,* 6: 165-218. John Spence, Sydney, 1922.

MARONG, T. The *Smilaceæ* of North and Central America. *Bull. Torr. Bot.* Club, 21: 419-443, 1894.

MARTIN, J. N. The Comparative Morphology of Some *Leguminosæ. Bot. Gaz.,* 58: 154-167, 1914.

MATHIAS, M. E. Studies in The *Umbelliferæ. Ann. No. Bot. Gard.,* 15: 91-108, 1928 *et seq.*

MC KOWN, L. W. Some Misunderstood *Piperaceæ. Bot. Gaz.* 85: 46-65, 1928.

MC NAIR, J. B. The Taxonomy of The Poison Ivy. *Field Mus. Nat. Hist., Bot. Ser.,* 4: 3: 55, 1925.

McNair, J. B. The Taxonomic and Climatic Distribution of Oils, Fats, and Waxes in Plants. *Amer. Jour. Bot.,* 16:832-841, 1929.

Melchior, H. Sind die *Violaceen* und die *Resedaceen* miteinander verwandt? *Ber. der Deutsch. Bot. Gesellsch.,* 45:171-179, 1927.

Müller, W. Beiträge zur Entwicklungsgeschichte der infloreszenzen der *Boragineen* und *Solaneen. Flora,* 94:385-419, 1905.

Munz, Philip A. Studies in The *Onagraceæ. Amer. Journ. Bot.,* 15:223-240, 1928. *Bot. Gaz.,* 85:235-270, 1928; *Amer. Journ. Bot.,* 16:246-257, 702-715, 1929; 17:358-370, 1930.

Nay, W. Serodiagnostische Verwandtschaftsforschungen innerhalb der *Rosales, Myrtifloræ, Umbelliferæ. Beitr. Biol. d. Pflanzen,* 15:5-37, 1927.

Needham, J. G. Scientific Names. *Science,* 71:26-28, 1930.

Nelson, C. I. and Birkeland, J. M. A Serological Ranking of Some Wheat Hybrids as an Aid in Selecting for Certain Genetic Characters. *Journ. Agric. Res.,* 38:169-181, 1929. (See Bibliography.)

Nicolas, G. Des Synanthies, à Propos du *Narcissus Tazeta* L. A paper dealing with the causes of synanthy, syncarpy, fasciation, etc. *Compt. Rend. Acad. Sci.,* Paris, 174:1126-1128, 1922.

Oliver, D. Observations on The Structure of The Stem in Certain Species of *Caryophyllaceæ* and *Plumbaginaceæ. Trans. Linn. Soc.,* 22:289, 1858.

Owen, F. V. Soy-Beans with Two Embryos. *Journ. Hered.,* 19:373-374, 1928.

Pace, Lula. *Parnassia* and Some Allied Genera. *Bot. Gaz.,* 54: 406-417, 1912.

Palla, E. Zur Kenntniss der Gattung *Scirpus. Bot. Jahrb.,* 10: 293-301, 1889.

Parkin, J. The Phylogenetic Classification of The Flowering Plants. Nature, 115:340-342, 1925.

Parkin, J. Diœcism in *Ranunculus acris. Nature,* 123:568-569, 1929.

Pax, F. Monographie der Gattung *Acer. Bot. Jahrb.,* 6:287-373, 1884; 7:177-263, 1886; 11:72-83 (1889), 1890; 16:393-404, 1893.

Pax, F. Prantl's Lehrbuch der Botanik. Engelmann, Berlin, 1916.

PENNELL, E. W. Notes on The *Scrophulariaceæ* of The Northwestern United States. *Bull. Torr. Bot. Club,* 55: 315-318, 1928.

PHIPPS, I. F. Heritable Characters of Maize. *Journ. Hered.,* 19: 9: 399-403, 1928.

PILGER, R. Über die Blütenstand und Aerchen der Bambuseen-Gattung *Guadua. Berl. Deutsch. Bot. Ges.,* 45: 562-570, 1927.

PIPER, C. V. North American Species of *Festuca. Contrib. U. S. Nat. Herb.,* 10, 1906.

PLOWMAN, A. B. The Comparative Anatomy and Phylogeny of The *Cyperaceæ. Ann. Bot.,* 20: 1-30, 1906.

PLOWMAN, A. B. Is The Boxelder a Maple? A Study of The Comparative Anatomy of *Negundo. Bot. Gaz.,* 60: 169-192, 1915.

READ, E. L. The Leaf Nectaries of *Gossypium. Bot. Gaz.,* 63: 229-231, 1917.

RIKLI, M. Beiträge zur vergleichende Anatomie der *Cyperaceæ.* Pringsheim's *Jahrb. für Wissensch. Bot.,* 27: 485-580, 1895.

ROBBINS, W. W. Botany of Crop Plants. 2nd ed. Blakiston, Philadelphia, 1924.

ROBINSON, B. L. Tropical Species of *Eupatorium. Proc. Amer. Acad. Arts and Sci.,* 54: 3-41, 1919.

ROBINSON, B. L. Tropical American *Compositæ,* Chiefly *Eupatorieæ.* I. *Contrib. Gray Herb.,* New Series, 60: 1-88, 1919, and *Proc. Amer. Acad. Arts and Sci.,* 55: 3-41, 1919; also *Contrib. Gray Herb., N. S.,* 61 : 3-30 and 61 : 30-80, 1920.

RYDBERG, P. A. Genera of North American *Fabaceæ. Amer. Journ. Bot.,* 15: 195-203, 1928.

SAEGER, A. The Flowering of The *Lemnaceæ. Bull. Torr. Bot. Club,* 56: 351-358, 1929.

SARGANT, ETHEL. On The Origin of The Monocotyledons. *Bot. Gaz.,* 37: 325-345, 1904.

SARGANT, ETHEL. On The Vascular System of Monocots. *New Phytologist,* 1: 107, 1902.

SARGANT, ETHEL. A Theory of The Origin of The Monocotyledons Founded on The Structure of Their Seedlings. *Ann. Bot.,* 17: 1-88, 1903.

SARGANT, ETHEL. The Reconstruction of A Race of Primitive Angiosperms. *Ann. Bot.,* 22: 121-186, 1908.

SARGENT, F. L. Corn Plants, Their Uses and Ways of Life. Houghton Mifflin, Boston, 1889.

SARGENT, F. L. Plants and Their Uses. Holt, New York, 1913.

SARGENT, O. H. Fragments of The Flower Biology of Westralian Plants. *Ann. Bot.*, 32:215-231, 1918.

SAUNDERS, EDITH. On Carpel Polymorphism. *Ann. Bot.*, 39:123-167, 1925 *et seq.*

SAUNDERS, EDITH. On a New View of The Nature of The Median Carpels of The *Cruciferæ*. *Amer. Journ. Bot.*, 16:122-137, 1929.

SCHILDER, P. Gedanken der Naturphilosophie. Springer, Vienna, 1928. See also *Bot. Centralbl., Neue Folge,* 155:321, 1928.

SCHMIDT, E. Untersuchungen über *Berberidaceen*. *Beihefte z. Bot., Centralbl.*, 45:329-396, 1928.

SCHNEIDER, CAMILLO. Notes on American Willows. *Bot. Gaz.*, 65:1-41, 1918; 66:318-353, 1918; 67:27-64, 1919.

SCHOUTE, J. C. Die Stelär Theorie. Groningen, 1902.

SCHRÖDINGER, R. Der Blütenbau der Zygomorphen *Ranunculaceen*, etc. *Abhandl. K. K. Zoöl.-Bot. Ges., Wien,* 4:1-63 (1909), 1906-1909.

SCHUMANN, K. Neue Untersuchungen über Blütenanschluss. Leipzig. 1890.

SCHUMANN, K. Æstivation der Blüten und ihre mechanischen Ursachen. *Ber. d. Deutsch. Bot. Ges.*, 4:553, 1886.

SCHUSTER, J. Über der Morphologie der Grasblüte. *Flora,* C 2: 212-266, 1910.

SEAGER, ALBERT. The Flowering of The *Lemnaceæ*. *Bull. Torr. Bot. Club,* 56:351-358, 1929.

SENN, G. Die Grundlagen des Hallierschen Angiospermensystem. *Beihefte z. Bot. Centralbl.*, 17:129-156, 1904.

SHERFF, E. E. North American Species of *Taraxacuin*. *Bot. Gaz.*, 70:339-358, 1920.

SHERFF, E. E. Studies in The Genus *Bidens*. *Bot. Gaz.*, 56:490-495, 1913; 59:301-316, 1915; 61:495-506, 1916; 64:21-41, 1917; 70:89-109, 1920; 76:144-166, 1923; 85:1-29, 1928.

SINGER, C. The Evolution of Biology. Univ. Press, Cambridge, 1929.

SINNOTT, E. W. and BAILEY, J. W. Investigations on The Phylogeny of Angiosperms. IV. *Ann. Bot.*, 28:547-600, 1914.

SMALL, JAMES. Origin and Development of The *Compositæ*. Wm. Wesley, London, 1924.

SMITH, C. P. Plurality of Seeds in Acorns of *Quercus Prinus*. *Rhodora,* 16:41, 1916.

SNOW, LETITIA. Contributions to The Knowledge of The Diaphragms of Water Plants. *Bot. Gaz.*, 58: 485-517, 1914.

STACE, ARTHUR J. Plant Odors. *Bot. Gaz.*, 12: 265-268, 1887.

STEVENS, N. E. Observations on Heterostylous Plants. *Bot. Gaz.*, 53: 277-308, 1912.

STOLL, W. Organization und Gestalt im Bereich der Blüte. Berlin, 1928.

STOUT, A. B. Types of Sterility in Plants and Their Significance in Horticulture. *Mem. Hort. Soc.*, N. Y., 3: 3-8, 1927.

STRASBURGER, E. *et al.* Textbook of Botany. 5th English edition. Macmillan, London, 1908.

SZYSZYLOWICZ, I. Zur Systematik der *Tiliaceen*. *Bot. Jahrb.*, 6: 427-457, 1885.

TANNERT, P. Entwicklung und Bau von Blüte und Frucht von *Avena sativa*. Dissertation. Zurich, 1905.

THOMSON, J. A. The Gospel of Evolution. Putnam, New York, 1926.

TISCHLER, G. Die *Berberidaceen* und *Podophyllaceen*. *Bot. Jahrb.*, 31: 596-732, 1902.

TRELEASE, WM. Winter Botany. Urbana, Illinois. 1919.

TROLL, W. Organization und Gestalt im Bericht der Blüte. Springer, Berlin, 1928.

ULBRICH, E. Über die Systematische Gliederung und Geographische Verbreitung der Gattung *Anemone: Verh. Bot. Ver. Prov. Brandenburg,* 48, and *Jahrb. für Syst,* 37: 172, 1905.

ULBRICH, E. Biologie der Früchte und Samen. Springer, Berlin, 1928.

ULINE, E. B. Eine Monographie der *Dioscoraceæ*. *Bot. Jahrb. f. Syst., Pflanzengeschichte, und Pflanzengeographie,* 25: 126-165, 1898. Reviewed in *Bot. Centralbl.*, 75: 134-138, 1898.

VAN TIEGHEM, P. L'Œuf des Plantes considerée comme Base de Leur Classification. *Ann. Sc. Nat. Bot.*, 8, Sér. T 14: 213, 1901.

VASEY, G. Grasses of The Southwest. *U. S. Dept. Agric., Div. Bot., Bull.* No. 12, 1890.

VASEY, G. Grasses of The Pacific Slope. *U. S. Dept. Agric., Div. Bot. Bull. No.* 13, 1892.

VASEY, G. Illustrations of North American Grasses. Vol. I, 1891; Vol. II, 1893. Gov't Printing Office, Washington, D. C.

VELENOVSKY, J. Vergleichende Studien über die Salix-Blüte. *Bot. Centralbl.,* 17: 123-128, 1904.

VICTORIN, FRÈRE MARIE-. Les Liliiflorés du Québec. *Contrib. du Lab. de Bot. de l'Univ. de Montréal,* No. 14: 1-202, 1929.

VICTORIN, FRÈRE MARIE-. Sur Quelques Composées Nouvelles, Rares, ou Critiques du Québec Oriental. *Contrib. du Lab. de Bot., de l'Univ. de Montréal,* No. 5: 79-96, 1925.

VICTORIN, FRÈRE MARIE-. Nouvelles Études sur Les Composées du Québec. *Contrib. du Lab. de Bot. de l'Univ. de Montréal,* No. 8, 20 pp., 1926.

VIERHAPPER, F. Zur Systematik der Gattung *Avena. Verh. Zoöl.-Bot. Ges.,* Wien, 56: 369, 1906.

VINES, S. H. Elementary Textbook of Botany. Macmillan, London, 1898.

VUILLEMIN, PAUL. L'Inflorescence. *Bull. Soc. Sci.,* Nancy, Series 4, 1: 57-67, 1921.

WALTHER, A. Einführung in die Mathematische Behandlung Naturwissenschaftlichen Fragen. Springer, Berlin, 1929.

WARBURG, O. Monographie der *Myristicaceen. Nova Acta. Abhandl. der Kaiserlichen Leopoldinischen Carolinischen Deutschen Akademie der Naturforschung,* Bd. 68, 1898. Reviewed in *Bot. Centralbl.,* 75: 79-84, 1898.

WATT, G. Dictionary of The Economic Products of India. 6 vols. Vol. 3, The Yams, *Dioscorea.* 1890.

WAUGH, F. A. A Conspectus of The Genus *Lilium. Got. Gaz.,* 27: 235-254, 341-360, 1890.

WEATHERWAX, P. The Development of The Spikelet of *Zea Mays. Bull. Torr. Bot. Club,* 54: 483-496, 1917.

WEATHERWAX, P. Cleistogamy in Two Species of *Danthonia. Bot. Gaz.,* 85: 104-109, 1928.

WEATHERWAX, P. The Morphology of The Spikelets of Six Genera of *Oryzeæ. Amer. Journ. Bot.,* 16: 547-555, 1929.

WERMUND, R. Untersuchungen über die Brauchbarkeit der Serodiagnostik für die Botanische Verwandtschaftsforschung. *Ebenda,* 16: 39-80, 1928.

WETTSTEIN, R. Handbuch der Systematischen Botanik. Leipzig and Vienna, 1924.

WIEGAND, K. A Synopsis of The Plant Kingdom. In Bailey's *Standard Cyclopedia of Horticulture,* Vol. 1: 1-147. Macmillan, New York, 1906.

WIELAND, G. R. The Origin of The Dicotyls. *Science,* 48: 18-21, 1918.

WILSON, E. H. The Lilies of Eastern Asia. Dulau, London, 1925.

WODEHOUSE, R. P. Pollen-Grains in The Identification and Classification of Plants. *Bull. Torr. Bot. Club,* 55: 449-462, 1928.

WODEHOUSE, R. P. The Origin of Symmetry-Patterns in Pollen-Grains. *Bull. Torr. Bot. Club,* 56: 339-350, 1929.

WOODSON, R. E. Studies in The *Apocynaceæ. Ann. Mo. Bot. Gard.,* 15: 341-378, 1928 *et seq.*

WORSDELL, W. C. The Evolution of The Vascular Tissue of Plants. *Bot. Gaz.,* 34: 216-223, 1902.

WORSDELL, W. C. The Origin of The Flower. *Science Progress,* 2: 255-262, 1907.

WORSDELL, W. C. A Study of The Vascular System in Certain Orders of The *Ranales. Ann. Bot.,* 22: 651-682, 1908.

ZARNACK, G. Untersuchungen über die Brauchbarkeit der Serodiagnostik für die Botanischen Verwandtschaftsforschung, Erlaütert an der Reihe der *Ranales. Ebenda,* 15: 180-227, 1927.

ZUDERALL, H. Über das Aufblühen der Gräser. *Sitzungsbericht der Kais. Akad. Wiss.,* Wien, 118, 1910.

REFERENCES ON TERATOLOGICAL PHENOMENA

ALLARD, H. A. Abnormalities in *Nicotiana. Bot. Gaz.,* 65: 175-185, 1918.

ANDREWS, F. M. Monstrosities in *Trillium. Proc. Indiana Acad. Sci.,* 37: 325-326, 1928.

ANDREWS, F. M. Modified *Narcissus* Flowers. *Proc. Indiana Acad. Sci.,* 37: 326-327, 1928.

CAMPBELL, R. S. A Case of Phyllody in *Yucca elata. Bot. Gaz.,* 88: 109-110, 1929.

COOKSON, ISABEL. Floral Abnormalities in The Genera *Eriostemon* and *Glossodia. Proc. Roy. Soc.* Victoria, 33: 32-38, 1921.

HOLDEN, H. S. Observations on The Anatomy of Teratological Seedlings. *Ann. Bot.,* 32: 513-530, 1918 *et seq.*

LANDE, H. H. and GATES, F. C. A Head of *Sorghum* with Greatly Proliferated Spikelets. *Bot. Gaz.,* 88: 440-450, 1929.

LIFE, A. C. An Abnormal *Ambrosia. Bot. Gaz.,* 38: 383-384, 1904.

MASTERS, M. T. Vegetable Teratology. Ray Soc., London, 1869.

STROHL, J. Missbildungen in der Thier- und Pflanzenreich. Gustav Fischer, Jena, 1929.

TONI, J. B. DE. Contribution to The Teratology of The Genus *Datura* L. *Journ. Linn. Soc. Bot.*, London, 47: 419-420, 1921.

WORSDELL, W. C. The Principles of Plant Teratology. DuLau, London, 1916.

YOUNCKER, T. G. Observations on The Teratology of The Genus *Cuscuta*. *Proc. Indiana Acad. Sci.*, 37: 335-338, 1928.

ECOLOGY, PHYTOGEOGRAPHY, NATURAL HISTORY

AARONSOHN, A. L. Über die in Palästina und Syrien Wildwachsenden Getreidearten. *Verhandl. K. K. Zool.-Bot. Ges.*, Wien, 59: 485-509 (1909), Jan. 20, 1910.

ABRAMS, LEROY. A Phytogeographic and Taxonomic Study of The Southern California Trees and Shrubs. *Bull. N. Y. Bot. Gard.*, 6: 303-485, 1910.

ABRAMS, LEROY. Origin and Geographical Affinities of The Flora of California. *Ecology*, 6: 1-6, 1925.

ADAMS, C. C. Southeastern United States as A Centre of Geographical Distribution of Flora and Fauna. *Biol. Bull.*, 3: 115-131, 1902.

ADAMS, C. C. Post-glacial Origin and Migrations of The Life of The Northeastern United States. *Journ. Geog.*, 1: 303-310, 352-357, 1902.

ADAMS, C. C. Post-glacial Dispersal of The North American Biota. *Biol. Bull.*, 9: 53-71, 1905.

AGASSIZ, A. Three Cruises of The *Blake*. 2 vols. Houghton Mifflin, Boston, 1888.

ANDRÉ, E. A Naturalist in The Guiana. Smith Elder, London, 1904.

ANONYMOUS. The Origin of The Potato. *Gard. Chron.*, 71: 37, 1922.

ANTEVS, ERNST. The Big Tree as A Climatic Measure. *Carneg. Inst. Wash. Yearbook*, 22: 299-301, 1924.

ARBER, AGNES. Water Plants. Cambridge, 1920.

ARESCHOUG, F. W. C. Der Einfluss des Klimas auf die Organization der Pflanzen Insbesonder auf die Anatomische Struktur der Blattorgane. *Bot. Jahrb.*, 2: 511, 1882.

AVEBURY, LORD (SIR JOHN LUBBOCK). Ants, Bees, and Wasps. New ed. Routledge & Sons, London, 1929.

BAILEY, I. W. Notes on Neotropical Ant-Plants. *Bot. Gaz.*, 74: 369-391, 1922.

BAKER, E. G. Aspen as A Temporary Forest Type. *Journ. For.*, 16: 294, 1916.

BALDWIN, E. G. The Honey Plant Regions of Indiana. *Amer. Bee Journ.*, 61: 51-53, 1921.

BALL, C. R. History and Distribution of *Sorghum*. *Bull. U. S. Dept. Agric.*, 1910.

BALL, J. On The Origin of The Flora of The European Alps. *Proc. Roy. Geog. Soc.*, London, 1879.

BATES, H. W. The Naturalist on The River Amazon. Murray, London, 1892.

BATESON, W. Materials for The Study of Variation, etc. Macmillan, New York, 1894.

BATESON, W. Mendel's Principles of Heredity. Putnam, New York, 1909.

BATHER, F. A. Evolution Through Adaptation. *Nature*, 123: 602-603, 1929.

BELT, T. A. A Naturalist in Nicaragua. London, 1888.

BERRY, E. W. The Origin and Distribution of The Family *Myrtaceæ*. *Bot. Gaz.*, 59: 484-490, 1915.

BERRY, E. W. A Note on The Age and Area Hypothesis. *Science*, 56: 539-540, 1917.

BERTSCH, K. Die ältesten Getreidereste Deutschlands. *Ber. d. Deutsch. Bot. Ges.*, 47: 121-125, 1929.

BEWS, J. W. The Southeast African Flora: Its Origin, Migrations, and Evolutionary Tendencies. *Ann. Bot.*, 36: 209-224, 1922.

BEWS, J. W. Plant Forms and Their Evolution in South Africa. Longmans, London, 1925.

Bews, J. W. Studies in The Ecological Evolution of the Angiosperms. Wheldon & Wesley, London, 1927. Reprint No. 16 of *New Phytologist*.

BLAKESLEE, A. F. A Paradise for Plant Lovers. The Biological Station at Alto da Serra, Brazil. *Scientific Monthly*, 25: 5-18, 1927.

Blakeslee, A. F. and Cartledge, J. L. Sterility of Pollen in *Datura*. *Mem. Hort. Soc.*, N. Y., 3: 305-312, 1927.

Blytt, A. On Variations of Climate in The Course of Time. *Videnskabs-Selskabets Forhandlinger,* No. 8. Oslo, 1886.

Bottomley, M. E. How The Ancient Water-Lily Became Modern. *Parks and Recreation,* 11: 404-409, 1928.

Boué, A. On The Changes Which Appear to Have Taken Place During Different Periods of The Earth's Formation, in The Climate of Our Globe, and in The Nature and Physical and Geographical Distribution of Animals and Plants. *Edinburgh Phil. Journ.,* 1: 88, 1826.

Bowman, I. Forest Physiography. Wiley, New York, 1914.

Boyson, V. F. The Falkland Islands. Clarendon Press, Oxford, 1924. Notes on Natural History.

Braun-Blanquet, J. Notions d'Élement et de Territoire Phytogéographiques. *Archives des Sciences Physiques et Naturelles,* 1: 497-512, 1919.

Braun-Blanquet, J. Pflanzensoziologie. Julius Springer, Berlin, 1928.

Braun-Blanquet, J. and Furrer, E. Sur l'Étude des Associations. *Bull. de la Société Languedocienne de Géographie,* Tome 36: 1-22, 1913.

Bray, W. L. The Plant Geography of North America. The Relation of The North American Flora to That of South America. *Science, N. S.,* 12: 709-716, 1900.

Brendel, F. Notes on The Flora of Florida. *Amer. Nat.,* 8: 449, 1874.

Britton, N. L. The Origins and Relations of The Floras and Faunas of The Antarctic and Adjacent Regions. *Science, N. S.,* 3: 310, 1896.

Brues, C. T. The Selection of Food Plants by Insects. *Amer. Nat.,* 54: 313-332, 1920.

Bunbury, C. J. F. Botanical Fragments. London, 1883. Author deals with the influence of the chemical composition of rocks on vegetation. Quotes from the second book of the *Georgics* to show that Virgil was not unfamiliar with this factor in plant distribution.

Bunning, E. Untersuchungen über die Seismoreaktione von Staubgefässen und Narben. *Zeitschr. für Botanik,* 21: 465-536, 1929.

Burgeff, H. Die Wurzelpilze der *Orchideen,* Ihre Kultur und Ihr Leben in der Pflanze. Jena, 1909.

BURTT, D. J. The Distribution and Origin of *Salix* in South Africa. *Ecology,* 10:62-86, 1922.

BUSH, B. F. Geographic Origin of *Cratægus viridis* L. *Rhodora,* 14:81-86, 1912.

BUTTERS, F. K. The Vegetation of The Selkirks. In "Mountaineering and Exploration in The Selkirks," by Howard Palmer, Appendix A: 352-362. Putnam, New York, 1914.

CALVERT, A. S. and P. P. A Year of Costa Rica Natural History. Macmillan, New York, 1917.

CAMPBELL, D. H. The Derivation of The Flora of Hawaii. *Leland Stanford Junior Publications,* Palo Alto, 1919.

CAMPBELL, E. G. Plant Relations In Brazos County, Texas, with Special Reference to Eastern and Western Types. *Ecology,* 6: 163-170, 1925.

CAMPBELL, R. S. Vegetative Succession in The *Prosopis* Sand Dunes of Southern New Mexico. *Ecology,* 10:392-398, 1929.

CANNON, W. A. Root Habits of Desert Plants. *Carneg. Inst. Wash., Publ.* No. 131, 1911.

CARPENTER, G. B. The Biology of Insects. Sidgwick & Jackson, London; Harcourt, Brace, New York, 1928.

CARVER, JONATHAN. Travels Through The Interior Parts of North America in The Years 1766, 1767, and 1768. Dublin, 1779.

CHRIST, H. Das Pflanzenleben der Schweitz. Zurich. 1879.

CHRYSLER, M. A. The Plant-Life of Maryland. *Maryland Weather Service, Special Bull.,* 5: 149-219, 1910.

CLEMENT, E. The Nonsymbiotic Germination of Orchid Seeds. *Orch. Rev.,* 34: 164, 1926; 34:68-75, 1929.

CLEMENTS, F. E. The Life History of Lodgepole Pine Burns. *U. S. For. Serv. Bull.,* 79, 1910.

CLEMENTS, F. E. Plant Succession. *Carneg. Inst. Wash., Publ.* No. 242, 1916.

CLEMENTS, F. E. Plant Indicators. *Carneg. Inst. Wash., Publ.* No. 290, 1920.

CLEMENTS, F. E. The Original Vegetation of Death Valley. *Carneg. Inst. Wash. Yearbook,* 22: 317, 1924.

CLEMENTS, F. E. and LONG, FRANCES. Experimental Pollination. An Outline of The Ecology of Flowers and Insects. *Carneg. Inst. Wash., Publ.* No. 336, 1923.

COCKAYNE, L. The Vegetation of New Zealand. Second edition, Vol. 14 in Engler and Drude, Die Vegetation der Erde. Engelmann, Leipzig, 1928.

COCKERELL, T. D. A. Aspects of The Madeira Flora. *Bot. Gaz.,* 85: 66-73, 1928.

COKER, W. C. The Distribution of The Venus's Fly-Trap, *Dionæa muscipula. Journ. Elisha Mitchell Sci. Soc.,* 43: 221-228, 1928.

COOK, M. T. and GLEASON, H. A. Ecological Survey of The Flora of Porto Rico. *Journ. Dept. Agric.,* Porto Rico, 12: 1-139, 1928.

COOPER, W. S. The Climax Forest of Isle Royale, Lake Superior, and Its Development. *Bot. Gaz.,* 55: 144, 115-140, 189-235, 1913.

COOPER, W. S. The Broad Sclerophyll Vegetation of California. *Carneg. Inst. Wash., Publ.* No. 319, 1922.

COOPER, W. S. Fundamentals of Vegetational Change. *Ecology,* 7: 391-419, 1926.

COVILLE, F. V. Botany of The Death Valley Expedition. *Contrib. U. S. Nat. Herb.,* Vol. 4, 1893.

COWLES, H. C. The Ecological Relations of Vegetations on The Sand Dunes of Lake Michigan. *Bot. Gaz.,* 27: 95-117, *et seq.,* 1899.

COWLES, H. C. The Physiographic Ecology of Chicago and Vicinity. *Bot. Gaz.,* 31: 73-108, 145-182, 1901.

COWLES, H. C. The Causes of Vegetation Cyclics. *Bot. Gaz.,* 51: 161-193, 1911.

CRAMPTON, C. B. The Geological Relations of Stable and Migratory Plant Formations. *Scottish Botanical Review,* 1: 1-16, 1912.

CROSSLAND, C. Note on The Dispersal of Mangrove Seedlings. *Ann. Bot.,* 17: 267-270, 1903.

CROWDER, R. V. Hawaii That Has All Things. *Nature Magazine,* 13: 363-371, 1929.

DACHNOWSKI, A. P. Peat Deposits and Their Evidence of Climatic Changes. *Bot. Gaz.,* 72: 57-89, 1821.

DAHL, K., LID, J., and MUNSTER, T. A Division of Norway into Biographical Sectional Areas. *Videnskaps-Selskapets Skrifter,* 1, 7, Oslo, 1924.

DARWIN, C. The Effect of Cross- and Self-Fertilization in The Vegetable Kingdom. Murray, London, 1876.

DARWIN, C. The Effect of Continental Glaciation on The Distribution of Plants. *Life and Letters,* 1: 71-72, 372, New York, 1902. *The Origin of Species,* 2: 152, New York, 1902.

DARWIN, C. The Formation of Vegetable Mould. Appleton, New York, 1881.

DARWIN, C. Insectivorous Plants. Appleton, New York, 1884.

DARWIN, C. Movements and Habits of Climbing Plants. 7th ed. rev. Appleton, New York, 1884.

DARWIN, C. Journal of Researches into The Natural History and Geology of The Countries Visited During The Voyage of H.M.S. *Beagle,* 'Round The World. Appleton, New York, 1887.

DARWIN, C. Variation in Animals and Plants under Domestication. Appleton, New York, 1900.

DARWIN, C. The Various Contrivances by Which Orchids Are Fertilized by Insects. 2nd ed. London. 1904.

DAVENPORT, E. Domesticated Animals and Plants. Ginn, Boston, 1910.

DAVEY, J. B. The Distribution of *Salix* in South Africa. *Ecology,* 10:62, 1922.

DEAN, C. C. Plants New and Rare in Indiana. *Proc. Indiana Acad. Sci.,* 37:321-323, 1928.

DE CANDOLLE, A. Géographic Botanique Raisonée. Paris, 1855.

DE CANDOLLE, E. Origin of Cultivated Plants. Appleton, New York, 1908.

DES CARS, A. A Treatise on Pruning Forest and Ornamental Trees. Transl. from the 7th French edition. Introduction by C. S. Sargent. Boston, *The Society,* 1900.

DE VRIES, H. L'Évolution des Étres Organisées par Sautes brusques. *Scientia,* No. 1, 1916.

DE VRIES, H. The Origin by Mutation of Endemic Plants of Ceylon. *Science,* 43:785-787, 1916.

DIELS, L. Pflanzengeographie. 2 Auflage. Samml. Göschen, 389, 1918.

DOUGLAS, DAVID. Journal, 1823-1827. Edited by W. Wilkes. Wesley, London, 1915.

DRUDE, O. Handbuch der Pflanzengeographie. Stuttgart, 1890.

DU RIETZ, G. E. *Gyrophora rigida* DR. in North America. A New Member of The West-Arctic Element in The Scandinavian Mountain Flora. *Svensk Bot. Tidskr.,* 22:278-281, 1928.

DURAND, E. and FLAHAULT, C. Les Limites de la Région Méditeranéene en France. *Bull. Soc. Bot. Fr.,* 33, xxiv-xxxiii. 1886.

DYER, W. T. Article on Distribution of Plants in *Encyclopedia Brittanica,* Ed. 9, Vol. 7, 1877.

Emerson, F. W. Subterranean Organs of Bog Plants. *Bot. Gaz.*, 72: 359-374, 1921.

Engler, A. and Drude, O. Die Vegetation der Erde. 15 vols. Engelmann, Leipzig, 1896 to date.

Engler, A. and Prantl, K. Die Natürlichen Pflanzenfamilien. Leipzig, 1899 to present.

Eriksson, J. V. Bälinge Mossars Utvecklingshistoria och Vegetation. *Svensk Bot. Tidskr.*, 6: 8-194, 1912.

Everett, T. H. Curious Insectivorous Plants. *Gard. Chron.*, 32: 14, 1928.

Fairchild, David. The Jungles of Panama. *Nat. Geogr. Mag.*, 41: 131-146, 1922.

Fernald, M. L. Expedition to Newfoundland. *Rhodora*, 13: 109-167, 1911.

Fernald, M. L. The Alpine Bearberries. *Rhodora*, 16: 21, 1916.

Fernald, M. L. The Content of The Floras of Eastern and Western Newfoundland. *Amer. Journ. Bot.*, 5: 237-247, 1918.

Fernald, M. L. Expedition to Nova Scotia. *Rhodora*, 23: 89-111, *et seq.*, 1921.

Fernald, M. L. The Geographic Distribution of Hybrids. *Science*, 54: 73-74, 1921.

Fernald, M. L. Isolation and Endemism in Northeastern America and Their Relation to The Age-and-Area Hypothesis. *Amer. Journ. Bot.*, 1: 558-572, 1924.

Fernald, M. L. Persistence of Plants in The Unglaciated Areas of Boreal America. *Mem. Amer. Acad. Arts and Sciences*, 15: 237-342, 1925.

Fernald, M. L. Botanizing in Newfoundland. *Contrib. Gray Herb.*, 74-78, N. S., 1925-1927.

Fernald, M. L. The Antiquity and Dispersal of Vascular Plants. *Quart. Rev. Biol.*, 1: 212-245, 1926.

Fernald, M. L. Unverified Geographical Ranges. *Science*, 68: 145-149, 1928.

Flint, W. F. The Distribution of Plants in New Hampshire. *Geology of New Hampshire*, 1: 393, 1874.

Floderus, Björn. Bidrag till Kännedomen om Novaja Semljas *Salices*. *Svensk Bot. Tidskr.*, 6: 387-426, 1912.

Forbes, H. O. A Naturalist's Wanderings in The Eastern Archipelago. London, 1885.

Forsaith, F. J. Pollen Sterility in Relation to Geographic Distribution of Some *Onagraceæ*. *Bot. Gaz.*, 62: 466-487, 1916.

FRIES, ROB. E. Die Vegetation des Bangweolo-Gebietes. *Svensk Bot. Tidskr.*, 7: 233-257, 1913.

FUCHS, T. Die Mediterranflora in Ihrer Abhängigkeit von der Bodenunterlage. *Sitzber. K. Ak. Wiss,* Wien, 76, 1887.

GANONG, W. F. The Vegetation of The Bay of Fundy Salt and Diked Marshes. *Bot. Gaz.,* 36: 161-186, 1903.

GASKILL, A. Why Prairies Are Treeless. *Proc. Soc. Amer. Foresters,* 1: 158-178, 1906.

GATES, F. C. and SEWELL, M. C. Tillage and Weeds. *Ecology,* 6: 138-142, 1925.

GLEASON, H. A. An Isolated Prairie Grove and Its Phytogeographical Significance. *Bot. Gaz.,* 53: 38-49, 1912.

GLEASON, H. A. The Vegetational History of The Middle West. *Ann. Assoc. Amer. Geog.,* 12: 39-85, 1922.

GLEASON, H. A. Studies in The Flora of Northern South America. *Bull. Torr. Bot. Club,* 51: 443-448, 1924; *et seq.*

GLENN, L. C. Denudation and Erosion in The Southern Appalachian Region and The Monongahela Basin. *U. S. Geol. Surv. Professional Paper* 72, 1911.

GORDON, A. R. Report on Hudson Bay Expedition. Dept. Marine and Fisheries. Ottawa, 1884-5.

GRAY, ASA. Observations Upon The Relations of The Japanese Flora to That of North America and Other Parts of The North Temperate Zone. *Mem. Amer. Acad. Sci.,* Vol. 6, 1850.

GRAY, ASA. Forest Geography and Archeology. *Amer. Journ. Sci. and Arts,* 16: 85-94, 183-196, 1878.

GRAY, ASA. Die Vegetation d. Rocky Mountain Gebietes und Ein Vergleich Derselben mit Andere Welttheile. *Bot. Jahrb.,* 2: 256, 1882.

GRAY, ASA. Characteristics of The North American Flora. *Journ. Sci.,* 28: 323, 1884.

GRAY, ASA. On East-American and East-Asiatic Floras. *Scientific Papers of Asa Gray,* 1: 125, 1889.

GREELY, A. W. Three Years' Arctic Service. An Account of The Lady Franklin Expedition. New York, 1886.

GREENE, E. L. Notes On The Flora of Santa Cruz Islands. *Bull. Calif. Acad. Sci.,* 2: 377-418, 1886-1887.

GUPPY, H. B. Observations of A Naturalist in The Pacific Between 1896 and 1899. Macmillan, New York, Vol. 1, Vanua Levu, Fiji, 1903; Vol. 2, Plant Dispersal, 1906.

Guppy, H. B. Studies in Seeds and Fruits. Williams & Norgate, London, 1912.

Guppy, H. B. Fruits, Seeds, and Currents. Williams & Norgate, London, 1917.

Guppy, H. B. Plant Distribution from The Standpoint of An Idealist. *Journ. Linn. Soc. Bot.,* 44: 439-472, 1919.

Hall, C. F. Narrative of The Second Arctic Expedition. Washington, 1879.

Hall, H. M. A Botanical Survey of San Jacinto Mountain. *Univ. Calif. Publ.,* 1: 1-140. Univ. Press, Berkeley, 1902.

Handlirsch, A. Über Relikte. *Verhandl. K. K. Zool.-Bot. Ges.,* Wien, 59: 183-207, 1909.

Harper, R. M. The Plant Population of Northern Lower Michigan. *Bull. Torr. Bot. Club,* 45: 23-42, 1918.

Harrison, J. W. H. The Colors of Primroses. *Nature,* 107: 359-360, 1921.

Harshberger, J. W. The Comparative Age of The Different Floristic Elements of Eastern North America. *Proc. Acad. Nat. Sci., Phila.,* 56: 601-615, 1904.

Harshberger, J. W. Original Centers Concerned in North American Plant Distribution. *Proc. Acad. Nat. Sci.* Phila., 47: 2, 1905.

Harshberger, J. W. Phytogeographic Survey of North America. Stechert, New York, 1911.

Hart, H. C. On The Botany of The British Polar Expedition, 1875-1876. *Journ. Bot.,* 18: 52-56, 70-79, 111-115, 141-145, 177-182, 204-208, 235-242, 303-306, 1880.

Hayes, C. W. and Campbell, M. R. The Relation of Biology to Physiography. *Science,* N. S., 12: 131-133, 1900.

Hendricks, H. V. Torsion Studies in Twining Plants. *Bot. Gaz.,* 68: 425-440, 1919.

Herklotz, G. A. C. Krakatau and The Thousand Islands. *Discovery,* 10: 376-380, 1929.

Hildebrand, F. R. Die Lebensdauer und Vegetationsweise der Pflanzen, Ihre Ursachen und Ihre Entwickelung. *Bot. Jahrb.,* 2: 52, 135, 1882.

Hilgard, E. W. Soils. Macmillan, New York, 1907.

Hitchcock, A. H. Floral Aspects of The Hawaiian Islands. *Ann. Rept. Smithsonian Inst.* for 1917: 449-462, 1919.

Hollick, A. Plant Distribution As a Factor in The Interpretation of Geological Phenomena, With Special Reference to Long

Island and Vicinity. *Trans. N. Y. Acad. Sci.,* 12: 189-202, 1893.

Holm, T. The Vegetation of The Alpine Region of The Rocky Mountains in Colorado. *Nat. Acad. Sci. Mem.,* No. 3. 16, 17, 1923.

Holttum, R. E. The Vegetation of Greenland. *Ecology,* 10: 87-108, 1922.

Hooker, J. D. The Distribution of North American Flora. *Amer. Naturalist,* 13: 155, 1879.

Hopkins, A. D. The Bioclimatic Zones of Continents with Proposed Designations and Classification. *Journ. Wash. Acad. Sci.,* 11: 227-229, 1921b.

Hopkins, A. D. The Bioclimatic Zones Determined by Meteorological Data. *U. S. Dept. Agric. Mon. Weath. Rev.,* 40: 299-300, 1921c.

Humboldt, Alexander von. De Distributione Geographica Plantarum. Paris, 1817.

Hutchinson, A. H. Limiting Factors in Relation to Specific Ranges of Forest Trees. *Bot. Gaz.,* 66: 465-493, 1918.

Iltis, Hugo, and Schulz, Bert. Totius Flora Photographica Arte Depicta (Flowers of The World). Transl. by W. C. Worsdell. Rudolph M. Roher, Brünn, 1928.

Irmscher, Edgar. Pflanzenverbreitung und Entwicklung der Kontinente. *Mitteil. Inst. Allg. Bot.,* Hamburg, 5: 17-235, 1923.

Jackson, V. W. Vegetational Succession in The Red River Basin (Manitoba). *Sci. Agric.,* 4: 301-307, 1924.

Jacobsen, W. C. Goat-Grass (*Ægilops triuncialis*), A Weed Pest of The Range. *Monthly Bull. Dept. Agric. State of California,* Jan., 1929.

Jenny-Lips, H. Vegetationsbedingungen und Pflanzengesellschaften auf Felsschutt. Phytosoziologische Untersuchungen in Den Glarner Alpen. *Beih. Bot. Centralbl.,* 46: 119-296, 1930.

Jewell, Minna E. and Brown, H. W. Studies on Northern Michigan Bog Lakes. *Ecology,* 10: 427-475, 1929.

Johansson, K. Hufvuddragen af Götlands Växttopografi och Växtgeografi. *Sv. Vet.-ak., Handl.,* 29: 1-270 (1896), 1897.

Johansson, K. Bidrag till Götlänska Pulsatillornas Naturhistoria. *Svensk Bot. Tidskr.,* 6: 1-40, 1912.

Jurica, H. S. A Study in The Grouping of Plants. *Bot. Gaz.,* 78: 326-334, 1924.

KATZ, N. J. Zur Kenntniss der Moore Nordeuropas. *Beih. Bot. Centralbl.*, Bd. 46, Hf. 2/3: 297-394, 1930.

KELLER, PAUL. Beiträge zur Kenntniss der Nacheiszeitlichen Waldentwicklung in der Ostschweiz. *Beih. Bot. Centralbl.*, 45: 189-219, 1928.

KELLER, PAUL. Postglaziale Waldperioden in Den Zentralalpen Graubündens. *Beih. Bot. Centralbl.*, 46: 395-489, 1930.

KENOYER, L. A. The Plant Life of British India. *Sci. Monthly*, 18: 48-65, 1924.

KENOYER, L. A. Plant Physiognomy. *Ecology*, 10: 409, 1929.

KERNER, P. and OLIVER, F. W. (Translator). Natural History of Plants. Holt, New York. Undated.

KIRKWOOD, J. E. Forest Distribution in The Rocky Mountains. *Montana State Univ. Stud., Ser.* No. 2: 15-180, 1922.

KLUGH, A. B. Flower-visiting Insects and The Reflection of Ultraviolet by Flowers. *Ecology*, 6: 179-181, 1925.

KNOLL, F. Üeber Abendschwärmer und Schwärmerblumen. *Ber. d. Deutsch. Bot. Ges.*, 21: 460-461, 1929.

KNOWLTON, F. H. Plants of the Past. University Press, Princeton, 1927.

KNUDSON, L. Nonsymbiotic Germination of Orchid Seeds. *Bot. Gaz.*, 73: 1-25, 1922; 77: 212-219, 1924; 79: 345-379, 1925.

KNUDSON, L. Nonsymbiotic Germination of Orchid Seeds. *New Phytologist*, 26: 328-336, 1927.

KNUTH, P. Beiträge zur Biologie der Blüten. *Bot. Centralbl.*, 70: 274-278, 1897, *et seq.*

KNUTH, P. Handbook of Flower-Pollination. Eng. Transl. Clarendon Press, Oxford, 1906-1909.

KRASAN, F. Beitrag zur Klärung einiger phytogeographischer Begriffe. *Bot. Johrb. Beibl.*, 31: 3-38, 1901.

KRAUSE, K. Über die Heimat der Kirsche un Ihr Vorkommen im Pontischen Kleinasien. *Die Wissenschaft*, 15: 425-428, 1927.

KRÜGER, P. Die Oberirdischen Vegetationsorgane der *Orchideen* in Ihren Beziehung zu Klima und Standort. *Flora*, 66: 435-443, 451-459, 467-477, 499-510, 515-522, 1883.

KURZ, H. Northern Aspect and Phenology of Tallahassee Red Hill Flora. *Bot. Gaz.*, 85: 83-89, 1928.

LAGARDE, R. V. Non-symbiotic Germination of Orchids. *Ann. Mo. Bot. Gard.*, 16: 499-514, 1929.

LAGORCE, JOHN O. The Fight at the Timber Line. *Nat. Geogr. Mag.*, 42: 165-196, 1922.

Lamson-Scribner, F. and Merrill, E. D. The Grasses of Alaska. *Contrib. U. S. Nat. Herb.*, 13, 1910.

La Pylaie. Voyage a l'Isle de Terre-Neuve, 5. 1925.

Lecoq, H. L'Étude sur la Géographie Botanique de l'Europe. 1854.

Livingston, B. E. Climatic Areas of The United States As Related to Plant Growth. *Proc. Amer. Philos. Soc.*, 52: 257-275, 1913.

Livingston, B. E. The Relation of Plants to Climatic Conditions. *Johns Hopkins Univ. Circular,* p. 153, March, 1917.

Livingston, B. E. and MacLean, F. T. A Living Climatological Instrument. *Science,* 43: 362-363, 1916.

Livingston, B. E. and Shreve, F. The Distribution of Vegetation in the United States, As Related to Climatic Conditions. *Carneg. Inst. Wash., Publ.* No. 284, 1921.

Lotsy, J. P. Die Endemischen Pflanzen von Ceylon und Die Mutationshypothese. *Biol. Centralbl.,* 36: 207-209, 1916.

Lovell, J. H. The Insect Visitors of The *Caprifoliaceæ*. *Amer. Naturalist,* 34: 37-51, 1900.

Lutz, H. J. The Vegetation of Heart's Content, A Virgin Forest in Northwestern Pennsylvania. *Ecology,* 11: 1-29, 1920.

Mac Dougal, D. T. Delta and Desert Vegetation. *Bot. Gaz.,* 38: 44-63, 1904.

Mac Dougal, D. T. The Botanical Features of North American Deserts. *Carneg. Inst. Wash., Publ.* No. 99, 1908.

Mac Dougal, D. T. Review of Willis's Age-and-Area Hypothesis. *Pl. World,* 193, 1914.

Mac Dougal, D. T. Growth in Trees. *Proc. Amer. Philos. Soc.,* 40: 7-14, 1921.

Mac Dougal, D. T. *et al.* The Salton Sea. A Study of The Geography, Geology, Floristics, and Ecology of A Desert Basin. *Carneg. Inst., Wash., Publ.* No. 193, 1914.

Mac Dougal, D. T. *et al.* Basis of Succulence in Plants. *Bot. Gaz.,* 67: 405-416, 1919.

Mac Dougal, D. T., and Spoehr, H. A. The Origination of Zerophytes. *Plant World,* 21: 245-249, 1918.

Macmillan, Conway. Observations on The Distribution of Plants Along The Shore of Lake of The Woods. *Minn. Bot. Stud.,* 1: 949-1023, 1897.

Macmillan, Conway. Minnesota Plant Life. Saint Paul. 1899.

MARSHALL, R. The Problem of The Wilderness. *Scientific Monthly,* 30: 141-148, 1930.

MASTERS, M. T. Vegetable Teratology. Ray Society, London, 1869.

MATSSON, L. P. R. Til Frågan om Rosornas Befruktning. *Svensk Bot. Tidskr.,* 6: 589-605, 1912.

Mc DOUGAL, W. B. The Classification of Symbiotic Phenomena. *Plant World,* 21: 250-256, 1918.

Mc DOUGAL, W. B. Symbiosis in A Deciduous Forest. *Bot. Gaz.,* 73: 200-212, 1922.

Mc LEAN, R. C. Studies in The Ecology of A Tropical Rain-Forest, With Special Reference to The Forests of South Brazil. *Journ. Ecol.,* 7: 5-54, 1919.

MARRIAM, C. H. Life Zones and Crop Zones in the United States. *U. S. Dept. Agric., Bull.* 10, 1898.

MERRILL, E. D. Distribution of The *Dipterocarpaceæ.* Origin and Relationships of The Philippine Flora, etc. *Philip. Journ. Sci.,* 23: 1-32, 1923.

METCALF, W. Growth of Eucalyptus in California Plantations. *Agric. Exp. Sta. Bull.* No. 380, Berkeley, 1924.

MEYEN, F. J. F. Grundriss der Pflanzengeographie. Berlin, 1836.

MOLISCH, H. Die Lebensdauer der Pflanze. Gustav Fischer, Jena, 1929.

MOSS, C. E. The Fundamental Units of Vegetation. *New Phytologist,* 9: 18-53, 1910.

MUELLER, H. The Fertilization of Flowers. Engl. Transl. Macmillan, London, 1883.

NANSEN, F. The First Crossing of Greenland. Longmans, London, 1919.

NELSON, E. W. Report Npon The Natural History Collections Made in Alaska Between The Years 1877 and 1882. *Signal Service, U. S. Army.* 1887.

NICHOLS, G. E. Methods in The Floristic Study of Vegetation. *Ecology,* 11: 127-135, 1930.

NORDHAGEN, ROLF. Planteveksten paa Froöene. *Det Kongelige Videnskapers-Selskaps Skrifter,* No. 8. Oslo, 1916.

ORTON, J. A. The Andes and The Amazon. Harpers, New York, 1875.

OSTENFELD, C. H. Flora Arctica. Copenhagen, 1902.

PALLAS, P. S. Reise durch Die Verschiedene Provinzen des Russischen Reichs. St. Petersburg, 1771-1776.

PALMER, E. J. Botanical Reconnaissance of Southern Illinois. *Journ. Arnold Arboretum,* 2: 129-153, 1921.

PAYSON, E. The Pollination of *Asclepias cryptoceras. Bot. Gaz.,* 61: 72-74, 1916.

PELLETT, F. C. Garden Plants Which Attract Bees. *Amer. Bee Journ.,* 61: 138, 1921.

PERRYCOSTE, F. H. The Colors of Primroses. *Nature,* 107: 459, 1921.

PETER, A. Über Spontane und Künstliche Gartenbastarde der Gattung *Hieracium. Bot. Jahrb.,* 6: 111-136, 1885.

POOL, R. J. The Vegetation of The Sandhills of Nebraska. *Minn. Bot. Stud.,* 4: 189-312, 1914.

POUND, R. and CLEMENTS, F. E. The Phytogeography of Nebraska. Lincoln, 1900.

PREBLES, E. A. A Biological Investigation of The Hudson Bay Region. North American Fauna. *U. S. Biol. Surv.,* Bull. No. 22, 1902.

PRINTZ, HENRIK. The Vegetation of The Siberian-Mongolian Frontiers. *Det Kongelige Norske Videnskapers Selskap.* Oslo, 1921.

RAE, JOHN. Narrative of An Expedition to The Shores of The Arctic Sea in 1846-1847. London, 1850.

RAMSBOTTOM, J. The Germination of Orchid Seeds. *Orchid Rev.,* 30: 197, 1922.

RAMSBOTTOM, J. Orchid Mycorrhiza. *Trans. British Mycol. Soc.,* 8: 28-61, 1923.

RAYNER, M. C. The Ecology of *Calluna vulgaris. New Phytol.,* 12: 59-77, 1913.

RAYNER, M. C. Mycorrhiza in The *Ericaceæ. Trans. British Mycol. Soc.,* 8: 61-66, 1923.

Report of The Committee on Floral Areas. *Rhodora,* 27: 56, 1925.

RETTIG, ERNST. Ameisenpflanzen-Pflanzenameisen. Gustav Fischer, Jena, 1904. Reviewed in *Bot. Centralbl.,* 96: 51-52, 1904.

Reviews of Russian papers on ecology and phytogeography in *Bot. Centralbl.,* 156, N. F. 14: 25-40, 1929.

RIDLEY, H. N. Endemic Plants. *Journ. Bot.,* 63: 182-183, 1925.

ROBERTS, EDITH. The Distribution of Beach Plants. *Bot. Gaz.,* 60: 406-411, 1915.

Robertson, C. Insect Relations of Certain Asclepiads. *Bot. Gaz.*, 12:207-216, 1887.

Robertson, C. Zygomorphy and Its Causes. *Bot. Gaz.*, 13:146-151, 1888.

Robertson, C. Flowers and Insects. *Umbelliferæ. Trans. St. Louis Acad. Sci.*, 5:449-460, 1890; *Asclepiadaceæ* to *Scrophulariaceæ, ibid.*, 5:569-598, 1891; *Labiatæ, ibid.*, 6:101, 1892; *Rosaceæ* and *Compositæ, ibid.*, 6:433, 1894.

Robertson, C. Philosophy of The Flower Seasons. *Amer. Nat.*, 29:97-117, 1895.

Robertson, C. Flowers and Insects. *Bot. Gaz.*, 15:79-84, 1890; 25:229-245, 1898.

Robertson, C. Ecological Adaptation and Ecological Selection. *Science*, 23:307-310, 1906.

Robertson, C. Flowers and Insects. Evolution of Entomophilous Flowers. *Bot. Gaz.*, 63:307-316, 1917.

Robertson, C. Pollination of *Asclepias cryptoceras. Bot. Gaz.*, 66:177, 1918.

Robertson, C. Proterandry and The Flight of Bees. *Entomol. News*, 29:340, 1918.

Robertson, C. Flower Seasons. *Scientific Monthly*, 14:201-203, 1922.

Robertson, C. Phenology of Entomophilous Flowers. *Ecology*, 5:393-407, 1924.

Robertson, C. Conditions of Natural Selection. *Science*, 59:363-364, 1924.

Robertson, C. Color Preference of Bees. *Bot. Gaz.*, 78:68-84, 1924.

Rübel, E. Ökologische Pflanzengeographie. In *Handwörterbuch der Naturwissenschaft*, Bd. 4:858-907, 1913.

Rübel, E. Geobotanische Untersuchungsmethoden. Gebrüder Borntraeger, Berlin, 1922.

Rydberg, P. A. Composition of The Rocky Mountain Flora. *Science, N. S.*, 12:870-873, 1900.

Rydberg, P. A. Phytogeographical Notes on The Rocky Mountain Region—IV. Forests of The Subalpine and Montane Zones. *Bull. Torr. Club*, 42:11-25, 1915.

Salisbury, E. J. The Oak-Hornbeam Woods of Herefordshire. *Journ. Ecol.*, 4:82-117, 1916.

Sampson, A. W. Climate and Plant Growth in Certain Vegetative Associations. *U. S. Dept. Agric.*, Bull. 700, 1919.

SAMPSON, H. C. An Ecological Survey of The Prairie Vegetation of Illinois. *Ill. Nat. Hist. Surv.*, 13: 523-577, 1921.

SAMUELSSON, GUNNAR. Studien über Die Entwicklungsgeschichte der Blüten Einiger Bicornes-Typen. Ein Beitrag zur Kenntniss der Systematischen Stellung der *Diapensiaceen* und *Empetraceen. Svensk. Bot. Tidskr.*, 7: 97-188, 1913.

SCHAFFNER, J. H. Origin of Timber Belts. From Observations in Kansas. *Bot. Gaz.*, 27: 392-393, 1899.

SCHARFF, R. F. Distribution and Origin of Life In America. Macmillan, New York, 1912.

SCHIMPER, A. F. W. Notizen uber Insectenfressenden Pflanzen. *Bot. Zeitung*, 40: 224, 1882.

SCHIMPER, A. F. W. Plant Geography Upon A Physical Basis. Engl. Transl. Clarendon Press, Oxford, 1903.

SCHOTTKY, E. Die Eichen des Extratropischen Ostasiens und Ihre Pflanzengeographische Bedeutung. Dissertation, Berlin, 1912. *Bot. Jahrb.*, 47: 617-708, 1912.

SCHOUW, J. F. Grundzüge der Allegemeine Pflanzengeographie. Berlin, 1823.

SCHRÖTER, C. Das Pflanzenleben der Alpen. Zurich, 1908.

SCHWEINFURTH, G. Die Entdeckung des Wilden Urweizens in Palästina. *Ber. d. Deutsch. Bot. Ges.*, 26a: 309-324, 1908.

SERNANDER, R. Sveriges Växtvärld i Forntid och Nutid. In "Sveriges Rike." J. F. Nyström, Stockholm, 1900.

SERNANDER, R. Die Schwedischen Torfmoore als Zeugen Postglacialer Klimaschwankungen. *Intern. Geologenkongress*, XI, Stockholm, 1910.

SETCHEL, W. A. Morphological and Phenological Notes on *Zostera marina*. *Univ. Calif. Publications (Botany)*, 14: 389-449, 1929.

SEWARD, A. C. Greenland: As It Is and As It Was. *Nature*, 123: 455-462, 1929.

SHANTZ, H. L. and MARBUT, C. F. Vegetations and Soils of Africa. *Amer. Geog. Soc. Research Series*, No. 13. New York, 1923.

SHREVE, F. A Map of The Vegetation of The United States. *Geogr. Rev.*, 3: 119-125, 1917.

SHREVE, F. Ecological Aspects of The Deserts of California. *Ecology*, 6: 93-103, 1925.

SHREVE, F. Changes in Desert Vegetation. *Ecology*, 10: 364-373, 1929.

SIMMONS, H. G. Har en Landbrygga öfver Nordatlanten funnits i Postglacial Tid? (Was There a Land Bridge across The Atlantic in Postglacial Time?). *Ymer*, 2: 150-155, 1905.

SIMMONS, H. G. Remarks About The Relations of The Floras of The Northern Atlantic, The Polar Sea, and The North Pacific. *Beih. Bot. Centralbl.*, 19, Abt. 2: 149-194, 1905.

SIMMONS, H. G. The Vascular Plants in The Flora of Ellesmereland. *Report Second Norw. Arctic Exped. in The Fram*, 1898-1903. *Videnskaps-Selskapets Forhandl.*, Oslo, 1906.

SIMMONS, H. G. A Survey of The Phytogeography of The Arctic Archipelago. Lund Universitets *Årskrift, N. F.*, Afd. 2, 9(19), 1913.

SINNOTT, E. W. The Pond Flora of Cape Cod. *Rhodora*, 14: 25, 1913.

SINNOTT, E. W. Comparative Rapidity of Evolution in Various Plant Types. *Amer. Naturalist*, 50: 466-478, 1916.

SINNOTT, E. W. The "Age-and-Area" Hypothesis and The Problem of Endemism. *Ann. Bot.*, 31: 209-216, 1917.

SINNOTT, E. W. The "Age-and-Area" Hypothesis of Willis. *Science*, 46:457-459, 1917.

SINNOTT, E. W. Age and Area and The History of Species. *Amer. Journ. Bot.*, 11: 573-578, 1924.

SINNOTT, E. W. and BAILEY, I. W. The Origin and Dispersal of Herbaceous Angiosperms. *Ann. Bot.*, 28: 547-600, 1914.

SKENE, M. The Biology of The Flowering Plants. Macmillan, London, 1924.

SKOTSBERG, V. C. A Botanical Survey of The Falkland Islands. Botanische Ergebnisse der Schwedischen Expedition Nach Patagonien, 1913. *Kgl. Svenska Vet.-Ak. Handl.*, 50: 1-129, 1913.

SMALL, J. The Age-and-Area Law. Science Progress, 12: 439-449, 1918.

SMILEY, F. J. The Alpine and Subalpine Vegetation of The Lake Tahoe Region. *Bot. Gaz.*, 59: 265-286, 1915.

SPALDING, V. M. Biological Relations of Certain Desert Shrubs. I. The Creosote Bush. *Bot. Gaz.*, 38: 122, 1904; II. *Bot. Gaz.*, 41: 262-282, 1906.

SPALDING, V. M. Distribution and Movements of Desert Plants. *Carneg. Inst. Wash., Publ.*, No. 113, 1909.

SPRUCE, R. Notes of A Botanist On The Amazon and Andes. Edited by Alfred Russell Wallace. London, 1908.

STALLARD, HARVEY. Secondary Succession in The Climax Formations of Northern Minnesota. *Ecology*, 10: 476-547, 1929.

ST. JOHN, HAROLD. A Botanical Exploration of The North Shore of The Gulf of St. Lawrence. *Canada Dept. Mines, Mem.*, 126, iii + 130 pp., 1922.

Stephens, Edith. Carnivorous Plants of The Cape Peninsula. *Journ. Bot. Soc. South Africa,* 9: 21-24. 1923.

Stone, W. The Plants of Southern New Jersey, With Special Reference to The Flora of The Pine Barrens and The Geographic Distribution of The Species. *Ann. Rep. N. J. State. Mus.,* Part 2: 21-828, Trenton. 1912.

Sverdrup, Otto. The Second Norwegian Polar Expedition in The *Fram. Roy Geogr. Journ.,* 22:38-56, 1903.

Tansley, A. G. The Classification of Vegetation and The Concept of Development. *Journ. Ecology,* 8: 118-149, 1920.

Tansley, A. G. Practical Plant Ecology. Dodd, Mead, New York, 1923.

Tansley, A. G. and Chipp, T. F. Aims and Methods in The Study of Vegetation. *Brit. Emp. Veg. Committee,* London, 1926.

Taylor, W. A Distributional and Ecological Study of Mt. Ranier, Washington. *Ecology,* 3: 214-236, 1922.

Taylor, W. P. and McGinnis, W. G. The Bio-Ecology of Forest and Range. *Scientific Monthly,* 27: 177-182, 1928.

Tengwall, T. A. De Sydliga Skandinaviska Fjällväxterna och Deras Invandringshistoria. *Sv. Bot. Tidskr.,* 7: 258-275, 1913.

Thomson, J. A. The Biology of The Seasons. Melrose, London, 1911.

Thomson, J. A. The Haunts of Life. Harcourt, Brace, London, 1922.

Thomson, J. A. Everyday Biology. Doran, New York, 1924.

Thomson, J. A. The Gospel of Evolution. Putnam, New York, 1926.

Toumey, J. W. The Vegetation of The Forest Floor. *Proc. Intern. Cong. Pl. Sci.,* 1: 575, 1929.

Transeau, E. N. On The Geographic Distribution and Ecological Relations of The Bog Plant Societies of Northern North America. *Bot. Gaz.,* 36: 401-420, 1903.

Transeau, E. N. Climatic Centers and Centers of Plant Distribution. *Mich. Acad. Sci. Ann. Rep.,* 7: 73-75, 1905.

Transeau, E. N. Forest Centers of Eastern America. *Amer. Naturalist,* 39: 875-889, 1905.

Troll, W. Über Antherenbau, Pollen, und Pollination von *Galanthus. Flora, N. F.,* 23: 321-343, 1928.

Turrill, W. B. The Plant-Life of The Balkan Peninsula. A Phytogeographic Study. University Press, Cambridge, 1929.

UPHOFF, J. C. T. Ecological Observations on Plants of The Marshes and Swamps of Central Cuba. *Ecology,* 5:363-371, 1924.

VAN DYKE, J. C. The Desert. Scribner's, New York, 1902.

VERDOORN, INEZ C. The Flowering Parasitic Plants Found in South Africa. *So. Afr. Journ. Nat. Hist.,* 4:221-228, 1923.

VESTAL, A. G. Preliminary Account of The Forests in Cumberland County, Illinois. *Trans. Ill. State Acad. Sci.,* 12:234-245, 1919.

VICTORIN, FRÈRE MARIE-. Esquisse Systematique et Écologique de La Flore Dendrologique d'Une Portion de La Rive Sud du Saint-Laurent, aux Environs de Longeuil, P. Q., *Rév. Trimestrielle Canad.,* 8:1-33, 1922.

VICTORIN, FRÈRE MARIE-. Études Floristiques sur La Région du Lac Saint-Jean. *Contrib. du Lab. Bot. de l'Univ. de Montréal,* No. 4:1-174, 1925.

VICTORIN, FRÈRE MARIE-. Deux Epibiotes Remarquables de La Minganie. *Contrib. du Lab. Bot. de l'Univ. de Montréal,* No. 12:163-175, 1928. Extract from *Trans. Roy. Soc. Can.,* Series 3, 22 (Sect. 5): 163-176, 1928.

VICTORIN, FRÈRE MARIE-. Le Dynamisme dans La Flore du Québec. *Contrib. du Lab. Bot. de l'Univ. de Montréal,* No. 13, 1929.

WAKEHAM, COM. W. Report on The Second Hudson Bay Expedition. *Dept. Marine and Fisheries,* Ottawa, Canada, 1897.

WALLACE, A. R. The Tropical Nature and Other Essays. Macmillan, London, 1878.

WALLACE, A. R. Natural Selection and Tropical Nature. Macmillan, London, 1891.

WALLACE, A. R. Darwinism: An Exposition of The Theory of Natural Selection. Macmillan, London, 1905.

WALLACE, A. R. The World of Life. Moffat, Yard, New York, 1911.

WALLACE, A. R. The Malay Archipelago. Harpers, New York, 1885.

WARMING, E. On The Vegetation of Tropical America. *Bot. Gaz.,* 27:1-18, 1899.

WARMING, E. Œcology of Plants. Engl. transl. Clarendon Press, Oxford, 1909.

WATERMAN, W. G. Ecological Problems from *Sphagnum.* Bogs of Illinois. *Ecology,* 7:255-272, 1926.

Waterman, W. G. Development of Root-Systems under Dune Conditions. *Bot. Gaz.*, 68: 22-53, 1919.

Waterman, W. G. Development of Plant Communities of A Sand Ridge in Michigan. *Bot. Gaz.*, 74: 1-31, 1921.

Waterman, W. G. Plant Communities of Alpine Park. *Bot. Gaz.*, 80: 188-202, 1925.

Weaver, J. E. Root Development in The Grassland Formation. *Carneg. Inst. Wash., Publ.* No. 292, 1921.

Weaver, J. E. and Clements, F. E. Plant Ecology. McGraw-Hill, New York, 1929.

Weaver, J. E. and Thiel, A. F. Ecological Studies in The Tension Zone Between Grassland and Woodland. *Rep. Bot. Surv. Neb., n. s.*, 1, 1917.

Westell, S. Nature in Field and Meadow. Sheldon Press, London, 1929.

Wherry, E. T. Northward Range-Extension of Some Southern Orchids in Relation to Soil Reaction. *Journ. Wash. Acad. Sci.*, 18: 212-216, 1928.

White, T. H. Mutation, Adaptation to Temperature Differences, and Geographical Distribution in Plants. *Verh. V. Internat. Kongr.-Vereb.*, 2: 1575-1586, 1928.

Wiegand, K. F. N. Some Studies Regarding the Biology of Buds on Twigs in Winter. *Bot. Gaz.*, 42: 373-424, 1906.

Willard, Daniel F. The Story of The Prairies. Rand-McNally, New York, 1903.

Wille, N. The Flora of Norway and Its Immigration. *Ann. Mo. Bot. Gard.*, 2: 59-108, 1915.

Williams, C. B. A Short Bioclimatic Study in The Egyptian Desert. *Egypt. Min. Agr. Tech., Sci. Serv.*, Bull. 29, 1923, 1924. Reviewed in *Ecology*, 7: 384-385, 1926, by M. T. Cook.

Willis, J. C. The Flora of The Hilltops in Ceylon. *Ann. Roy. Bot. Gard. Perad.*, 4: 131-138, 1908.

Willis, J. C. The Endemic Flora of Ceylon. *Philos. Trans., B.*, 206: 307-342, 1915. Correction in *Proc. Roy. Soc., B.*, 206: 307-342, 1916.

Willis, J. C. The Evolution of Species in Ceylon. *Ann. Bot.*, 30: 1-23, 1916.

Willis, J. C. The Distribution of Species in New Zealand. *Ann. Bot.*, 30: 437-457, 1916.

Willis, J. C. The Relative Age of Endemic Species, and Other Controversial Points. *Ann. Bot.*, 31: 189-208, 1917.

WILLIS, J. C. The Source and Distribution of The New Zealand Flora, With A Reply to Criticism. *Ann. Bot.,* 32: 339-367, 1918.

WILLIS, J. C. Age and Area. A Study in Geographic Distribution and The Origin of Species. University Press, Cambridge, 1922.

WILLIS, J. C. Age and Area. *Quart. Rev. Biol.,* 1: 553-571, 1926.

WOODARD, J. The Origin of The Prairies in Illinois. *Bot. Gaz.,* 77: 241-261, 1924.

WOODARD, J. The Distribution of Tree Vegetation. *Ecology,* 6: 150-156, 1925.

WORSDELL, W. C. The Principles of Plant Teratology. Dulau, London, 1916.

WYLIE, R. B. The Pollination of *Vallisneria spiralis*. *Bot. Gaz.,* 63: 135-145, 1917.

FORESTS AND FORESTRY

ANONYMOUS. Important Forest Trees of The Eastern United States. *U. S. Dept. Agric.,* Circ. No. 223, 1922.

ASHE, W. W. Forest Types of The Appalachians and White Mountains. *Journ. Elisha Mitchell Sci. Soc.,* 37: 183-198, 1922.

BESSEY, C. E. The Forests and Forest Trees of Nebraska. *Ann. Rep. State Bd. of Agric.* for 1889: 79-102.

BRANDIS, D. Indian Trees. London, 1906.

BROWN, J. E. Forests of Western Australia and Their Development. Perth, 1899.

BUNYARD, E. A. The Winter Study of Fruit Trees. *Journ. Roy. Hort. Soc.,* 47: 18-25, 1922.

BÜSGEN, M. and MÜNCH, E. The Structure and Life of Forest Trees. Translation of "Bau und Leben unserer Waldbäume" by T. Thomson. Chapman & Hall, London, 1929.

CAJANDER, A. K. Über Waldtypen. *Acta Forestalia Fennica.* Helsingfors, 1913.

CAJANDER, A. K. The Theory of Forest Types. *Acta Forestalia Fennica.* Helsingfors, 1926.

CAMUS, H. and A. Les Chénes d'Indo-Chine (The Oaks of Indo-China). *Ann. Sci. Nat. Bot.,* 3: 777-409, 1921.

CHEESEMAN, T. F. The Kauri Forest. *New Zeal. Journ. Sci. and Tech.,* 5: 166-168, 1922.

Clarke, D. A. The Commercial Forest Trees of Massachusetts, How You May Know Them. A Pocket Manual. State Forester, State House, Boston, 1908.

Clements, F. E. Plant Formations and Forest Types. *Proc. Soc. Amer. For.*, 4, 1909.

Craib, I. J. Some Aspects of Soil Moisture in The Forest. Yale Univ. *School* of *Forestry Bull.* 25: 1-61, 1929.

Durland, Wm. B. The Forests of The Dominican Republic. *Geogr. Rev.*, 12: 206-222, 1922.

Evelyn, J. *Silva:* or A Discourse on Forest-Trees, and The Propagation of Timber in His Majesty's Dominions, etc. York. 1786. First edition, 1664.

Fernow, B. E. Forests and Floods. *Garden and Forest*, 3: 9-10, 1890.

Fernow, B. E. Forest Influences. *U. S. Dept. Agric., Div. For.*, Bull. 7, 1893.

Garman, H. The Woody Plants of Kentucky. *Ky. Agric. Exp. Sta.*, Bull. No. 169, 1913.

Grant, Madison. Saving The Redwoods. *Nat. Geogr. Mag.*, 37: 519-536, 1920.

Greeley, W. B. Classification of Forest Types. *Proc. Soc. Amer. For.*, 8: 76, 1913.

Guthrie, J. D. Alaska's Interior Forests. *Journ. For.*, 20: 363-373, 1922.

Guthrie, J. D. Snapshots of European Forests. *Amer. For. and For. Life*, 34: 170-178, 1928.

Hanson, C. O. Forestry for Woodmen. Clarendon Press, Oxford, 1921.

Henry, A. Forests, Woods, and Trees in Relation to Hygiene. London, 1919.

Hoffman, J. V. Natural Reproduction from Seed in The Forest Floor. *Journ. Agric. Res.*, 2: 1-27, 1917.

Illick, J. S. The Birches. *Amer. For.*, 28: 355-363, 1922; The Beeches. *Amer. For.*, 28: 546-551, 1922; The White Oaks. *Amer. For.*, 28: 586-592, 1922.

Kindle, E. M. Notes on The Forests of Southeastern Labrador. *Geogr. Rev.*, 12: 57-71, 1922.

KOEHLER, ARTHUR. Guidebook for The Identification of Woods Used for Ties and Timbers. *U. S. Dept. Agric., For. Serv. Misc.* RL-1, October 15, 1917.

KOEHLER, ARTHUR. The Identification of True Mahogany, Certain So-Called Mahoganies, and Some Common Substitutes. *U. S. Dept. Agric.,* Bull. No. 1050, 1922.

MEAD, D. W. The Flow of Streams and The Factors Which Modify It, With Special Reference to Wisconsin Conditions. *Univ. Wis. Bull.* No. 425, 1911.

MAIDEN, J. H. The Forest Flora of New South Wales. Vol. 8, Part 6, pp. 71-80. A. J. Kent, Sydney, 1924 *et seq.*

METCALF, W. Growth of Eucalyptus in California Plantations. *Agric. Exp. Sta. Bull.* No. 380, Berkeley, 1924.

METCALF, W. The Chinese Elm—A Valuable Tree. *Amer. For. and For. Life,* 34: 229 *et seq.,* 1928.

MICHAUX, ANDRÉ. L'Histoire des Arbres Forestiers de l'Amerique Septentrionale. Paris. 1810.

MICHAUX, ANDRÉ. The North American Sylva. Translated from the French. Philadelphia. 1850.

MUNGER, T. T. Avalanches and Forest Cover in The Northern Cascades. *U. S. For. Circ.,* 173, 1911.

OSMORTON, A. E. Notes on The Forest Communities of The Gahrwall Himalayas. *Ecology,* 1: 129, 1922.

PALMER, ERNEST J. The Forest Flora of The Ozark Region. *Journ. Arnold Arboretum,* 2: 216-232, 1922.

PAMMEL, L. H. and KING, C. M. The Germination of Some Trees and Shrubs and Their Juvenile Forms. *Proc. Iowa Acad. Sci.,* 27: 75-80, 1920.

PEARSON, U. G. The Factors Controlling The Distribution of Forest Types. *Ecology,* 1: 139-159; 289-308, 1920.

PRATT, M. B. Shade and Ornamental Trees in California. Pp. 130, Pls. 137. *Calif. St. Bd. For.* Sacramento, 1922.

RING, E., and TOWER, G. E. Forest Trees of Maine. Forestry Dept., Bangor, 1908.

SARGENT, C. S. Report on The Forests of North America Exclusive of Mexico. *Dept. Interior, Census Office.* Washington, 1884.

SCOTT, CHAS. A. *et al.* Kansas Trees and Their Uses. *Kansas State Bd. Agric.,* Topeka, 1928.

SHANTZ, H. L. and ZON, R. Atlas of Agric. Classification and Map of the Vegetation of The United States. *Part 1, Sec. E., U. S. Dept. Agric.* Washington, 1924.

SUDWORTH, G. B. Miscellaneous Conifers of The Rocky Mountain Region. *U. S. Dept. Agric.,* Bull. No. 680, 1918.

TOUMEY, J. W. The Relation of Forests and Stream-Flow. *U. S. Dept. Agric. Yearbook,* 1903.

WALCOTT, C. D. Forest Reserves. *Nineteenth Ann. Rep., U. S. Geol. Surv.,* Part 5: 1-399, 1899.

WARD, R. D. The Influence of Forests Upon Climate. *Pop. Sci. Mon.,* 82: 313-331, 1913.

WHITFORD, H. H. and CRAIG, R. D. Forests of British Columbia. *Committee of Conservation,* Ottawa, 1918.

WILSON, E. H. Notes from Australasia. *Journ. Arnold Arboretum,* 2: 160-163, 1921. An account of the author's impressions of west Australian forests.

WILSON, E. H. The New Zealand Forests. *Journ. Arnold Arboretum,* 2: 232-236, 1922.

WOEIKOW, A. I. Der Einfluss der Wälder auf Das Klima. Petermanns *Mitteilungen,* 31: 81-87, 1885.

WOODARD, J. Distribution of Tree Vegetation. *Ecology,* 6: 150-156, 1925.

ZON, R. The Relation of Forests in The Atlantic Plain to The Humidity of the Central States and Prairie Region. *Proc. Soc. Amer. For.,* 8: 138-153, 1914.

ZON, R. Forests and Water in The Light of Scientific Investigation. *Appendix V, Final Report National Waterways Commission,* 1912. Gov't Printing Office (reprint), 1927.

GEOLOGICAL REFERENCES

CHAMBERLAIN, T. C. and SALISBURY, R. D. Geology. 3 vols. Holt, New York, 1905.

MILLER, W. J. An Introduction to Historical Geology, With Special Reference to North America. Van Nostrand, New York, 1920.

QUIRKE, T. T. Elements of Geology. Holt, New York, 1925.

SCHUCHERT, C. A Textbook of Geology. Wiley, New York, 1920.
SCOTT, W. B. Introduction to Geology. Macmillan, New York, 1902.

WILLARD, D. F. The Story of The Prairies. Rand-McNally, New York, 1903.

STANDARD REFERENCES ON PALEONTOLOGY

SCOTT, D. H. Studies in Fossil Botany. 2 vols. Black, London, 1920.

ZITTEL, K. Text-Book of Palæontology. 3 vols. Translation by Chas. R. Eastman. Macmillan, New York, 1902.

For the general reader—excellent and informational:

BERRY, E. W. Tree Ancestors. Williams & Wilkins, Baltimore, 1923.

KNOWLTON, F. H. Plants of The Past. University Press, Princeton, 1927.

HISTORY OF BOTANY

GARNSEY, HENRY and BALFOUR, I. B. Sachs' History of Botany. Clarendon Press, Oxford, 1890.
GEDDES, PATRICK. Chapters in Modern Botany. Scribner's, New York, 1893.
GRAY, ASA. Scientific Papers. Selected by C. S. Sargent. Houghton Mifflin, New York, 1889.
GREEN, J. R. A History of Botany, 1860-1900. Clarendon Press, Oxford, 1890.
GREENE, E. L. Landmarks of Botanical History, Part I. *Smithsonian Inst. Wash., Misc. Collections,* Part of Volume 54, 1909.

HARVEY-GIBSON, R. J. Outlines of The History of Botany. A. & C. Black, London, 1919.
HAWKES, ELLISON. Pioneers of Plant Study. Sheldon Press, London, 1928.

Lones, Thos. E. Aristotle's Researches in Natural Science. London, 1912.

Osborne, H. F. From The Greeks to Darwin. Macmillan, New York, 1913.

SOME BOTANICAL PERIODICALS IN ENGLISH

Addisonia. New York Botanical Garden, New York City.
American Botanist. Willard N. Clute, Joliet, Illinois.
American Journal of Botany. Brooklyn Botanic Garden, Brooklyn, N. Y.
Annals of Botany. University Press, Oxford.
Annals of The Missouri Botanical Garden. Saint Louis, Mo.
Botanical Gazette. University of Chicago Press, Chicago, Ill.
Bulletin of The Brooklyn Botanic Garden. Brooklyn, N. Y.
Bulletin of The Torrey Botanical Club. New York.
Curtis' Botanical Magazine. Bernard Quaritch, London.
Journal of Botany. Taylor & Francis, Fleet St., London.
Kew Bulletin. Kew Gardens, London.
Midland Naturalist (Ohio Journal of Science). Ohio State University, Columbus, Ohio.
Rhodora. Gray Herbarium, Cambridge, Mass.
Torreya. Torrey Botanical Club, New York City.

ECOLOGICAL JOURNALS

Ecology. Ecological Society of America, Brooklyn, N. Y.
Journal of Ecology. University Press, Cambridge.

TEXTBOOKS ON ECOLOGY

Mc Dougal, W. B. Plant Ecology. Lea & Febiger, Philadelphia, 1927.

Tansley, A. G. Practical Plant Ecology. Dodd, Mead, New York, 1923.

Tansley, A. G. and Chipp, T. F. Aims and Methods in The Study of Vegetation. The Crown Agent for The Colonies, London, 1926.

WEAVER, J. E. and CLEMENTS, F. E. Plant Ecology. McGraw-Hill, New York, 1929.

In the German:

BRAUN-BLANQUET, J. Pflanzensoziologie. Julius Springer, Berlin, 1928.

RÜBEL, EDUARD. Geobotanische Untersuchungen. Gebrüder Borntraeger, Berlin, 1922.

RECENT TEXTBOOKS ON SYSTEMATIC BOTANY

HITCHCOCK, A. S. Methods of Descriptive Systematic Botany. Wiley, New York, 1925.

POOL, R. J. Flowers and Flowering Plants. McGraw-Hill, New York, 1929.

SWINGLE, D. B. Textbook of Systematic Botany. McGraw-Hill, New York, 1928.

INDEXES

TOPICAL INDEX

NOTE: Bold face figures refer to main discussion of or points of special interest in connection with the subject.

INDEX TO FIGURES